D0831067
WITHDRAWN
FROM
STOCK

6186 9851

OPERATIONS AND PROCESS MANAGEMENT

Visit the *Operations and Process Management*, Third Edition Pearson eText at www.pearsoned.co.uk/slack to find interactive material to support your learning, including:

- Videos from the authors to introduce the subjects
- Animated figures with commentary to explain the concepts
- Active cases with audio to bring Case Studies to life
- Hints to the Applying the Principles exercises.

Use your Pearson eText to get the most from *Operations and Process Management*, Third Edition.

If your copy of the book did not contain an access card, you can purchase access online at www.pearsoned.co.uk/slack by clicking on the Pearson eText link and following the onscreen instructions.

OPERATIONS AND PROCESS MANAGEMENT

Principles and Practice for Strategic Impact

Third Edition

Nigel Slack

Alistair Brandon-Jones

Robert Johnston

Alan Betts

PEARSON

Harlow, England • London • New York • Boston • San Francisco • Toronto • Sydney
Auckland • Singapore • Hong Kong • Tokyo • Seoul • Taipei • New Delhi
Cape Town • São Paulo • Mexico City • Madrid • Amsterdam • Munich • Paris • Milan

Pearson Education Limited
Edinburgh Gate
Harlow
Essex CM20 2JE
England

and Associated Companies throughout the world

Visit us on the World Wide Web at:
www.pearson.com/uk

First published 2006
Second edition published 2009
Third edition published 2012

ISBN: 978-0-273-75187-8

British Library Cataloguing-in-Publication Data
A catalogue record for this book is available from the British Library

Library of Congress Cataloging-in-Publication Data
A catalog record for this book is available from the Library of Congress

10 9 8 7 6 5
16 15 14

Typeset in 9.25/12.5 pt Syntax by 73
Printed by CPI UK

Brief Contents

Contents

Supporting resources

Visit www.pearsoned.co.uk/slack to find valuable online resources.

Pearson eText

The Pearson eText contains interactive material to support your learning, including:

- Videos from the authors to introduce the subjects
- Animated figures with commentary to explain the concepts
- Active cases with audio to bring case studies to life
- Hints to the Applying the Principles exercises.

For instructors

- Detailed downloadable Instructor's Manual featuring teaching notes and guided solutions
- PowerPoint slides of all the figures and illustrations from the book.

Also: The Pearson eText also provides the following features:

- Search tool to help locate specific items of content
- Bookmarking to make it easy to return to specific content of interest
- The ability to annotate and highlight.

For more information please contact your local Pearson Education sales representative or visit www.pearsoned.co.uk/slack.

Guide to case studies

Chapter	Page	Case study	Region	Manufac-turing/ Service	Company size	Topics/ techniques
Chapter 1 Operations and processes	27	Design House Partnerships at Concept Design Services	Europe	M, S	Medium	Role of operations, process objectives, types of operation and process
Chapter 2 Operations strategy	61	McDonalds: half a century of growth	World	S	Large	Operations strategy, operations objectives, strategic fit
Chapter 3 Supply network design	92	Disneyland Resort Paris (abridged)	France	S	Large	Location strategy, service design, capacity, job design
Chapter 4 Process design 1 – positioning	132	McPherson Charles Solicitors	UK	S	Medium	Process design, job design. Process technology, process resourcing.
Chapter 5 Process design 2 – analysis	165	The Action Response Applications Processing Unit (ARAPU)	Africa, Asia, UK	S	Small	Process design, process mapping, balancing, Little's law
Chapter 6 Designing the innovation process	205	Developing 'Savory Rosticrisps' at Dreddo Dan's	World	M	Large	Product development, operations strategy, process performance.
Chapter 7 Supply chain management	244	Supplying fast fashion	World	S M	Large	Outsourcing, supply chain design, fast response
Chapter 8 Capacity management	273	Blackberry Hill Farm	UK	S M	Small	Capacity management, forecasting, cumulative production and demand plotting.
Chapter 9 Inventory management	308	supplies4medics.com	Europe	S	Medium	Inventory management, Inventory information systems, ABC analysis
Chapter 10 Resource planning and control	337	subText Studios Singapore	Singapore	S	Medium	Planning and control, Gantt charts, activity monitoring, controlling activities.
Chapter 11 Lean synchronisation	380	Implementing lean at CWHT	UK	S	Large	Improvement principles, lean philosophy, change management, public sector operations.
Chapter 12 Quality management	408	Turnround at the Preston plant	Canada	M	Medium	Improvement principles, statistical process control, process learning, operations capabilities
Chapter 13 Improvement	458	Geneva Construction and risk	World	S	Large	Improvement principles, six sigma, change management
Chapter 14 Risk and resilience	491	Slagelse Industrial Services (SIS)	Denmark	S M	Large	Risk, failure prevention, supplier selection, relationship management
Chapter 15 Project management	521	United Photonics Malaysia Sdn Bhd	Malaysia	S	Large	Project planning, project risk, project monitoring,

Preface

Why is operations and process management essential?

Because without effective operations and processes there can be no long-term success for any organisation. Because it is at the heart of what all organisations do; they create value through their productive resources. Because it is the essential link that connects broad long-term strategy and day-to-day ongoing activities. This is why operations and process management has been changing. It has always been exciting, and it has always been challenging, but now it has acquired a much more prominent profile. The 3rd Edition of this book reflects this in a number of ways.

It stresses the importance of operations and process management

Of course, it has always been important, but increasingly managers in all types of enterprise are accepting that operations management can make or break their businesses. Effective operations management can keep costs down, enhance the potential to improve revenue, promote an appropriate allocation of capital resources, and most important, develop the capabilities that provide for future competitive advantage.

It stresses the real strategic impact operations and process management

Operations are not always operational. The operations function also has a vital strategic dimension, and operations management is now expected to play a part in shaping strategic direction, not just responding to it.

It stresses that operations and process management matters to all sectors of the economy

At one time operations management was seen as being of most relevance to manufacturing and a few types of mass service businesses. Now the lessons are seen as applying to all types of enterprise; all types of service and manufacturing, large or small organisations, public or private, for-profit or not-for-profit.

It stresses that operations and process management is of interest to all managers

Perhaps most importantly, because operations management is accepted as being founded on the idea of managing process, and because managers in all functions of the business now accept that they spend much of their time managing processes, it is clear that to some extent, all managers are operations managers. The principles and practice of operations management are relevant to every manager.

It extends the scope of operations and process management

The obvious unit of analysis of operations management is the operations function itself – the collection of resources that produces products and services. But, if managers from other functions are to be included, operations management must also address itself to process management at a more generic level. Also, no operations can consider itself in isolation from its customers, suppliers, collaborators and competitors. It must see itself as part of the extended supply network. Operations management increasingly needs to work at all three levels of analysis – the individual process, the operation itself, and the supply network.

All this has implications for the way operations management is studied, especially at post experience and post-graduate levels, and the way operations management is practised. It has also very much shaped the way this book has been structured. In addition to covering all the important topics that make the subject so powerful, it places particular emphasis on the following:

- **Principles** – that is, the core ideas that describe how operations behave, how they can be managed, and how they can be improved. They are not immutable laws or prescriptions that dictate how operations *should* be managed, nor are they descriptions that simply explain or categorise issues.
- **Diagnosis** – an approach that questions and explores the fundamental drivers of operations performance. Aims to uncover or 'diagnose' the underlying trade-offs which operations need to overcome and the implications and consequences of the courses of action that could be taken.
- **Practice** – anyone with managerial experience, or who is approaching careers choices, understands the importance of developing practical knowledge and skills that can be applied in practice. This requires an approach, as well as frameworks and techniques, which can be adapted to take account of the complexity and ambiguity of operations, yet give guidance to identifying and implementing potential solutions.

Who should use this book?

This book is intended to provide an introduction to operations and process management for everyone who wishes to understand the nature, principles and practice of the subject. It is aimed primarily at those who have some management experience (although no prior academic knowledge of the area is assumed), or who are about to embark on a career in management. For example:

- *MBA students* should find that its practical discussions of operations management activities enhance their own experience.
- *Postgraduate students* on other specialist masters degrees should find that it provides them with a well-grounded and, at times, critical approach to the subject.
- *Executives* should find its diagnostic structure helps to provide an understandable route through the subject.

What makes this book distinctive?

It has a clear structure

The book is structured on a model of operations management that distinguishes between activities that contribute to the direction, design, delivery and development of operations and processes.

It has significant website resources

Significant supplementary material is fully integrated with the text in order to help 'time poor' experienced students customise their learning to their own needs. This combination of text and integrated supporting material will help to fulfil readers' demanding requirements. The web resources include:

- video-based introduction to each chapter
- active cases that allow the testing and exploration of principles in realistic contexts
- study guides that follow the flow of each chapter and include further examples and animated diagrams
- Excel spreadsheets and examples
- practice notes that provide a step-by-step guide to operations techniques
- hints on how to tackle the 'applying the principles' examples in the text
- 'test yourself' multiple-choice questions
- flashcards for you to test your knowledge of key terms and phrases.

It uses diagnostic logic chains

Every chapter follows a series of questions that forms a 'diagnostic logic' for the topic. These are the questions that anyone can ask to reveal the underlying state of their, or any other, operations. The questions provide an aid to diagnosing where and how an operation can be improved.

It is illustrations-based

Operations management is a practical subject and cannot be taught satisfactorily in a purely theoretical manner. Because of this, each chapter starts with two real-life examples of how the topic is treated in practice

It identifies key operations principles

Whenever a core idea of operations and process management is described in the text, a brief 'operations principle' summary is included in the margin. This helps to distil those essential points of the topic.

It includes critical commentaries

Not everyone agrees about what is the best approach to the various topics and issues within the subject. This is why we have, at the end of each chapter, included a 'critical commentary'. These are alternative views to the one being expressed in the main flow of the text. They do not necessarily represent our view, but they are worth debating.

Each chapter includes a summary checklist

Each chapter is summarised in the form of a list of checklist questions. These cover the essential questions that anyone should ask if they wish to understand the way their own, or any other, operation works. More importantly, they can also act as prompts for operations and process improvement.

Each chapter finishes with a case study

Every chapter includes a case study, relating real or realistic situations that require analysis, decision, or both. The cases have sufficient content to serve as the basis of case sessions in class, but are short enough to serve as illustrations for the less formal reader. As mentioned, further 'active cases' are offered on the website accompanying the book.

Each chapter includes an 'Applying the principles' section

Selected problems, short exercises and activities are included at the end of each chapter. These provide an opportunity to test out your understanding of the principles covered in the chapter.

Each chapter includes a 'Taking it further' section

A short annotated list of further reading and useful websites is provided which takes the topics in the chapter further, or treats some important related issues.

Instructor's manual and PowerPoint slides

Visit www.pearsoned.co.uk/slack to find valuable online resources. A dedicated new web-based instructor's manual is available to lecturers adopting this textbook. It includes teaching notes for all chapters, guided solutions for all case studies in the book, guided solutions for active cases and ideas for teaching with them. A set of PowerPoint slides featuring figures and illustrations from the main text is also available.

Guided tour of the book

2

Operations strategy

Introduction

In the long term, the major (and some would say, only) objective for operations and processes is to provide a business with some form of strategic advantage. That is why the management of a business's processes and operations and its intended overall strategy must be logically connected. Yet for many in business, the very idea of an 'operations strategy' is a contradiction in terms. After all, to be involved in the strategy process is the complete opposite of those day-to-day tasks and activities associated with being an operations manager. Nevertheless, it is also clear that operations can have a real strategic impact. Many *enduringly* remarkable enterprises, from Apple to Zara, use their operations resources to gain long-term strategic success. Such firms have found that it is the way they manage their operations that sets them apart from, and above, their competitors. Without a strong link with overall strategy, operations and processes will be without a coherent direction. And without direction they may finish up making internal decisions that either do not reflect strategy, or that conflict with each other, or both. So, although operations and process management is largely 'operational', it also has a strategic dimension that is vital if operations are to fulfil their potential to contribute to competitiveness. Figure 2.1 shows the position of the ideas described in this chapter in the general model of operations management.

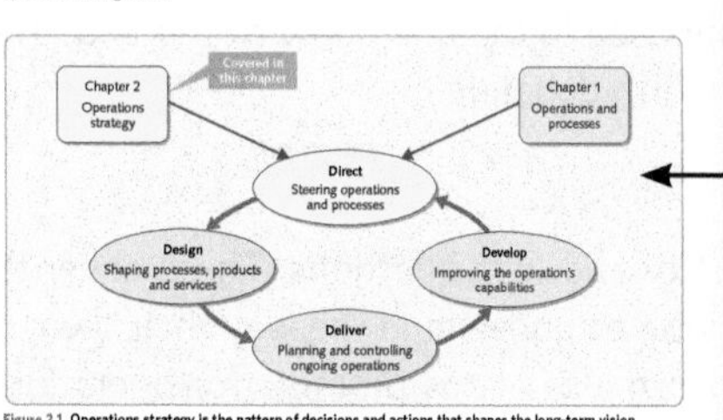

Figure 2.1 **Operations strategy is the pattern of decisions and actions that shapes the long-term vision, objectives and capabilities of the operation and its contribution to overall strategy**

Introduction – every chapter begins with a short introductory paragraph to set the context for the material covered in the chapter.

Clear structure – the book is structured on a model of operations management that distinguishes between activities that contribute to the *direction*, *design*, *delivery*, and *development* of operations and processes.

EXECUTIVE SUMMARY

Decision logic chain for operations strategy

What is operations strategy?
↓
Does the operation have a strategy?
↓
Does operations strategy make sense from the top and the bottom of the business? | Does operations strategy align market requirements with operations resources?
↓
Does operations strategy set an improvement path?

Each chapter is structured around a set of diagnostic questions. These questions suggest what you should ask in order to gain an understanding of the important issues of a topic, and, as a result, improve your decision making. An executive summary addressing these questions is provided below.

What is operations strategy?

Operations strategy is the pattern of decisions and actions that shapes the long-term vision, objectives and capabilities of the operation and its contribution to overall strategy. It is the way in which operations resources are developed over the long term to create sustainable competitive advantage for the business. Increasingly, many businesses are seeing their operations strategy as one of the best ways to differentiate themselves from competitors. Even in those companies that are marketing led (such as fast-moving consumer goods), an effective operations strategy can add value by allowing the exploitation of market positioning.

Does the operation have a strategy?

Strategies are always difficult to identify because they have no presence in themselves, but are identified by the pattern of decisions that they generate. Nevertheless one can identify what an operations strategy should do. First, it should provide a vision for how the operation's resources can contribute to the business as a whole. Second, it should define the exact meaning of the operation's performance objectives. Third, it should identify the broad decisions that will help the operation achieve its objectives. Finally, it should reconcile strategic decision with performance objectives.

Does operations strategy make sense from the top and the bottom of the business?

Operations strategy can been seen both as a top-down process that reflects corporate and business strategy through to a functional level, and as a bottom-up process that allows the experience and learning at an operational level to contribute to strategic

Decision logic chain – every chapter is organised around a series of questions that form a 'decision logic chain' for the topic. This problem-solving approach equips you with the questions to effectively assess the operations and process in your company and diagnose how they can be improved.

Executive summary – a summary answer to the diagnostic questions can be found at the start of each chapter.

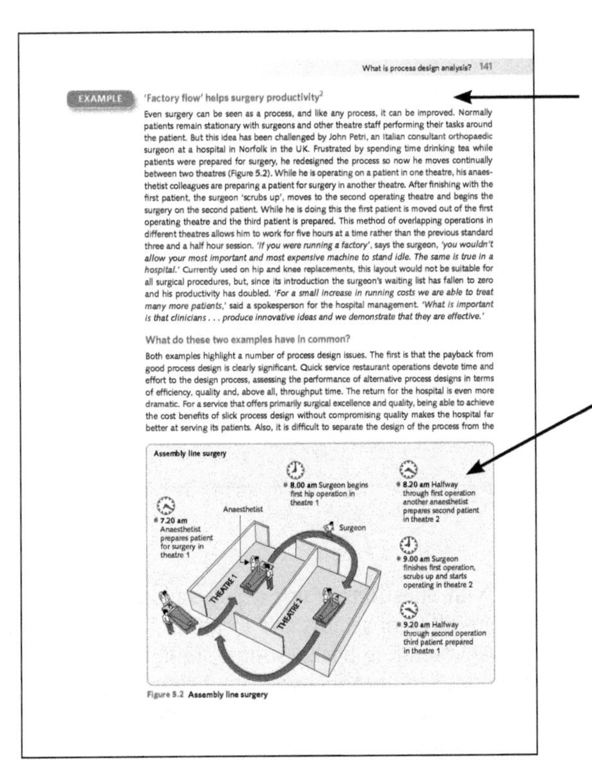
What is process design analysis? 141

EXAMPLE

'Factory flow' helps surgery productivity[2]

Even surgery can be seen as a process, and like any process, it can be improved. Normally patients remain stationary with surgeons and other theatre staff performing their tasks around the patient. But this idea has been challenged by John Petri, an Italian consultant orthopaedic surgeon at a hospital in Norfolk in the UK. Frustrated by spending time drinking tea while patients were prepared for surgery, he redesigned the process so now he moves continually between two theatres (Figure 5.2). While he is operating on a patient in one theatre, his anaesthetist colleagues are preparing a patient for surgery in another theatre. After finishing with the first patient, the surgeon 'scrubs up', moves to the second operating theatre and begins the surgery on the second patient. While he is doing this the first patient is moved out of the first operating theatre and the third patient is prepared. This method of overlapping operations in different theatres allows him to work for five hours at a time rather than the previous standard three and a half hour session. *'If you were running a factory'*, says the surgeon, *'you wouldn't allow your most important and most expensive machine to stand idle. The same is true in a hospital.'* Currently used on hip and knee replacements, this layout would not be suitable for all surgical procedures, but, since its introduction the surgeon's waiting list has fallen to zero and his productivity has doubled. *'For a small increase in running costs we are able to treat many more patients,'* said a spokesperson for the hospital management. *'What is important is that clinicians . . . produce innovative ideas and we demonstrate that they are effective.'*

What do these two examples have in common?

Both examples highlight a number of process design issues. The first is that the payback from good process design is clearly significant. Quick service restaurant operations devote time and effort to the design process, assessing the performance of alternative process designs in terms of efficiency, quality and, above all, throughput time. The return for the hospital is even more dramatic. For a service that offers primarily surgical excellence and quality, being able to achieve the cost benefits of slick process design without compromising quality makes the hospital far better at serving its patients. Also, it is difficult to separate the design of the process from the

Figure 5.2 Assembly line surgery

Examples – each chapter opens with two examples. A balance of service and manufacturing examples gives you a practical and wide-ranging understanding of operations and process management.

Figures and diagrams are featured throughout the text to highlight key points and clarify topics discussed.

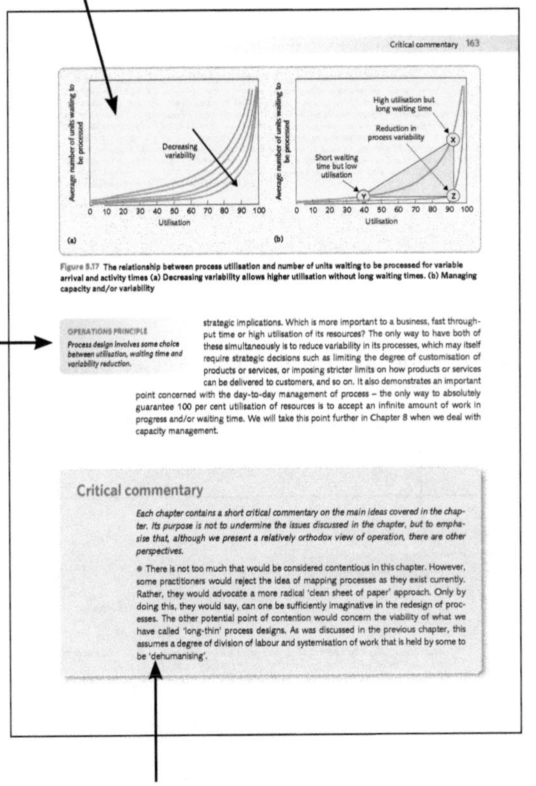
Critical commentary 163

Figure 5.17 The relationship between process utilisation and number of units waiting to be processed for variable arrival and activity times (a) Decreasing variability allows higher utilisation without long waiting times. (b) Managing capacity and/or variability

OPERATIONS PRINCIPLE

Process design involves some choice between utilisation, waiting time and variability reduction.

strategic implications. Which is more important to a business, fast throughput time or high utilisation of its resources? The only way to have both of these simultaneously is to reduce variability in its processes, which may itself require strategic decisions such as limiting the degree of customisation of products or services, or imposing stricter limits on how products or services can be delivered to customers, and so on. It also demonstrates an important point concerned with the day-to-day management of process – the only way to absolutely guarantee 100 per cent utilisation of resources is to accept an infinite amount of work in progress and/or waiting time. We will take this point further in Chapter 8 when we deal with capacity management.

Critical commentary

Each chapter contains a short critical commentary on the main ideas covered in the chapter. Its purpose is not to undermine the issues discussed in the chapter, but to emphasise that, although we present a relatively orthodox view of operation, there are other perspectives.

- There is not too much that would be considered contentious in this chapter. However, some practitioners would reject the idea of mapping processes as they exist currently. Rather, they would advocate a more radical 'clean sheet of paper' approach. Only by doing this, they would say, can one be sufficiently imaginative in the redesign of processes. The other potential point of contention would concern the viability of what we have called 'long-thin' process designs. As was discussed in the previous chapter, this assumes a degree of division of labour and systemisation of work that is held by some to be 'dehumanising'.

Operations principles – key concepts are distilled and presented in the margin of the text as a useful reference and reminder.

204 Chapter 6 • Designing the innovation process

SUMMARY CHECKLIST

This checklist comprises questions that can be usefully applied to any type of operations and reflect the major diagnostic questions used within the chapter.

- ☐ Is the importance of innovation as a contributor to achieving strategic impact fully understood?
- ☐ Is innovation really treated as a process?
- ☐ Is the innovation process itself designed with the same attention to detail as any other process?
- ☐ Are innovation objectives specified so as to give a clear priority between quality, speed, dependability, flexibility, cost and sustainability?
- ☐ Are the stages in the innovation process clearly defined?
- ☐ Are ideas and concepts for new offerings captured from all appropriate internal and external sources?
- ☐ Are potential offerings screened in a systematic manner in terms of their feasibility, acceptability and vulnerability?
- ☐ During preliminary design, have all possibilities for design standardisation, commonality and modularisation of design elements been explored?
- ☐ Has the concept of mass customisation been linked to the innovation process?
- ☐ Are potential offerings thoroughly evaluated and tested during the innovation process?
- ☐ Are the resources for developing innovation adequate?
- ☐ Is sufficient capacity devoted to the innovation process?
- ☐ Have all options for outsourcing parts of the innovation process been explored?
- ☐ Has the possibility of involving customers formally in the innovation process been explored?
- ☐ Are appropriate technologies such as CAD, extensions of CAD, and knowledge management, being used in the innovation process?
- ☐ Are the design of offering and the design of processes that create and deliver it considered together as one integrated process?
- ☐ Is overlapping (simultaneous) design of the stages in the innovation process used?
- ☐ Is senior management effort deployed early enough to ensure early resolution of design conflict?
- ☐ Does the organisational structure of the innovation process reflect the nature of the offering?
- ☐ Are some functions of the business more committed to innovating new service and product offerings than others?
- ☐ If so, have the barriers to cross-functional commitment been identified and addressed?

Critical commentaries provide an alternative viewpoint to the perspective presented in the main text.

Summary checklists present the essential questions that anyone should ask if they wish to understand how an operation works. They are located at the end of each chapter and reflect the major diagnostic questions used within the chapter.

Case study: Slagelse Industrial Services (SIS) 491

CASE STUDY

Slagelse Industrial Services (SIS)[11]

Slagelse Industrial Services (SIS) had become one of Europe's most respected die casters of zinc, aluminum and magnesium parts suppliers for hundreds of companies in many industries, especially automotive and defence. The company cast and engineered precision components by combining the most modern production technologies with precise tooling and craftsmanship. Slagelse Industrial Services (SIS) began life as a classic family firm run by Erik Paulsen, who opened a small manufacturing and die-casting business in his hometown of Slagelse, a town in east Denmark, about 100 km southwest of Copenhagen. He had successfully leveraged his skills and passion for craftsmanship over many years while serving a variety of different industrial and agricultural customers. His son, Anders had spent nearly ten years working as a production engineer for a large automotive parts supplier in the UK, but eventually returned to Slagelse to take over the family firm. Exploiting his experience in mass manufacturing, Anders spent years building the firm into a larger scale industrial component manufacturer but retained his father's commitment to quality and customer service. After 20 years he sold the firm to a UK-owned industrial conglomerate and within ten years it had doubled in size again and now employed in the region of 600 people and had a turnover approaching £200 million. Throughout this period the firm had continued to target their products into niche industrial markets where their emphasis upon product quality and dependability meant they were less vulnerable to price and cost pressures. However in 2009, in the midst of difficult economic times and widespread industrial restructuring, they had been encouraged to bid for higher volume, lower margin work. This process was not very successful but eventually culminated in a tender for the design and production of a core metallic element of a child's toy (a 'transforming' robot).

Interestingly, the client firm, Alden Toys, was also a major customer for other businesses owned by SIS's corporate parent. They were adopting a preferred supplier policy and intended to have only one or two purchase points for specific elements in their global toy business. They had a high degree of trust in the parent organisation and on visiting the SIS site were impressed by the firm's depth of experience and commitment to quality. In 2010, they selected SIS to complete the design and begin trial production.

Source: EM Clements Photography

'Some of us were really excited by the prospect . . . but you have to be a little worried when volumes are much greater than anything you've done before. I guess the risk seemed okay because in the basic process steps, in the type of product if you like, we were making something that felt very similar to what we'd been doing for many years.' (SIS Operations Manager)

'Well obviously we didn't know anything about the toy market but then again we didn't really know all that much about the auto industry or the defence sector or any of our traditional customers before we started serving them. Our key competitive advantage, our capabilities, call it what you will, they are all about keeping the customer happy, about meeting and sometimes exceeding specification.' (SIS Marketing Director)

The designers had received an outline product specification from Alden Toys during the bid process and some further technical detail afterwards. Upon receipt of this final brief, a team of engineers and managers confirmed that the product could and would be manufactured using a scaled-up version of current production processes. The key operational challenge appeared to be accessing sufficient (but not too much) capacity. Fortunately, for a variety of reasons, the parent company was very supportive of the project and promised to underwrite any sensible capital expenditure plans. Although this opinion of the production challenge was widely accepted throughout the firm (and shared by Alden Toys and SIS's parent group) it was left to one specific senior engineer to actually sign both the final bid and technical completion documentation. By

Applying the principles 207

(such as the purchase of test equipment, etc.). Monica was unsure whether the budget would be big enough.

'I know everyone in my position wants more money, but it is important not to under-fund a project like this. Increasing our development staff by 70 per cent is not really enough. In my opinion we need an increase of at least 90 per cent to make sure that we can launch when we want. This would need another $5m, spread over the next 20 months. We could get this by not building the pilot plant I suppose, but I am reluctant to give that up. It would mean begging for test capacity on other companies' plants, which is never satisfactory from a knowledge-building viewpoint. Also it would compromise security. Knowledge of what we were doing could easily leak to competitors. Alternatively we could subcontract more of the research which may be less expensive, especially in the long run, but I doubt if it would save the full $5m we need. More important, I am not sure that we should subcontract anything which would compromise safety, and increasing the amount of work we send out may do that. No, it's got to be the extra cash or the project could overrun. The profit projections for the Orlando products look great, (see Table 6.2), but delay or our inability to respond to competitor pressures would depress those figures significantly. Our competitors could get into the market only a little after us. Word has it that marketing's calculations indicate a delay of only six months could not only delay the profit stream by the six months but also cut it by up to 30 per cent.'

Table 6.2 Preliminary 'profit stream' projections for the Project Orlando offering, assuming launch in 24 months time

Time period*	1	2	3	4	5	6	7
Profit flow ($ million)	10	20	50	90	120	130	135

*six-month periods

Monica was keen to explain two issues to the management committee when it met to consider her request for extra funding. First, that there was a coherent and well thought-out strategy for the innovation project over the next two years. Second, that saving $5m on Project Orlando's budget would be a false economy.

QUESTIONS

1. How would you rank the innovation objectives for the project?
2. What are the key issues in resourcing this innovation process?
3. What are the main factors influencing the resourcing decisions?
4. What advice would you give Monica?

Case studies with questions which depict various-sized organisations in different sectors and locations, from both service and manufacturing backgrounds, are included at the end of every chapter. Apply your understanding of concepts and techniques by addressing operations-related problems in a real business context.

Applying the principles contain selected problems, short exercises and activities to test your understanding of the principles covered in each chapter.

Supplements covering quantitative topics appear at the end of relevant chapters.

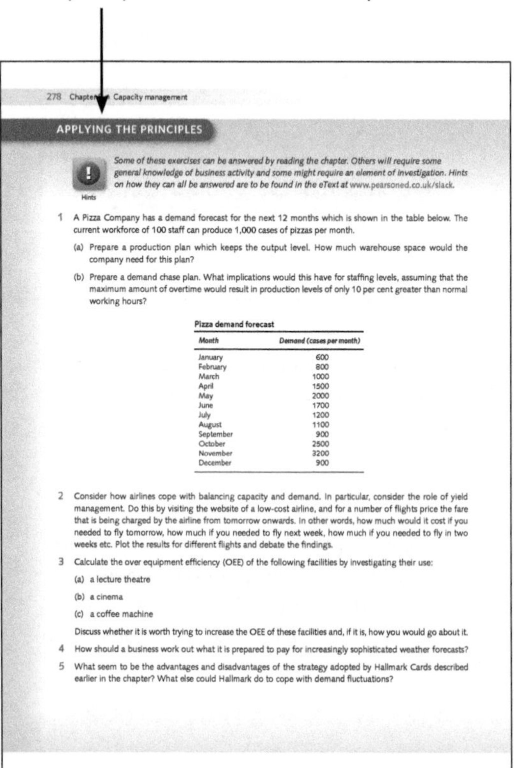

278 Chapter Capacity management

APPLYING THE PRINCIPLES

Some of these exercises can be answered by reading the chapter. Others will require some general knowledge of business activity and some might require an element of investigation. Hints on how they can all be answered are to be found in the eText at www.pearsoned.co.uk/slack.

Hints

1. A Pizza Company has a demand forecast for the next 12 months which is shown in the table below. The current workforce of 100 staff can produce 1,000 cases of pizzas per month.
   (a) Prepare a production plan which keeps the output level. How much warehouse space would the company need for this plan?
   (b) Prepare a demand chase plan. What implications would this have for staffing levels, assuming that the maximum amount of overtime would result in production levels of only 10 per cent greater than normal working hours?

Pizza demand forecast

Month	*Demand (cases per month)*
January	600
February	800
March	1000
April	1500
May	2000
June	1700
July	1200
August	1100
September	900
October	2500
November	3200
December	900

2. Consider how airlines cope with balancing capacity and demand. In particular, consider the role of yield management. Do this by visiting the website of a low-cost airline, and for a number of flights price the fare that is being charged by the airline from tomorrow onwards. In other words, how much would it cost if you needed to fly tomorrow, how much if you needed to fly next week, how much if you needed to fly in two weeks etc. Plot the results for different flights and debate the findings.
3. Calculate the over equipment efficiency (OEE) of the following facilities by investigating their use:
   (a) a lecture theatre
   (b) a cinema
   (c) a coffee machine

   Discuss whether it is worth trying to increase the OEE of these facilities and, if it is, how you would go about it.
4. How should a business work out what it is prepared to pay for increasingly sophisticated weather forecasts?
5. What seem to be the advantages and disadvantages of the strategy adopted by Hallmark Cards described earlier in the chapter? What else could Hallmark do to cope with demand fluctuations?

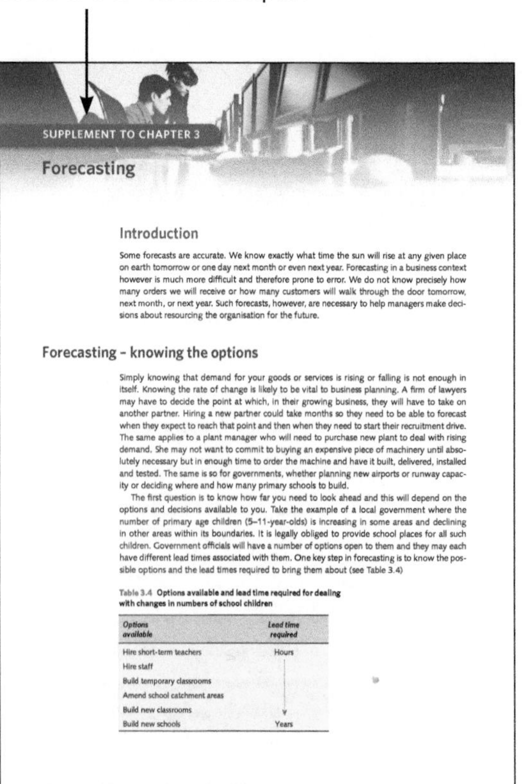

SUPPLEMENT TO CHAPTER 3

Forecasting

Introduction

Some forecasts are accurate. We know exactly what time the sun will rise at any given place on earth tomorrow or one day next month or even next year. Forecasting in a business context however is much more difficult and therefore prone to error. We do not know precisely how many orders we will receive or how many customers will walk through the door tomorrow, next month, or next year. Such forecasts, however, are necessary to help managers make decisions about resourcing the organisation for the future.

Forecasting – knowing the options

Simply knowing that demand for your goods or services is rising or falling is not enough in itself. Knowing the rate of change is likely to be vital to business planning. A firm of lawyers may have to decide the point at which, in their growing business, they will have to take on another partner. Hiring a new partner could take months so they need to be able to forecast when they expect to reach that point and then when they need to start their recruitment drive. The same applies to a plant manager who will need to purchase new plant to deal with rising demand. She may not want to commit to buying an expensive piece of machinery until absolutely necessary but in enough time to order the machine and have it built, delivered, installed and tested. The same is so for governments, whether planning new airports or runway capacity or deciding where and how many primary schools to build.

The first question is to know how far you need to look ahead and this will depend on the options and decisions available to you. Take the example of a local government where the number of primary age children (5–11-year-olds) is increasing in some areas and declining in other areas within its boundaries. It is legally obliged to provide school places for all such children. Government officials will have a number of options open to them and they may each have different lead times associated with them. One key step in forecasting is to know the possible options and the lead times required to bring them about (see Table 3.4)

Table 3.4 Options available and lead time required for dealing with changes in numbers of school children

Options available	*Lead time required*
Hire short-term teachers	Hours
Hire staff	↓
Build temporary classrooms	↓
Amend school catchment areas	↓
Build new classrooms	↓
Build new schools	Years

Guided tour of the Pearson eText

Operations and Process Management, Third Edition is available with a Pearson eText, which can be accessed at www.pearsoned.co.uk/slack. You can access the Pearson eText either by redeeming an access card or by purchasing access online.

Pearson eTexts gives you access to the text when and wherever you have access to the Internet.

eText pages look exactly like the printed text, offering powerful new functionality for students and instructors. Users can create notes, highlight text in different colours, create bookmarks, zoom, and view in single-page or two-page view.

Pearson eText allows for quick navigation to key parts of the eText using a table of contents and provides full-text search.

The **toolbar** allows you navigate through the eText, add notes or bookmarks, access the glossary, search for text, change settings and access the download manager.

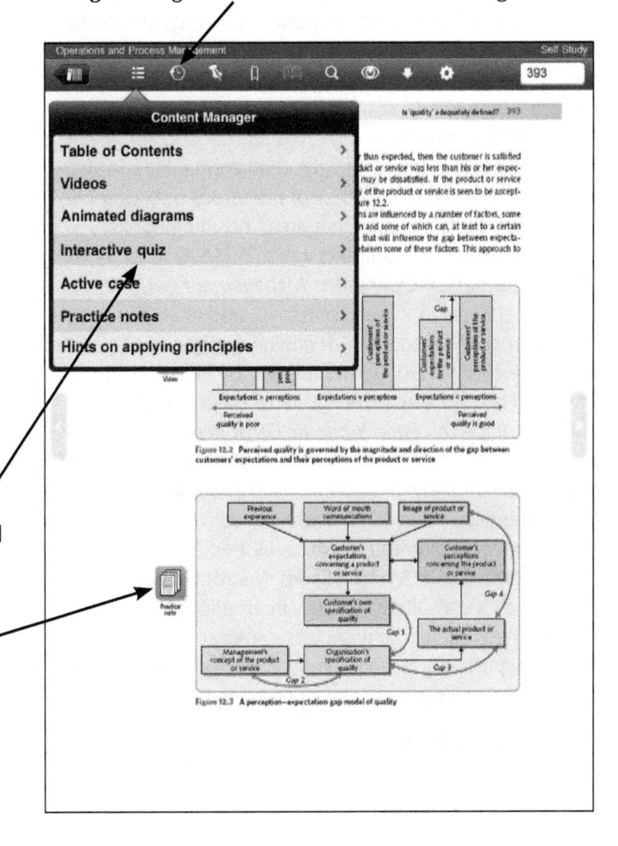

The table of contents and all of the associated media files can be accessed through the **Content Manager**.

Icons such as this Practice Note icon link directly to their associated media files.

The eText also offers links to associated media files wherever you see the following symbols:

Video – this icon appears at the start of every diagnostic question. The icon links to a short video of a member of the author team introducing the question and explaining how it can relate to your business.

Animated Diagram – this icon appears next to a selection of the diagrams in the text. The icon links to an animated version of the diagram with commentary providing an explanation of the concepts.

Practice Note – this icon appears when selected operations techniques are introduced. The icon links to a step-by-step guide which will show you how to apply the technique.

Hints – this icon appears at the start of the Applying the Principles box in every chapter. The icon links to hosts that will help you to solve the problems in this section.

Interactive Quiz – this icon appears at the end of every chapter. The icon links to a multiple-choice quiz that will test your understanding of the key concepts covered in that chapter.

Use your Pearson eText to get the most from *Operations and Process Management*, Third Edition.

About the authors

NIGEL SLACK is Professor of Operations Management and Strategy at Warwick University. Previously he was Professor of Service Engineering at Cambridge University, Professor of Manufacturing Strategy and Lucas Professor of Manufacturing System Engineering at Brunel University, University Lecturer in Management Studies at Oxford University and Fellow in Operations Management at Templeton College, Oxford. He worked initially as an industrial apprentice in the hand-tool industry and then as a production engineer and production manager in light engineering. He is a chartered engineer, and the author of numerous publications in the operations management area, including books, academic papers and chapters in books. Most recently, in 2011, *Essentials of Operations Management* (with Alistair Brandon-Jones and Robert Johnston), and in 2009, *Operations Management*, 6th edition (with Stuart Chambers and Robert Johnston), both published by Financial Times Prentice Hall. He is also the author of other books including, *The Blackwell Encyclopaedic Dictionary of Operations Management*, 2nd edition, published by Blackwell in 2005; *Operations Strategy, 3rd edition*, published by Financial Times Prentice Hall in 2011, and *Perspectives in Operations Management* (Volumes I to IV), published by Routledge in 2003, all three with Michael Lewis of Bath University. He also acts as a consultant to many international companies around the world in many sectors, especially financial services, transport, leisure and manufacturing. His research is in the process management and operations strategy areas.

ALISTAIR BRANDON-JONES is a Senior lecturer in Operation Management at the University of Bath School of Management, and a visiting lecturer at Warwick Medical School. Previously, he was a Teaching Fellow at Warwick Business School and also worked in a number of logistics and retail roles. He has a Bachelor's degree in Management Science and a Doctorate in Business from the University of Warwick and is widely published in leading operations and supply chain management journals. *Operations and Process Management* is his third text (he is co-author for *Essentials of Operations Management* with Nigel Slack and Robert Johnston, and *Quantitative Analysis in Operations Management* with Nigel Slack, both published by Financial Times Prentice Hall). Alistair is a leading authority on the role of users in the successful implementation of electronic procurement systems within and across organisations. He is also the UK-lead for the International Purchasing Survey, a collaboration of supply management researchers at leading European and American institutions. More recently, Alistair has been researching how organisations deal with service failure, both in terms of recovering the customer and in using failure for operational improvement. Alistair has consulting and executive development experience with organisations around the world, in various sectors including petrochemicals, health, financial services, manufacturing, defence, and government.

ROBERT JOHNSTON is Professor of Operations Management at Warwick Business School. He is the founding editor of the *International Journal of Service Industry Management* and he also serves on the editorial board of the *Journal of Operations Management* and the *International Journal of Tourism and Hospitality Research*. Before moving to academia Dr Johnston held several line management and senior management posts in a number of service organisations on both the public and private sectors. He continues to maintain close and active links with many large and small organisations through his research, management training and consultancy activities. As a specialist in service operations, his research interests include service design, service recovery, performance measurement and service quality. He is the author of *Service Operations Management*, 3rd edition, with Graham Clark, published by Financial Times Prentice Hall in 2009, and numerous other publications in the service and general operations management area.

ALAN BETTS is a freelance consultant and trainer working, primarily, with executives of service organisations to apply the principles of operations and process management. Following a career in financial services, Alan completed Institute of Personnel and Development qualifications and an MA in Human Resource Management and moved to the Operations Management group at Warwick Business School as a Senior Research Fellow. His chief interests are the development of innovative approaches to e-learning and m-learning, together with the coaching and development of managers and executives. Alan is a director of Bedford Falls Learning Limited, HT2 Limited and Capability Development Limited. He is a visiting Professor at the University of San Diego and is a Fellow of the Royal Society of Arts.

Acknowledgements

In preparing this book, the authors, as usual, unashamedly exploited their friends and colleagues. In particular we had invaluable help from a great and distinguished reviewer team and for colleagues who have provided valuable feedback on various aspects of the project. For their help we are particularly grateful to:

Pär Åhlström, Stockholm School of Economics, Sweden,
Malcolm Afferson, Sheffield Hallam University, UK
David Bamford, Manchester Business School, University of Manchester, UK
Umit Bitici, Strathclyde University, UK
Briony Boydell, Portsmouth University, UK
Paul Coughlan, Trinity College Dublin, Ireland
Marc Day, Reading University, UK
Stephen Disney, Cardiff University, UK
Des Doran, Kingston University, UK
Paul Forrester, Birmingham University, UK
Gino Franco, Derby University, UK
Abby Ghobadian, Reading University, UK
Roger Hall, Huddersfield University, UK
Ingjaldur Hannibalsson, University of Iceland, Iceland
Matthias Holweg, Cambridge University, UK
Koos Krabbendam, University of Twente, The Netherlands
Michael Lewis, Bath University, UK
Bart MacCarthy, Nottingham University, UK
John Maguire, Sunderland University, UK
Harvey Maylor, Cranfield University, UK
Duncan McFarlane, Cambridge University, UK
Ronnie Mcmillan, University of Strathclyde, UK
Andrea Masini, London Business School, UK
Phil Morgan, Oxford Brookes University, UK
Andy Neely, Cambridge University, UK
Steve New, Oxford University, UK
Peter Race, Reading University, UK
Alison Smart, Manchester Business School, University of Manchester, UK
Nigel Spinks, Reading University, UK
Martin Spring, Lancaster University, UK
Venu Venugopal, Nyenrode University, The Netherlands
Zoe Radnor, Cardiff University, UK
Jan de Vries, University of Groningen, The Netherlands
Graham Walder, Wolverhampton University, UK
Helen Walker, Cardiff University, UK
Maggie Zeng, Gloucestershire University

Our academic colleagues in the Operations Management Group at Warwick Business School and the School of Management at Bath University also helped, both by contributing ideas and by creating a lively and stimulating work environment. In particular Emma Brandon-Jones, Researcher in the School of Management at Bath University provided many useful ideas.

Mary Walton is co-ordinator to our group at Warwick Business School. Her continued efforts at keeping us organised (or as organised as we are capable of being) are always appreciated, but never more so than when we were engaged on this book.

The prerequisite for any book of this type is that it serves a real market need. We were privileged to receive advice from the Pearson team; some of the most insightful people in business education publishing, in particular, Rufus Curnow (Senior acquisitions editor), Joy Cash (Senior editor), Robert Sykes (Editorial assistant), Colin Reed (Senior designer – text), Michelle Morgan (Senior designer – cover), Caterina Pellegrino (Senior production controller), Mel Beard (Media producer), Katie Eyles (Media editor), Rachel Chiles (Media editor) and Najette Hunte (Marketing manager).

Our thanks also go to Janey Webb at Pearson Education. Although not directly involved with this edition, without her considerable effort, enthusiasm, common sense, and professional dedication, the first edition of this project would have been significantly impaired.

Finally, the manuscript, and much more besides, was organised and put together by Angela Slack. It was another heroic effort, which she undertook with (relatively) little complaint. To Angela – our thanks.

Picture credits

The publisher would like to thank the following for their kind permission to reproduce their photographs:

(Key: b-bottom; c-centre; l-left; r-right; t-top)

Air France: Media Library 316; **Alamy Images:** Alex Segre 380, Andrew Woodley 140, Chad Ehlers 92, David Pearson 61, Hemis 5br (f), Ian Dagnall 469, Jim Holden 266, Kevin Foy 241, Marka 5tl (a), Mediablitzimages (UK) Ltd 205, Mike Booth 408, Mira 183bl, North Wind Picture Archives 196, Peter Titmus 390, Tristar Photos 8, Vario Images GmbH & Co 7tl, World of Asia 428, Yang Yu 27; **Corbis:** Eleanor Bentall 395, Fancy / Veer 273, Jean Phillipe Arles 499, Keith Dannemiller 36, Michael Pole 5cr (d), Paul Drabble 287, William Taufic 521; **E M Clements Photography:** 491; **Exel:** 308; **Getty Images:** 165, 183tl, AFP 37, Bloomberg 215, 430, China Photos 500, Digital Vision 337, 458, Gaby Messina 251, Janet Kimber 132, Jim Lai / AFP 7cl, Keith Brofsky / Photodisc 286, Matt Mawson 5tr (b), Tosh Fumi Kitamura 75, Upper Cut Images 357bl; **Images courtesy of The Advertising Archives:** 83, 217; **Mutiara Beach Resort, Penang:** 252; **Nigel Slack:** 5cl (c), 317; **Photographers Direct/Ottmar Beirwagen:** 388; **PO Box 814:** 357tl; **Rex Features:** Geoffrey Robinson 113, Sipa Press 5bl (e), 373; **Science Photo Library Ltd:** Simon Fraser 470; **Space 4 Housing:** 114; **Wincanton Transport:** 73

Cover image: *Front:* **Getty Images:** Scott E Barbour

All other images © Pearson Education

Every effort has been made to trace the copyright holders and we apologise in advance for any unintentional omissions. We would be pleased to insert the appropriate acknowledgement in any subsequent edition of this publication.

1 Operations and processes

Introduction

Operations and process management is about how organisations create goods and services. Everything you wear, eat, use or read comes to you courtesy of the operations managers who organised its creation, as does every bank transaction, hospital visit and hotel stay. The people who created them may not always be called operations managers, but that is what they really are. Within the operations function of any enterprise, operations managers look after the processes that create products and services. But managers in other functions, such as Marketing, Sales and Finance, also manage processes. These processes often supply internal 'customers' with services such as marketing plans, sales forecasts, budgets, and so on. In fact, parts of all organisations are made up of processes. That is what this book is about – the tasks, issues and decisions that are necessary to manage processes effectively, both within the operations function, and in other parts of the business where effective process management is equally important. This is an introductory chapter, so we will examine some of the basic principles of operations and process management. The model that is developed to explain the subject is shown in Figure 1.1.

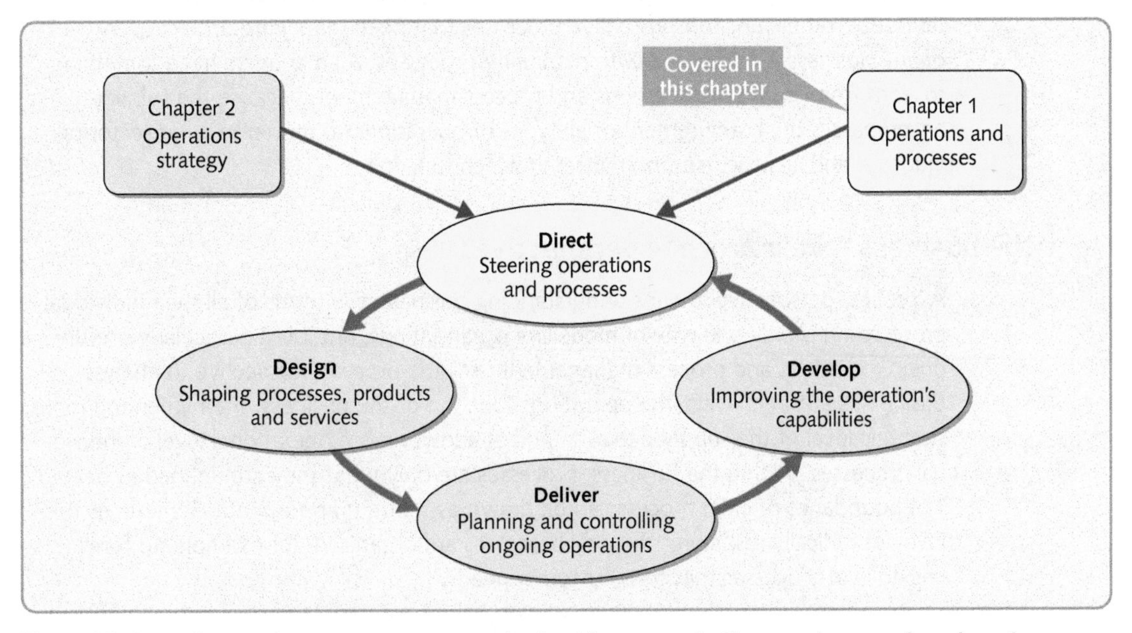

Figure 1.1 Operations and process management is about how organisations produce goods and services

EXECUTIVE SUMMARY

Decision logic chain for operations and processes

What is operations and process management?

↓

Does the business take a process perspective?

↓

Does operations and process management have a strategic impact?

↓

Should all processes be managed in the same way?

Each chapter is structured around a set of diagnostic questions. These questions suggest what you should ask in order to gain an understanding of the important issues of a topic, and, as a result, improve your decision making. An executive summary addressing these questions is provided below.

What is operations and process management?

The operations function is the part of the organisation that produces products or services. Every organisation has an operations function because every organisation produce some mixture of products and services. 'Operations' is not always called by that name, but whatever its name, it is always concerned with managing the core purpose of the business – producing some mix of products and services. Processes also produce products and services, but on a smaller scale. They are the component parts of operations. But, other functions also have processes that need managing. In fact *every* part of *any* business is concerned with managing processes. All managers have something to learn from studying operations and process management, because the subject encompasses the management of all types of operation, no matter in what sector or industry, and all processes, no matter in which function.

Does the business take a process perspective?

A 'process perspective' means understanding businesses in terms of all their individual processes. It is only one way of modelling organisations, but it is a particularly useful one. Operations and process management uses the process perspective to analyse businesses at three levels: the operations function of the business, the higher and more strategic level of the supply network, and at a lower more operational level of individual processes. Within the business, processes are only what they are defined as being. The boundaries of each process can be drawn as thought appropriate. Sometimes this involves radically reshaping the way processes are organised, for example, to form end-to-end processes that fulfil customer needs.

Does operations and process management have a strategic impact?

Operations and process management can make or break a business. When they are well managed, operations and processes can contribute to the strategic impact of the business in four ways: cost, revenue, investment and capabilities. Because the operations function has responsibility for much of a business's cost base, its first imperative is to keep costs under control. But also, through the way it provides service and quality, it should be looking to enhance the business's ability to generate revenue. Furthermore, as all failures are ultimately process failures, well-designed processes should have less chance of failing and more chance of recovering quickly from failure. Also, because operations are often the source of much investment, it should be aiming to get the best possible return on that investment. Finally, the operations function should be laying down the capabilities that will form the long-term basis for future competitiveness.

Should all processes be managed in the same way?

Not necessarily. Processes differ, particularly in what are known as the four Vs: volume, variety, variation and visibility. High-volume processes can exploit economies of scale and be systematised. High-variety processes require enough inbuilt flexibility to cope with the wide variety of activities expected of them. High-variation processes must be able to change their output levels to cope with highly variable and/or unpredictable levels of demand. High-visibility processes add value while the customer is 'present' in some way and therefore must be able to manage customers' perceptions of their activities. Generally, high volume together with low variety, variation and visibility facilitate low-cost processes, while low volume together with high levels of variety, variation and visibility all increase process costs. Yet in spite of these differences, operations managers use a common set of decisions and activities to manage them. These activities can be clustered under four groupings: directing the overall strategy of the operation; designing the operation's products, services and processes; planning and controlling process delivery; developing process performance.

Operations concerned with managing the core purpose of the business

DIAGNOSTIC QUESTION

What is operations and process management?

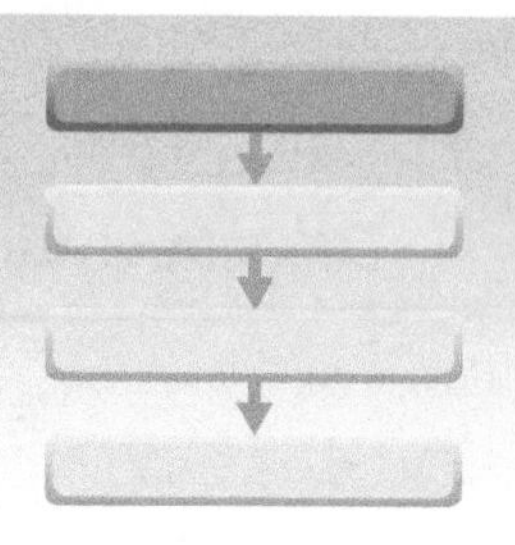

Operations and process management is the activity of managing the resources and processes that produce products and services. The core body of knowledge for the subject comes from 'operations management', which examines how the 'operations function' of a business produces products and services for external customers. We also use the shorter terms 'the operation' or 'operations', interchangeably with the 'operations function'. In some organisations an operations manager could be called by some other name, for example, a 'fleet manager' in a logistics company, an 'administrative manager' in a hospital, or a 'store manager' in a supermarket.

All business have 'operations', because all businesses produce products, services or a mixture of both. If you think that you don't have an operations function, you are wrong. If you think that your operations function is not important, you are also wrong. Look at the six businesses illustrated in Figure 1.2. There are two financial service companies, two manufacturing companies and two hotels. All of them have *operations functions* that produce the things that their customers are willing to pay for. Hotels produce accommodation services, financial services invest, store, move or sell us money and investment opportunities, and manufacturing businesses physically change the shape and the nature of materials to produce products. These businesses are from different sectors (banking, hospitality and manufacturing), but the main difference between their operations activities is not necessarily what one expects. There are often bigger differences *within* economic sectors than *between* them. All the three operations in the left-hand column provide value-for-money products and services and compete largely on cost. The three in the right-hand column provide more 'up-market' products and services that are more expensive to produce and compete on some combination of high specification and customisation. The implication of this is important. It means that the surface appearance of a business and its economic sector are less important to the way its operations should be managed than its intrinsic characteristics, such as the volume of its output, the variety of different products and services it needs to produce, and, above all, how it is trying to compete in its market.

OPERATIONS PRINCIPLE
All organisations have 'operations' that produce some mix of products and services.

OPERATIONS PRINCIPLE
The economic sector of an operation is less important in determining how it should be managed than its intrinsic characteristics.

Operations *and process* management

Within the operations shown in Figure 1.2, resources such as people, information systems, buildings and equipment will be organised into several individual 'processes'. A 'process' is an arrangement of resources that transforms inputs into outputs that satisfy (internal or external) customer needs. So, among other processes, banking operations contain account management processes, hotel operations contain room cleaning processes, furniture manufacturing operations contain assembly processes, and so on. The difference between *operations* and *processes* is one of scale, and therefore complexity. Both transform inputs into outputs, but processes are the smaller version. They are the component parts of operations, so the total operations function is made up of individual processes. But, within any business, the production of products and services is not confined to the operations function. For example, the marketing function 'produces' marketing plans and sales forecasts, the accounting function 'produces' budgets, the human resources function 'produces' development and recruitment

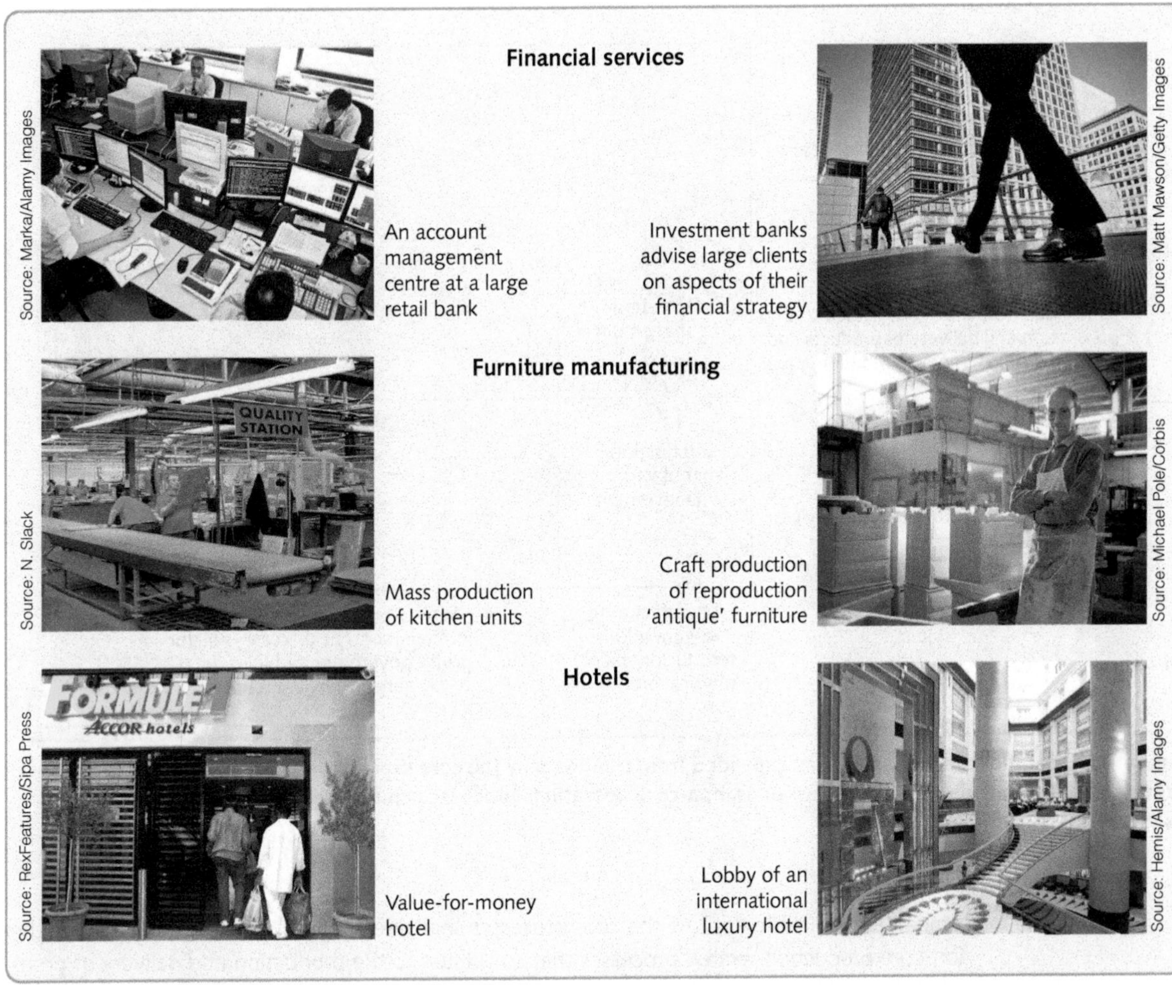

Figure 1.2 All types of business have 'operations' because all businesses produce some mix of products and services. The differences in the operations *within* a category of business are often greater than the differences *between* businesses

plans, and so on. In fact *every* part of *any* business is concerned with managing processes. So, 'operations and process management' is the term we use to encompass the management of all types of operation, no matter in what sector or industry, and all processes, no matter in which function of the business. The general truth is that processes are everywhere, and all types of manager have something to learn from studying operations and process management.

From 'production', to 'operations', to 'operations and process' management

Figure 1.3 illustrates how the scope of this subject has expanded. Originally, operations management was seen as very much associated with the manufacturing sector. In fact, it would have been called 'production' or 'manufacturing' management, and was concerned exclusively with the core business of producing physical products. Starting in the 1970s and 1980s, the term *operations management* became more common. It was used to reflect two trends. First, and most importantly, it was used to imply that many of the ideas, approaches and techniques traditionally used in the manufacturing sector could be equally applicable in the production of services. The second use of the term was to expand the scope of 'production' in manufacturing

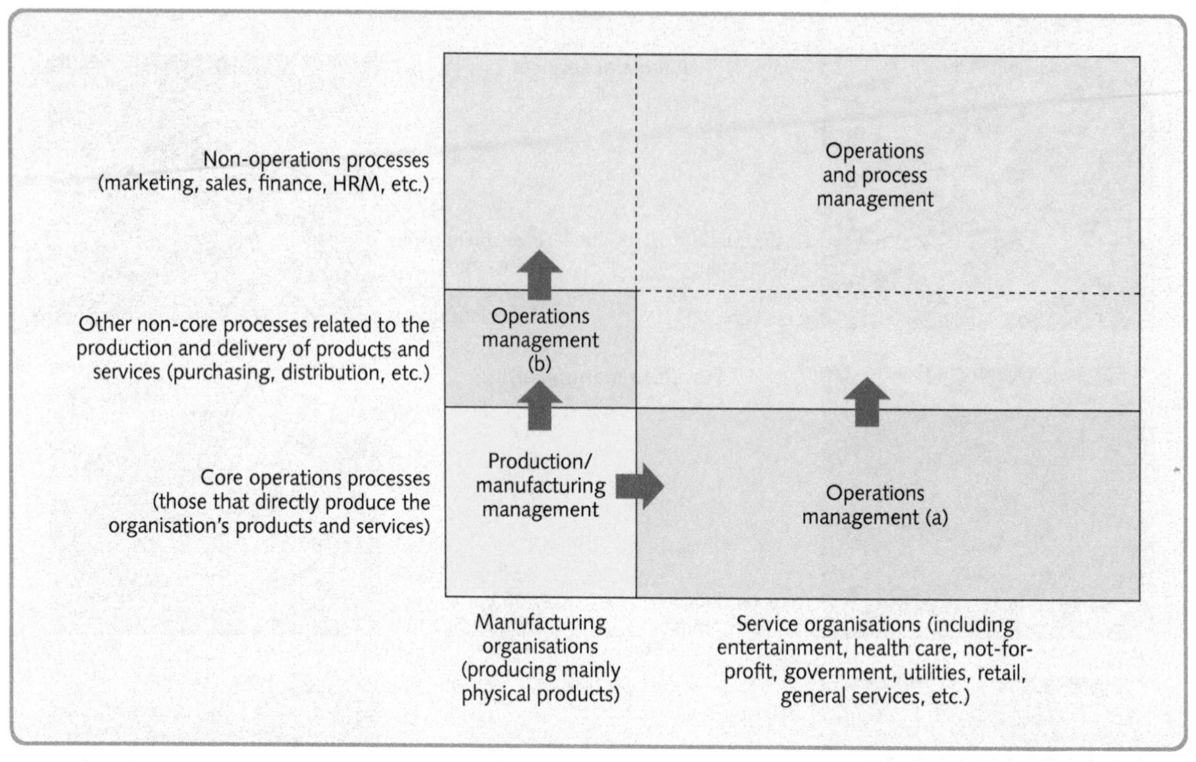

Figure 1.3 Operations management has expanded from treating only the core production processes in manufacturing organisations to include service organisations, non-core operations processes and processes in other functions such as marketing, finance and HRM

companies to include not just the core processes that directly produce products, but also the non-core production-related processes that contribute to the production and delivery of product. This would include such processes as purchasing, physical distribution, after-sales service, and so on. More recently the term *operations and process management* (or sometimes just *process management*) has been used to denote the shift in the scope of the subject to include the whole organisation. It is a far wider term than operations management because it applies to all parts of the organisation. This is very much how we treat the subject in this book. That is why it is called 'Operations and *Process* Management'. It includes the examination of the operations function in both manufacturing and service sectors and also the management of processes in non-operations functions.

Towards the beginning of all chapters we present two examples of individual businesses, or types of business, that illustrate the topic being examined in the chapter. Here we look at two businesses, one service company, and one manufacturing company, which have succeeded in large part because of their effective use of operations and process management principles.

EXAMPLE

IKEA[1]

It's the school holidays in an IKEA superstore in London, and parents with children in tow are crowding the isles and café area. Families push their way past the 'Anniversary Edition' of the 'Billy' bookcase (one of the firm's most popular products) proudly displayed by the entrance. Looking at the crowds it is easy to believe that IKEA is the most successful furniture retailer ever. With stores all over the world, they have managed to develop their own special way of

Source: Vario Images GmbH & Co./Alamy

Source: Jim Lai/AFP/Getty

selling furniture. Their stores' layout means customers often spend two hours in the store – far longer than in rival furniture retailers. IKEA's philosophy goes back to the original business, started in the 1950s in Sweden by Ingvar Kamprad. He built a showroom on the outskirts of Stockholm where land was cheap and simply set the furniture out as it would be in a domestic setting. Also, instead of moving the furniture from the warehouse to the showroom area, he asked customers themselves to pick the furniture up from the warehouse – still the basis of IKEA's process today.

The stores are all designed to facilitate the smooth flow of customers, from parking, moving through the store itself, to ordering and picking up goods. At the entrance to each store large noticeboards provide advice to shoppers who have not used the store before. For young children, there is a supervised children's play area, a small cinema, a parent and baby room and toilets, so parents can leave their children in the supervised play area for a time. Parents are recalled via the loudspeaker system if the child has any problems. IKEA 'allow customers to make up their minds in their own time' but 'information points' have staff who can help. All furniture carries a ticket with a code number which indicates its location in the warehouse. (For larger items customers go to the information desks for assistance.) There is also an area where smaller items are displayed, and can be picked directly. Customers then pass through the warehouse where they pick up the items viewed in the showroom. Finally, customers pay at the checkouts, where a ramped conveyor belt moves purchases up to the checkout staff. The exit area has service points, and a loading area that allows customers to bring their cars from the car park and load their purchases.

IKEA's success is founded on 'listening to its customers' and a disciplined (some would say, obsessive) elimination of waste in its processes; not just its retail processes, but also its design, distribution and administrative processes. Yet success brings its own problems and some customers became increasingly frustrated with overcrowding and long waiting times. In response, IKEA in the UK launched a programme to 'design out' the bottlenecks. The changes include:

- clearly marked in-store short cuts allowing customers who just want to visit one area to avoid having to go through all the preceding areas
- express checkout tills for customers with a bag only rather than a trolley
- extra 'help staff' at key points to help customers
- redesign of the car parks, making them easier to navigate
- dropping the ban on taking trolleys out to the car parks for loading (originally implemented to stop vehicles being damaged)
- a new warehouse system to stop popular product lines running out during the day
- more children's play areas.

IKEA spokeswoman Nicki Craddock said: *'We know people love our products but hate our shopping experience. We are being told that by customers every day, so we can't afford not to make changes. We realised a lot of people took offence at being herded like sheep on the long route around stores. Now if you know what you are looking for and just want to get in, grab it and get out, you can.'*

And the future? Martin Hansson, who runs the UK arm of IKEA, wants to see more emphasis on promoting their environmental agenda. *'We can be better at that. We're not good at showing how we handle waste and energy – it's a lost opportunity.'*

EXAMPLE

Operations at Virgin Atlantic[2]

The airline business is particularly difficult to get right. Few businesses can cause more customer frustration and few businesses can lose their owners so much money. This is because running an airline, and also running the infrastructure on which the airlines depend, is a hugely complex business, where the difference between success and failure really is how you manage your operations on a day-to-day basis. In this difficult business environment, one of the most successful airlines, and one whose reputation has grown because of the way it manages its operations, is Virgin Atlantic. Part of Sir Richard Branson's Virgin Group, Virgin Atlantic Airways was founded in 1984 and is owned 51 per cent by the Virgin Group and 49 per cent by Singapore Airlines. Now, the airline flies over 5,000,000 passengers to 30 destinations worldwide with a fleet of 38 aircraft and almost 10,000 employees.

Source: © Tristar Photos/Alamy

In many ways, it can be seen as being representative of the whole Virgin story – a small newcomer taking on the giant and complacent establishment while introducing better services and lower costs for passengers, yet also building a reputation for quality and innovative service development. The company's mission statement is 'to grow a profitable airline, that people love to fly and where people love to work', a commitment to service excellence that is reflected in the many awards they have won.

Virgin Atlantic's reputation includes a history of innovation in its service processes. It spent £100,000,000 installing its revolutionary new Upper Class suite that provides the longest and most comfortable flat bed and seat in airline history. It was also the first airline to offer business class passengers individual televisions. It now has one of the most advanced in-flight entertainment systems of any airline with over 300 hours of video content, 14 channels of audio, over 50 CDs, audio books and computer games on demand. The Upper Class area at London's Heathrow airport has a dedicated security channel exclusively for the use of Virgin Atlantic customers enabling business passengers to speed through the terminal moving from limousine to lounge in minutes.

Virgin Atlantic emphasise the practical steps they are taking to make their business as sustainable as possible, using the slogan 'we recycle exhaustively, especially our profits'. This refers to the pledge given by the company's chairman, Sir Richard Branson, to invest profits over the next ten years from the Virgin transport companies into projects to tackle climate change. *'We must rapidly wean ourselves off our dependence on coal and fossil fuels,'* Sir Richard said. *'The funds will be invested in schemes to develop new renewable energy technologies, through an investment unit called Virgin Fuels.'* Friends of the Earth welcomed Sir Richard's announcement, but the environmental pressure group also warned that the continued fast growth in air travel could not be maintained 'without causing climatic disaster'.

What do these two examples have in common?

All the operations managers in these two companies will be concerned with the same basic task – managing the processes that produce their products and services. And many, if not most, of the managers in each company who are called by some other title, will be concerned with managing their own processes that contribute to the success of their business. Although there will be differences between each company's operations and processes, such as the type of services they provide, the resources they use, and so on, the managers in each company will be making the same *type* of decisions, even if *what* they actually decide is different. The fact that both companies are successful because of their innovative and effective operations and processes also implies further commonality. First it means that they both understand the importance of taking a 'process perspective' in understanding their supply networks, running their operations, and managing all their individual processes. Without this they could not have sustained their strategic impact in the face of stiff market competition. Second, both businesses will expect their operations to make a contribution to their overall competitive strategy. Third, in achieving a strategic impact, they both will have come to understand the importance of managing *all* their individual processes throughout the business so that they too can all contribute to the businesses success.

DIAGNOSTIC QUESTION

Does the business take a process perspective?

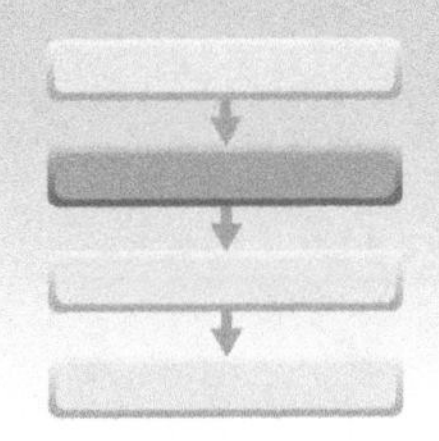

Video

If a business takes a process perspective, it understands that all parts of the business can be seen as processes, and that all processes can be managed using operations management principles. But it is also important to understand that a process perspective is not the only way of describing businesses, or any type of organisation. One could represent an organisation as a conventional 'organisational structure' that shows the reporting relationships between various departments or groups of resources. But even a little experience in any organisation shows that rarely, if ever, does this fully represent the way the organisation actually works. Alternatively one could describe an organisation through the way it makes decisions: how it balances conflicting criteria, weighs up risks, decides on actions and learns from its mistakes. On the other hand, one could describe the organisation's culture (its shared values, ideology, pattern of thinking and day-to-day rituals) or its power relationships (how it is governed, seeks consensus – or at least reconciliation – and so on. Or, and this is the significant point, one can represent the organisation as a collection of processes, interconnecting and (hopefully) all contributing to fulfilling its strategic aims. This is the perspective that we emphasise throughout this book. As we define it here, the process perspective analyses businesses as a collection of interrelated processes. Some of these processes will be within the operations function, and will contribute directly to the production of its products and services. Other processes will be in the other functions of the business, but will still need managing using similar principles to those within the operations function.

OPERATIONS PRINCIPLE

There are many valid approaches to describing organisations. The process perspective is a particularly valuable one.

None of these various perspectives gives a total picture. Each perspective adds something to our ability to understand and therefore more effectively manage a business. Nor are they mutually exclusive. A process perspective does not preclude understanding the influence of power relationships on how processes work, and so on. We use the process perspective here, not because it is the *only* useful and informative way of understanding businesses, but because it is the perspective that directly links the way we manage resources in a business with its strategic impact. Without effective process management, the best strategic plan can never become reality. The most appealing promises made to clients or customers will never be fulfilled. In addition, the process perspective has traditionally been undervalued. The subject of operations and process management has only recently come to be seen as universally applicable and, more importantly, universally valuable.

So, operations and process management is relevant to all parts of the business

If processes exist everywhere in the organisation, operations and process management will be a common responsibility of all managers irrespective of which function they are in. Each function will have its 'technical' knowledge of course. In Marketing, this includes the market

OPERATIONS PRINCIPLE

All parts of the business manage processes, so all parts of the business have an operations role and need to understand operations management.

expertise needed for designing and shaping marketing plans; in Finance, it includes the technical knowledge of financial reporting conventions. Yet each will also have an *operations* role that entails using its processes to produce plans, policies, reports and services. For example, the Marketing function has processes with inputs of market information, staff, computers, and so on. Its staff transform the information into outputs such as marketing plans, advertising campaigns and sales force organisation. In this sense, all functions are operations with their own collection of processes. The implications of this are very important. Because every manager in all parts of an organisation is, to some extent, an operations manager, they all should want to give good service to their customers, and they all will want to do this efficiently. So, operations management must be relevant for all functions, units and groups within the organisation. And the concepts, approaches and techniques of operations management can help to improve any process in any part of the organisation.

The 'input–transformation–output' model

OPERATIONS PRINCIPLE

All processes have inputs of transforming and transformed resources that they use to create products and services.

Central to understanding the processes perspective is the idea that all processes transform *inputs* into *outputs*. Figure 1.4 shows the *general transformation process model* that is used to describe the nature of processes. Put simply, processes take in a set of input resources, some of which are transformed into outputs of products and/or services and some of which do the transforming.

Process inputs

Transformed resource inputs are the resources that are changed in some way within a process. They are usually materials, information or customers. For example, one process in a bank prints statements of accounts for its customers. In doing so, it is processing materials. In the bank's branches, customers are processed by giving them advice regarding their financial affairs, cashing their cheques, etc. However, behind the scenes, most of the bank's processes are concerned with processing information about its customers' financial affairs. In fact, for the bank's operations function as a whole, its information transforming processes are probably the

OPERATIONS PRINCIPLE

Transformed resource inputs to a process are materials information or customers.

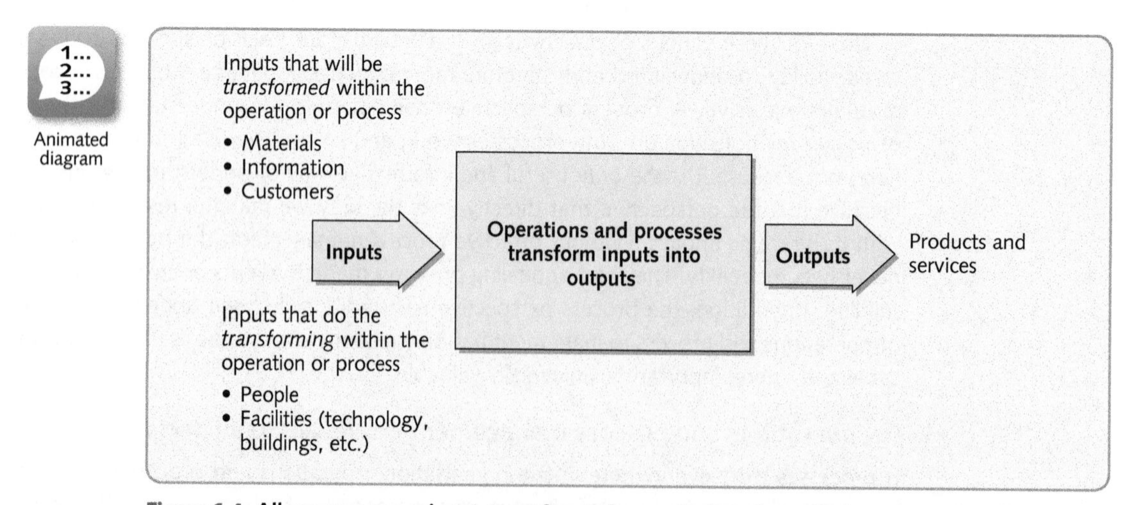

Figure 1.4 All processes are input–transformation–output systems that use 'transforming' resources to work on 'transformed' resources in order to produce products and services

most important. As customers, we may be unhappy with badly printed statements and we may even be unhappy if we are not treated appropriately in the bank. But if the bank makes errors in our financial transactions, we suffer in a far more fundamental way.

OPERATIONS PRINCIPLE
All processes have transforming resources of facilities (equipment, technology, etc.) and people.

There are two types of *transforming* resource that form the 'building blocks' of all processes. They are *facilities* – the buildings, equipment, plant and process technology of the operation, and *people* – who operate, maintain, plan and manage the operation.

The exact nature of both facilities and people will differ between processes. In a five-star hotel, facilities consist mainly of buildings, furniture and fittings. In a nuclear-powered aircraft carrier, its facilities are the nuclear generator, turbines and sophisticated electronic detection equipment. Although one operation is relatively 'low-technology' and the other 'high-technology', their processes all require effective, well-maintained facilities. Staff will also differ between processes. Most staff employed in a domestic appliance assembly process may not need a very high level of technical skill, whereas most staff employed by an accounting firm in an audit process are highly skilled in their own particular 'technical' skill (accounting). Yet although skills vary, all staff have a contribution to make to the effectiveness of their operation. An assembly worker who consistently misassembles refrigerators will dissatisfy customers and increase costs just as surely as an accountant who cannot add up.

Process outputs

All processes produce products and services, and although products and services are different, the distinction can be subtle. Perhaps the most obvious difference is in their respective tangibility. Products are usually tangible (you can physically touch a television set or a newspaper), services are usually intangible (you cannot touch consultancy advice or a haircut – although you may be able to see or feel the results). Also, services may have a shorter stored life. Products can usually be stored for a time, some food products only for a few days, and some buildings for thousands of years. But the life of a service is often much shorter. For example, the service of 'accommodation in a hotel room for tonight' will 'perish' if it is not sold before tonight – accommodation in the same room tomorrow is a different service.

The three levels of analysis

Operations and process management uses the process perspective to analyse businesses at three levels. The most obvious level is that of the business itself, or more specifically, the operations function of the business. The other functions of the business could also be treated at this level, but that would be beyond the scope of this book. And, while analysing the business at the level of the operation is important, for a more comprehensive assessment we also need to analyse the contribution of operations and process management at a higher and more strategic level (the level of its supply network) and at a lower more operational level (the level of the individual processes). These three levels of operations analysis are shown in Figure 1.5.

OPERATIONS PRINCIPLE
A process perspective can be used at three levels: the level of the operation itself, the level of the supply network and the level of individual processes.

The process perspective at the level of the operation

The operations part of a business is itself an input–transformation–output system, which transforms various inputs to produce (usually) a range of different products and services. Table 1.1 shows some operations described in terms of their main inputs, the purpose of their operations, and their outputs. Note how some of the inputs to the operation are transformed in some way while other inputs do the transforming. For example, an airline's aircraft, pilots, air crew, and ground crew are brought into the operation in order to act on passengers and cargo

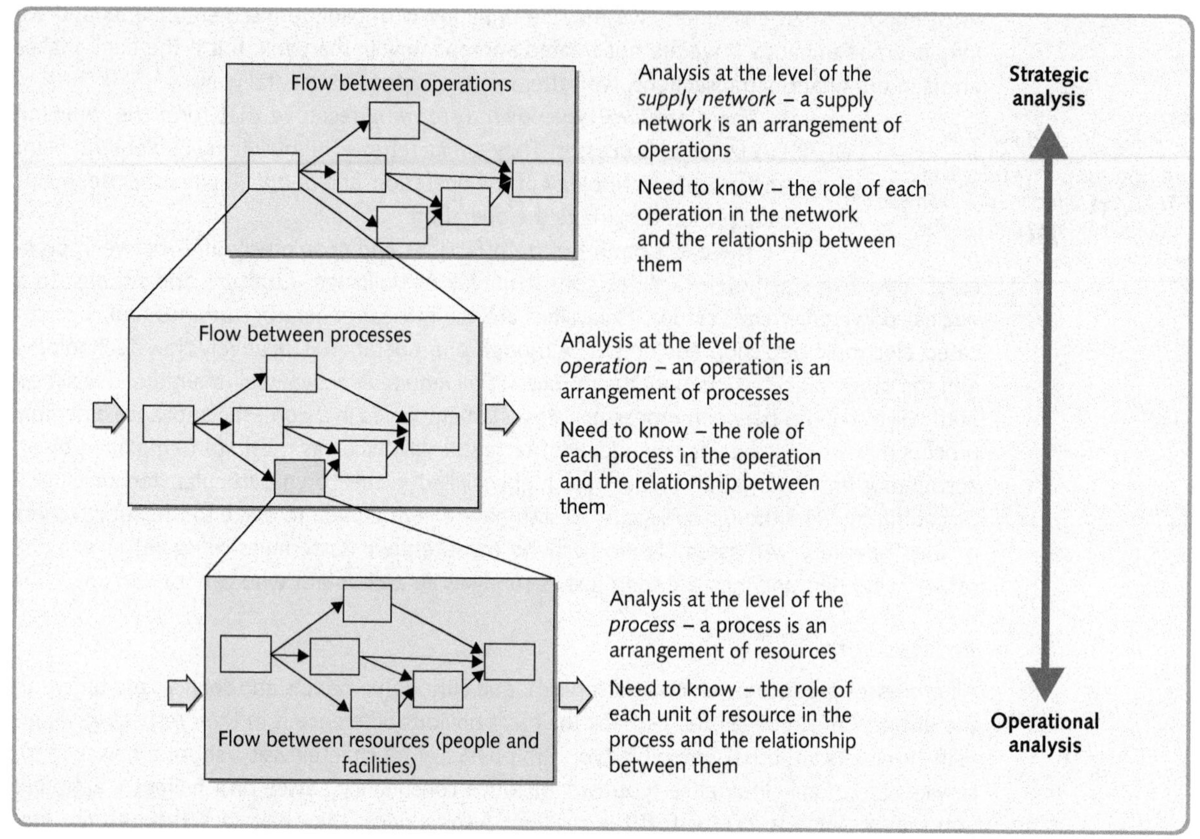

Figure 1.5 Operations and process management requires analysis at three levels: the supply network, the operation and the process

Table 1.1 Some operations described in terms of their inputs, purpose and outputs

Type of operation	*What are the operation's inputs?*	*What does the operation do?*	*What are operation's outputs?*
Airline	Aircraft Pilots and air crew Ground crew *Passengers* *Cargo*	Moves passengers and freight around the world	Transported passengers and freight
Department store	*Goods for sale* Staff sales Computerised registers *Customers*	Displays goods Gives sales advice Sells goods	Customers and goods 'Assembled' together
Police department	Police officers Computer systems *Information* *Public (law-abiding and criminal)*	Prevents crime Solves crime Apprehends criminals	Lawful society Public with feeling of security
Frozen food manufacturer	*Fresh food* Operators Food-processing equipment Freezers	Food preparation Freezes	Frozen food

Note: input resources that are transformed are printed in *italics*.

and change (transform) their location. Note also how in some operations customers themselves are inputs. (The airline, department store and police department are all like this.) This illustrates an important distinction between operations whose customers receive their outputs without seeing inside the operation, and those whose customers are inputs to the operation and therefore have some visibility of the operation's processes. Managing high visibility operations where the customer is inside the operation usually involves different set of requirements and skills to those whose customers never see inside the operation. (We will discuss this issue of visibility later in this chapter.)

Most operations produce both products and services

Some operations produce just products and others just services, but most operations produce a mixture of the two. Figure 1.6 shows a number of operations positioned in a spectrum from almost 'pure' goods producers to almost 'pure' service producers. Crude oil producers are concerned almost exclusively with the product which comes from their oil wells. So are aluminium smelters, but they might also produce some services such as technical advice. To an even greater extent, machine tool manufacturers produce services such as technical advice and applications engineering services as well as products. The services produced by restaurants are an essential part of what the customer is paying for. They both manufacture food and provide service. A computer systems services company may produce software 'products', but more so, it is providing an advice and customisation service to its customers. A management consultancy, although producing reports and documents, would see itself largely as a service provider. Finally, some pure services do not produce products at all. A psychotherapy clinic, for example, provides therapeutic treatment for its customers without any physical product.

OPERATIONS PRINCIPLE

Most operations produce a mixture of tangible products and intangible services.

Services and products are merging

Increasingly the distinction between services and products is both difficult to define and not particularly useful. Even the official statistics compiled by governments have difficulty in

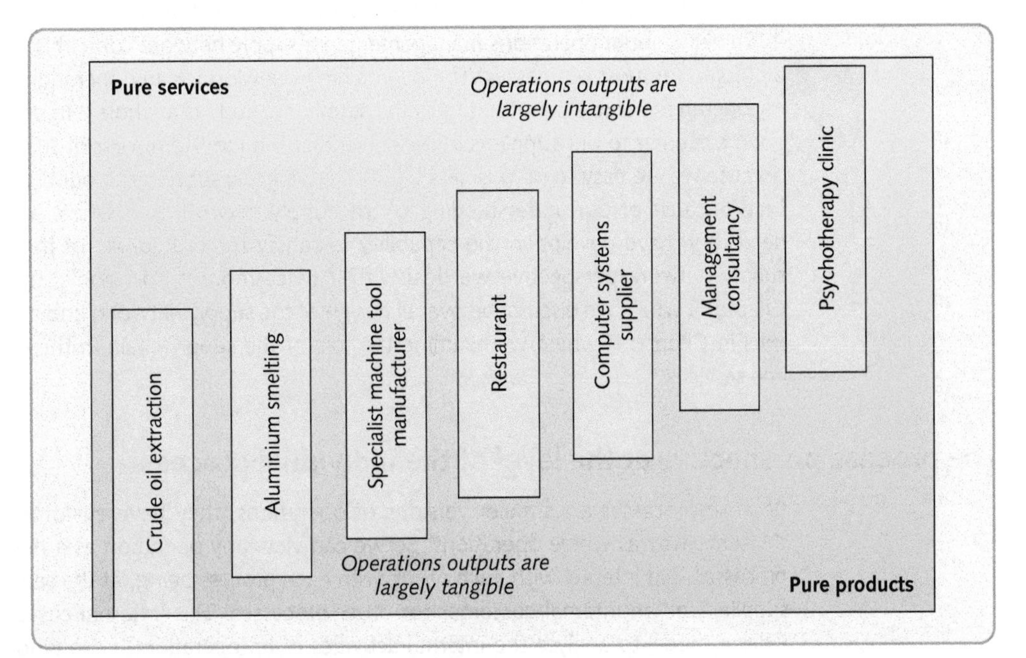

Figure 1.6 Relatively few operations produce either purely products or purely services. Most types of operation produce a mixture of goods and services

separating products and services. Software sold on a disk is classified as a product. The same software sold over the internet is a service. Some authorities see the essential purpose of all businesses, and therefore all operations, as being to 'serve customers'. Therefore, they argue, all operations are service providers who may (or may not) produce products as a means of serving their customers. Our approach in this book is close to this. We treat operations and process management as being important for all organisations. Whether they see themselves as manufacturers or service providers is very much a secondary issue.

The process perspective at the level of the supply network

Any operation can be viewed as part of a greater network of operations. It will have operations that supply it with the products and services it needs to make its own products and services, and unless it deals directly with the end consumer, it will supply customers who themselves may go on to supply their own customers. Moreover, any operation could have several suppliers, several customers and may be in competition with other operations producing similar services to those it produces itself. This collection of operations is called the supply network.

There are three important issues to understand about any operation's supply network. First, it can be complex. Operations may have a large number of customers and suppliers who themselves have large numbers of customers and suppliers. Also, the relationships between operations in the supply network can be subtle. One operation may be in direct competition with another in some markets while at the same time acting as collaborators or suppliers to each other in others. Second, theoretically the boundaries of any operation's supply chain can be very wide indeed. They could go back to the operation that digs raw material out of the ground and go forward to the ultimate reuse and/or disposal of a product. Sometimes it is necessary to do this (for example, when considering the environmental sustainability of products), but generally some kind of boundary to the network needs to be set so that more attention can be given to the most immediate operations in the network. Third, supply networks are always changing. Not only do operations sometimes lose customers and win others, or change their suppliers, they also may acquire operations that once were their customers or suppliers, or sell parts of their business, so converting them into customers or suppliers.

Thinking about operations management in a supply network context is a particularly important issue for most businesses. The overarching question for any operations manager is, 'Does my operation make a contribution to the supply network as a whole?' In other words, are we a good customer to our suppliers in the sense that the long-term cost of supply to us is reduced because we are easy to do business with? Are we good suppliers to our customers in the sense that, because of our understanding of the supply network as a whole, we understand their needs and have developed the capability to satisfy them. Because of the importance of the supply network perspective we deal with it twice more in this book; at a strategic level in Chapter 3 where we discuss the overall design of the supply network, and at a more operational level in Chapter 7 where we examine the role of the supply chain in the delivery of products and services.

The process perspective at the level of the individual process

Because processes are smaller versions of operations, they have customers and suppliers in the same way as whole operations. So we can view any operation as a network of individual processes that interact with each other, with each process being, at the same time, an internal supplier and an internal customer for other processes. This 'internal customer' concept provides a model to analyse the internal activities of an operation. If the whole operation is not working as it should, we may be able to trace the problem back along this internal network of customers and suppliers. It can also be a useful reminder to all parts of the operation that, by

Table 1.2 Some examples of processes in non-operations functions

Organisational function	*Some of its processes*	*Outputs from its process*	*Customer(s) for its outputs*
Marketing and sales	Planning process Forecasting process Order-taking process	Marketing plans Sales forecasts Confirmed orders	Senior management Sales staff, planners, operations Operations, finance
Finance and accounting	Budgeting process Capital approval processes Invoicing processes	Budget Capital request evaluations Invoices	Everyone Senior management, requestees External customers
Human resources management	Payroll processes Recruitment processes Training processes	Salary statements New hires Trained employees	Employees All other processes All other processes
Information technology	Systems review process Help desk process System implementation project processes	System evaluation Advice Implemented working systems and aftercare	All other processes All other processes All other processes

treating their internal customers with the same degree of care that they exercise on their external customers, the effectiveness of the whole operation can be improved. Again, remember that many of an organisation's processes are not operations processes, but are part of some other function. Table 1.2 illustrates just some of the processes that are contained within some of the more common non-operations functions, the outputs from these processes and their 'customers'.

There is an important implication of visualising each function of an organisation as being a network of processes. The diverse parts of a business are connected by the relationships between their various processes, and the organisational boundaries between each function and each part of the business is really a secondary issue. Firms are always reorganising the boundaries between processes. They frequently move responsibility for tasks between departments. The tasks and the processes change less often. Similarly, tasks and processes may move between various businesses; that is what outsourcing and the 'do or buy' decision is all about (see Chapter 3). In other words, not only can separate businesses be seen as networks of processes, whole supply networks can also. Who owns which processes and how the boundaries between them are organised are separate decisions.

OPERATIONS PRINCIPLE

Whole businesses, and even whole supply networks, can be viewed as networks of processes.

'End-to-end' business processes

Animated diagram

This separation of process boundaries and organisational boundaries also applies at a more micro level. This means that we can define what is inside a process in any way we want. The boundaries between processes, the activities that they perform, and the resources that they use, are all there because they have been designed in that way. It is common in organisations to find processes defined by the type of activity they engage in, for example, invoicing processes, product design processes, sales processes, warehousing processes, assembly processes, painting processes, etc. This can be convenient because it groups similar resources together, but it is only one way of drawing the boundaries between processes. Theoretically, in large organisations there must be an almost infinite number of ways that activities and resources could be collected together as distinct processes. One way of redefining the boundaries and responsibilities of processes is to consider the 'end-to-end' set of activities that satisfy defined

OPERATIONS PRINCIPLE

Processes are defined by how the organisation chooses to draw process boundaries.

customer needs. Think about the various ways in which a business satisfies its customers. Many different activities and resources will probably contribute to 'producing' each of its products and services. Some authorities recommend grouping the activities and resources together in an end-to-end manner to satisfy each defined customer need. This approach is closely identified with the 'business process engineering' (or re-engineering) movement (examined in Chapter 13). It calls for a radical rethink of process design that will probably involve taking activities and resources out of different functions and placing them together to meet customer needs. Remember though, designing processes around end-to-end customer needs is only one way (although often the sensible one) of designing processes.

EXAMPLE

The Programme and Video Division (PVD)

A broadcasting company has several divisions including several television and radio channels (entertainment and news), a 'general services' division that includes a specialist design workshop, and the 'Programme and Video Division' (PVD) that makes programmes and videos for a number of clients including the television and radio channels that are part of the same company. The original ideas for these programmes and videos usually come from the clients who commission them, although PVD itself does share in the creative input. The business is described at the three levels of analysis in Figure 1.7.

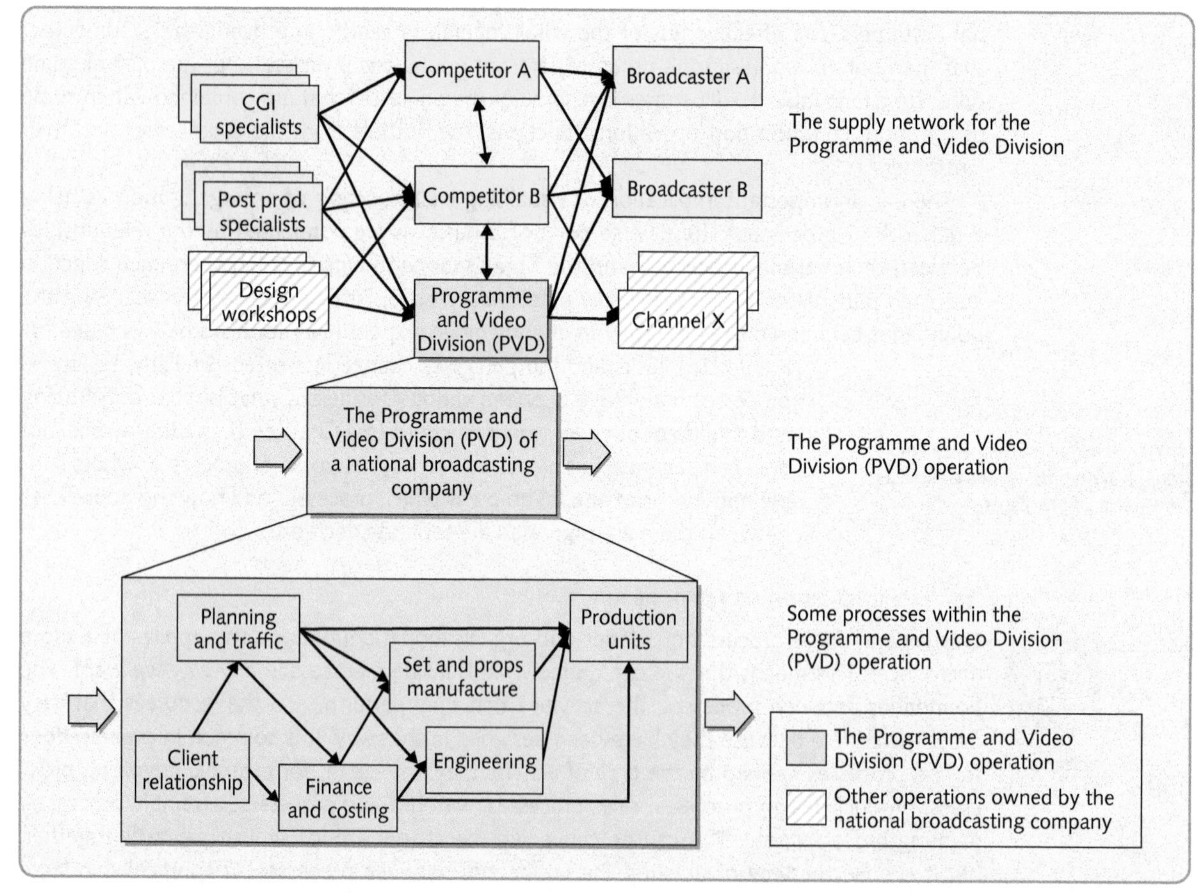

Figure 1.7 Operations and process management analysis for the Programme and Video Division (PVD) of a national broadcasting company at three levels: the supply network, the operation and individual processes

At the level of the operation – the division produces products in the form of tapes, discs and media files, but its real 'product' is the creativity and 'artistry' that is captured in the programmes. *'We provide a service,'* says the division's boss, *'that interprets the client's needs (and sometimes their ideas), and transforms them into appealing and appropriate shows. We can do this because of the skills, experience and creativity of our staff, and our state-of-the-art technology.'*

At the level of the supply network – the division has positioned itself to specialise in certain types of product, including children's programmes, wild life programmes and music videos. *'We did this so that we could develop a high level of expertise in a few relatively high margin areas. It also reduces our dependence on our own broadcasting channels. Having specialised in this way, we are better positioned to partner and do work for other programme makers who are our competitors in some other markets. Specialisation has also allowed us to outsource some activities such as computer graphic imaging (CGI) and post-production that are no longer worth keeping in-house. However, our design workshop became so successful that they were "spun out" as a division in their own right and now work for other companies as well as ourselves.'*

At the level of individual processes – many smaller processes contribute directly or indirectly to the production of programmes and videos, include the following:

- The planning and traffic department who act as the operations managers for the whole operation. They draw up schedules, allocate resources and 'project manage' each job through to completion.
- Workshops that manufacture some of the sets, scenery and props for the productions.
- Client liaison staff who liaise with potential customers, test out programme ideas and give information and advice to programme makers.
- An engineering department that cares for, modifies and designs technical equipment.
- Production units that organise and shoot the programmes and videos.
- The finance and costing department that estimates the likely cost of future projects, controls operational budgets, pays bills and invoices customers.

Creating end-to-end processes – PVD produces several products and services that fulfil customer needs. Each of these, to different extents, involves several of the existing departments within the company. For example, preparing a 'pitch' (a sales presentation that includes estimates of the time and cost involved in potential projects) mainly needs the contributions of Client relations and the Finance and costing departments, but also needs smaller contributions from other departments. Figure 1.8 illustrates the contribution of each department to each product or service. (No particular sequence is implied by Figure 1.8.) The contributions of each department may not all occur in the same order. Currently, all the division's processes are clustered into conventional departments defined by the type of activity they perform, engineering, client relationship, etc. A radical redesign of the operation could involve regrouping activities and resources into five 'business' processes that fulfil each of the five defined customer needs. This is shown diagrammatically by the dotted lines in Figure 1.8. It would involve the physical movement of resources (people and facilities) out of the current functional processes into the new end-to-end business processes. This is an example of how processes can be designed in ways that do not necessarily reflect conventional functional groupings.

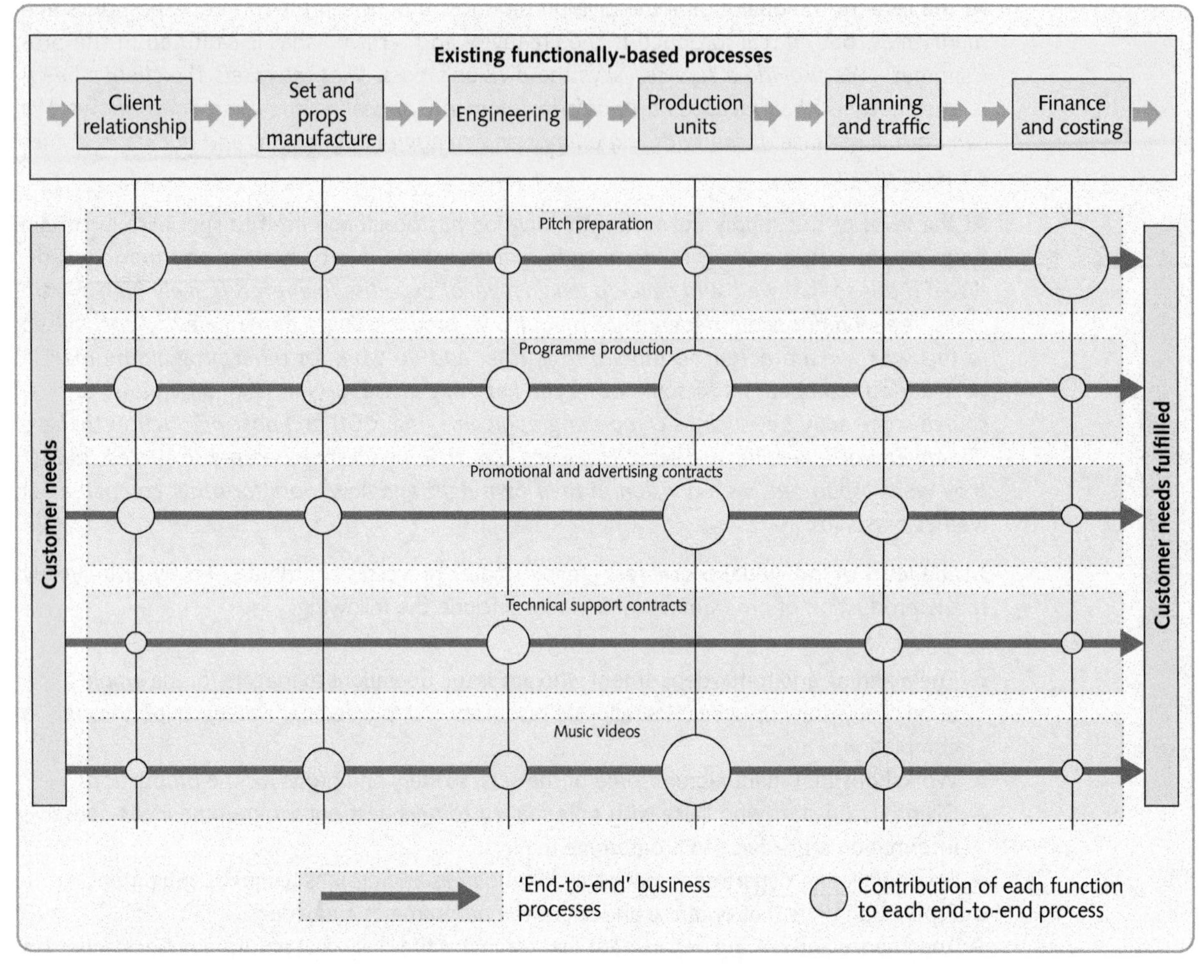

Figure 1.8 An example of how processes in the Programme and Video Division (PVD) could be reorganised around end-to-end business processes that fulfil defined customer needs

DIAGNOSTIC QUESTION

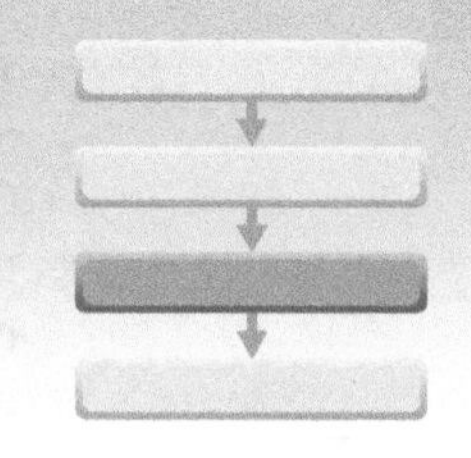

Does operations and process management have a strategic impact?

Video

One of the biggest mistakes a business can make is to confuse 'operations' with 'operational'. Operational is the opposite of strategic; it means detailed, localised, short-term, day-to-day. 'Operations', on the other hand, is the set of resources that produce products and services. Operations can be treated both at an operational *and a strategic level*. We shall examine some views of operations strategy in the next chapter. For now, we treat a fundamental question for any operation – does the way we manage operations and processes have a strategic impact? If a business does not fully appreciate the strategic impact that effective operations and process management can have, at the very least, it is missing an opportunity. The IKEA and Virgin Atlantic examples at the beginning of this chapter are just two of many businesses that have harnessed their operations to create strategic impact.

Operations and process management can make or break a business. Although for most businesses, the operations function represents the bulk of its assets and the majority of its people, the true value of the operation is more than 'bulk'. It can 'make' the businesses in the sense that it gives the ability to compete through both the short-term ability to respond to customers and the long-term capabilities that will keep it ahead of its competitors. But if an operations function cannot produce its products and services effectively, it could 'break' the business by handicapping its performance no matter how it positions and sells itself in its markets.

Cost, revenue, risk, investment and capability

The strategic importance of operations and process management is being increasingly recognised. When compared with only a few years ago, it attracts far more attention and, according to some reports, accounts for the largest share of all the money spent by businesses on consultancy advice. This may be partly because the area has been neglected in the past, but it also denotes an acceptance that it can have both short-term and long-term impact. This can be seen in the impact that operations and process management can have on the businesses' cost, revenue, risk, investment and capabilities.

- It can reduce the **costs** of producing products and services by being efficient. The more productive the operation is at transforming inputs into outputs, the lower will be the cost of producing a unit of output. Cost is never totally unimportant for any business, but generally the higher the cost of a product or service when compared to the price it commands in the market, the more important cost reduction will be as an operations objective. Even so, cost reduction is almost always treated as an important contribution that operations can make to the success of any business.
- It can increase **revenue** by increasing customer satisfaction through quality, service and innovation. Existing customers are more likely to be retained and new customers are more likely to be attracted to products and services if they are error-free and appropriately designed, if the operation is fast and responsive in meeting their needs and keeping its delivery promises, and if an operation can be flexible, both in customising its products and services and introducing new ones. It is operations that directly influence the quality, speed, dependability and flexibility of the business, all of which have a major impact on a company's ability to maximise its revenue.
- It can reduce the **risk** of operational failure, because well-designed and run operations should be less likely to fail. All failures can eventually be traced back to some kind of failure within a process. Furthermore, a well-designed process, if it does fail, should be able to recover faster and with less disruption (this is called *resilience*).
- It can ensure **effective investment** (capital employed) to produce its products and services. Eventually all businesses in the commercial world are judged by the return that they produce for their investors. This is a function of profit (the difference between costs and revenues) and the amount of money invested in the business's operations resources. We have already established that effective and efficient operations can reduce costs and increase revenue, but what is sometimes overlooked is the operation's role in reducing the investment required per unit of output. It does this by increasing the effective capacity of the operation and by being innovative in how it uses its physical resources.
- It can **build capabilities** that will form the basis for *future* innovation by building a solid base of operations skills and knowledge within the business. Every time an operation produces a product or a service it has the opportunity to accumulate knowledge about how that product or service is best produced. This accumulation of knowledge should be used as a basis for learning and improvement. If so, in the long term, capabilities can be built that will allow the operation to respond to future market challenges. Conversely, if an operations function is simply seen as the mechanical and routine fulfilment of customer requests, then it is difficult to build the knowledge base that will allow future innovation.

OPERATIONS PRINCIPLE

All operations should be expected to contribute to their business by controlling costs, increasing revenue, reducing risks, making investment more effective and growing long-term capabilities.

EXAMPLE

The programme and Video Division (PVD) continued

The PVD described earlier should be able to identify all four ways in which its operations and processes can have a strategic impact. The division is expected to generate reasonable returns by controlling its costs and being able to command relatively high fees. *'Sure, we need to keep our costs down. We always review our budgets for bought-in materials and services. Just as important, we measure the efficiency of all our processes, and we expect annual improvements in process efficiency to compensate for any increases in input costs.'* (Reducing costs.) *'Our services are in demand by customers because we are good to work with,'* says the division's Managing Director. *'We have the technical resources to do a really great job and we always give good service. Projects are completed on time and within budget. More importantly, our clients know that we can work with them to ensure a high level of programme creativity. That is why we can command reasonably high prices.'* (Increasing revenue.) *'Also, we have a robust set of processes that minimise the chances of projects failing to achieve success.'* (Reducing risk.) The division also has to justify its annual spend on equipment to its main board. *'We try and keep up to date with the new technology that can really make an impact on our programme making, but we always have to demonstrate how it will improve profitability.'* (Effective investment.) *'We also try to adapt new technology and integrate it into our creative processes in some way so that gives us some kind of advantage over our competitors.'* (Build capabilities.)

Operations management in not-for-profit organisations

Terms such as *competitive advantage*, *markets* and *business* that are used in this book are usually associated with companies in the for-profit sector. Yet operations management is also relevant to organisations whose purpose is not primarily to earn profits. Managing operations in an animal welfare charities, hospitals, research organisations or government departments is essentially the same as in commercial organisations. However, the strategic objectives of not-for-profit organisations may be more complex and involve a mixture of political, economic, social or environmental objectives. Consequently, there may be a greater chance of operations decisions being made under conditions of conflicting objectives. So, for example, it is the operations staff in a children's welfare department who have to face the conflict between the cost of providing extra social workers and the risk of a child not receiving adequate protection.

DIAGNOSTIC QUESTION

Should all processes be managed in the same way?

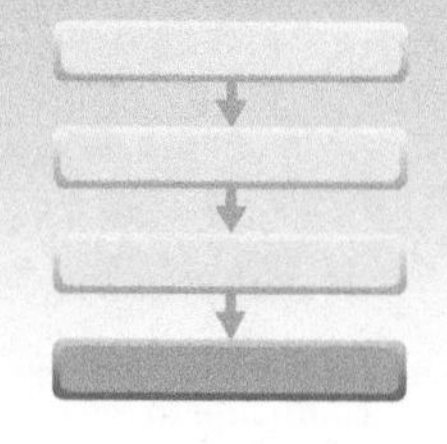

OPERATIONS PRINCIPLE

The way in which processes need to be managed is influenced by volume, variety, variation and visibility.

All processes differ in some way, so, to some extent, all processes will need to be managed differently. Some of the differences between processes are 'technical' in the sense that different products and services require different skills and technologies to produce them. However, processes also differ in terms of the nature of demand for their products or services. Four characteristics of demand in particular have a significant effect on how processes need to be managed:

Video

- the volume of the products and services produced
- the variety of the different products and services produced
- the variation in the demand for products and services
- the degree of visibility that customers have of the production of products and services.

The four Vs of processes

Volume

Animated diagram

Processes with a high volume of output will have a high degree of repeatability, and because tasks are repeated frequently it often makes sense for staff to specialise in the tasks they perform. This allows the systemisation of activities, where standard procedures may be codified and set down in a manual with instructions on how each part of the job should be performed. Also, because tasks are systemised and repeated, it is often worthwhile developing specialised technology that gives higher processing efficiencies. By contrast, low-volume processes with less repetition cannot specialise to the same degree. Staff are likely to perform a wide range of tasks, and while this may be more rewarding, it is less open to systemisation. Nor is it likely that efficient, high-throughput technology could be used. The implications of this are that high-volume processes have more opportunities to produce products or services at low unit cost. So, for example, the volume and standardisation of large fast-food restaurant chains, such as McDonald's or KFC, enables them to produce with greater efficiency than a small, local cafeteria or diner.

Variety

Processes that produce a high variety of products and services must engage in a wide range of different activities, changing relatively frequently between each activity. It must also contain a wide range of skills and technology sufficiently 'general purpose' to cope with the range of activities and sufficiently flexible to change between them. A high level of variety may also imply a relatively wide range of inputs to the process and the additional complexity of matching customer requirements to appropriate products or services. So, high variety processes are invariably more complex and costly than low variety ones. For example, a taxi company is usually prepared to pick up and drive customers almost anywhere (at a price); they may even take you by the route of your choice. There are an infinite number of potential routes (products) that it offers. But, its cost per kilometre travelled will be higher than a less customised form of transport such as a bus service.

Variation

Processes are generally easier to manage when they only have to cope with predictably constant demand. Resources can be geared to a level that is just capable of meeting demand. All activities can be planned in advance. By contrast, when demand is variable and/or unpredictable, resources will have to be adjusted over time. Worse still, when demand is unpredictable, extra resources will have to be designed into the process to provide a 'capacity cushion' that can absorb unexpected demand. So, for example, processes that manufacture high fashion garments will have to cope with the general seasonality of the garment market together with the uncertainty of whether particular styles may or may not prove popular. Operations that make conventional business suits are likely to have less fluctuation in demand over time, and be less prone to unexpected fluctuations. Because processes with lower variation do not need any extra safety capacity and can be planned in advance, they will generally have lower costs than those with higher variation.

Visibility

Process visibility is a slightly more difficult concept to envisage. It indicates how much of the processes are 'experienced' directly by customers, or how much the process is 'exposed' to its customers. Generally processes that act directly on customers (such as retail processes or health care process) will have more of their activities visible to their customers than those that act on materials and information. However, even material and information transforming processes may provide a degree of visibility to the customers. For example, parcel distribution operations provide internet-based 'track and trace' facilities to enable their customers to have visibility of where their packages are at any time. Low-visibility processes, if they communicate with their customers at all, do so using less immediate channels such as the telephone or the internet. Much of the process can be more 'factory-like'. The time lag between customer request and response could be measured in days rather than the near immediate response expected from high-visibility processes. This lag allows the activities in a low-visibility process to be performed when it is convenient to the operation, so achieving high utilisation. Also, because the customer interface needs managing, staff in high-visibility processes need customer contact skills that shape the customer's perception of process performance. For all these reasons, high-visibility processes tend to have higher costs than low-visibility processes.

Many operations have both high- and low-visibility processes. This serves to emphasise the difference that the degree of visibility makes. For example, in an airport, some of its processes are relatively visible to its customers (check-in desks, information desks, restaurants, passport control, security staff etc.). These staff operate in a high-visibility 'front-office' environment. Other processes in the airport have relatively little, if any, customer visibility (baggage handling processes, overnight freight operations, loading meals on to the aircraft, cleaning etc.). We rarely see these processes they perform the vital but low-visibility tasks, in the 'back-office' part of the operation.

The implications of the four Vs of processes

All four dimensions have implications for processing costs. Put simply, high volume, low variety, low variation and low visibility all help to keep processing costs down. Conversely, low volume, high variety, high variation and high customer contact generally carry some kind of cost penalty for the process. This is why the volume dimension is drawn with its 'low' end at the left, unlike the other dimensions, to keep all the 'low cost' implications on the right. Figure 1.9 summarises the implications of such positioning.

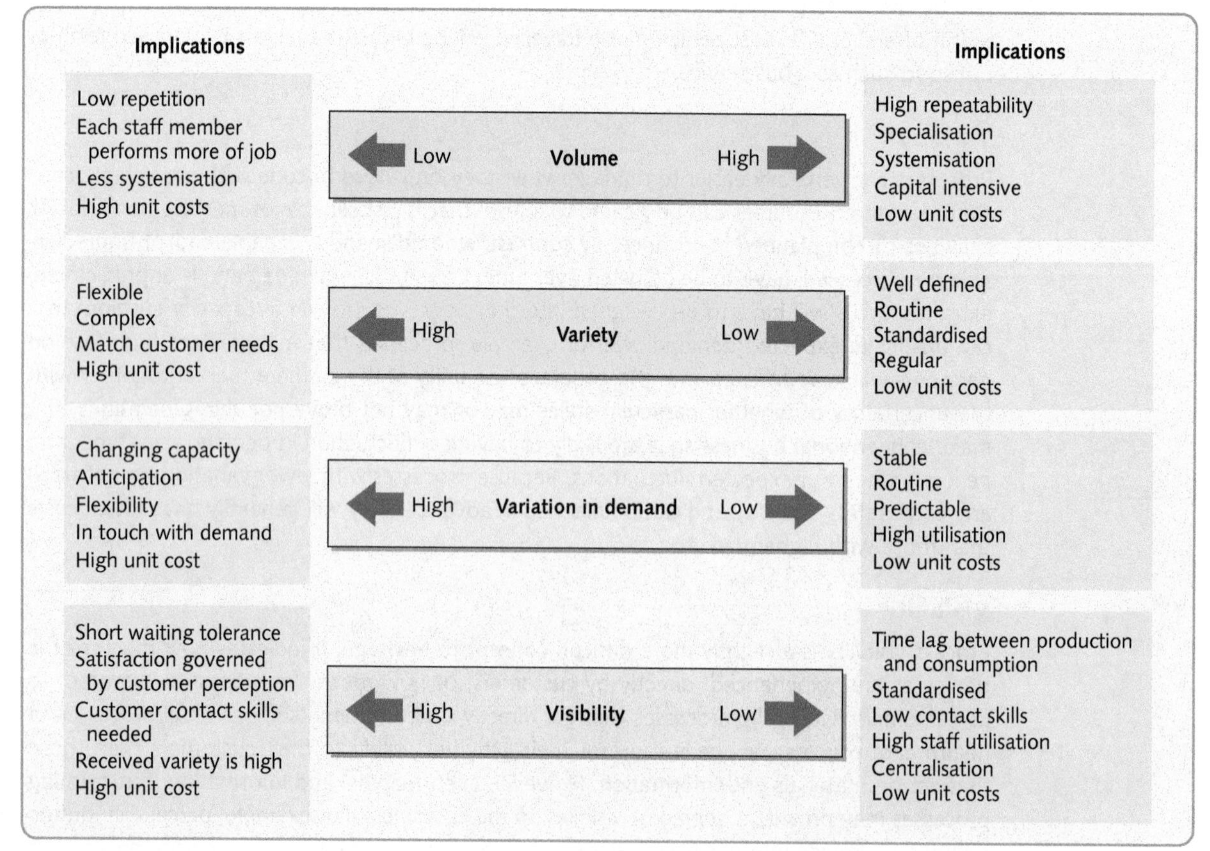

Figure 1.9 A typology of operations

Charting processes using the four Vs

In almost any operation, processes can be identified that have different positions on the four dimensions, and which therefore have different objectives and will need managing in different ways. To a large extent the position of a process on the four dimensions is determined by the demand of the market it is serving. However, most processes have some discretion in moving themselves on the dimensions. Look at the different positions on the visibility dimension that retail banks have adopted. At one time, using branch tellers was the only way customers could contact a bank. Now access to the bank's services could be through (in decreasing order of visibility) a personal banker, who visits your home or office, a conversation with a branch manager, the teller at the window, telephone contact through a call centre, internet banking services or an ATM cash machine. These other processes offer services that have been developed by banks to serve different market needs.

OPERATIONS PRINCIPLE

Operations and processes can (other things being equal) reduce their costs by increasing volume, reducing variety, reducing variation and reducing visibility.

Practice note

Figure 1.10 illustrates the different positions on the four Vs for some retail banking processes. Note that the personal banking/advice service is positioned at the high-cost end of the four Vs. For this reason, such services are often only offered to relatively wealthy customers that represent high profit opportunities for the bank. Note also that the more recent developments in retail banking, such as call centres, internet banking and ATMs, all represent a shift towards the low-cost end of the four Vs. New processes that exploit new technologies can often have a profound impact on the implications of each dimension. For example, internet banking, when compared with an ATM cash machine, offers a far higher variety of options for customers but, because the process is automated through its information technology, the cost of offering this variety is less than at a conventional branch or even a call centre.

A model of operations and process management

Managing operations and processes involves a whole range of separate decisions that will determine their overall purpose, structure and operating practices. These decisions can be grouped together in various ways. Look at other books on operations management and you will find many different ways of structuring operations decisions and therefore the subject as

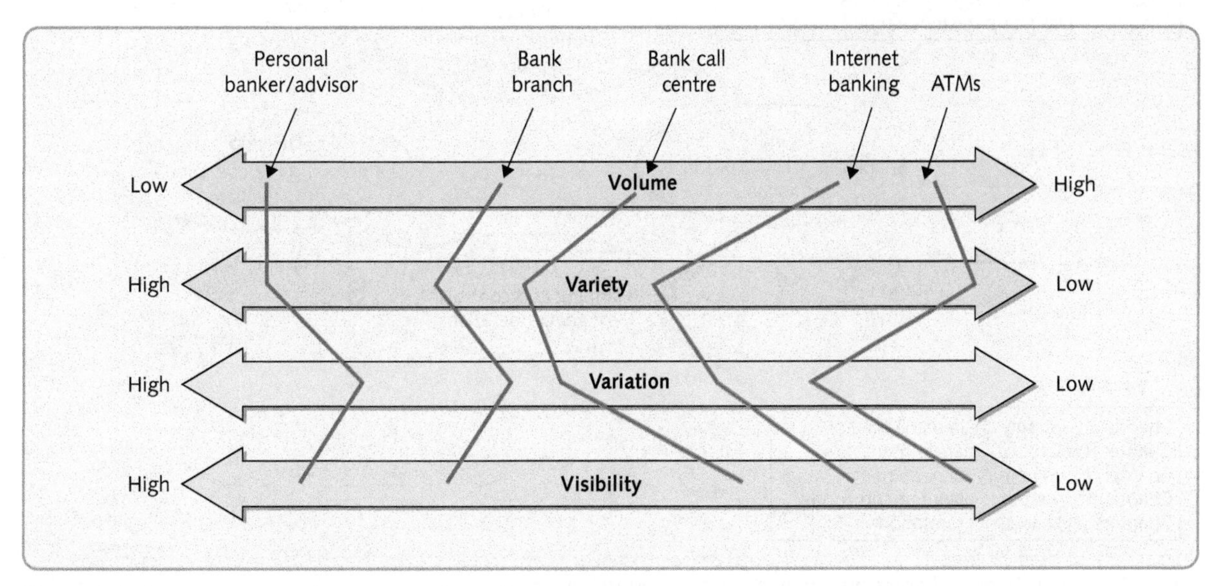

Figure 1.10 Four Vs analysis for some retail banking processes

OPERATIONS PRINCIPLE

Operations management activities can be grouped into four broad categories: directing the overall strategy of the operation; designing the operation's products, services and processes; planning and controlling delivery; developing process performance.

a whole. Here we have chosen to classify activities into four broad groups, relating to four broad activities. Although there are some overlaps between these four categories, they more or less follow a sequence that corresponds to the life cycle of operations and processes.

- **Directing** the overall strategy of the operation. A general understanding of operations and processes and their strategic purpose, together with an appreciation of how strategic purpose is translated into reality (direct), is a prerequisite to the detailed design of operations and process.
- **Designing** the operation's products, services and processes. Design is the activity of determining the physical form, shape and composition of operations and processes together with the products and services that they produce.
- Planning and control process **delivery**. After being designed, the delivery of products and services from suppliers and through the total operation to customers must be planned and controlled.
- **Developing** process performance. Increasingly it is recognised that operations and process managers cannot simply routinely deliver products and services in the same way that they always have done. They have a responsibility to develop the capabilities of their processes to improve process performance.

Video

We can now combine two ideas to develop the model of operations and process management that will be used throughout this book. The first is the idea that *operations* and the *processes* that make up both the operations and other business functions are transformation systems that take in inputs and use process resources to transform them into outputs. The second idea is that the resources, both in an organisation's operations as a whole and in its individual processes, need to be managed in terms of how they are *directed*, how they are *designed*, how *delivery* is planned and controlled and how they are *developed* and improved. Figure 1.11 shows how

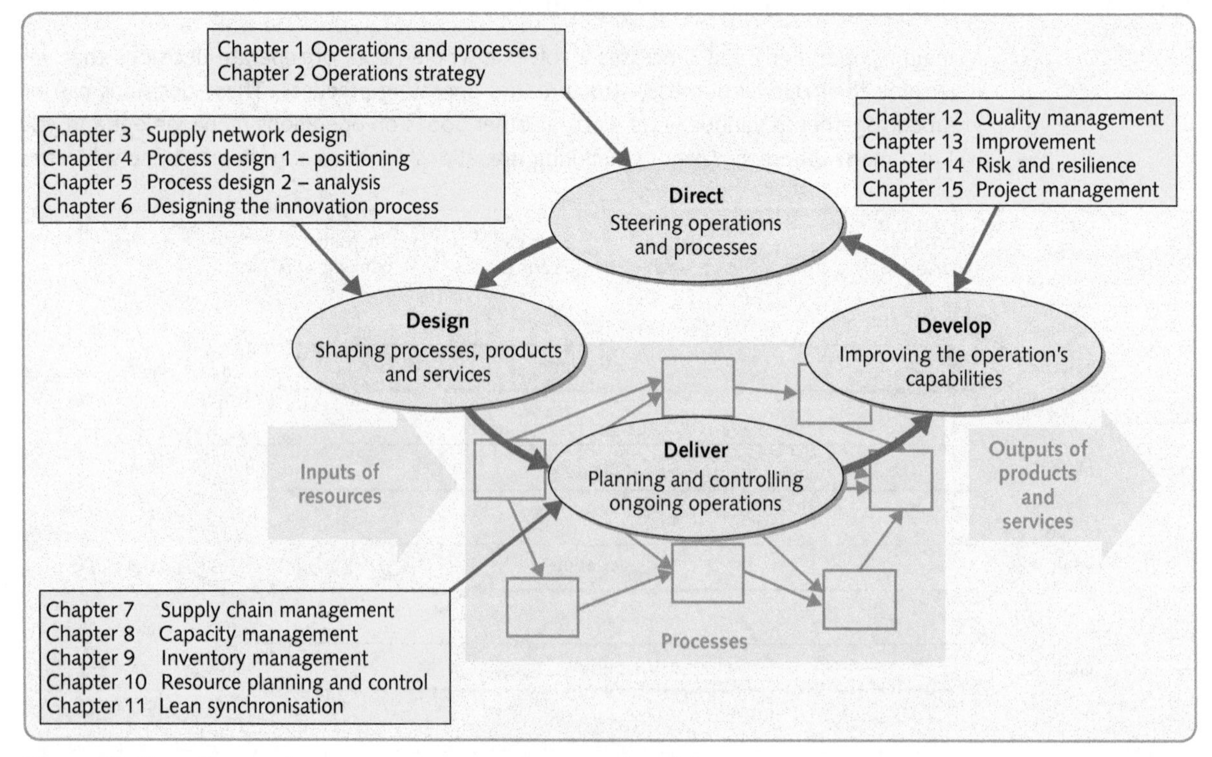

Figure 1.11 Operations and processes management: a general model

these two ideas go together. This book will use this model to examine the more important decisions that should be of interest to all managers of operations and processes.

Critical commentary

All chapters contain a short critical commentary on the main ideas covered in the chapter. Its purpose is not to undermine the issues discussed in the chapter, but to emphasise that, although we present a relatively orthodox view of operation, there are other perspectives.

- The central idea in this introductory chapter is that all organisations have operations (and other functions) that have processes that produce products and services, and that all these processes are essentially similar. However, some believe that by even trying to characterise organisations in this way (perhaps even by calling them 'processes') one loses or distorts their nature and depersonalises or takes the 'humanity' out of the way in which we think of the organisation. This point is often raised in not-for-profit organisations, especially by 'professional' staff. For example, the head of one European 'Medical Association' (a Doctors' Trade Union) criticised hospital authorities for expecting a 'sausage factory service based on productivity targets'. No matter how similar they appear on paper, it is argued, a hospital can never be viewed in the same as a factory. Even in commercial businesses, professionals, such as creative staff, often express discomfort at their expertise being described as a 'process'.

- To some extent these criticisms of taking such a process perspective are valid. How we describe organisations does say much about our underlying assumptions of what an 'organisation' is and how it is supposed to work. Notwithstanding the point we made earlier about how a purely process view can misleadingly imply that organisations are neat and controllable with unambiguous boundaries and lines of accountability, a process perspective can risk depicting the messy reality of organisations in a naïve manner. Yet, in our view it is a risk well worth taking.

SUMMARY CHECKLIST

This checklist comprises questions that can be usefully applied to any type of operations and reflect the major diagnostic questions used within the chapter.

- ☐ Is the operations function of the business clearly defined?
- ☐ Do operations managers realise that they are operations managers even if they are called by some other title?
- ☐ Do the non-operations functions within the business realise that they manage processes?
- ☐ Does everyone understand the inputs, activities and outputs of the processes of which they are part?
- ☐ Is the balance between products and services produced by the operations function well understood?
- ☐ Are future changes that may occur in the balance between products and services produced by the operation understood?
- ☐ What contribution are operations making towards reducing the cost of products and services?
- ☐ What contribution are operations making towards increasing the revenue from products and services?
- ☐ What contribution are operations making towards reducing the risks of failure and increasing the effectiveness of recovery?
- ☐ What contribution are operations making towards better use of capital employed?
- ☐ How are operations developing the capability for future innovation?
- ☐ Does the operation understand its position in the overall supply network?
- ☐ Does the operation contribute to the overall supply network?
- ☐ Are the individual processes that comprise the operations function defined and understood?
- ☐ Are individual processes aware of the internal customer and supplier concept?
- ☐ Do they use the internal customer and supplier concept to increase their contribution to the business as a whole?
- ☐ Do they use the ideas and principles of operations management to improve the performance of their processes?
- ☐ Has the concept of end-to-end business processes being examined and considered?
- ☐ Are the differences (in terms of volume, variety, variation and visibility) between processes understood?
- ☐ Are the volume, variety, variation and visibility characteristics of processes reflected in the way they are managed?

CASE STUDY

Design House Partnerships at Concept Design Services

'I can't believe how much we have changed in a relatively short time. From being an inward looking manufacturer, we became a customer focused "design and make" operation. Now we are an "integrated service provider". Most of our new business comes from the partnerships we have formed with design houses. In effect, we design products jointly with specialist design houses that have a well-known brand, and offer them a complete service of manufacturing and distribution. In many ways we are now a "business-to-business" company rather than a "business-to-consumer" company.' (Jim Thompson, CEO, Concept Design Services (CDS))

Concept Design Services (CDS) had become one of Europe's most profitable homeware businesses. It had moved in two stages from making precision plastic components, mainly in the aerospace sector, together with some cheap 'homeware' items such as buckets and dustpans, sold under the 'Focus' brand name, to making very high-quality (expensive) stylish homewares with a high 'design value' for well-known brands.

The first stage – from 'Focus' to 'Concept'

The initial move into higher margin homeware had been masterminded by Linda Fleet, CDS's Marketing Director. *'My previous experience in the decorative products industry had taught me the importance of fashion and product development, even in mundane products such as paint. Premium-priced colours and new textures would catch the popular imagination and would need supporting by promotion and editorial features in lifestyle magazines. The players who embraced this fashion element of the market were dramatically more profitable than those who simply provided standard ranges. Instinctively, I felt that this must also apply to homeware. We decided to develop a whole coordinated range of such items, and to open up a new distribution network for them to serve the more exclusive stores, kitchen equipment and specialty retailers. Within a year of launching our first new range of kitchen homeware under the 'Concept' brand name, we had over 3,000 retail outlets across Northern Europe with full point-of-sale display facilities and supported by press coverage and product placement on TV 'lifestyle'*

Source: Yang Yu/Alamy Images

programmes. Within two years 'Concept' products were providing over 75 per cent of our revenue and 90 per cent of our profits.' (The margin on Concept products is many times higher than for the Focus range. During this period the Focus (basic) range continued to be produced, but as a drastically reduced range.)

The second stage – from 'Concept' to 'design house partnerships'

Linda was also the driving force behind the move to design house partnerships. *'It started as a simple design collaboration between our design team and an Italian "design house".'* (Design houses are creative product designers who may, or may not, own a brand of their own, but rarely manufacture or distribute their products, relying on outsourcing to subcontractors.) *'It seemed a natural progression to them asking us to first manufacture and then distribute this and other of their designs. Over the next five years, we built up this business, so now we design (often jointly with the design house), manufacture and distribute products for several of the more prestigious European design houses. We think this sort of business is likely to grow. The design houses appreciate our ability to offer a full service. We can design products in conjunction with their own design staff and offer them a level of manufacturing expertise they can't get elsewhere. More significantly, we can offer a distribution service which is tailored to their needs. From the customer's point of view the distribution arrangements appear to belong to the design house itself. In fact they are based exclusively on our own call centre, warehouse and distribution resources.'*

The most successful collaboration was with Villessi, the Italian designers. Generally it was CDS's design expertise which was attractive to 'design house' partners. Not only did CDS employ professionally respected designers, they had also acquired a reputation for being able to translate difficult technical designs into manufacturable and saleable products. Design house partnerships usually involved relatively long lead times but produced unique products with very high margins, nearly always carrying the design house's brand.

Manufacturing operations

All manufacturing was carried out in a facility located 20 km from Head Office. Its moulding area housed large injection-moulding machines, most with robotic material handling capabilities. Products and components passed to the packing hall, where they were assembled and inspected. The newer more complex products often had to move from moulding to assembly and then back again for further moulding. All products followed the same broad process route but with more products needing several progressive moulding and assembly stages, there was an increase in 'process flow recycling' which was adding complexity. One idea was to devote a separate cell to the newer and more complex products until they had 'bedded in'. This cell could also be used for testing new moulds. However, it would need investment in extra capacity that would not always be fully utilised. After manufacture, products were packed and stored in the adjacent distribution centre.

'When we moved into making the higher margin 'Concept' products, we disposed of most of our older, small injection-moulding machines. Having all larger machines allowed us to use large multi-cavity moulds. This increased productivity by allowing us to produce several products, or components, each machine cycle. It also allowed us to use high-quality and complex moulds which, although cumbersome and more difficult to change over, were very efficient and gave a very high-quality product. For example, with the same labour we could make three items per minute on the old machines, and 18 items per minute on the modern ones using multi-moulds. That's a 600 per cent increase in productivity. We also achieved high-dimensional accuracy, excellent surface finish and extreme consistency of colour. We could do this because of our expertise derived from years making aerospace products. Also, by standardising on single large machines, any mould could fit any machine. This was an ideal situation from a planning perspective, as we were often asked to make small runs of Concept products at short notice.' (Grant Williams, CDS Operations Manager)

Increasing volume and a desire to reduce cost had resulted in CDS subcontracting much (but not all) of its Focus products to other (usually smaller) moulding companies. *'We would never do it with any complex or design house partner products, but it should allow us to reduce the cost of making basic products while releasing capacity for higher margin ones. However there have been quite a few "teething problems". Coordinating the production schedules is currently a problem, as is agreeing quality standards. To some extent it's our own fault. We didn't realise that subcontracting was a skill in its own right. And although we have got over some of the problems, we still do not have a satisfactory relationship with all of our subcontractors.'* (Grant Williams, CDS Operations Manager)

Planning and distribution services

The distribution services department of the company was regarded as being at the heart of the company's customer service drive. Its purpose was to integrate the efforts of design, manufacturing and sales by planning the flow of products from production, through the distribution centre, to the customer. Sandra White, the Planning Manager, reported to Linda Fleet and was responsible for the scheduling of all manufacturing and distribution, and for maintaining inventory levels for all the warehoused items. *'We try to stick to a preferred production sequence for each machine and mould so as to minimise set-up times by starting on a light colour, and progressing through a sequence to the darkest. We can change colours in 15 minutes, but because our moulds are large and technically complex, mould changes can take up to three hours. Good scheduling is important to maintain high plant utilisation. With a higher variety of complex products, batch sizes have reduced and it has brought down average utilisation. Often we can't stick to schedules. Short-term changes are inevitable in a fashion market. Certainly better forecasts would help . . . but even our own promotions are sometimes organised at such short notice that we often get caught with stockouts. New products in particular are difficult to forecast, especially when they are 'fashion' items and/or seasonal. Also, I have to schedule production time for new product mould trials; we normally allow 24 hours for the testing of each new mould received, and this has to be done on production machines. Even if we have urgent orders, the needs of the designers always have priority.'* (Sandra White)

Customer orders for Concept and design house partnership products were taken by the company's sales call centre located next to the warehouse. The individual orders would then be dispatched using the company's own fleet of medium and small distribution vehicles for UK orders, but using carriers for the Continental European market. A standard delivery timetable was used and an 'express delivery' service was offered for those customers prepared to pay a small delivery premium. However, a recent study had shown that almost 40 per cent of express deliveries were initiated by the company rather than customers. Typically this would be to fulfil deliveries of orders containing products out of stock at the time of ordering. The express delivery service was not required for Focus products because almost all deliveries were to five large customers. The size of each order was usually very large, with deliveries to customers' own distribution depots. However, although the organisation of Focus delivery was relatively straightforward, the consequences of failure were large. Missing a delivery meant upsetting a large customer.

Challenges for CDS

Although the company was financially successful and very well regarded in the homeware industry, there were a number of issues and challenges that it knew it would have to address. The first was the role of the design department and its influence over new product development. New product development had become particularly important to CDS, especially since they had formed alliances with design houses. This had led to substantial growth in both the size and the influence of the design department, which reported to Linda Fleet. *'Building up and retaining design expertise will be the key to our future. Most of our growth is going to come from the business which will be bought in through the creativity and flair of our designers. Those who can combine creativity with an understanding of our partners' business and design needs can now bring in substantial contracts. The existing business is important of course, but growth will come directly from these peoples' capabilities.'* (Linda Fleet)

But not everyone was so sanguine about the rise of the Design department. *'It is undeniable that relationships between the designers and other parts of the company have been under strain recently. I suppose it is, to some extent, inevitable. After all, they really do need the freedom to design as they wish. I can understand it when they get frustrated at some of the constraints which we have to work under in the manufacturing or distribution parts of the business. They also should be able to expect a professional level of service from us. Yet the truth is that they make most of the problems themselves. They sometimes don't seem to understand the consequences or implications of their design decisions or the promises they make to the design houses. More seriously, they don't really understand that we could actually help them do their job better if they cooperated a bit more. In fact, I now see some of our design house partners' designers more than I do our own designers. The Villessi designers are always in my factory and we have developed some really good relationships.'* (Grant Williams)

The second major issue concerned sales forecasting, and again there were two different views. Grant Williams was convinced that forecasts should be improved. *'Every Friday morning we devise a schedule of production and distribution for the following week. Yet, usually before Tuesday morning, it has had to be significantly changed because of unexpected orders coming in from our customers' weekend sales. This causes tremendous disruption to both manufacturing and distribution operations. If sales could be forecast more accurately we would achieve far higher utilisation, better customer service and, I believe, significant cost savings.'*

However, Linda Fleet saw things differently. *'Look, I do understand Grant's frustration, but after all, this is a fashion business. By definition it is impossible to forecast accurately. In terms of month-by-month sales volumes we are in fact pretty accurate, but trying to make a forecast for every week end every product is almost impossible to do accurately. Sorry, that's just the nature of the business we're in. In fact, although Grant complains about our lack of forecast accuracy, he always does a great job in responding to unexpected customer demand.'*

Jim Thompson, the Managing Director, summed up his view of the current situation. *'Particularly significant has been our alliances with the Italian and German design houses. In effect we are positioning ourselves as a complete service partner to the designers. We have a world-class design capability together with manufacturing, order processing, order taking and distribution services. These abilities allow us to develop genuinely equal partnerships which integrate us into the whole industry's activities.'*

Linda Fleet also saw an increasing role for collaborative arrangements. *'It may be that we are seeing a fundamental change in how we do business within our industry. We have always seen ourselves as primarily a company that satisfies consumer desires through the medium of providing good service to retailers. The new partnership arrangements put us more into the 'business-to-business' sector. I don't have any problem with this in principle, but I'm a little anxious as to how much it gets us into areas of business beyond our core expertise.'*

The final issue which was being debated within the company was longer term, and particularly important. *'The two big changes we have made in this company have both happened because we exploited a strength we already had within the company. Moving into Concept products was only possible because we brought our high-tech precision expertise that we had developed in the aerospace sector into the homeware sector where none of our new competitors could match our manufacturing excellence. Then, when we moved into design house partnerships we did so because we had a set of designers who could command respect from the world-class design houses with whom we formed partnerships. So what is the next move for us? Do we expand globally? We are strong in Europe but nowhere else in the world. Do we extend our design scope into other markets, such as furniture? If so, that would take us into areas where we have no manufacturing expertise. We are great at plastic injection moulding, but if we tried any other manufacturing processes, we would be no better than, and probably worse than, other firms with more experience. So what's the future for us?'* (Jim Thompson, CEO CDS)

QUESTIONS

1 Why is operations management important in CDS?

2 Draw a 'four Vs' profile for the company's products/services.

3 What would you recommend to the company if they asked you to advise them in improving their operations?

APPLYING THE PRINCIPLES

Hints

Some of these exercises can be answered by reading the chapter. Others will require some general knowledge of business activity and some might require an element of investigation. Hints on how they all can be answered are to be found in the eText at **www.pearsoned.co.uk/slack**.

1 Quentin Cakes make about 20,000 cakes per year in two sizes, both based on the same recipe. Sales peak at Christmas time when demand is about 50 per cent higher than in the more quiet summer period. Their customers (the stores who stock their products) order their cakes in advance through a simple internet-based ordering system. Knowing that they have some surplus capacity, one of their customers has approached them with two potential new orders.

(a) The *Custom Cake* option – this would involve making cakes in different sizes where consumers could specify a message or greeting to be 'iced' on top of the cake. The consumer would give the inscription to the store who would e-mail it through to the factory. The customer thought that demand would be around 1,000 cakes per year, mostly at celebration times such as Valentine's Day and Christmas.

(b) The *Individual Cake* option – this would involve Quentin Cakes introducing a new line of very small cakes intended for individual consumption. Demand for this individual-sized cake was forecast to be around 4,000 per year, with demand likely to be more evenly distributed throughout the year than their existing products.

The total revenue from both options is likely to be roughly the same and the company has only capacity to adopt one of the ideas. But which one should it be?

2 Described as having *'revolutionised the concept of sandwich making and eating'*, Pret A Manger opened their first shop in London in the mid 1980s. Now they have over 130 shops in UK, New York, Hong Kong and Tokyo. They say that their secret is to focus continually on quality, in all its activities. *'Many food retailers focus on extending the shelf life of their food, but that's of no interest to us. We maintain our edge by selling food that simply can't be beaten for freshness. At the end of the day, we give whatever we haven't sold to charity to help feed those who would otherwise go hungry.'* The first Pret A Manger shop had its own kitchen where fresh ingredients were delivered first thing every morning, and food was prepared throughout the day. Every Pret shop since has followed this model. The team members serving on the tills at lunchtime will have been making sandwiches in the kitchen that morning. They rejected the idea of a huge centralised sandwich factory even though it could significantly reduce costs. Pret also own and manage all their shops directly so that

they can ensure consistently high standards. '*We are determined never to forget that our hardworking people make all the difference. They are our heart and soul. When they care, our business is sound. If they cease to care, our business goes down the drain. We work hard at building great teams. We take our reward schemes and career opportunities very seriously. We don't work nights (generally), we wear jeans, we party!*'

(a) Do you think Pret A Manger fully understand the importance of their operations management?

(b) What evidence is there for this?

(c) What kind of operations management activities at Pret A Manger might come under the four headings of direct, design, deliver and develop?

3 Visit a furniture store (other than IKEA). Observe how the shop operates, for example, where customers go, how staff interact with them, how big it is, how the shop has chosen to use its space, what variety of products it offers, and so on. Talk with the staff and managers if you can. Think about how the shop that you have visited is different from IKEA. Then consider the question:

- What implications do the differences between IKEA and the shop you visited have for their operations management?

4 Write down five services that you have 'consumed' in the last week. Try and make these as varied as possible. Examples could include public transport, a bank, any shop or supermarket, attendance at an education course, a cinema, a restaurant etc.

For each of these services, ask yourself the following questions:

(a) Did the service meet your expectations? If so, what did the management of the service have to do well in order to satisfy your expectations? If not, where did they fail? Why might they have failed?

(b) If you were in charge of managing the delivery of these services what would you do to improve the service?

(c) If they wanted to, how could the service be delivered at a lower cost so that the service could reduce its prices?

(d) How do you think that the service copes when something goes wrong (such as a piece of technology breaking down)?

(e) Which other organisations might supply the service with products and services? (In other words, they are your 'supplier', but who are *their* suppliers)?

(f) How do you think the service copes with fluctuation of demand over the day, week, month or year?

These questions are just some of the issues which the operations managers in these services have to deal with. Think about the other issues they will have to manage in order to deliver the service effectively.

5 Find a copy of a financial newspaper (*Financial Times*, *Wall Street Journal*, *Economist* etc.) and identify one company that is described in the paper that day. What do you think would be the main operations issues for that company?

Notes on chapter

1 Thanks to Mike Shulver and Paul Walley for this information. Further sources include 'IKEA plans an end to stressful shopping' (2006), *London Evening Standard*, 24 April.

2 Source: Virgin Atlantic website.

TAKING IT FURTHER

Chase, R.B., Aquilano, N.J. and Jacobs, F.R. (2010) *Production and Operations Management: Manufacturing and Services* (11th Edition), McGraw-Hill. *There are many good general textbooks on operations management. This was one of the first and is still one of the best, though written very much for an American audience.*

Hall, J.M. and Johnson, M.E. (2009) 'When should a process be art', *Harvard Business Review*, March. *An article that provides a good discussion of how to deal with more creative processes.*

Johnston, R. and Clark, E. (2008) *Service Operations Management*, Financial Times Prentice Hall. *What can we say! A great treatment of service operations from the same stable as this textbook.*

Schmenner, R.W., van Wassenhove, L., Ketokivi, M., Heyl, J. and Lusch, R.F. (2009) 'Too much theory, not enough understanding', *Journal of Operations Management*, Vol. 27, 339–343. *An academic treatment of one aspect of the operations management literature.*

Slack, N. and Lewis, M. (eds) (2005) *The Blackwell Encyclopedic Dictionary of Operations Management* (2nd Edition). Blackwell Business, Oxford. *For those who like technical descriptions and definitions.*

Sousa, R. and Voss, C.A. (2008) 'Contingency research in operations management practices', *Journal of Operations Management*, Vol. 26, 697–713. *Another academic treatment, this time, of why context matters.*

USEFUL WEBSITES

www.opsman.org *Definitions, links and opinions on operations and process management.*

www.iomnet.org *The Institute of Operations Management site. One of the main professional bodies for the subject.*

www.poms.org *A US academic society for production and operations management. Academic, but some useful material, including a link to an encyclopedia of operations management terms.*

www.sussex.ac.uk/users/dt31/TOMI/ *One of the longest established portals for the subject. Useful for academics and students alike.*

www.ft.com *Useful for researching topics and companies.*

Interactive quiz

For further resources including examples, animated diagrams, self-test questions, Excel spreadsheets and video materials please explore the eText on the companion website at **www.pearsoned.co.uk/slack**.

2

Operations strategy

Introduction

In the long term, the major (and some would say, only) objective for operations and processes is to provide a business with some form of strategic advantage. That is why the management of a business's processes and operations and its intended overall strategy must be logically connected. Yet for many in business, the very idea of an 'operations strategy' is a contradiction in terms. After all, to be involved in the strategy process is the complete opposite of those day-to-day tasks and activities associated with being an operations manager. Nevertheless, it is also clear that operations can have a real strategic impact. Many *enduringly* remarkable enterprises, from Apple to Zara, use their operations resources to gain long-term strategic success. Such firms have found that it is the way they manage their operations that sets them apart from, and above, their competitors. Without a strong link with overall strategy, operations and processes will be without a coherent direction. And without direction they may finish up making internal decisions that either do not reflect strategy, or that conflict with each other, or both. So, although operations and process management is largely 'operational', it also has a strategic dimension that is vital if operations are to fulfil their potential to contribute to competitiveness. Figure 2.1 shows the position of the ideas described in this chapter in the general model of operations management.

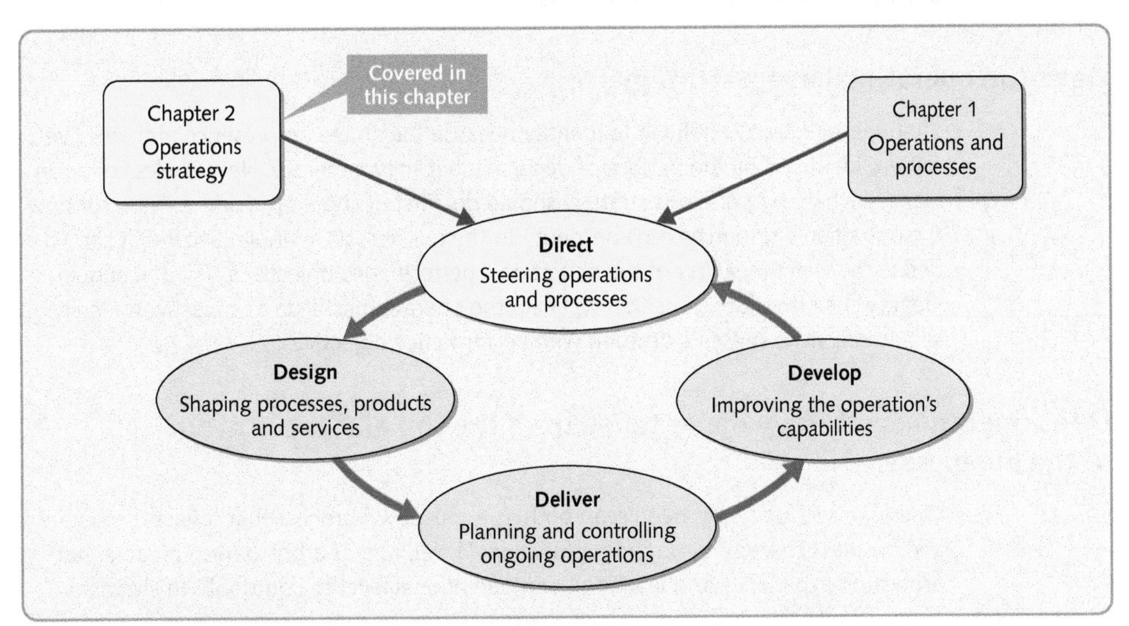

Figure 2.1 Operations strategy is the pattern of decisions and actions that shapes the long-term vision, objectives and capabilities of the operation and its contribution to overall strategy

EXECUTIVE SUMMARY

Each chapter is structured around a set of diagnostic questions. These questions suggest what you should ask in order to gain an understanding of the important issues of a topic, and, as a result, improve your decision making. An executive summary addressing these questions is provided below.

What is operations strategy?

Operations strategy is the pattern of decisions and actions that shapes the long-term vision, objectives and capabilities of the operation and its contribution to overall strategy. It is the way in which operations resources are developed over the long term to create sustainable competitive advantage for the business. Increasingly, many businesses are seeing their operations strategy as one of the best ways to differentiate themselves from competitors. Even in those companies that are marketing led (such as fast-moving consumer goods), an effective operations strategy can add value by allowing the exploitation of market positioning.

Does the operation have a strategy?

Strategies are always difficult to identify because they have no presence in themselves, but are identified by the pattern of decisions that they generate. Nevertheless one can identify what an operations strategy should do. First, it should provide a vision for how the operation's resources can contribute to the business as a whole. Second, it should define the exact meaning of the operation's performance objectives. Third, it should identify the broad decisions that will help the operation achieve its objectives. Finally, it should reconcile strategic decision with performance objectives.

Does operations strategy make sense from the top and the bottom of the business?

Operations strategy can been seen both as a top-down process that reflects corporate and business strategy through to a functional level, and as a bottom-up process that allows the experience and learning at an operational level to contribute to strategic thinking. Without both of these perspectives, operations strategy will be only partially effective. It should communicate both top to bottom *and* bottom to top throughout the hierarchical levels of the business.

Does operations strategy align market requirements with operations resources?

The most important short-term objective of operations strategy is to ensure that operations resources can satisfy market requirements. But this is not the only objective. In the longer term, operations strategy must build the capabilities within its resources that will allow the business to provide something to the market that its competitors find difficult to imitate or match. These two objectives are called the market requirements perspective and the operations resource capability perspective. The latter is very much influenced by the resource-based view (RBV) of the firm. The objective of operations strategy can be seen as achieving 'fit' between these two perspectives.

Does operations strategy set an improvement path?

The purpose of operations strategy is to improve the business's performance relative to its competitors' in the long term. It therefore must provide an indication of how this improvement is to take place. This is best addressed by considering the trade-offs between performance objectives in terms of the 'efficient frontier' model. This describes operations strategy as a combination of repositioning performance along an existing efficient frontier, and increasing overall operations effectiveness by overcoming trade-offs to expand the efficient frontier.

DIAGNOSTIC QUESTION

What is operations strategy?

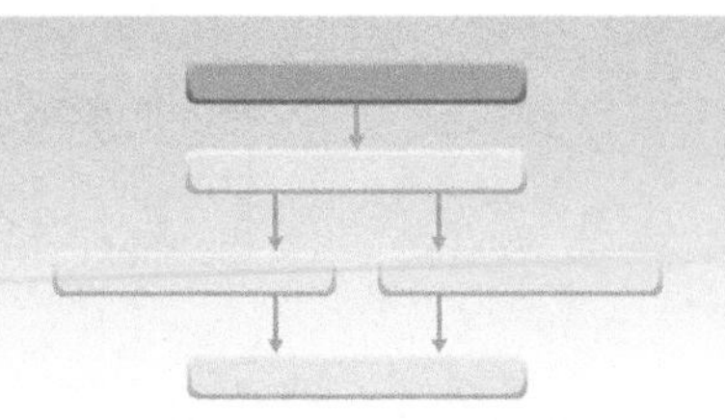

Video

Operations strategy is the pattern of decisions and actions that shapes the long-term vision, objectives and capabilities of the operation and its contribution to the overall strategy of the business[1]. The term 'operations strategy' sounds at first like a contradiction. How can 'operations', a subject that is generally concerned with the day-to-day creation and delivery of goods and services, be strategic? 'Strategy' is usually regarded as the opposite of those day-to-day routine activities. But, as we indicated previously, 'operations' is not the same as 'operational'. 'Operations' are the resources that create products and services. 'Operational' is the opposite of strategic, meaning day-to-day and detailed.

Perhaps more significantly, many of the businesses that seem to be especially competitively successful, and who appear to be sustaining their success into the longer term, have a clear and often inventive operations strategy. Just look at some of the high-profile companies quoted in this book, or that feature in the business press. It is not just that their operations strategy provides these companies with adequate support; it is their operations strategy that is the pivotal reason for their competitive superiority. Just as revealing, when companies stumble, it is often because they have either taken their eye off the operations ball, or failed to appreciate its importance in the first place. More generally, all enterprises, *and all parts of the enterprise*, need to prevent strategic decisions being frustrated by poor operational implementation. And this idea leads us to the second purpose of this chapter (and indeed the book as a whole): to show how, by using the principles of operations strategy, *all* parts of any business and *all* functions of a business can contribute effectively to the overall success of the business. So the idea of 'operations strategy' has two different but related meanings. The first is concerned with the operations function itself, and how it can contribute to strategic success. The second is concerned with how *any* function can develop its process and resources and establish its strategic role.

Look at these two examples of businesses with operations strategies that are clear and explicit and have contributed to their competitive success.

EXAMPLE

Flextronics[2]

Behind every well-known brand name in consumer electronics, much of the high-tech manufacturing which forms the heart of the product is probably done by companies few of us have heard of. Companies such as Ericsson and IBM are increasingly using Electronic Manufacturing Services (EMS) companies which specialise in providing the outsourced design, engineering, manufacturing and logistics operations for big brand names. Flextronics is one of the leading EMS providers in providing 'operational services' to technology companies. With over 70,000 employees spread throughout its facilities in 28 countries, it has a global presence that allows it the flexibility to serve customers in all the key markets throughout the world.

Source: Keith Dannemiller/Corbis

Flextronics manufacturing locations have to balance their customers' need for low costs (electronic goods are often sold in a fiercely competitive market) with their need for responsive and flexible service (electronics markets can also be volatile). Flextronics could have set up manufacturing plants close to its main customers in North America and Western Europe. This would certainly facilitate fast response and great service to customers; unfortunately these markets also tend to have high manufacturing costs. Flextronics' operations strategy must therefore achieve a balance between low costs and high levels of service. One way Flextronics achieves this is through its strategic location and supply network decisions, adopting what it calls its 'Industrial Park Strategy'. This involves finding locations which have relatively low manufacturing costs but are close to its major markets. It has established Industrial Parks in places such as Hungary, Poland, Brazil and Mexico (the Guadalajara Park in Mexico is shown in the illustration above). Flextronics own suppliers also are encouraged to locate within the Park to provide stability and further reduce response times.

EXAMPLE

Amazon[3]

As a publicly stated ambitious target it takes some beating. *'Amazon.com strives to be'*, it says, *'Earth's most customer-centric company.'* Founded by Jeff Bezos in 1995, the Amazon.com website started as a place to buy books, giving its customers what at the time was a unique customer experience. Bezos believed that only the internet could offer customers the convenience of browsing a selection of millions of book titles in a single sitting. During its first 30 days of business, Amazon.com fulfilled orders for customers in 45 countries – all shipped from Bezos' Seattle-area garage. And that initial success has been followed by continued growth that is based on a clear strategy of technological innovation. Among its many technological innovations for customers, Amazon.com offers a personalised shopping experience for each customer, book discovery through 'Search Inside The Book', convenient checkout using '1-Click® Shopping', and community features like Listmania and Wish Lists that help customers discover new products and make informed buying decisions. In addition, Amazon.com operates retail websites and offers programs that enable other retailers and to individual sellers to sell products on their websites. Now, many prominent retailers work with Amazon Services to power their e-commerce offerings from end-to-end, including technology services, merchandising, customer service and order fulfilment.

Source: AFP/Getty Images.

In the mid-2000s, Bezos, speaking at a number of public events about the company's plans, made its future strategy clearer. Although Amazon was generally seen as an internet book retailer and then a more general internet retailer, Bezos was promoting Amazon's 'utility computing' services. These provided cheap access to online computer storage, allowed program developers to rent computing capacity on Amazon systems, and connected firms with other firms who perform specialist tasks that are difficult to automate. The problem with online retailing, said Bezos, is its seasonality. At peak times, such as Christmas, Amazon has far more computing capacity than it needs for the rest of the year. At low points it may be using as little as 10 per cent of its total capacity. Hiring out that spare capacity is an obvious way to bring in extra revenue. In addition, Amazon soon had developed a search engine, a video download business, a service ('Fulfilment By Amazon') that allowed other companies to use Amazon's logistics capability including the handling of returned items, and a service that provided access to Amazon's 'back-end' technology.

A couple of years later, Amazon announced its EC2 (Elastic Compute Cloud) service that provides resizable computing capacity 'in the cloud'[4]. It is designed, say Amazon, to make web-scale computing easier for developers: *'Amazon EC2's simple web service interface allows you to obtain and configure capacity with minimal friction. It provides you with complete control of your computing resources and lets you run on Amazon's proven computing environment. Amazon EC2 reduces the time required to obtain and boot new server instances to minutes, allowing you to quickly scale capacity, both up and down, as your computing requirements change. Amazon EC2 changes the economics of computing by allowing you to pay only for capacity that you actually use. Amazon EC2 provides developers with the tools to build failure resilient applications and isolate themselves from common failure scenarios.'* Don't worry if you can't follow the technicalities of Amazon's statement, it is aimed at IT professionals. The important point is that it is a business-to-business service based on the company's core competence of leveraging its processes and technology that can make retail operations ultra efficient.

However, Amazon's apparent redefinition of its strategy was immediately criticised by some observers. 'Why not,' they said, 'stick to what you know, focus on your core competence of internet retailing?' Bezos's response was clear. 'We *are* sticking to our core competence; this is what we've been doing for the last 11 years. The only thing that's changed is that we are exposing it for (the benefit of) others.' At least for Jeff Bezos, Amazon is not so much an internet retailer as a provider of internet-based technology and logistics services.

What do these two examples have in common?

Neither of these companies suffered from any lack of clarity regarding what they wanted to do in the market. They are clear about what they are offering their customers (and they document it explicitly on their websites). They are also both equally clear in spelling out their operations strategy. Flextronics is willing to relocate whole operations in its commitment to responsive but low-cost customer service. Amazon redefined its market to sell its core operations-based capability. Without this type of clear 'top-down' strategy it is difficult to achieve clarity in the way individual processes should be managed. Yet both these companies are also known for the way they have perfected the operational-level processes to the extent that their learned expertise contributes to strategy in a 'bottom-up' manner. Also, in both these cases the requirements of their markets are clearly reflected in their operations' performance objectives (responsiveness and cost for Flextronics, and closely shadowing a changing market with its own internal technological and process capabilities in the case of Amazon). Similarly, both businesses have defined the way in which they achieve these objectives by strategically directing their operations resources (through its technological and process capabilities with Amazon, and through their location and supply chain decisions in the case of Flextronics). In other words, both have reconciled their market requirements with what their operations resource capabilities. These are the issues that we shall address in this chapter.

DIAGNOSTIC QUESTION

Does the operation have a strategy?

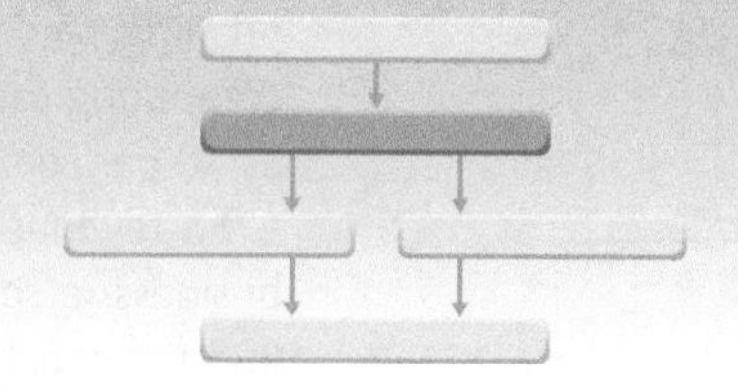

Video

There are some problems in asking this apparently simple question. In most operations management decisions you can see what you are dealing with. You can touch inventory, talk to people, programme machines, and so on. But strategy is different. You cannot see a strategy, feel it or touch it. Also, whereas the effects of most operations management decisions become evident relatively fast, it may be years before an operations strategy decision can be judged to be a success or not. Moreover, any 'strategy' is always more than a single decision. Operations strategy will be revealed in the total *pattern* of decisions that a business takes in developing its operations in the long term. Nevertheless, the question is an obvious starting point and one that must be addressed by all operations.

So, what should an operations strategy do? First, it should articulate a vision of how the businesses operations and processes can contribute to its overall strategy. This is something beyond the individual collection of decisions that will actually constitute the strategy. Second, it should translate market requirements into a message that will have some meaning within its operations. This means describing what customers want in terms of a clear and prioritised set of operations' *performance objectives*. Third, it should identify the broad decisions that will shape the operation's capabilities, and allow their long-term development so that they will provide the basis for the business's sustainable advantage. Finally, it should explain how its intended market requirements and its strategic operations decisions are to be reconciled.

An operations strategy should take significant stakeholders into account

All operations have stakeholders. They are the people and groups who have a legitimate interest in the operation's strategy. Some are internal (employees); others are external (customers, society or community groups, and a company's shareholders). External stakeholders may have a direct commercial relationship with the organisation, (suppliers and customers); others may not, (industry regulators). In not-for-profit operations, these stakeholder groups can overlap. So, voluntary workers in a charity may be employees, shareholders and customers all at once. However, in any kind of organisation, it is a responsibility of the operations function to understand the (often conflicting) objectives of its stakeholders and set it objectives accordingly. Yet, although all stakeholder groups, to different extents, will be interested in operations performance, they are likely to have very different views of which aspect of performance is important. Table 2.1 identifies typical stakeholder requirements. But stakeholder relationships are not just one way. It is also useful to consider what an individual organisation or business wants of the stakeholder groups themselves. Some of these requirements are illustrated in Table 2.1. The dilemma with using this wide range of stakeholders to judge performance is that organisations, particularly commercial companies, have to cope with the conflicting pressures of maximising profitability on one hand, with the expectation that they will manage in the interests of (all or part of) society in general with accountability and transparency. Even if a business wanted to reflect aspects of performance beyond its own immediate interests, how is it to do it?

OPERATIONS PRINCIPLE

Operations strategy should take significant stakeholders into account.

Corporate social responsibility (CSR) and the 'triple bottom line'

Strongly related to the stakeholder perspective of operations performance is that of corporate social responsibility (generally known as CSR). A direct link with the stakeholder concept is to be found in the definition used by Marks & Spencer, the UK-based retailer. *'Corporate Social Responsibility . . . is listening and responding to the needs of a company's stakeholders. This includes the requirements of sustainable development. We believe that building good relationships with employees, suppliers and wider society is the best guarantee of long-term success. This is the backbone of our approach to CSR.'* The issue of how broader social performance objectives can be included in operations management's activities is of increasing importance, both from an ethical and a commercial point of view. However, converting the CSR concept into operational reality presents considerable difficulty, although several attempts have been made. One such is the 'triple bottom line', an approach to value creation which attempts to integrate economic, environmental and

Table 2.1 Typical stakeholders' performance objectives

Stakeholder	*What stakeholders want from the operation*	*What the operation wants from stakeholders*
Shareholders	Return on investment Stability of earnings Liquidity of investment	Investment capital Long-term commitment
Directors/top management	Low/acceptable operating costs Secure revenue Well-targeted investment Low risk of failure Future innovation	Coherent, consistent, clear and achievable strategies Appropriate investment
Staff	Fair wages Good working conditions Safe work environment Personal and career development	Attendance Diligence/best efforts Honesty Engagement
Staff representative bodies (e.g. trade unions)	Conformance with national agreements Consultation	Understanding Fairness Assistance in problem solving
Suppliers (of materials, services, equipment etc.)	Early notice of requirements Long-term orders Fair price On-time payment	Integrity of delivery, quality and volume Innovation Responsiveness Progressive price reductions
Regulators (e.g. Financial regulators)	Conformance to regulations Feedback on effectiveness of regulations	Consistency of regulation Consistency of application of regulations Responsiveness to industry concerns
Government (local, national, regional)	Conformance to legal requirements Contribution to (local/national/regional) economy	Low/simple taxation Representation of local concerns Appropriate infrastructure
Lobby groups (e.g. environmental lobby groups)	Alignment of the organisation's activities with whatever the group are promoting	No unfair targeting Practical help in achieving aims (if the organisation wants to achieve them)
Society	Minimise negative effects from the operation (noise, traffic etc., and maximise positive effects (jobs, local sponsorship etc.)	Support for organisation's plans

social impacts. It first came to prominence in John Elkington's book, *Cannibals with Forks: the Triple Bottom Line of 21st Century Business*[5]. It advocated expanding the conventional financial reporting conventions to include ecological (sustainability) and social performance in addition to financial performance. So, for example, Holcim, the Swiss cement and aggregates group, have established a set of group-wide performance targets to achieve their triple bottom line business goals. But, before targets are met, the company aim to understand their current performance. They do this by establishing consistent measurement and reporting techniques, as well as implementing management systems to monitor progress toward their goals. Yet CSR-related performance measurement systems should not, say Holcim, be separate from the more conventional business systems. To work effectively, CSR performance systems are integrated into overall business processes and supported by appropriate training.

An operations strategy should articulate a vision for the operations contribution

The 'vision' for an operation is a clear statement of how operations intend to contribute value for the business. It is not a statement of what the operation wants to *achieve* (those are its objectives), but rather an idea of what it must *become* and what contribution it should make.

OPERATIONS PRINCIPLE

Operations strategy should articulate a 'vision' for the operations function contribution to overall strategy.

A common approach to summarising operations contribution is the Hayes and Wheelwright Four-Stage Model.[6] The model traces the progression of the operations function from what is the largely negative role of 'stage 1' operations to it becoming the central element of competitive strategy in excellent 'stage 4' operations. Figure 2.2 illustrates the four steps involved in moving from stage 1 to stage 4.

Stage 1: Internal neutrality

Video

This is the very poorest level of contribution by the operations function. The other functions regard it as holding them back from competing effectively. The operations function is inward looking and at best reactive with very little positive to contribute towards competitive success. Its goal is to be ignored. At least then it isn't holding the company back in any way. Certainly the rest of the organisation would not look to operations as the source of any originality, flair or competitive drive. Its vision is to be 'internally neutral', a position it attempts to achieve not by anything positive but by avoiding the bigger mistakes.

Stage 2: External neutrality

The first step of breaking out of stage 1 is for the operations function to begin comparing itself with similar companies or organisations in the outside market. This may not

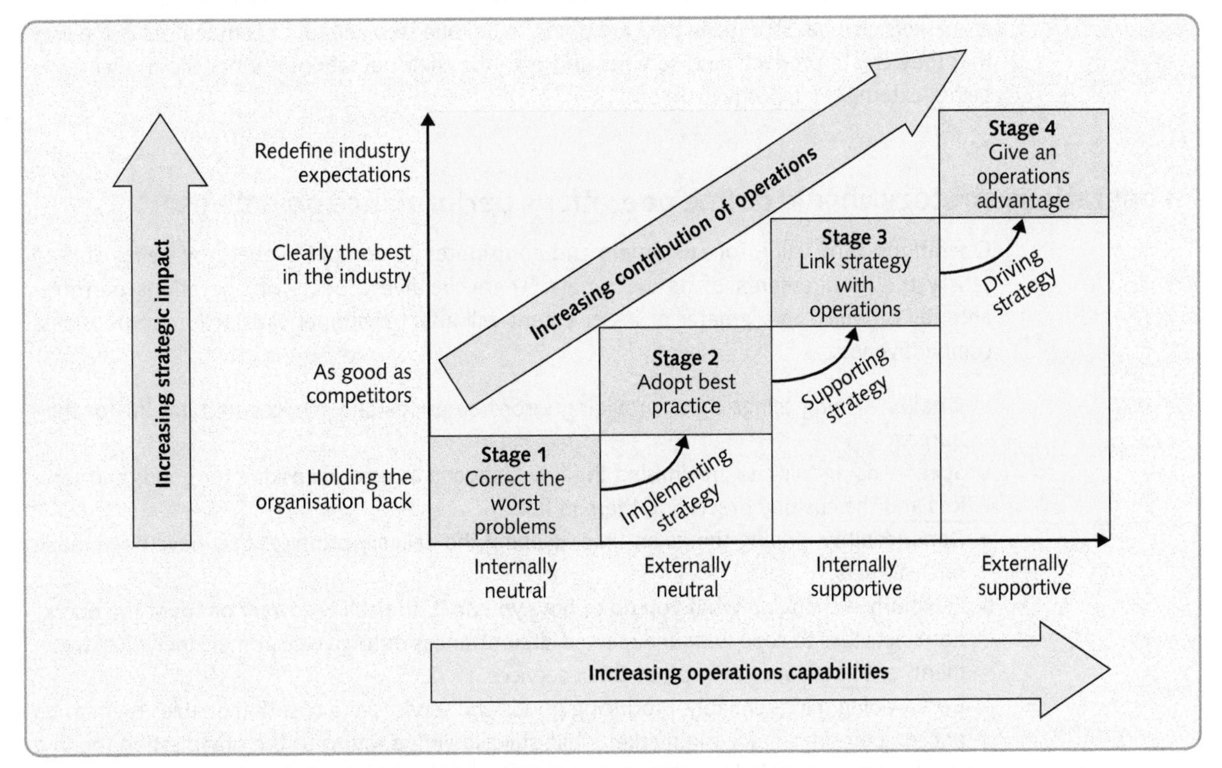

Figure 2.2 Hayes and Wheelwright's Four-Stage Model of operations contribution sees operations as moving from implementation of strategy, through to supporting strategy, and finally driving strategy

Practice note

immediately take it to the 'first division' of companies in the market, but at least it is measuring itself against its competitors' performance and trying to be 'appropriate', by adopting 'best practice' from them. Its vision is to become 'up to speed' or 'externally neutral' with similar businesses in its industry by adopting 'best practice' ideas and norms of performance from others.

Stage 3: Internally supportive

Stage 3 operations have probably reached the 'first division' in their market. They may not be better than their competitors on every aspect of operations performance but they are broadly up with the best. Yet, the vision of stage 3 operations is to be clearly and unambiguously the very best in the market. They may try to achieve this by gaining a clear view of the company's competitive or strategic goals and developing 'appropriate' operations resources to Excel in the areas in which the company needs to compete effectively. The operation is trying to be 'internally supportive' by providing a credible operations strategy.

Stage 4: Externally supportive

Stage 3 used to be taken as the limit of the operations function's contribution. Yet the model captures the growing importance of operations management by suggesting a further stage – stage 4. The difference between stages 3 and 4 is subtle, but important. A stage 4 company is one where the vision for the operations function is to provide *the* foundation for competitive success. Operations looks to the long term. It forecasts likely changes in markets and supply, and, over time, it develops the operations-based capabilities that will be required to compete in future market conditions. The operations function is becoming central to strategy-making. Stage 4 operations are creative and proactive. They are innovative and capable of adaptation as markets change. Essentially they are trying to be 'one step ahead' of competitors in the way that they create products and services and organise their operations – what the model terms being 'externally supportive'.

An operations strategy should define operations performance objectives

Operations adds value for customers and contributes to competitiveness by being able to satisfy the requirements of its customers. There are five aspects of operations performance, all of which to a greater or lesser extent will affect customer satisfaction and business competitiveness.

- **Quality** – doing things right, providing error-free goods and services that are 'fit for their purpose'.
- **Speed** – doing this fast, minimising the time between a customer asking for goods and services and the customer receiving them in full.
- **Dependability** – doing things on time, keeping the delivery promises that have been made to customers.
- **Flexibility** – changing what you do or how you do it, the ability to vary or adapt the operation's activities to cope with unexpected circumstances or to give customers individual treatment, or to introduce new products or services.
- **Cost** – doing things cheaply, producing goods and services at a cost that enables them to be priced appropriately for the market while still allowing a return to the organisation (or, in a not-for-profit organisation, that give good value to the tax payers or whoever is funding the operation).

The exact meaning of performance objectives is different in different operations

Different operations will have different views of what each of the performance objectives actually mean. Table 2.2 looks at how two operations, an insurance company and a steel plant, define each performance objective. For example, the insurance company sees quality as being at least as much about the manner in which their customers relate to their service as it does about the absence of technical errors. The steel plant, on the other hand, while not ignoring quality of service, primarily, emphasises product-related technical issues. Although, they are selecting from the same pool of factors which together constitute the generic performance objective, they will emphasise different elements.

OPERATIONS PRINCIPLE
Operations performance objectives can be grouped together as quality, speed, dependability, flexibility and cost.

Sometimes operations may choose to re-bundle elements using slightly different headings. For example, it is not uncommon in some service operations to refer to 'quality of service' as representing all the competitive factors we have listed under quality *and* speed *and* dependability (and sometimes aspects of flexibility). For example, information network operations use the term 'Quality of Service' (QoS) to describe their goal of providing guarantees on the ability of a network to deliver predictable results. This is often specified as including uptime (dependability), bandwidth provision (dependability and flexibility), latency or delay (speed of throughput), and error rate (quality). In practice, the issue is not so much one of universal definition but rather consistency within one, or a group of operations. At the very least it is important that individual companies have it clear in their own minds how each performance objective is to be defined.

OPERATIONS PRINCIPLE
The interpretation of the five performance objectives will differ between different operations.

Table 2.2 Aspects of performance objectives for two operations

Insurance company	*Performance objectives*	*Steel plant*
Aspects of each performance objective include . . .		*Aspects of each performance objective include . . .*
• Professionalism of staff • Friendliness of staff • Accuracy of information • Ability to change details in future	**Quality**	• Percentage of products conforming to their specification • Absolute specification of products • Usefulness of technical advice
• Time for call centre to respond • Prompt advice response • Fast quotation decisions • Fast response to claims	**Speed**	• Lead-time from enquiry to quotation • Lead-time from order to delivery • Lead-time for technical advice
• Reliability of original promise date • Customers kept informed	**Dependability**	• Percentage of deliveries 'on-time, in-full' • Customers kept informed of delivery dates
• Customisation of terms of insurance cover • Ability to cope with changes in circumstances, such as level of demand • Ability to handle wide variety of risks	**Flexibility**	• Range of sizes, gauges, coatings etc. possible • Rate of new product introduction • Ability to change quantity, composition and timing of an order
• Premium charged • Arrangement charges • 'No-claims' deals • 'Excess' charges	**Cost**	• Price of products • Price of technical advice • Discounts available • Payment terms

The relative priority of performance objectives differs between businesses

Not every operation will apply the same priorities to its performance objectives. Businesses that compete in different ways should want different things from their operations functions. In fact, there should be a clear logical connection between the competitive stance of a business and its operations objectives. So, a business that competes primarily on low prices and 'value for money' should be placing emphasis on operations objectives such as cost, productivity and efficiency; one that competes on a high degree of customisation of its services or products should be placing an emphasis on flexibility, and so on. Many successful companies understand the importance of making this connection between their message to customers and the operations performance objectives that they emphasise. For example:[7]

> OPERATIONS PRINCIPLE
>
> *The relative importance of the five performance objectives depends on how the business competes in its market.*

> *'Our management principle is the commitment to quality and reliability . . . to deliver safe and innovative products and services . . . and to improve the quality and reliability of our businesses.'*
> *(Komatsu)*

> *'The management team will . . . develop high quality, strongly differentiated consumer brands and service standards . . . use the benefits of the global nature and scale economies of the business to operate a highly efficient support infrastructure (with) . . . high quality and service standards which deliver an excellent guest experience . . .'*
> (InterContinental Hotels Group)

> *'A level of quality, durability and value that's truly superior in the market place . . . the principle that what is best for the customer is also best for the company . . . (our) . . . customers have learnt to expect a high level of service at all times – from initiating the order, to receiving help and advice, to speedy shipping and further follow-up where necessary . . . (our) . . . employees "go that extra mile". '*
> (Lands' End)

An operations strategy should identify the broad decisions that will help the operation achieve its objectives

Few businesses have the resources to pursue every single action that might improve their operations performance. So an operations strategy should indicate broadly how the operation might best achieve its performance objectives. For example, a business might specify that it will attempt to reduce its costs by aggressive outsourcing if its non-core business processes and by investing in more efficient technology. Or, it may declare that it intends to offer a more customised set of products or services through adopting a modular approach to its product or service design. The balance here is between a strategy that is overly restrictive in specifying how performance objectives are to be achieved, and one that is so open that it gives little guidance as to what ideas should be pursued.

There are several categorisations of operations strategy decisions. Any of them are valid if they capture the key decisions. Here we categorise operations strategy decisions in the same way we categorise operations management decisions; as applying to the activities of design, delivery and development. Table 2.3 illustrates some of the broad operations strategy decisions that fall within each category.

An operations strategy should reconcile strategic decisions to objectives

> OPERATIONS PRINCIPLE
>
> *An operation's strategy should articulate the relationship between operations objectives and the means of achieving them.*

We can now bring together two sets of ideas and, in doing so, we also bring together the two perspectives of (a) market requirements and (b) operations resources to form the two dimensions of a matrix. This *operations strategy* matrix is shown in Figure 2.3. It describes operations strategy as the intersection of a company's performance objectives and the strategic decisions that it makes. In fact there are several

Table 2.3 Some strategic decisions that may be addressed in an operations strategy

Strategic decisions concerned with the *design* of operations and processes	• How should the operation decide which products or services to develop and how to manage the development process? • Should the operation develop its products or services in-house or outsource the designs? • Should the operation outsource some of its activities, or take more activities in-house? • Should the operation expand by acquiring its suppliers or its customers? If so, which ones should it acquire? • How many geographically separate sites should the operation have? • Where should operations sites be located? • What activities and capacity should be allocated to each site? • What broad types of technology should the operation be using? • How should the operation be developing its people? • What role should the people who staff the operation play in its management?
Strategic decisions concerned with planning and controlling the *delivery* of products and services	• How should the operation forecast and monitor the demand for its products and services? • How should the operation adjust its activity levels in response to demand fluctuations? • How should the operation monitor and develop its relationship with its suppliers? • How much inventory should the operation have and where should it be located? • What approach and system should the operation use to plan its activities?
Strategic decisions concerned with the development of operations performance	• How should the operation's performance be measured and reported? • How should the operation ensure that its performance is reflected in its improvement priorities? • Who should be involved in the improvement process? • How fast should improvement in performance be? • How should the improvement process be managed? • How should the operation maintain its resources so as to prevent failure? • How should the operation ensure continuity if a failure occurs?

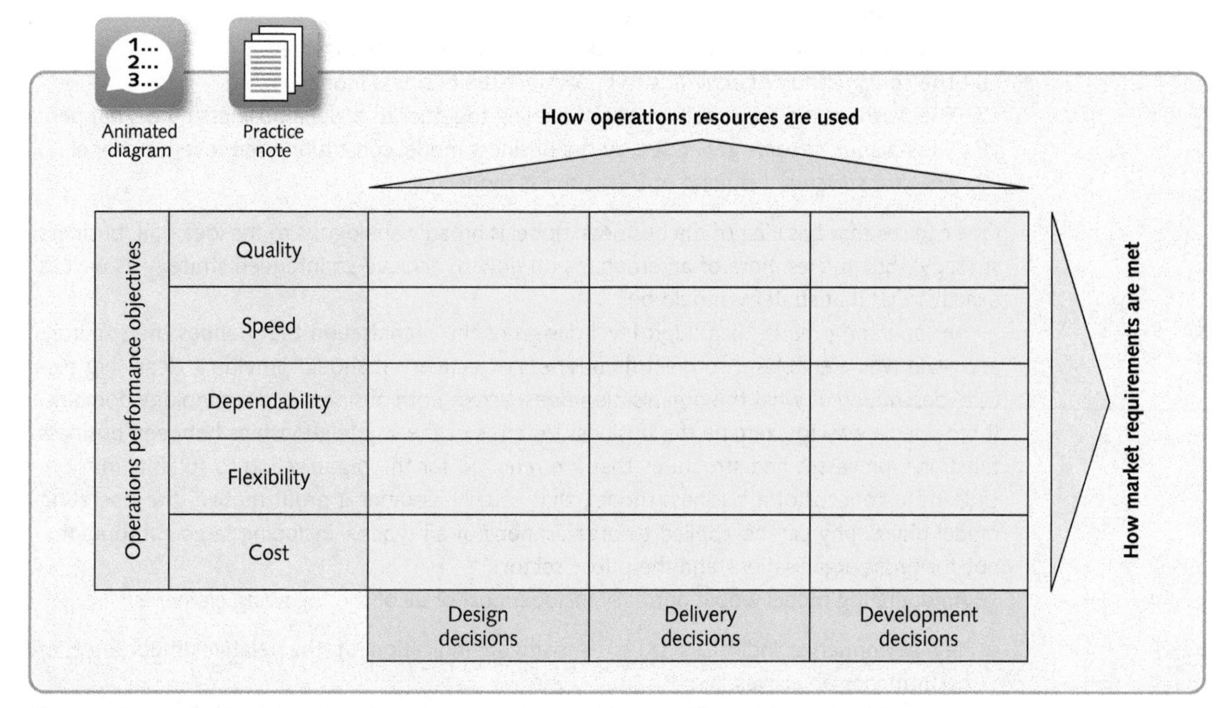

Figure 2.3 The operations strategy matrix defines operations strategy by the intersections of performance objectives and operations decisions

intersections between each performance objective and each decision area (however one wishes to define them). If a business thinks that it has an operations strategy, then it should have a coherent explanation for each of the cells in the matrix. That is, it should be able to explain and reconcile the intended links between each performance objective and each decision area. The process of reconciliation takes place between what is required from the operations function (performance objectives), and how the operation tries to achieve this through the set of choices made (and the capabilities that have been developed) in each decision area.

The concepts of the 'business model' and the 'operating model'

Two concepts have emerged over the last few years that are relevant to operations strategy (or at least the terms are new – one could argue that the ideas are far older). These are the concepts of the 'business model' and the 'operating model'.

Put simply, a 'business model' is the plan that is implemented by a company to generate revenue and make a profit. It includes the various parts and organisational functions of the business, as well as the revenues it generates and the expenses it incurs. In other words, what a company does and how they make money from doing it. More formally, it is 'a conceptual tool that contains a big set of elements and their relationships and allows [the expression of] the business logic of a specific firm. It is a description of the value a company offers to one or several segments of customers and of the architecture of the firm and its network of partners for creating, marketing and delivering this value and relationship capital, to generate profitable and sustainable revenue streams'[8].

One synthesis of literature[9] shows that business models have a number of common elements.

1. The *value proposition* of what is offered to the market.
2. The *target customer segments* addressed by the value proposition.
3. The communication and *distribution channels* to reach customers and offer the value proposition.
4. The *relationships* established with customers.
5. The *core capabilities* needed to make the business model possible.
6. The *configuration of activities* to implement the business model.
7. The *partners* and their motivations of coming together to make a business model happen.
8. The *revenue streams* generated by the business model constituting the revenue model.
9. The *cost structure* resulting of the business model.

One can see that this idea of the business model is broadly analogous to the idea of a 'business strategy', but implies more of an emphasis on *how* to achieve an intended strategy as well as exactly *what* that strategy should be.

An 'operating model' is a 'high-level design of the organisation that defines the structure and style which enables it to meet its business objectives'. It should provide a clear, 'big picture' description of what the organisation does, across both business and technology domains. It provides a way to examine the business in terms of the key relationships between business functions, processes and structures that are required for the organisation to fulfil its mission. Unlike the concept of a business model, that usually assumes a profit motive, the operating model philosophy can be applied to organisations of all types – including large corporations, not-for-profit organisations and the public sector.[10]

An operating model would normally include most or all of the following elements:

- Key performance indicators (KPIs) – with an indication of the relative importance of performance objectives.
- Core financial structure – Profit and Loss (P&L), new investments and cash flow.

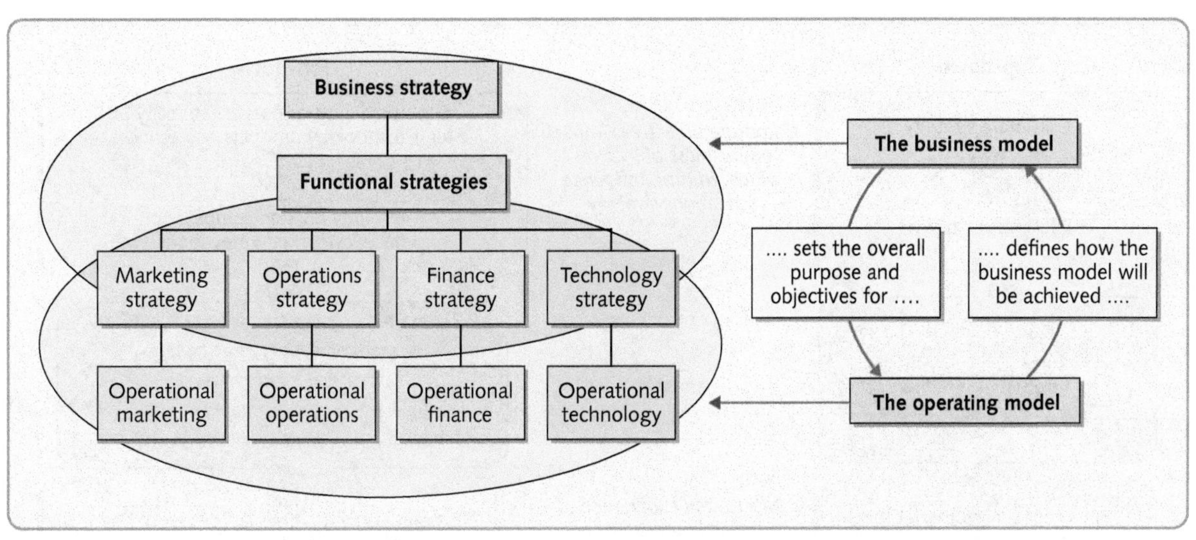

Figure 2.4 The relationship between the concepts of the 'business model' and the 'operating model'

- The nature of accountabilities for products, geographies, assets etc.
- The structure of the organisation – often expressed as capability areas rather than functional roles.
- Systems and technologies.
- Processes responsibilities and interactions.
- Key knowledge and competence.

Note two important characteristics of an operating model. First, it does not respect conventional functional boundaries as such. In some ways the concept of the operating model reflects the idea that we proposed in Chapter 1, namely that all managers are operations managers and all functions can be considered as operations because they comprise processes that deliver some kind of service. An operating model is like an operations strategy, but applied across all functions and domains of the organisation. Second, there are clear overlaps between the 'business model' and the 'operating model'. The main difference being that an operating model focuses more on how an overall business strategy is to be achieved. Operating models have an element of implied change or transformation of the organisation's resources and processes. Often the term 'target operating model' is used to describe the way the organisation should operate in the future if it is going to achieve its objectives and make a success of its business model. Figure 2.4 illustrates the relationship between business and operating models.

DIAGNOSTIC QUESTION

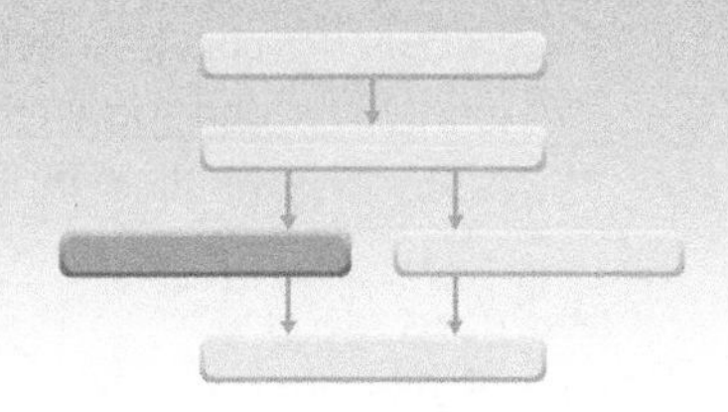

Does operations strategy make sense from the top and the bottom of the business?

The traditional view of operations strategy is that it is one of several *functional strategies* that are governed by decisions taken at the top of the organisational tree. In this view, operations strategy, together with marketing, human resources and other functional strategies, take their lead exclusively from the needs of the business as a whole. This is often called a 'top-down' perspective on operations strategy. An alternative view is that operations strategies emerge

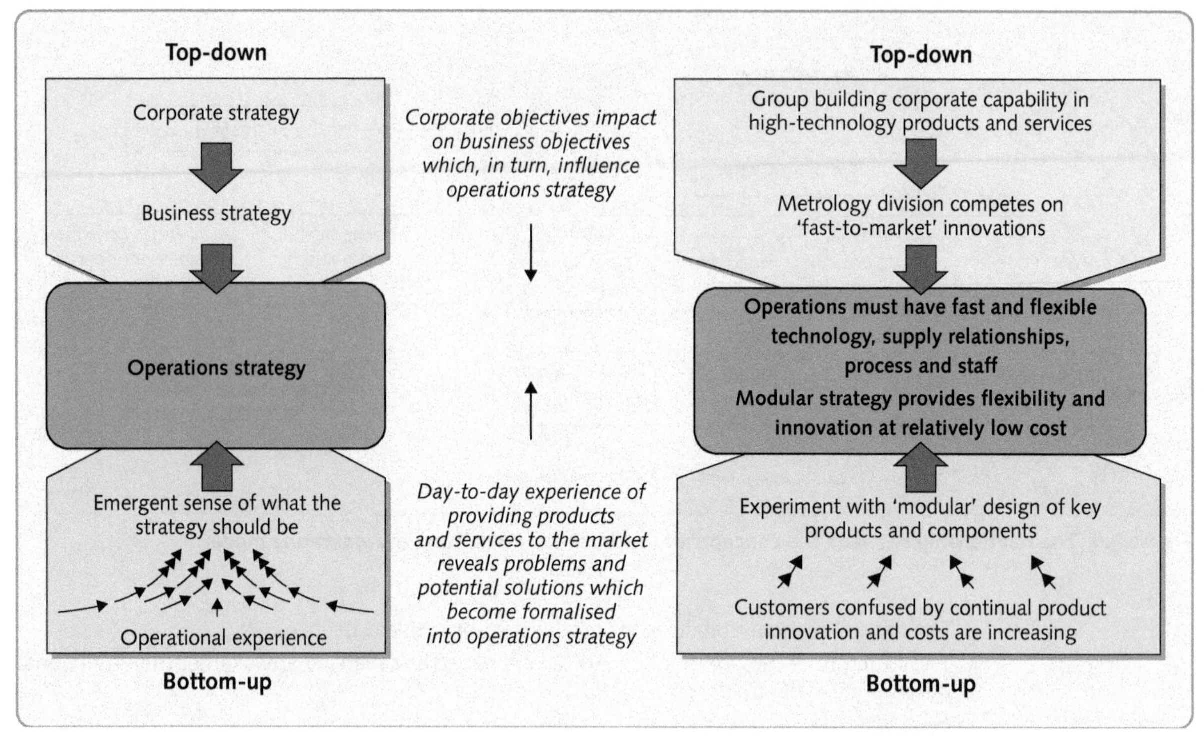

Figure 2.5 Top-down and bottom-up perspectives of strategy for the metrology company

over time from the operational level, as the business learns from the day-to-day experience of running processes (both operations and other processes). This is known as the 'emergent' or 'bottom-up' perspective on operations strategy. An operations strategy should reflect both of these perspectives. Any functional strategy, especially operations strategy, cannot afford to be in conflict with the business's overall strategy. Yet at the same time, any operation will be strongly influenced by its day-to-day experiences. Not only will operational issues set practical constraints on strategic direction, more significantly, day-to-day experiences can be exploited to provide an important contribution to strategic thinking. The left-hand side of Figure 2.5 illustrates this.

Top-down operations strategy should reflect the needs of the whole business

A top-down perspective often identifies three levels of strategy: corporate, business and functional. A corporate strategy should position the corporation in its global, economic, political and social environment. This will consist of decisions about what types of business the group wants to be in, what parts of the world it wants to operate in, how to allocate its cash between its various businesses, and so on. Each business unit within the corporate group will also need to put together its own business strategy which sets out its individual mission and objectives. This business strategy guides the business in relation to its customers, markets and competitors, and also defines its role within the corporate group of which it is a part. Similarly, within the business, functional strategies need to consider what part each function should play in contributing to the strategic objectives of the business. The operations, marketing, product/service development and other functions will all need to consider how best they should organise themselves to support the business's objectives.

> **OPERATIONS PRINCIPLE**
> *Operations strategies should reflect top-down corporate and/or business objectives.*

Bottom-up operations strategy should reflect operational reality

Although it is a convenient way of thinking about strategy, the top-down hierarchical model does not represent the way strategies are always formulated in practice. When any group is reviewing its corporate strategy, it will also take into account the circumstances, experiences and capabilities of the various businesses that form the group. Similarly, businesses, when reviewing their strategies, will consult the individual functions within the business about their constraints and capabilities. They may also incorporate the ideas which come from each function's day-to-day experience. In fact many strategic ideas emerge over time from operational experience rather than being originated exclusively at a senior level. Sometimes companies move in a particular strategic direction because the ongoing experience of providing products and services to customers at an operational level convinces them that it is the right thing to do. There may be no formal high-level decision making that examines alternative strategic options and chooses the one that provides the best way forward. Instead, a general consensus emerges from the operational experience. The 'high-level' strategic decision making, if it occurs at all, may simply confirm the consensus and provide the resources to make it happen effectively. This is sometimes called the concept of emergent strategies.[11] It sees strategies as often being formed in a relatively unstructured and fragmented manner to reflect the fact that the future is at least partially unknown and unpredictable.

OPERATIONS PRINCIPLE

Operations strategy should reflect bottom-up experience of operational reality.

This view of operations strategy reflects how things often happen, but at first glance it seems less useful in providing a guide for specific decision-making. Yet while emergent strategies are less easy to categorise, the principle governing a bottom-up perspective is clear: an operation's objectives and action should be shaped, at least partly, by the knowledge it gains from its day-to-day activities. The key virtues required for shaping strategy from the bottom up are an ability to learn from experience and a philosophy of continual and incremental improvement.

EXAMPLE

Flexibility in innovation

A metrology systems company develops integrated systems for large international clients in several industries. It is part of a group that includes several high-tech companies. It competes through a strategy of technical excellence and innovation together with an ability to advise and customise its systems to clients' needs. As part of this strategy it attempts to be the first in the market with every available new technical innovation. From a top-down perspective, its operations function, therefore, needs to be capable of coping with the changes which constant innovation will bring. It must develop processes that are flexible enough to develop and assemble novel components and systems. It must organise and train its staff to understand the way technology is developing so that they can put in place the necessary changes to the operation. It must develop relationships with its suppliers that will help them to respond quickly when supplying new components. Everything about the operation, its processes, staff, and its systems and procedures, must, in the short term, do nothing to inhibit and, in the long term, actively develop the company's competitive strategy of innovation.

However, over time, as its operations strategy develops, the business discovers that continual product and system innovation is having the effect of dramatically increasing its costs. And, although it does not compete on low prices, its rising costs were impacting profitability. Also there was some evidence that continual changes were confusing some customers. Partially in response to customer requests, the company's system designers started to work out a way of 'modularising' their system and product designs. This allowed one part of the system to be updated for those customers who valued the functionality the innovation could bring, without interfering with the overall design of the main body of the system. Over time, this approach becomes standard design practice within the company. Customers appreciated the extra customisation, and modularisation reduced operations costs. Note that this strategy emerged from the company's experience. No top-level board decision was ever taken to confirm this practice, but nevertheless it emerged as the way in which the company organises its design activity. The right-hand side of Figure 2.5 illustrates these top-down and bottom-up influences for the business.

DIAGNOSTIC QUESTION

Does operations strategy align market requirements with operations resources?

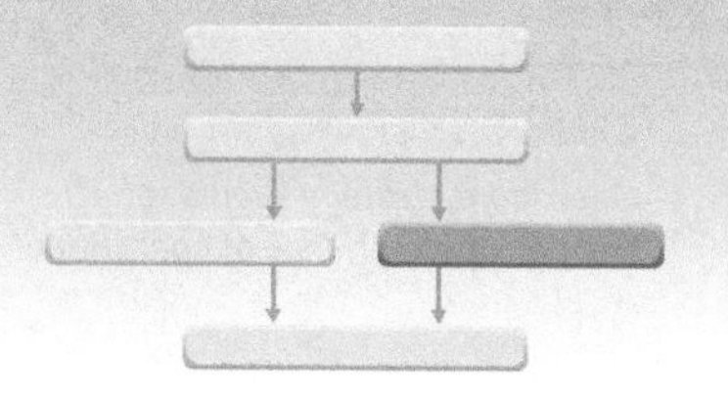

Video

Any operations strategy should reflect the intended market position of the business. Companies compete in different ways; some compete primarily on cost, others on the excellence of their products or services, others on high levels of customer service, and so on. The operations function must respond to this by providing the ability to perform in a manner that is appropriate for the intended market position. This is a market perspective on operations strategy. But, operations strategy must do more than simply meet the short-term needs of the market (important though this is). The processes and resources within operations also need to be developed in the long term to provide the business with a set of competencies or capabilities (we use the two words interchangeably). Capabilities in this context is the 'know-how' that is embedded within the business's resources and processes. These capabilities may be built up over time, as the result of the experiences of the operation, or they may be bought-in or acquired. If they are refined and integrated they can form the basis of the business's ability to offer unique and 'difficult to imitate' products and services to its customers. This idea of the basis of long-term competitive capabilities deriving from the operation's resources and processes is called the resource perspective on operations strategy.

Operations strategy should reflect market requirements

A particularly useful way of determining the relative importance of competitive factors is to distinguish between what have been termed 'order-winners' and 'qualifiers'. Figure 2.6 shows the difference between order-winning and qualifying objectives in terms of their utility, or worth, to the competitiveness of the organisation. The curves illustrate the relative amount of competitiveness (or attractiveness to customers) as the operation's performance varies.

OPERATIONS PRINCIPLE

Operations strategy should reflect the requirements of the business's markets.

- **Order-winners** are those things that directly and significantly contribute to winning business. They are regarded by customers as key reasons for purchasing the product or service. Raising performance in an order-winner will either result in more business or improve the chances of gaining more business. Order-winners show a steady and significant increase in their contribution to competitiveness as the operation gets better at providing them.
- **Qualifiers** may not be the major competitive determinants of success, but are important in another way. They are those aspects of competitiveness where the operation's performance has to be above a particular level just to be considered by the customer. Performance below this 'qualifying' level of performance may disqualify the operation from being considered by customers, but any further improvement above the qualifying level is unlikely to gain the company much competitive benefit. Qualifiers are those things that are generally expected by customers. Being great at them is unlikely to excite them, but being bad at them can disadvantage the competitive position of the operation.

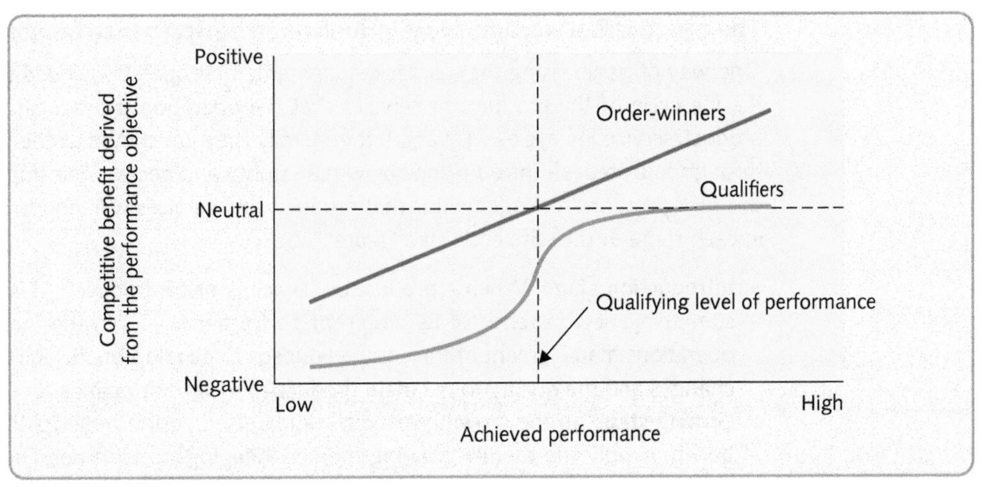

Figure 2.6 Order-winners and qualifiers. Order-winners gain more business the better you are. Qualifiers are the 'givens' of doing business

Different customer needs imply different objectives

If, as is likely, an operation produces goods or services for more than one customer group, it will need to determine the order-winners and qualifiers for each group. For example, Table 2.4 shows two 'product' groups in the banking industry. Here the distinction is drawn between the customers who are looking for banking services for their private and domestic needs and the corporate customers who need banking services for their (often large) businesses.

OPERATIONS PRINCIPLE

Different customer needs imply different priorities of performance objectives.

Table 2.4 Different banking services require different performance objectives

	Retail banking	Corporate banking
Products	Personal financial services, such as loans and credit cards	Special services for corporate customers
Customers	Individuals	Businesses
Product range	Medium but standardised, little need for special terms	Very wide range, many need to be customised
Design changes	Occasional	Continual
Delivery	Fast decisions	Dependable service
Quality	Means error-free transactions	Means close relationships
Volume per service type	Most service are high volume	Most services are low volume
Profit margins	Most are low to medium, some high	Medium to high
Order-winners	Price Accessibility Speed	Customisation Quality of service Reliability
Qualifiers	Quality Range	Speed Price
Performance objectives emphasised within the processes that produce each service	Cost Speed Quality	Flexibility Quality Dependability

The product/service life cycle influence on performance objectives

One way of generalising the market requirements that operations need to fulfil is to link it to the life cycle of the products or services that the operation is producing. The exact form of product/service life cycles will vary, but generally they are shown as the sales volume passing through four stages – introduction, growth, maturity and decline. The important implication of this for operations management is that products and services will require operations strategies in each stage of their life cycle (see Figure 2.7).

- **Introduction stage**. When a product or service is first introduced, it is likely to be offering something new in terms of its design or performance. Given the market uncertainty, the operations management of the company needs to develop the flexibility to cope with these changes and the quality to maintain product/service performance.
- **Growth stage**. In the growing market, standardised designs emerge that allows the operation to supply the rapidly growing market. Keeping up with demand through rapid and

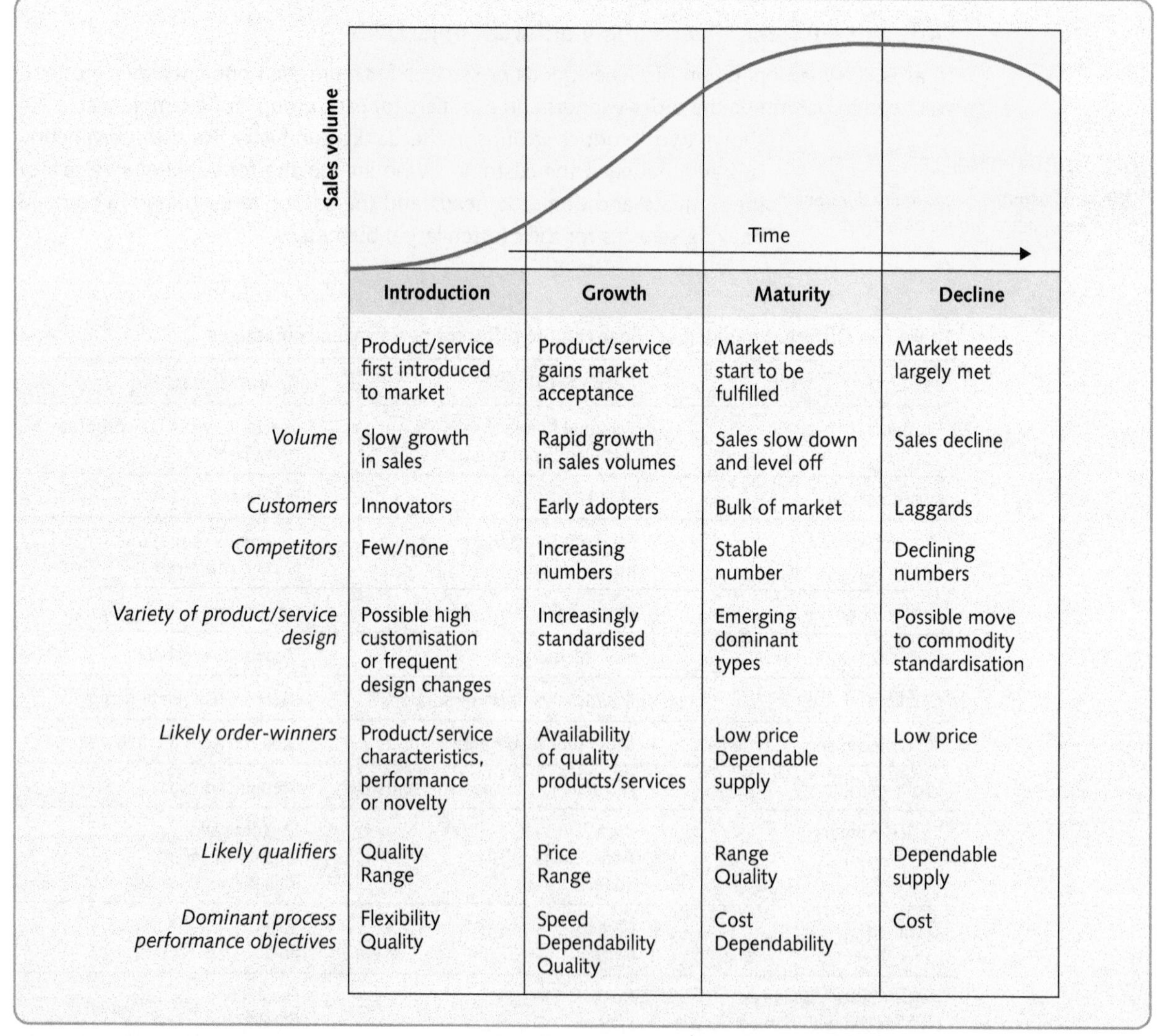

	Introduction	Growth	Maturity	Decline
	Product/service first introduced to market	Product/service gains market acceptance	Market needs start to be fulfilled	Market needs largely met
Volume	Slow growth in sales	Rapid growth in sales volumes	Sales slow down and level off	Sales decline
Customers	Innovators	Early adopters	Bulk of market	Laggards
Competitors	Few/none	Increasing numbers	Stable number	Declining numbers
Variety of product/service design	Possible high customisation or frequent design changes	Increasingly standardised	Emerging dominant types	Possible move to commodity standardisation
Likely order-winners	Product/service characteristics, performance or novelty	Availability of quality products/services	Low price Dependable supply	Low price
Likely qualifiers	Quality Range	Price Range	Range Quality	Dependable supply
Dominant process performance objectives	Flexibility Quality	Speed Dependability Quality	Cost Dependability	Cost

Figure 2.7 The effects of the product/service life cycle on the operation and its process performance objectives

dependable response and maintaining quality levels will help to keep market share as competition starts to increase.

- **Maturity stage**. Eventually demand starts to level off as the market becomes dominated by a few larger companies with standardised designs. Competition will probably emphasise price or value for money, so operations will be expected to get costs down in order to maintain profits, or to allow price cutting, or both. So, cost and productivity issues, together with dependable supply, are likely to be the operation's main concerns.
- **Decline stage**. After time, sales will decline. To the companies left there might be a residual market, but if capacity in the industry lags demand, the market will be dominated by price competition; therefore cost-cutting continues to be important.

Operations strategy should build operations capabilities

Building operations capabilities means understanding the existing resources and processes within the operation, starting with the simple questions, what do we have, and what can we do? However, trying to understand an operation by listing its resources alone is like trying to understand an automobile by listing its component parts. To understand an automobile we need to describe how the component parts form its internal mechanisms. Within the operation, the equivalents of these mechanisms are its *processes*. Yet, even a technical explanation of an automobile's mechanisms does not convey its style or 'personality'. Something more is needed to describe these. In the same way, an operation is not just the sum of its processes. It also has *intangible* resources. An operation's intangible resources include such things as:

OPERATIONS PRINCIPLE

The long-term objective of operation strategy is to build operations-based capabilities.

- its relationship with suppliers and the reputation it has with its customers
- its knowledge of and experience in handling its process technologies
- the way its staff can work together in new product and service development
- the way it integrates all its processes into a mutually supporting whole.

These intangible resources may not be as evident within an operation, but they are important and often have real value. And both tangible and intangible resources and processes shape its capabilities. The central issue for operations management, therefore, is to ensure that its pattern of strategic decisions really does develop appropriate capabilities.

The resource-based view

The idea that building operations capabilities should be an important objective of operations strategy is closely linked with the popularity of an approach to business strategy called the resource-based view (RBV) of the firm.[12] This holds that businesses with an 'above average' strategic performance are likely to have gained their sustainable competitive advantage because of their core competences (or capabilities). This means that the way an organisation inherits, or acquires, or develops its operations resources will, over the long term, have a significant impact on its strategic success. The RBV differs in its approach from the more traditional view of strategy which sees companies as seeking to protect their competitive advantage through their control of the market. For example, by creating *barriers to entry* through product differentiation, or making it difficult for customers to switch to competitors, or controlling the access to distribution channels (a major barrier to entry in gasoline retailing, for example, where oil companies own their own retail stations). By contrast, the RBV sees firms being able to protect their competitive advantage through *barriers to imitation*, that is, by building up 'difficult-to-imitate' resources. Some of these 'difficult-to-imitate' resources are particularly important, and can be classified as 'strategic' if they exhibit the following properties:

- *They are scarce.* Scarce resources, such as specialised production facilities, experienced engineers, proprietary software, etc. can underpin competitive advantage.
- *They are imperfectly mobile.* Some resources are difficult to move out of a firm. For example, resources that were developed in-house, or are based on the experience of the company's staff, or are interconnected with the other resources in the firm, cannot be traded easily.
- *They are imperfectly imitable and imperfectly substitutable.* It is not enough only to have resources that are unique and immobile. If a competitor can copy these resources or, replace them with alternative resources, then their value will quickly deteriorate. The more the resources are connected with process knowledge embedded deep within the firm, the more difficult they are for competitors to understand and to copy.

Reconciling market requirements and operations resource capabilities

The market requirements and the operations resource perspectives on operations strategy represent two sides of a strategic equation that all operations managers have to reconcile. On one hand, the operation must be able to meet the requirements of the market. On the other hand, it also needs to develop operations capabilities that make it able to do the things that customers find valuable but competitors find difficult to imitate. And ideally, there should be a reasonable degree of alignment, or 'fit' between the requirements of the market and the capabilities of operations resources. Figure 2.8 illustrates the concept of fit diagrammatically. The vertical dimension represents the nature of market requirements either because they reflect the intrinsic needs of customers or because their expectations have been shaped by the firm's marketing activity. This includes such factors as the strength of the brand or reputation, the degree of differentiation or the extent of market promises. Movement along the dimension indicates a broadly enhanced level of market 'performance'. The horizontal scale represents the nature of the firm's operations resources and processes. This includes such things as the performance of the operation in terms of its ability to achieve competitive objectives, the efficiency with which it uses its resources, and the ability of the firm's resources to underpin its business processes. Movement along the dimension broadly indicates an enhanced level of 'operations capability'.

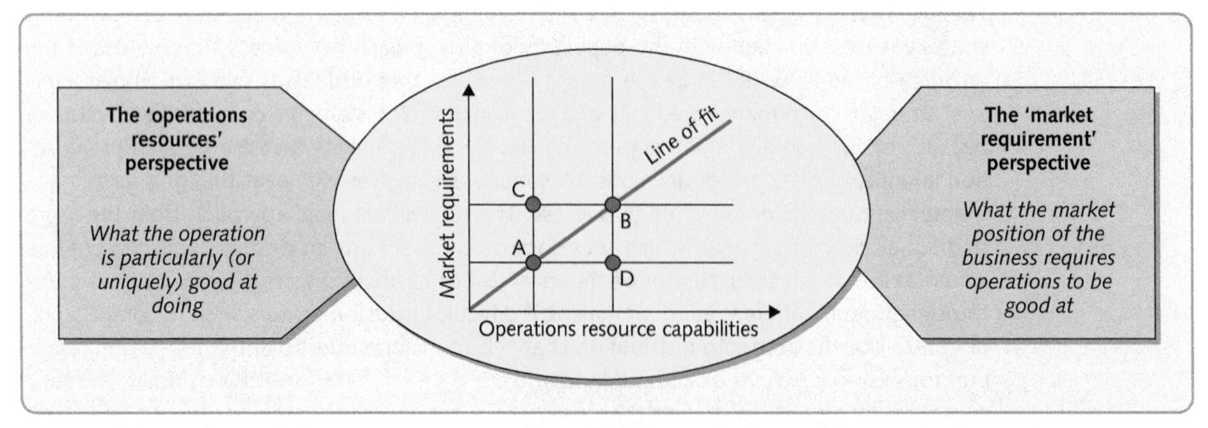

Figure 2.8 Operations strategy must attempt to achieve 'fit' between market requirements and operations resource capabilities

If market requirements and operations capability of an operation are aligned it would diagrammatically be positioned on the 'line of fit' in Figure 2.8. 'Fit' is to achieve an approximate balance between 'market requirements' and 'operations capability'. So when fit is achieved, firms' customers do not need, or expect, levels of operations capability that cannot be supplied. Nor does the operation have strengths that are either inappropriate for market needs or remain unexploited in the market.

An operation that has position A in Figure 2.8 has achieved 'fit' in so much as its operations capabilities are aligned with its market requirements, yet both are at a relatively low level. In other words, the market does not want much from the business, which is just as well because its operation is not capable of achieving much. An operation with position B has also achieved 'fit', but at a higher level. Other things being equal, this will be a more profitable position that position A. Positions C and D are out of alignment. Position C denotes an operation that does not have sufficient operations capability to satisfy what the market wants. Position D indicates an operation that has more operations capability than it is able to exploit in its markets. Generally, operations at C and D would wish to improve their operations capability (C), or reposition itself in its market (D) in order to get back into a position of fit.

DIAGNOSTIC QUESTION

Does operations strategy set an improvement path?

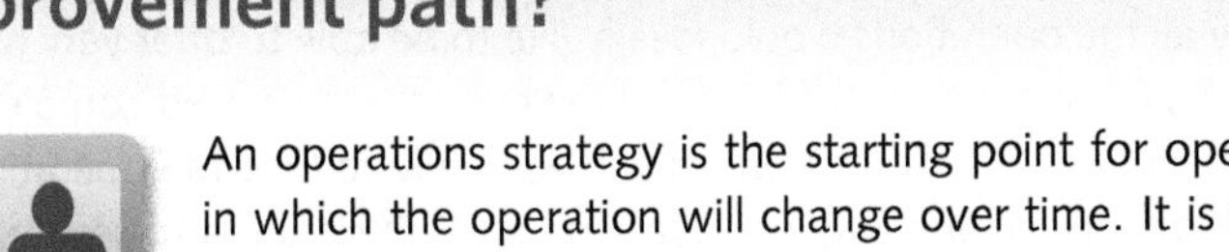

Video

An operations strategy is the starting point for operations improvement. It sets the direction in which the operation will change over time. It is implicit that the business will want operations to change for the better. Therefore, unless an operations strategy gives some idea as to how improvement will happen, it is not fulfilling its main purpose. This is best thought about in terms of how performance objectives, both in themselves and relative to each other, will change over time. To do this, we need to understand the concept of, and the arguments concerning, the trade-offs between performance objectives.

An operations strategy should guide the trade-offs between performance objectives

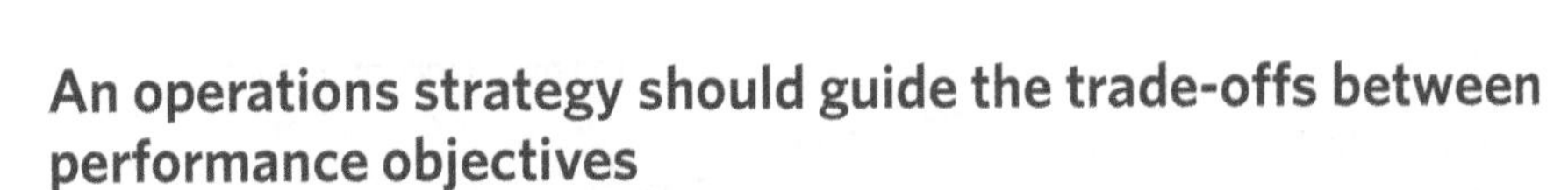

An operations strategy should address the relative priority of operation's performance objectives ('for us, speed of response is more important than cost efficiency, quality is more important than variety', and so on). To do this it must consider the possibility of improving its performance in one objective by sacrificing performance in another. So, for example, an operation might wish to improve its cost efficiencies by reducing the variety of products or services that it offers to its customers. Taken to its extreme, this 'trade-off' principle implies that improvement in one performance objective can *only* be gained at the expense of another. 'There is no such thing as a free lunch' could be taken as a summary of this approach to managing. Probably the best-known summary of the trade-off idea comes from Professor Wickham Skinner, the most influential of the originators of the strategic approach to operations, who said:[13]

> OPERATIONS PRINCIPLE
> *In the short term, operations cannot achieve outstanding performance in all its operations objectives.*

> *. . . most managers will readily admit that there are compromises or trade-offs to be made in designing an airplane or truck. In the case of an airplane, trade-offs would involve matters such*

as cruising speed, take-off and landing distances, initial cost, maintenance, fuel consumption, passenger comfort and cargo or passenger capacity. For instance, no one today can design a 500-passenger plane that can land on an aircraft carrier and also break the sound barrier. Much the same thing is true in . . . [operations].

But there is another view of the trade-offs between performance objectives. This sees the very idea of trade-offs as the enemy of operations improvement, and regards the acceptance that one type of performance can only be achieved at the expense of another as both limiting and unambitious. For any real improvement of total performance, it holds, the effect of trade-offs must be overcome in some way. In fact, overcoming trade-offs must be seen as the central objective of strategic operations improvement.

These two approaches to managing trade-offs result in two approaches to operations improvement. The first emphasises 'repositioning' performance objectives by trading-off improvements in some objectives for a reduction in performance in other. The other emphasises increasing the 'effectiveness' of the operation by overcoming trade-offs so that improvements in one or more aspects of performance can be achieved without any reduction in the performance of others. Most businesses at some time or other will adopt both approaches. This is best illustrated through the concept of the 'efficient frontier' of operations performance.

OPERATIONS PRINCIPLE
In the long term, a key objective of operations strategy is to improve all aspects of operations performance.

Trade-offs and the efficient frontier

Animated diagram

Figure 2.9(a) shows the relative performance of several companies in the same industry in terms of their cost efficiency and the variety of products or services that they offer to their customers. Presumably all the operations would ideally like to be able to offer very high variety while still having very high levels of cost efficiency. However, the increased complexity that a high variety of product or service offerings brings will generally reduce the operation's ability to operate efficiently. Conversely, one way of improving cost efficiency is to severely limit the variety on offer to customers. The spread of results in Figure 2.9(a) is typical of an exercise

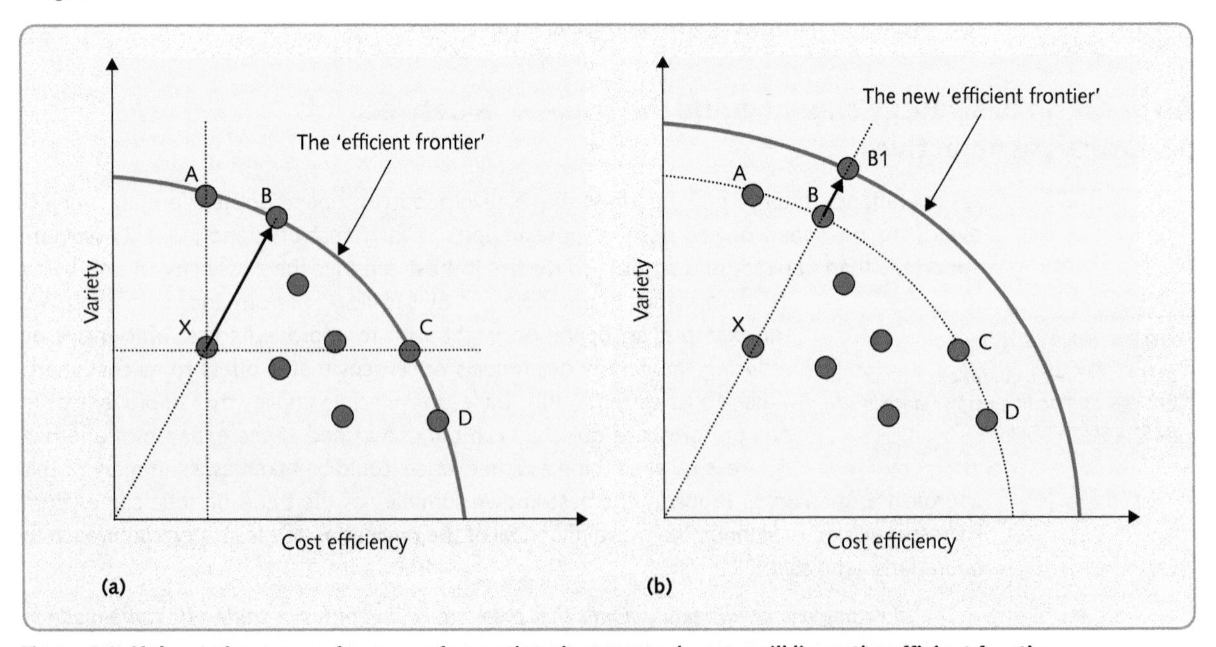

Figure 2.9 If the performance of a group of operations is compared, some will lie on the efficient frontier

such as this. Operations A, B, C, D all have chosen a different balance between variety and cost efficiency. But none is dominated by any other operation in the sense that another operation necessarily has 'superior' performance. Operation X however, has an inferior performance because operation A is able to offer higher variety at the same level of cost efficiency, and operation C offers the same variety but with better cost efficiency. The convex line on which operations A, B, C and D lie is known as the 'efficient frontier'. They may choose to position themselves differently (presumably because of different market strategies) but they cannot be criticised for being ineffective. Of course any of these operations that lie on the efficient frontier may come to believe that the balance they have chosen between variety and cost efficiency is inappropriate. In these circumstances they may choose to reposition themselves at some other point along the efficient frontier. By contrast, operation X has also chosen to balance variety and cost efficiency in a particular way but is not doing so effectively. Operation B has the same ratio between the two performance objectives but is achieving them more effectively. Operation X will generally have a strategy that emphasises increasing its effectiveness before considering any repositioning.

OPERATIONS PRINCIPLE

Operations that lie on the 'efficient frontier' have performance levels that dominate those which do not.

However, a strategy that emphasises increasing effectiveness is not confined to those operations that are dominated, such as operation X. Those with a position on the efficient frontier will generally also want to improve their operations effectiveness by overcoming the trade-off that is implicit in the efficient frontier curve. For example, suppose operation B in Figure 2.9(b) is the metrology company described earlier in this chapter. By adopting a modular product design strategy it improved both its variety and its cost efficiency simultaneously (and moved to position B1). What has happened is that operation B has adopted a particular operations practice (modular design) that has pushed out the efficient frontier. This distinction between positioning on the efficient frontier and increasing operations effectiveness to reach the frontier is an important one. Any operations strategy must make clear the extent to which it is expecting the operation to reposition itself in terms of its performance objectives and the extent to which it is expecting the operation to improve its effectiveness.

Improving operations effectiveness by using trade-offs

Improving the effectiveness of an operation by pushing out the efficient frontier requires different approaches depending on the original position of the operation on the frontier. For example, in Figure 2.10 operation P has an original position that offers a high level of variety at the expense of low cost efficiency. It has probably reached this position by adopting a series of operations practices that enable it to offer the variety even if these practices are intrinsically expensive. For example, it may have invested in general purpose technology and recruited employees with a wide range of skills. Improving variety even further may mean adopting even more extreme operations practices that emphasise variety. For instance, it may reorganise its processes so that each of its larger customers has a dedicated set of resources that understands the specific requirements of that customer and can organise itself to totally customise every product and service it produces. This will probably mean a further sacrifice of cost efficiency, but it allows an ever greater variety of products or services to be produced (P1). Similarly, operation Q may increase the effectiveness of its cost efficiency, by becoming even less able to offer any kind of variety (Q1). For both operations P and Q effectiveness is being improved through increasing the focus of the operation on one (or a very narrow set of) performance objectives and accepting an even further reduction in other aspects of performance.

The same principle of focus also applies to organisational units smaller than a whole operation. For example, individual processes may choose to position themselves on a highly focused set of performance objectives that match the market requirements of their own customers.

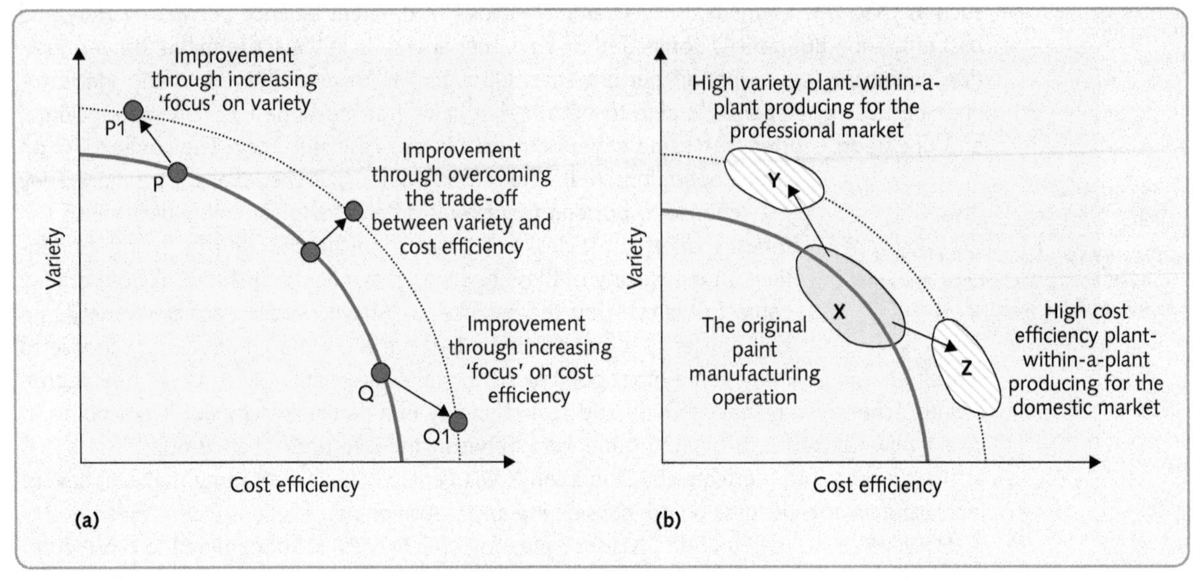

Figure 2.10 Operations 'focus' and the 'plant-within-a-plant' concept illustrated using the efficient frontier model

So, for example, a business that manufactures paint for interior decoration may serve two quite distinct markets. Some of its products are intended for domestic customers who are price sensitive but demand only a limited variety of colours and sizes. The other market is professional interior decorators who demand a very wide variety of colours and sizes but are less price sensitive. The business may choose to move from a position where all types of paint are made on the same processes (position X in Figure 2.10(b)) to one where it has two separate sets of processes (Y and Z); one that only makes paint for the domestic market and the other that only makes paint for the professional market. In effect, the business has segmented its operations processes to match the segmentation of the market. This is sometimes called the 'plant-within-a-plant' concept.

Improving operations effectiveness by overcoming trade-offs

This concept of highly focused operations is not universally seen as appropriate. Many companies attempt to give 'the best of both worlds' to their customers. At one time, for example, a high-quality, reliable and error-free automobile was inevitably an expensive automobile. Now, with few exceptions, we expect even budget-priced automobiles to be reliable and almost free of any defects. Auto manufacturers found that not only could they reduce the number of defects on their vehicles without necessarily incurring extra costs, but they could actually reduce costs by reducing errors in manufacture. If auto manufacturers had adopted a purely focused-based approach to improvement over the years, we may now only be able to purchase either very cheap, low-quality automobiles or very expensive, high-quality automobiles. So a permanent expansion of the efficient frontier is best achieved by overcoming trade-offs through improvements in operations practice.

OPERATIONS PRINCIPLE

An operation's strategy improvement path can be described in terms of repositioning and/or overcoming its performance trade-offs.

Even trade-offs that seem to be inevitable can be reduced to some extent. For example, one of the decisions that any supermarket manager has to make is how many checkout positions to open at any time. If too many checkouts are opened then there will be times when the checkout staff do not have any customers to serve and will be idle. The customers, however, will have excellent service in terms of little or no waiting time. Conversely, if too few checkouts are opened, the staff will be working all the time but customers will have to wait in long queues. There seems to be a direct trade-off between staff utilisation (and therefore cost) and customer waiting time (speed of service). Yet even the supermarket manager deciding how many checkouts to open can go some way to affecting the trade-off between customer waiting time and staff utilisation. The manager might, for example, allocate a number of 'core' staff to operate the checkouts but also arrange for those other staff who are performing other jobs in the supermarket to be trained and 'on-call' should demand suddenly increase. If the manager on duty sees a build-up of customers at the checkouts, these other staff could quickly be used to staff checkouts. By devising a flexible system of staff allocation, the manager can both improve customer service and keep staff utilisation high.

Critical commentary

Each chapter contains a short critical commentary on the main ideas covered in the chapter. Its purpose is not to undermine the issues discussed in the chapter, but to emphasise that, although we present a relatively orthodox view of operation, there are other perspectives.

- Starting any discussion of strategy from a stakeholder perspective is far from undisputed. Listen to Michael Jensen of Harvard Business School. '*At the economy-wide or social level, the issue is this: if we could dictate the criterion or objective function to be maximised by firms (and thus the performance criterion by which corporate executives choose among alternative policy options), what would it be? Or, to put the issue even more simply: how do we want the firms in our economy to measure their own performance? How do we want them to determine what is better versus worse?*' He also holds that using stakeholder perspectives gives undue weight to narrow special interests who want to use the organisation's resources for their own ends. The stakeholder perspective gives them a spurious legitimacy which '*undermines the foundations of value-seeking behaviour*'.[14]

- Similarly, the idea that operations strategy could ever become the driver of a business's overall strategy, and the associated concept of the resource-based view of the firm, are both problematic to some theorists. Business strategies, and functional strategies, were, for many years, seen as, first, market driven and, second, planned in a systematic and deliberative manner. So, it became almost axiomatic to see strategy as starting from a full understanding of market positioning. In fact, the main source of sustainable competitive advantage was seen as unequivocally associated with how a business positioned itself in its markets. Get the market proposition right and customers would respond by giving you business. Get it wrong and they would go to the competitors with a better offering. Strategy was seen as aligning the whole organisation to the market position that could achieve long-term profitable differentiation when compared to competitors. Functional strategies were simply a more detailed interpretation of this overall imperative. Furthermore, strategy must be something that could be planned and directed. If managers could not influence strategy, then how could business be anything other than a lottery?

- The idea that sustainable competitive advantage could come from the capabilities of one's resources was a clear threat to the established position. Furthermore, the idea that strategies emerged, sometimes haphazardly and unpredictably, over time rather than were deliberate decisions taken by senior managers was also seemingly counter-intuitive. Yet there is now considerable research evidence to support both these, once outrageous, propositions. The position we have taken in this chapter is one of blending some aspects of the traditional view with the more recent ideas. Nevertheless, it is important to understand that there are still different views on the very nature of strategic management.

SUMMARY CHECKLIST

This checklist comprises questions that can be usefully applied to any type of operations and reflect the major diagnostic questions used within the chapter.

- ☐ Does the operation have a fully articulated operations strategy?
- ☐ Does it include a vision for the role and contribution of the operations function?
- ☐ What position on the Hayes and Wheelwright stage 1 to 4 model are your operations?
- ☐ Are the operation's performance objectives fully articulated?
- ☐ Are the main strategic decisions that shape operations resources fully identified?
- ☐ Are the logical links established between what the market requires (in terms of performance objectives) and what capabilities an operations possesses (in terms of the major strategic decision areas)?
- ☐ What is the balance between top-down direction and bottom-up learning in formulating operations strategy?
- ☐ Is there a recognised process both for top-down and bottom-up communication on strategic issues?
- ☐ Are performance objectives understood in terms of whether they are order-winners or qualifiers?
- ☐ Have different parts of the operation (probably producing different products or services) their own relative priority of performance objectives that reflect their possibly different competitive positions?
- ☐ Is the idea of operations-based capabilities fully understood?
- ☐ What capabilities does the operation currently possess?
- ☐ Are these operations and/or resources scarce, imperfectly mobile, imperfectly imitable or imperfectly substitutable?
- ☐ If none of the above, are they really useful in terms of their strategic impact?
- ☐ Where would you put the operation in terms of Figure 2.8 that describes the broad fit between market requirements and operations resource capabilities?
- ☐ Have the key trade-offs for the operation been identified?
- ☐ What combination of repositioning in order to change the nature of trade-offs, and overcoming the trade-offs themselves, is going to be used to improve overall operations performance?

CASE STUDY

McDonald's: Half a Century of Growth[15]

It's loved and it's hated. It is a shining example of how good value food can be brought to a mass market. It is a symbol of everything that is wrong with 'industrialised', capitalist, bland, high-calorie and environmentally unfriendly commercialism. It is the best-known and most-loved fast food brand in the world with more than 32,000 restaurants in 117 countries, providing jobs for 1.7 million staff and feeding 60 million customers per day (yes, per day!). It is part of the homogenisation of individual national cultures, filling the world with bland, identical, 'cookie cutter', Americanised and soulless operations that dehumanise its staff by forcing them to follow rigid and over-defined procedures. But whether you see it as friend, foe, or a bit of both, McDonald's has revolutionised the food industry, affecting the lives both of the people who produce food and the people who eat it. It has also had its ups (mainly) and downs (occasionally). Yet, even in the toughest times it has always displayed remarkable resilience. Even after the economic turbulence of 2008, McDonald's reported an exceptional year of growth in 2009, posting sales increases and higher market share around the world – it was the sixth consecutive year of positive sales in every geographic region of their business.

Source: David Pearson/Alamy Images

Starting small

Central to the development of McDonald's is Ray Kroc, who by 1954 and at the age of 52 had been variously a piano player, a paper cup salesman and a multi-mixer salesman. He was surprised by a big order for eight multi-mixers from a restaurant in San Bernardino, California. When he visited the customer he found a small but successful restaurant run by two brothers Dick and Mac McDonald. They had opened their 'Bar-B-Que' restaurant 14 years earlier adopting the usual format at that time: customers would drive-in, choose from a large menu and be served by a 'car hop'. However, by the time Ray Kroc visited the brothers' operation it had changed to a self-service drive-in format with a limited menu of nine-items. He was amazed by the effectiveness of their operation. Focusing on a limited menu, including burgers, fries and beverages, had allowed them to analyse every step of the process of producing and serving their food. Ray Kroc was so overwhelmed by what he saw that he persuaded the brothers to adopt his vision of creating McDonald's restaurants all over the US, the first of which opened in Des Plaines, Illinois, in June 1955. However, later, Kroc and the McDonald brothers quarrelled, and Kroc brought the brothers out. Now with exclusive rights to the McDonald's name, the restaurants spread, and in five years there were 200 restaurants through the US. After ten years the company went public, the share price doubling in the first month. But through this and later expansion, Kroc insisted on maintaining the same principles that he had seen in the original operation. 'If I had a brick for every time I've repeated the phrase '*Quality, Service, Cleanliness and Value*', *I think I'd probably be able to bridge the Atlantic Ocean with them*.' (Ray Kroc)

Priority to the process

Ray Kroc had been attracted by the cleanliness, simplicity, efficiency and profitability of the McDonald brothers' operation. They had stripped fast food delivery down to its essence and eliminated needless effort to make a swift assembly line for a meal at reasonable prices. Kroc wanted to build a process that would become famous for food of consistently high quality using uniform methods of preparation. His burgers, buns, fries and beverages should taste just the same in Alaska as they did in Alabama. The answer was the 'Speedee Service System', a standardised process that prescribed exact preparation methods, specially designed equipment and strict product specifications. The emphasis on process standardisation meant that customers could be assured of identical levels of food and service quality every time they visited any store, anywhere. Operating procedures were specified in minute detail. In its first operations manual, which by 1991 had reached 750 pages, it prescribed specific cooking instructions such as temperatures, cooking times and portions to be followed rigorously. Similarly, operating procedures were defined to ensure the required customer experience, for example, no food items were to be held more than ten minutes in the transfer bin between being cooked and being served. Technology was also automated, and specially designed equipment helped to guarantee consistency using 'fool-proof' devices. For example, the ketchup was dispensed through a metered pump; specially designed 'clam shell' grills cooked both sides of each meat patty simultaneously for a pre-set time; and when it became clear that the metal tongs used by staff to fill French-fry containers were awkward to use efficiently, McDonald's engineers devised a simple V-shaped aluminium scoop that made the job faster and easier as well as presenting the fries in a more attractive alignment with their container.

For Kroc, the operating process was both his passion and the company's central philosophy. It was also the foundation of learning and improvement. The company's almost compulsive focus on process detail was not an end in itself. Rather it was to learn what contributed to consistent high-quality service in practice and what did not. Learning was always seen as important by McDonald's. In 1961, it founded 'Hamburger University', initially in the basement of a restaurant in Elk Grove Village, Illinois. It had a research and development laboratory to develop new cooking, freezing, storing and serving methods. Also franchisees and operators were trained in the analytical techniques necessary to run a successful McDonald's. It awarded degrees in 'Hamburgerology'. But learning was not just for headquarters. The company also formed a 'field service' unit to appraise and help its restaurants by sending field service consultants to review their performance on a number of 'dimensions' including cleanliness, queuing, food quality and customer service. As Ray Kroc, said, *'We take the hamburger business more seriously than anyone else. What sets McDonald's apart is the passion that we and our suppliers share around producing and delivering the highest-quality beef patties. Rigorous food safety and quality standards and practices are in place and executed at the highest levels every day.'*

No story illustrates the company's philosophy of learning and improvement better than its adoption of frozen fries. French-fried potatoes had always been important. Initially, the company tried observing the temperature levels and cooking methods that produced the best fries. The problem was that the temperature during the cooking process was very much influenced by the temperature of the potatoes when they were placed into the cooking vat. So, unless the temperature of the potatoes before they were cooked were also controlled (not very practical) it was difficult to specify the exact time and temperature that would produce perfect fries. But McDonald's researchers have perseverance. They discovered that, irrespective of the temperature of the raw potatoes, fries were always at their best when the oil temperature in the cooking vat increased by three degrees above the low temperature point after they were put in the vat. So by monitoring the temperature of the vat, perfect fries could be produced every time. But that was not the end of the story. The ideal potato for fries was the Idaho Russet, which was seasonal and not available in the summer months, when an alternative (inferior) potato was used. One grower, who, at the time, supplied a fifth of McDonald's potatoes, suggested that he could put Idaho Russets into cold storage for supplying during the summer period. Notwithstanding investment in cold storage facilities, all the stored potatoes rotted. Not to be beaten, he offered another suggestion. Why don't McDonald's consider switching to frozen potatoes? This was no trivial decision and the company was initially cautious about meddling with such an important menu item. However there were other advantages in using frozen potatoes. Supplying fresh potatoes in perfect condition to McDonald's rapidly expanding chain was increasingly difficult. Frozen potatoes could actually increase the quality of the company's fries if a method of satisfactorily cooking them could be found. Once again McDonald's developers came to the rescue. They developed a method of air drying the raw fries, quick frying, and then freezing them. The supplier, who was a relatively small and local suppler when he first suggested storing Idaho Russets grew its business to supply around half of McDonald's US business.

Throughout their rapid expansion a significant danger facing McDonald's was losing control of their operating system. They avoided this, partly by always focusing on four areas – improving the product, establishing strong supplier relationships, creating (largely customised) equipment, and developing franchise holders. But it was also their strict control of the menu which provided a platform of stability. Although their competitors offered a relatively wide variety of menu items, McDonald's limited theirs to ten items. This allowed uniform standards to be established, which in turn encouraged specialisation. As one of McDonald's senior managers at the time stressed, *'It wasn't because we were smarter. The fact that we were selling just ten items [and] had a facility that was small, and used a limited number of suppliers created an ideal environment.'* Capacity growth (through additional stores) was also managed carefully. Well-utilised stores were important to franchise holders, so franchise opportunities were located only where they would not seriously undercut existing stores. Ray Kroc used the company plane to spot from the air the best locations and road junctions for new restaurant branches.

Securing supply

McDonald's says that it has been the strength of the alignment between the company, its franchisees and its suppliers (collectively referred to as the System) that has been the explanation for its success. Expanding the McDonald's chain, especially in the early years, meant persuading both franchisees and suppliers to buy into the company's vision. 'Working', as Ray Kroc put it, *'not for McDonald's, but for themselves, together with McDonald's.'* He promoted the slogan, *'In business for yourself, but not by yourself.'* But when they started suppliers proved problematic. McDonald's approached the major food suppliers, such as Kraft and Heinz, but without much success. Large and established suppliers were reluctant to conform to McDonald's requirements, preferring to focus on retail sales. It was the relatively small companies who were willing to risk supplying what seemed then to be a risky venture. Yet as McDonald's grew, so did its suppliers. Also, McDonald's relationship with its suppliers was seen as less adversarial than with some other customers. One supplier is quoted as saying, *'Other chains would walk away from you for half a cent. McDonald's was more concerned with getting quality. McDonald's always treated me with respect even when they became much bigger and didn't have to.'* Furthermore, suppliers were always seen as a source if innovation. For example, one of McDonald's meat suppliers, Keystone Foods, developed a novel quick-freezing process that captured the fresh taste and texture of beef patties. This meant that every patty could retain its consistent quality until it hit the grill. Keystone shared its technology with other McDonald's meat suppliers for McDonald's, and today the process is an industry standard. Yet, although innovative and close, supplier relationships are also rigorously controlled. Unlike some competitors who simply accepted what suppliers provided, complaining only when supplies were not up to standard, McDonald's routinely analysed its supplier's products.

Fostering franchisees

McDonald's revenues consist of sales by company-operated restaurants and fees from restaurants operated by franchisees. McDonald's view themselves primarily as a franchisor and believe franchising is . . . *'important to delivering great, locally-relevant customer experiences and driving profitability'*. However, they also believe that directly operating restaurants is essential to providing the company with real operations experience. In 2009, of the 32,478 restaurants in 117 countries, 26,216 were operated by franchisees and 6,262 were operated by the company. Where McDonald's was different to other franchise operations was in their relationships. Some restaurant chains concentrated on recruiting franchisees that may then be ignored. McDonalds, on the other hand expected its franchisees to contribute their experiences for the benefit of all, Ray Kroc's original concept was that franchisees would make money before the company did, so he made sure that the revenues that went to McDonald's came from the success of the restaurants themselves rather from initial franchise fees.

Initiating innovation

Ideas for new menu items have often come from franchisees. For example, Lou Groen, a Cincinnati franchise holder had noticed that in Lent (a 40-day period when some Christians give up eating red meat on Fridays and instead eat only fish or no meat at all) some customers avoided the traditional hamburger. He went to Ray Kroc, with his idea for a 'Filet-o-Fish', a steamed bun with a shot of tartar sauce, a fish fillet, and cheese on the bottom bun. But Kroc wanted to push his own meatless sandwich, called the hula burger, a cold bun with a piece of pineapple and cheese. Groen and Kroc competed on a Lenten Friday to see whose sandwich would sell more. Kroc's hula burger failed, selling only six sandwiches all day while Groen sold 350 Filet-o-Fish. Similarly, the Egg McMuffin was introduced by franchisee Herb Peterson, who wanted to attract customers into his McDonalds stores all through the day, not just at lunch and dinner. He came up with idea for the signature McDonald's breakfast item because he was reputedly *'very partial to eggs Benedict and wanted to create something similar'*.

Other innovations came from the company itself. By the beginning of the 1980s, poultry was becoming more fashionable to eat and sales of beef were sagging. Fred Turner, then the Chairman of McDonald's had an idea for a new meal – a chicken finger-food without bones, about the size of a thumb. After six months of research, the food technicians and scientists managed to reconstitute shreds of white chicken meat into small portions which could be breaded, fried, frozen then reheated. Test-marketing the new product was positive, and in 1983 they were launched under the name Chicken McNuggets. These were so successful that within a month McDonald's became the second largest purchaser of chicken in the USA. By 1992, Americans were eating more chicken than beef.

Other innovations came as a reaction to market conditions. Criticised by nutritionists who worried about calorie-rich burgers and shareholders who were alarmed by flattening sales, McDonald's launching its biggest menu revolution in 30 years in 2003 when it entered the prepared salad market. They offered a choice of dressings for their grilled chicken salad with Caesar dressing (and croutons) or the lighter option of a drizzle of balsamic dressing. Likewise, recent moves towards coffee sales were prompted by the ever-growing trend set by big coffee shops like Starbucks. McCafé, a coffee-house-style food and drink chain, owned by McDonald's, had expanded to about 1,300 stores worldwide by 2011.

Problematic periods

The period from the early 1990s to the mid 2000s was difficult for parts of the McDonald's Empire. Although growth in many parts of the world continued, in some developed markets, the company's hitherto rapid growth stalled. Partly this was due to changes in food fashion, nutritional concerns and demographic changes. Partly it was because competitors were learning to either emulate McDonald's operating system, or focus on one aspect of the traditional 'quick service' offering, such as speed of service, range of menu items, (perceived) quality of food, or price. Burger King, promoted itself on its 'flame-grilled' quality. Wendy's offered a fuller service level. Taco Bell undercut McDonald's prices with their 'value pricing' promotions. Drive-through specialists such as Sonic speeded up service times. But it was not only competitors that were a threat to McDonald's growth. So called 'fast food' was developing a poor reputation in some quarters, and as its iconic brand, McDonald's was taking much of the heat. Similarly the company became a lightning rod for other questionable aspects of modern life that it was held to promote, from cultural imperialism, low-skilled jobs, abuse of animals, the use of hormone-enhanced beef, to an attack on traditional (French) values (in France). A French farmer called Jose Bové (who was briefly imprisoned) got other farmers to drive their tractors through, and wreck, a half-built McDonald's. When he was tried, 40,000 people rallied outside the courthouse.

The Chief Executive of McDonald's in the UK, Jill McDonald (yes, really!), said that some past difficulties were self-induced. They included a refusal to face criticisms and a reluctance to acknowledge the need for change. *'I think by the end of 1990s we were just not as close to the customer as we needed to be, we were given a hard time in the press and we lost our confidence. We needed to reconnect, and make changes that would disrupt people's view of McDonald's.'* Investing in its people also needed to be re-emphasised. *'We invest about £35m a year in training people. We have become much more of an educator than an employer of people'*. Nor does she accept the idea of 'McJobs' (meaning boring, poorly paid, often temporary jobs with few prospects). *'That whole McJob thing makes me so angry. It's snobbish. We are the biggest employer of young people in Britain. Many join us without qualifications. They want a better life, and getting qualifications is something they genuinely value.'*

Surviving strategies

Yet, in spite of its difficult period, the company has not only survived, but through the late 2000s has thrived. In 2009, McDonald's results showed that in the US, sales and market share both grew for the seventh consecutive year with new products such as McCafé premium coffees, the premium Angus Third Pounder, smoothies and frappes, together with more convenient locations, extended hours, efficient drive thru service and value-oriented promotions. In the UK, changes to the stores' decor have also helped stimulate growth. Jill McDonald's views are not untypical of other regions: *'We have probably changed more in the past four years than the past 30: more chicken, 100% breast meat, snack wraps, more coffee – lattes and cappuccinos, ethically sourced, not at rip-off prices. That really connected with customers. We sold 100m cups last year.'*

Senior managers put their recent growth down to the decision in 2003 to reinvent McDonald's by becoming 'better, not just bigger' and implementing its 'Plan to Win'. This focused on 'restaurant execution', with the goal of . . . 'improving the overall experience for our customers'. It provided a common framework for their global business yet allowed for local adaptation. Multiple improvement initiatives were based on its 'five key drivers of exceptional customer experiences' (People, Products, Place, Price and Promotion). But what of McDonald's famous standardisation? During its early growth no franchise holder could deviate from the 700+ page McDonald's operations manual known as 'the Bible'. Now things are different, at least partly because different regions have developed their own products. In India, the 'Maharaja Mac' is made of mutton, and the vegetarian options contain no meat or eggs. Similarly, McDonald's in Pakistan offers three spicy 'McMaza meals'. Even in the US things have changed. In at least one location in Indiana, there's now a McDonald's with a full service 'Diner' inside, where waitresses serve 100 combinations of food, on china – a far cry from Ray Kroc's vision of stripping out choice to save time and money.

QUESTIONS

1. How has the competitive situation that McDonald's faces changed since it was founded (in its current form) in the 1950s?
2. How have McDonald's operations activities, in terms of its design, delivery and development, influenced its operations performance objectives?
3. Draw an operations strategy matrix for McDonald's.
4. Search the internet site of Intel, the best known microchip manufacturer, and identify what appear to be its main elements in its operations strategy.

APPLYING THE PRINCIPLES

Hints

Some of these exercises can be answered by reading the chapter. Others will require some general knowledge of business activity and some might require an element of investigation. Hints on how they can all be answered are to be found in the eText at www.pearsoned.co.uk/slack.

1 The environmental services department of a city has two recycling services – newspaper collection (NC) and general recycling (GR). The NC service is a door-to-door collection service which, at a fixed time every week, collects old newspapers which householders have placed in reusable plastic bags at their gate. An empty bag is left for the householders to use for the next collection. The value of the newspapers collected is relatively small, the service is offered mainly for reasons of environmental responsibility. By contrast the GR service is more commercial. Companies and private individuals can request a collection of materials to be disposed of, either using the telephone or the internet. The GR service guarantees to collect the material within 24 hours unless the customer prefers to specify a more convenient time. Any kind of material can be collected and a charge is made depending on the volume of material. This service makes a small profit because the revenue both from customer charges and from some of the more valuable recycled materials exceeds the operation's running costs.

   How would you describe the differences between the performance objectives of the two services?

2 *'It is about four years now since we specialised in the small-to-medium firms' market. Before that we also used to provide legal services for anyone who walked in the door. So now we have built up our legal skills in many areas of corporate and business law. However, within the firm, I think we could focus our activities even more. There seem to be two types of assignment that we are given. About 40 per cent of our work is relatively routine. Typically these assignments are to do with things like property purchase and debt collection. Both these activities involve a relatively standard set of steps which can be automated or carried out by staff without full legal qualifications. Of course, a fully qualified lawyer is needed to make some decisions, however most work is fairly routine. Customers expect us to be relatively inexpensive and fast in delivering the service. Nor do they expect us to make simple errors in our documentation; in fact if we did this too often we would lose business. Fortunately our customers know that they are buying a standard service and don't expect it to be customised in any way. The problem here is that specialist agencies have been emerging over the last few years and they are starting to undercut us on price. Yet I still feel that we can operate profitably in this market and anyway, we still need these capabilities to serve our other clients. The other 60 per cent of our work is for clients who require far more specialist services, such as assignments involving company merger deals or major company restructuring. These assignments are complex, large, take longer, and require significant legal skill and judgement. It is vital that clients respect and trust the advice we give them across a wide range of legal specialisms. Of course they assume that we will not be slow or unreliable in preparing advice, but mainly it's trust in our legal judgement which is important to the client. This is popular work with our lawyers. It is both interesting and very profitable.'*

   'The help I need from you is in deciding whether to create two separate parts to our business; one to deal with routine services and the other to deal with specialist services. What I may do is appoint a senior "Operations Partner" to manage each part of the business, but if I do, what aspects of operations performance should they be aiming to Excel at?'

   (Managing Partner, Branton Legal Services)

3 Revisit the descriptions of Flexitronics and Amazon at the beginning of the chapter. Think about the following issues:

   (a) What are the key performance objectives for Flexitronics and the retail (Amazon.com) part of Amazon?

   (b) What are the key design, delivery, and development decisions that will need to be taken by each business?

   (c) Why does the 'capabilities-based' strategy described by Jeff Bezos present such a challenge?

Notes on chapter

1 For a more thorough explanation, *see* Slack, N. and Lewis, M. (2001) *Operations Strategy*, Financial Times Prentice Hall.

2 Source: company website (www.flextronics.com) and press releases.

3 Sources include: 'Clouds under the hammer: processing capacity is becoming a tradable commodity' (2010), *The Economist* 11 March; 'Lifting the bonnet' (2006),*The Economist*, 7 October; company website.

4 Cloud computing is a general term for the provision of internet-based access to computing capacity so that shared resources, such as software, can be provided on-demand.

5 Elkington, J. (1998), *Cannibals with Forks: the Triple Bottom Line of 21st Century Business*, New Society Publishers.

6 Hayes, R.H. and Wheelwright, S.C. (1984) *Restoring our Competitive Edge*, John Wiley.

7 All quotes taken from each company's website.

8 Osterwalder, A., Pigneur, Y. and Tucci, C. (2005) 'Clarifying business models: origins, present and future of the concept, *CAIS*, Vol. 15, p. 751–75.

9 Osterwalder, A (2005) 'What is a business model?', http://business-model-design.blogspot.com/2005/11/what-is-business-model.html.

10 Based on the definitions developed by Cap Gemin.i.

11 Mintzberg, H. and Waters, J.A. (1995) 'Of strategies: deliberate and emergent', *Strategic Management Journal*, July/Sept.

12 For a full explanation of this concept see Slack, N. and Lewis, M. (2011) *Operations Strategy* (3rd Edition), Financial Times Prentice Hall.

13 A point made initially by Skinner in Skinner, W. (1985) *Manufacturing: The Formidable Competitive Weapon*, John Wiley.

14 From a speech by Michael Jensen of Harvard Business School.

15 Sources include: Kroc, R.A. (1977) *Grinding it Out: The Making of McDonald's*, St. Martin's Press; Love, J. (1995) *McDonald's: Behind the Golden Arches*, Random House Publishing Group; www.aboutmcdonalds.com (2009); Davidson, A. (2011) 'So Mrs McDonald, would you like fries with that?' *Sunday Times*, 13 February; McDonalds Annual Report (2009); Upton, D. (1992) *McDonald's Corporation Case Study*, Harvard Business School.

TAKING IT FURTHER

Johnson, G., Whittington R. and Scholes, K. (2011) *Exploring Strategy* (9th Edition), Financial Times Prentice Hall.

Boyer, K. and McDermot, C. (1999) 'Strategic consensus in operations strategy', *Journal of Operations Management*, Vol. 17, Issue 3, March. *Academic, but interesting.*

Hamel, G. and Prahalad, C.K. (1993) 'Strategy as stretch and leverage', *Harvard Business Review*, Vol. 71, Nos 2 and 3. *This article is typical of some of the (relatively) recent ideas influencing operations strategy.*

Hayes, R. (2006) 'Operations, strategy, and technology: pursuing the competitive edge', *Strategic Direction*, Vol. 22 Issue 7, Emerald Group Publishing Limited. *A summary of the subject from one of (if not* **the***) leading academics in the area.*

Hayes, R.H. and Pisano, G.P. (1994) 'Beyond world class: the new manufacturing strategy', *Harvard Business Review*, Vol. 72, No 1. *Same as above.*

Hill, T. (2006) *Manufacturing Operations Strategy* (3rd Edn Texts and Cases), Palgrave Macmillan; *The descendant of the first non-US book to have a real impact in the area.*

Slack, N. and Lewis, M. (2011) *Operations Strategy (3rd Edn)*, Financial Times Prentice Hall. *What can we say – just brilliant!*

USEFUL WEBSITES

www.opsman.org *Definitions, links and opinions on operations and process management.*

www.cranfield,ac.uk/som *Look for the 'Best factory awards' link. Manufacturing, but interesting.*

www.worldbank.org *Global issues. Useful for international operations strategy research.*

www.weforum.org *Global issues, including some operations strategy ones.*

www.ft.com *Great for industry and company examples.*

Interactive quiz

For further resources including examples, animated diagrams, self-test questions, Excel spreadsheets and video materials please explore the eText on the companion website at **www.pearsoned.co.uk/slack**.

3 Supply network design

Introduction

Every business or organisation is part of a larger and interconnected network of other businesses and organisations. This is the supply network. It includes suppliers and customers, suppliers' suppliers and customers' customers, and so on. At a strategic level, operations managers are involved in influencing the nature or 'design' of the network, and the role of their operation in it. This chapter treats some of these strategic design decisions in the context of supply networks. It forms the context for 'supply chain management', the more operational aspects managing the individual 'trends' or chains through the supply network. We shall discuss supply chain management in Chapter 7, and because many supply network decisions require an estimate of future demand, this chapter includes a supplement on forecasting (see Figure 3.1).

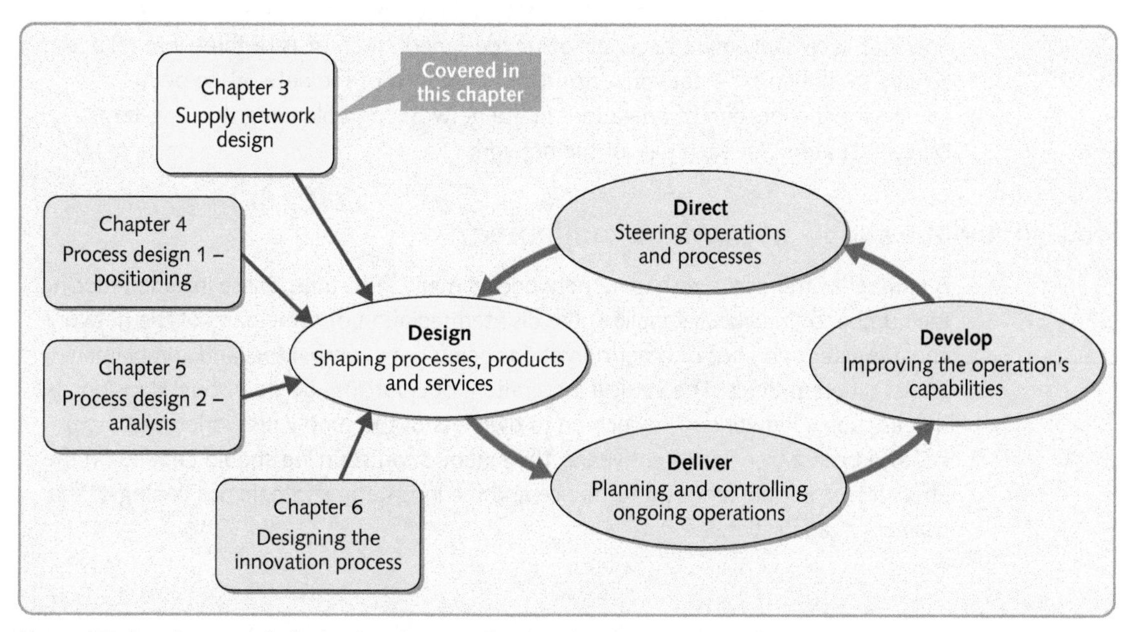

Figure 3.1 Supply network design involves configuring the shape and capabilities of the supply network

EXECUTIVE SUMMARY

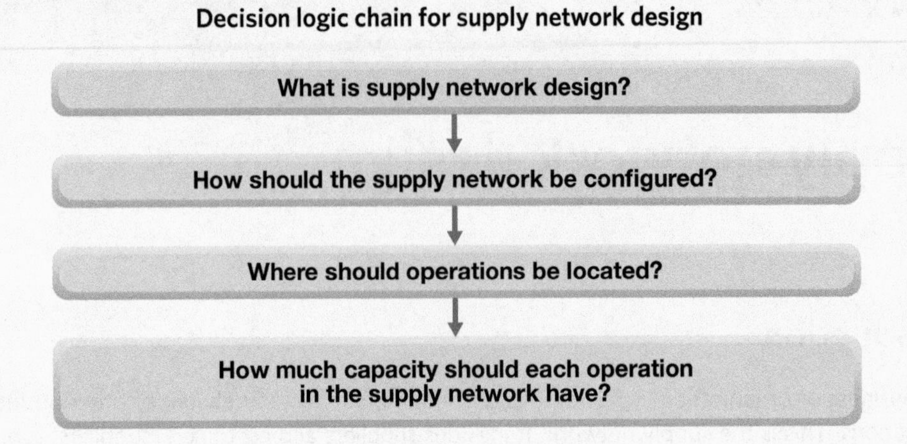

Each chapter is structured around a set of diagnostic questions. These questions suggest what you should ask in order to gain an understanding of the important issues of a topic, and, as a result, improve your decision making. An executive summary addressing these questions is provided below.

What is supply network design?

Supply network design involves configuring the shape and capabilities of the supply network. The supply network includes the chains of suppliers providing inputs to the operation, the chain of customers who receive outputs from the operation, and sometimes other operations who may at times compete and other times cooperate. It is a complex task that is different to 'design' in its conventional sense, because one does not necessarily own the assets being 'designed'. It consists of three interrelated activities – shaping the network (including how much of the network to own), influencing the location of operations in the network, and planning the long-term capacity strategy for each part of the network.

How should the supply network be configured?

A number of trends are reshaping networks in many industries. These include reducing the number of individual suppliers, the disintermediation of some parts of the network, and a greater tolerance of other operations being both competitors and complementors at different times. The vertical integration, outsourcing, or 'do or buy' decision also shapes supply networks. The decision to own less of the supply network (outsource) has also been a trend in recent years. The extent of outsourcing should depend on the effect it has on operations performance and the long-term strategic positioning of the business's capabilities.

Where should operations be located?

Location means the geographical positioning of an operation. It is important because it can affect costs, revenues, customer service and capital investment. Many businesses never consider relocation, but even those who see no immediate need to relocate may find benefits from relocation. Those businesses who actively investigate relocation often do so either because of changes in demand or changes in supply. The process of evaluating alternative locations involves identifying alternative location options, usually reduced to a list of representative locations, and evaluating each option against a set of (hopefully) rational criteria, usually involving consideration of capital requirements, market factors, cost factors, future flexibility and risk.

How much capacity should each operation in the supply network have?

This will depend on demand at any point in time. Capacity will need to be changed in the long term; long-term demand changes, either in advance of demand changes (capacity leading) or after demand changes (capacity lagging). However, the concept of the economy of scale will always be important. Economies of scale derive from both capital and operating efficiencies that derive from large-scale processes. However, after a certain level of capacity, most operations start to suffer diseconomies of scale. The economies of scale curves for operations of different sizes put together give an idea of the optimal capacity size for a particular type of operation. However, this is only an indication. In reality, economy of scale will depend on factors that include risk and strategic positioning.

DIAGNOSTIC QUESTION

What is supply network design?

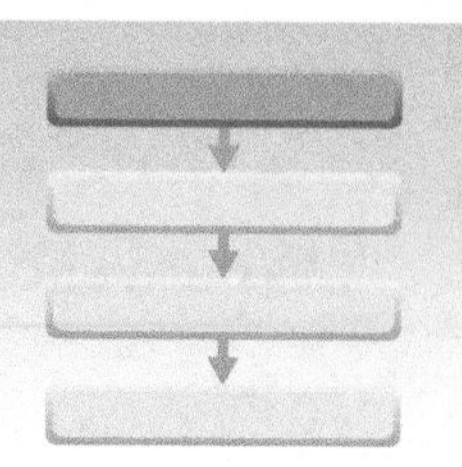

Video

Supply network design involves configuring the shape and capabilities of one's own and other operations with which the business interacts. These include all its suppliers and their suppliers, all its customers and their customers. It may also include other businesses who could be competitors under some circumstances. It is the most strategic of all design activities and in many ways is not the same type of activity as smaller scale design. In process design, decisions can be made with a high degree of confidence that they will be enacted as intended. In supply network design not only is the task intrinsically more complex, but there is a further crucial difference; most of the network that is being 'designed' may not be under the direct control of the 'designers'. Suppliers, customers and others in the network are independent operations. They will naturally pursue what they see as their own best interests, which may not coincide with one's own. The 'design' in supply network 'design' is being used to mean 'influencing' and 'negotiating' rather than design in its conventional sense. However, design is a useful perspective on the tasks involved in the strategic configuration of supply networks because it conveys the idea of understanding the individual components in the network, their characteristics, and the relationships between them.

Terminology is important when describing supply networks. On the 'supply side' of the 'focal' operation (the operation from whose perspective the network is being drawn) is a group of operations that directly supply the operation; these are often called first-tier suppliers. They are supplied by second-tier suppliers. However, some second-tier suppliers may also supply an operation directly, thus missing out a link in the network. Similarly, on the demand side of the network, 'first-tier' customers are the main customer group for the operation. These in turn supply 'second-tier' customers, although again the operation may at times supply second-tier customers directly. The suppliers and customers who have direct contact with an operation are called its immediate supply network. Figure 3.2 illustrates this.

Animated diagram

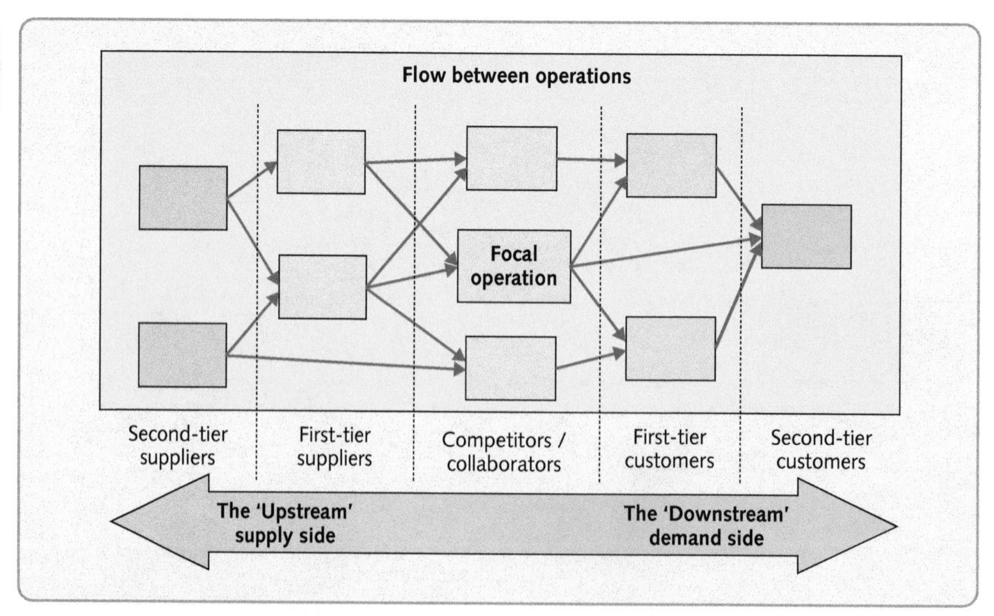

Figure 3.2 Supply network terminology

Flow through the network

Materials, parts, information, ideas and sometimes people all flow through the network of customer–supplier relationships formed by all these operations. Also, along with the forward flow of transformed resources (materials, information and customers) in the network, each customer–supplier linkage will feed back orders and information. For example, when stocks run low, retailers place orders with distributors who likewise place orders with the manufacturer, who will in turn place orders with its suppliers, who will replenish their own stocks from their own suppliers. So flow is a two-way process with items flowing one way and information flowing the other.

It is not only manufacturers who are part of a supply network. The flow of physical materials may be easier to visualise, but service operations also have suppliers and customers who themselves have their own suppliers and customers. One way to visualise the supply networks of some service operations is to consider the downstream flow of information that passes between operations. Most financial service supply networks can be thought about like this. However, not all service supply networks deal primarily in information. For example, property companies that own and/or run shopping malls have suppliers who provided security services, cleaning services, maintenance services, and so on. These first-tier suppliers will themselves receive service from recruitment agencies, consultants, etc. First-tier customers of the shopping mall are the retailers who lease retail space within the mall, who themselves serve retail customers. This is a supply network like any other. What is being exchanged between operations is the quality, speed, dependability, flexibility and cost of the services each operation supplies to its customers. In other words, there is a flow of 'operations performance' through the network. And although visualising the flow of 'performance' through supply networks is an abstract approach to visualising supply networks, it is a unifying concept. Broadly speaking, all types of supply network exist to facilitate the flow of 'operations performance'.

Note that this chapter deals exclusively with decision at the level of the supply network rather than at the level of 'the operation' or 'the process'. Yet networks of processes exist within operations, and networks of resources exist within processes. The issues of network configuration, location and capacity at these more operational levels are discussed in Chapters 4 and 5 (for layout and flow) and Chapter 8 (for capacity management).

Here are two examples of organisations that have, in their own way, made 'design' decisions concerning their supply networks that have influenced how we think about the issue.

EXAMPLE

Wincanton benefits from the logistics outsourcing trend[1]

Source: Wincanton Transport

There is a certain irony in the fact that it is the major players in the day-to-day business of supply chain management who have benefited the most from one of the most important strategic supply decisions – what should firms outsource and what should they do themselves? For many of today's most successful companies, the answer to that is question is 'logistics'. 'Why', runs the logic, 'should a business who's expertise is in retailing, or making cornflakes, or online bookselling, be running its own trucks? Why not leave it to the specialists in logistics to provide what has become known as 'third-party logistics (3PL)?' And few logistics companies have benefited more from this logic than Wincanton, the second largest contract logistics business in Europe, who manages the integrated transport of goods by road, rail, barge, sea and air through its 28,000 employees, 6,500 vehicles, 25 locomotives and a total warehousing space of 2.8 million square metres. Wincanton's customers rely on them every day to deliver goods dependably and in a cost-effective manner to them and, in turn, to their own customers. They also expect Wincanton to provide expertise and solutions to the challenges that can occur at every stage of the supply chain.

So why have companies such as Wincanton and other third-party logistics providers become such an important part of today's supply networks? It certainly is not a new phenomenon. The growth of third-party logistics companies began over thirty years ago when businesses began to look for innovative ways in which they could become leaner by farming out their logistics operations in order to focus on their core business. Particularly influential at this time was FedEx which offered its overnight delivery service and, by doing, changed the expectations of business-to-business firms. From that point on, firms like Wincanton which could offer businesses the opportunity of using just-in-time principles (see Chapter 11), saving warehousing space and reducing overall costs had a convincing case to propose to prospective customers. And as companies saw the benefits of outsourcing their delivery and warehousing operations, the number of companies offering third-party logistics services began to increase and began to offer an ever-increasing number of services. This predictably led to greater competition between third-party logistics firms, which in turn led to larger savings for the companies who employed them. Notwithstanding this increased competition, Wincanton succeeded by focusing on its 'One Wincanton' programme which helps it to differentiate itself from the competition. It focuses on four 'operating principles', which are operational excellence, customer intimacy, product leadership and value. '*The results*,' says Conor Whelan, Group IT Director at Wincanton plc, '*are solutions that allow customers to focus on their core business. Warehousing solutions are tailored to requirements on a dedicated, shared-user or networked basis, while transport is provided utilising road, rail, air and intermodal systems. One Wincanton is a group-wide initiative aimed at creating a stronger, more consistent brand and a consistent set of standards and procedures across the growing organisation. We want to help our employees provide customers with the same experience wherever they do business with us. It will also encourage greater internal collaboration, development of leadership skills, communication and understanding. One of the key communications tools we will use to facilitate 'One Wincanton' is our new intranet portal, 'One Place', which will encourage greater collaboration and knowledge sharing across the Group as well as providing information about our brand and the 'One Wincanton' campaign*'.

The other important factor in winning business in the third-party logistics market is providing innovative solutions. But when Wincanton asked their customers how they defined innovation, they got a wide variety of answers. '*It's hard to pin down*', says Whelan. '*We believe that we're continually providing our customers with innovative solutions, but it's what they expect of us so they often regard it as business as usual. Traditionally a warehouse or transport environment* [was] *manually driven,* [now] *we're increasingly seeing the introduction of voice-activated solutions that increase operational efficiency by providing a "hands free" environment*'.

EXAMPLE

Dell develops its supply network[2]

When he was a student at the University of Texas at Austin, Michael Dell's side-line of buying unused stock of PCs from local dealers, adding components, and re-selling the now higher-specification machines to local businesses was so successful he quit university and founded a computer company which was to revolutionise the industry's supply network management. But his fledgling company was just too small to make their own components. Better, he figured

Source: Toshifumi Kitamura/Getty Images

to learn how to manage a network of committed specialist component manufacturers and take the best of what was available in the market. Dell says that his commitment to outsourcing was always done for the most positive of reasons. '*We focus on how we can coordinate our activities to create the most value for customers*.' Yet Dell still faced a cost disadvantage against its far bigger competitors, so they decided to sell their computers direct to their customers, bypassing retailers. This allowed the company to cut out the retailer's (often considerable) margin, which in turn allowed Dell to offer lower prices. Dell also realised that cutting out the link in the supply network between them and the customer also provided them with significant learning opportunities by offering an opportunity to get to know their customers' needs far more intimately. This allowed them to forecast based on the thousands of customer contact calls every hour. It also allowed them to talk with customers about what they really want from their machines. Most importantly it allowed Dell to learn how to run its supply chain so that products could move through the supply chain to the end customer in a fast and efficient manner, reducing Dell's level of inventory and giving Dell a significant cost advantage.

However, what is right at one time may become a liability later on. Two decades later Dell's growth started to slow down. The irony of this is that, what had been one of the company's main advantages, its direct sales model using the internet and its market power to squeeze price reductions from suppliers, were starting to be seen as disadvantages. Although the market had changed, Dell's operating model had not. Some commentators questioned Dell's size. How could a $56 billion company remain lean, sharp and alert? Other commentators pointed out that Dell's rivals had also now learnt to run efficient supply chains ('*Getting a 20 year competitive advantage from your knowledge of how to run supply chains isn't too bad*.') However, one of the main factors was seen as the shift in the nature of the market itself. Sales of PCs to business users had become largely a commodity business with wafer-thin margins, and this part of the market was growing slowly compared to the sale of computers to individuals. Selling computers to individuals provided slightly better margins than the corporate market, but they increasingly wanted up-to-date computers with a high-design value, and most significantly, they wanted to see, touch and feel the products before buying them. This was clearly a problem for a company like Dell who had spent 20 years investing in its telephone and, later, internet-based sales channels. What all commentators agreed on was that in the fast-moving and cut-throat computer business, where market requirements could change overnight, operations resources must constantly develop appropriate new capabilities. And Dell, like other PC makers, have had to struggle to remain competitive in a rapidly changing market.

After Michael Dell returned as the boss of the company in 2007, he revamped its operations. Dell switched some of its focus to consumers and the developing world. He also conceded that the company had missed out on the boom in supplying computers to home users – who make up just 15 per cent of its revenues – because it was focused on supplying businesses. '*Let's say you wanted to buy a Dell computer in a store nine months ago – you'd have searched a long time and not found one. Now we have over 10,000 stores that sell our products.*' He rejected the idea that design was not important to his company, though he accepted that it had not been a top priority when all the focus was on business customers. '*As we've gone to the consumer we've been paying quite a bit more attention to design, fashion, colours, textures and materials.*' Yet challenges remain. First, the growing popularity of smart phones, tablet computers, and other mobile devices has had an impact on laptop sales. Second, the growth of cloud computing, which allows companies to store and process vast amounts of data in huge warehouses of servers. Hosting companies are claiming that companies would be better off renting capacity 'in the cloud' than buying their own servers from suppliers such as Dell.

What do these two examples have in common?

At first glance these two examples seem to be about different things. Wincanton has fought in an increasingly competitive market to prove that it can better any in-house logistics operation. Dell started by reconfiguring its own supply network through necessity, made a virtue of necessity, and struggled to adapt as the nature of the market changed. Yet both organisations are making decisions that will shape their, and others', supply networks in some way. Wincanton (and others in the third-party logistics market) have become the norm in industries once dominated by in-house operations. Dell has shown how even the most revolutionary supply network approach cannot remain appropriate forever and needs to adapt to competitive pressures. Between them the two examples illustrate the essentially long-term nature of how supply networks are configured. Both companies have developed their strategies over a number of decades. Note also that the developments in both Wincanton and Dell rely on assumptions regarding the level and nature of future demand. The supplement to this chapter explores forecasting in more detail. In Chapter 7, we will cover the more operational day-to-day issues of managing operations networks.

DIAGNOSTIC QUESTION

How should the supply network be configured?

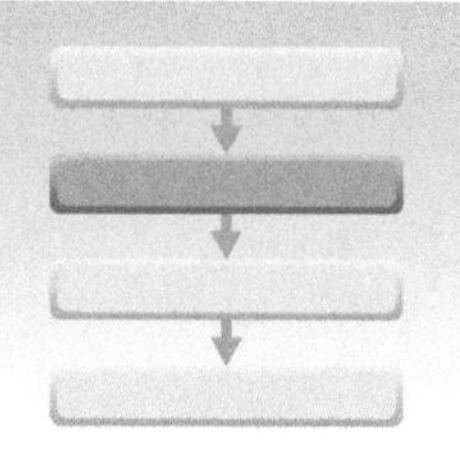

Video

An operation may want to use its influence to manage network behaviour by reconfiguring the network so as to change the scope of the activities performed in each operation and the nature of the relationships between them. The most common example of network reconfiguration has come in the attempts made over the last few years by many companies to reduce the number of suppliers with whom they have direct contact. The complexity of dealing with many hundreds of suppliers may both be expensive for an operation and (sometimes more important) prevent the operation from developing close relationships with suppliers. It is not easy to be close to hundreds of different suppliers. This has led many companies to reconfigure their supply side network to make it simpler and more orderly. It has also meant that some suppliers have become increasingly important to their customers.

OPERATIONS PRINCIPLE
Reducing the number of suppliers can reduce transaction costs and enrich supply relationships.

For example, take the front part of a car, the bit with the bumper, radiator grill, fog lights, side-lights, badge and so on.[3] At one time each of these components came from different specialist suppliers. Now the whole of this 'module' may come from one 'system supplier'. Traditional car makers are getting smaller and are relying on systems suppliers such as TRW in the US, Bosch in Germany, and Magna in Canada to provide them with whole chunks of car. Some of these system suppliers are global players who rival the car makers themselves in scope and reach. Cost pressures have forced car makers to let their suppliers take more responsibility for engineering and pre-assembly. This also means them working with fewer suppliers. For example, Ford Europe's old Escort model took parts from around 700 direct suppliers, while the replacement Focus model used only 210. Future models may have fewer than 100 direct suppliers. This can also make joint development easier. For example, Volvo paired up with one supplier (Autoliv) to develop safety systems incorporating side air bags. In return, Volvo got exclusive rights to use the systems for the first year. A smaller number of system suppliers also make it easier to update components. While a car maker may not find it economic to change its seating systems more than once every seven or eight years, a specialist supplier could have several alternative types of seat in parallel development at any one time.

Disintermediation

Another trend in some supply networks is that of companies within a network bypassing customers or suppliers to make contact directly with customers' customers or suppliers' suppliers. 'Cutting out the middle men' in this way is called *disintermediation*. An obvious example of this is the way the internet has allowed some suppliers to 'disintermediate' traditional retailers in supplying goods and services to consumers. So, for example, many services in the travel industry that used to be sold through retail outlets (travel agents) are now also available direct from the suppliers. The option of purchasing the individual components of a vacation through the websites of the airline, hotel, car-hire company etc., is now easier for consumers. Of course, they may still wish to purchase an 'assembled' product from retail travel agents which can have the advantage of convenience. Nevertheless the process of disintermediation has developed new linkages in the supply network.

Coopetition

One approach to thinking about supply networks is called the 'value net' for a company. It sees any business as being surrounded by four types of players; suppliers, customers, competitors and complementors. Complementors enable your products or services to be valued more by customers because they can also have the complementor's products or services, as opposed to when they have yours alone. Competitors are the opposite; they make customers value your product or service less when they can have their product or service, rather than yours alone. Competitors can also be complementors and vice versa. For example, adjacent restaurants may see themselves as competitors for customers' business. A customer standing outside and wanting a meal will choose between the two of them. Yet in another way they are complementors. Would that customer have come to this part of town unless there was more than one restaurant to choose from? Restaurants, theatres, art galleries, and tourist attractions generally, all cluster together in a form of cooperation to increase the total size of their joint market. It is important to distinguish between the way companies cooperate in increasing the total size of a market and the way in which they then compete for a share of that market.

Customers and suppliers have 'symmetric' roles. Historically, insufficient emphasis has been put on the role of the supplier. Harnessing the value of suppliers is just as important as listening to the needs of customers. Destroying value in a supplier in order to create it in a customer does not increase the value of the network as a whole. For example, pressurising suppliers because customers are pressurising you will not add long-term value. In the long term it creates value for the total network to find ways of increasing value for suppliers and well as customers. All the players in the network, whether they be customers, suppliers, competitors or complementors, can be both friends and enemies at different times. This is not 'unusual' or 'aberrant' behaviour. It is the way things are. The term used to capture this idea is 'coopetition'.[4]

Insource or outsource? Do or buy? The vertical integration decision

No single business does everything that is required to produce its products and services. Bakers do not grow wheat or even mill it into flour. Banks do not usually do their own credit checking, they retain the services of specialist credit checking agencies that have the specialised information systems and expertise to do it better. This process is called outsourcing and has become an important issue for most businesses. This is because, although most companies have always outsourced some of their activities, a larger proportion of direct activities are now being bought from suppliers. Also, many indirect processes are now being outsourced. This is often referred to as business process outsourcing (BPO). Financial service companies in particular are starting to outsource some of their more routine back-office processes. In a similar way, many processes within the Human Resource function, from simply payroll services through to more complex training and development processes, are being outsourced to specialist companies. The processes may still be physically located where they were before, but the staff and technology are managed by the outsourcing service provider. The reason for doing this is often primarily to reduce cost. However, there can sometimes also be significant gains in the quality and flexibility of service offered. *'People talk a lot about looking beyond cost cutting when it comes to outsourcing companies' human resource functions'*, says Jim Madden, CEO of Exult, the California-based specialist outsourcing company. *'I don't believe any company will sign up for this* [outsourcing] *without cost reduction being part of it, but for the clients whose Human Resource functions we manage, such as BP, and Bank of America, it is not just about saving money.'*

OPERATIONS PRINCIPLE

Vertical integration can be characterised by its direction, extent and balance among stages.

The outsourcing debate is just part of a far larger issue which will shape the fundamental nature of any business. Namely, what should the scope of the business be? In other words, what should it do itself and what should it buy in? This is often referred to as the 'do or buy', decision, when individual components or activities are being considered, or 'vertical integration', when it is the ownership of whole operations that are being decided. Vertical integration is the extent to which an organisation owns the network of which it is a part. It usually involves an organisation assessing the wisdom of acquiring suppliers or customers. Vertical integration can be defined in terms of three factors (Figure 3.3).[5]

- **The direction of vertical integration.** Should an operation expand by buying one of its suppliers or by buying one of its customers? The strategy of expanding on the supply side of the network is sometimes called backward or upstream vertical integration, and expanding on the demand side is sometimes called forward or downstream vertical integration.
- **The extent of vertical integration.** How far should an operation take the extent of its vertical integration? Some organisations deliberately choose not to integrate far, if at all, from their original part of the network. Alternatively, some organisations choose to become very vertically integrated.
- **The balance among stages.** This is not strictly about the ownership of the network, but rather the exclusivity of the relationship between operations. A totally balanced network relationship is one where one operation produces only for the next stage in the network and totally satisfies its requirements. Less than full balance allows each operation to sell its output to other companies or to buy in some of its supplies from other companies. Fully

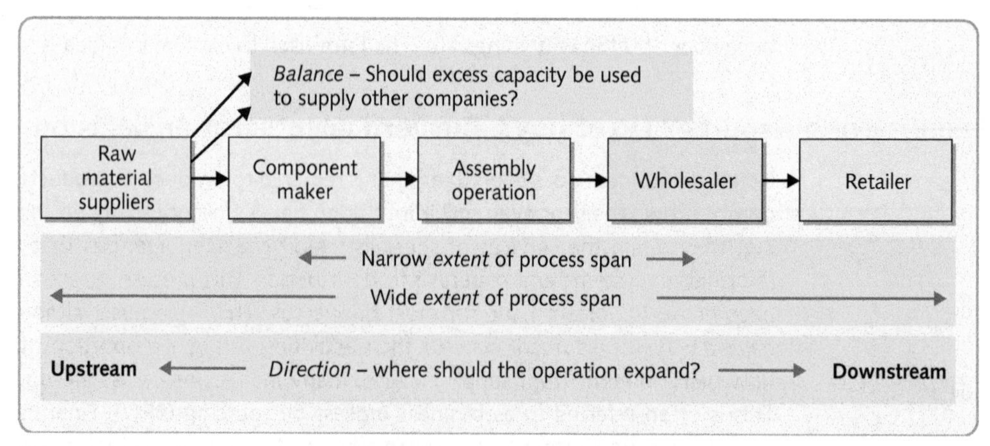

Figure 3.3 The direction, extent and balance of vertical integration

balanced networks have the virtue of simplicity and also allow each operation to focus on the requirements of the next stage along in the network. Having to supply other organisations, perhaps with slightly different requirements, might serve to distract from what is needed by their (owned) primary customer. However, a totally self-sufficient network is sometimes not feasible, nor is it necessarily desirable.

How any business positions itself in its supply network is a function of what it sees as its particular areas of expertise and where it feels it can be most profitable. However, these two factors do not always coincide. Where a business has expertise may not always be the most profitable part of the supply network. This can be a driver of vertical integration. For example, Ghana's largest cooperative, Kuapa Kokoo, owns 45 per cent of Divine, a chocolate company that produces in Germany. This is because, although chocolate sales globally are around $75 billion a year, traditionally the growers of cocoa (from which chocolate is made) have received only around $4 billion a year from the sale of cocoa beans. By taking a share of Divine, Kuapa Kokoo can capture a little more of the revenue further down the supply network.[6]

Making the outsourcing/vertical integration decision

Whether it is referred to as 'do or buy', vertical integration or no vertical integration, inhouse or outsourced supply, the choice facing operations is rarely simple. Organisations in different circumstances with different objectives are likely to take different decisions. Yet the question itself is relatively simple, even if the decision itself is not: 'Does inhouse or outsourced supply in a particular set of circumstances give the appropriate performance objectives that it requires to compete more effectively in its markets?' For example, if the main performance objectives for an operation are dependable delivery and meeting short-term changes in customers' delivery requirements, the key question should be: 'How does inhouse or outsourcing give better dependability and delivery flexibility performance?' This means judging two sets of opposing factors – those that give the potential to improve performance, and those that work against this potential being realised. Table 3.1 summarises some arguments for inhouse supply and outsourcing in terms of each performance objective.

OPERATIONS PRINCIPLE

Assessing the advisability of outsourcing should include how it impacts on relevant performance objectives.

Deciding whether to outsource

Although the effect of outsourcing on the operation's performance objective is important, there are other factors that companies take into account when deciding if outsourcing an activity is a sensible option. For example, if an activity has long-term strategic importance to a company, it is unlikely to outsource it. A retailer might choose to keep the design and development of its website inhouse, even though specialists could perform the activity at less cost, because it plans to move into web-based retailing at some point in the future. Nor would a company usually outsource an activity where it had specialist skills or knowledge. For example, a company making laser printers may have built up specialist knowledge in the production of sophisticated laser drives. This capability may allow it to introduce product or process innovations in the future. It would be foolish to 'give away' such capability. After these two more strategic factors have been considered the company's operations performance can be taken into account. Obviously if its operation's performance is already too superior to any potential supplier, it would be unlikely to outsource the activity. Also, even if its performance was currently below that of potential suppliers, it may not outsource the activity if it feels that it could significantly improve its performance. Figure 3.4 illustrates this decision logic.

OPERATIONS PRINCIPLE

Assessing the advisability of outsourcing should include consideration of the strategic importance of the activity and the operation's relative performance.

Table 3.1 How inhouse and outsourced supply may affect an operation's performance objective

Performance objective	*'Do it yourself': Inhouse supply*	*'Buy it in': outsourced supply*
Quality	The origins of any quality problems are usually easier to trace, and improvement can be more immediate, but there can be some risk of complacency.	Supplier may have specialised knowledge and more experience and motivation through market pressures, but communication of quality problems can be more difficult.
Speed	Can mean closer synchronisation of schedules which speeds up the throughput of materials and information, but if the operation also has external customers, internal customers may receive low priority.	Speed of response can be built into the supply contract where commercial pressures will encourage good performance, but there may be significant transport/ delivery delays.
Dependability	Easier communications can help dependable delivery, but as with speed, if the operation also has external customers, internal customers may receive low priority.	Late delivery penalties in the supply contract can encourage good delivery performance, but distance and organisational barriers may inhibit in communication.
Flexibility	Closeness to the real needs of a business can alert the inhouse operation that some kind of change is required in its operations, but the ability to respond may be limited by the scale and scope of internal operations.	Outsource suppliers are likely to be larger and have wider capabilities than inhouse suppliers with more ability to respond to changes, but they may have to balance the conflicting needs of different customers.
Cost	Inhouse operations do not have to make the margin required by outside suppliers so the business can capture the profits which would otherwise be given to the supplier, but relatively low volumes may mean that it is difficult to gain economies of scale or the benefits of process innovation.	Probably the main reason why outsourcing is so popular. Outsourced companies can achieve economies of scale and they are motivated to reduce their own costs because it directly impacts on their profits, but extra costs of communication and coordination with an external supplier need to be taken into account.

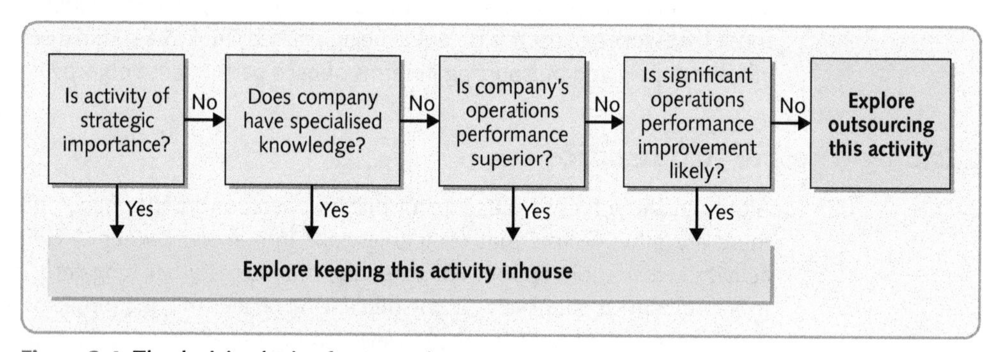

Figure 3.4 The decision logic of outsourcing

Outsourcing and offshoring

Two supply network strategies that are often confused are those of outsourcing and offshoring. Outsourcing means deciding to buy in products or services rather than perform the activities inhouse. Offshoring means obtaining products and services from operations that are based outside one's own country. Of course, one may both outsource and offshore as illustrated in Figure 3.5. Offshoring is very closely related to outsourcing and the motives for each may be similar. Offshoring to a lower-cost region of the world is usually done to reduce an operation's overall costs, as is outsourcing to a supplier who has greater expertise or scale or both.[7]

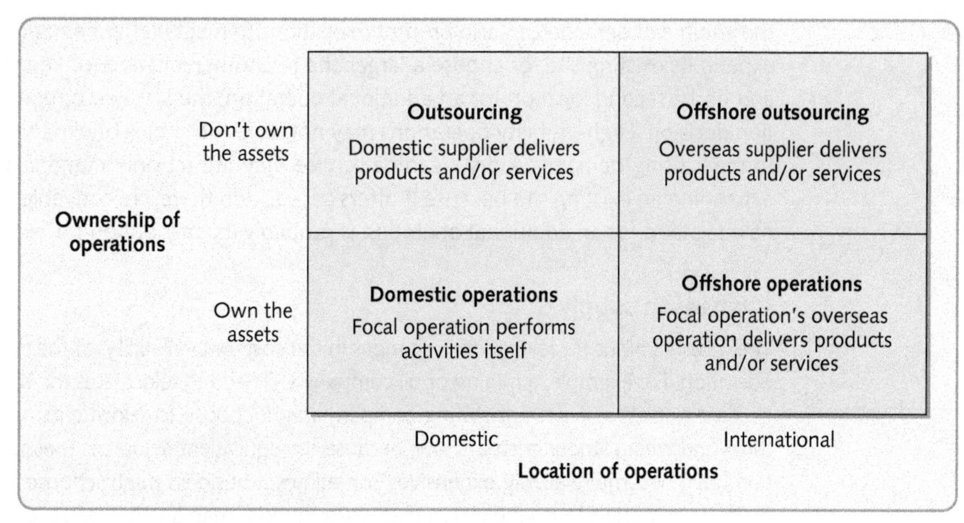

Figure 3.5 Offshoring and outsourcing are related but different

DIAGNOSTIC QUESTION

Where should operations be located?

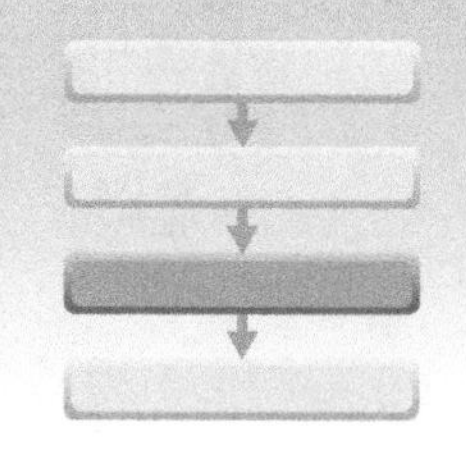

Location is the geographical positioning of an operation. It can be an important decision because it usually has an effect on an operation's costs as well as its ability to serve its customers (and therefore its revenues). So getting location wrong can have a significant impact on profits. In retailing, a difference in location of a few metres can make the difference between profit and loss. Mislocating a fire service station can slow down the average journey time of the fire crews in getting to the fires. Locating a factory where there is difficulty attracting labour with appropriate skills may damage the effectiveness of the factory's operations, and so on. The other reason why location decisions are important is that, once taken, they are difficult to undo. The costs of moving an operation from one site to another can be hugely expensive and the risks of inconveniencing customers very high. No operation wants to move very often.

Why relocate?

OPERATIONS PRINCIPLE

An operation should only change its location if the benefits of moving outweigh the costs of operating in the new location plus the cost of the move itself.

Not all operations can logically justify their location. Some are where they are for historical reasons. Yet even the operations that are 'there because they're there' are implicitly making a decision not to move. Presumably their assumption is that the cost and disruption involved in changing location would outweigh any potential benefits of a new location. Two stimuli often cause organisations to make location decisions: changes in demand for goods and services and changes in supply of inputs to the operation.

Changes in demand

A change in location may be prompted by customer demand shifting. For example, as garment manufacture moved to Asia, suppliers of zips, threads etc. started to follow them. Changes in the volume of demand can also prompt relocation. To meet higher demand, an operation could expand its existing site, or choose a larger site in another location, or keep its existing location and find a second location for an additional operation; the last two options will involve a location decision. High-visibility operations may not have the choice of expanding on the same site to meet rising demand. A dry cleaning service may attract only marginally more business by expanding an existing site because it offers a local, and therefore convenient, service. Finding a new location for an additional operation is probably its only option for expansion.

Changes in supply

The other stimulus for relocation is changes in the cost, or availability, of the supply of inputs to the operation. For example, a mining or oil company will need to relocate as the minerals it is extracting become depleted. A manufacturing company might choose to relocate its operations to a part of the world where labour costs are low, because the equivalent resources (people) in its original location have become relatively expensive. Sometimes a business might choose to relocate to release funds if the value of the land it occupies is worth more than an alternative, equally good, location.

Evaluating potential changes in location

Practice note

Evaluating possible locations is almost always a complex task because the number of location options, the criteria against which they could be evaluated, and the comparative rarity of a single location that clearly dominates all others, make the decision strategically sensitive. Furthermore, the decision often involves high levels of uncertainty. Neither the relocation activity itself, nor the operating characteristics of the new site could be as assumed when the decision was originally made. Because of this, it is useful to be systematic in terms of (a) identifying alternative options, and (b) evaluating each option against a set of rational criteria.

Identify alternative location options

The first relocation option to consider is not to. Sometimes relocation is inevitable, but often staying put is a viable option. Even if seeking a new location seems the obvious way forward, it is worth evaluating the 'do nothing' option, if only to provide a 'base case' against which to compare other options. In addition to the 'do nothing' option there should be a number of alternative location options. It is a mistake to consider only one location, but seeking out possible locations can be a time-consuming activity. Increasingly for larger companies, the whole world offers possible locations. While it has always been possible to manufacture in one part of the world in order to sell in another, until recently non-manufacturing operations were assumed to be confined to their home market. But no longer; the operational skills (as well as the brand image) of many service operations are transferable across national boundaries. Hotel, fast food, retailers and professional services all make location decisions on an international stage. Similarly, information-processing operations can now locate outside their immediate home base, thanks to virtually seamless telecommunications networks. If a financial services, or any other, business sees a cost advantage in locating part of its back-office operations in a part of the world where the 'cost per transaction' is lower, it can do so.

The implication of the globalisation of the location decision is to increase both the number of options and the degree of uncertainty in their relative merits. The sheer number of possibilities makes the location decision impossible to 'optimise'. Unless every single option is explored, no one best choice is possible. Rather, the process of identifying location options usually involves selecting a limited number of sites that represent different attributes. For example, a distribution centre, while always needing to be close to transport links, could be located in any of several regions and could either be close to population centres, or in a more rural location. The options may be chosen to reflect a range of both these factors. However, this assumes that the 'supply' of location options is relatively large, which is not always the case. For example, in many retail location decisions, there are a limited number of High Street locations that become available at any point in time. Often, a retailer will wait until a feasible location becomes available and then decide whether to either take up that option or wait and take the chance that a better location becomes available soon. In effect, the location decision here is a sequence of 'take or wait' decisions. Nevertheless, in making the decision to take or wait, a similar set of criteria to the more common location problem may be appropriate.

Set location evaluation criteria

Although the criteria against which alternative locations can be evaluated will depend on circumstances, the following five broad categories are typical:

Capital requirements. The capital or leasing cost of a site is usually a significant factor. This will probably be a function of the location of the site and its characteristics. For example, the shape of the site and its soil composition can limit the nature of any buildings erected there. Access to the site is also likely to be important, as are the availability of utilities, etc. In addition the cost of the move itself may depend on which site is eventually chosen.

Market factors. Location can affect how the market, either in general, or as individual customers, perceive an operation. Locating a general hospital in the middle of the countryside may have many advantages for its staff, but it clearly would be very inconvenient for its customers. Likewise, restaurants, stores, banks, petrol filling stations and many other high-visibility operations, must all evaluate how alternative locations will determine their image and the level of service they can give. The same arguments apply to labour markets. Location may affect the attractiveness of the operation in terms of staff recruitment and retention. For example, 'science parks' are usually located close to universities because they hope to attract companies who are interested in using the skills available at the university, but not all locations necessarily have appropriate skills available immediately. Staff at a remote call centre in the western islands of Scotland, used to a calm and tranquil life, were stunned by the aggressive nature of many callers to the call centre, some being reduced to tears by bullying customers. They had to be given assertiveness training by the call centre management.

Source: The Advertising Archives

The Swedish company Vin & Spirit was bought by Pernod Ricard which immediately gave assurances that Absolut Vodka would continue to be made in Swedan. Making such a quintessential Swedish product as Absolut Vodka anywhere else would seriously harm the brand.

Cost factors. Two major categories of cost are affected by location. The first is the costs of producing products or services. For example, labour costs can vary between different areas in any country, but are likely to be a far more significant factor when international comparisons are made, when they can exert a major influence on the location decision, especially in some industries such as clothing, where labour costs as a proportion of total costs are relatively high. Other cost factors, known as community factors derive from the social, political and economic environment of its site. These include such factors as local tax rates, capital movement restrictions, government financial assistance, political stability, local attitudes to 'inward investment', language, local amenities (schools, theatres, shops, etc.), the availability of support services, the history of labour relations and behaviour, environmental restrictions and planning procedures. The second category of costs relate to both the cost of transporting inputs from their source to the location of the operation and the cost of transporting products and services from the location to customers. Whereas almost all operations are concerned to some extent with the former, not all operations are concerned with the latter, either because customers come to them (for example, hotels), or because their services can be 'transported' at virtually no cost (for example, some technology help desks). For supply networks that process physical items, however, transportation costs can be very significant.

Future flexibility. Because operations rarely change their location, any new location must be capable of being acceptable, not only under current circumstances, but also under possible future circumstances. The problem is that no one knows exactly what the future holds. Nevertheless, especially in uncertain environments, any evaluation of alternative locations should include some kind of scenario planning that considers the robustness of each in coping with a range of possible futures. Two types of the flexibility of any location could be evaluated. The most common is to consider the potential of the location for expansion to cope with increased activity levels. The second is the ability to adapt to changes in input or output factors. For example, suppliers or customers may themselves relocate in the future. If so, could the location still operate economically?

Risk factors. Closely related to the concept of future flexibility, is the idea of evaluating the risk factors associated with possible locations. Again, the risk criterion can be divided into 'transition risk' and 'long-term risk'. Transition risk is simply the risk that something goes wrong during the relocation process. Some possible locations might be intrinsically more difficult to move to than others. For example, moving to an already congested location could pose higher risks to being able to move as planned than moving to a more accessible location. Long-term risks could again include damaging changes in input factors such as exchange rates or labour costs, but can also include more fundamental security risks to staff or property.

DIAGNOSTIC QUESTION

How much capacity should each operation in the supply network have?

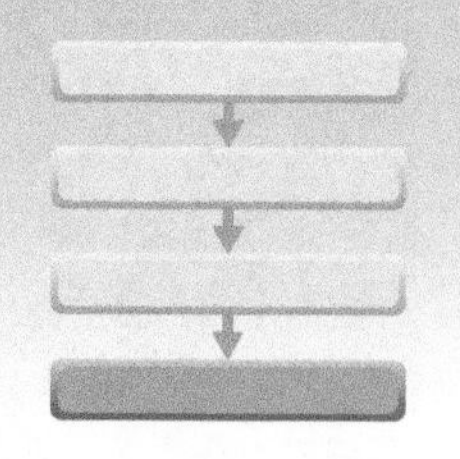

Video

The design of supply networks also includes defining the capacity of each operation in the network. Unless the capacity of individual operations reflects the needs of the network as a whole, it will either limit flow through the network or be the cause of capacity under-utilisation, both of which will in the long term reduce the effectiveness of the network as a whole. Here we shall treat capacity in a general long-term sense. Shorter-term aspects of capacity management are treated in Chapter 8. But whether over the short or long term, demand forecasts are one of the main inputs to capacity management, which is why forecasting is treated in the supplement to this chapter.

The optimum capacity level

Most organisations need to decide on the size (in terms of capacity) of each of their facilities. A chain of truck service centres, for example, might operate centres which have various capacities. The effective cost of running each centre will depend on the average service bay occupancy. Low occupancy because of few customers will result in a high cost per customer served because the fixed costs of the operation are being shared between few customers. As demand, and therefore service bay occupancy, increases, the cost per customer will reduce. However, operating at very high levels of capacity utilisation (occupancy levels close to capacity) can mean longer customer waiting times and reduce customer service. This effect is described in more detail in Chapter 5. There may also be less obvious cost penalties of operating centres at levels close to its nominal capacity. For example, long periods of overtime may reduce productivity levels as well as costing more in extra payments to staff: utilising bays at

OPERATIONS PRINCIPLE

All types of operation exhibit economy of scale effects, where operating costs reduce as the scale of capacity increases.

Animated diagram

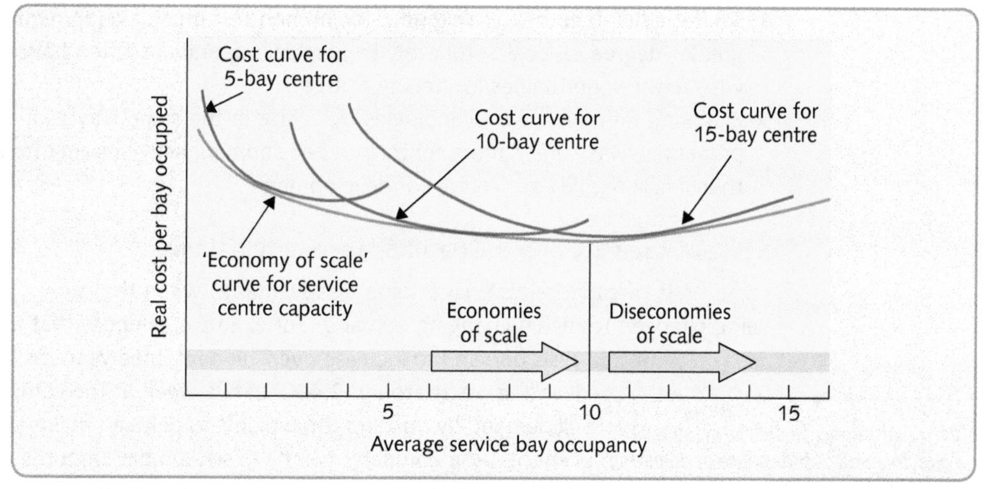

Figure 3.6 Unit cost curves for individual truck service centres of varying capacities

very high utilisation reduces maintenance and cleaning time that may increase breakdowns, reduce effective life, and so on. This usually means that average costs start to increase after a point which will often be lower than the theoretical capacity of the operation.

The blue curves in Figure 3.6 shows this effect for the service centres of 5-, 10-, and 15-bay capacity. As the nominal capacity of the centres increases, the lowest cost point at first reduces. This is because the fixed costs of any operation do not increase proportionately as its capacity increases. A 10-bay centre has less than twice the fixed costs of a 5-bay centre. Also the capital costs of constructing the operations do not increase proportionately to their capacity. A 10-bay centre costs less to build than twice the cost of a 5-bay centre. These two factors, taken together, are often referred to as economies of scale. However, above a certain size, the lowest cost point may increase. This occurs because of what are called diseconomies of scale, two of which are particularly important. First, complexity costs increase as size increases. The communications and coordination effort necessary to manage an operation tends to increase faster than capacity. Although not seen as a direct cost, this can nevertheless be very significant. Second, a larger centre is more likely to be partially underutilised because demand within a fixed location will be limited. The equivalent in operations that process physical items is transportation costs. For example, if a manufacturer supplies the whole of its European market from one major plant in Denmark, all supplies may have to be brought in from several countries to the single plant and all products shipped from there throughout Europe.

OPERATIONS PRINCIPLE

Diseconomies of scale increase operating costs above a certain level of capacity resulting in a minimum cost level of capacity.

Being small may have advantages

Although large-scale capacity operations will usually have a cost advantage over smaller units, there are also potentially significant advantages that can be exploited by small-scale operations. One significant research study showed that small-scale operations can provide significant advantages in the following four areas.[8]

- They allow businesses to locate near to 'hot spots' that can tap into local knowledge networks. Often larger companies centralise their research and development efforts, losing touch with where innovative ideas area generated.
- Responding rapidly to regional customer needs and trends by basing more and smaller units of capacity close to local markets.

- Taking advantage of the potential for human resource development by allowing staff a greater degree of local autonomy. Larger-scale operations often have longer career paths with fewer opportunities for 'taking charge'.
- Exploring radically new technologies by acting in the same way as a smaller, more entrepreneurial rival. Larger, more centralised development activities are often more bureaucratic than smaller-scale agile centres of development.

Scale of capacity and the demand–capacity balance

Large units of capacity also have some disadvantages when the capacity of the operation is being changed to match changing demand. For example, suppose that a manufacturer forecasts demand to increase over the next three years, as shown in Figure 3.7, levelling off at around 2,400 units a week. If the company seeks to satisfy all demand by building three plants as demand builds up, each of 800 units capacity, the company will have substantial amounts of over-capacity for much of the period when demand is increasing. Over-capacity means low capacity utilisation, which in turn means higher unit costs. If the company builds smaller plants, say 400-unit plants, there will still be over-capacity but to a lesser extent, which means higher capacity utilisation and possibly lower costs.

> **OPERATIONS PRINCIPLE**
> *Changing capacity using large units of capacity reduces the chance of achieving demand–capacity balance.*

The timing of capacity change

Changing the capacity of an operation is not just a matter of deciding on the best size of a capacity increment. The operation also needs to decide when to bring 'on-stream' new capacity. For example, Figure 3.8 shows the forecast demand for the new manufacturing operation. It has been decided to build 400-unit-per-week plants in order to meet the growth in demand for its new product. In deciding when the new plants are to be introduced the company must choose a position somewhere between two extreme strategies:

> **OPERATIONS PRINCIPLE**
> *Capacity-leading strategies increase opportunities to meet demand.*

- Capacity *leads* demand – timing the introduction of capacity in such a way that there is always sufficient capacity to meet forecast demand.
- Capacity *lags* demand – timing the introduction of capacity so that demand is always equal to or greater than capacity.

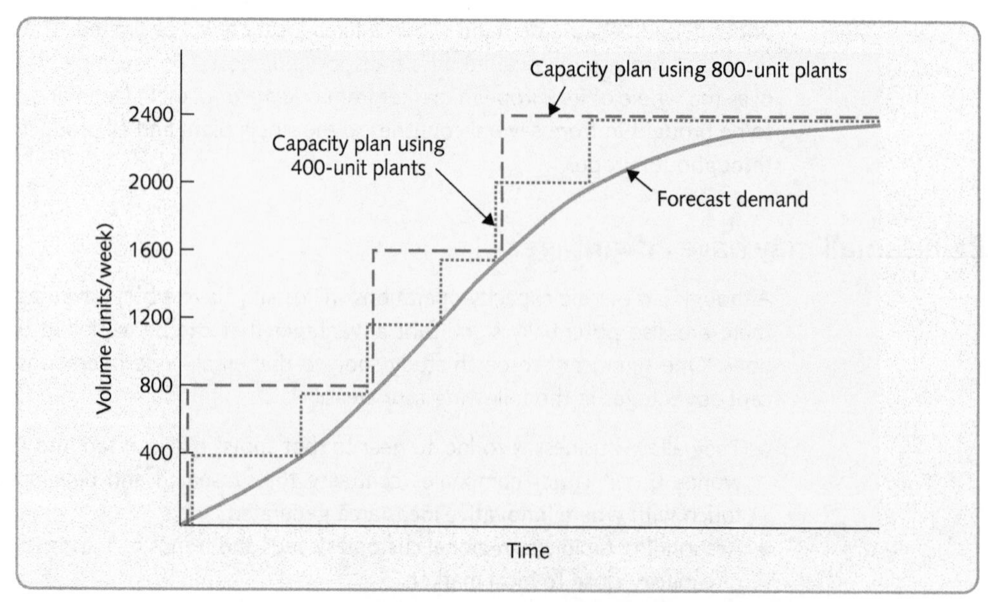

Figure 3.7 The scale of capacity increments affects the utilisation of capacity

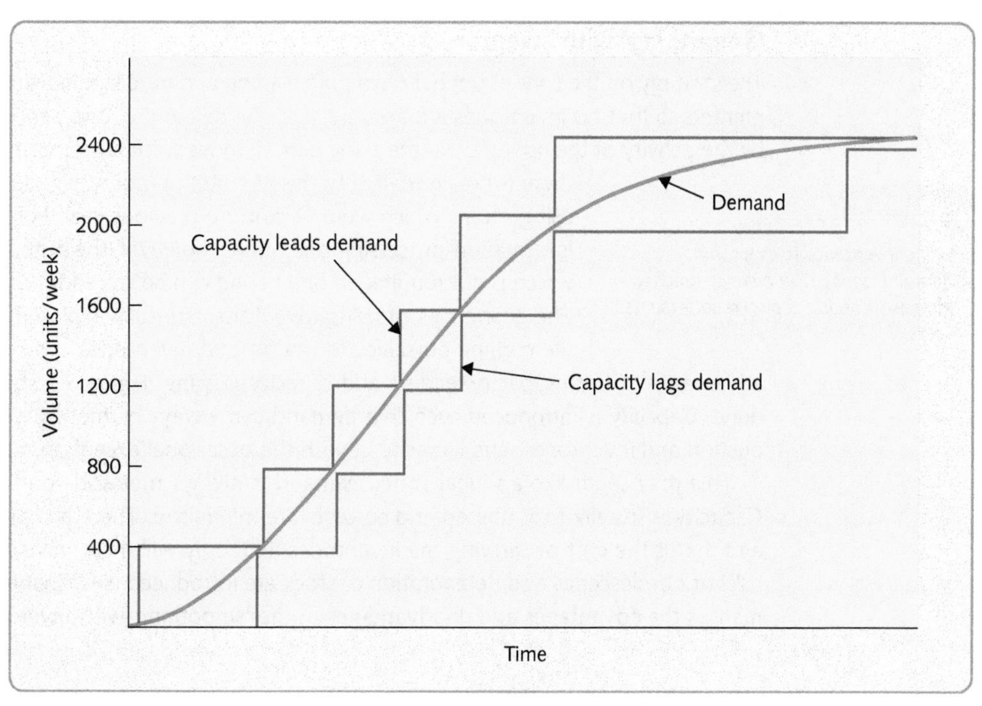

Figure 3.8 Capacity-leading and capacity-lagging strategies

Figure 3.8 shows these two extreme strategies, although in practice the company is likely to choose a position somewhere between the two. Each strategy has its own advantages and disadvantages. These are shown in Table 3.2. The actual approach taken by any company will depend on how it views these advantages and disadvantages. For example, if the company's access to funds for capital expenditure is limited, it is likely to find the delayed capital expenditure requirement of the capacity-lagging strategy relatively attractive.

OPERATIONS PRINCIPLE
Capacity-lagging strategies increase capacity utilisation.

Table 3.2 The arguments for and against pure leading and pure lagging strategies of capacity timing

Advantages	*Disadvantages*
Capacity-leading strategies	
• Always sufficient capacity to meet demand, therefore revenue is maximised and customers satisfied • Most of the time there is a 'capacity cushion' which can absorb extra demand if forecasts are pessimistic • Any critical start-up problems with new plants are less likely to affect supply to customers	• Utilisation of the plants is always relatively low, therefore costs will high • Risk of even greater (or even permanent) over-capacity if demand dos not reach forecast levels • Capital spending on plant early
Capacity-lagging strategies	
• Always sufficient demand to keep the plants working at full capacity, therefore unit costs are minimised • Over-capacity problems are minimised if forecasts are optimistic • Capital spending on the plants is delayed	• Insufficient capacity to meet demand fully, therefore reduced revenue and dissatisfied customers • No ability to exploit short-term increases in demand • Under-supply position even worse if there are start-up problems with the new plants

'Smoothing' with inventory

The strategy on the continuum between pure leading and pure lagging strategies can be implemented so that no inventories are accumulated. All demand in one period is satisfied (or not) by the activity of the operation in the same period. Indeed, for customer-processing operations there is no alternative to this. A hotel cannot satisfy demand in one year by using rooms which were vacant the previous year. For some materials- and information-processing operations, however, the output from the operation which is not required in one period can be stored for use in the next period. The economies of using inventories are fully explored in Chapter 10. Here we confine ourselves to noting that inventories can be used to obtain the advantages of both capacity-leading and capacity-lagging. Figure 3.9 shows how this can be done. Capacity is introduced such that demand can always be met by a combination of production and inventories, and capacity is, with the occasional exception, fully utilised.

OPERATIONS PRINCIPLE

Using inventories to overcome demand-capacity imbalance tends to increase working capital requirements.

This may seem like an ideal state. Demand is always met and so revenue is maximised. Capacity is usually fully utilised and so costs are minimised. There is a price to pay, however, and that is the cost of carrying the inventories. Not only will these have to be funded but the risks of obsolescence and deterioration of stock are introduced (see Chapter 9). Table 3.3 summarises the advantages and disadvantages of the 'smoothing-with-inventory' strategy.

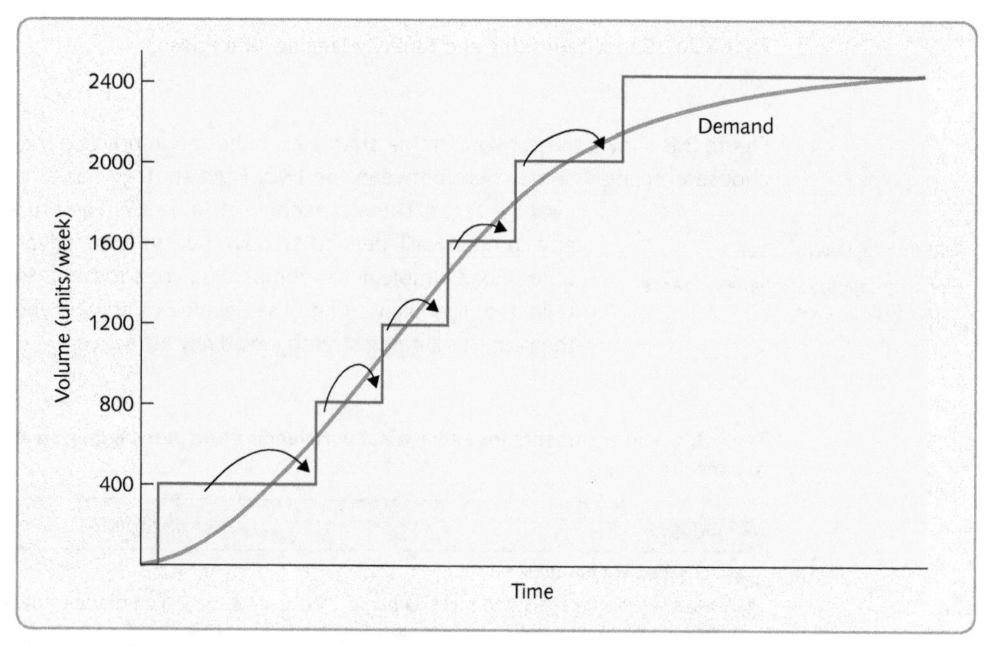

Figure 3.9 Smoothing with inventory means using the excess capacity of one period to produce inventory with which to supply the under-capacity of another period

Table 3.3 The advantages and disadvantages of a smoothing-with-inventory strategy

Advantages	*Disadvantages*
• All demand is satisfied, therefore customers are satisfied and revenue is maximised • Utilisation of capacity is high and therefore costs are low • Very short-term surges in demand can be met from inventories	• The cost of inventories in terms of working capital requirements can be high. This is especially serious at a time when the company requires funds for its capital expansion • Risk of product deterioration and obsolescence

Break-even analysis of capacity expansion

An alternative view of capacity expansion can be gained by examining the cost implications of adding increments of capacity on a break-even basis. Figure 3.10 shows how increasing capacity can move an operation from profitability to loss. Each additional unit of capacity results in a fixed-cost break, which is a further lump of expenditure which will have to be incurred before any further activity can be undertaken in the operation.

The operation is therefore unlikely to be profitable at very low levels of output. Eventually, assuming that prices are greater than marginal costs, revenue will exceed total costs. However, the level of profitability at the point where the output level is equal to the capacity of the operation may not be sufficient to absorb all the extra fixed costs of a further increment in capacity. This could make the operation unprofitable in some stages of its expansion.

WORKED EXAMPLE

A specialist graphics company is investing in new systems that will enable it to make high-quality images for its clients. Demand for these images is forecast to be around 100,000 images in year 1 and 220,000 images in year 2. The maximum capacity of each system is 100,000 images per year. They have a fixed cost of €200,000 per year and a variable cost of processing of €1 per image. The company believes they will be able to charge an average of €4 per image. What profit are they likely to make in the first and second years?

Year 1 demand = 100,000 images; therefore company will need one machine

Cost of producing images = fixed cost for one machine + (variable cost × 100,000)
= €200,000 + (€1 × 100,000)
= €300,000

Revenue = demand × price
= 100,000 × €4
= €400,000

Therefore profit = €400,000 − €300,000
= €100,000

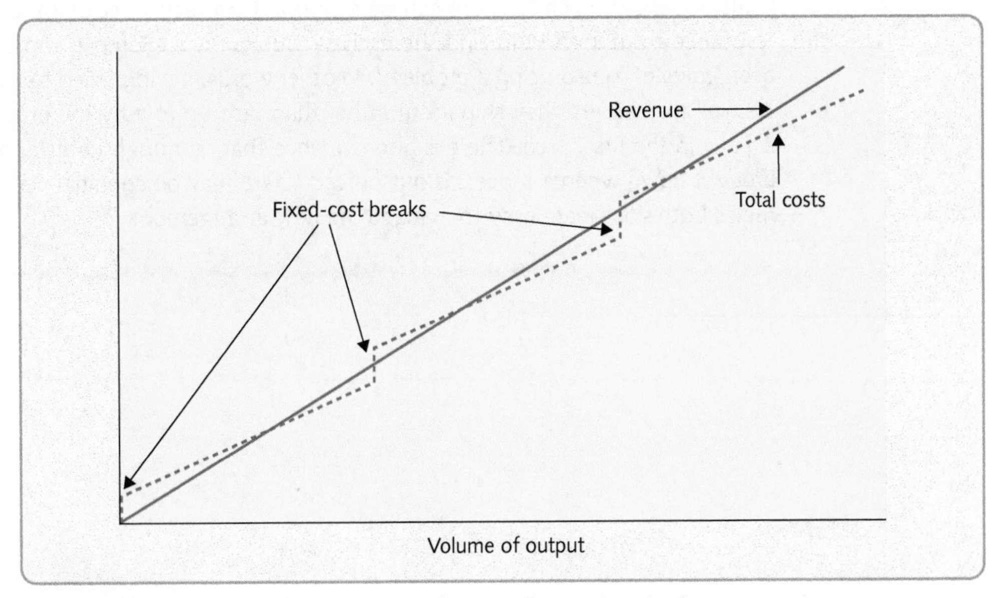

Figure 3.10 Incurring fixed costs repeatedly can raise total costs above revenue over some ranges of output

$$
\begin{aligned}
\text{Year 2 demand} &= \text{220,000 units; therefore company will need three machines}\\
\text{Cost of producing images} &= \text{fixed cost for three machines} + (\text{variable cost} \times 220{,}000)\\
&= (3 \times €200{,}000) + (€1 \times 220{,}000)\\
&= €820{,}000\\
\text{Revenue} &= \text{demand} \times \text{price}\\
&= 220{,}000 \times €4\\
&= €880{,}000\\
\text{Therefore profit} &= €880{,}000 - €820{,}000\\
&= €60{,}000
\end{aligned}
$$

Note: the profit in the second year will be lower because of the extra fixed costs associated with the investment in the two extra machines.

Critical commentary

Each chapter contains a short critical commentary on the main ideas covered in the chapter. Its purpose is not to undermine the issues discussed in the chapter, but to emphasise that, although we present a relatively orthodox view of operation, there are other perspectives.

- Probably the most controversial issue in supply network design is that of outsourcing. In many instances there has been fierce opposition to companies outsourcing some of their processes. Trade unions often point out that the only reason that outsourcing companies can do the job at lower cost is that they either reduce salaries, reduce working conditions, or both. Furthermore, they say, flexibility is only achieved by reducing job security. Employees who were once part of a large and secure corporation could find themselves as far less secure employees of a less benevolent employer with a philosophy of permanent cost-cutting. Even some proponents of outsourcing are quick to point out the problems. There can be significant obstacles, including understandable resistance from staff who find themselves 'outsourced'. Some companies have also been guilty of 'outsourcing a problem'. In other words, having failed to manage a process well themselves, they ship it out rather than face up to why the process was problematic in the first place. There is also evidence that, although long-term costs can be brought down when a process is outsourced, there may be an initial period when costs rise as both sides learn how to manage the new arrangement.

SUMMARY CHECKLIST

This checklist comprises questions that can be usefully applied to any type of operations and reflect the major diagnostic questions used within the chapter.

- ☐ Is the operation fully aware of all its first- and second-tier suppliers' and customers' capabilities and requirements?
- ☐ Are capabilities of suppliers and requirements of customers understood in terms of all aspects of operations performance?
- ☐ Does the operation have a view on how it would like to see its supply network develop over time?
- ☐ Have the benefits of reducing the number of individual suppliers been explored?
- ☐ Are any parts of the supply network likely to become disintermediated, and have the implications of this been considered?
- ☐ Does the operation have an approach to how its treats others in the supply network who might be both complementors and competitors?
- ☐ Is the vertical integration/outsourcing issue always under review for possible benefits?
- ☐ Is outsourcing (or bringing back inhouse) evaluated in terms of all the operation's performance objectives?
- ☐ Is there a rational set of criteria used for deciding whether to outsource?
- ☐ Is the relocation decision ever considered?
- ☐ Have factors such as changes in demand or supply that may prompt a relocation been considered?
- ☐ If considering a relocation, are alternative locations always evaluated against each other and against a 'do nothing' option?
- ☐ Are sufficient location options being considered?
- ☐ Do location evaluation criteria include capital, market, cost, flexibility, and risk factors?
- ☐ Is the optimum economy of scale for the different types of operation with the business periodically assessed?
- ☐ Are the various strategies for timing changes in capacity always evaluated in terms of their advantages and disadvantages?
- ☐ Are the fixed cost breaks of capacity increase understood, and are they taken into account when increasing or decreasing capacity?

CASE STUDY

Disneyland Resort Paris (abridged)[9]

In August 2006, the company behind Disneyland Resort Paris reported a 13 per cent rise in revenues, saying that it was making encouraging progress with new rides aimed at getting more visitors. '*I am pleased with year-to-date revenues and especially with the third quarter's, as well as with the success of the opening of Buzz Lightyear Laser Blast, the first step of our multi-year investment programme. These results reflect the Group's strategy of increasing growth through innovative marketing and sales efforts as well as a multi-year investment programme. This performance is encouraging as we enter into the important summer months*,' said Chairman and Chief Executive, Karl L. Holz. Revenue for the quarter ending 30 June rose to €286.6 million ($362 million) from €254 million a year earlier. The results helped to boost overall profits at Disney Company and the company's stock price soared.

Yet it hadn't always been like that. The 14-year history of Disneyland Paris had more ups and downs than any of its rollercoasters. The company had hauled itself back from what some commentators had claimed was the brink of bankruptcy in 2005. In fact, from 12 April 1992, when Euro Disney opened, through to this more optimistic report, the resort had been subject simultaneously to wildly optimistic forecasts and widespread criticism and ridicule. An essay on one critical internet site (called 'An Ugly American in Paris') summarised the whole venture in this way: '*When Disney decided to expand its hugely successful theme park operations to Europe, it brought American management styles, American cultural tastes, American labor practices, and American marketing pizzazz to Europe. Then, when the French stayed away in droves, it accused them of cultural snobbery.*'

The 'Magic' of Disney

Since its founding in 1923, The Walt Disney Company had striven to remain faithful in its commitment to '*producing unparalleled entertainment experiences based on its rich legacy of quality, creative content and exceptional storytelling*'. It did this through four major business divisions: Studio Entertainment, Parks and Resorts, Consumer Products and Media Networks. Each segment consisted of integrated businesses that worked together to '*maximise exposure and growth worldwide*'.

In the Parks and Resorts division, according to the company's description, customers could experience the '*magic of Disney's beloved characters*'. It was founded in 1952, when Walt Disney formed what is now known as Walt Disney Imagineering to build Disneyland in Anaheim, California. By 2006, Walt Disney Parks and Resorts operated or licensed 11 theme parks at five Disney destinations around the world. They were: Disneyland Resort, California; Walt Disney World Resort, Florida; Tokyo Disney Resort; Disneyland Resort Paris; and the latest park, Hong Kong Disneyland. In addition, the division operated 35 resort hotels, two luxury cruise ships and a wide variety of other entertainment offerings. But in the history of the Walt Disney Company, perhaps none of its ventures had proved to be as challenging as its Paris resort.

Source: Chad Ehlers/Alamy Images

Service delivery at Disney resorts and parks

The core values of the Disney Company and, arguably, the reason for its success originated in the views and personality of Walt Disney, the company's founder. He had what some called an obsessive focus on creating images, products and experiences for customers that epitomised fun, imagination and service. Through the 'magic' of legendary fairytale and story characters, customers could escape the cares of the real world. Different areas of each Disney Park are themed, often around various 'lands' such as Frontierland, Fantasyland, Tomorrowland and Adventureland. Each land contains attractions and rides, most of which are designed to be acceptable to a wide range of ages. Very few rides are 'scary' when compared with those in many other entertainment parks. The architectural styles, décor, food, souvenirs and cast costumes are all designed to reflect the theme of the 'land', as are the films and shows.

Although there were some regional differences, all the theme parks followed the same basic setup. Over the years, Disney had built up a reputation for imaginative rides. Its 'imagineers' had years of experience in using 'auto animatronics' to help recreate and reinforce the essence of the theme. The terminology used by the company reinforced its philosophy of consistent entertainment. Employees, even those working 'back stage', were called 'cast members'. They did not wear uniforms but 'costumes', and rather than being given a job they were 'cast in a role'. All park visitors were called 'guests'.

Disney employees were generally relatively young, often of school or college age. Most were paid hourly on tasks that could be repetitive even though they usually involved constant contact with customers. Yet employees were still expected to maintain a high level of courtesy and work performance. All cast members were expected to conform to strict dress and grooming standards. Applicants to become cast members were screened for qualities such as how well they responded to questions, how well they listened to their peers, how they smiled and used body language, and whether they had an 'appropriate attitude'.

All Disney parks had gained a reputation for their obsession with delivering a high level of service and experience through attention to operations detail. To ensure that their strict service standards were met they had developed a number of specific operations policies:

- All parks employed effective queue-management techniques such as providing information and entertainment for visitors.
- Visitors (guests) were seen as having a role within the park. They were not merely spectators or passengers on the rides, they were considered to be participants in a play. Their needs and desires were analysed and met through frequent interactions with staff (cast members). In this way they could be drawn into the illusion that they were actually part of the fantasy.
- Disney's stated goal was to exceed its customers' expectations every day.
- Service delivery was mapped and continuously refined in the light of customer feedback.
- The staff induction programme emphasised the company's quality-assurance procedures and service standards. These were based on the four principles of safety, courtesy, show and efficiency.
- Parks were kept fanatically clean.
- The same Disney character never appeared twice within sight – how could there be two Mickeys?
- Staff were taught that customer perceptions were the key to customer delight but also were extremely fragile. Negative perceptions can be established after only one negative experience.
- Disney University was the company's inhouse development and learning facility with departments in each of the company's sites. The University trained Disney's employees in strict service standards as well as providing the skills to operate new rides as they were developed.
- Staff recognition programmes attempted to identify outstanding service delivery performance as well as *'energy, enthusiasm, commitment and pride'*.
- All parks contained phones connected to a central question hot-line for employees to find the answer to any question posed by customers.

Tokyo Disneyland

Tokyo Disneyland, opened in 1982, was owned and operated by the Oriental Land Company. Disney had designed the park and advised on how it should be run and it was considered a great success. Japanese customers revealed a significant appetite for American themes and brands, and already had a good knowledge of Disney characters. Feedback was extremely positive, with visitors commenting on the cleanliness of the park and the courtesy and efficiency of staff members. Visitors also appreciated the Disney souvenirs because giving gifts is deeply embedded in the Japanese culture. The success of the Tokyo Park was explained by one American living in Japan. *'Young Japanese are very clean-cut. They respond well to Disney's clean-cut image, and I am sure they had no trouble filling positions. Also, young Japanese are generally comfortable wearing uniforms, obeying their bosses, and being part of a team. These are part of the Disney formula. Also, Tokyo is very crowded and Japanese here are used to crowds and waiting in line. They are very patient. And above all, Japanese are always very polite to strangers.'*

Disneyland Paris

By 2006, Disneyland Paris consisted of three parks: the Disney village, Disneyland Paris itself and the Disney Studio Park. The Village was comprised of stores and restaurants, Disneyland Paris was the main theme park and Disney Studio Park had a more general movie-making theme. At the time of the European park's opening, more than 2,000,000 Europeans visited the US Disney parks, accounting for 5 per cent of the total visitors. The company's brand was strong and it had over half a century of translating the Disney brand into reality. The name 'Disney' had become synonymous with wholesome family entertainment that combined childhood innocence with high-tech 'Imagineering'.

Alternative locations

Initially, as well as France, Germany, Britain, Italy and Spain were all considered as possible locations, though

Germany, Britain and Italy were soon discarded from the list of potential sites. The decision soon came to a straight contest between the Alicante area of Spain, which had a similar climate to Florida for a large part of the year, and the Marne-la-Vallée area just outside Paris. Certainly, winning the contest to host the new park was important for all the potential host countries – the new park promised to generate more than 30,000 jobs.

The major advantage of locating in Spain was the weather. However, the eventual decision to locate near Paris was thought to have been driven by a number of factors that weighed more heavily with Disney executives. These included the following:

- There was a site available just outside Paris which was both large enough and flat enough to accommodate the park.
- The proposed location put the park within a two-hour drive for 17,000,000 people, a four-hour drive for 68,000,000 people, a six-hour drive for 110,000,000 people and a two-hour flight for a further 310,000,000 or so.
- The site also had potentially good transport links. The Euro Tunnel that was to connect England with France was due to open in 1994. In addition, the French autoroutes network and the high-speed TGV network could both be extended to connect the site with the rest of Europe.
- Paris was already a highly attractive vacation destination and France generally attracted around 50,000,000 tourists each year.
- Europeans generally take significantly more holidays each year than Americans (five weeks of vacation as opposed to two or three weeks).
- Market research indicated that 85 per cent of people in France would welcome a Disney park in their country.
- Both national and local government in France were prepared to give significant financial incentives (as were the Spanish authorities), including an offer to invest in local infrastructure, reduce the rate of value-added tax on goods sold in the park, provide subsidised loans and value the land artificially low to help reduce taxes. Moreover, the French government was prepared to expropriate land from local farmers to smooth the planning and construction process.

Early concerns that the park would not have the same sunny, happy feel in a climate cooler than Florida were allayed by the spectacular success of Disneyland Tokyo in a location with a similar climate to Paris.

Construction was starting on the 2000-hectare site in August 1988. But from the announcement that the park would be built in France, it was subject to a wave of criticism. One critic called the project a *'cultural Chernobyl'* because of how it might affect French cultural values. Another described it as *'a horror made of cardboard, plastic, and appalling colours; a construction of hardened chewing-gum and idiot folk lore taken straight out of comic books written for obese Americans'*. However, as some commentators noted, the cultural arguments and anti-Americanism of the French intellectual elite did not seem to reflect the behaviour of most French people, who *'eat at McDonald's, wear Gap clothing, and flock to American movies'*.

Designing Disneyland Resort Paris

Phase 1 of the Euro Disney Park was designed to have 29 rides and attractions, a championship golf course together with many restaurants, shops, live shows and parades as well as six hotels. Although the park was designed to fit in with Disney's traditional appearance and values, a number of changes were made to accommodate what were thought to be the preferences of European visitors. For example, market research indicated that Europeans would respond to a 'wild west' image of America. Therefore, both rides and hotel designs were made to emphasise this theme. Disney was also keen to diffuse criticism, especially from French left-wing intellectuals and politicians, that the design of the park would be too 'Americanised' and would become a vehicle for American 'cultural imperialism'. To counter charges of American imperialism, Disney gave the park a flavour that stressed the European heritage of many of the Disney characters and increased the sense of beauty and fantasy. It was, after all, competing against Paris's exuberant architecture and sights. For example, Discoveryland featured storylines from Jules Verne, the French author. Snow White (and her dwarfs) was located in a Bavarian village. Cinderella was located in a French inn. Even Peter Pan was made to appear more 'English Edwardian' than in the original US designs.

Because of concerns about the popularity of American 'fast food', Euro Disney introduced more variety into its restaurants and snack bars, featuring foods from around the world. In a bold publicity move, Disney invited a number of top Paris chefs to visit and taste the food. Some anxiety was also expressed concerning the different 'eating behaviour' between Americans and Europeans. Whereas Americans preferred to 'graze', eating snacks and fast meals throughout the day, Europeans generally preferred to sit down and eat at traditional meal times. This would have a very significant impact on peak demand levels in dining facilities. A further concern was that in Europe (especially French) visitors would be intolerant of long queues. To overcome this, extra diversions such as films and entertainments were planned for visitors as they waited in line for a ride.

Before the opening of the park, Euro Disney had to recruit and train between 12,000 and 14,000 permanent and around 5000 temporary staff. All these new employees were required to undergo extensive training to prepare them to achieve Disney's high standard of customer service as well as to understand operational routines and safety procedures. Originally, the company's objective was to hire 45 per cent of its employees from France, 30 per cent from other European countries and 15 per cent from outside of Europe. However, this proved difficult and when the park opened around 70 per cent of employees were French. Most cast members were paid around 15 per cent above the French minimum wage.

An information centre was opened in December 1990 to show the public what Disney was constructing. The 'casting centre' was opened on 1 September 1991 to recruit the 'cast members' needed to staff the park's attractions. But the hiring process did not go smoothly. In particular, Disney's grooming requirements that insisted on a 'neat' dress code, a ban on facial hair, set standards for hair and finger nails, and an insistence on 'appropriate undergarments' proved controversial. Both the French press and trade unions strongly objected to the grooming requirements, claiming they were excessive and much stricter than was generally held to be reasonable in France. Nevertheless, the company refused to modify its grooming standards. Accommodating staff also proved to be a problem when the large influx of employees swamped the available housing in the area. Disney had to build its own apartments as well as rent rooms in local homes just to accommodate its employees. Nevertheless, notwithstanding all the difficulties, Disney did succeed in recruiting and training all its cast members before the opening.

The park opens

The park opened to employees for testing during late March 1992, during which time the main sponsors and their families were invited to visit the new park, but the opening was not helped by strikes on the commuter trains leading to the park, staff unrest, threatened security problems (a terrorist bomb had exploded the night before the opening) and protests in surrounding villages which demonstrated against the noise and disruption from the park. The opening day crowds, expected to be 500,000, failed to materialise and at close of the first day only 50,000 people had passed through the gates.

Disney had expected the French to make up a larger proportion of visiting guests than they did in the early days. The poor turnout may have been partly due to protests from French locals who feared their culture would be damaged by Euro Disney. Also, all Disney parks had traditionally been alcohol-free and to begin with Euro Disney was no different. However, this was extremely unpopular, particularly with French visitors who like to have a glass of wine or beer with their food. Whatever the cause, the low initial attendance was very disappointing for the Disney Company.

It was reported that in the first nine weeks of operation, approximately 1000 employees left Euro Disney, about one half of them 'voluntarily'. The reasons cited varied. Some blamed the hectic pace of work and the long hours that Disney expected. Others mentioned the 'chaotic' conditions in the first few weeks of the park opening. Even Disney conceded that conditions had been tough immediately after the park opened. Some leavers blamed Disney's apparent difficulty in understanding 'how Europeans work'. *'We can't just be told what to do, we ask questions and don't all think the same.'* Some visitors who had experience of the American parks commented that the standards of service were noticeably below what would be acceptable in America. There were reports that some cast members were failing to meet Disney's normal service standard. *'Even on opening weekend some clearly couldn't care less. My overwhelming impression . . . was that they were out of their depth. There is much more to being a cast member than endlessly saying 'Bonjour'. Apart from having a detailed knowledge of the site, Euro Disney staff have the anxiety of not knowing in what language they are going to be addressed. Many were struggling.'*

It was also noticeable that different nationalities exhibited different types of behaviour when visiting the park. Some nationalities always used the waste bins while others were more likely to drop litter on the floor. Most noticeable were differences in queuing behaviour. Northern Europeans tend to be disciplined and content to wait for rides in an orderly manner. By contrast some southern European visitors *'seem to have made an Olympic event out of getting to the ticket taker first'*. Nevertheless, not all reactions were negative. European newspapers quoted plenty of positive reaction from visitors, especially children. Euro Disney was so different from the existing European theme parks, with immediately recognisable characters and a wide variety of attractions. Families who could not afford to travel to the United States could now interact with Disney characters and *'sample the experience at far less cost'*.

The next 15 years

By August 1992, estimates of annual attendance figures were being drastically cut from 11,000,000 to just over 9,000,000. Euro Disney's misfortunes were further compounded in late 1992 when a European recession caused property prices to drop sharply and interest payments on its large start-up loans forced the company to admit serious financial difficulties. Also, the cheap dollar resulted in more people taking their holidays in Florida at Walt Disney

World. While at the first anniversary of the Paris park's opening, in April 1993, Sleeping Beauty's Castle was decorated as a giant birthday cake to celebrate the occasion, further problems were approaching. After criticism for having too few rides, the rollercoaster 'Indiana Jones and the Temple of Peril' was opened in July. This was the first Disney rollercoaster that included a 360-degree loop, but just a few weeks after opening emergency brakes locked on during a ride, causing some guest injuries. The ride was temporarily shut down for investigation. Also in 1993, the proposed Euro Disney phase 2 was shelved due to financial problems, which meant Disney MGM Studios Europe and 13,000 hotel rooms would not be built to the original 1995 deadline agreed upon by the Walt Disney Company. However, Discovery Mountain, one of the planned phase 2 attractions, did get approval.

By the start of 1994, rumours were circulating that the park was on the verge of bankruptcy. Emergency crisis talks were held between the banks and backers, with things coming to a head during March when Disney offered the banks an ultimatum. It would provide sufficient capital for the park to continue to operate until the end of the month, but unless the banks agreed to restructure the park's $1 billion debt, the Walt Disney company would close the park and walk away from the whole European venture, leaving the banks with a bankrupt theme park and a massive expanse of virtually worthless real estate. Michael Eisner, Disney's CEO, announced that Disney was planning to pull the plug on the venture at the end of March 1994 unless the banks were prepared to restructure the loans. The banks agreed to Disney's demands.

In May 1994, the connection between London and Marne-la-Vallée was completed, along with a TGV link, providing a connection between several major European cities. By August the park was starting to find its feet at last, and all of its hotels were fully booked during the peak holiday season. Also, in October, the park's name was officially changed from Euro Disney to 'Disneyland Paris' in order to *'show that the resort now was named much more like its counterparts in California and Tokyo'*. The end-of-year figures for 1994 showed encouraging signs despite a 10 per cent fall in attendance caused by the bad publicity over the earlier financial problems.

For the next few years new rides continued to be introduced. 1995 saw the opening of the new rollercoaster, 'Space Mountain de la Terre à la Lune', and Disneyland Paris announced its first annual operating profit in November that year, helped by the opening of Space Mountain in June. In 1997 the five-year celebrations included parties, a new parade with Quasimodo and all the characters from the latest Disney blockbusting classic 'The Hunchback of Notre Dame', the 'YEAR TO BE HERE' marketing campaign, the resort's first Halloween celebration and a new Christmas parade. A new attraction was added in 1999, 'Honey I Shrunk The Audience', making the audience the size of a bug while being invited to 'Inventor of the Year Award Ceremony'. However, the planned Christmas and New Year celebrations were disrupted when a freak storm caused havoc, destroying the Mickey Mouse glass statue that had just been installed for the Lighting Ceremony and many other attractions. Also damaged were trees next to the castle, the top of which developed a pronounced lean, as did many street signs and lamp posts.

Disney's 'Fastpass' system was introduced in 2000, a new service that allowed guests to use their entry passes to obtain a ticket for certain attractions to gain direct entry without queuing. Two new attractions were also opened, 'Indiana Jones et le Temple du péril' and 'Tarzan le Recontre' starring a cast of acrobats along with Tarzan, Jane and all their Jungle friends with music from the movie in different European languages. In 2001 the 'ImagiNations Parade' was replaced by the 'Wonderful World of Disney Parade' which received some criticism for being 'less than spectacular', with only eight parade floats. Also Disney's 'California Adventure' was opened in California. The resort's tenth anniversary saw the opening of the new Walt Disney Studios Park attraction, based on a similar attraction in Florida that had proved to be a success.

André Lacroix from Burger King was appointed as CEO of Disneyland Resort Paris in 2003, to *'take on the challenge of a failing Disney park in Europe and turn it around'*. Increasing investment, he refurbished whole sections of the park and introduced the Jungle Book Carnival in February to increase attendance during the slow months. By 2004, attendance had improved but the company announced that it was still losing money. And even the positive news of 2006, although generally well received, still left questions unanswered. As one commentator put it: *'Would Disney, the stockholders, the banks, or even the French government make the same decision to go ahead if they could wind the clock back to 1987? Is this a story of a fundamentally flawed concept, or was it just mishandled?'*

QUESTIONS

1 What markets are the Disney resorts and parks aiming for?

2 Was Disney's choice of the Paris site a mistake?

3 What aspects of its parks' design did Disney change when it constructed Euro Disney?

4 What did Disney not change when it constructed Euro Disney?

5 What were Disney's main mistakes from the conception of the Paris resort through to 2006?

APPLYING THE PRINCIPLES

Hints

Some of these exercises can be answered by reading the chapter. Others will require some general knowledge of business activity and some might require an element of investigation. Hints on how they can all be answered are to be found in the eText at www.pearsoned.co.uk/slack.

1 Visit sites on the internet that offer (legal) downloadable music using MP3 or other compression formats. Consider the music business supply chain, a) for the recordings of a well-known popular music artist, and b) for a less well-known (or even largely unknown) artist struggling to gain recognition. How might the transmission of music over the internet affect each of these artists' sales? What implications does electronic music transmission have for record shops?

2 Visit the websites of companies that are in the paper manufacturing/pulp production/packaging industries. Assess the extent to which the companies you have investigated are vertically integrated in the paper supply chain that stretches from foresting through to the production of packaging materials.

3 Many developing nations are challenging the dominance of more traditional Western locations, notably Silicon Valley, for high-tech research and manufacturing. Two examples are Bangalore in India and Shanghai in China. Make a list of all the factors you would recommend a multinational corporation to take into account in assessing the advantages and disadvantages and risks of locating in developing countries. Use this list to compare Bangalore and China for a multi-national computer corporation:

   (a) siting its research and development facility;

   (b) siting a new manufacturing facility.

4 Tesco.com is now the world's largest and most profitable online grocery retailer. In 1996 Tesco.com was alone in developing a 'store-based' supply network strategy. This means that it used is existing stores to assemble customer orders which were placed online. Tesco staff would simply be given print-outs of customer orders and then walk round the store picking items off the shelves. The groceries would then be delivered by a local fleet of Tesco vans to customers. By contrast, many new e-grocery entrants and some existing supermarkets pursued a 'warehouse' supply network strategy of building new, large, totally automated and dedicated regional warehouses. Because forecasts for online demand were so high, they believed that the economies of scale of dedicate warehouses would be worth the investment. In the late 1990s Tesco came under criticism for being over-cautious and in 1999 reviewed its strategy. They concluded that its store-based strategy was correct and persevered. The most famous of the pure e-grocery companies was called WebVan. At the height of the dot-com phenomenon WebVan Group went public with a first-day market capitalisation of $7.6 billion. By 2001, having burnt its way through $1.2 billion in capital before filing for bankruptcy, WebVan Group went bust, letting go all of its workers and auctioning off everything from warehouse equipment to software.

   (a) Draw the different supply network strategies for Tesco and companies like WebVan.

   (b) What do you think the economy of scale curves for the Tesco operation and the WebVan operation would look like relative to each other?

   (c) Why do you think WebVan went bust and Tesco was so successful?

Notes on chapter

1 Sources include: 'Wincanton moves to more agile operations' (2009) IBM CIO programme; 'Wincanton standardises pan-European operations' (2008), http://www.logisticsit.com/, 24 June; IBC Ltd. Wincanton company website.

2 Source: 'Rebooting their systems' (2011), *The Economist* (2011), 10 March; 'For whom the Dell tolls' (2006), *The Economist*, 13 May13; Cellan-Jones, R. (2008), 'Dell aims to reclaim global lead', *BBC Business*, 14 April.

3 Source: Zwick, S. (1999) 'World Cars', *Time* magazine, 22 February.

4 Brandenburger, A.M. and Nalebuff, B.J. (1996) *Co-Opetion*, Doubleday, New York.

5 Hayes, R.H. and Wheelwright, S.C. (1994) *Restoring our Competitive Edge*, John Wiley.

6 'Thinking outside the box' (2007), *The Economist*, 7 April.

7 Bacon, G., Machan, I. and Dnyse, J. (2008) 'Offshore challenges: manufacturing', *The Institute of Electrical Engineers*, January.

8 Pil, F.K. and Holweg, M. (2003), 'Exploring scale: the advantages of thinking small', *MIT Sloan Management Review*, Winter.

9 This case was prepared using published sources of information. It does not reflect the views of the Walt Disney Company, which should not be held responsible for the accuracy or interpretation of any of the information or views contained in the case. It is not intended to illustrate either good or bad management practice.

TAKING IT FURTHER

Chopra, S. and Meindl, P. (2009) *Supply Chain Management: Strategy, planning and operations* (4th edn), Prentice Hall, NJ. *A good textbook that covers both strategic and operations issues.*

Cousins, P., Lamming R., Dr Lawson B., Dr Squire B. (2008) *Strategic Supply Management: Principles, Theories and Practice*, Financial Times Prentice Hall. *A modern perspective on the subject from some of the leading European academics in the area.*

Dell, M. (with Fredman C.) (1999) *Direct from Dell: Strategies that revolutionized an Industry*, Harper Business. *Michael Dell explains how his supply network strategy (and other decisions) had such an impact on the industry. Interesting and readable, but not a critical analysis!*

Gunasekarana, A., Laib, K. and Cheng, T.C.E. (2008) 'Responsive supply chain: a competitive strategy in a networked economy', *Omega*, Volume 36, Issue 4.

Srai, J.S. Gregory, M. (2008) 'A supply network configuration perspective on international supply chain development', *International Journal of Operations & Production Management*, Vol. 28 Issue: 5.

Quinn, J.B. (1999) 'Strategic outsourcing: leveraging knowledge capabilities', *Sloan Management Review*, Summer. *A bit academic, but a good discussion on the importance of 'knowledge' in the outsourcing decision.*

USEFUL WEBSITES

www.opsman.org *Definitions, links and opinions on operations and process management.*

www.locationstrategies.com *Exactly what the title implies. Good industry discussion.*

www.conway.com *American location selection site. You can get a flavour of how location decisions are made.*

www.transparency.org *A leading site for international business (inducing location) that fights corruption.*

www.intel.com *More details on its 'Copy Exactly' strategy and other capacity strategy issues.*

www.outsourcing.com *Site of the Institute of Outsourcing. Some good case studies.*

www.bath.ac.uk/crisps *A centre for research in strategic purchasing and supply. Some interesting papers.*

www.outsourcing.co.uk *Site of the UK National Outsourcing Association. Some interesting reports, news items, etc.*

Interactive quiz

For further resources including examples, animated diagrams, self-test questions, Excel spreadsheets and video materials please explore the eText on the companion website at **www.pearsoned.co.uk/slack**.

SUPPLEMENT TO CHAPTER 3

Forecasting

Introduction

Some forecasts are accurate. We know exactly what time the sun will rise at any given place on earth tomorrow or one day next month or even next year. Forecasting in a business context however is much more difficult and therefore prone to error. We do not know precisely how many orders we will receive or how many customers will walk through the door tomorrow, next month, or next year. Such forecasts, however, are necessary to help managers make decisions about resourcing the organisation for the future.

Forecasting – knowing the options

Simply knowing that demand for your goods or services is rising or falling is not enough in itself. Knowing the rate of change is likely to be vital to business planning. A firm of lawyers may have to decide the point at which, in their growing business, they will have to take on another partner. Hiring a new partner could take months so they need to be able to forecast when they expect to reach that point and then when they need to start their recruitment drive. The same applies to a plant manager who will need to purchase new plant to deal with rising demand. She may not want to commit to buying an expensive piece of machinery until absolutely necessary but in enough time to order the machine and have it built, delivered, installed and tested. The same is so for governments, whether planning new airports or runway capacity or deciding where and how many primary schools to build.

The first question is to know how far you need to look ahead and this will depend on the options and decisions available to you. Take the example of a local government where the number of primary age children (5–11-year-olds) is increasing in some areas and declining in other areas within its boundaries. It is legally obliged to provide school places for all such children. Government officials will have a number of options open to them and they may each have different lead times associated with them. One key step in forecasting is to know the possible options and the lead times required to bring them about (see Table 3.4)

Table 3.4 Options available and lead time required for dealing with changes in numbers of school children

Options available	*Lead time required*
Hire short-term teachers	Hours
Hire staff	↓
Build temporary classrooms	
Amend school catchment areas	
Build new classrooms	
Build new schools	Years

Individual schools can hire (or lay off) short-term (supply) teachers from a pool not only to cover for absent teachers but also to provide short-term capacity while teachers are hired to deal with increases in demand. Acquiring (or dismissing) such temporary cover may only require a few hours' notice. (This is often referred to as short-term capacity management.)

Hiring new (or laying off existing) staff is another option but both of these may take months to complete (medium-term capacity management).

A shortage of accommodation may be fixed in the short to medium term by hiring or buying temporary classrooms. It may only take a couple of weeks to hire such a building and equip it ready for use.

It may be possible to amend catchment areas between schools to try to balance an increasing population in one area against a declining population in another. Such changes may require lengthy consultation processes.

In the longer term, new classrooms or even new schools may have to be built. The planning, consultation, approval, commissioning, tendering, building and equipping process may take one to five years depending on the scale of the new build (long-term capacity planning – see Chapter 6).

Knowing the range of options, managers can then decide the time scale for their forecasts; indeed several forecasts might be needed for the short term, medium term and long term.

In essence forecasting is simple

In essence forecasting is easy. To know how many children may turn up in a local school tomorrow you can use the number that turned up today. In the long term, in order to forecast how many primary-age children will turn up at a school in five years' time one need simply look at the birth statistics for the current year for the school's catchment area – see Figure 3.11

However, such simple extrapolation techniques are prone to error and indeed such approaches have resulted in some local governments committing themselves to building schools which, five or six years later when complete, had few children, and other schools were bursting at the seams with temporary classrooms and temporary teachers, often resulting in falling morale and declining educational standards. The reason why such simple approaches are prone to problems is that there are many contextual variables (see Figure 3.12) which will have a potentially significant impact on, for example, the school population five years hence. For example, one minor factor in developed countries, though a major factor in developing countries, might be the death rate in children between birth and five years of age. This may be dependant upon location with a slightly higher mortality rate in the poorer areas compared to the more affluent areas. Another more significant factor is immigration and emigration as people move into or out of the local area. This will be affected by housing stock and housing developments and the ebb and flow of jobs in the area and the changing economic prosperity in the area.

One key factor which has an impact on the birth rate in an area is the amount and type of the housing stock. City centre tenement buildings tend to have a higher proportion of children per dwelling, for example, than suburban semi-detached houses. So not only will existing

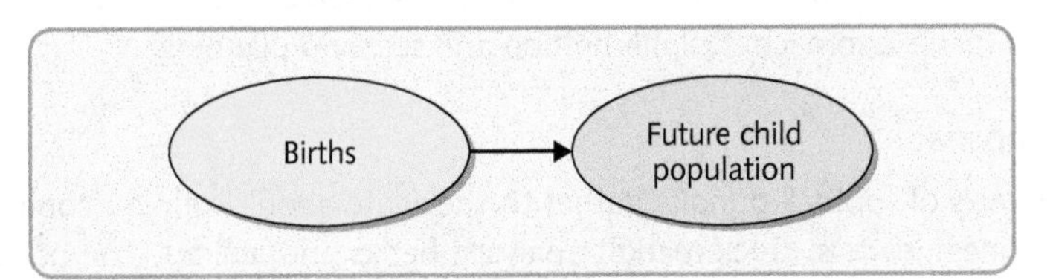

Figure 3.11 Simple prediction of future child population

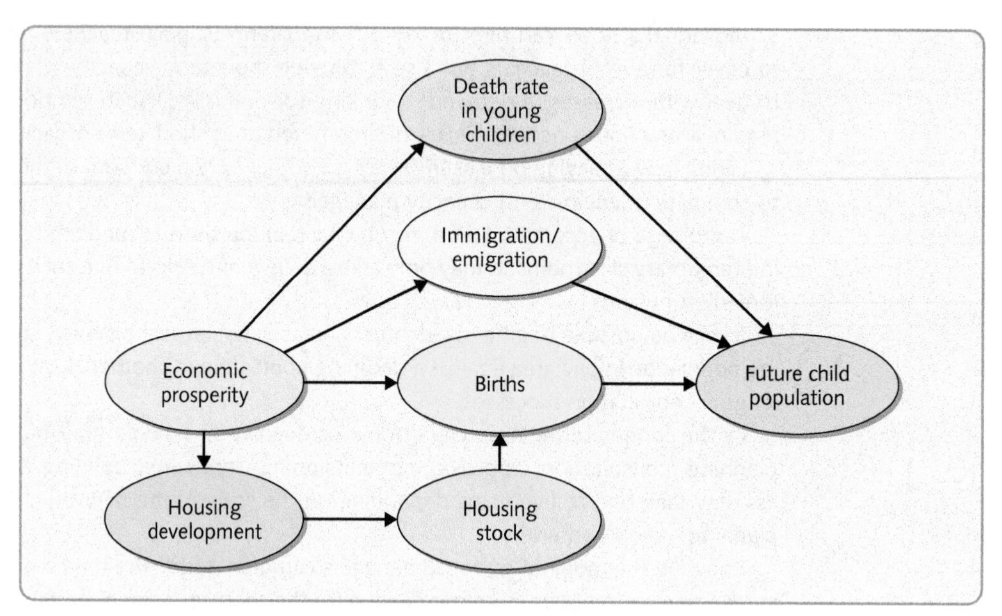

Figure 3.12 Some of the key causal variables in predicting child populations

housing stock have an impact on the child population but so also will the type of housing developments under construction, planned and proposed.

Approaches to forecasting

There are two main approaches to forecasting. Managers sometimes use qualitative methods based on opinions, past experience even best guesses. There is also a range of qualitative forecasting techniques available to help managers evaluate trends, causal relationships and make predictions about the future. Also, quantitative forecasting techniques can be used to model data. Although no approach or technique will result in an accurate forecast a combination of qualitative and quantitative approaches can be used to great effect by bringing together expert judgements and predictive models.

Qualitative methods

Imagine you were asked to forecast the outcome of a forthcoming football match. Simply looking at the teams' performance over the last few weeks and extrapolating from it is unlikely to yield the right result. Like many business decisions, the outcome will depend on many other factors. In this case the strength of the opposition, their recent form, injuries to players on both sides, the match location and even the weather will have an influence on the outcome. A qualitative approach involves collecting and appraising judgements, options, even best guesses, as well as past performance from 'experts' to make a prediction. There are several ways this can be done: a panel approach, Delphi method and scenario planning.

Panel approach

Just as panels of football pundits gather to speculate about likely outcomes, so too do politicians, business leaders, stock market analysts, banks and airlines. The panel acts like a focus group, allowing everyone to talk openly and freely. Although there is the great advantage of several brains being better than one, it can be difficult to reach a consensus, or sometimes the views of the loudest or highest status may emerge (the bandwagon effect). Although more reliable than one person's views, the panel approach still has the weakness that everybody, even the experts, can get it wrong.

Delphi method

Perhaps the best known approach to generating forecasts using experts is the Delphi method.[1] This is a more formal method which attempts to reduce the influences from procedures of face-to-face meetings. It employs a questionnaire, e-mailed or posted to the experts. The replies are analysed and summarised and returned, anonymously, to all the experts. The experts are then asked to reconsider their original response in the light of the replies and arguments put forward by the other experts. This process is repeated several more times to conclude either with a consensus or at least a narrower range of decisions. One refinement of this approach is to allocate weights to the individuals and their suggestions based on, for example, their experience, their past success in forecasting, other people's views of their abilities. The obvious problems associated with this method include constructing an appropriate questionnaire, selecting an appropriate panel of experts and trying to deal with their inherent biases.

Scenario planning

One method for dealing with situations of even greater uncertainty is scenario planning. This is usually applied to long-range forecasting, again using a panel. The panel members are usually asked to devise a range of future scenarios. Each scenario can then be discussed and the inherent risks considered. Unlike the Delphi method, scenario planning is not necessarily concerned with arriving at a consensus but looking at the possible range of options and putting plans in place to try to avoid the ones that are least desired and taking action to follow the most desired.

Quantitative methods

There are two main approaches to quantitative forecasting; time series analysis and causal modelling techniques Time series examine the pattern of past behaviour of a single phenomenon over time, taking into account reasons for variation in the trend in order to use the analysis to forecast the phenomenon's future behaviour. Causal modelling is an approach which describes and evaluates the complex cause effect relationships between the key variables (such as in Figure 3.12).

Time series analysis

Simple time series plot a variable over time, then by removing underlying variations with assignable causes use extrapolation techniques to predict future behaviour. The key weakness with this approach is that it simply looks at past behaviour to predict the future, ignoring causal variables which are taken into account in other methods such causal modelling or qualitative techniques. For example, suppose a company is attempting to predict the future sales of a product. The past three years' sales, quarter by quarter, are shown in Figure 3.13(a). This series of past sales may be analysed to indicate future sales. For instance, underlying the series might be a linear upward trend in sales. If this is taken out of the data, as in Figure 3.13(b), we are left with a cyclical seasonal variation. The mean deviation of each quarter from the trend line can now be taken out, to give the average seasonality deviation. What remains is the random variation about the trends and seasonality lines, Figure 3.13(c). Future sales may now be predicted as lying within a band about a projection of the trend, plus the seasonality. The width of the band will be a function of the degree of random variation.

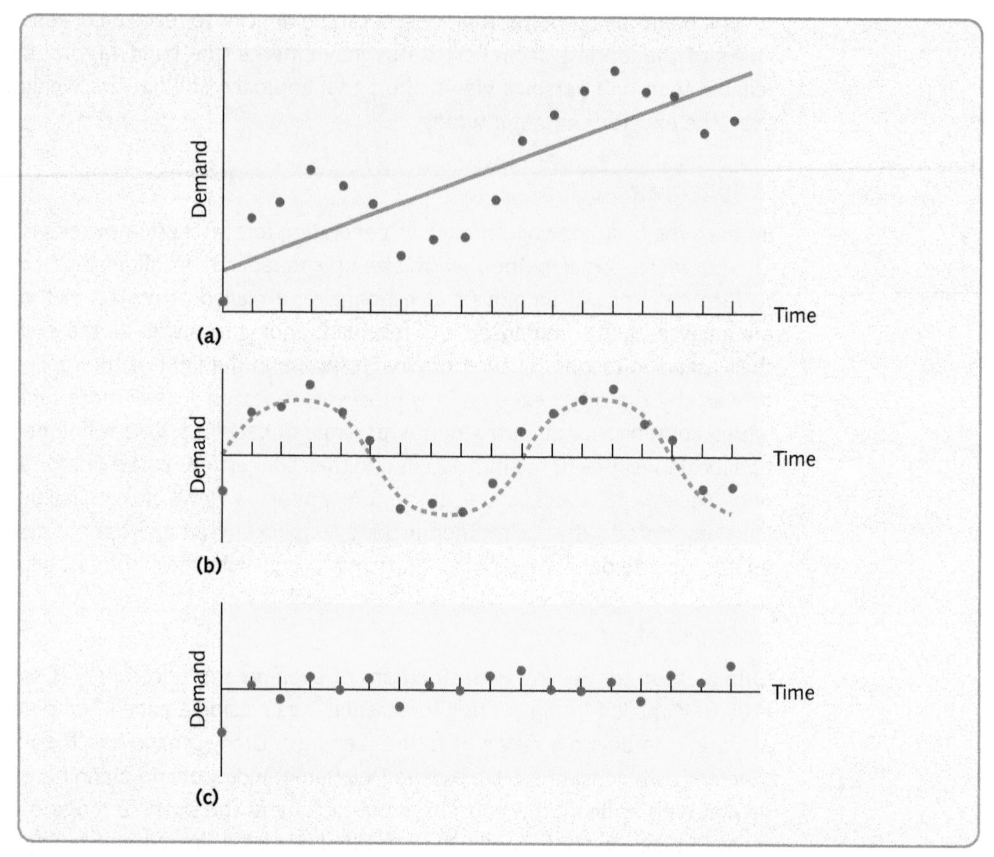

Figure 3.13 Time series analysis with (a) trend, (b) seasonality and (c) random variation

Forecasting unassignable variations

The random variations which remain after taking out trend and seasonal effects are without any known or assignable cause. This does not mean that they do not have a cause, however, just that we do not know what it is. Nevertheless, some attempt can be made to forecast it, if only on the basis that future events will, in some way, be based on past events. We will examine two of the more common approaches to forecasting which are based on projecting forward from past behaviour. These are:

- moving-average forecasting
- exponentially smoothed forecasting.

The moving-average approach to forecasting takes the previous *n* periods' actual demand figures, calculates the average demand over the *n* periods, and uses this average as a forecast for the next period's demand. Any data older than the *n* periods plays no part in the next period's forecast. The value of *n* can be set at any level, but is usually in the range 4 to 7.

EXAMPLE

Eurospeed parcels

Table 3.5 shows the weekly demand for Eurospeed, a European-wide parcel delivery company. It measures demand, on a weekly basis, in terms of the number of parcels which it is given to deliver (irrespective of the size of each parcel). Each week, the next week's demand is forecast by taking the moving average of the previous four weeks' actual demand. Thus if the forecast demand for week t is F_t and the actual demand for week t is A_t, then:

$$F_t = \frac{1}{4}(A_{t-4} + A_{t-3} + A_{t-2} + A_{t-1})$$

Table 3.5 **Moving-average forecast calculated over a four-week period**

Week	*Actual demand (thousands)*	*Forecast*
20	63.3	
21	62.5	
22	67.8	
23	66.0	
24	67.2	64.9
25	69.9	65.9
26	65.6	67.7
27	71.1	66.3
28	68.8	67.3
29	68.4	68.9
30	70.3	68.5
31	72.5	69.7
32	66.7	70.0
33	68.3	69.5
34	67.0	69.5
35		68.6

For example, the forecast for week 35:

$$F_{35} = (72.5 + 66.7 + 68.3 + 67.0)/4$$
$$= 68.8$$

Exponential smoothing

There are two significant drawbacks to the moving-average approach to forecasting. First, in its basic form, it gives equal weight to all the previous *n* periods which are used in the calculations (although this can be overcome by assigning different weights to each of the *n* periods). Second, and more important, it does not use data from beyond the *n* periods over which the moving average is calculated. Both these problems are overcome by exponential smoothing, which is also somewhat easier to calculate. The exponential-smoothing approach forecasts demand in the next period by taking into account the actual demand in the current period and the forecast which was previously made for the current period. It does so according to the formula:

$$F_t = \alpha A_{t-1} + (1 - x)F_{t-1}$$

where α = the smoothing constant.

The smoothing constant α is, in effect, the weight which is given to the last (and therefore assumed to be most important) piece of information available to the forecaster. However, the other expression in the formula includes the forecast for the current period which included the previous period's actual demand, and so on. In this way all previous data has a (diminishing) effect on the next forecast.

EXAMPLE

Eurospeed parcel using exponential smoothing

Table 3.6 shows the data for Eurospeed's parcels forecasts using this exponential-smoothing method, where $\alpha = 0.2$. For example, the forecast for week 35 is:

$$F_{35} = (0.2 \times 67.0) + (0.8 \times 68.3) = 68.4$$

Table 3.6 **Exponentially smoothed forecast calculated with smoothing constant $\alpha = 0.2$**

Week (t)	*Actual demand (thousands) (A)*	*Forecast demand (F)*
20	63.3	60.00
21	62.5	60.66
22	67.8	60.03
23	66.0	61.58
24	67.2	62.83
25	69.9	63.70
26	65.6	64.94
27	71.1	65.07
28	68.8	66.28
29	68.4	66.78
30	70.3	67.12
31	72.5	67.75
32	66.7	68.70
33	68.3	68.30
34	67.0	68.30
35		68.04

The value of α governs the balance between the responsiveness of the forecasts to changes in demand, and the stability of the forecasts. The closer α is to 0 the more forecasts will be dampened by previous forecasts (not very sensitive but stable). Figure 3.14 shows the Eurospeed volume data plotted for a four-week moving average, exponential smoothing with $\alpha = 0.2$ and exponential smoothing with $\alpha = 0.3$.

Causal models

Causal models often employ complex techniques to understand the strength of relationships between the network of variables and the impact they have on each other. Simple regression

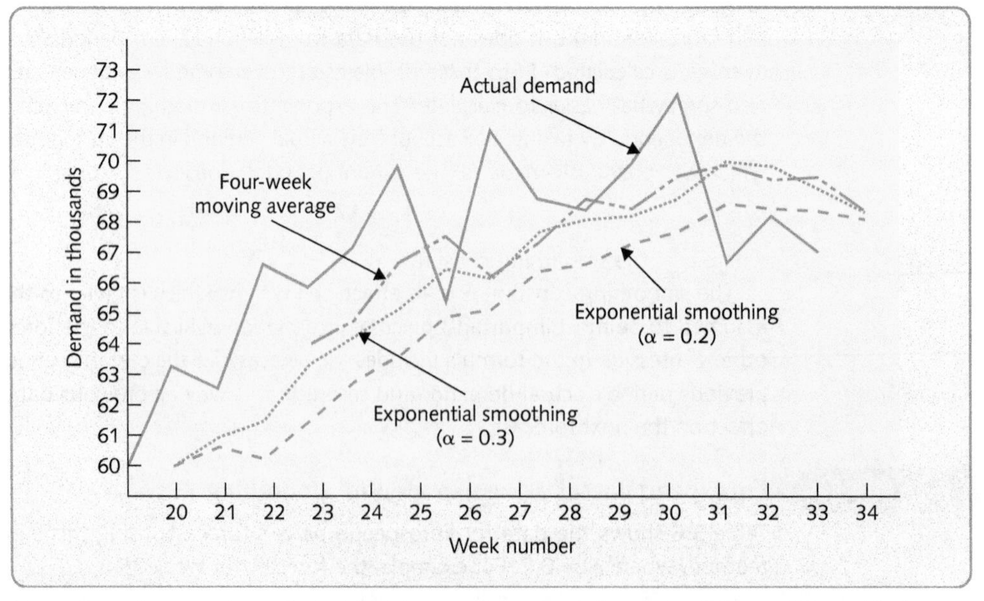

Figure 3.14 A comparison of a moving-average forecast and exponential smoothing with the smoothing constant $\alpha = 0.2$ and 0.3

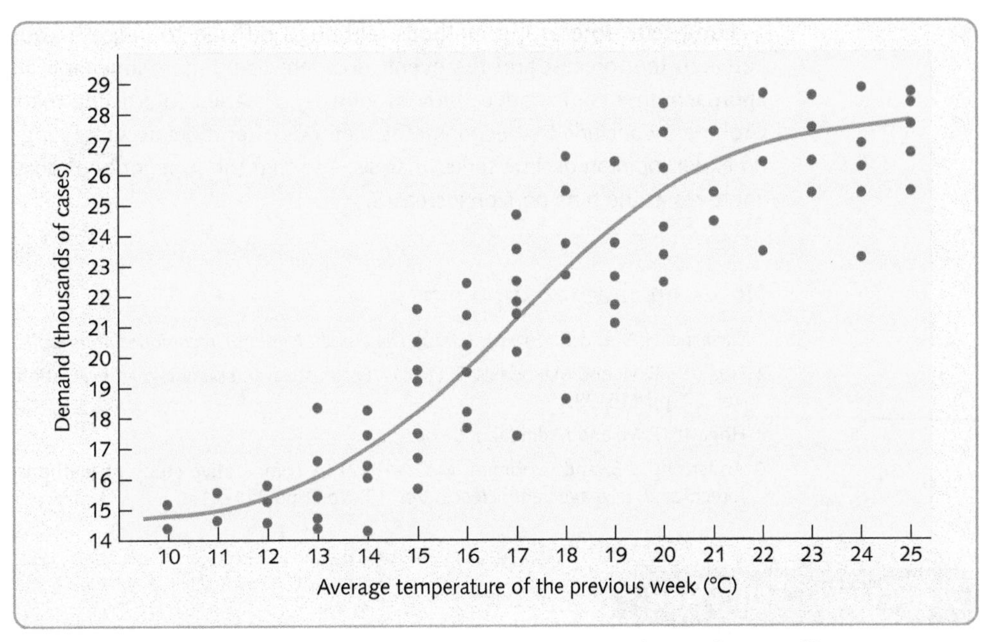

Figure 3.15 Regression line showing the relationship between the previous week's average temperature and demand

models try to determine the 'best fit' expression between two variables. For example, suppose an ice-cream company is trying to forecast its future sales. After examining previous demand, it figures that the main influence on demand at the factory is the average temperature of the previous week. To understand this relationship, the company plots demand against the previous week's temperatures. This is shown in Figure 3.15. Using this graph, the company can make a reasonable prediction of demand, once the average temperature is known, provided that the other conditions prevailing in the market are reasonably stable. If they are not, then these other factors which have an influence on demand will need to be included in the regression model, which becomes increasingly complex.

These more complex networks comprise many variables and relationships, each with their own set of assumptions and limitations. While developing such models and assessing the importance of each of the factors and understanding the network of interrelationships is beyond the scope of this text, many techniques are available to help managers undertake this more complex modelling and also feedback data into the model to further refine and develop it, in particular structural equation modelling.

The performance of forecasting models

Forecasting models are widely used in management decision-making, and indeed most decisions require a forecast of some kind, yet the performance of this type of model is far from impressive. Hogarth and Makridakis,[2] in a comprehensive review of the applied management and finance literature, show that the record of forecasters using both judgement and sophisticated mathematical methods is not good. What they do suggest, however, is that certain forecasting techniques perform better under certain circumstances. In short-term forecasting there is:

> . . . considerable inertia in most economic and natural phenomena. Thus the present states of any variables are predictive of the short-term future (i.e. three months or less). Rather simple mechanistic methods, such as those used in time series forecasts, can often make accurate short-term forecasts and even out-perform more theoretically elegant and elaborate approaches used in econometric forecasting.[3]

Long-term forecasting methods, although difficult to judge because of the time lapse between the forecast and the event, do seem to be more amenable to an objective causal approach. In a comparative study of long-term market forecasting methods, Armstrong and Grohman[4] conclude that econometric methods offer more accurate long-range forecasts than do expert opinion or time series analysis, and that the superiority of objective causal methods improves as the time horizon increases.

Notes on chapter supplement

1 Linstone, H.A. and Turoof, M. (1975) *The Delphi Method: Techniques and Applications*, Addison-Wesley.

2 Hogarth, R.M. and Makridakis, S. (1981) 'Forecasting and planning: an evaluation', *Management Science*, Vol. 27, pp 115–38.

3 Hogarth, R.M. and Makridakis, S., *op. cit.*

4 Armstrong, J.S. and Grohman, M.C. (1972) 'A comparative study of methods for long-range market forecasting', *Management Science*, Vol. 19, No 2, pp 211–21.

TAKING IT FURTHER

Hoyle R.H. (ed), *Structural Equation Modeling*, Sage, Thousand Oaks, California, 1995. *For the specialist.*

Morlidge S. and Player S. (2010) *Future Ready: How to Master Business Forecasting*, John Wiley & Sons. *A good, not too technical, treatment.*

4

Process design 1 - positioning

Introduction

Processes are everywhere. They are the building blocks of all operations, and their design will affect the performance of the whole operation and, eventually, the contribution it makes to its supply network. No one, in any function or part of the business, can fully contribute to its competitiveness if the processes in which they work are poorly designed and ineffective. It is not surprising then that process design has become such a popular topic in the management press and among consultants. This chapter is the first of two that examine the design of processes. Chapter 4 is primarily concerned with how processes and the resources they contain must reflect the volume and variety requirements placed on them. The next chapter examines the more detailed and analytical aspects of process analysis (see Figure 4.1).

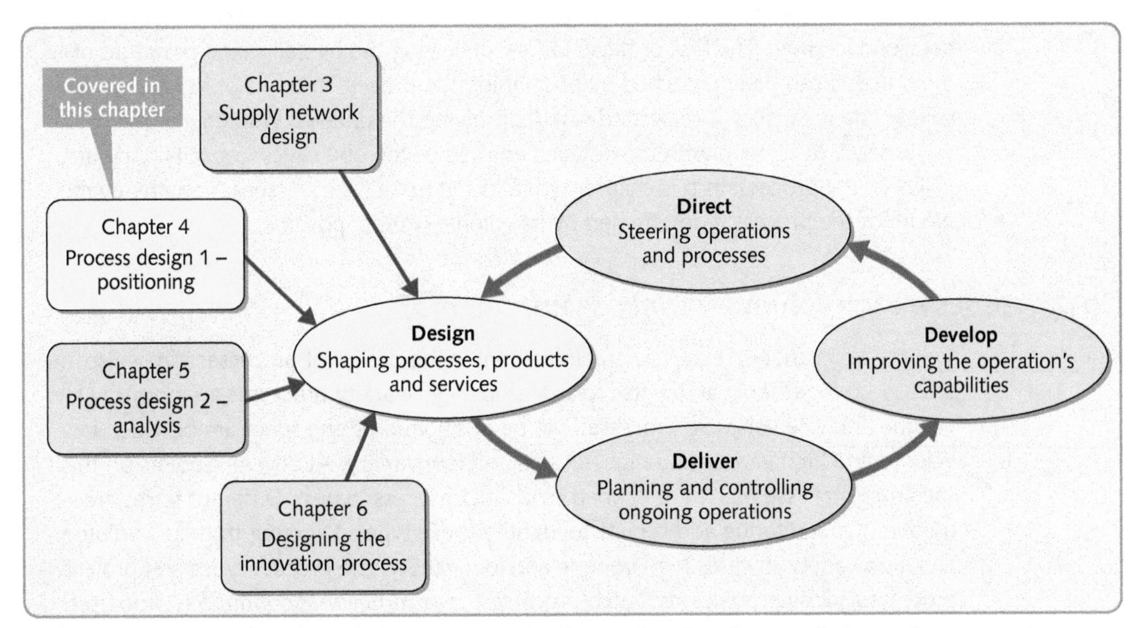

Figure 4.1 Process design positioning is concerned with ensuring that the overall shape of processes is appropriate to their volume–variety position

EXECUTIVE SUMMARY

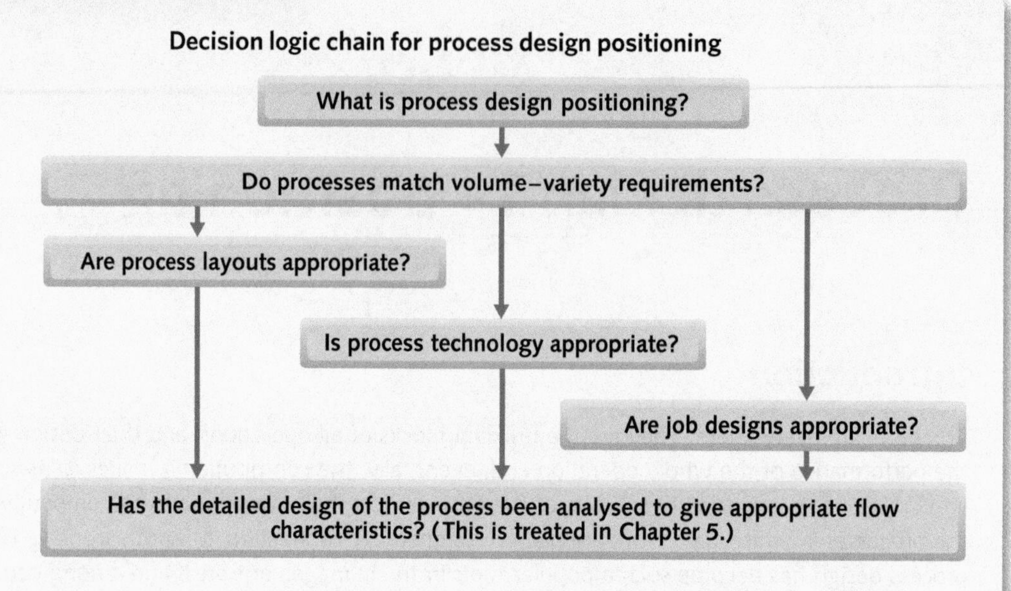

Each chapter is structured around a set of diagnostic questions. These questions suggest what you should ask in order to gain an understanding of the important issues of a topic, and, as a result, improve your decision making. An executive summary addressing these questions is provided below.

What is process design positioning?

Process design is concerned with conceiving the overall shape of processes and their detailed workings. The first of these tasks (conceiving the overall shape or nature of the process) can be approached by positioning the process in terms of its volume and variety characteristics. The second task (conceiving the detailed workings of the process) is more concerned with the detailed analysis of the objectives, capacity and variability of the process. In this chapter we treat the first of these issues: how the overall nature of the process is determined by its volume–variety position.

Do processes match volume–variety requirements?

Volume and variety are particularly influential in the design of processes. They also tend to go together in an inverse relationship. High-variety processes are normally low volume and vice versa. So processes can be positioned on the spectrum between low volume and high variety and high volume and low variety. At different points on this spectrum processes can be described as distinct process 'types'. Different terms are used in manufacturing and service to identify these types. Working from low volume and high variety towards high volume and low variety, the process types are: project processes; jobbing processes; batch processes; mass processes; continuous processes. The same sequence in service types are known as: professional services; service shops; mass services. Whatever terminology is used, the overall design of the process must fit its volume–variety position. This is usually summarised in the form of the 'product–process' matrix.

Are process layouts appropriate?

There are different ways in which the different resources within a process (people and technology) can be arranged relative to each other. But however this is done it should reflect the process's volume–variety position. Again, there are pure 'types' of layout that correspond with the different volume–variety positions. These are, fixed position layout, functional layout, cell layout, and product layout. Many layouts are hybrids of these pure types, but the type chosen is influenced by the volume and variety characteristics of the process.

Is process technology appropriate?

Process technologies are the machine's equipment and devices that help processes transform materials and information and customers. It is different to product technology that is embedded with the product or service itself. Again, product technology should reflect volume and variety. In particular, the degree of automation in the technology, the scale and/or scalability of the technology, and the coupling and/or connectivity of the technology should be appropriate to volume and variety. Generally, low volume and high variety requires relatively unautomated, general purpose, small-scale and flexible technologies. By contrast, high-volume and low-variety processes require automated, dedicated and large-scale technologies that are sometimes relatively inflexible.

Are job designs appropriate?

Job design is about how people carry out their tasks within a process. It is particularly important because it governs people's expectations and perceptions of their contribution to the organisation as well as being a major factor in shaping the culture of the organisation. Some aspects of job design are common to all processes irrespective of their volume and variety position. These are such things as ensuring the safety of everyone affected by the process, ensuring a firm ethical stance, and upholding an appropriate work/life balance. However, other aspects of job design are influenced by volume and variety. In particular, the extent of division of labour, the degree to which jobs are defined, and the way in which job commitment is encouraged. Broadly, high-variety and low-volume processes require broad, relatively undefined jobs with decision-making discretion. Such jobs tend to have intrinsic job commitment. By contrast, high-volume and low-variety processes tend to require jobs that are relatively narrow in scope and closely defined with relatively little decision-making discretion. This means some deliberative action is needed in the design of the job (such as job enrichment) in order to help maintain commitment to the job.

DIAGNOSTIC QUESTION

What is process design positioning?

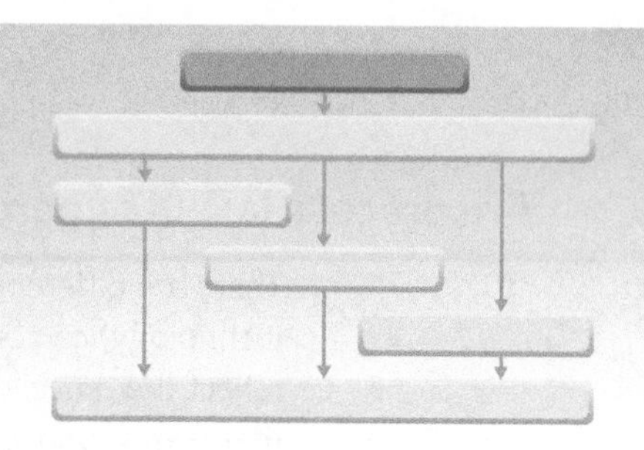

To 'design' is to conceive the looks, arrangement and workings of something *before it is constructed*. In that sense it is a conceptual exercise. Yet it is one which must deliver a solution that will work in practice. Design is also an activity that can be approached at different levels of detail. One may envisage the general shape and intention of something before getting down to defining its details. However, it is often only through getting to grips with the detail of a design that the feasibility of its overall shape can be assessed. So it is with designing processes. First, one must consider the overall shape and nature of the process. The most common way of doing this is by positioning it according to its volume and variety characteristics. Second, one must analyse the details of the process in order to ensure that it fulfils its objectives effectively. But don't think of this as a simple sequential process. There may be aspects concerned with the broad positioning of the process that will need to be modified following its more detailed analysis.

In this chapter we discuss the more general approach to process design by showing how a process's position on the volume–variety scale will influence its layout, technology, and the design of its jobs. In the next chapter we will discuss the more detailed aspects of process design, in particular its objectives, current configuration, capacity, and variability. This is illustrated in Figure 4.2.

EXAMPLE

Tesco's store flow processes[1]

Successful supermarkets, like Tesco, know that the design of their stores has a huge impact on profitability. They must maximise their revenue per square metre and minimise the costs of operating the store, while keeping customers happy. At a basic level, supermarkets have to get the amount of space allocated to the different areas right. Tesco's 'One in front' campaign, for example, tries to avoid long waiting times by opening additional tills if more than one

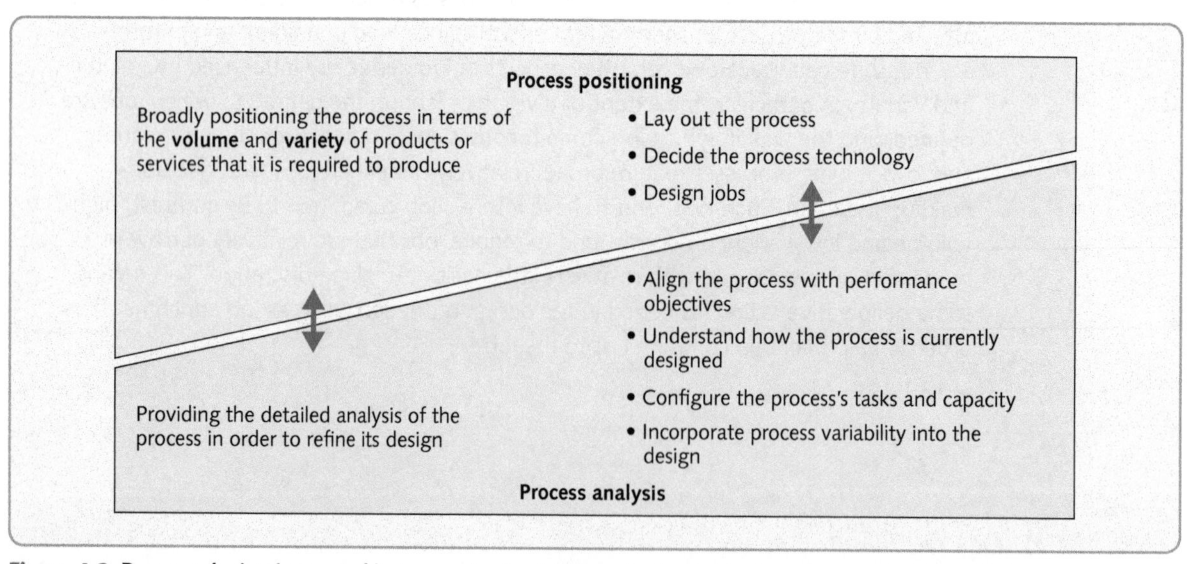

Figure 4.2 Process design is treated in two parts: Positioning, that sets the broad characteristics of the design; and Analysis, that refines the details of the design

Source: Geoffrey Robinson/Rex Features

customer is waiting at a checkout. Tesco also uses technology to understand exactly how customers flow through their stores. The 'Smartlane' system from Irisys, a specialist in intelligent infrared technologies, counts the number and type of customers entering the store (in family or other groups known as 'shopping units') tracks their movement using infra red sensors, and predicts the likely demand at the checkouts up to an hour in advance.

The circulation of customers through the store must be right and the right layout can make customers buy more. Some supermarkets put their entrance on the left-hand side of a building with a layout designed to take customers in a clockwise direction around the store. Aisles are made wide to ensure a relatively slow flow of trolleys so that customers pay more attention to the products on display (and buy more). However, wide aisles can come at the expense of reduced shelf space that would allow a wider range of products to be stocked.

The actual location of all the products is a critical decision, directly affecting the convenience to customers, their level of spontaneous purchase, and the cost of filling the shelves. Although the majority of supermarket sales are packaged tinned or frozen goods, the displays of fruit and vegetables are usually located adjacent to the main entrance, as a signal of freshness and wholesomeness, providing an attractive and welcoming point of entry. Basic products that figure on most people's shopping lists, such as flour, sugar and bread, may be located at the back of the store and apart from each other so that customers have to pass higher-margin items as they search. High-margin items are usually put at eye level on shelves (where they are more likely to be seen) and low-margin products lower down or higher up. Some customers also go a few paces up an aisle before they start looking for what they need. Some supermarkets call the shelves occupying the first metre of an aisle 'dead space', not a place to put impulse-bought goods. But the prime site in a supermarket is the 'gondola-end', the shelves at the end of the aisle. Moving products to this location can increase sales 200 or 300 per cent. Not surprising that suppliers are willing to pay for their products to be located here. The supermarkets themselves are keen to point out that, although they obviously lay out their stores with customers' buying behaviour in mind, it is counterproductive to be too manipulative. Nor are some commonly held beliefs about supermarket layout always true. They deny that they periodically change the location of food stuffs in order to jolt customers out of their habitual shopping patterns so that they are more attentive to other products and end up buying more. Occasionally layouts are changed they say, but mainly to accommodate changing tastes and new ranges.

EXAMPLE

Space4 housing processes[2]

You don't usually build a house this way. It's more like the way you would expect an automobile to be made. Nevertheless, Space4's huge building in Birmingham (UK) contains what some believe could be the future of house building. It is a production line whose 90 operators, many of whom have automobile assembly experience, are capable of producing the timber-framed panels that form the shell of the new homes at a rate of a house every hour. The automated, state-of-the-art electronic systems within the production process control all facets of the operation, ensuring that scheduling and operations are timely and accurate. There is a direct link between the Computer Aided Design (CAD) systems that design the houses and the manufacturing processes that make them, reducing the time between design and manufacture. The machinery itself incorporates automatic predictive and preventative maintenance routines that minimise the chances of unexpected breakdowns. But not everything about the process relies on automation. Because of their previous automobile assembly experience, staff are used to the just-in-time high-efficiency culture of modern mass production. After production, the completed panels are stacked in 3-metre-high piles and are then fork lifted into trucks where they are dispatched to building sites across the UK. Once the panels arrive at the building site, the construction workforce can assemble the

Source: Space4 Housing

exterior of a 112 square metres (average size) new home in a single day. Because the external structure of a house can be built in a few hours, and enclosed in a weatherproof covering, staff working on the internal fittings of the house, such as plumbers and electricians can have a secure and dry environment in which to work, irrespective of external conditions. Furthermore, the automated production process uses a type of high-precision technology which means there are fewer mistakes in the construction process on site. This means that the approval process from the local regulatory authority takes less time. This process, says Space4, speeds up the total building time from 12–14 weeks to 8–10 weeks.

Space4 is a division of Persimmon, who are the UK's largest house builder. It believes that Space4's production process could produce 5,000 homes a year, which would account for about 50 per cent of the Group's total sales. It is optimistic that Space4, with its mass production process can give it an edge over rivals. Also driving the adoption of mass processing were the tough new energy targets. Managing Director of Space4, Craig Hagan said, '*To make the new, improved, standards as part of the drive towards zero-carbon, homes need to be energy efficient and insulated. Our success has been down to a combination of the new energy standards favouring our product and the fact we work so quickly.*' Space4 also claims many other advantages for its process. It combines accurate quality conformance, flexibility, a construction routine that uses considerably less water than conventional building methods, which significantly reduces the drying out period and surface cracking, and reduced site generated waste, requiring fewer skips and providing tidier safer site. When built, the homes also have good thermal insulation. The panels from the factory are injected with a special resin mixture that creates a foam that keeps heat in and energy bills down. It is also a process that can produce at high volume while allowing the company to respond quickly to volatile economic conditions, because the process has the ability to change volumes relatively easily. In fact it was partly this last point that convinced Persimmon that this type of house building process had such a bright future. When the company inherited Space4 after its takeover of a smaller rival, in 2005, Persimmon had its doubts about the division. At first it regarded Space4 as a 'non-core' operation and its future appeared to be hanging in the balance. But three years later the housing market crashed and house building volumes in Britain fell sharply. Rather than building a house every hour, Space4 built only 1,000 homes in the entire year. Both during this period of low demand, and later, when volumes recovered, the production process was able to manufacture with reasonable efficiency. Such volume flexibility and the popularity of its energy efficient timber-frame construction convinced Persimmon that Space4 is a key part of its future.

What do these two examples have in common?

Both the operations described in the two examples have something to tell us about the influence of the volume and variety of demand on the way processes are designed. The Space4 way of producing homes is very different to the traditional method. By manufacturing panels in relatively high volume in its factory, it can use automated technology and people who have mass production experience to drive down its costs. The effects of variety, which would normally increase costs, are minimised by using flexible technology. The supermarket also has high volume, but what of its variety? Looked at one way, every customer is different and therefore variety will be extremely high. Yet, because of the way the process is designed, partly to encourage similarity of flow and partly by making customers 'customise' their own service, variety is made to seem very low. For example, checkout processes deal only with two 'products': customers with baskets and customers with trolleys. The important point is that in both cases the volume and variety of demand (or how it is interpreted by the operation) has a profound effect on the design of the processes. Just think about the design of a small convenience store, or a small handyman/building company. Because they have different volume/variety characteristics, their processes would be designed in very different ways.

DIAGNOSTIC QUESTION

Do processes match volume–variety requirements?

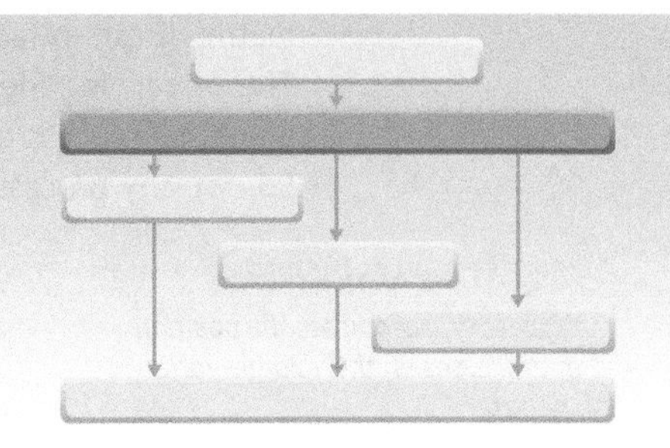

Video

Two factors are particularly important in process design: the *volume* and *variety* of the products and services that it processes. Moreover, volume and variety are related in so much as low-volume operations processes often have a high variety of products and services, and high-volume operations processes often have a narrow variety of products and services. So, we can *position* processes on a continuum from those that operate under conditions of low volume and high variety, through to those that operate under conditions of high volume and low variety. The volume–variety position of a process influences almost every aspect of its design. Processes with different volume–variety positions will be arranged in different ways, have different flow characteristics, and have different technology and jobs. So, a first step in process design is to understand how volume and variety shape process characteristics, and to check whether processes have been configured in a manner that is appropriate for their volume–variety position. Think about the volume–variety characteristics of the following examples.

> **OPERATIONS PRINCIPLE**
> *The design of any process should be governed by the volume and variety it is required to produce.*

The 'product-process' matrix

The most common method of illustrating the relationship between a process's volume–variety position and its design characteristics is shown in Figure 4.3. Often called the 'product–process'

Animated diagram

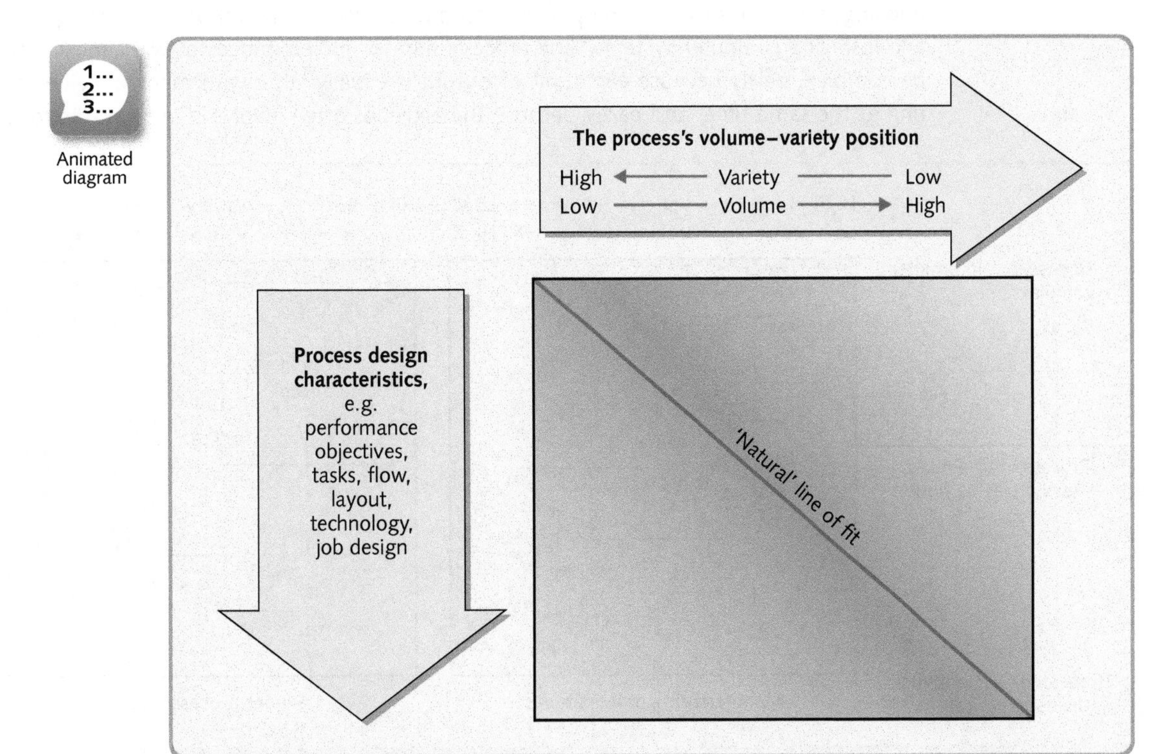

Figure 4.3 The elements of process design are strongly influenced by the volume–variety requirements placed on the process

matrix, it can in fact be used for any type of process whether producing products or services. The underlying idea of the product–process matrix is that many of the more important elements of process design are strongly related to the volume–variety position of the process. So, for any process, the tasks that it undertakes, the flow of items through the process, the layout of its resources, the technology it uses, and the design of jobs, are all strongly influenced by its volume–variety position. This means that most processes should lie close to the diagonal of the matrix that represents the 'fit' between the process and its volume–variety position. This is called the 'natural' diagonal.[3]

OPERATIONS PRINCIPLE

Process types indicate the position of processes on the volume-variety spectrum.

Process types

Processes that inhabit different points on the diagonal of the product–process matrix are sometimes referred to as 'process types'. Each process type implies differences in the set of tasks performed by the process and the way in which materials, or information or customers, flow through the process. Different terms are sometimes used to identify process types, depending on whether they are predominantly manufacturing or service processes, and there is some variation in how the names are used. This is especially so in service process types. It is not uncommon to find manufacturing terms used also to describe service processes. Perhaps most importantly, there is some degree of overlap between process types. The different process types are shown in Figure 4.4.

Project processes

Project processes are those which deal with discrete, usually highly customised products. Often the timescale of making the product is relatively long, as is the interval between the completions of each product. The activities involved in the process may be ill-defined and uncertain, sometimes changing during the process itself. Examples include advertising agencies, shipbuilding, most construction companies, and movie production companies, drilling oil wells and installing computer systems. Any process map for project processes will almost certainly be complex, partly because each unit of output is usually large with many activities occurring at the same time, and partly because the activities often involve significant discretion to

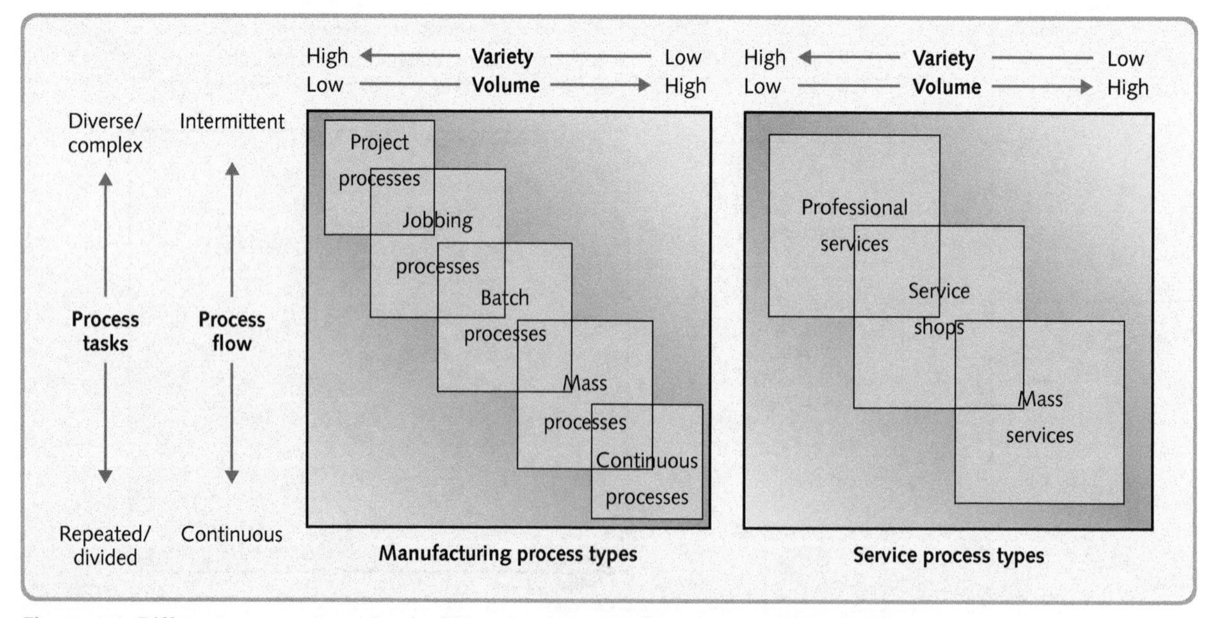

Figure 4.4 Different process types imply different volume–variety characteristics for the process

act according to professional judgement. In fact a process map for a whole project would be extremely complex, so rarely would a project be mapped, but small parts may be.

Jobbing processes

Jobbing processes also deal with very high variety and low volumes, but whereas in project processes each project has resources devoted more or less exclusively to it, in jobbing processes each 'product' has to share the operation's resources with many others. The process will work on a series of products but, although all the products will require the same kind of attention, each will differ in its exact needs. Examples of jobbing processes include many precision engineers such as specialist toolmakers, furniture restorers, 'make-to-measure' tailors, and the printer who produces tickets for the local social event. Jobbing processes produce more and usually smaller items than project processes but, like project processes, the degree of repetition is low. Many jobs could be 'one-offs'. Again, any process map for a jobbing process could be relatively complex for similar reasons to project processes. Although jobbing processes sometimes involve considerable skill, they are usually more unpredictable than project processes.

Batch processes

Batch processes can look like jobbing processes, but without the degree of variety normally associated with jobbing. As the name implies, batch processes usually produce more than one 'product' at a time. So each part of the operation has periods when it is repeating itself, at least while the 'batch' is being processed. The size of the batch could be just two or three, in which case the batch process would differ little from jobbing, especially if each batch is a totally novel product. Conversely, if the batches are large, and especially if the products are familiar to the operation, batch processes can be fairly repetitive. Because of this, the batch type of process can be found over a wider range of volume and variety levels than other process types. Examples of batch processes include machine tool manufacturing, the production of some special gourmet frozen foods, the manufacture of most of the component parts that go into mass-produced assemblies such as automobiles, and the production of most clothing. Batch process maps may look straightforward, especially if different products take similar routes through the process with relatively standard activities being performed at each stage.

Mass processes

Mass processes produce in high volume, usually with narrow effective variety. An automobile plant, for example, might produce several thousand variants of car if every option of engine size, colour, and equipment is taken into account. Yet its effective variety is low because the different variants do not affect the basic process of production. The activities in the automobile plant, like all mass processes, are essentially repetitive and largely predictable. In addition to the automobile plant, examples of mass processes include consumer durable manufacturers, most food processes such as a frozen pizza manufacturer, beer bottling plants and CD production. Process maps for this type of process will be straightforward sequences of activities.

Continuous processes

Continuous processes are one step beyond mass processes insomuch as they operate at even higher volume and often have even lower variety. Sometimes they are literally continuous in that their products are inseparable, being produced in an endless flow. Continuous processes are often associated with relatively inflexible, capital-intensive technologies with highly predictable flow. Examples of continuous processes include petrochemical refineries, electricity utilities, steel making and internet server farms. Like mass processes, process maps will show few elements of discretion, and although products may be stored during the process, the predominant characteristic of most continuous processes is of smooth flow from one part of the process to another.

Professional services

Professional services are high-variety, low-volume processes, where customers may spend a considerable time in the service process. Such services usually provide high levels of customisation, so contact staff are given considerable discretion. They tend to be people-based rather than equipment-based, with emphasis placed on the process (how the service is delivered) as much as the 'product' (what is delivered). Examples include management consultants, lawyers' practices, architects, doctors' surgeries, auditors, health and safety inspectors and some computer field service operations. Where process maps are used they are likely to be drawn predominantly at a high level. Consultants for example frequently use a predetermined set of broad stages, starting with understanding the real nature of the problem through to the implementation of their recommended solutions. This high level process map guides the nature and sequence of the consultants' activities.

Service shops

Service shops are characterised by levels of customer contact, customisation, volume of customers and staff discretion that position them between the extremes of professional and mass services (see next section). Service is provided via a mix of front- and back-office activities. Service shops include banks, high street shops, holiday tour operators, car rental companies, schools, most restaurants, hotels and travel agents. For example, an equipment hire and sales organisation may have a range of equipment displayed in front-office outlets, while back-office operations look after purchasing and administration. The front-office staff have some technical training and can advise customers during the process of selling the product. Essentially the customer is buying a fairly standardised 'product' but will be influenced by the process of the sale which is customised to the individual customer's needs.

Mass services

Mass services have many customer transactions and little customisation. Such services are often predominantly equipment-based and 'product' oriented, with most value added in the back-office, sometimes with comparatively little judgement needed by front-office staff who may have a closely defined job and follow set procedures. Mass services include supermarkets, national rail networks, airports and many call centres. For example, airlines move a large number of passengers on their networks. Passengers pick a journey from the range offered. The airline can advise passengers on the quickest or cheapest way to get from A to B, but they cannot 'customise' the service by putting on special flights for them.

Moving off the natural diagonal

Animated diagram

A process lying on the natural diagonal of the matrix shown in Figure 4.3 will normally have lower operating costs than one with the same volume–variety position that lies off the diagonal. This is because the diagonal represents the most appropriate process design for any volume–variety position. Processes that are on the right of the 'natural' diagonal would normally be associated with lower volumes and higher variety. This means that they are likely to be more flexible than seems to be warranted by their actual volume–variety position. That is, they are not taking advantage of their ability to standardise their activities. Because of this, their costs are likely to be higher than they would be with a process that was closer to the diagonal. Conversely, processes that are on the left of the diagonal have adopted a position that would normally be used for higher-volume and lower-variety processes. Processes will therefore be 'over-standardised' and probably too inflexible for their volume–variety position. This lack of flexibility can also lead to high costs because the process will not be able to change from one activity to another as readily as a more flexible process. One note of caution regarding this

OPERATIONS PRINCIPLE

Moving off the 'natural diagonal' of the product–process matrix will incur excess cost.

idea: although logically coherent, it is a conceptual model rather than something that can be 'scaled'. Although it is intuitively obvious that deviating from the diagonal increases costs, the precise amount by which costs will increase is very difficult to determine.

Nevertheless, a first step in examining the design of an existing process is to check if it is on the natural diagonal of the product–process matrix. The volume–variety position of the process may have changed without any corresponding change in its design. Alternatively, design changes may have been introduced without considering their suitability for the processes volume–variety position.

EXAMPLE

Meter installation

The 'meter installation' unit of a water utility company installed and repaired water meters. The nature of each installation job could vary significantly because the metering requirements of each customer varied and because meters had to be fitted into very different water pipe systems. When a customer requested an installation or repair, a supervisor would survey the customer's water system and transfer the results of the survey to the installation team of skilled plumbers. An appointment would then be made for a plumber to visit the customer's location and install or repair the meter on the agreed appointment date. The company decided to install for free a new 'standard' remote-reading meter that would replace the wide range of existing meters and could be read automatically using the customer's telephone line. This would save meter-reading costs. It also meant a significant increase in work for the unit and more skilled plumbing staff were recruited. The new meter was designed to make installation easier by including universal quick-fit joints that reduced pipe cutting and jointing during installation. As a pilot, it was also decided to prioritise those customers with the oldest meters and conduct trials of how the new meter worked in practice. All other aspects of the installation process were left as they were.

The pilot was not a success. Customers with older meters were distributed throughout the company's area, so staff could not service several customers in one area and had to travel relatively long distances between customers. Also, because customers had not initiated the visit themselves, they were more likely to have forgotten the appointment, in which case plumbers had to return to their base and try to find other work to do. The costs of installation were proving to be far higher than forecast and the plumbers were frustrated at the waste of their time and the now relatively standardised installation job. The company decided to change its process. Rather than replace the oldest meters which were spread around its region, it targeted smaller geographic areas to limit travelling time. It also cut out the survey stage of the process because, using the new meter, 98 per cent of installations could be fitted in one visit, minimising disruption to the customer and the number of missed appointments. Just as significantly, fully qualified plumbers were often not needed, so installation could be performed by less expensive labour.

This example is illustrated in Figure 4.5. The initial position of the installation process is at point A. The installation unit were required to repair and install a wide variety of meters into a very wide variety of water systems. This needed a survey stage to assess the nature of the job and the use of skilled labour to cope with the complex tasks. The installation of the new type of meter changed the volume–variety position for the process by reducing the variety the jobs tackled by the process and increasing the volume it had to cope with. However the process was not changed. By choosing a wide geographic area to service, retaining the unnecessary survey stage and hiring over skilled staff, the company was still defining itself as a high-variety, low-volume 'jobbing' process. The design of the process was appropriate for its old volume–variety position, but not the new one. In effect it had moved to point B in Figure 4.5. It was off the diagonal, with unnecessary flexibility and high operating costs. Redesigning the process to take advantage of the reduced variety and complexity of the job (position C on Figure 4.5) allowed installation to be performed far more efficiently.

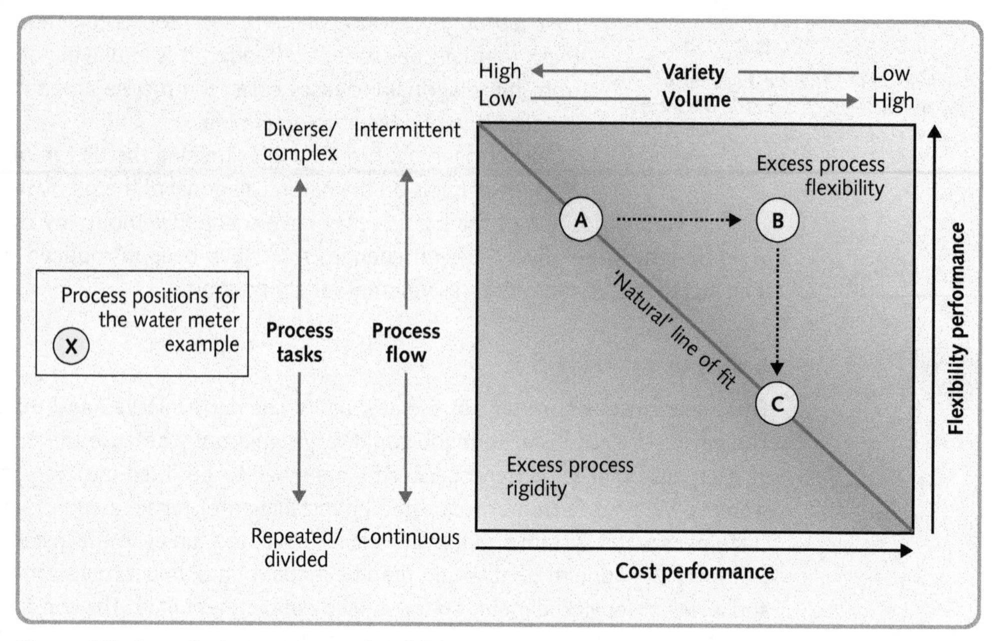

Figure 4.5 A product-process matrix with process positions from the water meter example

Layout, technology and design

If movement down the natural diagonal of the product–process matrix changes the nature of a process, then the key elements of its design will also change. At this broad level, these 'key elements' of the design are the two 'ingredients' that make up processes, technology, and people, and the way in which these ingredients are arranged within the process relative to each other. This latter aspect is usually called *layout*. In the remainder of the chapter, we start by discussing layout and then the design decisions that relate to process technology and the jobs that the people within the process undertake.

DIAGNOSTIC QUESTION

Are process layouts appropriate?

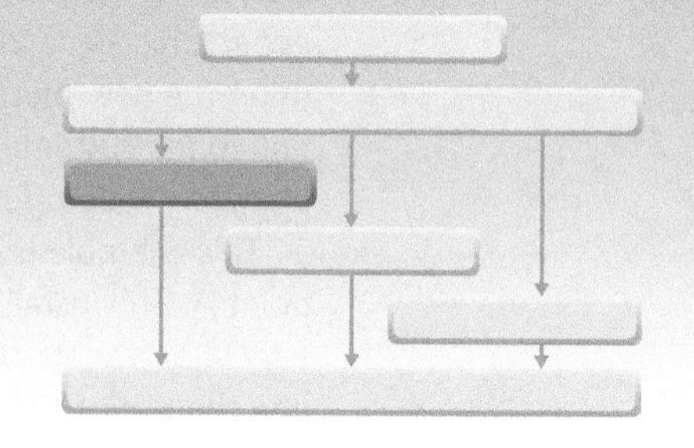

Video

There is little point in having a well-sequenced process if in reality its activities are physically located in a way that involves excessive movement of materials, information or customers. Usually the objective of the layout decision is to minimise movement, but, especially in information transforming processes, where distance is largely irrelevant, other criteria may dominate. For example, it may be more important to layout processes that similar activities or resources are grouped together. So, an international bank may group its foreign exchange dealers together to encourage communication and discussion between them, even though the 'trades' they make are processed in an entirely different location. Some high-visibility processes may fix their layout to emphasise the behaviour of the customers who are being processed.

Layout should reflect volume and variety

Again, the layout of a process is determined partly by its volume and variety characteristics. When volume is very low and variety is relatively high, 'flow' may not be a major issue. For example, in telecommunications satellite manufacture each product is different, and because products 'flow' through the operation very infrequently, it is not worth arranging facilities to minimise the flow of parts through the operation. With higher volume and lower variety, flow becomes a far more important issue. If variety is still high, however, an entirely flow-dominated arrangement is difficult because there will be different flow patterns. For example, a library will arrange its different categories of books and its other services partly to minimise the average distance its customers have to 'flow' through the operation. But, because its customers' needs vary, it will arrange its layout to satisfy the majority of its customers (but perhaps inconvenience a minority). When the variety of products or services reduces to the point where a distinct 'category' with similar requirements becomes evident but variety is still not small, appropriate resources could be grouped into a separate cell. When variety is relatively small and volume is high, flow can become regularised and resources can be positioned to address the (similar) needs of the products or services, as in a classic flow line.

> **OPERATIONS PRINCIPLE**
>
> *Resources in low-volume, high-variety processes should be arranged to cope with irregular flow.*

> **OPERATIONS PRINCIPLE**
>
> *Resources in high-volume, low-variety processes should be arranged to cope with smooth, regular flow.*

Most practical layouts are derived from only four *basic layout types* that correspond to different positions on the volume–variety spectrum. These are illustrated diagrammatically in Figure 4.6 and are as follows:

Fixed-position layout

Fixed-position layout is in some ways a contradiction in terms, since the transformed resources do not move between the transforming resources. Instead of materials, information

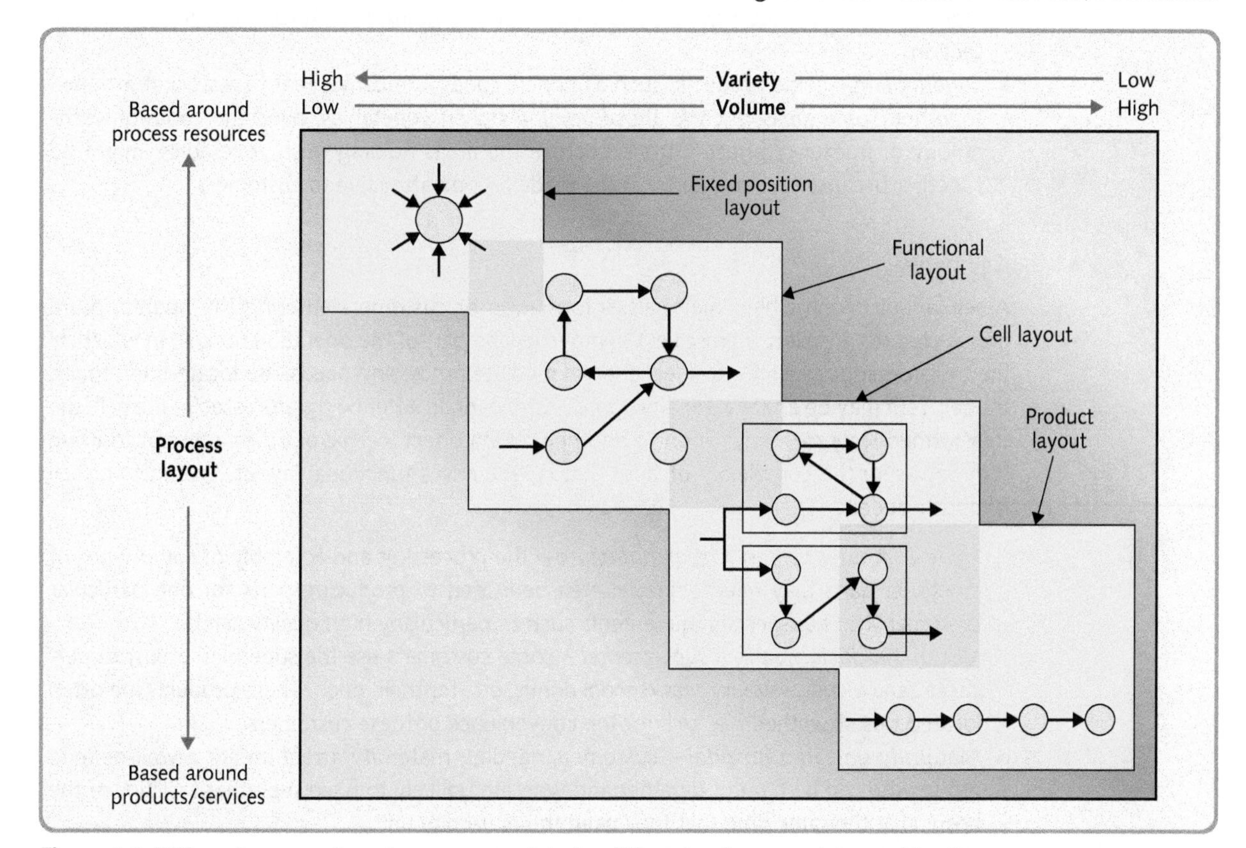

Figure 4.6 Different process layouts are appropriate for different volume–variety combinations

or customers flowing through an operation, the recipient of the processing is stationary and the equipment, machinery, plant and people who do the processing move as necessary. This could be because the product or the recipient of the service is too large to be moved conveniently, or it might be too delicate to move, or perhaps it could object to being moved; for example:

- *Power generator construction* – the product is too large to move.
- *Open-heart surgery* – patients are too delicate to move.
- *High-class restaurant* – customers would object to being moved to where food is prepared.

Functional layout

Practice note

Functional layout is so called because the functional needs and convenience of the transforming resources which constitute the processes dominate the layout decision. (Confusingly, functional layout can also be called 'process layout'.) In functional layout, similar activities or resources (or those with similar needs) are located together. This may be because it is convenient to group them together, or that their utilisation can be improved. It means that when materials, information or customers flow through the operation, they will take a route from activity to activity according to their needs. Usually this makes the flow pattern in the operation complex. Examples of process layouts include:

- *Hospital* – some processes (e.g. radiography equipment and laboratories) are required by several types of patient.
- *Machining the parts for aircraft engines* – some processes (e.g. heat treatment) need specialist support (heat and fume extraction); some processes (e.g. machining centres) require the same technical support from specialists; some processes (e.g. grinding machines) get high machine utilisation as all parts which need grinding pass through a single grinding section.
- *Supermarket* – some products, such as tinned goods, are convenient to restock if grouped together. Some areas, such as those holding frozen vegetables, need the common technology of freezer cabinets. Others, such as the areas holding fresh vegetables, might be together because that way they can be made to look attractive to customers.

Cell layout

A cell layout is one where materials, information or customers entering the operation are preselected (or preselect themselves) to move to one part of the operation (or cell) in which all the transforming resources, to meet their immediate processing needs, are located. Internally, the cell itself may be arranged in any appropriate manner. After being processed in the cell, the transformed resources may go on to another cell. In effect, cell layout is an attempt to bring some order to the complexity of flow that characterises functional layout. Examples of cell layouts include:

- *Some computer component manufacture* – the processing and assembly of some types of computer parts may need a special area dedicated to producing parts for one particular customer who has special requirements such as particularly high quality levels.
- *'Lunch' products area in a supermarket* – some customers use the supermarket just to purchase sandwiches, savoury snacks, cool drinks, etc. for their lunch. These products are often located close together in a 'cell' for the convenience of these customers.
- *Maternity unit in a hospital* – customers needing maternity attention are a well-defined group who can be treated together and who are unlikely to need the other facilities of the hospital at the same time that they need the maternity unit.

Product layout

Product layout involves locating people and equipment entirely for the convenience of the transformed resources. Each product, piece of information or customer follows a pre-arranged route in which the sequence of required activities corresponds to the sequence in which the processes have been located. The transformed resources 'flow' along a 'line'. This is why this type of layout is sometimes called flow or line layout. Flow is clear, predictable and therefore relatively easy to control. It is the high volume and standardised requirements of the product or service which allows product layouts. Examples of product layout include:

- *Automobile assembly* – almost all variants of the same model require the same sequence of processes.
- *Self-service cafeteria* – generally the sequence of customer requirements (starter, main course, dessert, drink) is common to all customers, but layout also helps control customer flow.

Layout selection

Getting the process layout right is important, if only because of the cost, difficulty and disruption of making any layout change. It is not an activity many businesses would want to repeat very often. Also, an inappropriate layout could mean that extra cost is incurred *every time* an item is processed. But more than this, an effective layout gives clarity and transparency to the flow of items through a process. There is no better way of emphasising that everyone's activities are really part of an overall process than by making the flow between activities evident to everyone.

One of the main influences on which type of layout will be appropriate, is the nature of the process itself, as summarised in its 'process type'. There is often some confusion between process types and layout types. Layout types are not the same as process types. Process types were described earlier in the chapter and indicate a broad approach to the organisation and operation of a process. Layout is a narrower concept but is very clearly linked to process type. Just as process type is governed by volume and variety, so is layout. But for any given process type there are usually at least two alternative layouts. Table 4.1 summarises the alternative layouts for particular process types. Which of these is selected, or whether some hybrid layout is chosen, depends on the relative importance of the performance objectives of the process, especially cost and flexibility. Table 4.2 summarises.

Table 4.1 Alternative layout types for each process type

<table>
<tr><th>Manufacturing process type</th><th colspan="2">Potential layout types</th><th>Service process type</th></tr>
<tr><td>Project</td><td>Fixed-position layout
Functional layout</td><td rowspan="2">Fixed-position layout
Functional layout
Cell layout</td><td rowspan="2">Professional service</td></tr>
<tr><td>Jobbing</td><td>Functional layout
Cell layout</td></tr>
<tr><td>Batch</td><td>Functional layout
Cell layout</td><td rowspan="2">Functional layout
Cell layout</td><td rowspan="2">Service shop</td></tr>
<tr><td>Mass</td><td>Cell layout
Product layout</td></tr>
<tr><td>Continuous</td><td>Product layout</td><td>Cell layout
Product layout</td><td>Mass service</td></tr>
</table>

Table 4.2 The advantages and disadvantages of the basic layout types

	Advantages	*Disadvantages*
Fixed-position	• Very high mix and product flexibility • Product or customer not moved or disturbed • High variety of tasks for staff	• High unit costs • Scheduling of space and activities can be difficult • Can mean much movement of plant and staff
Functional	• High mix and product flexibility • Relatively robust if in the case of disruptions • Relatively easy supervision of equipment or plant	• Low facilities utilisation • Can have very high work-in-progress or customer queueing • Complex flow can be difficult to control
Cell	• Can give a good compromise between cost and flexibility for relatively high-variety operations • Fast throughput • Group work can result in good motivation	• Can be costly to rearrange existing layout • Can need more plant and equipment • Can give lower plant utilisation
Product	• Low unit costs for high volume • Gives opportunities for specialisation of equipment • Materials or customer movement is convenient	• Can have low mix flexibility • Not very robust if there is disruption • Work can be very repetitive

DIAGNOSTIC QUESTION

Is process technology appropriate?

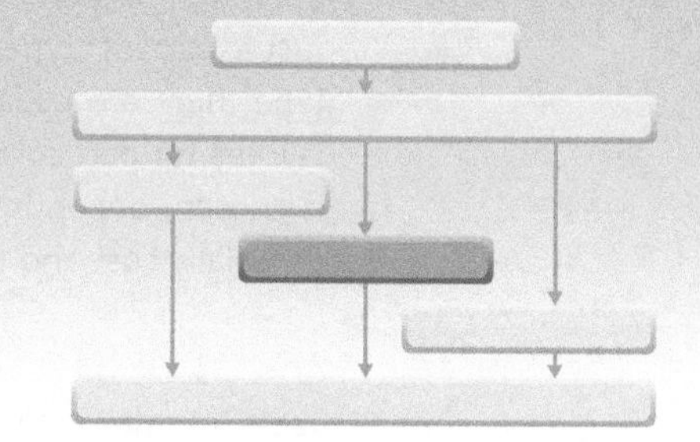

Video

Process technologies are the machines, equipment and devices that help processes 'transform' materials and information and customers. It is a particularly important issue because few operations have been unaffected by the advances in process technology over the last two decades, and the pace of technological development is not slowing down. However, it is important to distinguish between *process technology* (the machines and devices that help to *create* products and services) and *product technology* (the technology that is embedded within the product or service and creates its specification or functionality. Some process technology, although not used for the actual creation of goods and services, nonetheless plays a key role in *facilitating* their creation. For example, the information technology systems that run planning and control activities can be used to help managers and operators run the processes. Sometimes this type of technology is called *indirect* process technology, and it is becoming increasingly important. Many businesses spend more on the computer systems that run their processes than they do on the direct process technology that creates their products and services.

Process technology should reflect volume and variety

Again, different process technologies will be appropriate for different parts of the volume–variety continuum. High-variety, low-volume processes generally require process technology that is *general purpose,* because it can perform the wide range of processing activities that high variety demands. High-volume, low-variety processes can use technology that is more *dedicated* to its narrower range of processing requirements. Within the spectrum from general purpose to dedicated process technologies, three dimensions in particular tend to vary with volume and variety. The first is the extent to which the process technology carries out activities or makes decisions for itself, that is, its degree of 'automation'. The second is the capacity of the technology to process work, that is, its 'scale' or 'scalability'. The third is the extent to which it is integrated with other technologies, that is, its degree of 'coupling' or 'connectivity'. Figure 4.7 illustrates these three dimensions of process technology.[5]

OPERATIONS PRINCIPLE

Process technology in high-volume, low-variety processes is relatively automated, large scale and closely coupled when compared to that in low-volume, high-variety processes.

The degree of automation of the technology

To some extent, all technology needs human intervention. It may be minimal, for example the periodic maintenance interventions in a petrochemical refinery. Conversely, the person who operates the technology may be the entire 'brains' of the process, for example the surgeon using keyhole surgery techniques. Generally, processes that have high variety and low volume will employ process technology with lower degrees of automation than those with higher volume and lower variety. For example, investment banks trade in highly complex and sophisticated financial 'derivatives', often customised to the needs of individual clients, and each may be worth millions of dollars. The back-office of the bank

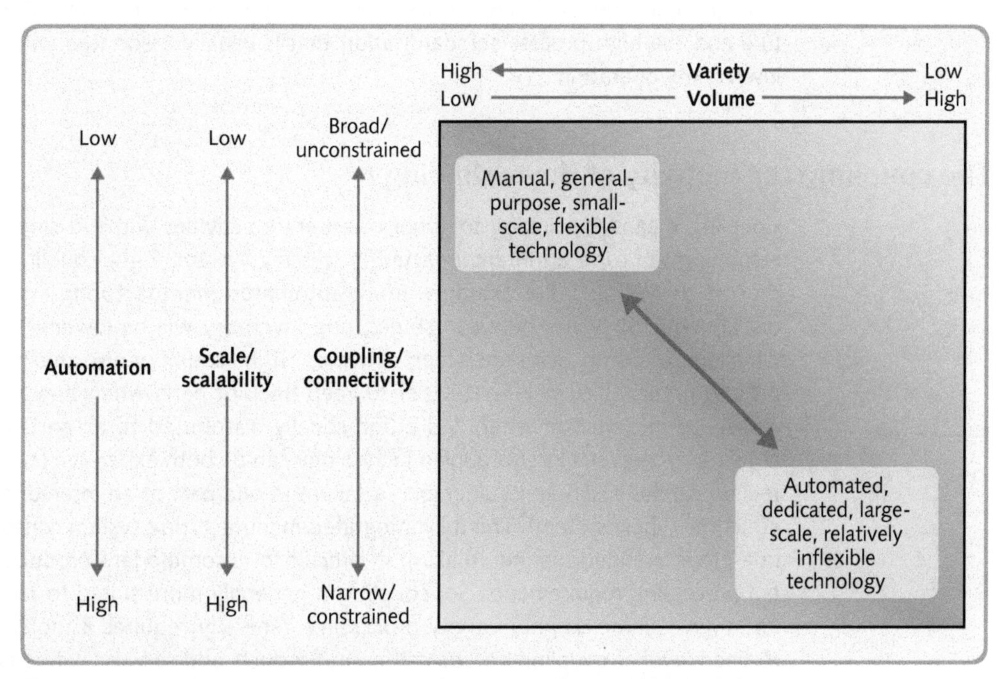

Figure 4.7 Different process technologies are important for different volume–variety combinations

has to process these deals to make sure that payments are made on time, documents are exchanged, and so on. Much of this processing will be done using relatively general purpose technology such as spreadsheets. Skilled back-office staff are making the decisions rather than the technology. Contrast this with a higher-volume, low-variety products, such as straightforward equity (stock) trades. Most of these products are simple and straightforward and are processed in very high volume of several thousand per day by 'automated' technology.

The scale/scalability of the technology

There is usually some discretion as to the scale of individual units of technology. For example, the duplicating department of a large office complex may decide to invest in a single, very large, fast copier, or alternatively in several smaller, slower copiers distributed around the operation's various processes. An airline may purchase one or two wide-bodied aircraft or a larger number of smaller aircraft. The advantage of large-scale technologies is that they can usually process items cheaper than small-scale technologies, but usually need high volume and can cope only with low variety. By contrast, the virtues of smaller-scale technology are often the nimbleness and flexibility that is suited to high-variety, lower-volume processing. For example, four small machines can between them produce four different products simultaneously (albeit slowly), whereas a single large machine with four times the output can produce only one product at a time (albeit faster). Small-scale technologies are also more robust. Suppose the choice is between three small machines and one larger one. In the first case, if one machine breaks down, a third of the capacity is lost, but in the second, capacity is reduced to zero.

The equivalent to scale for some types of information processing technology is *scalability*. By scalability we mean the ability to shift to a different level of useful capacity quickly, and cost-effectively. Scalability is similar to absolute scale insomuch as it is influenced by the same volume–variety characteristics. IT scalability relies on consistent IT platform architecture and the high process standardisation that is usually associated with high-volume and low-variety operations.

The coupling/connectivity of the technology

Coupling means the linking together of separate activities within a single piece of process technology to form an interconnected processing system. Tight coupling usually gives fast process throughput. For example, in an automated manufacturing system products flow quickly without delays between stages, and inventory will be lower – it can't accumulate when there are no 'gaps' between activities. Tight coupling also means that flow is simple and predictable, making it easier to keep track of parts when they pass through fewer stages, or information when it is automatically distributed to all parts of an information network. However, closely coupled technology can be both expensive (each connection may require capital costs) and vulnerable (a failure in one part of an interconnected system can affect the whole system). The fully integrated manufacturing system constrains parts to flow in a predetermined manner, making it difficult to accommodate products with very different processing requirements. So, coupling is generally more suited to relatively low variety and high volume. Higher-variety processing generally requires a more open and unconstrained level of coupling because different products and services will require a wider range of processing activities.

DIAGNOSTIC QUESTION

Are job designs appropriate?

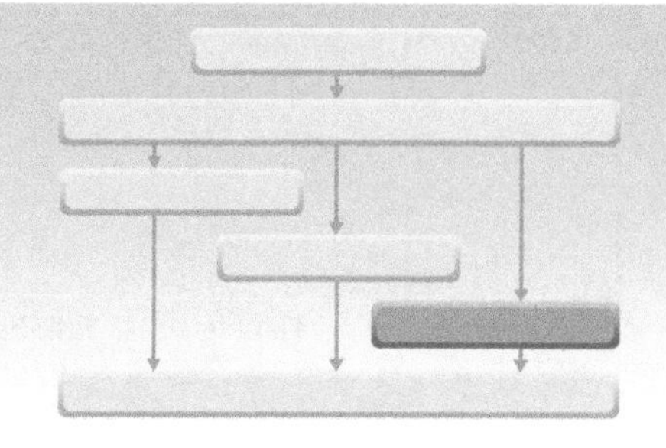

Job design is about how people carry out their tasks within a process. It defines the way they go about their working lives. It positions the expectations of what is required of them, and it influences their perceptions of how they contribute to the organisation. It also defines their activities in relation to their work colleagues and it channels the flow of communication between different parts of the operation. But, of most importance, it helps to develop the culture of the organisation – its shared values, beliefs and assumptions. Inappropriately designed jobs can destroy the potential of a process to fulfill its objectives, no matter how appropriate its layout or process technology. So jobs must be designed to fit the nature of the process. However, before considering this, it is important to accept that some aspects of job design are common to all processes, irrespective of what they do or how they do it. Consider the following:

- *Safety*. The primary and universal objective of job design is to ensure that all staff performing any task within a process are protected against the possibility of physical or mental harm.
- *Ethical issues*. No individual should be asked to perform any task that is either illegal or (within limits) conflicts with strongly held ethical beliefs.
- *Work/life balance*. All jobs should be structured so as to promote a healthy balance between time spent at work and time away from work.

Note that all these objectives of job design are also likely to improve overall process performance. However, the imperative to follow such objectives for their own sake transcends conventional criteria.

Job design should reflect volume and variety

As with other aspects of process design, the nature and challenges of job design are governed largely by the volume–variety characteristics of a process. An architect designing major construction projects will perform a wide range of very different, often creative and complex tasks, many of which are not defined at the start of the process, and most of which have the potential to give the architect significant job satisfaction. By contrast, someone in the architect's accounts office keying in invoice details has a job that is repetitive, has little variation, is tightly defined and that cannot rely on the intrinsic interest of the task itself to maintain job commitment. These two jobs will have different characteristics because they are part of processes with different volume and variety positions. Three aspects of job design in particular are affected by the volume–variety characteristics of a process: how tasks are to be allocated to each person in the process; the degree of job definition; the methods used to maintain job commitment. Figure 4.8 illustrates this.

> **OPERATIONS PRINCIPLE**
> *Job designs in high-volume, low-variety processes are relatively closely defined with little decision-making discretion and needing action to help commitment when compared to those in low-volume, high-variety processes.*

How should tasks be allocated? – the division of labour

The most obvious aspect of any individual's job is how big it is, that is, how many of the tasks within any process are allocated to an individual. Should a single individual perform all the

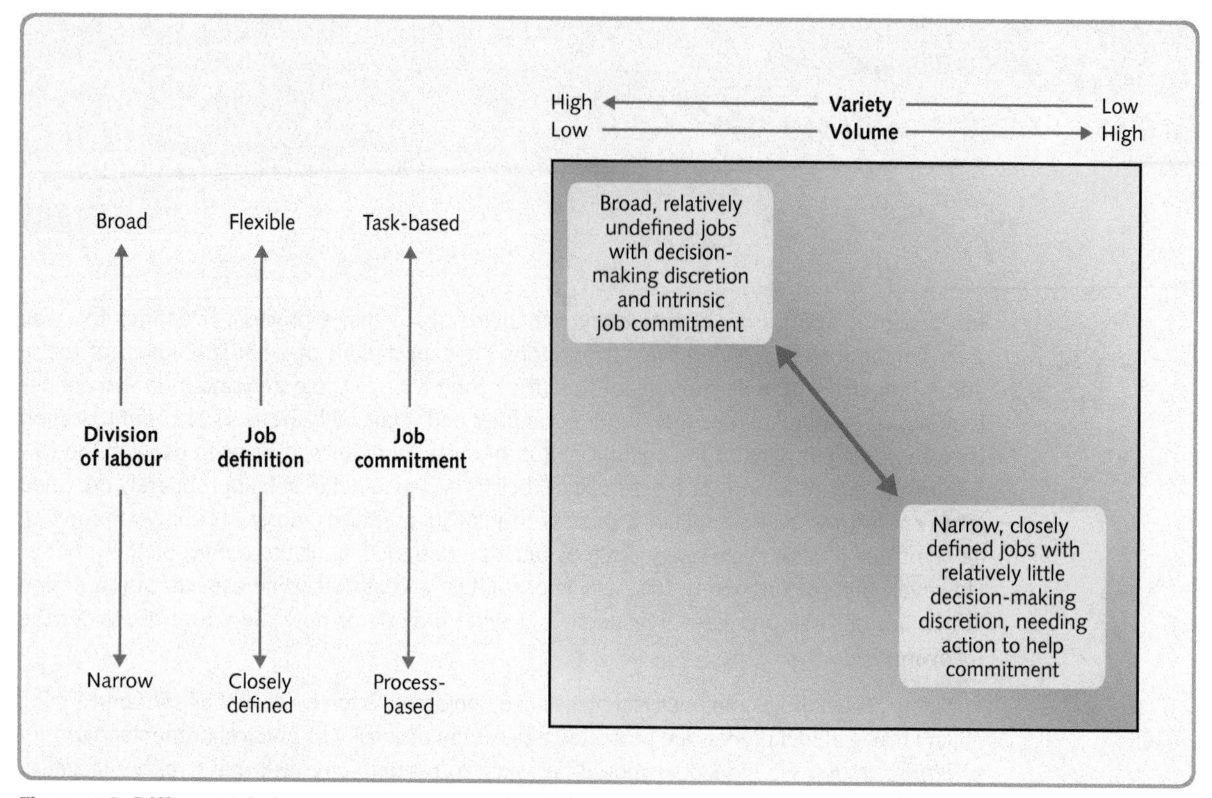

Figure 4.8 Different job designs are appropriate for different volume—variety combinations

Practice note

process? Alternatively, should separate individuals or teams perform each task? Separating tasks into smaller parts between individuals is called the *division of labour*. Perhaps its epitome is the assembly line, where products move along a single path and are built up by operators continually repeating a single task. This is the predominant model of job design in most high volume–low variety processes. For such processes there are some *real advantages* in division-of-labour principles:

- *It promotes faster learning*. It is obviously easier to learn how to do a relatively short and simple task than a long and complex one, so new members of staff can be quickly trained and assigned to their tasks.
- *Automation becomes easier*. Substituting technology for labour is considerably easier for short and simple tasks than for long and complex ones.
- *Non-productive work is reduced*. In large, complex tasks the proportion of time between individual value adding elements can be very high, for example in manufacturing, picking up tools and materials, putting them down again and generally searching and positioning.

There are also *serious drawbacks* to highly divided jobs:

- *It is monotonous*. Repeating the same task eight hours a day and five days a week is not fulfilling. This may lead to an increased likelihood of absenteeism, staff turnover, and error rates.
- *It can cause physical injury*. The continued repetition of a very narrow range of movements can, in extreme cases, lead to physical injury. The over-use of some parts of the body (especially the arms, hands and wrists) can result in pain and a reduction in physical capability, called repetitive strain injury (RSI).
- *It can mean low flexibility*. Dividing a task up into many small parts often gives the job design a rigidity which is difficult to change under changing circumstances. For example, if an assembly line has been designed to make one particular product but then has to change

to manufacture a quite different product, the whole line will need redesigning. This will probably involve changing every operator's set of tasks.

- *It can mean poor robustness*. Highly divided jobs imply items passing between several stages. If one of these stages is not working correctly, for example because some equipment is faulty, the whole operation is affected. On the other hand, if each person is performing the whole of the job, any problems will only affect that one person's output.

To what degree should jobs be defined?

Jobs in high-variety processes are difficult to define in anything but the most general terms. Such jobs may require tacit knowledge gained over time and through experience and often require individuals to exercise significant discretion in what they do and how they do it. Some degree of job definition is usually possible and advisable, but it may be stated in terms of the 'outcome' from the task rather than in terms of the activities within the task. For example, the architect's job may be defined in terms of *'achieving overall coordination, taking responsibility for articulating the overall vision of the project, ensuring stakeholders are comfortable with the process, etc.'* By contrast, a process with less variety and higher volume is likely to be defined more closely, with the exact nature of each activity defined and individual staff trained to follow a job step-by-step.

How should job commitment be encouraged?

Many factors may influence job commitment. An individual's job history and expectations, relationships with co-workers, personal circumstances, can all be important. So are the volume and variety characteristics of the process by defining the possible ways in which commitment can be enhanced. In high-variety processes, especially those with a high degree of staff discretion, job commitment is likely to come from the *intrinsic nature of the task* itself. Exercising skill and decision-making, for example, can bring its own satisfaction. Of course commitment can be enhanced through extra responsibility, flexibility in working times and so on, but the main motivator is the job itself. By contrast, low-variety, high-volume jobs, especially those designed with a high division of labour and little discretion can be highly alienating. Such jobs have relatively little intrinsic task satisfaction. It has to be *'designed into' the process* by emphasising the satisfaction to be gained from the performance of the process overall. A number of job design approaches have been suggested for achieving this in processes involving relatively repetitive work.

- **Job enlargement** involves allocating a larger number of tasks to individuals, usually by combining tasks that are broadly of the same type as those in the original job. This may not involve more demanding or fulfilling tasks, but it may provide a more complete and therefore slightly more meaningful job. If nothing else, people performing an enlarged job will not repeat themselves as often. For example, suppose that the manufacture of a product has traditionally been split up on an assembly-line basis into ten equal and sequential jobs. If that job is then redesigned so as to form two parallel assembly lines of five people, each operator would have twice the number of tasks to perform.
- **Job enrichment** like job enlargement, increases the number of tasks in a job, but also implies allocating tasks that involve more decision making, or greater autonomy, and therefore greater control over the job. These could include the maintenance of, and adjustments to, any process technology used, the planning and control of activities within the job, or the monitoring of quality levels. The effect is both to reduce repetition in the job *and* to increase personal development opportunities. So, in the assembly-line example, each operator, a job could also be allocated responsibility for carrying out routine maintenance and such tasks as record-keeping and managing the supply of materials.

Video

- **Job rotation** means moving individuals periodically between different sets of tasks to provide some variety in their activities. When successful, job rotation can increase skill flexibility and make a small contribution to reducing monotony. However, it is not always viewed as beneficial either by management (because it can disrupt the smooth flow of work) or by the people performing the jobs (because it can interfere with their rhythm of work).
- **Empowerment** means enhancing individuals' ability, and sometimes authority, to change how they do their jobs. Some technologically constrained processes, such as those in chemical plants may limit the extent that staff can dilute their highly standardised task methods without consultation. Other less defined processes to empowerment may go much further.
- **Team-working** is closely linked to the empowerment. Team-based work organisation (sometimes called self-managed work teams), is where staff, often with overlapping skills, collectively perform a defined task and have some discretion over how they perform the task. The team may control such things as task allocation between members, scheduling work, quality measurement and improvement, and sometimes even the hiring of staff. Groups are described as 'teams' when the virtues of working together are being emphasised and a shared set of objectives and responsibilities is assumed.

Critical commentary

Each chapter contains a short critical commentary on the main ideas covered in the chapter. Its purpose is not to undermine the issues discussed in the chapter, but to emphasise that, although we present a relatively orthodox view of operation, there are other perspectives.

- There could be three sets of criticisms prompted by the material covered in this chapter. The first relates to the separation of process design into two parts – positioning and analysis. It can reasonably be argued that this separation is artificial insomuch as (as is admitted at the beginning of this chapter) the two approaches are very much interrelated. An alternative way of thinking about the topic would be to consider all aspects of the arrangement of resources together. This would include the issues of layout that have been discussed in this chapter together with the more detailed process mapping issues described in Chapter 5. The second criticism would challenge the core assumption of the chapter – that many significant process design decisions are influenced primarily by volume and variety. Whereas it is conventional to relate layout and (to a slightly lesser extent) process technology to volume–variety positioning, it is less conventional to do so for issues of job design. Some would argue that the vast majority of job design decisions will not vary significantly with volume and variety. The final criticism is also related to job design. Some academics would argue that our treatment of job design is over-influenced by the discredited (in their eyes) principles of the 'scientific' management movement that grew into 'work study' and 'time and motion' management.

SUMMARY CHECKLIST

This checklist comprises questions that can be usefully applied to any type of operations and reflect the major diagnostic questions used within the chapter.

- [ ] Do processes match volume–variety requirements?
- [ ] Are 'process types' understood and do they match volume–variety requirements?
- [ ] Can processes be positioned on the 'diagonal' of the product–process matrix?
- [ ] Are the consequences of moving away from the 'diagonal' of the product–process matrix understood?
- [ ] Are the implications of choosing an appropriate layout, especially the balance between process flexibility and low processing costs, understood?
- [ ] Are the process layouts appropriate?
- [ ] Which of the four basic layout types that correspond to different positions on the volume–variety spectrum is appropriate for each process?
- [ ] Is process technology appropriate?
- [ ] Is the effect of the three dimensions of process technology (the degree of automation, the scale/scalability, and the coupling/connectivity of the technology) understood?
- [ ] Are job designs appropriate?
- [ ] Does job design ensure the imperative to design jobs that are safe, ethical and promote adequate work/life balance?
- [ ] Is the extent of division of labour in each process appropriate for its volume–variety characteristics?
- [ ] Is the extent of job definition in each process appropriate for its volume–variety characteristics?
- [ ] Are job commitment mechanisms in each process appropriate for its volume–variety characteristics?

CASE STUDY

McPherson Charles Solicitors

Grace Whelan, Managing Partner of McPherson Charles, welcomed the three solicitors into the meeting room. She outlined the agenda, essentially their thoughts and input into the rolling three-year plan. McPherson Charles, based in Bristol in the west of England, had grown rapidly to be one of the biggest law firms in the region; with 21 partners and around 400 staff, it was an ambitious partnership aiming to maintain its impressive growth record. The firm was managed through 15 teams each headed by a partner. The meeting was intended to be the first stage in 'Plans for the Future', a programme to improve the effectiveness of the firm's operations. The three partners attending the meeting with Grace were Simon Reece (Family Law), Kate Hutchinson (Property) and Hazel Lewis (Litigation). Grace asked for ideas on what the firm should prioritise if it was to improve its performance further.

Simon Reece kicked things off: '*I think the first thing we need to agree on is that, for a professional service firm like ourselves, the quality of our people will always be the most important issue. We need to be absolutely confident that our staff not only have the best possible understanding of their own branch of the law, but also have the necessary client relationship skills to consolidate our business position with increasingly demanding clients.*' Hazel was not so sure. '*Of course I agree that the quality of our staff is an important issue, but that has always been true. What is new is the help we can get from some serious investment in technology and software. Just getting our systems and processes right would, I am sure, save us a lot of time and effort, and of course, reduce our cost base.*'

'*I really don't think spending more money on technology is the answer Hazel.*' Simon continued. '*We need more time to really understand our clients and being process and IT focused just doesn't work for us, we need another way of managing. The key is increasing revenue not penny pinching about costs, and to do that we need to really concentrate on relationship skills. Family law is like walking through a minefield, you can easily offend clients who are, almost by definition, in a highly emotional state. I think we need to make sure that senior members of staff with experience of managing client relationships pass on their knowledge to those who are less experienced.*'

'*I disagree, Simon.*' It was Kate now. '*Our clients really are increasingly cost-conscious and if we don't deliver*

Source: Janet Kimber/Getty Images

value for money word will spread very fast and our business will dry up. Much of the time we over-engineer our services. Why should we use highly qualified and expensive lawyers for every single task? I am convinced that, with slick systems and enhanced training, non-qualified people could do much of our work.' Grace knew that solicitors liked nothing better than disagreeing – it was what they did best – and she knew that this was going to be another long meeting.

In very simple terms, these are the type of activities that each team was engaged in.

Simon Reece, Family Law

'*We are called the 'family law' team but basically what we do is to help people through the trauma of divorce, separation and break up. Our biggest "high value" clients come to us because of word-of-mouth recommendation. Last year we had 89 of these "high value" clients and they all valued the personal touch that we were able to give them, getting to know them well and spending time with them to understand the, often "hidden" aspects of their case. These interviews cannot be rushed. These clients tend to be wealthy people and we will often have to drop everything and go off half way round the world to meet and discuss their situation. There are no standard procedures, every client is different, and everyone has to be treated as an individual. So we have a team of individuals who rise to the challenge each time and give great service. Of course, not all our clients are the super-rich. About a third of our annual family law income comes from about 750 relatively routine divorce and counselling cases. This work is a lot less interesting and I try to make sure that all my team have a mix of interesting and routine work over the year. I encourage them to exercise and develop their professional judgement. They are empowered to deal with any issues themselves or call in myself or one of the more senior members of the team for advice if appropriate. It is important to give this kind of responsibility to them so that they see themselves as part of a team. We are also the only part of the firm that has adopted an open-plan office arrangement centred around our specialist library of family case law.'*

Hazel Lewis, Litigation

'The litigation team provides a key service for our commercial client base. Our primary work consists of handling bulk collections of debt. The group's has 17 clients of which 5 provide 85 per cent of total volume. We work closely with the accounts departments of the client companies and have developed a semi-automatic approach to debt collection. Staff input data received from their clients into the system, from that point everything progresses through a pre-defined process, letters are produced, queries responded to and eventually debts collected, ultimately through court proceedings if necessary. Work tends to come in batches from clients and varies according to economic conditions, time of year and client sales activities. At the moment things are fairly steady, we had 872 new cases last week. The details of each case are sent over by the client; our people input the data onto our screens and set up a standard diary system for sending letters out. Some people respond quickly to the first letter and often the case is closed within a week or so; other people ignore letters and eventually we initiate court proceedings. We know exactly what is required for court dealings and have a pretty good process to make sure all the right documentation is available on the day.'

Kate Hutchinson, Property

'We are really growing fast and are building up an excellent reputation locally for being fast, friendly and giving value for money. Most of our work is "domestic", acting for individuals buying or selling their home, or their second home. Each client is allocated to a solicitor who becomes their main point of contact. But, given that we can have up to a hundred domestic clients a week, most of the work is carried out by the rest of the team behind the scenes. There is a relatively standard process to domestic property sales and purchases and we think that now we are pretty efficient at managing these standard jobs. Our process has four stages, one dealing with land registry searches, one liaising with banks who are providing the mortgage finance, one who makes sure surveys are completed and one division that finalises the whole process to completion. We believe that this degree of specialisation can help us achieve the efficiencies that are becoming important as the market gets more competitive. Increasingly we are also getting more complex "special" jobs. These are things like "volume re-mortgage" arrangements; rather complex "one-off" jobs, where a mortgage lender transfers a complex set of loan assets to another lender. "Special" jobs are always more complex than the domestic work and sometimes there are times when fast completion is particularly important and that can throw us a bit. The firm has recently formed partnerships with two large speculative builders, so we are getting into special "plot sales". All these "specials" do involve a lot of work and can occupy several members of the team for a time. We are now getting up to 25 of these "specials" each week, and they can be somewhat different to our normal work, but we try to follow roughly the same process with them as the normal domestic jobs.'

Are each team's processes appropriate?

Grace was concerned. The three teams obviously had to cope with very different volumes of work and variety of activities. It was also clear that each team had developed different approaches to managing their processes. The question that she needed to address was whether each team's approach was appropriate for the demands placed upon it.

QUESTIONS

1. What are the individual 'services' offered by each of the three teams?
2. Where would you place each service in a scale that goes from relatively low volume, relatively high variety, to relatively high volume, relatively low variety?
3. How would you describe each teams' process in terms of its layout, the technology (if any) it uses, and the jobs of its staff?
4. Use the above information to draw a product–process matrix. What does it indicate?

APPLYING THE PRINCIPLES

Hints

Some of these exercises can be answered by reading the chapter. Others will require some general knowledge of business activity and some might require an element of investigation. Hints on how they can all be answered are to be found in the eText at **www.pearsoned.co.uk/slack.**

1 Visit a branch of a retail bank and consider the following questions:

(a) What categories of service does the bank seem to offer?

(b) To what extent does the bank design separate processes for each of its types of service?

(c) What are the different process design objectives for each category of service?

2 Revisit the example at the beginning this of the chapter that examines some of the principles behind supermarket layout. Then visit a supermarket and observe people's behaviour. You may wish to try and observe which areas they move slowly past and which areas they seem to move past without paying attention to the products. (You may have to exercise some discretion when doing this; people generally don't like to be stalked round the supermarket too obviously.) Try and verify, as far as you can, some of the principles that were outlined in the box. If you were to redesign the supermarket what would you recommend? What do you think, therefore, are the main criteria considered in the design of a supermarket layout? What competitive or environmental changes may result in the need to change store layouts in the future? Will these apply to all supermarkets, or only certain ones, such as town centre stores?

3 Consider a retail bank. It has many services that it could offer through its branch network, using telephone-based call centres, or using an internet-based service. Choose a service that (like the bank's services) could be delivered in different ways. For example, you could choose education courses (that can be delivered full-time, part-time, distance learning, or e-learning, etc.), or a library (using a fixed facility, a mobile service, internet-based service, etc.), or any other similar service. Evaluate each alternative delivery process in terms of its effect on its customers' experiences.

4 Many universities and colleges around the world are under increasing pressure to reduce the cost per student of their activities. How do you think technology could help operations such as universities to keep their costs down but their quality of education high?

5 Robot-type technology is starting to play a part in some medical surgical procedures. What would it take before you were willing to subject yourself to a robot doctor?

6 Security devices are becoming increasingly high-tech. Most offices and similar buildings have simple security devices such as 'swipe cards' that admit only authorised people to the premises. Other technologies are becoming more common (although perhaps more in movies than in reality) such as finger print, iris and face scanning. Explore websites that deal with advanced security technology and gain some understanding of their state of development, advantages and disadvantages. Use this understanding to design a security system for an office building with which you a familiar. Remember that any system must allow access to legitimate users of the building (at least to obtain information for which they have clearance) and yet provide maximum security against any unauthorised access to areas and/or information.

Notes on chapter

1 Sources: Paul Walley – our colleague in the Operations Management Group at Warwick Business School; Martin, P. (2000), 'How supermarkets make a meal of you', *Sunday Times*, 4 November.

2 Sources include: Davey, J. (2010) 'Today we built a house every hour', *Sunday Times*, 31 January; Persimmon company website; Brown, G. (2010), 'Space4growth gives boost to former car sector workers', *Birmingham Post*, 27 August.

3 Hayes, R.H. and Wheelwright, S.C. (1984) *Restoring our Competitive Edge*, John Wiley.

4 Example provided by Professor Michael Lewis, Bath University.

5 Typology derived from Slack, N. and Lewis, M.A. (2001) *Operations Strategy*, Financial Times Prentice Hall.

TAKING IT FURTHER

Harvard Business Review (2011), Improving Business Processes (Harvard Pocket Mentor), Harvard Business School Press.

Hammer, M. (1990) 'Reengineering work: don't automate, obliterate', *Harvard Business Review*, July–August. *This is the paper that launched the whole idea of business processes and process management in general to a wider managerial audience. Slightly dated but worth reading.*

Harrington H.J. (2011) *Streamlined Process Improvement*, McGraw Hill Professional.

Hopp, W.J. and Spearman, M.L. (2001) *Factory Physics* (2nd edn), McGraw Hill. *Very technical so don't bother with it if you aren't prepared to get into the maths. However, some fascinating analysis, especially concerning Little's law.*

Ramaswamy, R. (1996) *Design and Management of Service Processes*, Addison-Wesley Longman. *A relatively technical approach to process design in a service environment.*

van der Aalst, W.M.P., ter Hofstede, A.H.M., and Weske, M. (eds) (2003) 'International Conference on Business Process Management (BPM 2003)', *Lecture Notes in Computer Science*, Vol. 2678, Springer-Verlag, Berlin. *Comprehensive academic survey.*

USEFUL WEBSITES

www.opsman.org *Definitions, links and opinions on operations and process management.*

www.bpmi.org *Site of the Business Process Management Initiative. Some good resources including papers and articles.*

www.bptrends.com *News site for trends in business process management generally. Some interesting articles.*

www.bls.gov/oes/ *US Department of Labor employment statistics.*

www.iienet.org *The American Institute of Industrial Engineers site. They are an important professional body for process design and related topics.*

www.waria.com *A Workflow and Reengineering association website. Some useful topics.*

Interactive quiz

For further resources including examples, animated diagrams, self-test questions, Excel spreadsheets and video materials please explore the eText on the companion website at **www.pearsoned.co.uk/slack**.

Process design 2 - analysis

Introduction

The previous chapter set the broad parameters for process design; in particular it showed how volume and variety shape the positioning of the process in terms of layout, process technology and the design of jobs. But this is only the beginning of process design. Within these broad parameters there are many, more detailed decisions to be made that will dictate the way materials, information and customers flow through the process. Do not dismiss these detailed design decisions as merely the 'technicalities' of process design. They are important because they determine the actual performance of the process in practice and eventually its contribution to the performance of the whole business (see Figure 5.1).

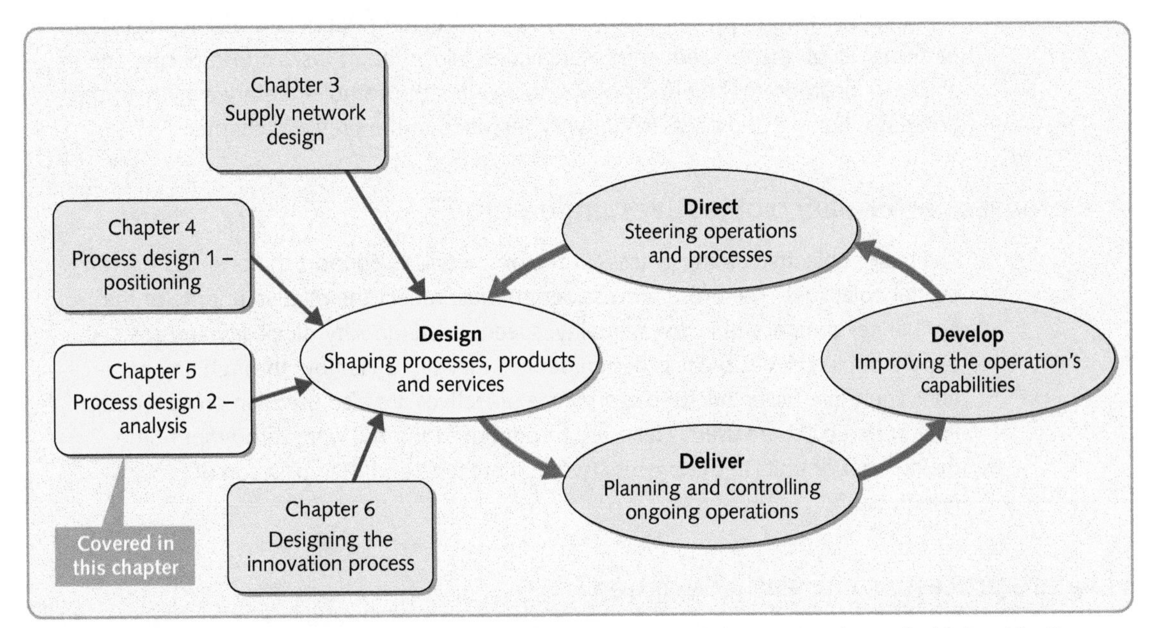

Figure 5.1 Process design analysis involves calculating the details of the process, in particular its objectives, sequence of activities, allocation of tasks and capacity and its ability to incorporate the effects of variability

EXECUTIVE SUMMARY

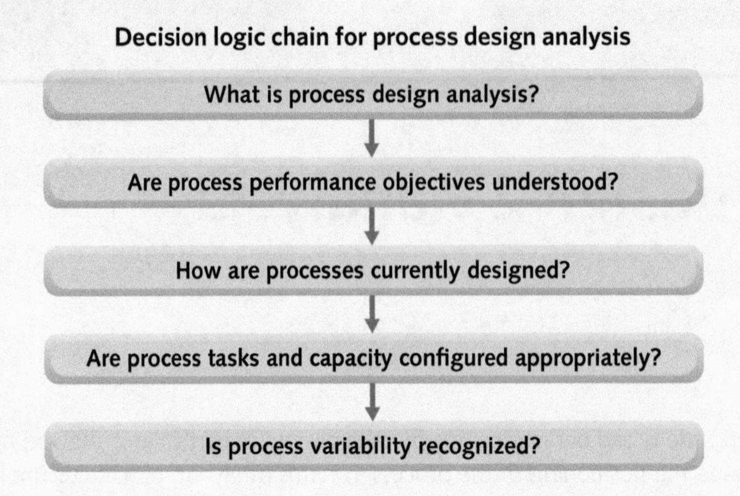

Each chapter is structured around a set of diagnostic questions. These questions suggest what you should ask in order to gain an understanding of the important issues of a topic, and, as a result, improve your decision making. An executive summary addressing these questions is provided below.

What is process design analysis?

The analysis stage of process design involves calculating the details of the process, in particular its objectives, sequence of activities, allocation of tasks and capacity, and its ability to incorporate the effects of variability. It is the complementary activity to the broad positioning of processes that was described in the previous chapter.

Are process performance objectives understood?

The major objective of any process in the business is to support the business's overall objectives. Therefore process design must reflect the relative priority of the normal performance objectives: quality, speed, dependability, flexibility and cost. At a more detailed level, process design defines the way units flow through an operation. Therefore more 'micro' performance objectives are also useful in process design. Four in particular are used. These are throughput (or flow) rate, throughput time, the number of units in the process (work in process) and the utilisation of process resources.

How are processes currently designed?

Much process design is in fact redesign, and a useful starting point is to fully understand how the current process operates. The most effective way of doing this is to map the process in some way. This can be done at different levels using slightly different mapping techniques. Sometimes it is useful to define the degree of visibility for different parts of the process, indicating how much of the process is transparent to customers.

Are process tasks and capacity configured appropriately?

This is a complex question with several distinct parts. First, it is necessary to understand the task precedence to be incorporated in the process. This defines what activities must occur before others. Second, it is necessary to examine how alternative process design options can incorporate series and parallel configuration. These are sometimes called 'long-thin' and 'short-fat' arrangements. Third, cycle time and process capacity must be calculated. This can help to allocate work evenly between the stages of the process (called balancing). Fourth, the relationship between throughput, cycle time and work in process must be established. This is done using a simple but extremely powerful relationship known as Little's law (throughput time = work in process × cycle time).

Is process variability recognised?

In reality, processes have to cope with variability, both in terms of time and the tasks that are performed within the process. This variability can have very significant effects on process behaviour, usually to reduce process efficiency. Queueing theory can be used to understand this effect. In particular, the relationship between process utilisation and the number of units waiting to processed (or throughput time) is important to understand.

DIAGNOSTIC QUESTION

What is process design analysis?

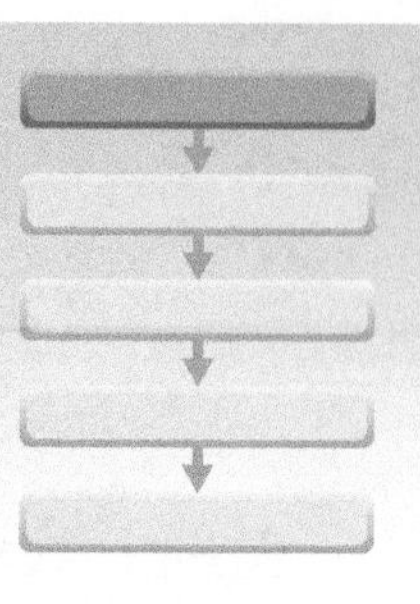

Video

To 'design' is to conceive the looks, arrangement and workings of something *before it is constructed*. Process design should be treated at two levels – the broad, aggregated level and the more detailed level. The previous chapter took a broad approach by relating process design to the volume–variety position of the process. That will have identified the broad process type, and given some guidance as to the layout, process technology and job designs to be used within the process. This chapter takes a more detailed view. However, in working out the details of a process design it is sometimes necessary to revisit the overall broad assumptions under which it is being designed. This is why the detailed analysis of process design covered in this chapter should always be thought through in the context of the broader process positioning issues covered in Chapter 4. The following two examples illustrate processes whose detailed design is important in determining their effectiveness.

EXAMPLE

Processes for even faster food[1]

The quick service restaurant (QSR) industry reckon that the very first drive-through dates back to 1928 when Royce Hailey first promoted the drive-through service at his Pig Stand restaurant in Los Angeles. Customers would simply drive by the back door of the restaurant where the chef would come out and deliver the restaurant's famous 'barbequed pig' sandwiches. Today, drive-through processes are slicker and faster. They are also more common: in 1975, McDonald's did not have any drive-throughs, and now more than 90 per cent of its US restaurants incorporate a drive-through process. In fact, 80 per cent of recent fast-food growth has come through the growing number of drive-throughs. Says one industry specialist: *'There are a growing number of customers for whom fast-food is not fast enough. They want to cut waiting time to the very minimum without even getting out of their car. Meeting their needs depends on how smooth we can get the process.'*

Source: Andrew Woodley/Alamy Images

The competition to design the fastest and most reliable drive-through process is fierce. Starbuck's drive-throughs have strategically placed cameras at the order boards so that servers can recognise regular customers and start making their order even before it's placed. Burger King has experimented with sophisticated sound systems, simpler menu boards and see-through food bags to ensure greater accuracy (no point in being fast if you don't deliver what the customer ordered). These details matter. McDonalds reckon that their sales increase one per cent for every six seconds saved at a drive-through, while a single Burger King restaurant calculated that its takings increased by $15,000 a year each time it reduced queuing time by one second. Menu items must be easy to read and understand. Designing 'combo meals' (burger, fries and a cola), for example, saves time at the ordering stage. But not everyone is thrilled by the boom in drive-throughs. People living in the vicinity may complain of the extra traffic they attract and the unhealthy image of fast-food combined with a process that does not even make customers get out of their car, is, for some, a step too far.

EXAMPLE

'Factory flow' helps surgery productivity[2]

Even surgery can be seen as a process, and like any process, it can be improved. Normally patients remain stationary with surgeons and other theatre staff performing their tasks around the patient. But this idea has been challenged by John Petri, an Italian consultant orthopaedic surgeon at a hospital in Norfolk in the UK. Frustrated by spending time drinking tea while patients were prepared for surgery, he redesigned the process so now he moves continually between two theatres (Figure 5.2). While he is operating on a patient in one theatre, his anaesthetist colleagues are preparing a patient for surgery in another theatre. After finishing with the first patient, the surgeon 'scrubs up', moves to the second operating theatre and begins the surgery on the second patient. While he is doing this the first patient is moved out of the first operating theatre and the third patient is prepared. This method of overlapping operations in different theatres allows him to work for five hours at a time rather than the previous standard three and a half hour session. *'If you were running a factory'*, says the surgeon, *'you wouldn't allow your most important and most expensive machine to stand idle. The same is true in a hospital.'* Currently used on hip and knee replacements, this layout would not be suitable for all surgical procedures, but, since its introduction the surgeon's waiting list has fallen to zero and his productivity has doubled. *'For a small increase in running costs we are able to treat many more patients,'* said a spokesperson for the hospital management. *'What is important is that clinicians . . . produce innovative ideas and we demonstrate that they are effective.'*

What do these two examples have in common?

Both examples highlight a number of process design issues. The first is that the payback from good process design is clearly significant. Quick service restaurant operations devote time and effort to the design process, assessing the performance of alternative process designs in terms of efficiency, quality and, above all, throughput time. The return for the hospital is even more dramatic. For a service that offers primarily surgical excellence and quality, being able to achieve the cost benefits of slick process design without compromising quality makes the hospital far better at serving its patients. Also, it is difficult to separate the design of the process from the

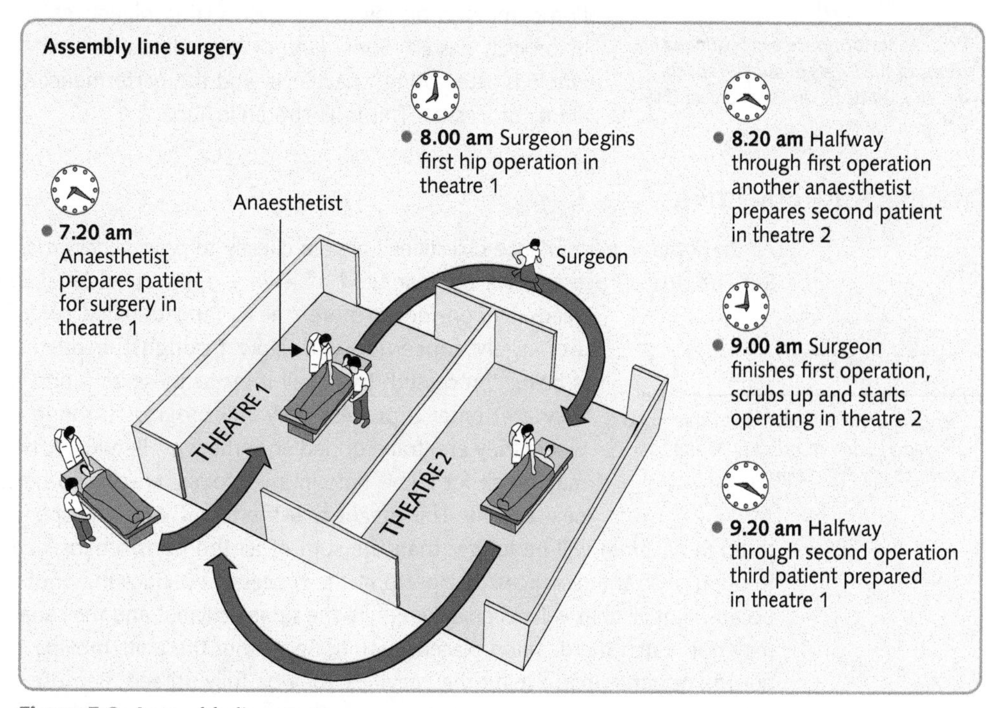

Figure 5.2 Assembly line surgery

design of the product or service that it produces. The 'combo meal' is designed with the constraints and capabilities of the drive-through process in mind, and even the design of the surgical procedure may need to be adapted marginally to facilitate the new process. An important point here is that, in both cases, processes are designed to be appropriate for the market they are serving. Different market strategies may require different process designs. So a good starting point for any operation is to understand the direct relationship between strategic and process performance objectives. But, notwithstanding this relatively strategic starting point for process design, both examples illustrate the importance of not being afraid to analyse processes at a very detailed level. This may include thoroughly understanding current processes so that any improvement can be based on the reality of what happens in practice. It will certainly involve allocating the tasks and associated capacity very carefully to appropriate parts of the process. And, for most processes, it will also involve a design that is capable of taking into consideration the variability that exists in most human tasks. These are the topics covered in this chapter.

OPERATIONS PRINCIPLE

Processes should always be designed to reflect customer and/or market requirements.

DIAGNOSTIC QUESTION

Are process performance objectives understood?

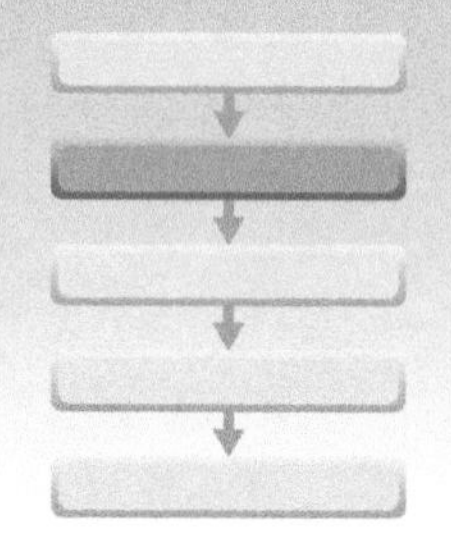

Video

The whole point of process design is to make sure that the performance of the process is appropriate for whatever it is trying to achieve. For example, if an operation competed primarily on its ability to respond quickly to customer requests, its processes would need to be designed to give fast throughput times. This would minimise the time between customers requesting a product or service and their receiving it. Similarly, if an operation competed on low price, cost-related objectives are likely to dominate its process design. Some kind of logic should link what the operation as a whole is attempting to achieve, and the performance objectives of its individual processes. This is illustrated in Table 5.1.

OPERATIONS PRINCIPLE

Process performance can be judged in terms of the levels of quality, speed, dependability, flexibility and cost they achieve.

Process flow objectives

All the strategic performance objectives translate directly to process design as shown in Table 5.1. But, because processes will be managed at a very operational level, process design also needs to consider a more 'micro' and detailed set of objectives. These are largely concerned with flow through the process. When whatever is being 'processed' (we shall refer to these as 'units' irrespective of what they are) enter a process they will progress through a series of activities where they are 'transformed' in some way. Between these activities the units may dwell for some time in inventories, waiting to be transformed by the next activity. This means that the time that a unit spends in the process (its throughput time) will be longer than the sum of all the transforming activities that it passes through. Also the resources that perform the process's activities may not be used all the time because not all units will necessarily require the same activities and the capacity of each resource may not match the demand placed upon it. So neither the units moving through the process, nor the resources performing the activities, may be fully utilised. Because of this, the way that units leave the process is unlikely to be exactly the same as the way they arrive at the process.

OPERATIONS PRINCIPLE

Process flow objectives should include throughput rate, throughput time, work-in-progress and resource utilisation; all of which are interrelated.

Table 5.1 The impact of strategic performance objectives on process design objectives and performance

Strategic performance objective	*Typical process design objectives*	*Some benefits of good process design*
Quality	• Provide appropriate resources, capable of achieving the specification of product of services • Error-free processing	• Products and service produced 'on-specification' • Less recycling and wasted effort within the process
Speed	• Minimum throughput time • Output rate appropriate for demand	• Short customer waiting time • Low in-process inventory
Dependability	• Provide dependable process resources • Reliable process output timing and volume	• On-time deliveries of products and services • Less disruption, confusion and rescheduling within the process
Flexibility	• Provide resources with an appropriate range of capabilities • Change easily between processing states (what, how, or how much is being processed?)	• Ability to process a wide range of products and services • Low cost/fast product and service change • Low cost/fast volume and timing changes • Ability to cope with unexpected events (e.g. supply or a processing failure)
Cost	• Appropriate capacity to meet demand • Eliminate process waste in terms of: – excess capacity – excess process capability – in-process delays – in-process errors – inappropriate process inputs	• Low processing costs • Low resource costs (capital costs) • Low delay/inventory costs (working capital costs)

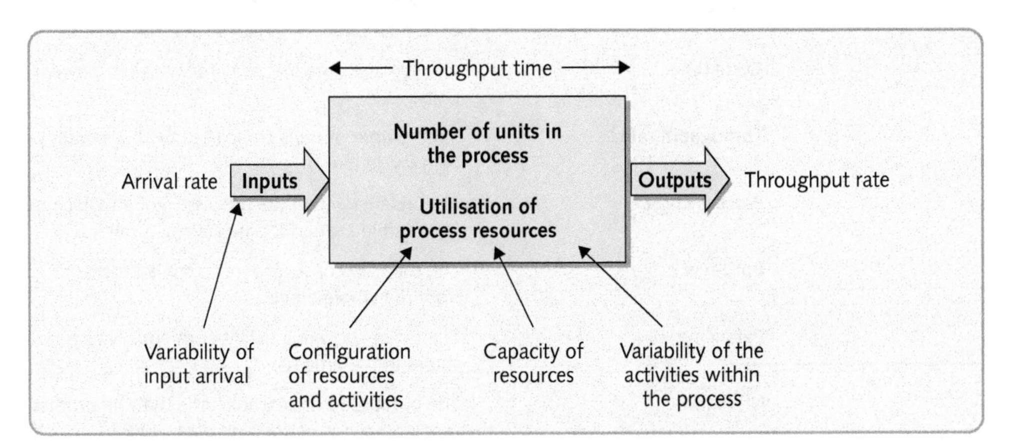

Figure 5.3 'Micro' process performance objectives and process design factors

Figure 5.3 illustrates some of the 'micro' performance flow objectives that describe process flow performance and the process design factors that influence them. The flow objectives are:

- Throughput rate (or flow rate) is the rate at which units emerge from the process, i.e. the number of units passing through the process per unit of time.

- Throughput time is the average elapsed time taken for inputs to move through the process and become outputs.
- The number of units in the process (also called the 'work in process', or 'in-process inventory'), as an average over a period of time.
- The utilisation of process resources is the proportion of available time that the resources within the process are performing useful work.

The design factors that will influence the flow objectives are:

- the variability of input arrival to the process
- the configuration of the resources and activities within the process
- the capacity of the resources at each point in the process
- the variability of the activities within the process.

As we examine each of these design factors, we will be using a number of terms that, although commonly used within process design, need some explanation. These terms will be described in the course of the chapter, but for reference, Table 5.2 summarises them.

Table 5.2 Some common process design terms[3]

Term	*Definition*
Process task	The sum of all the activities that must be performed by the process.
Work content of the process task	The total amount of work within the process task measured in time units.
Activity	A discrete amount of work within the overall process task.
Work content of an activity	The amount of work within an activity measured in time units.
Precedence relationship	The relationship between activities expressed in terms of their dependencies, i.e. whether individual activities must be performed before other activities can be started.
Cycle time	The average time that the process takes between completions of units.
Throughput rate	The number of units completed by the process per unit of time (= 1/cycle time).
Process stage	A work area within the process through which units flow; it may be responsible for performing several activities.
Bottleneck	The capacity constraining stage in a process; it governs the output from the whole process.
Balancing	The act of allocating activities as equally as possible between stages in the process.
Utilisation	The proportion of available time that the process, or part of the process, spends performing useful work.
Starving	Under-utilisation of a stage within a process caused by inadequate supply from the previous stage.
Blocking	The inability of a stage in the process to work because the inventory prior to the subsequent stage is full.
Throughput time	The elapsed time between a unit entering the process and it leaving the process.
Queue time	The time a unit spends waiting to be processed.

DIAGNOSTIC QUESTION

How are processes currently designed?

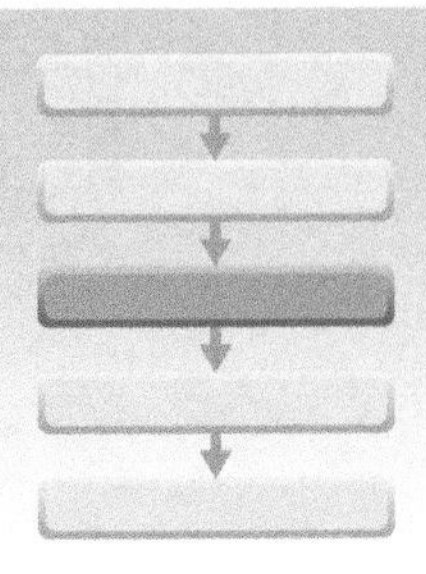

Video

Existing processes are not always sufficiently well defined or described. Sometimes this is because they have developed over time without ever being formally recorded, or they may have been changed (perhaps improved) informally by the individuals who work in the process. But processes that are not formally defined can be interpreted in different ways leading to confusion and inhibiting improvement. So, it is important to have some recorded visual descriptor of a process that can be agreed by all those who are involved in it. This is where process mapping comes in.

OPERATIONS PRINCIPLE

Process mapping is needed to expose the reality of process behaviour.

Process mapping

Animated diagram

Process mapping (or 'process blueprinting' as it is sometimes called) at its most basic level involves describing processes in terms of how the activities within the process relate to each other. There are many, broadly similar, techniques that can be used for process mapping. However, all the techniques have two main features:

- they identify the different types of activity that take place during the process
- they show the flow of materials or people or information through the process (or, put another way, the sequence of activities that materials, people or information are subjected to).

Different process mapping symbols are sometimes used to represent different types of activity. They can be arranged in order, and in series or in parallel, to describe any process. And although there is no universal set of symbols used all over the world, some are relatively common. Most derive either from the early days of 'scientific' management around a century ago or, more recently, from information system flowcharting. Figure 5.4 shows some of these symbols.

EXAMPLE

Theatre lighting operation

Figure 5.5 shows one of the processes used in a theatre lighting operation. The company hires out lighting and stage effects equipment to theatrical companies and event organisers. Customers' calls are routed to the store technician. After discussing their requirements, the technician checks the equipment availability file to see if the equipment can be supplied from the company's own stock on the required dates. If the equipment cannot be supplied, in-house customers may be asked whether they want the company to try and obtain it from other possible suppliers. This offer depends on how busy and how helpful individual technicians are. Sometimes customers decline the offer and a 'Guide to Customers' leaflet is sent to the customer. If the customer does want a search, the technician will call potential suppliers in an attempt to find available equipment. If this is not successful the customer is informed, but if suitable equipment is located it is reserved for delivery to the company's site. If equipment can be supplied from the company's own stores, it is reserved on the

Practice note

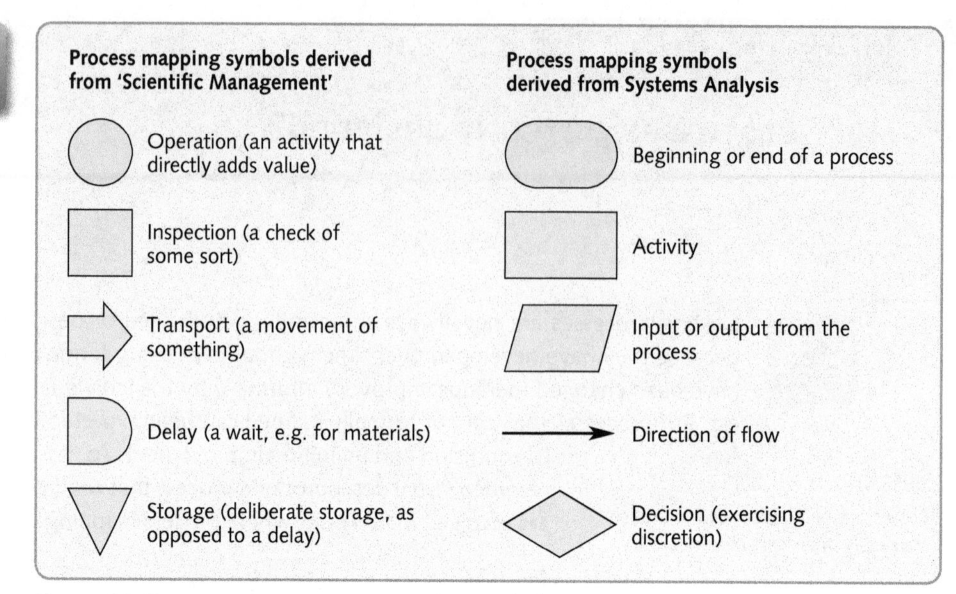

Figure 5.4 Some common process mapping symbols

equipment availability file and the day before it is required a 'kit wagon' is taken to the store where all the required equipment is assembled, taken back to the workshop checked, and if any equipment is faulty it is repaired at this point. After that it is packed in special cases and delivered to the customer.

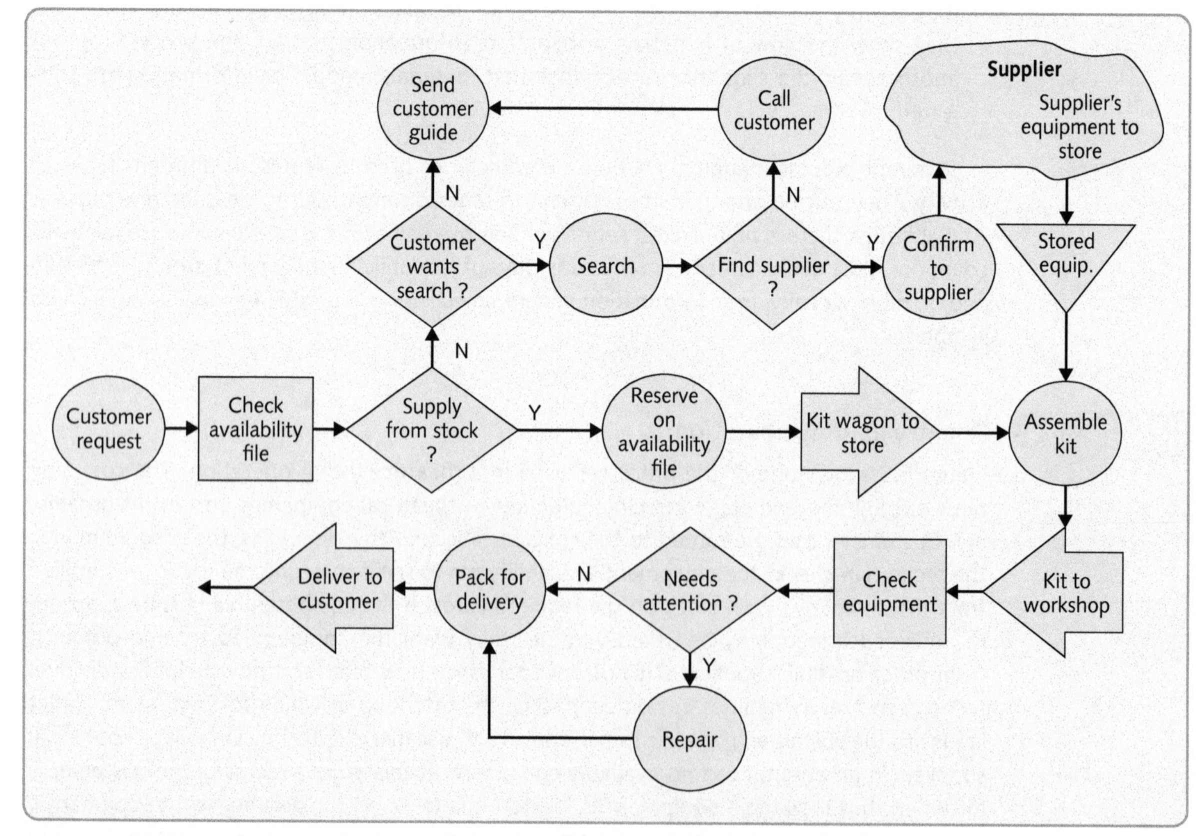

Figure 5.5 Process map for 'enquire to delivery' process at stage lighting operation

Different levels of process mapping

For a large process, drawing process maps at this level of detail can be complex. This is why processes are often mapped at a more aggregated level, called high level process mapping, before more detailed maps are drawn. Figure 5.5 illustrates this for the total *'supply and install lighting'* process in the stage lighting operation. At the highest level the process can be drawn simply as an input–transformation–output process with materials and customers as its input resources and lighting services as outputs. No details of how inputs are transformed into outputs are included. At a slightly lower or more detailed level, what is sometimes called an outline process map (or chart) identifies the sequence of activities but only in a general way. So the process of *'enquire to delivery'* that is shown in detail in Figure 5.5 is here reduced to a single activity. At the more detailed level, all the activities are shown in a 'detailed process map' (the activities within the process 'install and test' are shown).

Although not shown in Figure 5.6, an even more micro set of process activities could be mapped within each of the detailed process activities. Such a micro detailed process map could specify every single motion involved in each activity. Some quick service restaurants, for example, do exactly that. In the lighting hire company example most activities would not be mapped in any more detail than that shown in Figure 5.6. Some activities, such as 'return to base', are probably too straightforward to be worth mapping any further. Other activities, such as 'rectify faulty equipment', may rely on the

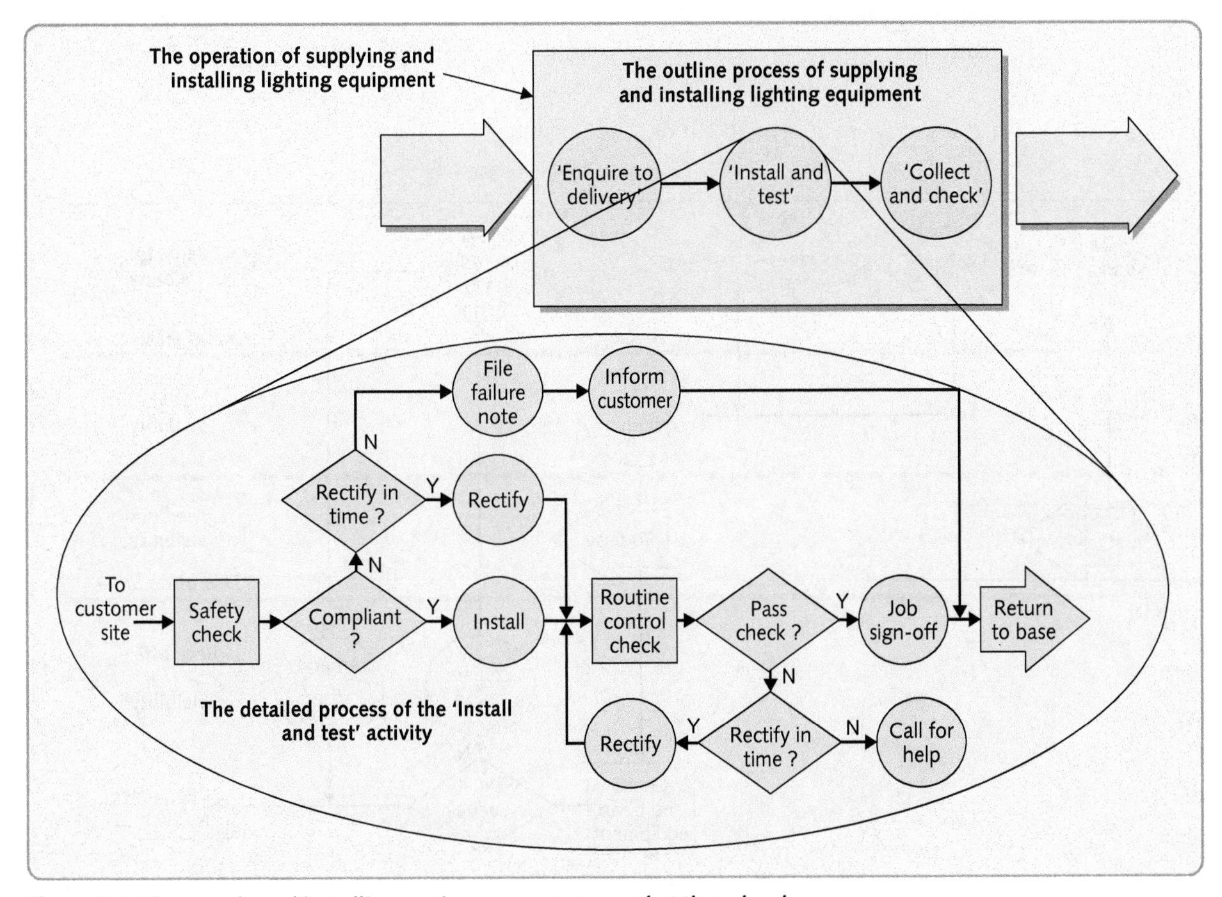

Figure 5.6 The 'supply and install' operations process mapped at three levels

technician's skills and discretion to the extent that the activity has too much variation and is too complex to map in detail. Some activities, however, may need mapping in more detail to ensure quality or to protect the company's interests. For example, the activity of safety checking the customer's site to ensure that it is compliant with safety regulations will need specifying in some detail to ensure that the company can prove it exercised its legal responsibilities.

Process visibility

It is sometimes useful to map such processes in a way that makes the degree of visibility of each part of the process obvious.[4] This allows those parts of the process with high visibility to be designed so that they enhance the customer's perception of the process. Figure 5.7 shows yet another part of the lighting equipment company's operation: 'the collect and check' process. The process is mapped to show the visibility of each activity to the customer. Here four levels of visibility are used. There is no hard and fast rule about this; many processes simply distinguish between those activities that the customer *could* see and those that they couldn't. The boundary between these two categories is often called the 'line of visibility'. In Figure 5.7, three categories of visibility are shown. At the very highest level of visibility, above the 'line of interaction', are those activities that involve direct interaction between the lighting company's staff and the customer. Other activities take place at the customer's site or in the presence of the customer but involve less or no direct interaction. Yet further activities (the two transport activities in this case) have some degree of visibility because they take place away from the company's base and are visible to potential customers, but are not visible to the immediate customer.

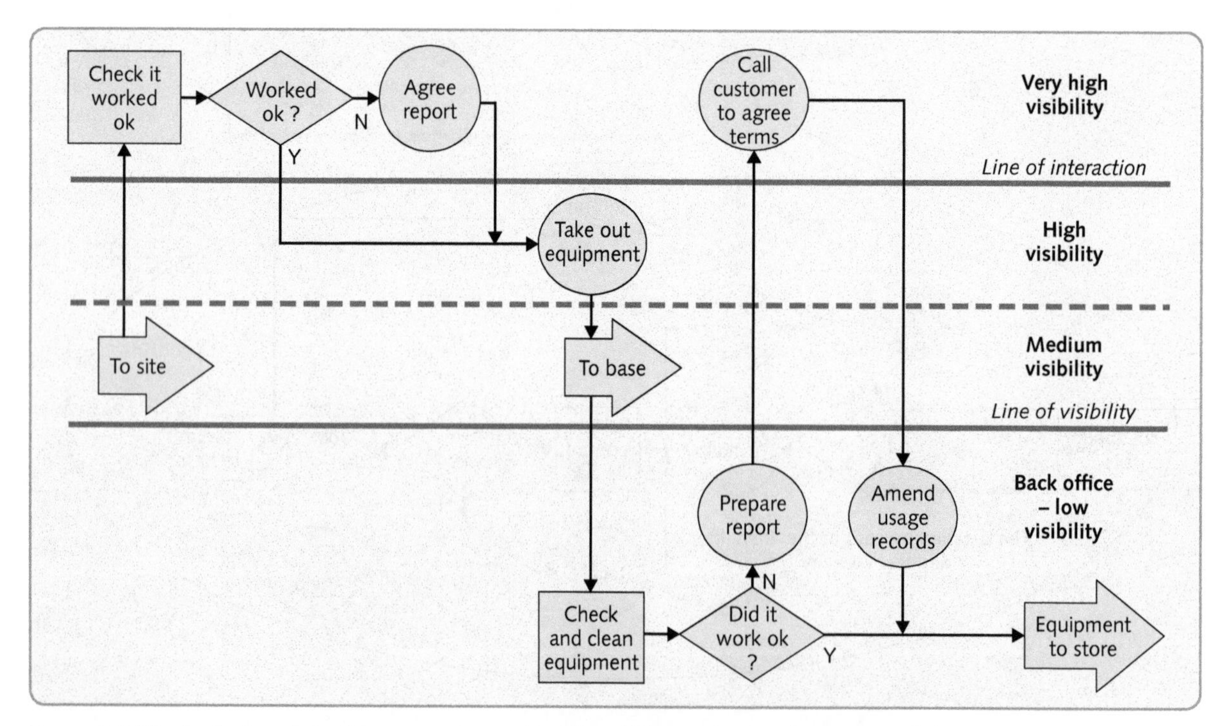

Figure 5.7 The 'collect and check' process mapped to show different levels of process visibility

DIAGNOSTIC QUESTION

Are process tasks and capacity configured appropriately?

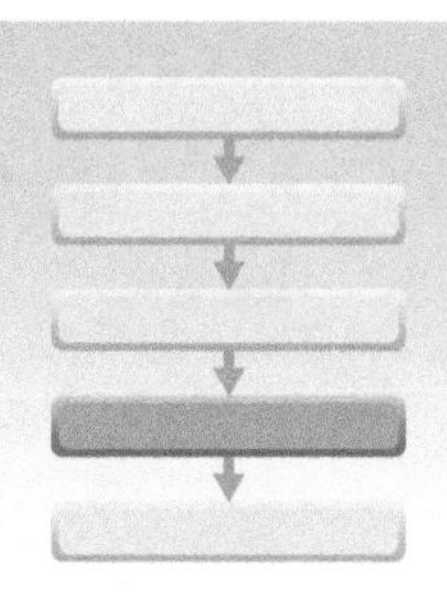

Video

Process maps show how the activities of any particular process are currently arranged and help to suggest how they can be reconfigured. But there are also some general issues that must be understood before processes can be analysed. These relate to how the total task can be divided up within the process and determine how capacity is allocated. This, in turn, determines the flow through the process.

Getting to grips with process capacity means understanding the following issues.

- task precedence
- series and parallel configurations
- cycle time and process flow
- process balancing
- throughput, cycle time and work in process.

Task precedence

OPERATIONS PRINCIPLE

Process design must respect task precedence.

Any process redesign needs to preserve the inherent precedence of activities within the overall task. Task 'precedence' defines what activities must occur before others, because of the nature of the task itself. At its simplest level task precedence is defined by:

- the individual activities that comprise the total process task
- the relationship between these individual activities.

Task precedence is usually described by using a 'precedence diagram', which, in addition to the above, also includes the following information:

- the time necessary to perform the total task (sometimes known as the 'work content' of the task)
- the time necessary to perform each of the individual activities within the task.

EXAMPLE

Computer repair service centre

A repair service centre receives faulty or damaged computers sent in by customers, repairs them and dispatches them back to the customer. Each computer is subject to the same set of tests and repair activities, and although the time taken to repair each computer will depend on the results of the tests, there is relatively little variation between individual computers.

Table 5.3 defines the process task of testing and repairing the computers in terms of the seven activities that comprise the total task, the relationship between the activities in terms of each activity's 'immediate predecessor', and the time necessary to perform each activity. Figure 5.8 shows the relationship between the activities graphically. This kind of illustration is called the 'precedence diagram' for the process task. It is useful because it indicates how activities *cannot* be sequenced in the eventual process design. For example, the process cannot perform activity 'b' before activity 'a' is completed. However, it does not determine how a process *can* be designed. Yet once the task has been analysed in this way, activities can be arranged to form the process's general configuration.

Table 5.3 Process task details for the 'computer test and repair' task

Activity code	*Activity name*	*Immediate predecessor*	*Activity time (minutes)*
a	Preliminary test 1	–	5
b	Preliminary test 2	a	6
c	Dismantle	b	4
d	Test and repair 1	c	8
e	Test and repair 2	c	6
f	Test and repair 3	c	4
g	Clean/replace casing elements	d,e,f	10

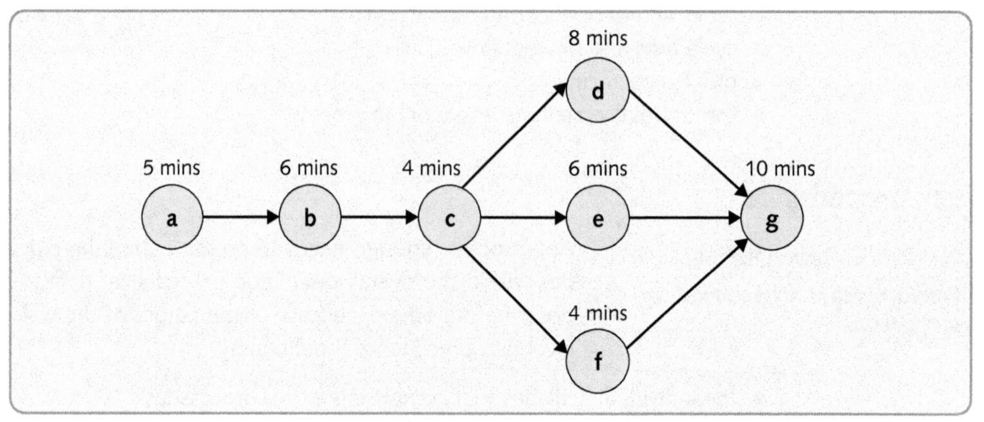

Figure 5.8 Precedence diagram showing the relationship between activities for the 'computer test and repair' task

Series and parallel configurations

Animated diagram

At its simplest level, the general configuration of a process involves deciding the extent to which activities are arranged sequentially and the extent to which they are arranged in parallel.

For example, the task illustrated in Figure 5.8 involves seven activities that in total take 43 minutes. Demand is such that the process must be able to complete the test and repair task at the rate of one every 12 minutes in order to meet demand. One possible process design is to arrange the seven activities in a series arrangement of stages. The first question to address is, how many stages would this type of series arrangement require? This can be calculated by dividing the total work content of the task by the required cycle time. In this case, number of stages = 43 minutes/12 minutes = 3.58 stages.

Given the practical difficulties of having a fraction of a stage, this effectively means that the process needs four stages. The next issue is to allocate activities to each stage. Because the output from the whole process will be limited by the stage with most work (the sum of its allocated activities), each stage can have activities allocated to it up to a maximum allocation of twelve minutes. Figure 5.9 illustrates how this could be achieved. The longest stage

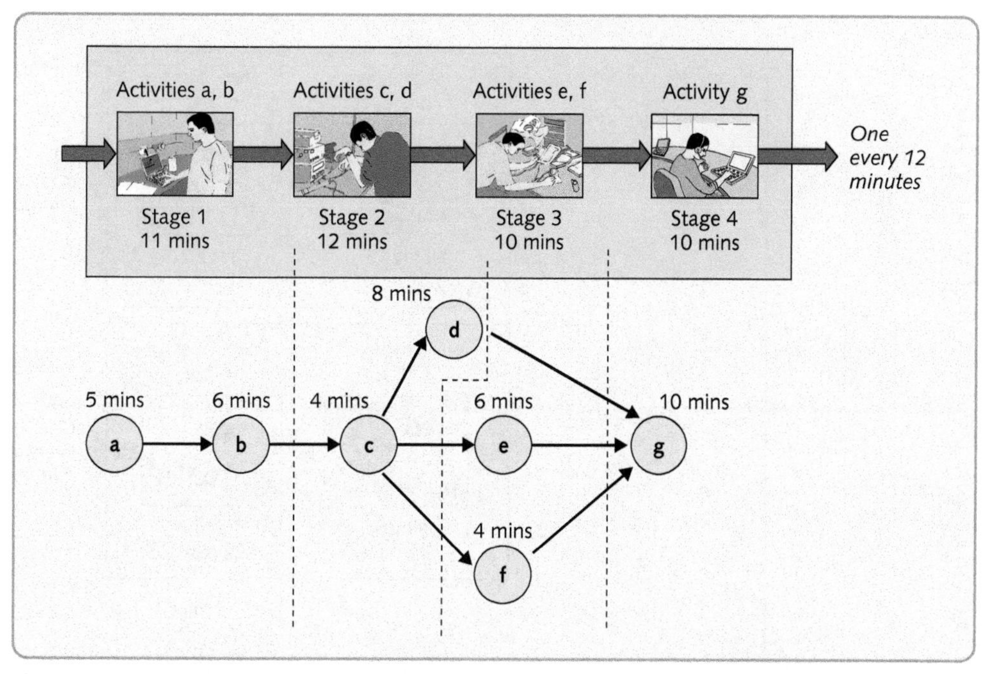

Figure 5.9 'Long-thin' arrangement of stages for the 'computer test and repair' task

(stage 2 in this case) will limit the output of the total process to one computer every twelve minutes and the other stages will be relatively under loaded.

However, there are other ways of allocating tasks to each stage, and involving the parallel arrangement of activities, that could achieve a similar output rate. For example, the four stages could be arranged as two parallel 'shorter' arrangements with each stage performing approximately half of the activities in the total tasks. This is illustrated in Figure 5.10. It involves two two-stage arrangements, with stage 1 being allocated four activities that amount to 21 minutes of work and the second stage being allocated three activities that amount to 22 minutes of work. So, each arrangement will produce one repaired computer every 22 minutes (governed by the stage with the most work). This means that the two arrangements together will produce two repaired computers every 22 minutes, an average of one repaired computer every eleven minutes.

Loading each stage with more work and arranging the stages in parallel can be taken further. Figure 5.11 illustrates an arrangement where the whole test and repair task is performed at individual stages, all of which are arranged in parallel. Here, each stage will produce one repaired computer every 43 minutes and so together will produce four repaired computers every 43 minutes, an average output rate of one repaired computer every 10.75 minutes.

This simple example represents an important process design issue. Should activities in a process be arranged predominately in a single series 'long-thin' configuration, or, predominately in several 'short-fat' parallel configurations, or somewhere in between? (Note that 'long' means the number of stages and 'fat' means the amount of work allocated to each stage.) Most processes will adopt a combination of series and parallel configurations, and in any particular situation there are usually technical constraints which limit either how 'long and thin' or how 'short and fat' the process can be. But there is usually a real choice to be made with a range of possible options. The advantages of each extreme of the long-thin to short-fat spectrum are very different and help to explain why different arrangements are adopted.

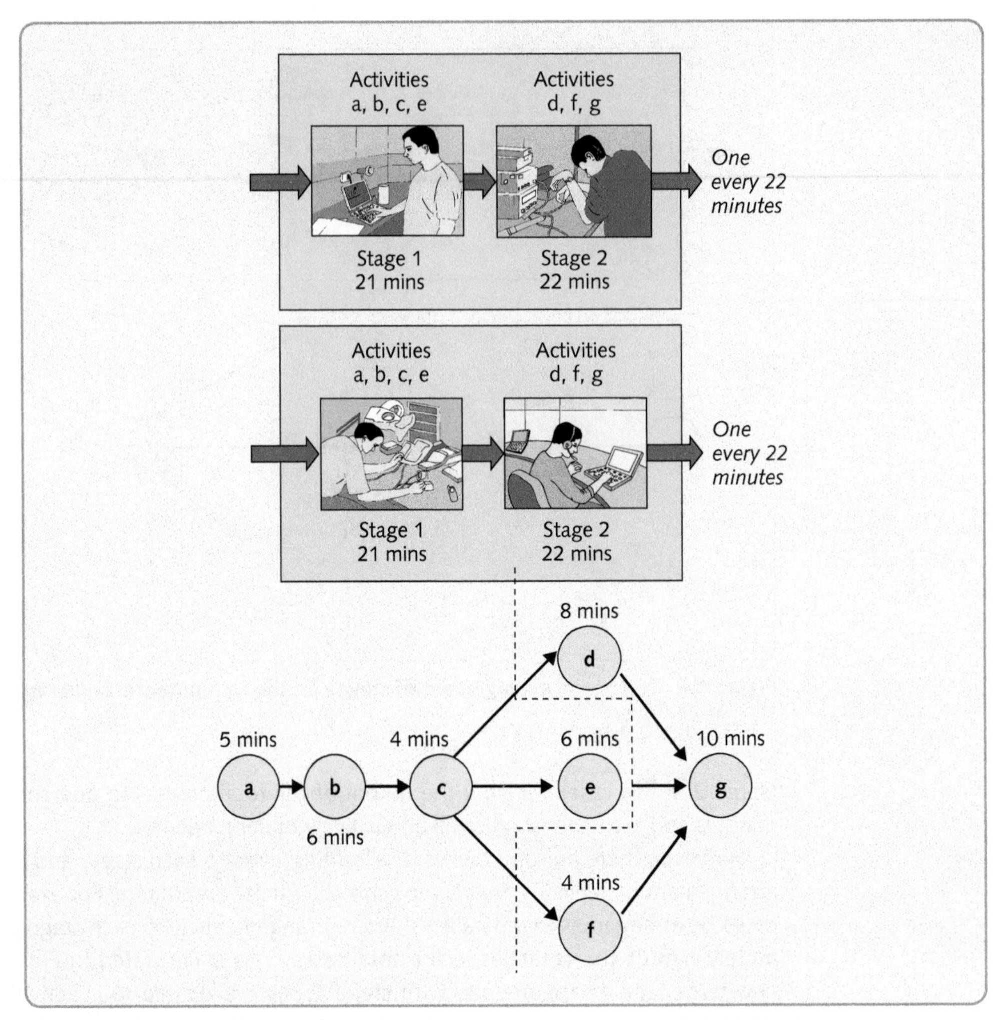

Figure 5.10 Intermediate configurations for the 'computer test and repair' task

The advantages of the series dominated (long-thin) configuration include:

- *A more controlled flow* through the process that is relatively easy to manage.
- *Simple materials handling* – especially if a product being manufactured is heavy, large or difficult to move.
- *Lower capital requirements*. If a specialist piece of equipment is needed for one element in the job, only one piece of equipment would need to be purchased; on short-fat arrangements every stage would need one.
- *More efficient operation*. If each stage is only performing a small part of the total job, the person at the stage may have a higher proportion of direct productive work as opposed to the non-productive parts of the job, such as picking up tools and materials.

The advantages of the parallel dominated (short-fat) configuration include:

- *Higher mix flexibility*. If the process needs to produce several types of product or service, each stage could specialise in different types.
- *Higher volume flexibility*. As volume varies, stages can simply be closed down or started up as required.

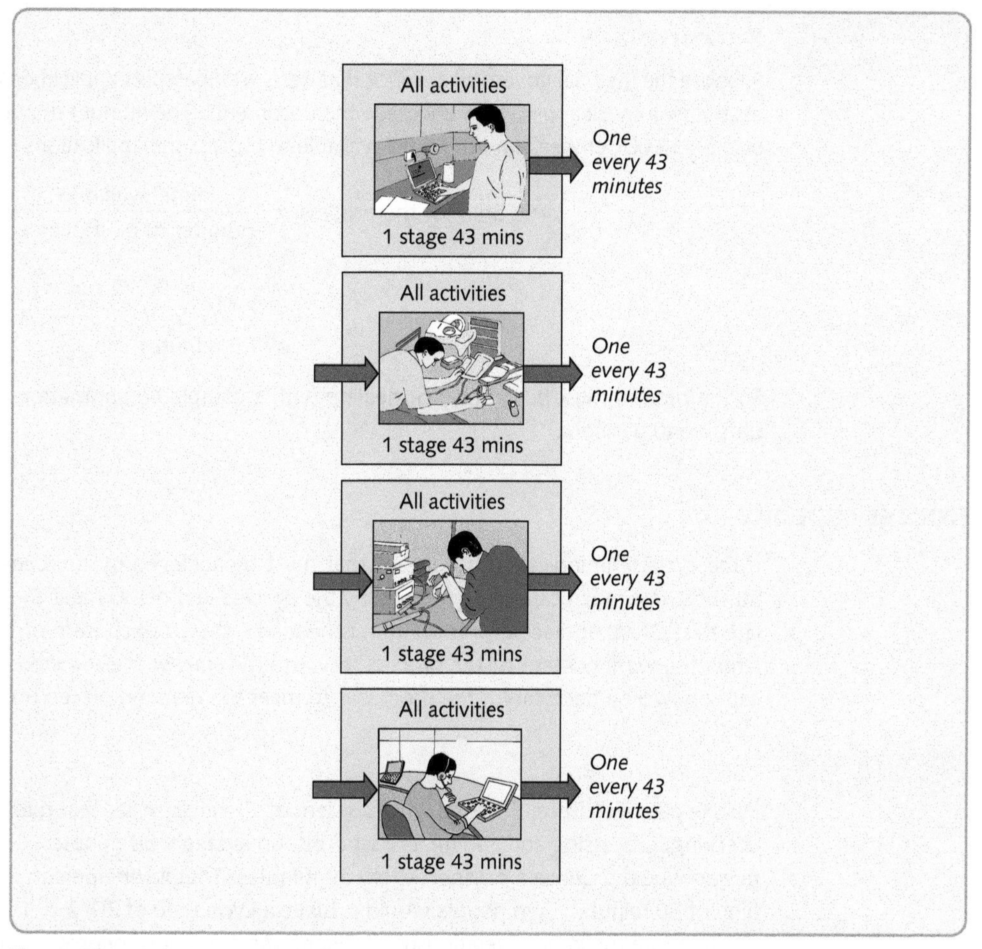

Figure 5.11 The 'short-fat' arrangement of stages for the 'computer test and repair' task

- *Higher robustness*. If one stage breaks down or ceases operation in some way, the other parallel stages are unaffected; a long-thin arrangement would cease operating completely.
- *Less monotonous work*. In the computer repair example, the staff in the short-fat arrangement are repeating their tasks only every 43 minutes; in the long-thin arrangement it is every 12 minutes.

Cycle time and process capacity

The cycle time of a process is the time between completed units emerging from it. Cycle time is a vital factor in process design and has a significant influence on most of the other detailed design decisions. It is usually one of the first things to be calculated because it can be used both to represent the demand placed on a process and the process's capacity. The cycle time also sets the pace or 'drum beat' of the process. However the process is designed it must be able to meet its required cycle time. It is calculated by considering the likely demand for the products or services over a period and the amount of production time available in that period.

OPERATIONS PRINCIPLE

Process analysis derives from an understanding of the required process cycle time.

EXAMPLE

Passport office

Suppose the regional government office that deals with passport applications is designing a process that will check applications and issue the documents. The number of applications to be processed is 1,600 per week and the time available to process the applications is 40 hours per week.

$$\text{cycle time for the process} = \frac{\text{time available}}{\text{number to be processed}}$$
$$= \frac{40}{1600} = 0.025 \text{ hours}$$
$$= 1.5 \text{ minutes}$$

So the process must be capable of dealing with a completed application once every 1.5 minutes, or 40 per hour.

Process capacity

If the cycle time indicates the output that must be achieved by a process, the next decision must be how much capacity is needed by the process in order to meet the cycle time. To calculate this, a further piece of information is needed – the work content of the process task. The larger the work content of the process task and the smaller the required cycle time, the more capacity will be necessary if the process is to meet the demand placed upon it.

EXAMPLE

Passport office

For the passport office, the total work content of all the activities that make up the total task of checking, processing and issuing a passport is, on average, 30 minutes. So, a process with one person would produce a passport every 30 minutes. That is, one person would achieve a cycle time of 30 minutes. Two people would achieve a cycle time of 30/2 = 15 minutes, and so on.

Therefore the general relationship between the number of people in the process (its capacity in this simple case) and the cycle time of the process is:

$$\frac{\text{work content}}{N} = \text{cycle time}$$

where N = the number of people in the process.
Therefore, in this case:

$$N = \frac{30}{\text{cycle time}}$$

In this case:

$$N = \frac{30}{1.5} = 20 \text{ people}$$

So, the capacity that this process needs if it is to meet demand is 20 people.

Process balancing

Video

Balancing a process involves attempting to allocate activities to each stage as equally as possible. Because the cycle time of the whole process is limited by the longest allocation of activity times to an individual stage, the more equally work is allocated, the less time will be 'wasted' at the other stages in the process. In practice, it is nearly always impossible to achieve perfect balance so some degree of imbalance in work allocation between stages will occur. The effectiveness of the balancing activity is measured by balancing loss. This is the time wasted

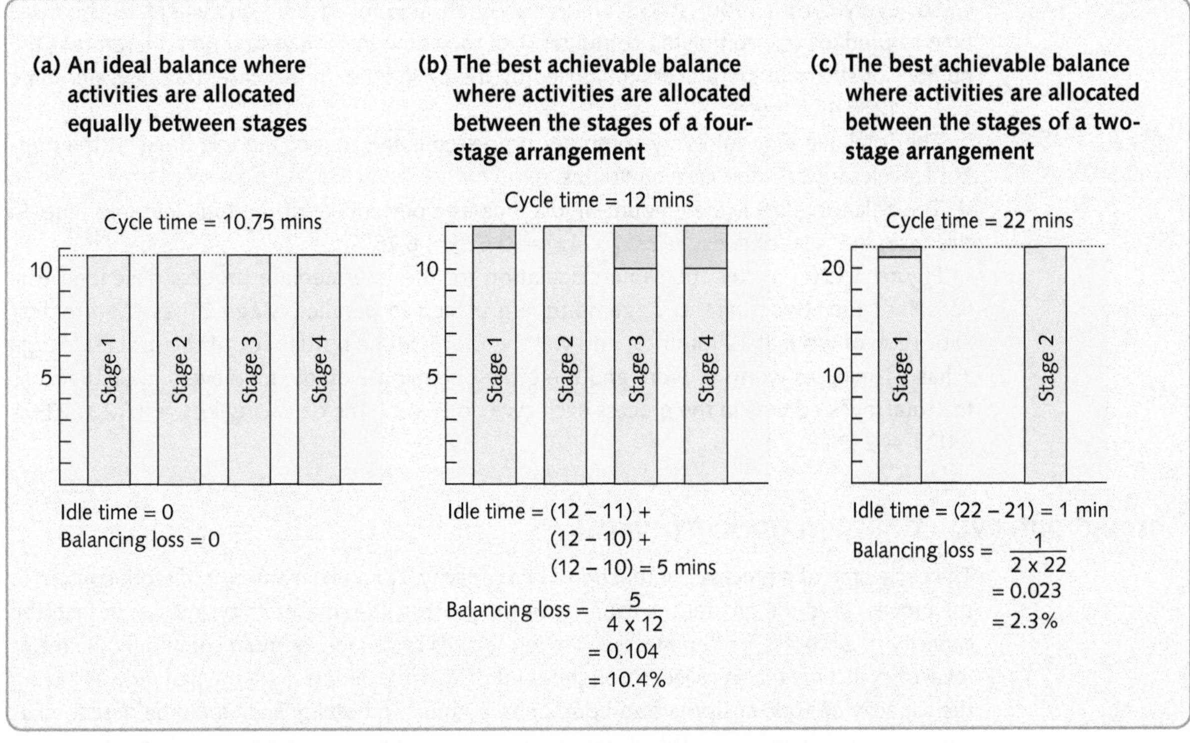

Figure 5.12 Balancing loss is that proportion of the time invested in processing the product or service that is not used productively. For the 'computer test and repair' process, (a) is the theoretical perfect balance, (b) is the best balance for four stages, and (c) is the best balance for two stages

OPERATIONS PRINCIPLE

Allocating work equally to each stage in a process (balancing) smoothes flow and avoids bottlenecks.

through the unequal allocation of activities as a percentage of the total time invested in processing. This is illustrated in Figure 5.12. Here the computer test and repair task is used to illustrate balancing loss for the 'long-thin' arrangement of four sequential stages and the 'intermediate' arrangement of two parallel two-stage arrangements.

Practice note

Figure 5.12(a) shows the ideal allocation of activities with each stage perfectly balanced. Here exactly a quarter of the total work content (10.75 minutes) has been allocated to each of the four stages. Every 10.75 minutes each stage performs its activities and passes a computer on to the next stage, or out of the process in the case of stage 4. No stage suffers any idle time and, because the stages are perfectly balanced, balancing loss = 0. In fact, because of the actual times of each activity, it is not possible to equally allocate work to each stage. Figure 5.12(b) shows the best allocation of activities. Most work is allocated to stage 2, so that stage will dictate the cycle time of the whole process. Stage 1 has only 11 minutes of work and so will be idle for (12 – 11) = 1 minute every cycle (or alternatively will keep processing one computer every 11 minutes and the build-up of inventory between stage 1 and stage 2 would grow to infinity). Similarly, stages 3 and 4 have idle time, in this case both have (12 – 10) = 2 minutes idle time. They can only process one computer every 12 minutes because stage 2 will only pass forward a computer to them every 12 minutes. So, they are being starved of work for 2 minutes every 12 minutes. In practice, stages that are not the bottlenecks stage may not actually be idle for a period of time every cycle. Rather they will slow down the pace of work to match the time of the bottleneck stage. Nevertheless, this still is effective idle time because under conditions of perfect balance they could be performing useful work.

So, every cycle all four stages are investing an amount of time equivalent to the cycle time to produce one completed computer. The total amount of invested time therefore is the number of stages in the process multiplied by the cycle time. In this case, total invested time $= 4 \times 12 = 48$ minutes.

The total idle time for every computer processed is the sum of the idle times at the non-bottleneck stages, in this case 5 minutes.

The balancing loss is the amount of idle time as a percentage of the total invested time. In this case, the balancing loss $= 3/(4 \times 12) = 0.625 = 6.25\%$.

Figure 5.12(c) makes the same calculation for the intermediate process described earlier. Here too, two-stage arrangements are placed in parallel. Stage 2 has the greatest allocation of work at 22 minutes, and will therefore be the bottleneck of the process. Stage 1 has 21 minutes worth of work and therefore one minute of idle time every cycle. Because the total invested time in the process each cycle $= 2 \times 22$, the balancing loss $= 1/(2 \times 22) = 0.023 = 2.3\%$.

Throughput, cycle time and work in process

The cycle time of a process is a function of its capacity. For a given amount of work content in the process task, the greater the capacity of the process, the smaller its cycle time. In fact, the capacity of a process is often measured in terms of its cycle time, or more commonly the reciprocal of cycle time that is called 'throughput rate'. So, for example, a theme park ride as having the capacity of 1000 customers an hour, or an automated bottling line would be described as having a capacity of 100 bottles a minute, and so on. However, a high level of capacity (short cycle time and fast throughput rate) does not necessarily mean that material, information or customers can move quickly through the process. This will depend on how many other units are contained within the process. If there is a large number of units within the process they may have to wait in 'work in process' inventories for part of the time they are within the process (throughput time).

Little's law

The mathematical relationship that relates cycle time to work in process and throughput time is called Little's law.[5] It is simple, but very useful, and it works for any stable process. Little's law can be stated as:

$$\text{throughput time} = \text{work in process} \times \text{cycle time}$$

Or:

$$\text{work in process} = \text{throughput time} \times (1/\text{cycle time})$$

That is:

$$\text{work in process} = \text{throughput time} \times \text{throughput rate}$$

For example, in the case of the computer test and repair process with four stages:

cycle time = 12 minutes (loading on the bottleneck station)

work in process = 4 units (one at each stage of the process assuming there is no space for inventory to build up between stages)

Therefore:

$$\text{throughput time} = \text{work in process} \times \text{cycle time}$$
$$= 12 \times 4 = 48 \text{ minutes}$$

OPERATIONS PRINCIPLE

Little's law states that throughput time = work in process × cycle time.

Similarly, for the example of the passport office, suppose the office has a 'clear desk' policy that means that all desks must be clear of work by the end of the day. How many applications should be loaded onto the process in the morning in order to ensure that every one is completed and desks are clear by the end of the day?

From before:

cycle time = 1.5 minutes, and assuming a 7.5-hour (450-minute) working day

From Little's law:

$$\text{throughput time} = \text{work in process} \times \text{cycle time}$$

$$450 \text{ minutes} = \text{work in process} \times 1.5$$

Therefore:

$$\text{work in process} = 450/1.5 = 300$$

So, 300 applications can be loaded onto the process in the morning and be cleared by the end of the working day.

EXAMPLE

Little's law at a seminar

Mike was totally confident in his judgement: *'You'll never get them back in time,'* he said. *'They aren't just wasting time, the process won't allow them to all have their coffee and get back for 11 o'clock.'* Looking outside the lecture theatre, Mike and his colleague Dick were watching the 20 business men who were attending the seminar queuing to be served coffee and biscuits. The time was 10.45 and Dick knew that unless they were all back in the lecture theatre at 11 o'clock there was no hope of finishing his presentation before lunch. *'I'm not sure why you're so pessimistic,'* said Dick. *'They seem to be interested in what I have to say and I think they will want to get back to hear how operations management will change their lives.'* Mike shook his head. *'I'm not questioning their motivation,'* he said. *'I'm questioning the ability of the process out there to get through them all in time. I have been timing how long it takes to serve the coffee and biscuits. Each coffee is being made fresh and the time between the server asking each customer what they want and them walking away with their coffee and biscuits is taking 48 seconds. Remember that, according to Little's law, throughput equals work in process multiplied by cycle time. If the work in process is the 20 managers in the queue and cycle time is 48 seconds, the total throughput time is going to 20 multiplied by 0.8 minutes which equals 16 minutes. Add to that sufficient time for the last person to drink their coffee and you must expect a total throughput time of a bit over 20 minutes. You just haven't allowed long enough for the process.'* Dick was impressed. *'Err . . . what did you say that law was called again?'* *'Little's law,'* said Mike.

EXAMPLE

Little's law at an IT support unit

Every year it was the same. All the workstations in the building had to be renovated (tested, new software installed, etc.) and there was only one week in which to do it. The one week fell in the middle of the August vacation period when the renovation process would cause minimum disruption to normal working. Last year the company's 500 workstations had all been renovated within one working week (40 hours). Each renovation last year took on average 2 hours and 25 technicians had completed the process within the week. This year there would be 530 workstations to renovate but the company's IT support unit had devised a faster testing and renovation routine that would only take on average 1½ hours instead of 2 hours. How many technicians will be needed this year to complete the renovation processes within the week?

Last year:

$$\begin{aligned}\text{work in progress (WIP)} &= 500 \text{ workstations}\\ \text{time available}(T_t) &= 40 \text{ hours}\\ \text{average time to renovate} &= 2 \text{ hours}\\ \text{therefore throughput rate}(T_r) &= \tfrac{1}{2} \text{ hour per technician}\\ &= 0.5\,N\\ \text{where } N &= \text{number of technicians}\end{aligned}$$

Little's law:

$$\begin{aligned}\text{WIP} &= T_t \times T_r\\ 500 &= 40 \times 0.5\,N\\ N &= \frac{500}{40 \times 0.5} = 25 \text{ technicians}\end{aligned}$$

This year:

$$\begin{aligned}\text{work in progress (WIP)} &= 530 \text{ workstations}\\ \text{time available} &= 40 \text{ hours}\\ \text{average time to renovate} &= 1.5 \text{ hours}\\ \text{throughput rate}(T_r) &= 1/1.5 \text{ per technician}\\ &= 0.67\,N\\ \text{where } N &= \text{number of technicians}\end{aligned}$$

Little's law:

$$\begin{aligned}\text{WIP} &= T_t \times T_r\\ 530 &= 40 \times 0.67\,N\\ N &= \frac{530}{40 \times 0.67}\\ &= 19.88 \text{ technicians}\end{aligned}$$

DIAGNOSTIC QUESTION

Is process variability recognised?

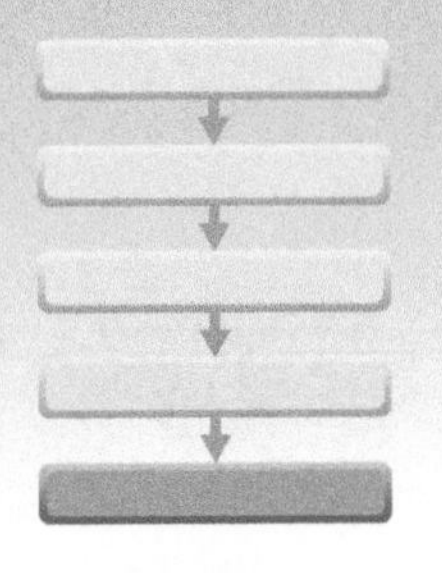

Video

So far in our treatment of process analysis we have assumed that there is no significant variability either in the demand to which the process is expected to respond, or in the time taken for the process to perform its various activities. Clearly, this is not the case in reality. So, it is important to look at the variability that can affect processes and take account of it. However, do not dismiss the deterministic analysis we have been examining up to this point. At worst it provides a good first approximation to analysing processes, while at best, the relationships that we have discussed do hold for average performance values.

Sources of variability in processes

There are many reasons why variability occurs in processes. A few of these possible sources of variation are listed below:

- The late (or early) arrival of material, information or customers at a stage within the process.
- The temporary malfunction or breakdown of process technology within a stage of the process.
- The necessity for recycling 'misprocessed' materials, information or customers to an earlier stage in the process.
- The misrouting of material, information or customers within the process that then needs to be redirected.
- Each product or service being processed might be different, for example, different models of automobile going down the same line.
- Products or services, although essentially the same, might require slightly different treatment. For instance, in the computer test and repair process, the time of some activities will vary depending on the results of the diagnostic checks.
- With any human activity there are slight variations in the physical coordination and effort on the part of the person performing the task that result in variation in activity times, even of routine activities.

All these sources of variation within a process will interact with each other, but result in two fundamental types of variability:

- Variability in the demand for processing at an individual stage within the process, usually expressed in terms of variation in the inter-arrival times of units to be processed.
- Variation in the time taken to perform the activities (i.e. process a unit) at each stage.

Activity time variability

> **OPERATIONS PRINCIPLE**
> *Variability in a process acts to reduce its efficiency.*

The effects of variability within a process will depend on whether the movements of units between stages, and hence the inter-arrival times of units at stages, are synchronised or not. For example, consider the computer test and repair process described previously. Figure 5.13 shows the average activity time at each stage of the process, but also the variability around the average time. Suppose that it was decided to synchronise the flow between the four stages by using an indexing conveyor or a simple traffic lights system that ensured all movement between

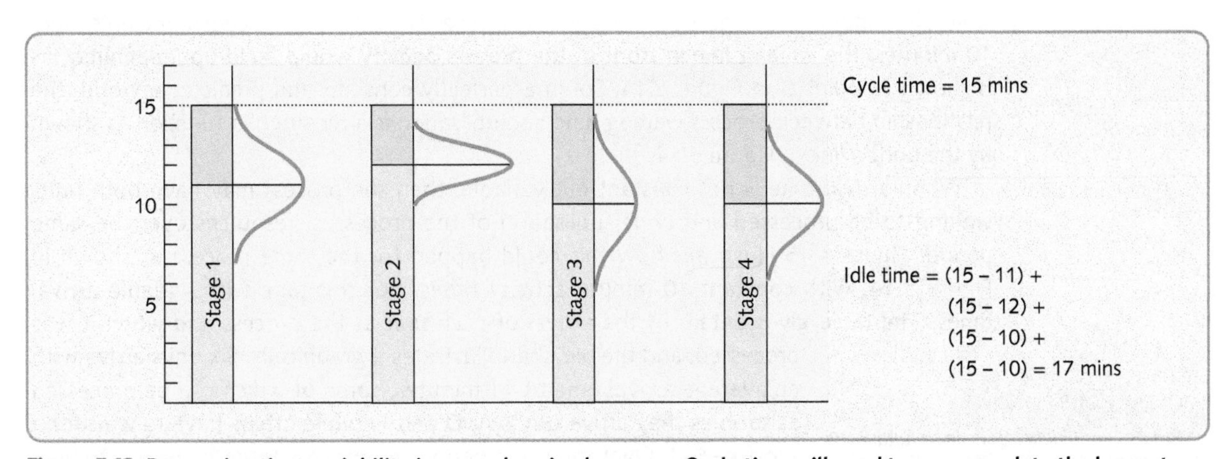

Figure 5.13 Processing time variability in a synchronised process. Cycle time will need to accommodate the longest activity time at any of the stages

the stages happened simultaneously. The interval between each synchronised movement would have to be set at an interval that would allow all stages to have finished their activities irrespective of whether they had experienced a particularly fast or particularly slow activity time. In this case, from Figure 5.13 that synchronised indexing time would have to be set at 15 minutes. This then becomes the effective cycle time of the process. Note that the effective bottleneck stage is now stage 1 rather than stage 2. Although stage 2 has the longer average activity time (12 minutes), activity 1 with an average activity time of 11 minutes has a degree of variability that results in a maximum activity of 15 minutes. Note also that every stage will experience some degree of idle time, the average idle time at each station being the cycle time minus the average activity time at that station. This reduction in the efficiency of the process is only partly a result of its imbalance. The extra lost time is as result of activity time variability.

This type of effect is not at all uncommon. For example, automobiles are assembled using a moving belt assembly line whose speed is set to achieve a cycle time that can accommodate activity time variability. However, a more common arrangement, especially when processing information or customers, is to move units between stages as soon as the activities performed by each stage are complete. Here, units move through the process in an unsynchronised manner rather than having to wait for an imposed movement time. This means that each stage may spend less time waiting to move their unit forward, but it does introduce more variation in the demand placed on subsequent stations. When movement was synchronised, the inter-arrival time of units at each stage was fixed at the cycle time. Without synchronisation, the inter-arrival time at each stage will itself be variable.

Arrival time variability

To understand the effect of arrival variability on process performance it is first useful to examine what happens to process performance in a very simple process as arrival time changes under conditions of no variability. For example, the simple process shown in Figure 5.14 is comprised of one stage that performs exactly 10 minutes of work. Units arrive at the process at a constant and predictable rate. If the arrival rate is one unit every 30 minutes, then the process will be utilised for only 33.33 per cent of the time, and the units will never have to wait to be processed. This is shown as point A on Figure 5.14. If the arrival rate increases to one arrival every 20 minutes, the utilisation increases to 50 per cent, and again the units will not have to wait to be processed. This is point B on Figure 5.14. If the arrival rate increases to one arrival every 10 minutes, the process is now fully utilised, but, because a unit arrives just as the previous one has finished being processed, no unit has to wait. This is point C on Figure 5.14. However, if the arrival rate ever exceeded one unit every 10 minutes, the waiting line in front of the process activity would build up indefinitely, as is shown as point D in Figure 5.14. So, in a perfectly constant and predictable world, the relationship between process waiting time and utilisation is a rectangular function as shown by the dotted line in Figure 5.14.

When arrival time is not constant but variable, then the process may have both units waiting to be processed and under-utilisation of the processes' resources over the same period. Figure 5.15 illustrates how this could happen for the same process as shown in Figure 5.14, with constant 10-minute activity times, but this time with variable arrival times. The table gives details of the arrival of each unit at the process and when it was processed, and the bar chart illustrates it graphically. Six units arrive with an average arrival time of 11 minutes, some of which can be processed as soon as they arrive (units A, D and F) while others have to wait for a short period. Over the same period the process three times has to wait for work.

OPERATIONS PRINCIPLE

Process variability results in simultaneous waiting and resource under utilisation.

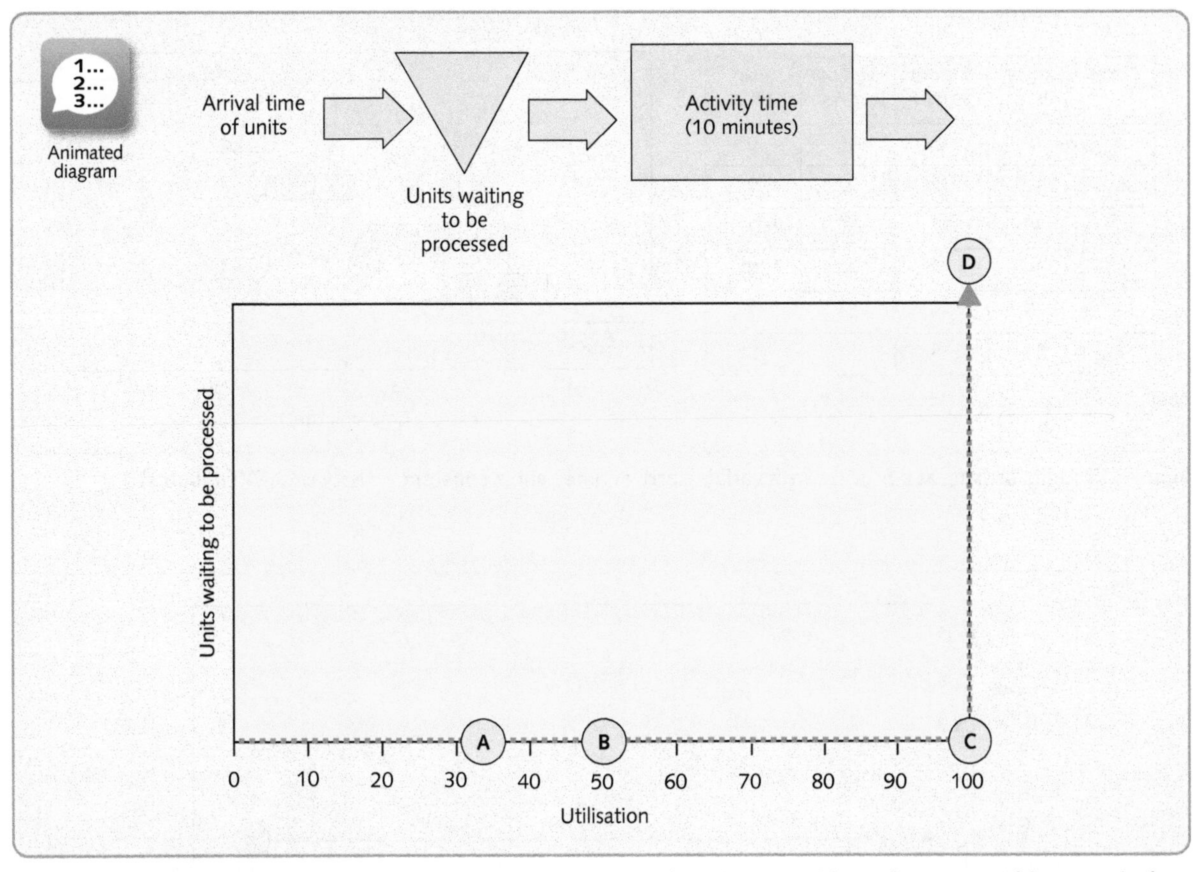

Figure 5.14 The relationship between process utilisation and number of units waiting to be processed for no arrival time or activity time variability

During the observed period:

$$\begin{aligned}
\text{time when a single unit was waiting} &= 3 \text{ minutes} \\
\text{elapsed time for processing the six units} &= 65 \text{ minutes} \\
\text{average number of units waiting} &= 3/65 \\
&= 0.046 \text{ units} \\
\text{process idle time} &= 5 \text{ minutes} \\
\text{So, process idle percentage} &= 5 \times 100/65 \\
&= 7.7\% \\
\text{Therefore, process utilisation} &= 92.3\%
\end{aligned}$$

This point is shown as point X in Figure 5.16. If the average arrival time were to be changed with the same variability, the dotted line in Figure 5.16 would show the relationship between average waiting time and process utilisation. The closer the process moves to 100 per cent utilisation the higher the average waiting time will become. Or, to put it another way, the only way to guarantee very low waiting times for the units is to suffer low process utilisation.

When both arrival times and activity times are variable, this effect is even more pronounced. And the greater the variability, the more the waiting time utilisation deviates from the simple rectangular function of the 'no variability' conditions that was shown in Figure 5.14. A set

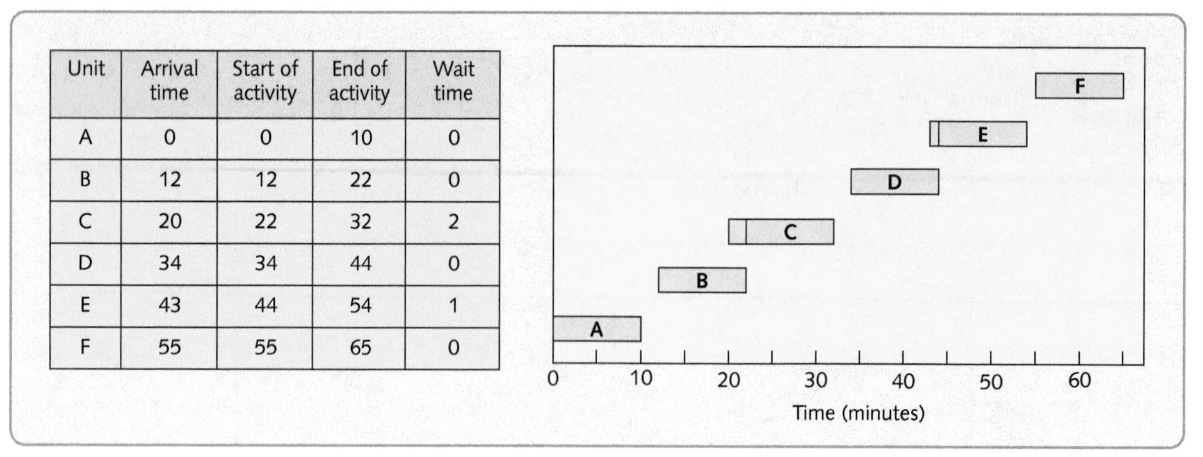

Unit	Arrival time	Start of activity	End of activity	Wait time
A	0	0	10	0
B	12	12	22	0
C	20	22	32	2
D	34	34	44	0
E	43	44	54	1
F	55	55	65	0

Figure 5.15 Units arriving at a process with variable arrival times and a constant activity time (10 minutes)

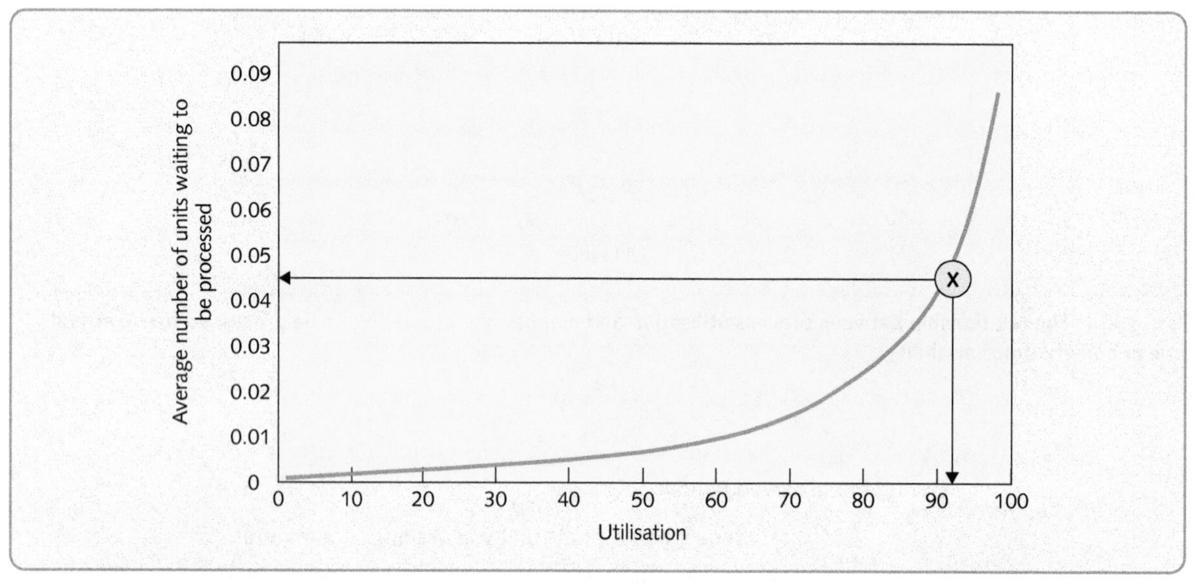

Figure 5.16 The relationship between process utilisation and number of units waiting to be processed for the variable arrival times in the example

Animated diagram

of curves for a typical process is shown in Figure 5.17(a). This phenomenon has important implications for the design of processes. In effect, it presents three options to process designers wishing to improve the waiting time or utilisation performance of their processes, as shown in Figure 5.17(b). Either:

- Accept long average waiting times and achieve high utilisation (point X).
- Accept low utilisation and achieve short average waiting times (point Y).
- Reduce the variability in arrival times, activity times, or both, and achieve higher utilisation and short waiting times (point Z).

To analyse processes with both inter-arrival and activity time variability, queuing or 'waiting line' analysis can be used. This is treated in the supplement to this chapter. But, do not dismiss the relationship shown in Figures 5.16 and 5.17 as some minor technical phenomenon. It is far more than this. It identifies an important choice in process design that could have

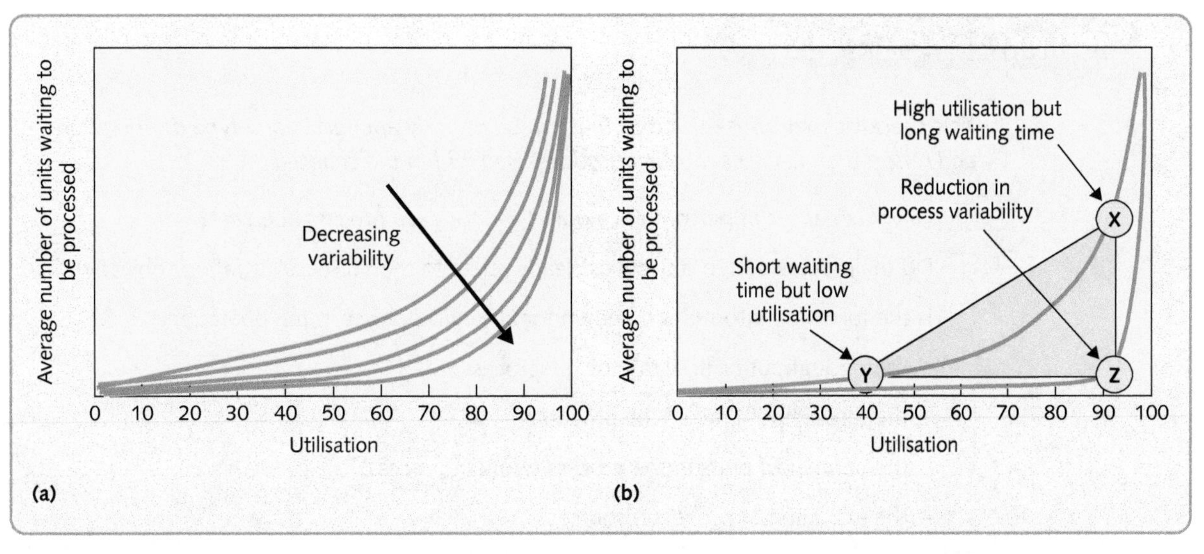

Figure 5.17 The relationship between process utilisation and number of units waiting to be processed for variable arrival and activity times (a) Decreasing variability allows higher utilisation without long waiting times. (b) Managing capacity and/or variability

OPERATIONS PRINCIPLE

Process design involves some choice between utilisation, waiting time and variability reduction.

strategic implications. Which is more important to a business, fast throughput time or high utilisation of its resources? The only way to have both of these simultaneously is to reduce variability in its processes, which may itself require strategic decisions such as limiting the degree of customisation of products or services, or imposing stricter limits on how products or services can be delivered to customers, and so on. It also demonstrates an important point concerned with the day-to-day management of process – the only way to absolutely guarantee 100 per cent utilisation of resources is to accept an infinite amount of work in progress and/or waiting time. We will take this point further in Chapter 8 when we deal with capacity management.

Critical commentary

Each chapter contains a short critical commentary on the main ideas covered in the chapter. Its purpose is not to undermine the issues discussed in the chapter, but to emphasise that, although we present a relatively orthodox view of operation, there are other perspectives.

- There is not too much that would be considered contentious in this chapter. However, some practitioners would reject the idea of mapping processes as they exist currently. Rather, they would advocate a more radical 'clean sheet of paper' approach. Only by doing this, they would say, can one be sufficiently imaginative in the redesign of processes. The other potential point of contention would concern the viability of what we have called 'long-thin' process designs. As was discussed in the previous chapter, this assumes a degree of division of labour and systemisation of work that is held by some to be 'dehumanising'.

SUMMARY CHECKLIST

This checklist comprises questions that can be usefully applied to any type of operations and reflect the major diagnostic questions used within the chapter.

- ☐ Have a clear set of performance objectives for each process been set?
- ☐ Do the process design objectives clearly relate to the business's strategic objectives?
- ☐ Is the following information known for all key processes in the operation:
  - the throughput or flow rate of the process?
  - the throughput time of the process?
  - the number of units in the process (work in process)?
  - the utilisation of process resources?
- ☐ Are processes documented using process mapping techniques?
- ☐ Are formal process descriptions followed in practice?
- ☐ If not, should the process descriptions be changed or should existing process descriptions be enforced?
- ☐ Is it necessary for process descriptions to include the degree of visibility at each stage of the process?
- ☐ Are the details of task precedence known for each process?
- ☐ Have the advantages and disadvantages of series and parallel configurations been explored?
- ☐ Is the process balanced? If not, can the bottleneck stages be redesigned to achieve better balance?
- ☐ Is the relationships between throughput, cycle time and 'work in process' understood (Little's law)?
- ☐ Are the sources of process variability recognised?
- ☐ Has the effect of variability been recognised in the design of the process?

CASE STUDY

The Action Response Applications Processing Unit (ARAPU)

Introduction

Action Response is a London-based charity dedicated to providing fast responses to critical situations throughout the world. It was founded by Susan N'tini, its Chief Executive, to provide relatively short-term aid for small projects until they could obtain funding from larger donors. The charity receives requests for cash aid usually from an intermediary charity and looks to process the request quickly, providing funds where and when they are needed. *'Give a man a fish and you feed him today; teach him to fish and you feed him for life. It's an old saying and it makes sense but, and this is where Action Response comes in, he might starve while he's training to catch fish.'* (Susan N'tini)

Source: Getty Images

Nevertheless, Susan does have some worries. She faces two issues in particular. First, she is receiving complaints that funds are not getting through quickly enough. Second, the costs of running the operation are starting to spiral. She explains. *'We are becoming a victim of our own success. We have striven to provide greater accessibility to our funds; people can access application forms via the internet, by post and by phone. But we are in danger of losing what we stand for. It is taking longer to get the money to where it is needed and our costs are going up. We are in danger of failing on one of our key objectives: to minimise the proportion of our turnover that is spent on administration. At the same time, we always need to be aware of the risk of bad publicity through making the wrong decisions. If we don't check applications thoroughly, funds may go to the "wrong" place and if the newspapers gets hold of the story we would run a real risk of losing the goodwill, and therefore the funds, from our many supporters.'*

Susan held regular meetings with key stakeholders. One charity that handled a large number of applications for people in Nigeria told her of frequent complaints about the delays over the processing of the applications. A second charity representative complained that when he telephoned to find out the status of an application, the ARAPU staff did not seem to know where it was or how long it might be before it was complete. Furthermore, he felt that this lack of information was eroding his relationship with his own clients, some of whom were losing faith in him as a result. *'Trust is so important in the relationship,'* he explained.

Some of Susan's colleagues, while broadly agreeing with her anxieties over the organisation's responsiveness and efficiency, took a slightly different perspective. *'One of the really good things about Action Response is that we are more flexible than most charities. If they need support until one of the larger charities can step in, then we will always consider a request for aid. I would not like to see any move towards high process efficiency harming our ability to be open-minded and consider requests that might seem a little unusual at first.'* (Jacqueline Horton, Applications Assessor)

Others saw the charity as performing an important counselling role. *'Remember that we have gained a lot of experience in this kind of short-term aid. We are often the first people that are in a position to advise on how to apply for larger and longer term funding. If we developed this aspect of our work we would again be fulfilling a need that is not adequately supplied at the moment.'* (Stephen Nyquist, Applications Assessor)

The Action Response Applications Processing Unit (ARAPU)

Potential aid recipients, or the intermediary charities representing them, apply for funds using a standard form. These forms can be downloaded from the internet or requested via a special help line. Sometimes the application will come directly from an individual community leader, but more usually it will come via an intermediary charity that is can help the applicant to complete the form. The application is sent to ARAPU, usually by fax or post (some were submitted online, but few communities have this facility).

ARAPU employs seven applications assessors with support staff who are responsible for data entry, coding, filing and 'completing' (staff who prepare payment, or explain why no aid can be given). In addition, a board of non-paid trustees meets every Thursday, to approve the assessors' decisions. The unit's IT system maintained records of all transactions, providing an update on the number of applications received, approved, declined, and payments allocated. These reports identified that the unit received about 300 new applications per week and responded to about the same number (the unit operates a 35-hour week). But while the unit's financial targets were being met, the trend indicated that cost per application was increasing. The target for the turnaround of an application, from receipt of application to response was 20 days, and although this was not measured formally, it was generally assumed that turnaround time was longer than this. Accuracy had never been an issue as all files were thoroughly assessed to ensure that all the relevant data was collected before the applications were processed. Productivity seemed high and there was always plenty of work waiting for processing at each section with the exception that the 'completers' were sometimes waiting for work to come from the committee on a Thursday. Susan had conducted an inspection of all sections' in-trays that had revealed a rather shocking total of about 2000 files waiting within the process, not counting those waiting for further information.

Processing applications

The processing of applications is a lengthy procedure requiring careful examination by applications assessors trained to make well-founded assessments in line with the charity's guidelines and values. Incoming applications are opened by one of the four 'receipt' clerks who check that all the necessary forms have been included in the application. The receipt clerks take about ten minutes per application. These are then sent to the coding staff in batches twice a day. The five coding clerks allocate a unique identifier to each application and key the information on the application into the system. The coding stage takes about 20 minutes for each application. Files are then sent to the senior applications assessors' secretary's desk. As assessors become available, the secretary provides the next job in the line to the assessor.

About 100 of the cases seen by the assessors each week are put aside after only 10 minutes 'scanning' because further information is needed. The assessor returns these files to the secretaries, who write to the applicant (usually via the intermediate charity) requesting additional information, and return the file to the 'receipt' clerks who 'store' the file until the further information eventually arrives (usually between one and eight weeks). When it does arrive, the file enters the process and progresses through the same stages again. Of the applications that require no further information, around half (150) are accepted and half (150) declined. On average, those applications that were not 'recycled' took around 60 minutes to assess.

All the applications, whether approved or declined, are stored prior to ratification. Every Thursday, the Committee of Trustees meets to formally approve the applications assessors' decisions. The committee's role is to sample the decisions to ensure that the guidelines of the charity are upheld. In addition they will review any particularly unusual cases highlighted by the applications assessors. Once approved by the committee, the files are then taken to the completion officers. There are three 'decline' officers whose main responsibility is to compile a suitable response to the applicant, pointing out why the application failed and offering, if possible, to provide helpful advice. An experienced declines officer takes about 30 minutes to finalise the file and write a suitable letter. Successful files are passed to the four 'payment' officers where again the file is completed, letters (mainly standard letters) are created and payment instructions are given to the bank. This usually takes around 50 minutes, including dealing with any queries from the bank about payment details. Finally the paperwork itself is sent, with the rest of the file, to two 'dispatch' clerks who complete the documents and mail them to the applicant. The dispatch activity takes, on average, ten minutes for each application.

The feeling among the staff was generally good. When Susan consulted the team they said their work was clear and routine, but their life was made difficult by charities that rang in expecting them to be able to tell them the status of an application they had submitted. It could take them hours, sometimes days, to find any individual file. Indeed two of the 'receipt' clerks now were working almost full-time on this activity. They also said that charities frequently complained that decision making seemed slow.

QUESTIONS

1 What objectives should the ARAPU process be trying to achieve?
2 What is the main problem with the current ARAPU processes?
3 How could the ARAPU process be improved?

APPLYING THE PRINCIPLES

Hints

Some of these exercises can be answered by reading the chapter. Others will require some general knowledge of business activity and some might require an element of investigation. Hints on how they can all be answered are to be found in the eText at www.pearsoned.co.uk/slack.

1 Choose a process with which you are familiar. For example, a process at work, registration for a university course, joining a video rental shop service, enrolling at a sports club or gym, registering at a library, obtaining a car parking permit, etc. Map the process that you went through, from your perspective, using the process mapping symbols explained in this chapter. Try to map what the 'back-office' process might be (that is the part of the process that is vital to achieving its objective, but you can't see). You may have to speculate on this but you could talk with someone who knows the process. How might the process be improved from your (the customer) perspective and from the perspective of the operation itself?

2 *'It is a real problem for us,'* said Angnyeta Larson. *'We now have only ten working days between all the expense claims coming from the departmental coordinators and authorising payments on the next month's payroll. This really is not long enough and we are already having problems during peak times.'* Angnyeta was the Department Head of the internal financial control department of a metropolitan authority in southern Sweden. Part of her department's responsibilities included checking and processing expense claims from staff throughout the metropolitan authority and authorising payment to the salaries payroll section. She had 12 staff who were trained to check expense claims and all of them were devoted full-time to processing the claims in the two weeks (ten working days) prior to the deadline for informing the salaries section. The number of claims submitted over the year averaged around 3,200, but this could vary between 1,000 during the quiet summer months up to 4,300 in peak months. Processing claims involved checking receipts, checking that claims met with the strict financial allowances for different types of expenditure, checking all calculations, obtaining more data from the claimant if necessary, and (eventually) sending an approval notification to salaries. The total processing time took on average 20 minutes per claim.

   (a) How many staff does the process need on average, for the lowest demand and for the highest demand?

   (b) If a more automated process involving electronic submission of claims could reduce the average processing time to 15 minutes, what effect would this have on the required staffing levels?

   (c) If department coordinators could be persuaded to submit their batched claims earlier (not always possible for all departments) so that the average time between submission of the claims to the finance department and the deadline for informing salaries section was increased to 15 working days, what effect would this have?

3 The headquarters of a major creative agency offered a service to all its global subsidiaries that included the preparation of a budget estimate that was submitted to potential clients when making a 'pitch' for new work. This service had been offered previously only to a few of the groups subsidiary companies. Now that it was to be offered worldwide, it was deemed appropriate to organise the process of compiling budget estimates on a more systematic basis. It was estimated that the worldwide demand for this service would be around 20 budget estimates per week, and that, on average, the staff who would put together these

estimates would be working a 35-hour week. The elements within the total task of compiling a budget estimate are shown in the following table.

The elements within the total task of compiling a budget estimate

Element	*Time (minutes)*	*What element (s) must be done prior to this one?*
A – obtain time estimate from creatives	20	None
B – obtain account handler's deadlines	15	None
C – obtain production artwork estimate	80	None
D – preliminary budget calculations	65	A, B, and C
E – check on client budget	20	D
F – check on resource availability and adjust estimate	80	D
G – complete final budget estimate	80	E and F

(a) What is the required cycle time for this process?

(b) How many people will the process require to meet the anticipated demand of 20 estimates per week?

(c) Assuming that the process is to be designed on a 'long-thin' basis, what elements would each stage be responsible for completing? And what would be the balancing loss for this process?

(d) Assuming that instead of the 'long-thin' design, two parallel processes are to be designed, each with half the number of stations of the 'long-thin' design. What now would be the balancing loss?

4 A company has decided to manufacture a general-purpose 'smoothing plane', a tool which smoothes and shapes wood. Its engineers estimated the time it would take to perform each element in the assembly process. The marketing department also estimated that the likely demand for the new product would be 98,000 units. The marketing department was not totally confident of its forecast. However, *'a substantial proportion of demand is likely to be export sales, which we find difficult to predict. But whatever demand does turn out to be, we will have to react quickly to meet it. The more we enter these parts of the market, the more we are into impulse buying and the more sales we lose if we don't supply.'*

An idea of the assembly task can be gained from the following table, which gives the 'standard time' for each element of the assembly task.

Standard times for each element of assembly task in standard minutes (SM)

Press elements	
Assemble poke	0.12 mins
Fit poke to front	0.10 mins
Rivet adjusting lever to front	0.15 mins
Press adjusting nut screw to front	0.08 mins
Bench elements	
Fit adjusting nut to front	0.15 mins
Fit frog screw to front	0.05 mins
Fit knob to base	0.15 mins
Fit handle to base	0.17 mins
Fit front assembly to base	0.15 mins
Assemble blade unit	0.08 mins
Final assembly	0.20 mins
Packing element	
Make box, wrap plane, pack	0.20 mins
Total time	**1.60 mins**

All elements must be performed sequentially in the order listed.

The standard costing system at the company involves adding a 150 per cent overhead charge to the direct labour cost of manufacturing the product, and the product would retail for the equivalent of around €35 in Europe where most retailers will sell this type of product for about 70–100 per cent more than they buy it from the manufacturer.

(a) How many people will be needed to assemble this product?

(b) Design a process for the assembly operation (to include the fly press work), including the tasks to be performed at each part of the system.

(c) How might the process design need to be adjusted as demand for this and similar products builds up?

Notes on chapter

1 Source: Horovitz, A. (2002), 'Fast food world ways drive-thru is the way to go', *USA Today*, 3 April.

2 Car-Brown, J. (2005), 'French factory surgeon cuts NHS queues', *The Sunday Times* 23 October.

3 Not everyone uses exactly the same terminology in this area. For example, some publications use the term 'cycle time' to refer to what we have called 'throughput rate'.

4 The concept of visibility is explained in Shostack, G.L. (1984) 'Designing services that deliver', *Harvard Business Review*, January–February, pp 133–39.

5 Little's law is best explained in Hopp, W.J. and Spearman, M.L. (2001), *Factory Physics* (2nd Edition), McGrawHill, New York.

TAKING IT FURTHER

Harvard Business Review (2011), Improving Business Processes (Harvard Pocket Mentor), **Harvard Business School Press.**

Hammer, M. (1990) 'Reengineering work: don't automate, obliterate', *Harvard Business Review*, July-August. *This is the paper that launched the whole idea of business processes and process management in general to a wider managerial audience. Slightly dated but worth reading.*

Harrington H.J. (2011) *Streamlined Process Improvement*, McGraw Hill Professional.

Hopp, W.J. and Spearman, M.L. (2001) *Factory Physics* (2nd edn), McGraw Hill. *Very technical so don't bother with it if you aren't prepared to get into the maths. However, some fascinating analysis, especially concerning Little's law.*

Ramaswamy, R. (1996) *Design and Management of Service Processes*, Addison-Wesley Longman. *A relatively technical approach to process design in a service environment.*

van der Aalst , W.M.P., ter Hofstede, A.H.M., and Weske, M. (eds) (2003) 'International Conference on business Process Management (BPM 2003), *Lecture Notes in Computer Science*, Vol. 2678, Springer-Verlag, Berlin. ***Comprehensive academic survey.***

USEFUL WEBSITES

www.opsman.org *Definitions, links and opinions on operations and process management.*

www.bpmi.org *Site of the Business Process Management Initiative. Some good resources including papers and articles.*

www.bptrends.com *News site for trends in business process management generally. Some interesting articles.*

www.bls.gov/oes/ *US Department of Labor employment statistics.*

www.iienet.org *The American Institute of Industrial Engineers site. They are an important professional body for process design and related topics.*

www.waria.com *A Workflow and Reengineering association website. Some useful topics.*

Interactive quiz

For further resources including examples, animated diagrams, self-test questions, Excel spreadsheets and video materials please explore the eText on the companion website at **www.pearsoned.co.uk/slack**.

SUPPLEMENT TO CHAPTER 5

Queuing analysis

Introduction

Queuing analysis (in many parts of the world it is called 'waiting line' analysis) is often explained purely in terms of customers being processed through service operations. This is misleading. Although queuing analysis can be particularly important in service operations, especially where customer really do 'queue' for service, the approach is useful in any kind of operation. Figure 5.18 shows the general form of queuing analysis.

The general form of queuing analysis

Customers arrive according to some probability distribution and wait to be processed (unless part of the operation is idle); when they have reached the front of the queue, they are processed by one of the n parallel 'servers' (their processing time also being described by a probability distribution), after which they leave the operation. There are many examples of this kind of system. Table 5.4 illustrates some of these. All of these examples can be described by a common set of elements that define their queuing behaviour.

- *The source of customers*, sometimes called the calling population, is the source of supply of customers. In queue management, 'customers' are not always human. 'Customers' could for example be trucks arriving at a weighbridge, orders arriving to be processed or machines waiting to be serviced, etc.
- *The arrival rate* is the rate at which customers needing to be served arrive at the server or servers. Rarely do customers arrive at a steady and predictable rate; usually there is variability in their arrival rate. Because of this it is necessary to describe arrival rates in terms of probability distributions.

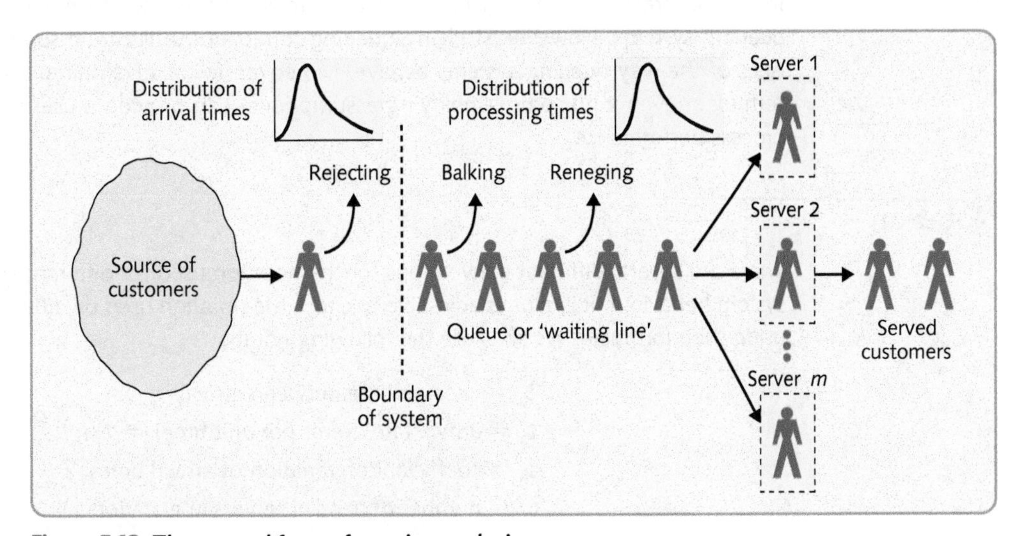

Figure 5.18 The general form of queuing analysis

Table 5.4 Examples of processes that can be analysed using queuing analysis

Operation	*Arrivals*	*Processing capacity*
Bank	Customers	Tellers
Supermarket	Shoppers	Checkouts
Hospital clinic	Patients	Doctors
Graphic artist	Commissions	Artists
Custom cake decorators	Orders	Cake decorators
Ambulance service	Emergencies	Ambulances with crews
Telephone switchboard	Calls	Telephonists
Maintenance department	Breakdowns	Maintenance staff

- *The queue*. Customers waiting to be served form the queue or waiting line itself. If there is relatively little limit on how many customers can queue at any time, we can assume that, for all practical purposes, an infinite queue is possible. Sometimes, however, there is a limit to how many customers can be in the queue at any one time.
- *Queue discipline*. This is the set of rules that determines the order in which customers waiting in the queue are served. Most simple queues, such as those in a shop, use a *first-come-first-served* queue discipline.
- *Servers*. A server is the facility that processes the customers in the queue. In any queuing system there may be any number of servers configured in different ways. In Figure 5.19 servers are configured in parallel, but some systems may have servers in a series arrangement. There is also likely to be variation in how long it takes to process each customer. Therefore processing time, like arrival time, is usually described by a probability distribution.

Calculating queue behaviour

Management scientists have developed formulae which can predict the steady-state behaviour of different types of queuing systems. Unfortunately, many of these formulae are extremely complicated, especially for complex queuing systems, and are beyond the scope of this book. In fact, in practice, computer programs are almost always used to predict the behaviour of queuing systems. However, studying queuing formulae can illustrate some useful characteristics of the way queuing systems behave. Moreover, for relatively simple systems, using the formulae (even with some simplifying assumptions) can provide a useful approximation to process performance.

Notation

There are several different conventions for the notation used for different aspects of queuing system behaviour. It is always advisable to check the notation used by different authors before using their formulae. We shall use the following notation.

$$t_a = \text{average time between arrivals}$$
$$r_a = \text{arrival rate (items per unit time)} = 1/t_a$$
$$c_a = \text{coefficient of variation of arrival times}$$
$$m = \text{number of parallel servers at a station}$$

t_e = mean processing time
r_e = processing rate (items per unit time) = m/t_e
c_e = coefficient of variation of process time
u = utilisation of station = r_a/r_e = $(r_a t_e)/m$
WIP = average work in process (number of items) in the queue
WIP = expected work in process (number of times) in the queue
W_q = expected waiting time in the queue
W = expected waiting time in the system (queue time + processing time)

Variability

The concept of variability is central to understanding the behaviour of queues. If there were no variability there would be no need for queues to occur because the capacity of a process could be relatively easily adjusted to match demand. For example, suppose one member of staff (a server) serves customers at a bank counter who always arrive exactly every five minutes (i.e. 12 per hour). Also suppose that every customer takes exactly five minutes to be served, then because,

(a) the arrival rate is ≤ processing rate, and
(b) there is no variation

no customer need ever wait because the next customer will arrive when, or before, the previous customer leaves. That is, $WIP_q = 0$.

Also, in this case, the server is working all the time, again because exactly as one customer leaves, the next one is arriving. That is, $u = 1$.

Even with more than one server, the same may apply. For example, if the arrival time at the counter is five minutes (12 per hour) and the processing time for each customer is now always exactly 10 minutes (6 per hour), the counter would need two servers ($m = 2$), and because,

(a) arrival rate is ≤ processing rate × m, and
(b) there is no variation,

again, $WIP_q = 0$, and $u = 1$.

Of course, it is convenient (but unusual) if arrival rate/processing rate is a whole number. When this is not the case (for this simple example with no variation):

$$\text{Utilisation} = \text{processing rate}/(\text{arrival rate} \times m)$$

For example, if arrival rate, $r_a = 5$ minutes, processing rate $r_e = 8$ minutes, and number of servers $m = 2$,
then,

$$\text{utilisation } u = 8/(5 \times 2) = 0.8 \text{ or } 80\%$$

Incorporating variability

The previous examples were not realistic because they assumed no variation in arrival or processing times. We also need to take into account the variation around these means. To do that, we need to use a probability distribution. Figure 5.19 contrasts two processes with different arrival distributions. The units arriving are shown as people, but they could be jobs arriving at a machine, trucks needing servicing or any other uncertain event. The top example shows low variation in arrival time where customers arrive in a relatively predictable manner. The bottom example has the same average number of customer arriving but this time they arrive

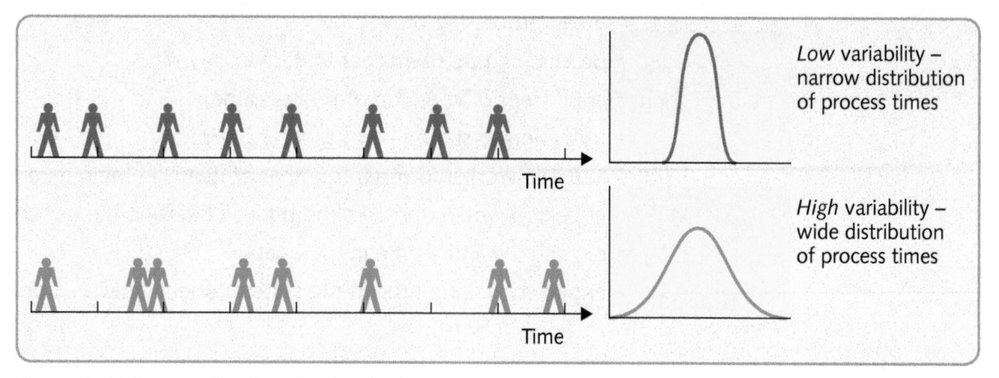

Figure 5.19 Low and high arrival variation

unpredictably with sometimes long gaps between arrivals and at other times two or three customers arriving close together. We could do a similar analysis to describe processing times.

In Figure 5.19 high arrival variation has a distribution with a wider spread (called 'dispersion') than the distribution describing lower variability. Statistically the usual measure for indicating the spread of a distribution is its standard deviation, σ. But variation does not only depend on standard deviation. For example, a distribution of arrival times may have a standard deviation of 2 minutes. This could indicate very little variation when the average arrival time is 60 minutes. But it would mean a very high degree of variation when the average arrival time is 3 minutes. Therefore to normalise standard deviation, it is divided by the mean of its distribution. This measure is called the coefficient of variation of the distribution. So,

$$c_a = \text{coefficient of variation of arrival times} = \sigma_a / t_a$$

$$c_e = \text{coefficient of variation of processing times} = \sigma_e / t_e$$

Incorporating Little's law

Little's law (described earlier in the chapter) describes the relationship between the cycle time, the work in process and the throughput time of the process. It was denoted by the following simple relationship:

$$\text{work in process} = \text{cycle time} \times \text{throughput time}$$

or

$$\text{WIP} = \text{C} \times \text{T}$$

Little's law can help to understand queuing behaviour. Consider the queue in front of station.

$$\begin{aligned}\text{work in process in the queue} = {} & \text{arrival rate at the queue (equivalent to cycle time)} \\ & \times \text{waiting time in the queue (equivalent to throughput time)}\end{aligned}$$

$$\text{WIP}_q = r_a \times W_q$$

and

$$\begin{aligned}\text{waiting time in the whole system} = {} & \text{the waiting time in the queue} \\ & + \text{the average process time at the station}\end{aligned}$$

$$W = W_q + t_e$$

We will use this relationship later to investigate queuing behaviour.

Types of queuing system

Conventionally, queuing systems are characterised by four parameters.

A = the distribution of arrival times (or more properly inter-arrival times, the elapsed times between arrivals)
B = the distribution of process times
m = the number of servers at each station
b = the maximum number of items allowed in the system

The most common distributions used to describe A or B are either:

- the exponential (or Markovian) distribution denoted by M.
- the general (for example normal) distribution denoted by G.

So, for example, an M/G/1/5 queuing system would indicate a system with exponentially distributed arrivals, process times described by a general distribution such as a normal distribution, with one server and a maximum number of five items allowed in the system. This type of notation is called Kendall's Notation.

Queueing theory can help us investigate any type of queuing system, but in order to simplify the mathematics, we shall here deal only with the two most common situations. Namely:

- M/M/m – the exponential arrival and processing times with m servers and no maximum limit to the queue.
- G/G/m – general arrival and processing distributions with m servers and no limit to the queue.

And first we will start by looking at the simple case when M = 1.

For M/M/1 queuing systems

The formulae for this type of system are as follows.

$$\text{WIP} = \frac{u}{1 - u}$$

Using Little's law:

$$\text{WIP} = \text{cycle time} \times \text{throughput time}$$
$$\text{throughput} = \text{WIP/cycle time}$$

Then:

$$\text{throughput time} = \frac{u}{1 - u} \times \frac{1}{r_a} = \frac{t_a}{1 - u}$$

and since, throughput time in the queue = total throughput time − average processing time:

$$\begin{aligned} W_q &= W - t_e \\ &= \frac{t_e}{1 - u} - t_e \\ &= \frac{t_e - t_e(1 - u)}{1 - u} = \frac{t_e - t_e - ut_e}{1 - u} \\ &= \frac{u}{(1 - u)}t_e \end{aligned}$$

Again, using Little's law:

$$\text{WIP}_q = r_a \times W_q = \frac{u}{1 - u}t_e r_a$$

and since,

$$u = \frac{r_a}{r_e} = r_a t_e$$

$$r_a = \frac{u}{t_e}$$

then:

$$\mathrm{WIP}_q = \frac{u}{1 - u} \times t_e \times \frac{u}{t_e}$$

$$= \frac{u^2}{(1 - u)}$$

For M/M/m systems

When there are *m* servers at a station the formula for waiting time in the queue (and therefore all other formulae) needs to be modified. Again, we will not derive these formulae but just state them.

$$W_q = \frac{u^{\sqrt{2(m+1)}-1}}{m(1 - u)} t_e$$

from which the other formulae can be derived as before.

For G/G/1 systems

The assumption of exponential arrival and processing times is convenient as far as the mathematical derivations of various formulae are concerned. However, in practice, process times in particular are rarely truly exponential. This is why it is important to have some idea of how G/G/1 and G/G/M queues behave. However, exact mathematical relationships are not possible with such distributions. Therefore some kind of approximation is needed. The one here is in common use, and although it is not always accurate, it is for practical purposes. For G/G/1 systems the formula for waiting time in the queue is as follows:

$$W_q = \left(\frac{c_a^2 + c_e^2}{2}\right)\left(\frac{u}{1 - u}\right)t_e$$

There are two points to make about this equation. The first is that it is exactly the same as the equivalent equation for an M/M/1 system but with a factor to take account of the variability of the arrival and process times. The second is that this formula is sometimes known as the VUT formula because it describes the waiting time in a queue as a function of:

V = the variability in the queuing system
U = the utilisation of the queuing system (that is demand versus capacity),

and

T = the processing times at the station.

In other words, we can reach the intuitive conclusion that queuing time will increase as variability, utilisation or processing time increase.

For G/G/m systems

The same modification applies to queuing systems using general equations and *m* servers. The formula for waiting time in the queue is now as follows.

$$W_q = \left(\frac{c_a^2 + c_e^2}{2}\right)\left(\frac{u^{\sqrt{2(m+1)}-1}}{m(1 - u)}\right)t_e$$

EXAMPLE

'I can't understand it. We have worked out our capacity figures and I am sure that one member of staff should be able to cope with the demand. We know that customers arrive at a rate of around six per hour and we also know that any trained member of staff can process them at a rate of eight per hour. So why is the queue so large and the wait so long? Have at look at what is going on there please.'

Sarah knew that it was probably the variation, both in customers arriving and in how long it took each of them to be processed, that was causing the problem. Over a two-day period when she was told that demand was more or less normal, she timed the exact arrival times and processing times of every customer. Her results were as follows.

$$\text{coefficient of variation, } c_a \text{ of customer arrivals} = 1$$
$$\text{coefficient of variation, } c_e \text{ of processing time} = 3.5$$
$$\text{average arrival rate of customers, } r_a = 6 \text{ per hour}$$
$$\text{therefore, average inter-arrival time} = 10 \text{ minutes}$$
$$\text{average processing rate, } r_e = 8 \text{ per hour}$$
$$\text{therefore, average processing time} = 7.5 \text{ minutes}$$
$$\text{therefore, utilisation of the single server, } u = 6/8 = 0.75$$

Using the waiting time formula for a G/G/1 queuing system:

$$W_q = \left(\frac{1 + 12.25}{2}\right)\left(\frac{0.75}{1 - 0.75}\right) \times 7.5$$
$$= 6.625 \times 3 \times 7.5 = 149.06 \text{ mins}$$
$$= 2.48 \text{ hours}$$

Also,

$$\text{WIP}_q = \text{cycle time} \times \text{throughput time}$$
$$\text{WIP}_q = 6 \times 2.48 = 14.88$$

So, Sarah had found out that the average wait that customers could expect was 2.48 hours and that there would be an average of 14.88 people in the queue.

'Ok, so I see that it's the very high variation in the processing time that is causing the queue to build up. How about investing in a new computer system that would standardise processing time to a greater degree? I have been talking with our technical people and they reckon that, if we invested in a new system, we could cut the coefficient of variation of processing time down to 1.5. What kind of a different would this make?'

Under these conditions with $c_e = 1.5$

$$W_q = \left(\frac{1 + 2.25}{2}\right)\left(\frac{0.75}{1 - 0.75}\right) \times 7.5$$
$$= 1.625 \times 3 \times 7.5 = 36.56 \text{ mins}$$
$$= 0.61 \text{ hours}$$

Therefore:

$$\text{WIP}_q = 6 \times 0.61 = 3.66$$

In other words, reducing the variation of the process time has reduced average queuing time from 2.48 hours down to 0.61 hours and has reduced the expected number of people in the queue from 14.88 down to 3.66.

EXAMPLE

A bank wishes to decide how many staff to schedule during its lunch period. During this period customers arrive at a rate of nine per hour and the enquiries that customers have (such as opening new accounts, arranging loans, etc.) take on average 15 minutes to deal with. The bank manager feels that four staff should be on duty during this period but wants to make sure that the customers do not wait more than three minutes on average before they are served. The manager has been told by his small daughter that the distributions that describe both arrival and processing times are likely to be exponential. Therefore, $r_a = 9$ per hour, so $t_a = 6.67$ minutes; and $r_e = 4$ per hour, so $t_e = 15$ minutes. The proposed number of servers, $m = 4$, therefore the utilisation of the system, $u = 9/(4 \times 4) = 0.5625$.

From the formula for waiting time for a M/M/m system,

$$W_q = \frac{u^{\sqrt{2(m+1)}-1}}{m(1 - u)} t_e$$

$$W_q = \frac{0.5625^{\sqrt{10}-1}}{4(1 - 0.5625)} \times 0.25$$

$$= \frac{0.5625^{2.162}}{1.75} \times 0.25$$

$$= 0.042 \text{ hours}$$

$$= 2.52 \text{ minutes}$$

Therefore the average waiting time with four servers would be 2.52 minutes. That is well within the manager's acceptable waiting tolerance.

TAKING IT FURTHER

Hopp, W.J. and Spearman, M.L. (2001) ***Factory Physics*** **(2nd edn), McGraw Hill.** *Very technical so don't bother with it if you aren't prepared to get into the maths. However, some fascinating analysis, especially concerning Little's law.*

6

Designing the innovation process

Introduction

Customers want innovation. They want service and product offerings to be continually updated, altered and modified. Some changes are small – incremental adaptations to existing ways of doing things. Others are radical – major departures from anything that has gone before. The innovation activity is about successfully delivering change in its many different forms. Being good at innovation has always been important. What has changed in recent years is the sheer speed and scale of innovation in industries all over the world. Innovation processes are also increasingly complex, with inputs from different individuals and departments within an organisation, and increasingly from a wide variety of external sources. This makes the design and management of the innovation process more critical than ever before. This chapter examines what is meant by innovation and why it matters, the general stages involved in bringing service and product offerings from concept to launch, and some of the resourcing and organisational considerations for innovation (see Figure 6.1)

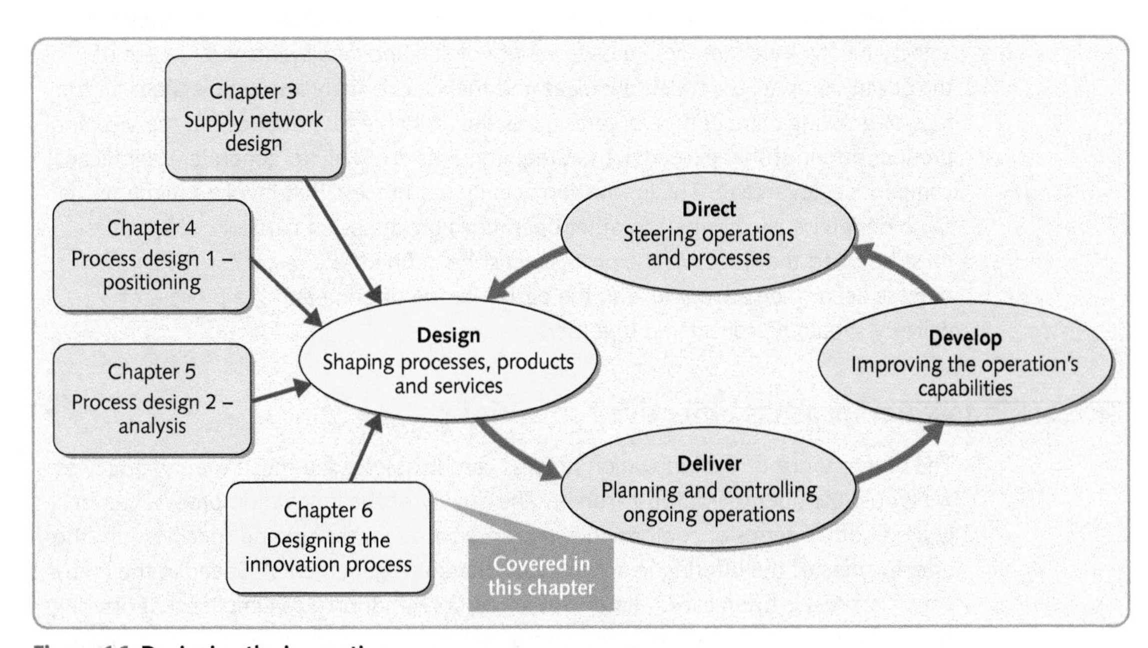

Figure 6.1 Designing the innovation process

EXECUTIVE SUMMARY

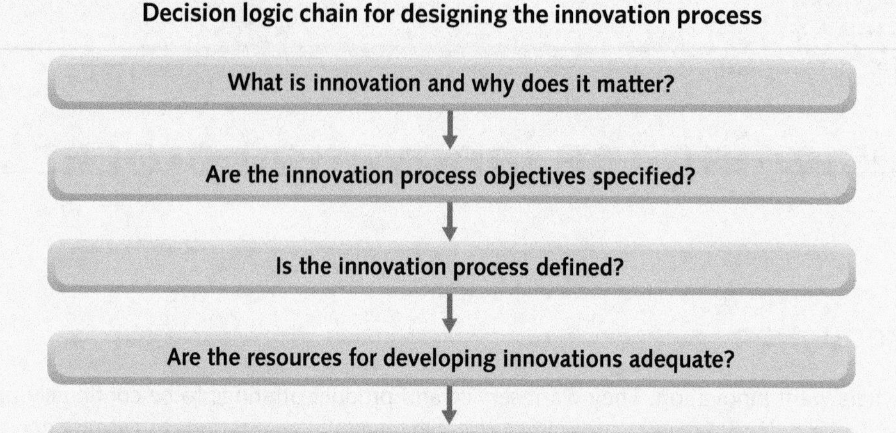

Each chapter is structured around a set of diagnostic questions. These questions suggest what you should ask in order to gain an understanding of the important issues of a topic, and, as a result, improve your decision making. An executive summary addressing these questions is provided below.

What is innovation and why does it matter?

Innovation is all about doing things differently. This includes changing what is offered to customers and the way these offerings are created and delivered. The innovation activity involves working with a wide variety of stakeholders inside and outside of the organisation to generate new ideas and then successfully implement them. In the face of growing competition, organisations that are able to introduce new service and product offerings that exceed market requirements are likely to generate a significant competitive advantage. The innovation activity is a process that involves many of the same design issues common to other operations processes. In particular, objectives must be clear; the stages of the process must be defined; the resources within the process need to be adequate; and the design of the offering and the process of delivery should be considered together.

Are the innovation process objectives specified?

The performance of the innovation process can be assessed in much the same way as we would consider the outputs from it. The quality of the innovation process can be judged both in terms of conformance (no errors in the offering) and specification (the effectiveness of the offering in achieving its market requirements). Speed in the innovation process is often called 'time-to-market' (TTM). Short TTM implies that offerings can be frequently introduced to the market achieving strategic impact. Dependability in the innovation process means meeting launch delivery dates. In turn, this often requires that the innovation process is sufficiently flexible to cope with disruptions. The cost of the innovation process can be thought of both as the amount of budget that is necessary to develop a new offering, and as the impact of the innovation on cost of delivering the service or product on an ongoing basis. Finally, the sustainability of the innovation process considers the impact on broad stakeholder objectives encompassed within the 'triple bottom line' – people, planet and profit.

Is the innovation process defined?

To create a fully specified service or product offering, potential designs tend to pass through a set of stages in the innovation process. Almost all stage models start with a general idea or 'concept' and progress through to a fully defined specification for the offering incorporating various service and product components. In between these two states, the offering may pass through stages such as concept generation, concept screening, preliminary design (including consideration of standardisation, commonality, modularisation and mass customisation), evaluation and improvement, prototyping and final design.

Are the resources for developing innovations adequate?

To be effective, the innovation process needs to be resourced adequately. The detailed principles of process design that were discussed in Chapters 4 and 5 are clearly applicable when developing new services or product offerings. However, because the innovation activity is often an operation in its own right, there are some more strategic issues to consider – how much capacity to devote to innovation, how much of the innovation activity to outsource, and what kinds of technology to use in the development of new offerings.

Is the design of the offering and of the process simultaneous?

The outputs from the innovation, in the form of new service and product offerings, are important inputs into the processes that create and deliver them on an ongoing basis. Therefore, it is often best to consider these in parallel rather than (as more traditionally done) in sequence. Merging the innovation process for new service and product offerings with the processes that create them is sometimes called simultaneous (or interactive) design. Its key benefit is the reduction in the time taken for the whole innovation activity. In particular, four simultaneous design factors can be identified that promote fast time-to-market. These are: routinely integrating the design of the product–service offering and the design of the process used to create and deliver them; overlapping the stages in the innovation process; the early deployment of strategic decision making to resolve design conflict; an organisational structure that reflects the nature of the offering.

DIAGNOSTIC QUESTION

What is innovation and why does it matter?

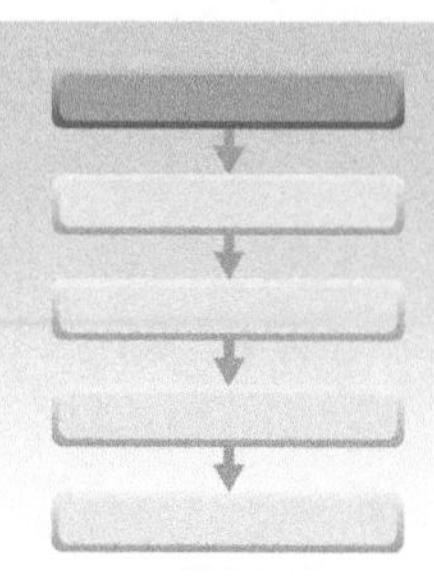

Innovation is all about doing things differently. This includes changing what is offered to customers as well as changing the way these offerings are created and delivered. The innovation activity involves working with customers, suppliers and internal functions to generate new ideas and then successfully implement them. Contributions are needed from: those who understand market requirements; those who understand the technical aspects of the offering; those with access to cost and investment information; those who can protect intellectual property; and, most importantly, the operations people responsible for the innovation's delivery.

The aim of the innovation process is firstly to create offerings that exceed customers' expectations in terms of quality, speed, dependability, flexibility, cost and sustainability, and secondly, to ensure that competitors find these offerings hard to imitate, substitute or gain access to. With increasingly demanding customers, higher levels of competition, and shorter product–service life cycles seen in many markets worldwide, organisations who can master the art of innovation will generate significant competitive advantage. In addition, if an organisation is able to deliver consistently innovations ahead of its competitors, it is more likely to set industry standards which others are often forced to follow.

Like any other process, the innovation activity can and should be managed – after all, successful outcomes, in the form of service and product offerings, are significantly reliant on an effectively designed innovation process. While innovation projects clearly vary, there is sufficient commonality in them to be able to model the process and seek to improve it. Figure 6.2 shows the innovation activity as a process, with inputs and outputs, as in any other process. Inputs come both from within the organisation – employees, R&D, Operations, Marketing, Human Resources, and Finance – and from outside it – market research, customers, lead users, suppliers, competitors, collaborators and wider stakeholders. The outputs (or outcomes) of the innovation process include: the concept or value proposition; the offering or package; and the process of delivery.

The following two examples are both innovation processes. For a company like Microsoft, the design of software is much more important than its subsequent manufacture. Similarly, pharmaceutical offerings go through a long and rigorous innovation process to deliver new drugs into a market.

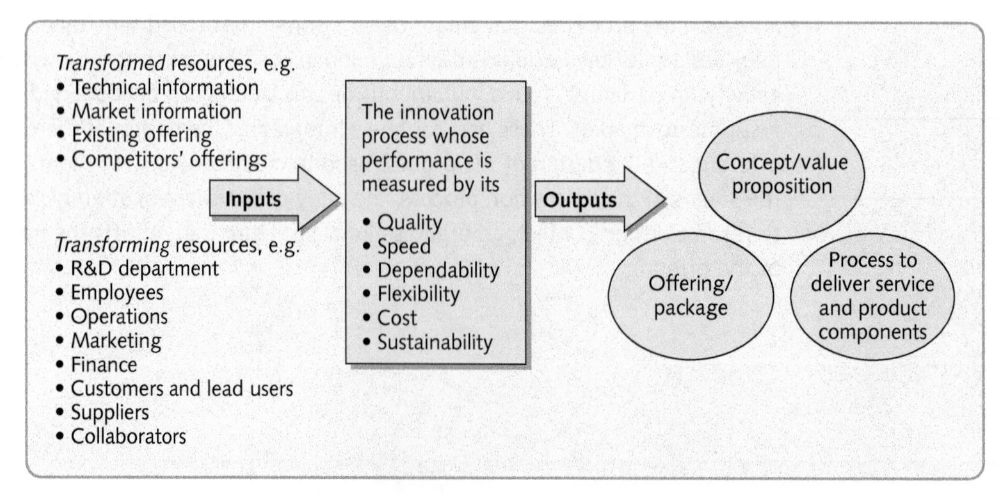

Figure 6.2 The innovation activity as a process

EXAMPLE

Innovation at Microsoft[1]

Software development is a highly challenging task given the unpredictable nature of user requirements and underlying technologies. Furthermore, changes to one part of a software product will almost certainly affect other parts. Above all, software is big and getting bigger. Some Microsoft products from the early 1980s had less than 100,000 lines of code. The first Windows NT in 1993 had about 4.5 million lines of code. Now, software can have tens of millions of lines of code with hundreds of developers. Microsoft has developed two main clusters of ideas for coping with such difficult innovation projects.

Source: Getty Images

The first concerns the company's approach to conceptualising the overall innovation task and allocating resources to its various stages. It includes the strict prioritisation of individual features within the software, the most important being developed first. Brief vision statements for each part of the project help to define the more detailed functional specifications needed to determine resources, but not so detailed as to prevent redefinition as the project progresses. A modular design also helps in making incremental changes. Microsoft likes to 'fix' project resources early in the innovation process, limiting the number of people and the time available for each part of the design process to encourage strict prioritisation. It also focuses the team's creativity towards achieving a 'working' (if not perfect) version of the product which can be made ready for market testing.

The second set of ideas concerns the company's approach to managing day-by-day innovation. It is an attempt to ensure design discipline and order without inhibiting the developers' creativity. Big projects do need clearly defined design phases and task allocation, but over-specifying a project structure can discourage innovation. So Microsoft uses relatively small teams of 300–400 developers who synchronise their decisions very frequently. These synchronisations are called 'builds', and always occur at a fixed time, often every day. Developers submit their work at the fixed 'build' time. Frequent synchronisation allows individual teams to be creative and to change their objectives in line with changing circumstances but, at the same time, ensures a shared discipline. Even so, it is necessary to stabilise the design periodically through the use of intermediate 'milestones'.

EXAMPLE

Novartis[2]

There are few industries where innovation is more important than in pharmaceuticals. Designing new drugs needs budgets of several billion dollars. To Novartis, one of the most respected companies in the pharmaceutical sector, managing the design of new drugs is arguably its most important activity. *'We want to build the most competitive, most innovative research organisation with the most promising pipeline in the industry to cure and help people around the world,'* says the company. *'We want to discover, develop and successfully market innovative products to cure diseases, to ease suffering and to enhance the quality of life.'* Of course the company also wants to do this while being able to provide a return to their shareholders, and with R&D expenditures averaging around 20 per cent of sales revenue for most pharmaceutical companies, managing the innovation process is a 'make or break' activity. Drug innovation consists of several, sometimes overlapping, stages. First is the 'drug discovery' phase. Naturally occurring or synthetic chemical compounds are investigated in the laboratory to explore their potential for further development. This process is no longer based on trial and error. Using a technique known as combinatorial chemistry, thousands

Source: Mira/Alamy Images

of compounds are produced and tested automatically. When promising candidates for further design have been identified, 'pre-clinical' testing begins, where further laboratory tests investigate the pharmacological characteristics of potential new drugs. Issues such as efficacy and toxicity are investigated, and first thought is given to how the drug could be manufactured should it ever go into production. Of 10,000 candidates that are screened during the drug discovery stage, only about 250 will be selected to go through to the pre-clinical phase, which further cuts the number of candidates down to around five. These five go on to the first of three stages of 'clinical trials', a process that all drugs must undergo before they can be considered for market approval. Phase one of clinical trials starts by testing the drugs on healthy volunteers, then on patients who have the disease that the drug is intended to treat. Phase two attempts to establish appropriate scales for measuring the effectiveness of the drug, while phase three completes the process of quantifying the effectiveness of the drug and checks for any significant side-effects. Generally clinical trials are carefully regulated and monitored by government agencies charged with giving approval for the drug to be marketed. Typically, for every five candidates entering the clinical testing phase, only one is approved to be sold into the market. In total the whole process can easily take up to 15 years. This means that having plenty of potentially marketable drugs in the innovation pipeline is vital for any pharmaceutical company.

What do these two examples have in common?

The nature of innovation will inevitably differ between organisations – the technical issues will be different, as often will the timescales involved. For example, computing products made in volume now may not even exist in two or three years' time. There are also differences between the design of offerings or packages with high or low levels of customer interaction. For offerings dominated by physical goods, customers judge attributes such as its functionality, its aesthetics, and so on. For offerings where the customer is an active participant (i.e. co-creates value with the firm), judgement is made not only on what is received but also on the process that they were put through in receiving it.

Yet notwithstanding these differences, the innovation *process* is essentially very similar irrespective of what is being offered. More importantly, it is a process that is seen as increasingly important. An inability to manage the innovation process by either Novartis or Microsoft would obviously have serious consequences. Conversely, good design that incorporates creativity within a well-ordered innovation process can generate huge competitive advantage. As such, the innovation process must be: designed; be clear about its process objectives; have clearly defined stages; have adequate resources; consider both the design of the offering and the process that will deliver it.

OPERATIONS PRINCIPLE

The innovation activity is a process that can be managed using the same principles as other processes.

DIAGNOSTIC QUESTION

Are the innovation process objectives specified?

The performance of the innovation process can be assessed in much the same way as we would consider the outputs from it, namely in terms of quality, speed, dependability, flexibility, cost and sustainability. These performance objectives have just as much relevance for innovation as they do for the ongoing delivery of offerings once they are introduced to the market.

OPERATIONS PRINCIPLE

Innovation processes can be judged in terms of their levels of quality, speed, dependability, flexibility, cost and sustainability.

What is the quality of the innovation process?

Design quality is not always easy to define precisely, especially if customers are relatively satisfied with existing service and product offerings. Many software companies talk about the 'I don't know what I want, but I'll know when I see it' syndrome, meaning that only when customers use the software are they in a position to articulate what they do or don't require. Nevertheless, it is possible to distinguish high and low quality designs (although this is easier to do with hindsight) by judging them in terms of their ability to meet market requirements. In doing this, the distinction between the specification quality and the conformance quality of designs is important. No business would want a design process that was indifferent to 'errors' in its designs, yet some are more tolerant than others. For example, neither Microsoft nor Novartis would want to produce products with errors in them. Both would suffer significant reputational damage if they did. Yet with Novartis the potential for harm is particularly high because their products directly affect our health. This is why the authorities insist on such a prolonged and thorough design process. Although withdrawing a drug from the market is unusual, it does occasionally occur. Far more frequent are the 'product recalls' that are relatively common in, say, the automotive industry. Many of these are design related and the result of 'conformance' failures in the design process. The 'specification' quality of design is different. It means the degree of functionality, or experience, or aesthetics, or whatever the product or service is primarily competing on. Some businesses require product or service designs that are relatively basic (although free from errors), while others require designs that are clearly special in terms of the customer response they hope to elicit.

What is the speed of the innovation process?

The speed of innovation matters more to some industries than others. For example, innovation in construction and aerospace happen at a much slower pace than in clothing or microelectronics. However, rapid innovation or 'time-based competition' has become the norm for an increasing number of industries. Sometimes this is the result of fast-changing consumer fashion. Sometimes it is forced by a rapidly changing technology base. Telecoms, for example, are updated frequently because their underlying technology is constantly improving. Sometimes both of these pressures are evident, as in many internet-based services. No matter what the motivation, fast design brings a number of advantages:

- *Early market launch*. An ability to innovate speedily means that service and product offerings can be introduced to the market earlier and thus earn revenue for longer, and may command price premiums.
- *Starting design late*. Alternatively, starting the design process later may have advantages, especially where either the nature of customer demand or the availability of technology is uncertain and dynamic. Fast design allows design decisions to be made closer to the time when service and product offerings are introduced to the market.
- *Frequent market stimulation*. Rapid innovations allow frequent new or updated offerings to be introduced into the market.

What is the dependability of the innovation process?

Rapid innovation processes that cannot be relied on to deliver dependably are, in reality, not fast at all. Design schedule slippage can extend design times, but worse, a lack of dependability adds to the uncertainty surrounding the innovation process. Conversely, processes that are dependable minimise design uncertainty. Unexpected technical difficulties, such as suppliers who themselves do not deliver solutions on time, customers or markets that change during the innovation process itself, and so on, all contribute to an uncertain and ambiguous design environment. Professional project management (see Chapter 15) of the innovation process can help to reduce uncertainty and prevent (or give early warning of) missed deadlines, process bottlenecks and resource shortages. However, external disturbances to the innovation process will remain. These may be minimised through close liaison with suppliers and market or environmental monitoring. Nevertheless, unexpected disruptions will always occur and the more innovative the design, the more likely they are to occur. This is why flexibility within the innovation process is one of the most important ways in which dependable delivery of new service and product offerings can be ensured.

What is the flexibility of the innovation process?

Flexibility in the innovation process is the ability to cope with external or internal change. The most common reason for external change is because markets, or specific customers, change their requirements. Although flexibility may not be needed in relatively predictable markets, it is clearly valuable in more fast-moving and volatile markets, where one's own customers and markets change, or where the designs of competitors' offerings dictate a matching or leapfrogging move. Internal changes include the emergence of superior technical solutions. In addition, the increasing complexity and interconnectedness of service and product components in an offering may require flexibility. A bank, for example, may bundle together a number of separate services for one particular segment of its market. Privileged account holders may obtain special deposit rates, premium credit cards, insurance offers, travel facilities, and so on, together in the same package. Changing one aspect of this package may require changes to be made in other elements. So extending the credit card benefits to include extra travel insurance may also mean the redesign of the separate insurance element of the package. One way of measuring innovation flexibility is to compare the cost of modifying a design in response to such changes against the consequences to profitability if no changes are made. The lower the cost of modifying an offering in response to a given change, the higher the level of flexibility.

What is the cost of the innovation process?

The cost of innovation is usually analysed in a similar way to the ongoing cost of delivering offerings to customers. These cost factors are split up into three categories: the cost of buying the inputs to the process; the cost of providing the labour in the process; and the other general overhead costs of running the process. In most inhouse innovation processes, the latter two costs outweigh the former.

One way of thinking about the effect of the other innovation performance objectives on cost is shown in Figure 6.3. Whether caused by quality errors, an intrinsically slow innovation process, a lack of project dependability, or delays caused through inflexibility, the end result is that the design is late. Delayed completion of the design results in both more expenditure on the design and delayed (and probably reduced) revenue. The combination of these effects usually means that the financial break-even point for a new offering is delayed far more than the original delay in its launch.

What is the sustainability of the innovation process?

The sustainability of an innovation is the extent to which it benefits the 'triple bottom line' – people, planet and profit. When organisations carry out their innovation activities, they must consider their objectives in relation to the triple bottom line, some of which may come from strategic objectives and others from external pressures, such as new legislation or changing social attitudes. The innovation process is very important in ultimately impacting on the ethical,

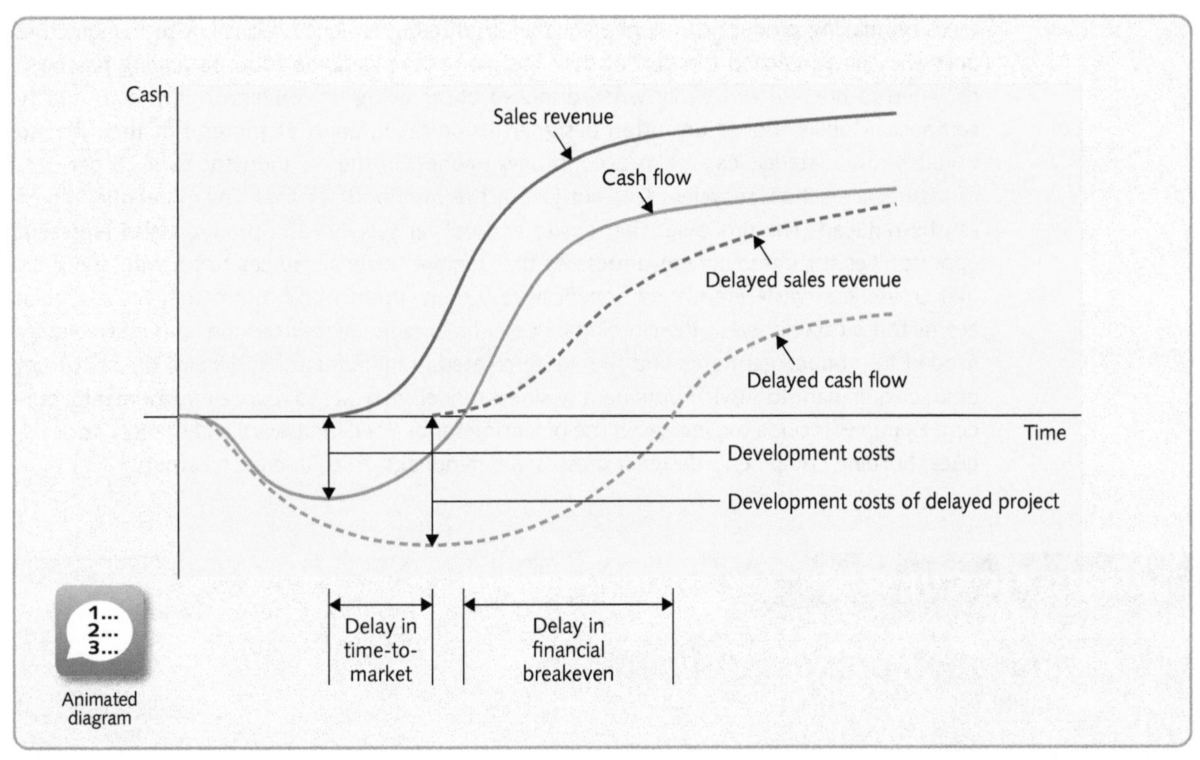

Figure 6.3 Delay in time-to-market of new innovations not only reduces and delays revenues, it also increases the costs of development; the combination of both of these effects usually delays the financial break-even point far more than the delay in the launch

environmental and economic wellbeing of stakeholders. Therefore, it is worth reflecting on a number of initiatives aimed at generating more sustainable innovations.

Some innovation activity is particularly focused on the ethical dimension of sustainability. For example, some banks have moved to offer customers ethical investments which seek to maximise social benefit as well as financial returns. Such investments tend to avoid businesses involved in weaponry, gambling, alcohol and tobacco, for example, and favour those promoting worker education, environmental stewardship and consumer protection. Other examples of ethically focused innovations include the development of 'Fair Trade' products such as bananas, tea, coffee, flowers, chocolate, cotton and handicrafts; clothing manufacturers establishing ethical trading initiatives with suppliers; supermarkets ensuring animal welfare for meat and dairy, and paying fair prices for vegetables; and online companies establishing customer complaint charters.

Innovation may also focus on the environmental dimension of sustainability. Considering the components of a product offering, changing materials in the design can significantly reduce the environmental burden. Examples include the use of: organic cotton or bamboo in clothing; wood or paper from managed forests used in garden furniture, stationery, and flooring; recycled materials for carrier bags; and natural dyes in clothing, curtains and upholstery. Other innovations may be more focused on the use stage of an offering. The MacBook Air, for example, has an advanced power management system that reduces the electricity requirements when being used. In the detergent industry, Persil and Ariel have developed products that allow clothes to be washed at much lower temperatures. Michelin recently developed energy saving tyres that are reported to last 6,000 miles longer than normal tyres and reduce fuel consumption. Architecture firms are increasingly designing houses that can operate with minimal energy or use sustainable sources of energy such as solar panels. Some innovations focus on making product components within an offering easier to recycle or re-manufacture once they have reached the end of their life. For example, some food packaging has been designed to break down easily when disposed of, allowing its conversion into high-quality compost. Mobile phones are often designed to be taken apart at the end of their life, so valuable raw materials can be re-used in new phones. In the car industry, over 75 per cent of materials are now recycled. It is not just in the offering itself that environmental impact can be reduced. The processes that create and deliver services and products also represent opportunities for improvement. Processes that require fewer resources to operate and ones that create less waste are clearly beneficial to the environmental bottom line. For example, eco factories established in the clothing industry have radically reduced the amount of energy needed to produce garments and the waste created in manufacturing. Finally, the shift from product-dominant to service-dominant business models may act to reduce environmental burden. Examples include the increased use of leasing for business hardware and printers, for cars, bikes, building equipment, designer dresses and handbags, suits, and even carpets.[3]

DIAGNOSTIC QUESTION

Is the innovation process defined?

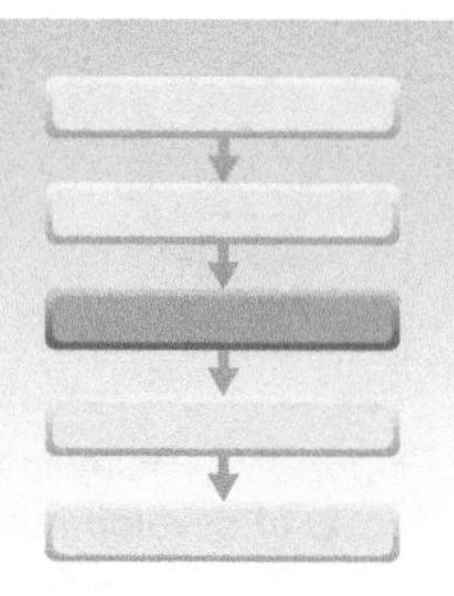

Video

To create a fully specified service or product offering, potential designs must pass through several stages. A typical innovation process is shown in Figure 6.4. Although these exact stages are not used by all companies and there are often re-work loops involved, there is still considerable similarity between the stages and sequence of the innovation process. Furthermore, they all have the same underlying principle: that over time an original idea, or 'concept', is refined and made progressively more detailed until it contains sufficient information to be turned into an actual service, product or process. At each stage in this

OPERATIONS PRINCIPLE

Innovation processes involve a number of stages that move an innovation from a concept to a fully specified state.

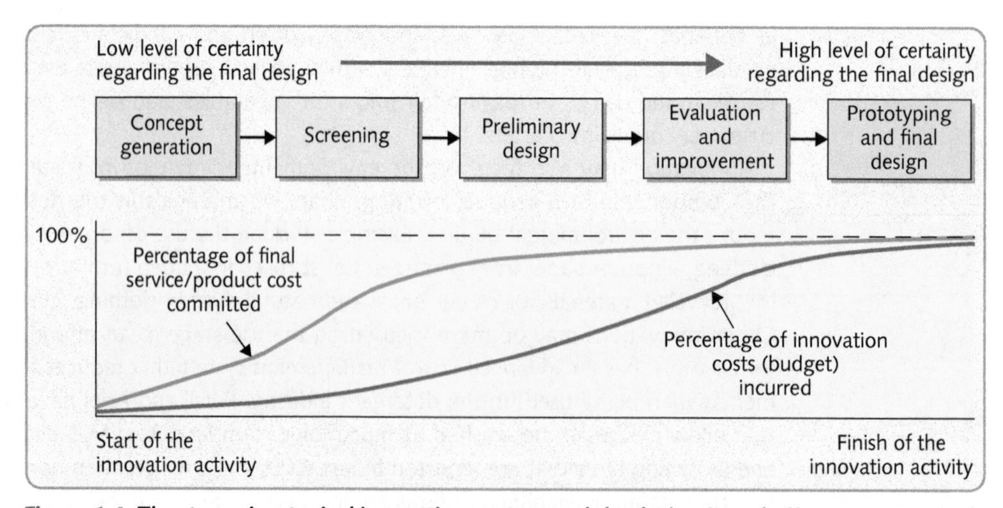

Figure 6.4 The stages in a typical innovation process and the design funnel effect – progressively reducing the number of possibilities until the final design is reached

progression the level of certainty regarding the final design increases as design options are discarded. The final design will not be evident until the very end of the process. Yet relatively early, many of the decisions that will affect the eventual cost of delivery will have been made. For example, choosing to make a mobile telephone's case out of a magnesium alloy will be a relatively early decision that may take little investigation. Yet this decision, although accounting for a small part of the total design budget, may have gone a long way to determining the final cost of the phone. The difference between the 'budget spend' of the innovation process and the actual costs committed by the innovation process are also shown in Figure 6.4.

Concept generation

Concept generation is all about ideas and ideas can come from anywhere! Within organisations, ideas are often expected to emerge from the research and development (R&D) or market research departments. However, this ignores the huge potential of other internal sources of innovation. Front-line service providers, in particular, are able to provide deep insights into what customers require based on informal interactions. For example, the 3M Corporation has been highly successful in generating new innovations by introducing formal incentives to encourage employee engagement. Similarly, while many customer complaints are dealt with at a relatively operational level, they have the potential to act as a useful source of customer opinion within the innovation process.

Outside of the firm, lead users can provide valuable inputs into the innovation process. Lead users are more demanding than the general market, are early adopters of new innovations, and have sufficient understanding to be able to co-develop innovations with a firm. In the other direction, supplier involvement in the innovations process is increasingly popular given its potential to improve the quality of products and services, minimise time-to-market, and spread the cost and risks of innovation. Some firms have gone a step further by outsourcing their research and development function altogether. In Europe and North America, it is now common for up to 30 per cent of R&D budgets to be outsourced.

A quasi form of outsourcing innovation is through the use of 'Skunkworks', a small team who are taken out of their normal working environment and granted the freedom to innovate with minimal managerial constraints. Since its original conception by the Lockheed aircraft company to encourage more radical innovations in high-speed fighter planes, many other companies have looked to use Skunkworks in their innovations processes. For example, Motorola's mobile phone 'Razr' was designed and developed in a special laboratory that the company set up, well away from its main research and development site in Illinois.

Competitors are another useful source of innovation. Reverse engineering can help isolate the key features of an offering which are worth emulating. Some (more back office) aspects of services may be difficult to reverse engineer, but consumer testing can enable educated guesses to be made about how these services have been created. Sometimes competitors choose to collaborate in their innovation efforts, in order to gain mutual benefits for intellectual property. For example, Toyota, Peugeot and Citroën recently collaborated on the 'B-Zero' project, sharing the development costs for their small city cars – the Aygo, Peugot 107 and Citroën C1. The three versions are essentially the same car with slight restyling and are produced in a new factory built in the Czech Republic. Finally, open-sourcing is increasingly popular in software innovation as a way of improving existing technologies. Programmers from different organisations are able to obtain and modify open source code which is then distributed back to the open-source community. A notable example is the development of Mozilla and Mozilla Firefox web browsers based on the open-source code of Netscape Navigator.

Concept screening

Concept screening is the first stage of implementation where potential innovations are considered for further development. It is not possible to translate all concepts into viable product–service packages. For example, DuPont estimates that the ratio of concepts to marketable offerings is around 250 to 1. In the pharmaceuticals industry (see the Novartis case earlier), the ratio is closer to 10,000 to 1. So, organisations need to be selective! The purpose of concept-screening is to take initial concepts and evaluate them for their feasibility (can we do it?), acceptability (do we want to do it?) and vulnerability (what are the risks of doing it?). Concepts may have to pass through many different screens, and several functions might be involved. Table 6.1 gives typical feasibility, acceptability and vulnerability questions for marketing, operations and finance functions.

During concept-screening a key issue to consider is deciding how big the innovation should be and where it should focus – innovation to the customer offering as opposed to innovation to the process of delivery. The vast majority of innovation is continuous or incremental in nature. Here the emphasis is on steady improvement to existing offerings and to the processes that deliver them. This kind of approach to innovation is very much reflected in the lean and total quality management perspectives. On the other hand, some innovation is discontinuous and involves radical change that is 'new to the world'. Discontinuous innovation is relatively rare – perhaps five to ten per cent of all innovations could be classified as such – but creates major challenges for existing players within a market. This is because organisations are often unwilling to disrupt current modes of working in the face of a barely emerging market, but by the time the threat has emerged more fully it may be too late to respond. Clayton Christensen refers to this problem as the 'Innovator's Dilemma', which supports the ideas of the renowned economist Joseph Schumpeter's that innovation should be a process of 'creative destruction'.

Preliminary design

Having generated one or more appropriate concepts, the next stage is to create preliminary designs. For service-dominant offerings this may involve documentation in the form of job instructions or 'service blueprints'. For product-dominant offerings, preliminary design involves defining product specifications (McDonald's has over 50 specifications for the potatoes used for its fries!) and the bill of materials, which details all the components needed for a single product. At this stage, there are significant opportunities to reduce cost through design simplification. The best innovations are often the simplest. Designers can adopt a number of approaches to reduce design complexity.

OPERATIONS PRINCIPLE

A key innovation objective should be the simplification of the design through standardisation, commonality, modularisation and mass customisation.

Table 6.1 Some typical evaluation questions for marketing, operations and finance

Evaluation criteria	*Marketing*	*Operations*	*Finance*
Feasibility	Is the market likely to be big enough?	Do we have the capabilities to deliver it?	Do we have access to finance to develop and launch it?
Acceptability	How much market share could it gain?	How much will we have to reorganise our activities to deliver it?	How much financial return will there be on our investment?
Vulnerability	What is the risk of it failing in the marketplace?	What is the risk of us being unable to deliver it acceptably?	How much money could we lose if things do not go to plan?

Standardisation

This is an attempt to overcome the cost of high variety by standardising offerings, usually by restricting variety. Examples include fast-food restaurants, discount supermarkets or telephone-based insurance companies. Similarly, although everybody's body shape is different, garment manufacturers produce clothes in a limited number of sizes. The range of sizes is chosen to give a reasonable fit for most, but not all, body shapes. Controlling variety is an important issue for most businesses, which all face the danger of allowing variety to grow excessively. Many organisations have significantly improved their profitability by careful variety reduction, often by assessing the real profit or contribution of each service or product.

Commonality

Common elements are used to simplify design complexity. If different services and products can draw on common components, the easier it is to deliver them. An example of this is Airbus, the European aircraft maker, which designed its new generation of aircraft with a high degree of commonality through the introduction of fly-by-wire technology. This meant that ten aircraft models featured virtually identical flight decks, common systems and similar handling characteristics. The advantages of commonality for the airline operators include a much shorter training time for pilots and engineers when they move from one aircraft to another. This offers pilots the possibility of flying a wide range of routes from short haul to ultra-long haul and leads to greater efficiencies because common maintenance procedures can be designed with maintenance teams capable of servicing any aircraft in the same family. Also, when up to 90 per cent of all parts are common within a range of aircraft, there is a reduced need to carry a wide range of spare parts. Similarly, Hewlett-Packard and Black & Decker have used platforms to reduce innovation costs, while Volkswagen's A5 platform is shared by nearly 20 different VW, Skoda, Seat, and Audi models.

Modularisation

This involves designing standardised 'sub-components' of an offering which can be put together in different ways. For example, the package holiday industry can assemble holidays to meet a specific customer requirement, from pre-designed and purchased air travel, accommodation, insurance, and so on. Similarly, in education there is an increasing use of modular courses which allow 'customers' choice but permit each module to have economical volumes of students. Dell uses the same logic for products, drawing together interchangeable subassemblies, that are manufactured in high volumes (a therefore lower cost), and assembled in a wide variety of combinations.

Mass customisation[4]

Flexibility in design can allow the ability to offer different things to different customers. Normally high variety means high cost, but some companies have developed their flexibility in such a way that customised offerings are produced using high-volume processes and thus costs are minimised. This approach is called mass customisation. For example, Paris Miki, an up-market eyewear retailer which has the largest number of eyewear stores in the world, uses its own 'Mikissimes Design System' to capture a digital image of the customer and analyse facial characteristics. Together with a list of customers' personal preferences, the system then recommends a particular design and displays it on the image of the customer's face. In consultation with the optician the customer can adjust shapes and sizes until the final design is chosen. Within the store the frames are assembled from a range of pre-manufactured components and the lenses ground and fitted to the frames. The whole process takes around an hour.

EXAMPLE

Customising for kids

A major challenge facing global programme makers is achieving economies of scale which come as a result of high-volume production while allowing customisation of programmes to suit different markets. 'Art Attack!' made for the Disney Channel, a children's TV channel shown around the world, has used the concept of mass customisation to meet this challenge. Typically, over 200 episodes of the show are made in six different language versions. About 60 per cent of each show is common across all versions. Shots without speaking or where the presenter's face is not visible are shot separately. For example, if a simple cardboard model is being made, all versions will share the scenes where the presenter's hands only are visible. Commentary in the appropriate language is over-dubbed onto the scenes which are edited seamlessly with other shots of the appropriate presenter. The final product will have the head and shoulders of Brazilian, French, Italian, German or Spanish presenters flawlessly mixed with the same pair of hands constructing the model. The result is that local viewers in each market see what appears to be a highly customised show, while the cost of making each episode is about one third of producing truly separate programmes for each market.

Design evaluation and improvement

The purpose of this stage in the design activity is to take the preliminary design and see if it can be improved before the service–product offering is tested in the market. There are a number of techniques that can be employed at this stage to evaluate and improve the preliminary design. Perhaps the best known is quality function deployment (QFD). The key purpose of QFD is to try to ensure that the eventual innovation actually meets the needs of its customers. It is a technique that was developed in Japan at Mitsubishi's Kobe shipyard and used extensively by Toyota, the motor vehicle manufacturer, and its suppliers. It is also known as the 'house of quality' (because of its shape) and the 'voice of the customer' (because of its purpose). The technique tries to capture what the customer needs and how it might be achieved. Figure 6.5

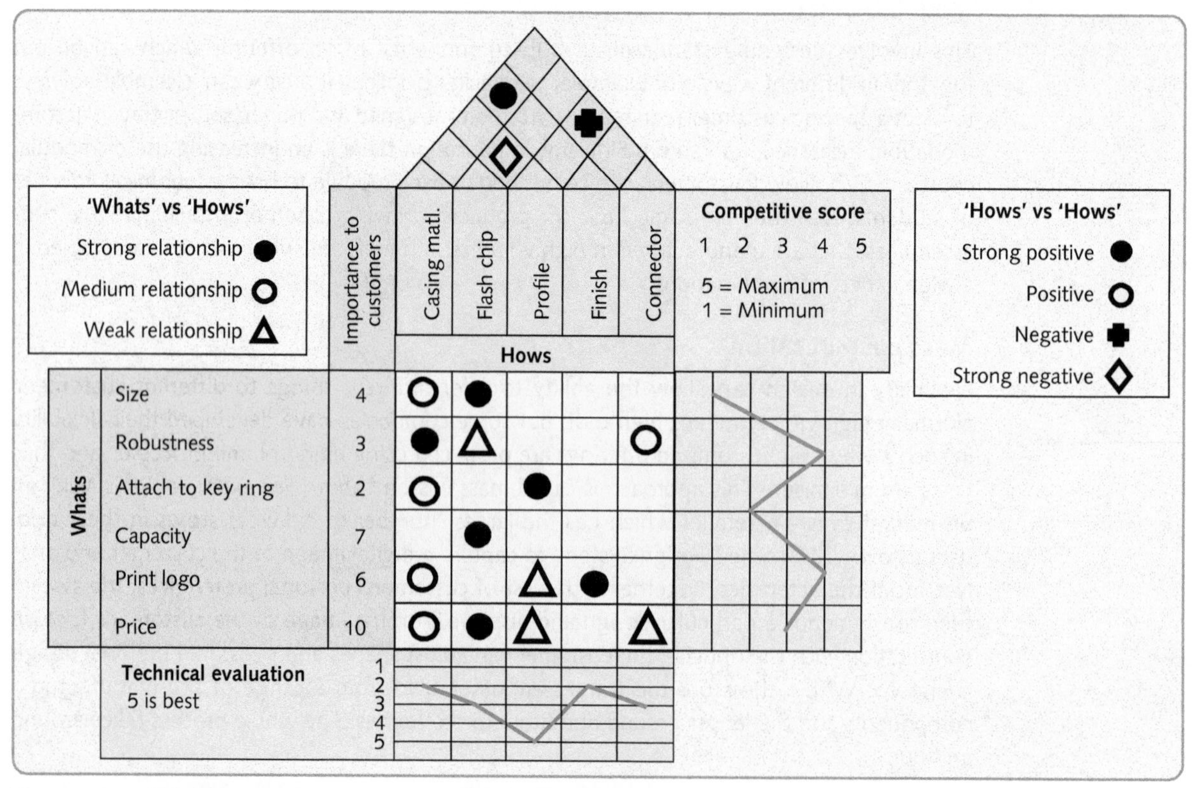

Figure 6.5 A QFD matrix for a promotional USB data storage pen

shows a simple QFD matrix used in the design of a promotional USB data storage pen. It is a formal articulation of how designers see the relationship between the requirements of the customer and the design characteristics of the offering.

It is at this stage in the process when both creativity and persistence are needed to move from a potentially good idea to a workable design. One product has commemorated the persistence of its design engineers in its company name. Back in 1953, the Rocket Chemical Company set out to create a rust-prevention solvent and degreaser to be used in the aerospace industry. Working in their lab in San Diego, California, it took them 40 attempts to get the water displacing formula worked out. So that is what they called the product. WD-40 literally stands for Water Displacement, fortieth attempt. Originally used to protect the outer skin of the Atlas Missile from rust and corrosion, the product worked so well that employees kept taking cans home to use for domestic purposes. Soon after, the product was launched with great success into the consumer market.

In fact, it's not just persistence that is important in the innovation process – failure itself can be beneficial if organisations can spot potential. Sometimes, when a design fails, it represents an opportunity to re-think the concept itself. For example, Pritt Stick, the world's first glue stick, was originally intended to be a super-glue, but product-testing proved unsatisfactory. So, Henkel, changed the product concept and successfully marketed the product as 'the non-sticky sticky stuff'!

EXAMPLE

Viagra - a success from a failure[5]

Sildenafil was developed by a group of chemists working for the pharmaceutical giant Pfizer's. The product was originally intended to help individuals with hypertension (high blood pressure) and angina. However, clinical trials proved somewhat unsuccessful, though doctors noted a side-effect of penile erections. Seeing the potential of this 'failed' innovation process, Pfizer marketed the drug, Viagra, for erectile dysfunction. In just two years, sales of Viagra had topped $1billion and the product has dominated the market ever since.

Prototyping and final design

At around this stage in the innovation activity it is necessary to turn the improved design into a prototype so that it can be tested. Product prototypes include everything from clay models to computer simulations. For more service-dominant offerings, prototyping often involves the actual implementation of the service on a pilot basis. For example, a retailer may organise piloting new services packages in a small number of stores in order to test customers' reaction to them. It may also be possible to 'virtually' prototype in much the same way as a product. This is a familiar idea in some industries such as magazine publishing, where images and text can be rearranged and subjected to scrutiny prior to them existing in any physical form, allowing them to be amended right up to the point of production. Although the means of prototyping may vary, the principle is always the same: do whatever one can to test out the innovation prior to delivery. Doing so minimises the likelihood of launching an unsuccessful innovation into the market and the financial and reputational risks that come with such a scenario.

DIAGNOSTIC QUESTION

Are the resources for developing innovations adequate?

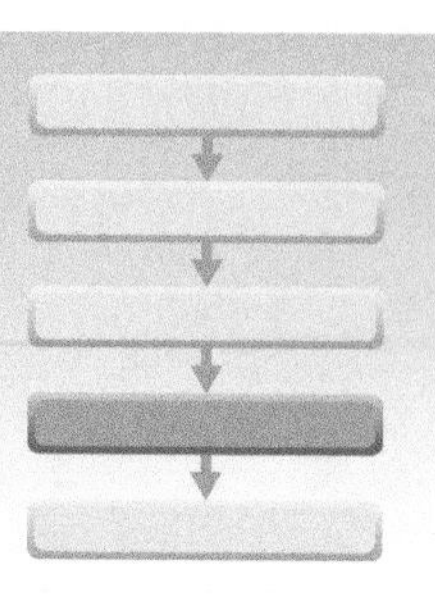

For any process to operate effectively, it must be appropriately designed and resourced. Innovation processes are no different. The detailed principles of process design that were discussed in Chapters 4 and 5 are of course applicable to innovation processes. However, because innovation processes are often an operation within the business in their own right, there are some more strategic issues to consider – how much capacity to devote to innovation, how much of the innovation activity to outsource, and what kinds of technology to use in the innovation process.

OPERATIONS PRINCIPLE

For innovation processes to be effective they must be adequately resourced.

Is there sufficient innovation capacity?

In general, capacity management involves deciding on the appropriate level of capacity needed by a process and how it can be adjusted to respond to changes in demand. In the case of innovation, demand is the number of new designs needed by the business. The chief difficulty is that, even in very large companies, the rate of new innovation is not constant. This means that product and service design processes are subjected to uneven internal 'demand' for designs, possibly with several new offerings being introduced to the market close together, while at other times little innovation is needed. This poses a resourcing problem because the capacity of an innovation activity is often difficult to flex. The expertise necessary for innovation is embedded within designers, technologists, market analysts, and so on. It may be possible to hire some expertise as and when it is needed, but much design resource is, in effect, fixed.

Such a combination of varying demand and relatively fixed design capacity means some organisations are reluctant to invest in innovation processes because they see it as an underutilised resource. This may lead to a vicious cycle in which companies fail to invest in innovation resources because many skilled design staff cannot simply be hired in the short term. This leads to innovation projects overrunning or failing to deliver appropriate solutions. This in turn may lead to the company losing business or otherwise suffering in the marketplace, which makes the company even less willing to invest in innovation resources. This issue relates to the relationship between capacity utilisation and throughput time that was discussed in Chapter 4. Either the business must accept relatively low utilisation of its innovation resources if it wants to maintain fast time-to-market, or, if it wants to maintain high levels of design resource utilisation, it must accept longer design times, or it must try and reduce the variability in the process in some way. Reducing variability may mean introducing new designs at fixed periods, for example, every year. See Figure 6.6.

Should all innovation activities be done in-house?

Just as there are supply networks that produce services and products, there are also supply networks of knowledge that connect suppliers and customers in the innovation process. This network is sometimes called the 'design (or development) network'. Innovation processes can adopt any position on a continuum of varying degrees of design engagement with suppliers, from retaining all the innovation capabilities in-house, to outsourcing all its innovation work.

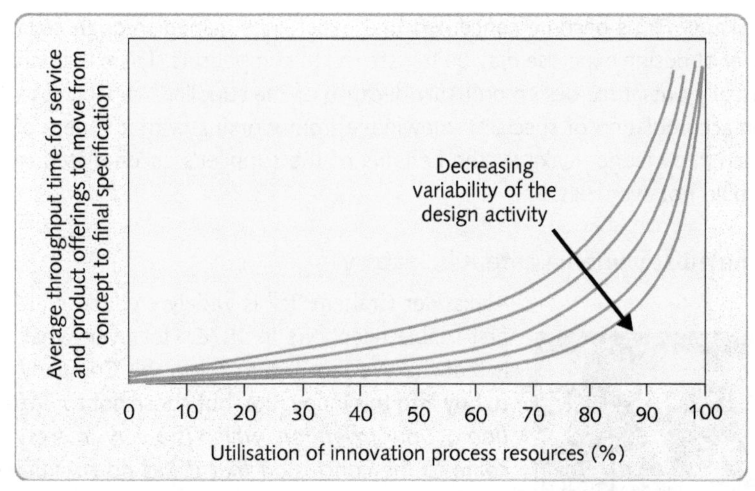

Figure 6.6 The relationship between innovation resource utilisation, innovation process throughput time, and innovation process variability

Between these extremes are varying degrees of internal and external capability. Figure 6.7 shows some of the more important factors that will vary depending on where an innovation process is on the continuum. Resources will be easy to control if they are kept in-house because they are closely aligned with the company's normal organisational structures, but control should be relatively loose because of the extra trust present in working with familiar colleagues. Outsourced design often involves greater control, with the use of contract with penalty clause for delay commonly used.

The overall cost of in-house versus outsourced innovation will vary, depending on the firm and the project. An important difference, however, is that external innovations tend to be regarded as a variable cost. The more external resources are used, the more these variable costs will be. In-house innovation is more of a fixed cost. Indeed a shift to outsourcing may occur because fixed design costs are viewed as too great. From an open innovation perspective[6] it is argued that firms should be willing to buy in or license inventions for other organisations, rather than relying solely on innovations generated internally. Similarly, it may be beneficial to give access to under-used proprietary innovations through joint ventures, licensing or spin-offs. However, a major inhibitor to open innovation is the fear of knowledge

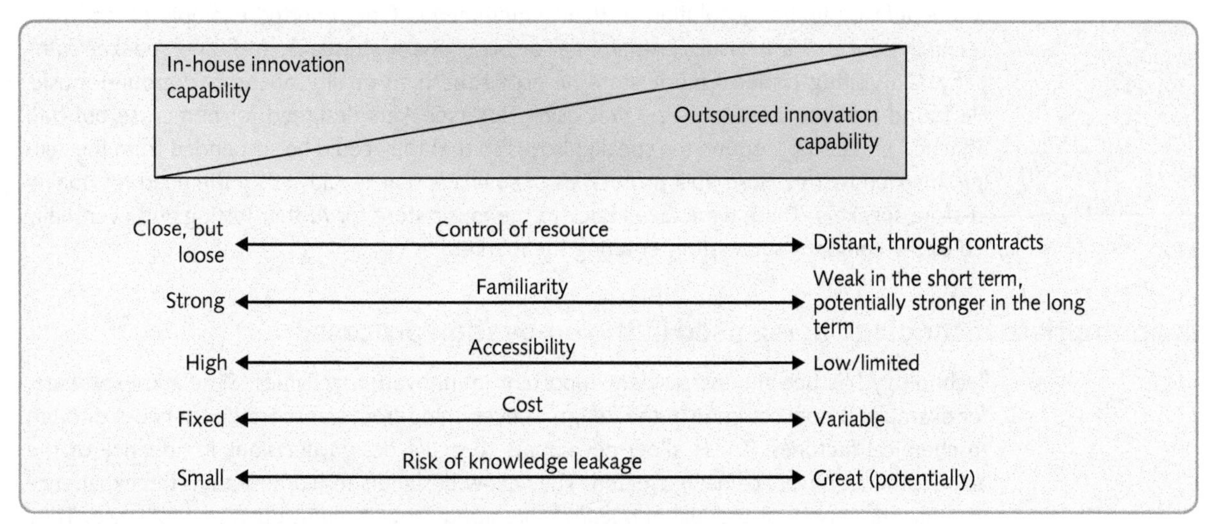

Figure 6.7 Some implications of the in-house–outsourced continuum

leakage. Firms become concerned that experience gained through collaboration with a supplier of design expertise may be transferred to competitors. There is a paradox here. Businesses usually outsource design primarily because of the supplier's capabilities which are themselves an accumulation of specialist knowledge from working with a variety of customers. Without such knowledge 'leakage' the benefits of the supplier's accumulated innovation capabilities would not even exist.

EXAMPLE

The most valuable patent in history

Source: North Wind Picture Archives/Alamy Images

Alexander Graham Bell is widely credited with the invention of the first usable telephone in 1876. However, what is less widely known is that the Western Union Telegraph Company had the opportunity to buy into this innovation but chose not to: *'After careful consideration of your invention, which is a very interesting novelty, we have come to the conclusion that it has no commercial possibilities . . . we see no future for an electrical toy . . .'* Within four years, 50,000 telephones had been sold in the USA and within 20 years the figure had reached 5 million. The patent, 174–465, is recognised as the most valuable in history and essentially created the telephone industry that we know today. This example illustrates the importance of drawing innovation from outside of the organisation. Perhaps one of the reasons why Western Union were so reluctant to invest in Bell's innovation was that is wasn't their idea! This so-called 'not-in-house' syndrome is surprisingly common – take for example Hoover's rejection of (but subsequent attempts to imitate) Dyson's radical innovations in the vacuum cleaner market.

Involving customers in the innovation activity

Few people know the merits and limitations of service and product offerings better than the customers who use them, which make them an obvious source of innovation. Different types of customer have the potential to provide different types of information. New users can pinpoint more attractive features of an offering while those who switch to a competitor can reveal its problems. A particularly interesting group of customers are the so called 'lead users', who have requirements of a service or product well ahead of the rest of the market, and who will benefit by finding a solution to their requirements. One reported example of lead user research concerns a new design manager at Bose, the high quality hi-fi and speaker company. On visiting his local music store he noted the high quality of the background music. He found that the store manager was using Bose speakers designed for home use but had attached metal strips around the speaker boxes so that they could be suspended from the ceiling. Inspired by this, Bose built prototypes of speakers that would satisfy the need for quality in-store speakers. These were taken back to the music store for further testing and eventually led to the company successfully entering the market.

Is appropriate technology being used in the innovation process?

Technology has become increasingly important in innovation activities. Simulation software, for example, is now common in the design of everything from transportation services through to chemical factories. These allow developers to make design decisions in advance of the actual product or service being created. They allow designers to work through the experience of using the service or product and learn more about how it might operate in practice. They can explore possibilities, gain insights and, most important, they can explore the consequences of their decisions. Innovation technologies are particularly useful when the design task is highly complex, because they allow developers to reduce their own uncertainty of how services or products will work in practice. Technologies also consolidate information on what is happening in the innovation process, thus presenting a more comprehensive vision within the organisation.

Computer-aided design (CAD)

The best-known innovation technology is computer-aided design (CAD). CAD systems store and categorise component information and allow designs to be built up on screen, often performing basic engineering calculations to test the appropriateness of proposed design solutions. They provide the computer-aided ability to create a modified product drawing and allow conventionally used shapes to be added swiftly to the computer-based representation of a product. Designs created on screen can be saved and retrieved for later use, which enables a library of standardised parts and components to be built up. Not only can this dramatically increase the productivity of the innovation process, it also aids the standardisation of parts in the design activity. For complex projects, the absolute size and interrelatedness of different aspects of the work requires sophisticated CAD systems to be successful. The powerful CAD system used on Boeing's 777 aircraft development project was credited with success in being able to involve its customers in the design, allowing more product configuration flexibility (such as the proportion of seats in each class, etc.) while still bringing this huge project to successful completion.

Extensions to Computer-aided design

Design for manufacture and assembly (DFMA) software is an extension of CAD, which allows those involved in the innovation process to integrate their designs prior to manufacture. For example, DFMA could be used to see how a new gun turret design will fit onto armoured vehicles during production, before either the turret or the vehicle has been made physically. 3-D object modelling is a rapid prototyping technique aimed at reducing the time taken to create physical models of products. Designs from CAD are created using machines that build models by layering up extremely thin (usually 0.05-mm) layers of photopolymer resin. An alternative to 3-D modelling is the use of virtual reality technologies. Here CAD information is converted into virtual images which can be viewed using 3-D glasses. This form of technology is more interactive than traditional CAD as designers (or customers) can 'walk around' a design and get a better sense of what it looks and feels like to be in. This means virtual reality technology is especially useful for designs that customers would be inside, such as sports venues, aeroplanes, buildings and amusement parks, for example.

Knowledge management technologies

In many professional service firms, such as management consultancies, design involves the evaluation of concepts and frameworks which can be used in client organisations to diagnose problems, analyse performance and construct possible solutions. They may include ideas of industry best practice, benchmarks of performance within an industry, and ideas which can be transported across industry boundaries. However, management consulting firms are often geographically dispersed and staff are often away from the offices, spending most of their time in client organisations. This creates a risk for such companies of 'reinventing the wheel' continually. Most consultancy companies attempt to tackle this risk by using knowledge management routines based on their intranet capabilities. This allows consultants to put their experience into a common pool, contact other staff within the company who have skills relevant to a current assignment, and identify previous similar assignments. In this way information is integrated into the ongoing knowledge innovation process within the company and can be tapped by those charged with developing new innovations.

DIAGNOSTIC QUESTION

Is the design of the offering and of the process simultaneous?

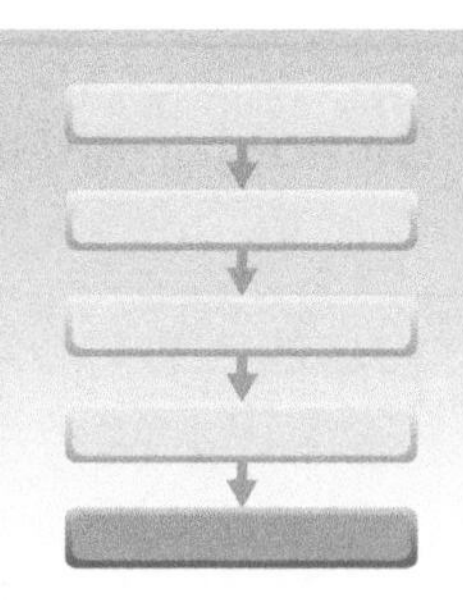

The outputs from the innovation, in the form of new service and product offerings, are important inputs into the processes that create and deliver them on an ongoing basis. This is why it is a mistake to separate the design of services and products from the design of the processes that will deliver them. Merging the innovation process for new service and product offerings with the processes that create them is sometimes called simultaneous (or interactive) design. Its key benefit is to reduce the elapsed time taken for the whole innovation activity. As noted earlier, reducing time-to-market (TTM) gives increased competitive advantage. For example, if it takes a business two years to develop an offering with a given set of resources, it can only introduce new offerings every two years. If its rival can develop offerings in one year, it can introduce new innovations into the market twice as often. This means the rival does not have to make such radical improvements in performance each time it introduces a new offering because it is introducing them more frequently. The factors that can significantly reduce time to market for innovations include the following:

OPERATIONS PRINCIPLE

Effective simultaneous innovation reduces time-to-market.

- integrating the design of the product–service offering and the design of the process used to create and deliver them
- overlapping the stages in the innovation process
- an early deployment of strategic decision making and resolution of design conflict
- an organisational structure that reflects the nature of the offering.

Integrating the design of the offering and the design of the process

What looks good as an elegant offering on paper may prove difficult to create and deliver on an ongoing basis. Conversely, a process designed for one set of services or products may be incapable of creating different ones. It clearly makes sense to design offerings and operations processes together. For service-dominated offerings, organisations have little choice but to do this because the process of delivery is usually *part* of the offering. However, it is useful to integrate the design of the offering and the process regardless of the kind of organisation. The fact that many businesses do not do this is only partly because of their ignorance or incompetence. There are real barriers to doing it. First, the timescales involved can be very different. Offerings may be modified, or even redesigned, relatively frequently. The processes that will be used to create and deliver an offering may be far too expensive to modify every time the offering changes. Second, the people involved with the innovation on one hand, and ongoing process design on the other, are likely to be organisationally separate. Finally, it is sometimes not possible to design an ongoing process for the creation and delivery of services and products until they are fully defined.

Yet none of these barriers is insurmountable. Although it may not be possible to change ongoing processes every time there is an alteration to the offering, they can be designed to cope with a range of potential services and products. The fact that design staff and operations staff are often organisationally separate can also be overcome. Even if it is not sensible to merge the two functions, there are communication and organisational mechanisms to encourage the two functions to work together. Even the claim that ongoing processes cannot be designed until they know the nature of the offering is not entirely true. There can be sufficient clues emerging from innovation activities for process design staff to consider how they might modify ongoing processes. This fundamental principle of simultaneous design is considered next.

Overlapping the stages of the innovation process

We have described the innovation process as a set of individual, predetermined stages, with one stage being completed before the next one commences. This step-by-step, or sequential, approach has been commonly applied in many organisations. It has some advantages. It is easy to manage and control innovation processes organised in this way because each stage is clearly defined. In addition, each stage is completed before the next stage is begun, so each stage can focus its skills and expertise on a limited set of tasks. The main problem of the sequential approach is that it is both time-consuming and costly. When each stage is separate, with a clearly defined set of tasks, any difficulties encountered during the design at one stage might necessitate the design being halted while responsibility moves back to the previous stage. This sequential approach is shown in Figure 6.8(a).

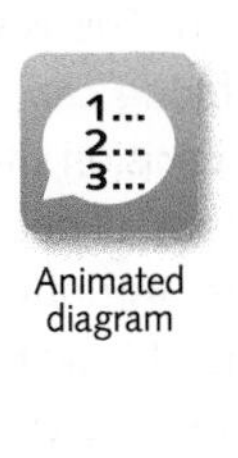

Animated diagram

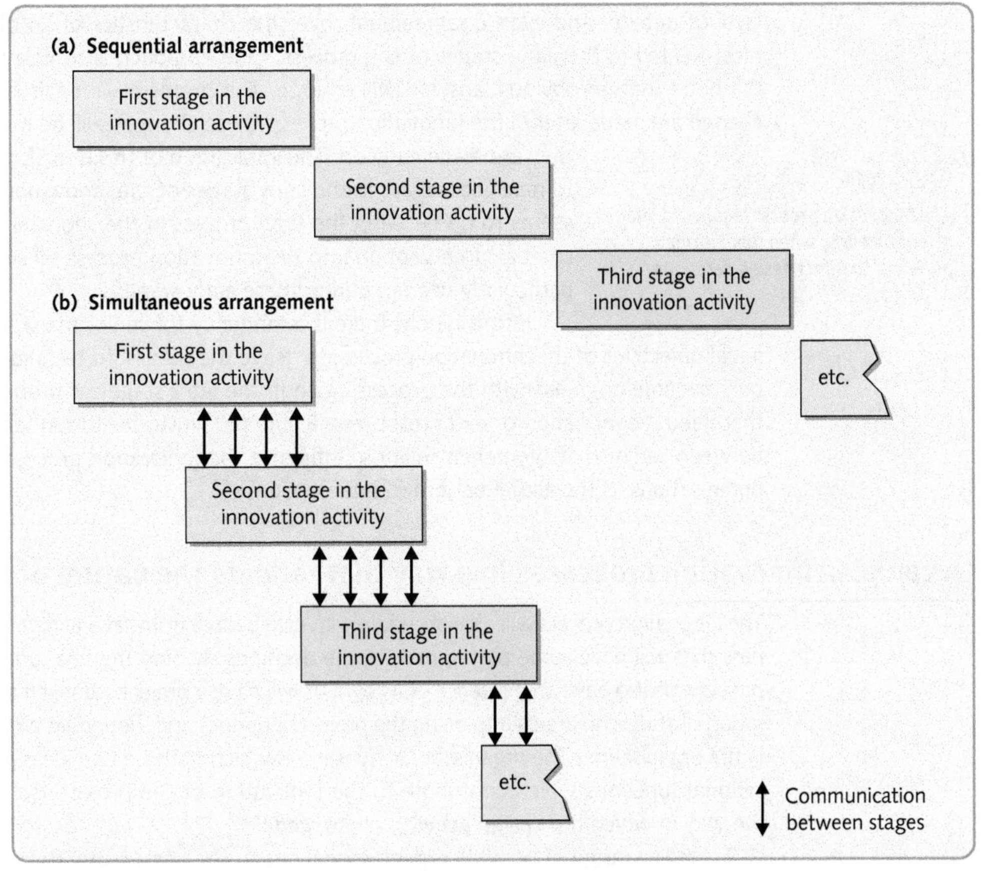

Figure 6.8 (a) Sequential arrangement of the stages in the innovation activity; (b) simultaneous arrangement of the stages in the innovation activity

Often there is really little need to wait until the absolute finalisation of one stage before starting the next[7]. For example, perhaps while generating the concept, the evaluation activity of screening and selection could be started. It is likely that some concepts could be judged as 'non-starters' relatively early on in the process of idea generation. Similarly, during the screening stage, it is likely that some aspects of the design will become obvious before the phase is finally complete. Therefore, the preliminary work on these parts of the design could be commenced at that point. This principle can be taken right through all the stages, one stage commencing before the previous one has finished, so there is simultaneous or concurrent work on the stages (see Figure 6.8(b)).

We can link this idea with the idea of uncertainty reduction, discussed earlier, when we made the point that uncertainty reduces as the design progresses. This also applies to each stage of innovation. If this is the case, then there must be some degree of certainty which the next stage can take as its starting point prior to the end of the previous stage. In other words, designers can be continually reacting to a series of decisions and clues which are given to them by those working on the preceding stage. However, this can only work if there is effective communication between each pair of stages.

Deploying strategic intervention and resolving conflicts early

Animated diagram

A design decision, once made, need not irrevocably shape the final offering. All decisions can be changed, but it becomes increasingly difficult to do so as the innovation process progresses. At the same time, early decisions are often the most difficult to make because of the high level of uncertainty surrounding what may or may not work as a final design. This is why the level of debate, and even disagreement, over the characteristics of an offering can be at its most heated in the early stages of the process. One approach is to delay decision making in the hope that an obvious 'answer' will emerge. The problem with this is that, if decisions to change are made later in the innovation process, these changes will be more disruptive than if they are made early on. The implication of this is, first, that it is worth trying to reach consensus in the early stages of the innovation process even if this seems to be delaying the total process in the short term, and second, that strategic intervention into the innovation process by senior management is particularly needed during these early stages.

OPERATIONS PRINCIPLE

The innovation process requires strategic attention early, when there is most potential to affect design decisions.

Unfortunately, there is a tendency for senior managers, after setting the initial objectives of the innovation process, to 'leave the details' to technical experts. They may only become engaged with the process again in the later stages as problems start to emerge that need reconciliation or extra resources. Figure 6.9 illustrates this in terms of the mismatch between senior management's ability to influence the innovation process and what, in many organisations, is the actual pattern of intervention.

Organising innovation processes in a way that reflects the nature of the offering

The innovation process will almost certainly involve people from several different areas of the business that will have some part in making the decisions shaping the final offering. Yet any design project will also have an existence of its own. It will have a project name, an individual manager or group of staff who are championing the project, a budget and, hopefully, a clear strategic purpose in the organisation. The organisational question is which of these two ideas – the various organisational functions which contribute to the innovation or the project itself – should dominate the way in which the design activity is managed?

There is a range of possible organisational structures – from pure functional to pure project forms. In a purely functional organisation, all staff associated with the innovation project are based unambiguously in their functional groups. There is no project-based group at all. They

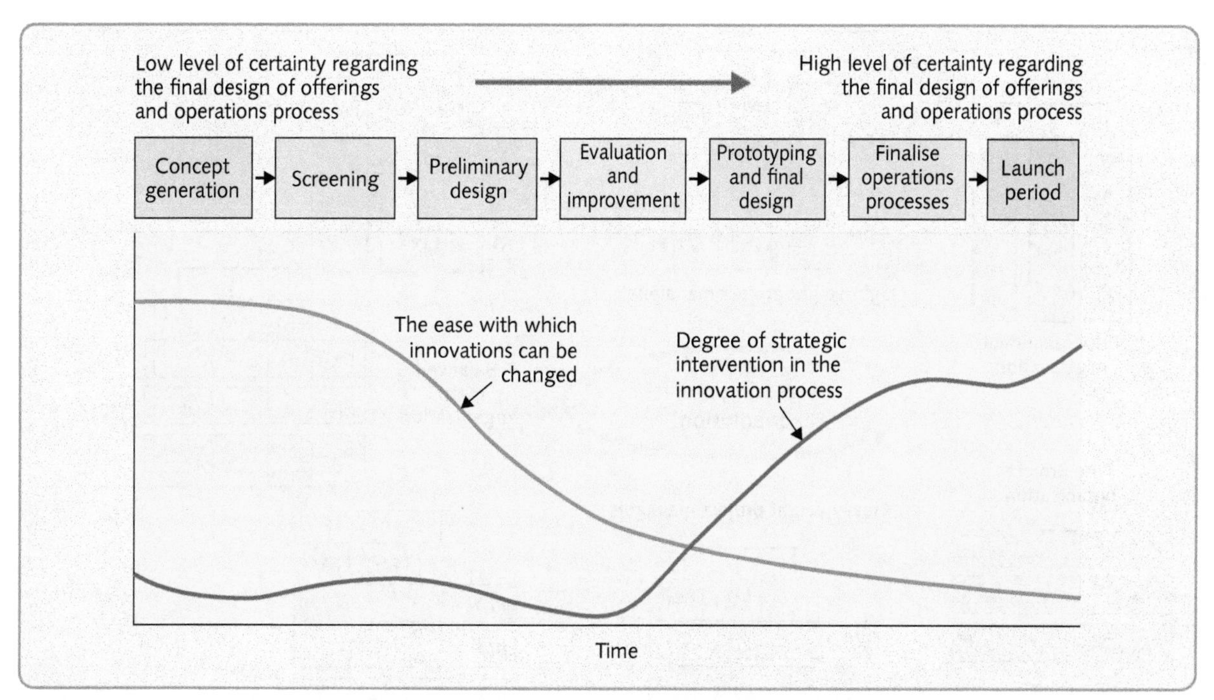

Figure 6.9 The degree of strategic intervention in the innovation process is often dictated by the need to resolve outstanding conflicts rather than the needs of the innovation process itself

may be working full-time on the project but all communication and liaison are carried out through their functional manager. At the other extreme, all the individual members of staff from each function involved in an innovation project could be moved out of their functions and perhaps even co-located in a task force dedicated solely to the project. The task force could be led by a project manager who might hold the entire budget allocated to the innovation project. Not all members of the task force necessarily have to stay in the team throughout the design period, but a substantial core might see the project through from start to finish. Some members of a design team may even be from other companies. In between these two extremes there are various types of matrix organisation with varying emphasis on these two aspects of the organisation[8] (see Figure 6.10).

- *Functional organisation.* The innovation project is divided into segments and assigned to relevant functional areas and/or groups within functional areas. The project is co-coordinated by functional and senior management.
- *Functional matrix* (or *lightweight* project manager). A person is formally designated to oversee the project across different functional areas. This person may have limited authority over the functional staff involved and serves primarily to plan and coordinate the project. Functional managers retain primary responsibility for their specific segments of the project.
- *Balanced matrix.* A person is assigned to oversee the project and interacts on an equal basis with functional managers. This person and the functional managers work together to direct innovation activities and approve technical and operational decisions.
- *Project matrix* (or *heavyweight* project manager). A manager is assigned to oversee the project and is responsible for its completion. Functional managers' involvement is limited to assigning personnel as needed and providing advisory expertise.
- *Project team* (or *tiger* team). A manager is given responsibility of a project team composed of a core group of personnel from several functional areas assigned on a full-time basis. The functional managers have no formal involvement.

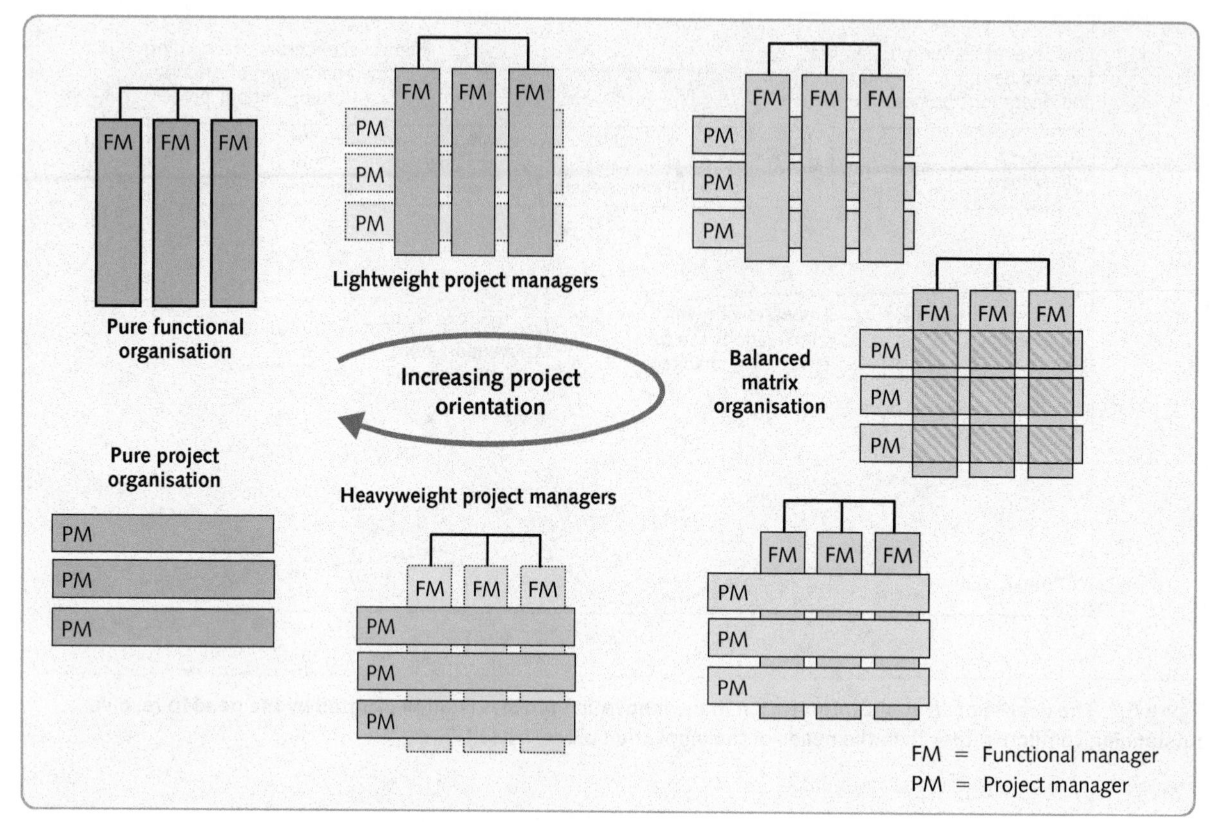

Figure 6.10 Organisation structures for innovation processes

Although there is no clear 'winner' among the alternative organisational structures, there is increasing support for structures towards the project rather than functional end of the continuum. Some authorities argue that heavyweight project manager structures and dedicated project teams are the most efficient forms of organisation in driving competitiveness, shorter lead times and technical efficiency. Other studies, although sometimes more equivocal, have shown that, in terms of the best total outcome from the development process, structures from balanced matrix through to project teams can all give high success rates.

Perhaps of more interest is the suitability of the alternative structures for different types of innovation. Matrix structures are generally deemed to be appropriate for both simple and highly complex projects. Dedicated project teams, on the other hand, are seen as appropriate for projects with a high degree of uncertainty, where their flexibility becomes valuable. Functionally based design structures, with resources clustered around a functional specialism, help the development of technical knowledge. Some organisations do manage to capture the deep technological and skills development advantages of functional structures, while at the same time co-coordinating between the functions so as to ensure satisfactory delivery of new service and product ideas. Perhaps the best known of these organisations is Toyota. It has a strong functionally based organisation to develop their offerings. It adopts highly formalised development procedures to communicate between functions and places strict limits on the use of cross-functional teams. But what is really different is their approach to devising an organisational structure for innovation which is appropriate for them. The argument which most companies have adopted to justify cross-functional project teams goes something like this: *'Problems with communication between traditional functions have been the main reasons for failing to deliver new innovation ideas to specification, on time and to budget. Therefore let us break down the walls between the functions and organise resources around the individual development projects. This will ensure good communication and a market-oriented culture'.* Toyota and similar companies on the other hand, have taken a different approach. Their argument goes something like this: *'The problem with cross-functional teams is that they can dissipate the carefully nurtured knowledge that exists within specialist functions. The real problem is how to retain this knowledge on which our future innovation depends, while overcoming some of the traditional functional barriers which have inhibited communication between the functions. The solution is not to destroy the function but to devise the organisational mechanisms to ensure close control and integrative leadership which will make the functional organisation work.'*[9]

Critical commentary

Each chapter contains a short critical commentary on the main ideas covered in the chapter. Its purpose is not to undermine the issues discussed in the chapter, but to emphasise that, although we present a relatively orthodox view of operation, there are other perspectives.

- The whole process-based approach to innovation could be interpreted as implying that all new offerings are created in response to a clear and articulated customer need. While this is usually the case, especially for services and products that are similar to (but presumably better than) their predecessors, more radical innovations are often brought about by the innovation itself creating demand. Customers don't usually know that they need something radical. For example, in the late 1970s people were not asking for micro-processors, they did not even know what they were. They were improvised by an engineer in the USA for a Japanese customer who made calculators. Only later did they become the enabling technology for the PC and after that the innumerable devices that now dominate our lives. Similarly, fly-by-wire, digital cameras, Maersk's super-slow container ships, sushi on conveyor belts, and the iPad, are all examples of innovations that have been 'pushed' by firms rather than 'pulled' by pre-existing customer demand.

- Nor does everyone agree with the dominant rational model[10] in which possible design options are progressively reduced stage by stage through the optimisation of known constraints and objectives. For some this neat model of the innovation, which underlies much business and engineering design literature, fails to accurately reflect the creativity, arguments and chaos that sometimes characterise real innovation projects. First, they argue, managers do not start out with an infinite number of options. No one could process that amount of information – and anyway, designers often have some set solutions in their mind, looking for an opportunity to be used. Second, the number of options being considered often increases as time goes by. This may actually be a good thing, especially if the activity was unimaginatively specified in the first place. Third, the real process of innovation involves cycling back, often many times, as potential solutions raise fresh questions or become dead ends, and as requirements and constraints evolve. In summary, the idea of the design funnel does not describe the process of innovation nor does it necessarily even describe what should happen. The action-centric or co-evolution[11] perspective of innovation represents the antithesis of the rational model. It posits that offerings are designed through a combination of emotion and creativity; that the process by which this is done is generally improvised; and the sequencing of stages is not universal in innovation processes.

SUMMARY CHECKLIST

This checklist comprises questions that can be usefully applied to any type of operations and reflect the major diagnostic questions used within the chapter.

- ☐ Is the importance of innovation as a contributor to achieving strategic impact fully understood?
- ☐ Is innovation really treated as a process?
- ☐ Is the innovation process itself designed with the same attention to detail as any other process?
- ☐ Are innovation objectives specified so as to give a clear priority between quality, speed, dependability, flexibility, cost and sustainability?
- ☐ Are the stages in the innovation process clearly defined?
- ☐ Are ideas and concepts for new offerings captured from all appropriate internal and external sources?
- ☐ Are potential offerings screened in a systematic manner in terms of their feasibility, acceptability and vulnerability?
- ☐ During preliminary design, have all possibilities for design standardisation, commonality and modularisation of design elements been explored?
- ☐ Has the concept of mass customisation been linked to the innovation process?
- ☐ Are potential offerings thoroughly evaluated and tested during the innovation process?
- ☐ Are the resources for developing innovation adequate?
- ☐ Is sufficient capacity devoted to the innovation process?
- ☐ Have all options for outsourcing parts of the innovation process been explored?
- ☐ Has the possibility of involving customers formally in the innovation process been explored?
- ☐ Are appropriate technologies such as CAD, extensions of CAD, and knowledge management, being used in the innovation process?
- ☐ Are the design of offering and the design of processes that create and deliver it considered together as one integrated process?
- ☐ Is overlapping (simultaneous) design of the stages in the innovation process used?
- ☐ Is senior management effort deployed early enough to ensure early resolution of design conflict?
- ☐ Does the organisational structure of the innovation process reflect the nature of the offering?
- ☐ Are some functions of the business more committed to innovating new service and product offerings than others?
- ☐ If so, have the barriers to cross-functional commitment been identified and addressed?

CASE STUDY

Developing 'Savory Rosti' crisps' at Dreddo Dan's

'Most people see the snack market as dynamic and innovative, but actually it is surprisingly conservative. Most of what passes for innovation is in fact tinkering with our marketing approach, things like special offers, promotion tie-ins and so on. We occasionally put new packs round our existing products and even more occasionally we introduce new flavours in existing ranges. Rarely though does anyone in this industry introduce something radically different. That is why 'Project Orlando' is both exciting and scary.'

Monica Allen, the Technical Vice-President of PJT's snack division, was commenting on a new product to be marketed under PJT's best-known brand 'Dreddo Dan's Surfer Snacks'. The Dreddo Dan's brand made use of surfing and outdoor 'action-oriented youth' imagery, but in fact was aimed at a slightly older generation who, although aspiring to such a lifestyle, had more discretionary spend for the premium snacks in which the brand specialised. Current products marketed under the brand included both fried and baked snacks in a range of exotic flavours. The project, internally known as Project Orlando, was a baked product that had been 'in development' for almost three years but had hitherto been seen very much as a long-term development, with no guarantee of it ever making it through to market launch. PJT had several of these long-term projects running at any time. They were allocated a development budget, but usually no dedicated resources were associated with the project. Less than half of these long-term projects ever even reached the stage of being test marketed. Around 20 per cent never got past the concept stage, and less than 20 per cent ever went into production. However, the company viewed the development effort put into these 'failed' products as being worthwhile because it often led to 'spin-off' developments and ideas that could be used elsewhere. Up to this point 'Orlando' had been seen as unlikely ever to reach the test marketing stage, but that had now changed dramatically.

'Orlando' was a concept for a range of snack foods, described within the company as 'savoury potato cookies'. Essentially they were 4-cm discs of crisp, fried potato with a soft dairy cheese-like filling. The idea of incorporating dairy fillings in snacks had been discussed within the industry for some time but the problems of manufacturing

Source: Mediablitzimages (UK) Ltd/Alamy Images

such a product were formidable. Keeping the product crisp on the outside yet soft in the middle, while at the same time ensuring microbiological safety, would not be easy. Moreover such a product would have to be capable of being stored at ambient temperatures, maintain its physical robustness and have a shelf life of at least three months.

Bringing Orlando products to market involved overcoming three types of technical problem. First, the formulation and ingredient mix for the product had to maintain the required texture, yet be capable of being baked on the company's existing baking lines. The risk of developing entirely new production technology for the offering was considered too great. Second, extruding the mixture into baking moulds while maintaining microbiological integrity (dairy products are difficult to handle) would require new extrusion technology. Third, the product would need to be packaged in a material that both reflected its brand image but also kept the product fresh through its shelf life. Existing packaging materials were unlikely to provide sufficient shelf life. The first of these problems had, more or less, been solved in PJT's development laboratories. The second two problems now seemed less formidable because of a number of recent technological breakthroughs made by equipment suppliers and packaging manufacturers. This had convinced the company that Orlando was worth significant investment and it had been given priority development status by the company's board. Even so, it was not expected to come to the market for another two years and was seen by some as potentially the most important new product development in the company's history.

The project team

Immediately after the board's decision, Monica had accepted responsibility to move the development forward. She decided to put together a dedicated project team to oversee the development. *'It is important to have representatives from all relevant parts of the company. Although the team will carry out much of the work themselves, they will still need the cooperation and the resources of their own departments. So, as well as being part of the team, they are also gateways to expertise around the company.'* The team consisted of representatives from marketing, the development kitchens (laboratories), PGT's technology centre (a development facility that served the whole group, not just the snack division), packaging engineers, and representatives from the division's two manufacturing plants. All but the manufacturing representatives were allocated to the project team on a full-time basis. Unfortunately, manufacturing had no one who had sufficient process knowledge and who could be spared from their day-to-day activities.

Development objectives

Monica had tried to set the objectives for the project in her opening remarks to the project team when they had first come together. *'We have a real chance here to develop an offering that not only will have major market impact, but will also give us a sustainable competitive advantage. We need to make this project work in such a way that competitors will find it difficult to copy what we do. The formulation is a real success for our development people, and as long as we figure out how to use the new extrusion method and packaging material, we should be difficult to beat. The success of Orlando in the marketplace will depend on our ability to operationalise and integrate the various technical solutions that we now have access to. The main problem with this type of offering is that it will be expensive to develop and yet, once our competitors realise what we are doing, they will come in fast to try and out-innovate us. Whatever else we do we must ensure that there is sufficient flexibility in the project to allow us to respond quickly when competitors follow us into the market with their own 'me-too' products. We are not racing against the clock to get this to market, but once we do make a decision to launch we will have to move fast and hit the launch date reliably. Perhaps most important, we must ensure that the crisps are 200 per cent safe. We have no experience in dealing with the microbiological testing which dairy-based food manufacture requires. Other divisions of PJT do have this experience and I guess we will be relying heavily on them.'* (Monica Allen)

Monica, who had been tasked with managing the, now much expanded, development process had already drawn up a list of key decisions she would have to take.

- *How to resource the innovation project.* The division had a small development staff, some of whom had been working on Project Orlando, but a project of this size would require extra staff amounting to about twice the current number of people dedicated to the innovation process.
- *Whether to invest in a pilot plant.* The process technology required for the new project would be unlike any of the division's current technology. Similar technology was used by some companies in the frozen food industry and one option would be to carry out trials at these (non-competitor) companies' sites. Alternatively, the Orlando team could build their own pilot plant which would enable them to experiment in-house. As well as the significant expense involved, this would raise the problem of whether any process innovations would work when scaled-up to full size. However, it would be far more convenient for the project team and allow them to 'make their mistakes' in private.
- *How much development to outsource.* Because of the size of the project, Monica had considered outsourcing some of the innovation activities. Other divisions within the company may be able to undertake some of the development work and there were also specialist consultancies that operated in the food-processing industries. The division had never used any of these consultancies before but other divisions had occasionally done so.
- *How to organise the innovation activities.* Currently the small development function had been organised around loose functional specialisms. Monica wondered whether this project warranted the creation of a separate department independent of the current structure. This might signal the importance of this innovation project to the whole division.

Fixing the budget

The budget to develop Project Orlando through to launch had been set at $30 million. This made provision to increase the size of the existing innovation team by 70 per cent over a 20-month period (for launch two years from now). It also included enough funding to build a pilot plant which would allow the team the flexibility to develop responses to potential competitor reaction after the launch. So, of the $30 million around $18 million was for extra staff and contracted-out innovation work, $7.5 million for the pilot plant and $4.5 million for one-off costs (such as the purchase of test equipment, etc.). Monica was unsure whether the budget would be big enough.

'I know everyone in my position wants more money, but it is important not to under-fund a project like this. Increasing our development staff by 70 per cent is not really enough. In my opinion we need an increase of at least 90 per cent to make sure that we can launch when we want. This would need another $5m, spread over the next 20 months. We could get this by not building the pilot plant I suppose, but I am reluctant to give that up. It would mean begging for test capacity on other companies' plants, which is never satisfactory from a knowledge-building viewpoint. Also it would compromise security. Knowledge of what we were doing could easily leak to competitors. Alternatively we could subcontract more of the research which may be less expensive, especially in the long run, but I doubt if it would save the full $5m we need. More important, I am not sure that we should subcontract anything which would compromise safety, and increasing the amount of work we send out may do that. No, it's got to be the extra cash or the project could overrun. The profit projections for the Orlando products look great, (see Table 6.2), but delay or our inability to respond to competitor pressures would depress those figures significantly. Our competitors could get into the market only a little after us. Word has is that marketing's calculations indicate a delay of only six months could not only delay the profit stream by the six months but also cut it by up to 30 per cent.'

Table 6.2 Preliminary 'profit stream' projections for the Project Orlando offering, assuming launch in 24 months time

Time period*	1	2	3	4	5	6	7
Profit flow ($ million)	10	20	50	90	120	130	135

*six-month periods

Monica was keen to explain two issues to the management committee when it met to consider her request for extra funding. First, that there was a coherent and well thought-out strategy for the innovation project over the next two years. Second, that saving $5m on Project Orlando's budget would be a false economy.

QUESTIONS

1 How would you rank the innovation objectives for the project?

2 What are the key issues in resourcing this innovation process?

3 What are the main factors influencing the resourcing decisions?

4 What advice would you give Monica?

APPLYING THE PRINCIPLES

Hints

Some of these exercises can be answered by reading the chapter. Others will require some general knowledge of business activity and some might require an element of investigation. Hints on how they can all be answered are to be found in the eText at www.pearsoned.co.uk/slack.

1 One offering where a very wide range of product types is valued by customers is that of domestic paint. Most people like to express their creativity in the choice of paints and other home decorating products that they use in their homes. Clearly, offering a wide range of paint must have serious cost implications for the companies which manufacture, distribute and sell the product. Visit a store that sells paint and get an idea of the range of products available on the market. How do you think paint manufacturers and retailers could innovate so as to increase variety but minimise costs?

2 Innovation becomes particularly important at the interface between offerings and the people that use them. Consider two types of website:

(a) those which are trying to sell something such as Amazon.com

(b) those which are primarily concerned with giving information, for example reuters.com or nytimes.com

What constitutes good innovation for these two types of website? Find examples of particularly good and particularly poor web design and explain the issues you've considered in making the distinction between them.

3 Visit the website of the UK's Design Council (www.design-council.org.uk). There you will find examples of innovations in many fields. Look through these examples and find one which you think represents excellence in innovation and one which you don't like (for example, because it seems trivial, or may not be practical, or for which there is no market, etc.). Prepare a case supporting your view of why one is good and the other bad. In doing this, derive a checklist of questions which could be used to assess the worth of any innovation idea.

4 How can the innovation of quick-service restaurant (fast-food) offerings be improved from the point of view of environmental sustainability? Visit two or three fast-food outlets and compare their approach to environmentally sensitive designs.

Notes on chapter

1 Sources: company website; Cusumano, M.A. (1997) 'How Microsoft make large teams work like small teams', *Sloan Management Review*, Fall.

2 Sources: Novartis company website; Rowberg, R.E (2001), 'Pharmaceutical research and design: a description and analysis of the process', *CRS Report for Congress*, 2 April.

3 See http://www.interfaceglobal.com for more information on this service.

4 Mass customisation was first fully articulated in Pine, B.J. (1993), *Mass Customization: The New Frontier in Business Competition*, Harvard Buseinss School Press, Boston, MA.

5 Sources: Phfizer company website; http://www.bnet.com/drug-business.

6 Open innovation is a term promoted by Henry Chesborough at the University of California, Berkley.

7 Wheelwright, S.C. and Clark, K.B. (1995), *Leading Product Development*, The Free Press, New York.

8 This idea is based on one presented by Hayes, Wheelwright and Clark, in Hayes, R.H., Wheelwright, S.C. and Clark, K.B. (1988) *Dynamic Manufacturing*, The Free Press, New York.

9 Sobek, D.K. II, Liker J.K. and Ward, A.K. (1998), 'Another look at how Toyota integrates product development', *Harvard Business Review* July–August.

10 This term is coined by Pahl, G. and Beitz, W. in (1996) *Engineering Design: A Systematic Approach*, Springer Verlag, London.

11 From Dorst, K. and Cross, N. (2001), 'Creativity in the design process: co-evolution of problem-solution', *Design Studies* (22), September, pp 425–37.

TAKING IT FURTHER

Bangle, C. (2001) 'The ultimate creativity machine: how BMW turns art into profit', *Harvard Business Review*, January, pp 47–55. *A good description of how good aesthetic design translates into business success.*

Baxter, M. (1995) *Product Design*, Chapman and Hall. *Presents a structured framework for innovation which will be of interest to practising managers.*

Blackburn, J.D. (ed) (1991) 'Time based competition: the next battle ground in American manufacturing', Urwin, Homewood, Ill. *A good summary of why interactive innovation gives fast time-to-market and why this is important.*

Bruce, M. and Bessant, J. (2002) *Design In Business: Strategic Innovation through Design*, Financial Times Prentice Hall and The Design Council. *Probably one of the best overviews of innovation in a business context available today.*

Chesbrough, H.W. (2003). *Open Innovation: The New Imperative for Creating and Profiting from Technology*, Harvard Business School Press, Boston. *A good overview of the reasons for open innovation and the challenges of making it work.*

Christensen, C.M. (1997), *The Innovator's Dilemma*, Harvard Business School Press, Boston, MA. *Ground-breaking booking on disruptive innovation.*

Dyson, J. (1997) *Against the Odds: An Autobiography*, Orion Business Books, London. *One of Europe's most famous designers gives his philosophy.*

Lowe, A. and Ridgway, K. (2000) 'A user's guide to quality function deployment', *Engineering Management Journal*, June. *A good overview of QFD explained in straightforward non-technical language.*

Mckeown, M. (2008), *The Truth About Innovation*, Prentice Hall. *A practical guide to innovation, including lots of examples of success and failure.*

USEFUL WEBSITES

www.cfsd.org.uk *The Centre for Sustainable Design. Some useful resources, but obviously largely confined to sustainability issues.*

www.conceptcar.co.uk *A site devoted to automotive innovation. Fun if you like new car designs!*

www.betterproductdesign.net *A site that acts as a resource for good innovation practice. Set up by Cambridge University and the Royal College of Art. Some good material that supports all aspects of innovation.*

www.designcouncil.org.uk *Site of the UK's Design Council. One of the best sites in the world for innovation related issues.*

Interactive quiz

For further resources including examples, animated diagrams, self-test questions, Excel spreadsheets and video materials please explore the eText on the companion website at **www.pearsoned.co.uk/slack**.

7 Supply chain management

Introduction

An operation's ability to deliver products or services to customers is fundamentally influenced by how its supply chains are managed. Chapter 3 treated the strategic design of supply networks. This chapter considers the planning and control activity for the individual supply chains in the network. Supply chain management is the overarching operations management activity that dictates an operation's *delivery* performance because it controls the flow of products and services from suppliers right through to the end customer. That is why it is the first chapter dealing with the planning and control of delivery. But planning and controlling delivery is a much larger topic, and includes, capacity management (Chapter 8), inventory management (Chapter 9), resource planning and control (Chapter 10), and lean synchronisation (Chapter 11). Figure 7.1 illustrates the supply–demand linkage treated in this chapter.

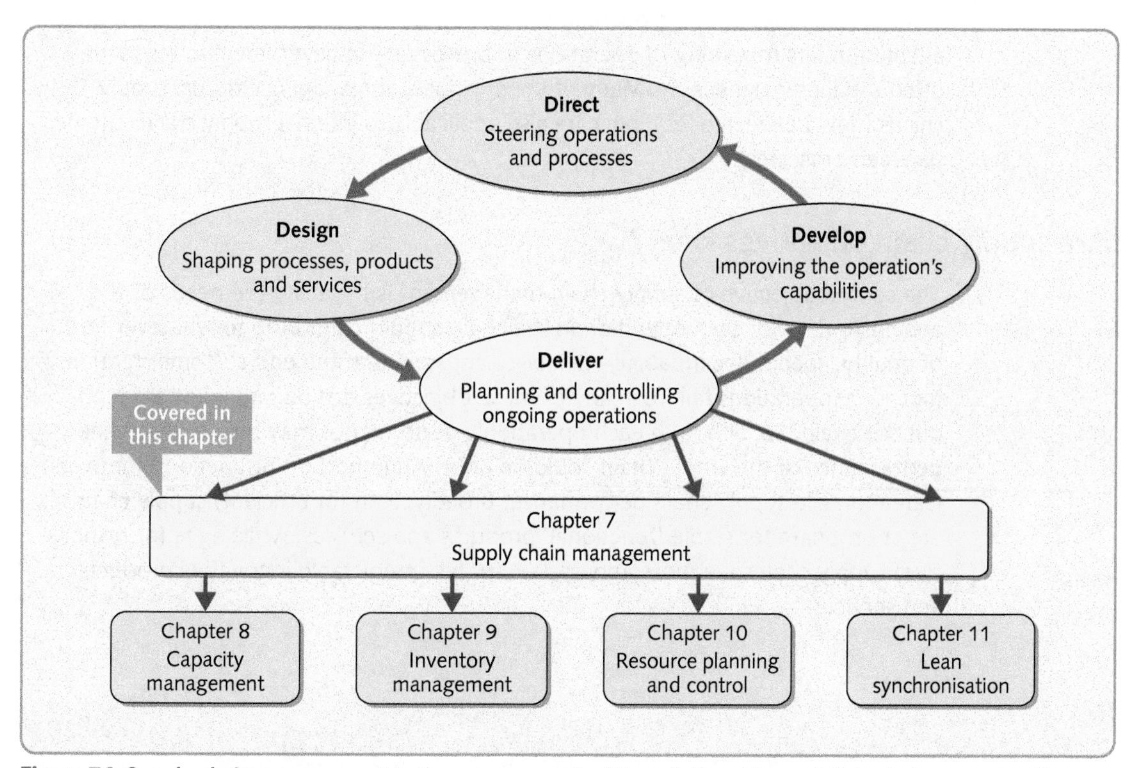

Figure 7.1 Supply chain management is the management of the relationships and flows between operations and processes; it is the topic that integrates all the issues concerning the delivery of products and services

EXECUTIVE SUMMARY

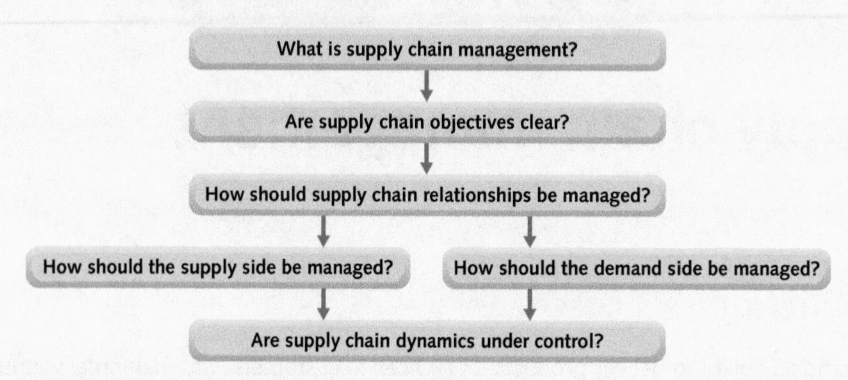

Each chapter is structured around a set of diagnostic questions. These questions suggest what you should ask in order to gain an understanding of the important issues of a topic, and, as a result, improve your decision making. An executive summary addressing these questions is provided below.

What is supply chain management?

Supply chain management is the management of relationships and flows between operations and processes. Technically, it is different from supply network management, which looks at all the operations or processes in a network. Supply chain management refers to a string of operations or processes. However, the two terms are often used interchangeably. Many of the principles of managing external supply chains (flow between operations) are also applicable to internal supply chains (flow between processes).

Are supply chain objectives clear?

The central objective of supply chain management is to satisfy the needs of the end customer. So, each operation in the chain should contribute to whatever mix of quality, speed, dependability, flexibility and cost that the end customer requires. Individual operations failure in any of these objectives can be multiplied throughout the chain. So, although each operation's performance may be adequate, the performance of the whole chain could be poor. An important distinction is between lean and agile supply chain performance. Broadly, lean (or efficient) supply chains are appropriate for stable 'functional' products and services, while agile (or responsive) supply chains are more appropriate for less predictable innovative products and services.

How should supply chain relationships be managed?

Supply chain relationships can be described on a spectrum from market-based, contractual, 'arms-length' relationships, through to close and long-term partnership relationships. Each has its advantages and disadvantages. Developing relationships involves assessing which relationship will provide the best potential for developing overall performance. However, the types of relationships adopted may be dictated by the structure of the market itself. If the number of potential suppliers is small, there are few opportunities to use market mechanisms to gain any kind of advantage.

How should the supply side managed?

Managing supply side relationships involves three main activities: selecting appropriate suppliers; planning and controlling on-going supply activity; supplier development. Supplier selection involves trading off different supplier attributes, often using scoring assessment methods. Managing ongoing supply involves clarifying supply expectations, often using service level agreements to manage the supply relationships. Supplier development can benefit both suppliers and customers, especially in partnership relationships. Very often barriers are the mismatches in perception between customers and suppliers.

How should the demand side managed?

This will depend partly on whether demand is dependent on some known factor and therefore predictable, or independent of any known factor and therefore less predictable. Approaches such as materials requirements planning (MRP) are used in the former case, while approaches such as inventory management are used in the latter case. The increasing outsourcing of physical distribution and the use of new tracking technologies, such as RFID, have brought efficiencies to the movement of physical goods and customer service. But customer service may be improved even more if suppliers take on responsibility for customer development, i.e. helping customers to help themselves.

Are supply chain dynamics under control?

Supply chains have a dynamic of their own that is often called the *bull-whip* effect. It means that relatively small changes at the demand end of the chain increasingly amplify into large disturbances as they move upstream. Three methods can be used to reduce this effect. Information sharing can prevent over-reaction to immediate stimuli and give a better view of the whole chain. Channel alignment through standardised planning and control methods allows for easier coordination of the whole chain. Improving the operational efficiency of each part of the chain prevents local errors multiplying to affect the whole chain.

DIAGNOSTIC QUESTION

What is supply chain management?

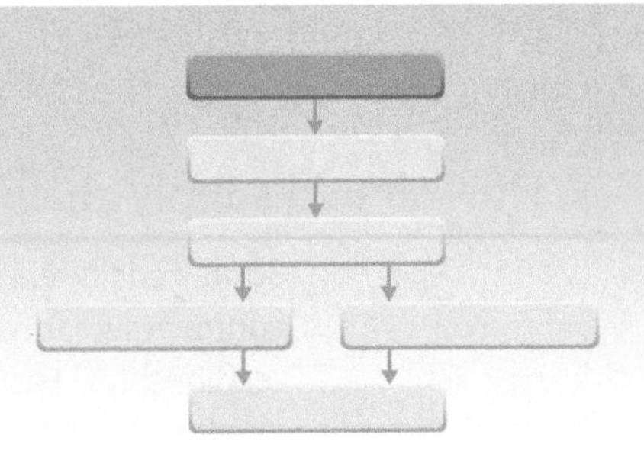

Video

Supply chain management (SCM) is the management of the relationships and flows between the 'string' of operations and processes that produce value in the form of products and services to the ultimate consumer. It is a holistic approach to managing across the boundaries of companies and of processes. Technically, supply *chains* are different to supply *networks*. A supply network is *all* the operations that are linked together so as to provide goods and services through to end customers. In large supply networks there can be many hundreds of supply chains of linked operations passing through a single operation. The same distinction holds within operations. Internal supply network, and supply chain management concerns flow between processes or departments (see Figure 7.2). Confusingly, the terms supply network and supply chain management are often used interchangeably.

OPERATIONS PRINCIPLE
The supply chain concept applies to the internal relationships between processes as well as the external relationships between operations.

It is worth emphasising again that the supply chain concept applies to internal process networks as well as external supply networks. Many of the ideas discussed in the context of the 'operation-to-operation' supply chain, also apply to the 'process-to-process' internal supply chain. It is also worth noting that the 'flows' in supply chains are not restricted to the downstream flow of products and services from suppliers through to customers. Although the most obvious failure in supply chain management occurs when downstream flow fails to meet customer requirements, the root cause may be a failure in the upstream flow of information. Modern supply chain management is as much concerned with managing information

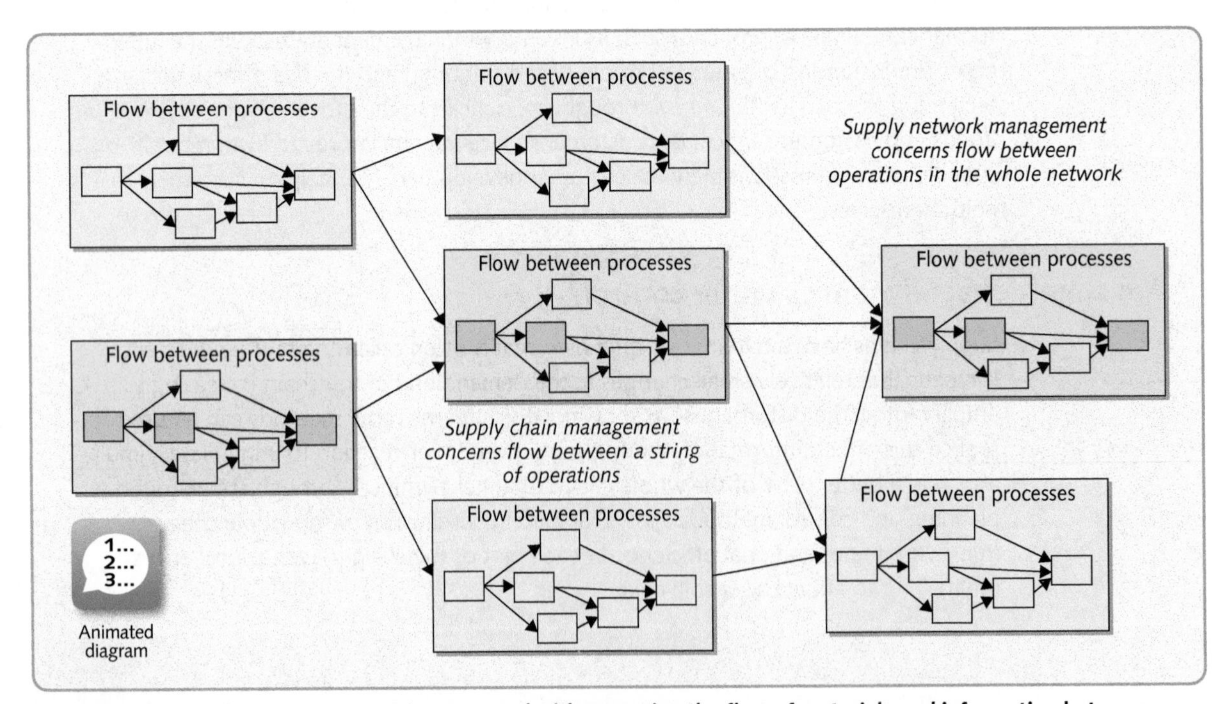

Figure 7.2 Supply chain management is concerned with managing the flow of materials and information between a string of operations that form the strands or 'chains' of a supply network

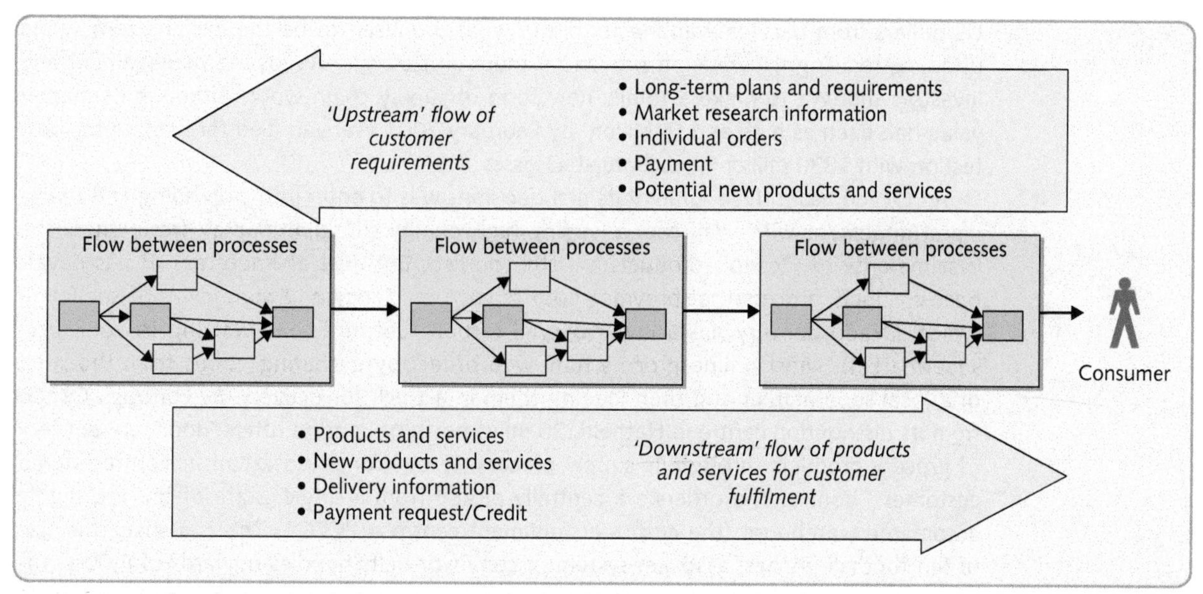

Figure 7.3 Supply chain management is concerned with the flow of information as well as the flow of products and services

flows (upstream and downstream) as it is with managing the flow of products and services (see Figure 7.3). In fact, the supply chain concept could be broadened further. There may not be any flow of physical items at all. Many 'flows' from suppliers to customers are non-tangible. Banks supply information about your cash (or lack of it). Media operations supply entertainment, education and information. Consultants supply advice, reassurance and knowledge. Police services supply a feeling of security, and so on.

OPERATIONS PRINCIPLE

The supply chain concept applies to non-physical flow between operations and processes as well as physical flows.

The following two examples of supply chain management illustrate some of the issues that are discussed in this chapter.

EXAMPLE

Ocado[1]

Source: Bloomberg/Getty Images

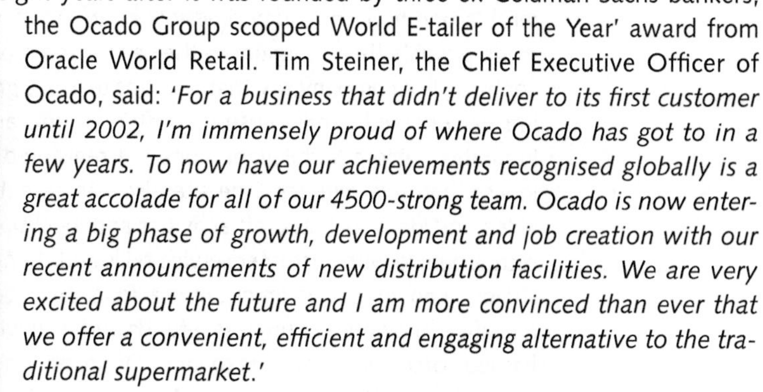

It was a proud moment for a business that many had forecast would never survive 12 months. In 2010, eight years after it was founded by three ex-Goldman Sachs bankers, the Ocado Group scooped World E-tailer of the Year' award from Oracle World Retail. Tim Steiner, the Chief Executive Officer of Ocado, said: *'For a business that didn't deliver to its first customer until 2002, I'm immensely proud of where Ocado has got to in a few years. To now have our achievements recognised globally is a great accolade for all of our 4500-strong team. Ocado is now entering a big phase of growth, development and job creation with our recent announcements of new distribution facilities. We are very excited about the future and I am more convinced than ever that we offer a convenient, efficient and engaging alternative to the traditional supermarket.'*

By 2010, Ocado was the only dedicated online supermarket in the UK and the largest dedicated online supermarket by turnover in the world. What it had succeeded in doing in the UK was to reshape the final 'business-to-consumer' configuration of the traditional food supply chain. It had become one of the most successful online grocers in the world. But it was not the first. Back in 1999 an internet grocer called Webvan erupted on to the scene in California. It gained considerable publicity and more than a billion

US dollars from backers wanting to join in what promised to be the exciting new world of online retailing. However, it proved far more difficult than Webvan's management and investors thought to make a totally new form of supply chain work. Although its market value had been as high as $15 billion, by February 2001 Webvan filed for bankruptcy protection with $830 million in accumulated losses.

Yet Ocado has thrived. One of its first decisions was to enter into a branding and sourcing arrangement with Waitrose, a leading high-quality UK supermarket, from where the vast majority of Ocado's products are still sourced. But, just as important, it has developed a supply process that provides both relative efficiency and high levels of service (a typical Ocado delivery has a lower overall carbon footprint than walking to your local supermarket). Most online grocers fulfil web orders by gathering goods from the shelf of a local supermarket and then loading them in a truck for delivery. By contrast, Ocado from its distribution centre in Hatfield, 20 miles north of London offers 'doorstep' delivery of grocery products through its supply process, to over one-and-a-half million registered customers' homes. The orders are centrally picked from a single, state-of-the-art, highly automated warehouse (the customer fulfilment centre or 'CFC'). This is a space the size of ten football pitches, a 15-km system of conveyor belts handles upwards of 8,000 grocery containers an hour, which are then shipped to homes, mainly in the southern part of the UK. Ocado operates what it calls its 'hub and spoke' supply system; with its central CFC (hub) serving regional (spoke) distribution points. In contrast with traditional 'bricks-and-mortar' supermarkets (Ocado has no 'physical' shops), it delivers direct to customers from its distribution centre rather than from stores. The largely automated picking process, which was developed by its own software engineers, allows the company to pick and prepare groceries for delivery up to seven times faster than its rivals. Although as many as one million separate items are picked for individual customer orders every day, there are fewer than 80 mistakes.

Making its deliveries of more than 21,000 different products from a central location means it can carry more items than smaller local stores which are more likely to run out of stock. Also fresh or perishable items that are prepared centrally will have more 'shelf life'. Ocado's food waste, at 0.3 per cent of sales, is the lowest in the industry. The structural advantage of this supply arrangement means that 99 per cent of all orders are fulfilled accurately. Just as important as the physical distribution to the customers' door is the ease of use of the company's website (Ocado.com) and the convenience of booking a delivery slot. Ocado offer reliable one-hour, next day timeslots in an industry where two-hour timeslots prevail. This is made possible thanks, again, to the centralised model and world-class processes, systems and controls. The company say that its website is designed to be simple to use and intuitive. Smart lists personalised to each customer offer prompts and ideas so that the absence of any in-store inspiration becomes irrelevant. For a pre-registered customer, a weekly shop can be completed in less than five minutes. The site also has an extensive range of recipes including some as video and ideas such as craft activities and lunchbox fillers. Ocado makes a conscious effort to recruit people with customer service skills and then train them as drivers rather than vice versa. Drivers, known as Customer Service Team Members, are paid well above the industry norm and are empowered to process refunds and deal with customer concerns on the doorstep. This has led some commentators to label Ocado 'the new Amazon'. *'Not so,'* say others. *'In some ways it's actually more complex than Amazon's operation. Amazon built a dominant brand in the US, the world's biggest market, by selling books and CDs, which essentially you just stick in an envelope and put a stamp on. That is not the same as having a highly automated warehouse with expensive machines and a huge fleet of delivery vans taking the goods to every house.'*

EXAMPLE

Burberry links the catwalk with the consumer[2]

Back in 1914, Thomas Burberry began producing trench coats for use by soldiers during World War I. Since then Burberry has become one of the best known luxury brands in the world. The British heritage is important. Even now Burberry trench coats still have the traditional 'D-rings', initially designed to hold hand grenades, and the collars are still sewn on by hand in a factory in Castleford in the north of England. The journey from a slightly old-fashioned maker of traditional rainwear into a luxury brand dynamo was a marketing, design and branding success story. Yet the operations side of the business did not fully reflect the company's status as a global fashion brand. Not only was the Castleford factory inefficient, the company's supply chain was confusing, complex and chaotic. *'It was,'* said Andy Janowski, Burberry's Chief Operating Officer, *'still jumbled under the hood, beset by late shipments and an onerous process that required 432 different steps to get from sketch to customer.'*

Source: The Advertising Archives

Burberry's Chief Executive, Rose Marie Bravo, had led the transformation of the Burberry brand, but it was her successor, Angela Ahrendts, who decided to tackle the supply chain that was draining cash from the company's finances and, because it was slow and unresponsive, holding back the company's brand strength. But Burberry was not alone. Supply inefficiency was (and is) common in the luxury-fashion industry, where slow manufacturing and late deliveries were normal. One of the main problems was the traditional supply chain configuration in the high-fashion industry where most sales are achieved through a complex arrangement of wholesalers, franchisees or licensing partners. This resulted in a number of problems for luxury brands. Perhaps most significantly, each independent stage in the supply chain had to make its own profit, and although margins are relatively high in the luxury goods market, it still left a smaller share for Burberry. The other operations between Burberry and its end customer also made it difficult to respond to market changes and emerging trends. However the multi-stage supply chain does have some advantages. Wholesalers and retailers carry inventory which acts as a buffer between Burberry's production schedules and the fluctuations in the market. Conversely, viewed another way, inventory can cover up numerous forecasting and supply flaws, and keeping close to its market is exactly what a fashion label must be good at. So Burberry elected to become a more retail-focused business. It sold more merchandise directly to customers, both through the internet and its own expanding network of stores. These included buying out its long-time franchise partner in China to operate the 50 stores there directly.

The supply side of Burberry's supply chain has also been reorganised. It consolidated its manufacturing operations on the Castleford site, reorganising its production processes to enhance efficiency and responsiveness. It also trimmed its manufacturing suppliers from 300 to 90, cut its network of 26 warehouses to three global distribution hubs and reduced 31 logistic hauliers down to three, all of which required a significant investment in new information technology systems. One innovation was to connect its designers and manufacturers on Skype to save time collaborating on new designs. The changes were effective. The average cost of making a garment at Castleford decreased by between 10 and 15 per cent within two years. In addition, delivery reliability improved. Before, deliveries could be as much as three months late. Andy Janowski holds that it was the overhaul of the supply chain that made it possible for Burberry to do things that would have been impossible in the past. For example, Burberry launched its 'Runway to Reality' initiative, in which it streamed its semi-annual fashion show live on the internet and simultaneously sold the designs coming down the catwalk online.

What do these two examples have in common?

The first lesson from these two companies is that they both take supply chain management seriously. In fact, more than that, they both understand that, no matter how good individual operations or processes are, a business's overall performance is a function of the whole chain of which it is a part. That is why both of these companies put so much effort into managing the whole chain. This does not mean that both companies adopt the same, or even similar, approaches to supply chain management. Each has a slightly different set of priorities. Ocado does not so much *have* a supply chain, it *is* a supply chain in the sense that its entire purpose is supply. Moreover it is making a business of supply using a relatively new concept (internet-based customer ordering) where other businesses have failed. It therefore needs to keep its costs under strict control if it is to avoid the fate of its predecessors like Webvan. Burberry, on the other hand, needs a responsive supply chain so that its fashion ideas can have up-to-the-minute impact. It needs agility from its supply chain so that it can keep close to its market. So, although they are very different businesses with different supply chain objectives, the commonality between them is that both see the way they configure and manage their supply chains as a source of innovation. Both are doing something different from their competitors; both are innovating through their supply chain management; both have a clear idea of what they want to be and realise the importance of understanding customers as the starting point of successful supply chain management. Although the examples emphasise Ocado's downstream customer relationships and Burberry's total supply chain relationships, the common theme is the importance of investing in a supply perspective. In addition, both companies have invested in mechanisms for communicating along the supply chain and coordinating material and information flows. The rest of this chapter is structured around these three main issues: clarifying supply chain objectives; supply chain relationships, both with suppliers and with customers; and controlling and coordinating flow.

DIAGNOSTIC QUESTION

Are supply chain objectives clear?

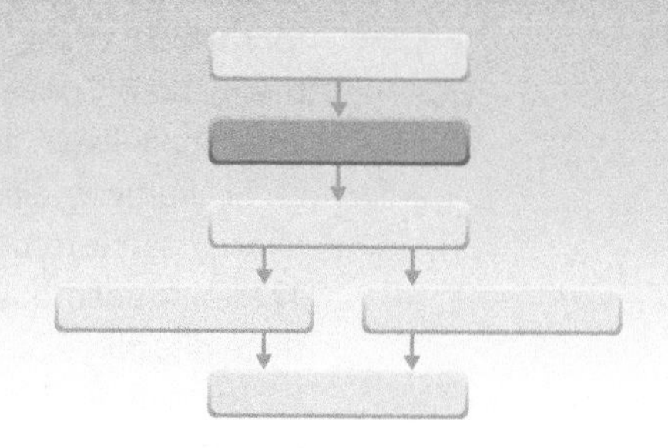

Video

All supply chain management shares one common, and central, objective – to satisfy the end customer. All stages in a chain must eventually include consideration of the final customer, no matter how far an individual operation is from the end customer. When a customer decides to make a purchase, he or she triggers action back along the whole chain. All the businesses in the supply chain pass on portions of that end customer's money to each other, each retaining a margin for the value it has added. Each operation in the chain should be satisfying its own customer, but also making sure that eventually the end customer is satisfied.

For a demonstration of how end customer perceptions of supply satisfaction can be very different from that of a single operation, examine the customer 'decision tree' in Figure 7.4. It charts the hypothetical progress of 100 customers requiring service (or products) from a business (for example, a printer requiring paper from an industrial paper stockist). Supply performance, as seen by the core operation (the warehouse), is represented by the shaded part of the diagram. It has received 20 orders, 18 of which were 'produced' (shipped to customers) as promised (on time, and in full). However, originally 100 customers may have requested service,

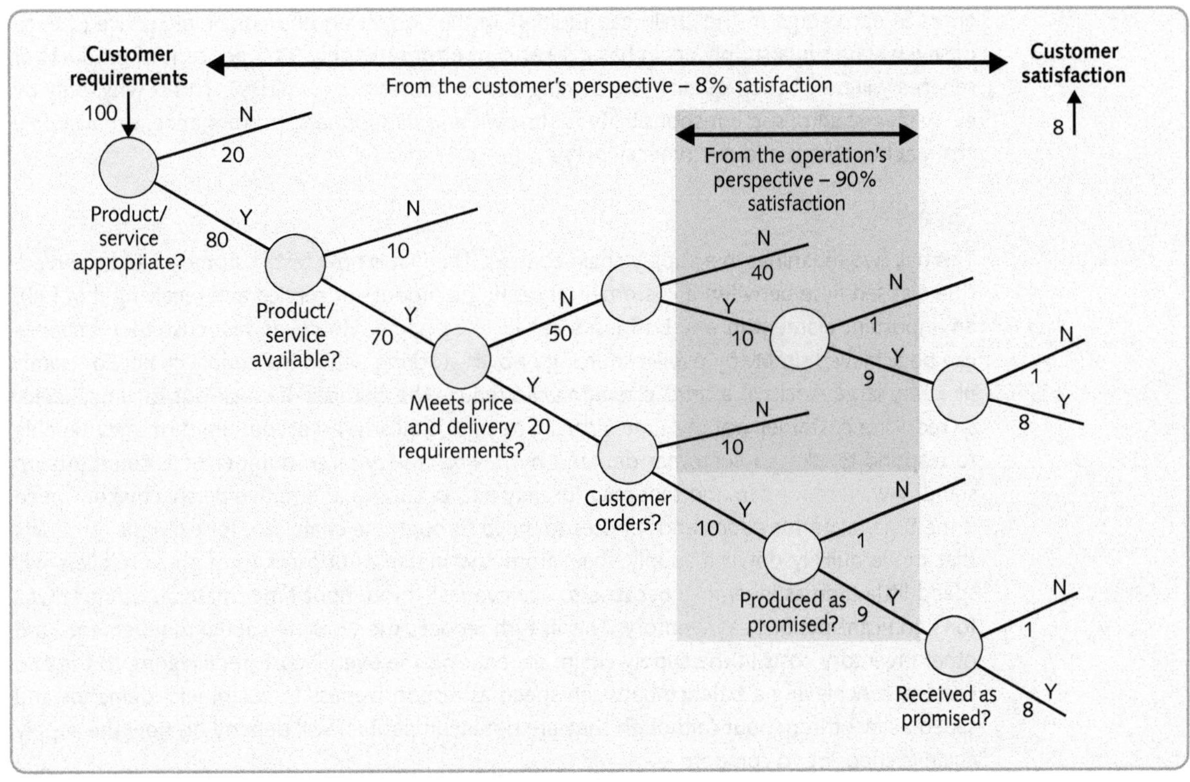

Figure 7.4 Taking a customer perspective of supply performance can lead to very different conclusions

> **OPERATIONS PRINCIPLE**
> *The performance of an operation in a supply chain does not necessarily reflect the performance of the whole supply chain.*

20 of who found the business did not have appropriate products (did not stock the right paper), 10 of whom could not be served because the products were not available (out of stock), 50 of whom were not satisfied with the price and/or delivery (of whom 10 placed an order notwithstanding). Of the 20 orders received, 18 were produced as promised (shipped) but two were not received as promised (delayed or damaged in transport). So what seems a 90 per cent supply performance is in fact an 8 per cent performance from the customer's perspective.

This is just one operation in a whole network. Include the cumulative effect of similar reductions in performance for all the operations in a chain, and the probability that the end customer is adequately served could become remote. The point here is not that all supply chains have unsatisfactory supply performances (although most supply chains have considerable potential for improvement). Rather it is that the performance both of the supply chain as a whole, and its constituent operations, should be judged in terms of how all end customer needs are satisfied.

Supply chain objectives

The objective of supply chain management is to meet the requirements of end customers by supplying appropriate products and services when they are needed at a competitive cost. Doing this requires the supply chain to achieve appropriate levels of the five operations performance objectives: quality, speed, dependability, flexibility and cost.

Quality

The quality of a product or service when it reaches the customer is a function of the quality performance of every operation in the chain that supplied it. The implication of this is that errors in each stage of the chain can multiply in their effect on end customer service (if each of seven stages in a supply chain has a 1 per cent error rate, only 93.2 per cent of products or services will be of good quality on reaching the end customer (i.e. 0.99). This is why, only by every stage taking some responsibility for its own *and its suppliers'* performance, can a supply chain achieve high end customer quality.

Speed

This has two meanings in a supply chain context. The first is how fast customers can be served, (the elapsed time between a customer requesting a product or service and receiving it in full), an important element in any business's ability to compete. However, fast customer response can be achieved simply by over-resourcing or over-stocking within the supply chain. For example, very large stocks in a retail operation can reduce the chances of stock-out to almost zero, so reducing customer waiting time virtually to zero. Similarly, an accounting firm may be able to respond quickly to customer demand by having a very large number of accountants on standby waiting for demand that may (or may not) occur. An alternative perspective on speed is the time taken for goods and services to move through the chain. So, for example, products that move quickly down a supply chain from raw material suppliers through to retailers will spend little time as inventory because to achieve fast throughput time, material cannot dwell for significant periods as inventory. This in turn reduces the working capital requirements and other inventory costs in the supply chain, so reducing the overall cost of delivering to the end customer. Achieving a balance between speed as responsiveness to customers' demands and speed as fast throughput (although they are not incompatible) will depend on how the supply chain is choosing to compete.

Dependability

Dependability in a supply chain context is similar to speed insomuch as one can almost guarantee 'on-time' delivery by keeping excessive resources, such as inventory, within the chain. However, dependability of throughput time is a much more desirable aim because it reduces uncertainty within the chain. If the individual operations in a chain do not deliver as promised on time, there will be a tendency for customers to over-order, or order early, in order to provide some kind of insurance against late delivery. The same argument applies if there is uncertainty regarding the *quantity* of products or services delivered. This is why delivery dependability is often measured as 'on time, in full' in supply chains.

Flexibility

In a supply chain context, this is usually taken to mean the chain's ability to cope with changes and disturbances. Very often this is referred to as supply chain agility. The concept of agility includes previously discussed issues such as focusing on the end customer and ensuring fast throughput and responsiveness to customer needs. But, in addition, agile supply chains are sufficiently flexible to cope with changes, either in the nature of customer demand or in the supply capabilities of operations within the chain.

Cost

In addition to the costs incurred within each operation to transform its inputs into outputs, the supply chain as a whole incurs additional costs that derive from each operation in a chain doing business with each other. These transaction costs may include such things as the costs of finding appropriate suppliers, setting up contractual agreements, monitoring supply performance, transporting products between operations, holding inventories, and so on. Many of the recent developments in supply chain management, such as partnership agreements or reducing the number of suppliers, are an attempt to minimise transaction costs.

Should supply chains be lean or agile?

Video

A distinction is often drawn between supply chains that are managed to emphasise supply chain efficiency (lean supply chains), and those that emphasise supply chain responsiveness and flexibility (agile supply chains). These two modes of managing supply chains are reflected in an idea proposed by Professor Marshall Fisher of Wharton Business School: that supply chains serving different markets should be managed in different ways. Even companies that have seemingly similar products or services, in fact, may compete in different ways with different products.[3] For example, shoe manufacturers may produce classics that change little over the years, as well as fashion shoes that last only one season. Chocolate manufacturers have stable lines that have been sold for 50 years, but also product 'specials' associated with an event or film release, the latter selling only for a matter of months. Hospitals have routine 'standardised' surgical procedures, such as cataract removal, but also have to provide emergency post-trauma surgery. Demand for the former products will be relatively stable and predictable, but demand for the latter will be far more uncertain. Also, the profit margin commanded by the innovative product will probably be higher than that of the more functional product. However, the price (and therefore the margin) of the innovative product may drop rapidly once it has become unfashionable in the market.

The supply chain policies that are seen to be appropriate for functional products and innovative products are termed efficient (or lean), and responsive (or agile) supply chain policies, respectively. Efficient supply chain policies include keeping inventories low, especially in the downstream parts of the network, so as to maintain fast throughput and reduce the amount of working capital tied up in the inventory. What inventory there is in the network is concentrated mainly in the manufacturing operation, where it can keep utilisation high and therefore manufacturing costs low. Information must flow quickly up and down the chain from retail outlets back up to the manufacturer so that schedules can be given the maximum amount of time to adjust efficiently. The chain is then managed to make sure that products flow as quickly as possible down the chain to replenish what few stocks are kept downstream.

OPERATIONS PRINCIPLE
Supply chains with different end objectives need managing differently.

By contrast, responsive supply chain policy stresses high service levels and responsive supply to the end customer. The inventory in the network will be deployed as closely as possible to the customer. In this way, the chain can still supply even when dramatic changes occur in customer demand. Fast throughput from the upstream parts of the chain will still be needed to replenish downstream stocks. But those downstream stocks are needed to ensure high levels of availability to end customers. Figure 7.5 illustrates how the different supply chain policies match the different market requirements implied by functional and innovative products.

OPERATIONS PRINCIPLE
'Functional' products require lean supply chain management; 'innovative' products require agile supply chain management.

The SCOR model[4]

The Supply Chain Operations Reference Model (SCOR) is a broad, but highly structured and systematic, framework to supply chain improvement that has been developed by the Supply Chain Council (SCC), a global non-profit consortium. The framework uses a methodology, and diagnostic and benchmarking tools that are increasingly widely accepted for evaluating and comparing supply chain activities and their performance. Just as important, the SCOR model allows its users to improve, and communicate supply chain management practices within and between all interested parties in their supply chain by using a standard language and a set of structured definitions. The SCC also provides a benchmarking database by which companies can compare their supply chain performance to others in their industries and training classes. Companies that have used the model include BP AstraZeneca, Shell, SAP AG, Siemens AG and Bayer.

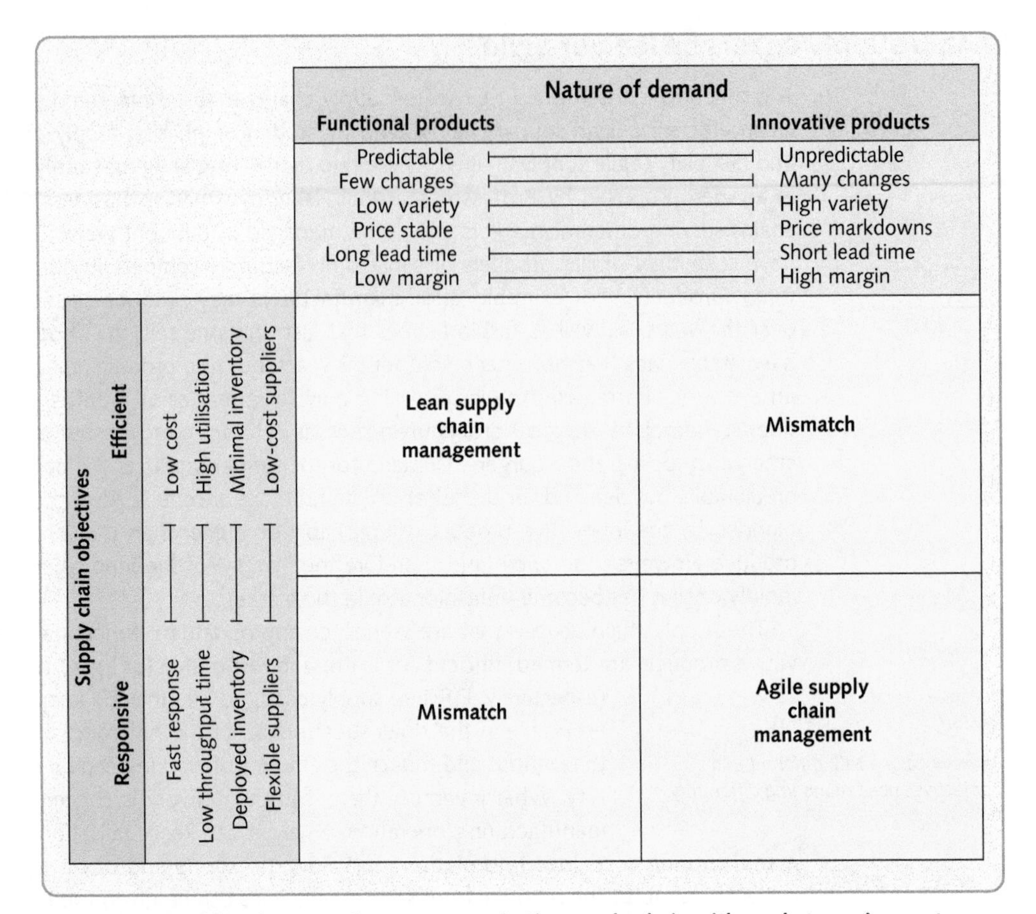

Figure 7.5 Matching the operations resources in the supply chain with market requirements

Source: Adapted from Fisher, M.C. (1997) 'What is the right supply chain for your product?' *Harvard Business Review*, March–April, pp 105–116.

The model uses three well-known individual techniques turned into an integrated approach. These are:

- business process modelling
- benchmarking performance
- best practice analysis.

Business process modelling

SCOR does not represent organisations or functions, but rather processes. Each basic 'link' in the supply chain is made up of five types of process, each process being a 'supplier–customer' relationship (see Figure 7.6).

- 'Source', is the procurement, delivery, receipt and transfer of raw material items, subassemblies, product and or services.
- 'Make, is the transformation process of adding value to products and services through mixing production operations processes.
- 'Deliver' processes perform all customer-facing order management and fulfilment activities including outbound logistics.
- 'Plan' processes manage each of these customer–supplier links and balance the activity of the supply chain. They are the supply and demand reconciliation process, which includes prioritisation when needed.

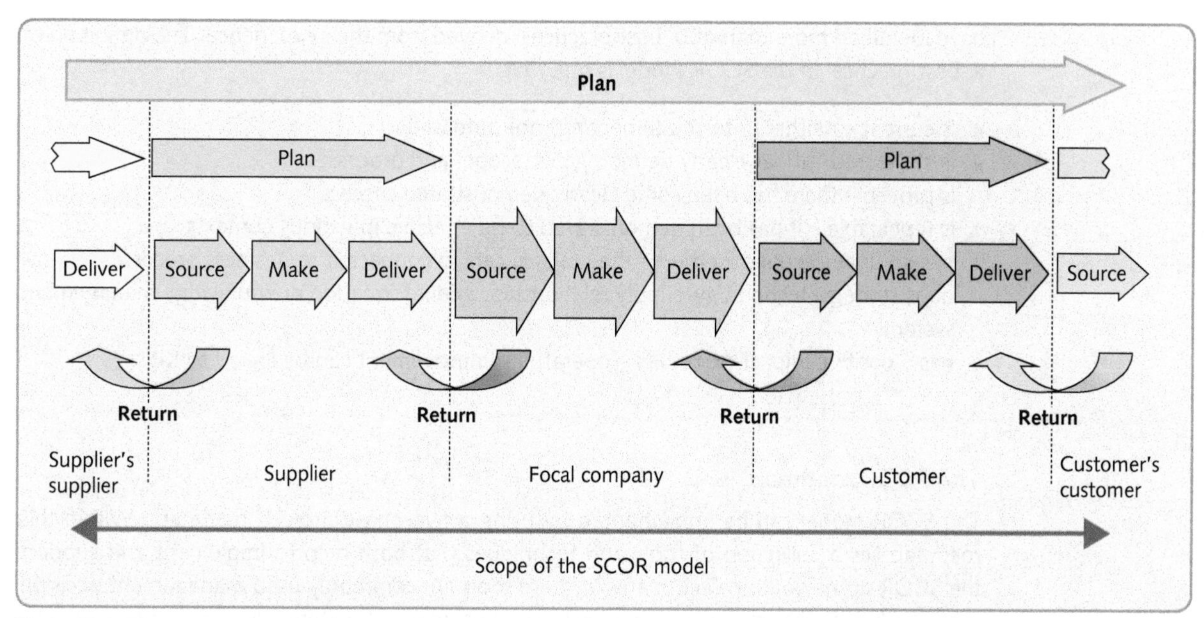

Figure 7.6 The structure of supply chains implicit in the SCOR model showing the relationship between Plan, Source, Make, Deliver and Return elements of the model

- 'Return' processes look after the reverse logistics flow of moving material back from end customers upstream in the supply chain because of product defects or post-delivery customer support.

All these processes are modelled at increasingly detailed levels (see Chapter 4 for a description of the different levels of process analysis). The first level (level 1) identifies the five processes and allows managers to set the scope of the business issues. The SCC advocates the idea that 'if it isn't broken, don't model it'. If no problem has been identified in a particular area, it will be of no significant help to map processes in any further level of detail. More detailed modelling (level 2) identifies which type of supply chain configuration the company operates, for example make-to-stock, make-to-order or engineer-to-order environment. Yet more detailed process modelling (level 3) is then done in terms of the company's ability to compete successfully in its chosen markets.

Benchmarking performance

Performance metrics in the SCOR model are also structured by level, as is process analysis. Level 1 metrics are the yardsticks by which an organisation can measure how successful it is in achieving its desired positioning within the competitive environment, as measured by the performance of a particular supply chain. These level 1 metrics are the Key Performance Indicators (KPIs) of the chain and are created from lower-level diagnostic metrics (called level 2 and level 3 metrics) which are calculated on the performance of lower level processes. Some metrics do not 'roll up' to level 1; these are intended to diagnose variations in performance against plan.

Best practice analysis

Best practice analysis follows the benchmarking activity that should have measured the performance of the supply chain processes and identified the main performance gaps. Best practice analysis identifies the activities that need to be performed to close the gaps. SCC members have identified more than 400 'best practices' derived from their experience. The definition of a 'best practice' in the SCOR model is one that:

- is current – neither untested (emerging) nor outdated
- is structured – it has clearly defined goals, scope, and processes
- is proven – there has been some clearly demonstrated success
- is repeatable – it has been demonstrated to be effective in various contexts
- has an unambiguous method – the practice can be connected to business processes, operations strategy, technology, supply relationships and information or knowledge management systems
- has a positive impact on results – operations improvement can be linked to KPIs.

The SCOR roadmap

The SCOR model can be implemented by using a five-phase project 'roadmap'. Within this roadmap lies a collection of tools and techniques that both help to implement and support the SCOR framework. In fact, many of these tools are commonly used management decision tools, such as Pareto charts, cause–effect diagrams, maps of material flow, brainstorming, etc. The roadmap has five stages as follows:

Phase 1: Discover. Involves supply-chain definition and prioritisation where a 'Project Charter' sets the scope for the project. This identifies logic groupings of supply chains within the scope of the project. The priorities, based on a weighted rating method, determine which supply chains should be dealt with first. This phase also identifies the resources that are required. identified and secured through business process owners/actors.

Phase 2: Analyse. Using data from benchmarking and competitive analysis, the appropriate level of performance metrics are identified, that will define the strategic requirements of each supply chain.

Phase 3: Material flow design. In this phase the project teams have their first go at creating a common understanding of how processes can be developed. The current state of processes are identified and an initial analysis attempts to see where there are opportunities for improvement.

Phase 4: Work and information flow design. The project teams collect and analyse the work involved in all relevant processes (plan, source, make, deliver and return) and map the productivity and yield of all transactions.

Phase 5: Implementation planning. This is the final preparation phase for communicating the findings of the project. Its purpose is to transfer the knowledge of the SCOR team(s) to individual implementation or deployment teams.

Benefits of the SCOR model

Claimed benefits from using the SCOR model include: improved process understanding and performance; improved supply chain performance; increased customer satisfaction and retention; a decrease in required capital; better profitability and return on investment; increased productivity. And, although most of these results could arguably be expected when any company starts focusing on business processes improvements, SCOR proponents argue that using the model gives an above average and supply focused improvement.

DIAGNOSTIC QUESTION

How should supply chain relationships be managed?

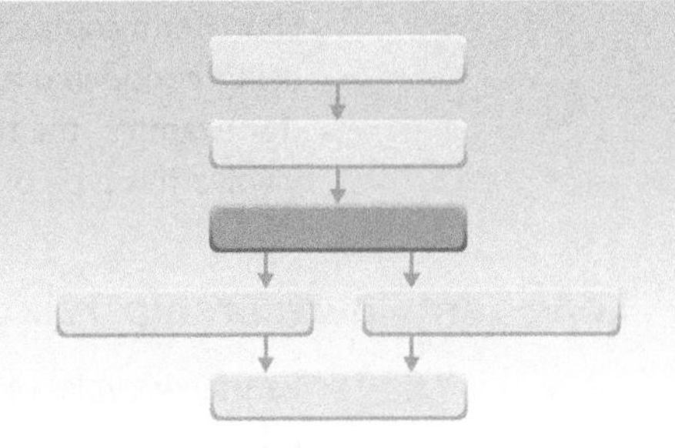

Video

The 'relationship' between operations in a supply chain is the basis on which the exchange of products, services, information and money is conducted. Managing supply chains is about managing relationships, because relationships influence the smooth flow between operations and processes. Different forms of relationship will be appropriate in different circumstances. An obvious but important factor in determining the importance of relationships to any operation is the extent to which they outsource their activities. In Chapter 3 we distinguished between non-vertically integrated operations that outsource almost all activities, and vertically integrated operations that outsource almost nothing. Only extremely vertically integrated businesses are able to ignore the question of how to manage customer–supplier relationships (because they do everything themselves). Initially, we can examine this question by describing two contrasting 'pure' arrangements – pure *contractual*, market-based, transactional relationships, and close, longer-term, pure *partnership* relationships. However, it is better to think of these as the two basic ingredients of any supply arrangement. Whatever arrangement with its suppliers a firm chooses to take; it can be described by the balance between contracts and partnerships.

Contract-based 'transactional' relationships

Contract-based, transactional relationships involve purchasing goods and services in a 'pure' market fashion, often seeking the 'best' supplier every time it is necessary to make a purchase. Each transaction effectively becomes a separate decision. The relationship may be short-term, with no guarantee of further trading between the parties once the goods or services are delivered and payment is made.[5] The *advantages* of contract-based 'transactional' relationships are usually seen as follows:

- They maintain competition between alternative suppliers. This promotes a constant drive between suppliers to provide best value.
- A supplier specialising in a small number of products or services, but supplying them to many customers, can gain natural economies of scale, enabling the supplier to offer the products and services at a lower price than if customers performed the activities themselves on a smaller scale.
- There is inherent flexibility in outsourced supplies. If demand changes, customers can simply change the number and type of suppliers, a faster and cheaper alternative to redirecting internal activities.
- Innovations can be exploited no matter where they originate. Specialist suppliers are more likely to come up with innovations that can be acquired faster and cheaper than developing them in-house.

There are, however, *disadvantages* in buying in a totally contractual manner:

- Suppliers owe little loyalty to customers. If supply is difficult, there is no guarantee of receiving supply.
- Choosing who to buy from takes time and effort. Gathering sufficient information and making decisions continually are, in themselves, activities that need to be resourced.

Short-term contractual relationships of this type may be appropriate when new companies are being considered as more regular suppliers, or when purchases are one-off or very irregular. (for example, the replacement of all the windows in a company's office block would typically involve this type of competitive-tendering market relationship).

Long-term 'partnership' relationships

Partnership relationships in supply chains are sometimes seen as a compromise between vertical integration on the one hand (owning the resources which supply you) and transactional relationships on the other. Partnership relationships are defined as:[6] '*. . . relatively enduring inter-firm cooperative agreements, involving flows and linkages that use resources and/or governance structures from autonomous organisations, for the joint accomplishment of individual goals linked to the corporate mission of each sponsoring firm.*' This means that suppliers and customers are expected to cooperate, even to the extent of sharing skills and resources, to achieve joint benefits beyond those they could have achieved by acting alone. At the heart of the concept of partnership lies the issue of the *closeness* of the relationship. Partnerships are close relationships, the degree of which is influenced by a number of factors, as follows:

- *Sharing success* – both partners jointly benefit from the cooperation rather than manoeuvring to maximise their own individual contribution.
- *Long-term expectations* – relatively long-term commitments, but not necessarily permanent ones.
- *Multiple points of contact* – communication is not restricted to formal channels, but may take place between many individuals in both organisations.
- *Joint learning* – a relationship commitment to learn from each other's experience.
- *Few relationships* – a commitment on the part of both parties to limit the number of customers or suppliers with whom they do business.
- *Joint coordination of activities* – fewer relationships allow joint coordination of activities such as the flow of materials or service, payment, and so on.
- *Information transparency* – confidence is built through information exchange between the partners.
- *Joint problem solving* – jointly approaching problems can increase closeness over time.
- *Trust* – probably the key element in partnership relationships. In this context, trust means the willingness of one party to relate to the other on the understanding that the relationship will be beneficial to both, even though that cannot be guaranteed. Trust is widely held to be both the key issue in successful partnerships, but also, by far the most difficult element to develop and maintain.

Which type of relationship?

It is very unlikely that any business will find it sensible to engage exclusively in one type of relationship or another. Most businesses will have a portfolio of, possibly, widely differing relationships. Also, there are degrees to which any particular relationship can be managed on a transactional or partnership basis. The real question is: where, on the spectrum from transactional to partnership, should each relationship be positioned? And, while there is no simple formula for choosing the 'ideal' form of relationship in each case, there are some important factors that can sway the decision. The most obvious issue will concern how a business intends to compete in its marketplace. If price is the main competitive factor then the relationship could be determined by which approach offers the highest potential savings.

OPERATIONS PRINCIPLE

All supply chain relationships can be described by the balance between their 'contractual' and 'partnership' elements.

On one hand, market-based contractual relationships could minimise the actual price paid for purchased products and services, while partnerships could minimise the transaction costs of doing business. If a business is competing primarily on product or service innovation, the type of relationship may depend on where innovation is likely to happen. If innovation depends on close collaboration between supplier and customer, partnership relationships are needed. On the other hand, if suppliers are busily competing to out-do each other in terms of their innovations, and especially if the market is turbulent and fast growing (as with many software and internet-based industries), then it may be preferable to retain the freedom to change suppliers quickly using market mechanisms. However, if markets are very turbulent, partnership relationships may reduce the risks of being unable to secure supply.

The main differences between the two ends of this relationship spectrum concerns whether a customer sees advantage in long-term or short-term relationships. Contractual relationships can be either long or short term, but there is no guarantee of anything beyond the immediate contract. They are appropriate when short-term benefits are important. Many relationships and many businesses are best served by concentrating on the short term (especially if, without short-term success, there is no long term). Partnership relationships are by definition long-term. There is a commitment to work together over time to gain mutual advantage. The concept of mutuality is important here. A supplier does not become a 'partner' merely by being called one. True partnership implies mutual benefit, and often mutual sacrifice. Partnership means giving up some freedom of action in order to gain something more beneficial over the long term. If it is not in the culture of a business to give up some freedom of action, it is very unlikely to ever make a success of partnerships. Opportunities to develop relationships can be limited by the structure of the market itself. If the number of potential suppliers is small, there may be few opportunities to use market mechanisms to gain any kind of supply advantage and it would probably be sensible to develop a close relationship with at least one supplier. On the other hand, if there are many potential suppliers, and especially if it is easy to judge the capabilities of the suppliers, contractual relationships are likely to be best.

OPERATIONS PRINCIPLE

True 'partnership' relationships involve mutual sacrifice as well as mutual benefit.

DIAGNOSTIC QUESTION

How should the supply side be managed?

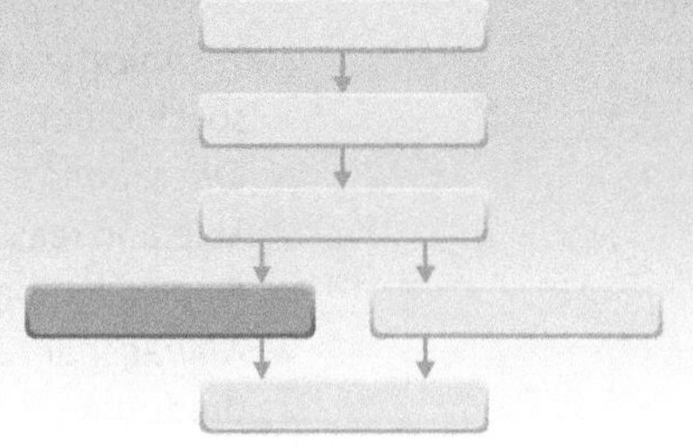

Video

The ability of any process or operation to produce outputs is dependent on the inputs it receives, so good supply management is a necessary (but not sufficient) condition for effective operations management in general. It involves three main activities: selecting appropriate suppliers; planning and controlling the ongoing supply activity; developing and improving suppliers' capabilities. All three activities are usually the responsibility of the purchasing or procurement function within the business. Purchasing should provide a vital link between the operation itself and its suppliers. They should understand the requirements of all the processes within their own operation and also the capabilities of the suppliers who could potentially provide products and services for the operation.

Supplier selection

Choosing appropriate suppliers should involve trading off alternative attributes. Rarely are potential suppliers so clearly superior to their competitors that the decision is self-evident. Most businesses find it best to adopt some kind of supplier 'scoring' or assessment procedure. This should be capable of rating alternative suppliers in terms of factors such as the following:

- range of products or services provided
- quality of products or services
- responsiveness
- dependability of supply
- delivery and volume flexibility
- total cost of being supplied
- ability to supply in the required quantity.

In addition, there are likely to be less quantifiable or longer-term factors that will need taking into consideration. These may include the following:

- potential for innovation
- ease of doing business
- willingness to share risk
- long-term commitment to supply
- ability to transfer knowledge as well as products and services.

Choosing suppliers should involve evaluating the relative importance of all these factors. So, for example, a business might choose a supplier who, although more expensive than alternative suppliers, has an excellent reputation for on time delivery, because that is more appropriate to the way the business competes itself, or because the high level of supply dependability allows the business to hold lower stock levels, which may even save costs overall. Other trade-offs may be more difficult to calculate. For example, a potential supplier may have high levels of technical capability, but may be financially weak, with a small but finite risk of going out of business. Other suppliers may have little track record of supplying the products or services required, but show the managerial talent and energy for potential customers to view developing a supply relationship as an investment in future capability. But to make sensible trade-offs it is important to assess four basic capabilities:

> **OPERATIONS PRINCIPLE**
> *Supplier selection should reflect overall supply chain objectives.*

- *Technical capability* – the product or service knowledge to supply to high levels of specification.
- *Operations capability* – the process knowledge to ensure consistent, responsive, dependable and reasonable cost supply.
- *Financial capability* – the financial strength to fund the business in the short and long terms.
- *Managerial capability* – the management talent and energy to develop supply potential in the future.

Single- or multi-sourcing

A closely linked decision is whether to source each individual product or service from one, or more than one, supplier (single-sourcing or multi-sourcing). Some of the advantages and disadvantages of single- and multi-sourcing are shown in Table 7.1.

It may seem as though companies who multi-source do so exclusively for their own short-term benefit. However, this is not always the case: multi-sourcing can have an altruistic motive, or at least one that brings benefits to both supplier and purchaser in the long term. For example, Robert Bosch GmbH, the German automotive components manufacturer and distributor,

Table 7.1 Advantages and disadvantages of single- and multi-sourcing

	Single-sourcing	*Multi-sourcing*
Advantages	Potentially better quality because of more supplier quality assurance possibilities Strong relationships that are more durable Greater dependency encourages more commitment and effort Better communication Easier to cooperate on new product/service development More economies of scale Higher confidentiality	Purchaser can drive price down by competitive tendering Can switch sources in case of supply failure Wide sources of knowledge and expertise to tap
Disadvantages	More vulnerable to disruption if a failure to supply occurs Individual supplier more affected by volume fluctuations Supplier might exert upward pressure on prices if no alternative supplier is available	Difficult to encourage commitment supplier Less easy to develop effective supplier quality assurance More effort needed to communicate Suppliers less likely to invest in new processes More difficult to obtain economies of scale

at one time required that sub-contractors do no more than 20 per cent of their total business with them.[7] This was to prevent suppliers becoming too dependent on them. The purchasing organisation could then change volumes up and down without pushing the supplier into bankruptcy. However, despite these perceived advantages, there has been a trend for purchasing functions to reduce their supplier base in terms of numbers of companies supplying any one part or service, mainly because it reduces the costs of transacting business.

Purchasing, the internet and e-commerce

For some years, electronic means have been used by businesses to confirm purchased orders and ensure payment to suppliers. The rapid development of the internet, however, opened up the potential for far more fundamental changes in purchasing behaviour. Partly this was as the result of supplier information made available through the internet. Previously, a purchaser of industrial components may have been predisposed to return to suppliers who had been used before. There was inertia in the purchasing process because of the costs of seeking out new suppliers. By making it easier to search for alternative suppliers, the internet changes the economics of the search process and offers the potential for wider searches. It also changed the economics of scale in purchasing. Purchasers requiring relatively low volumes find it easier to group together in order to create orders of sufficient size to warrant lower prices. In fact, the influence of the internet on purchasing behaviour is not confined to *e-commerce*. Usually e-commerce is taken to mean the trade that actually takes place over the internet. This is usually assumed to be a buyer visiting the seller's website, placing an order for parts and making a payment (also through the site). But the internet is also an important source of purchasing information. For every 1 per cent of business transacted directly via the internet, there may be 5 or 6 per cent of business that, at some point, involved it, probably with potential buyers using it to compare prices or obtain technical information.

One increasingly common use of internet technology in purchasing (or e-procurement as it is sometimes known) is for large companies, or groups of companies, to link their e-commerce systems into a common 'exchange'. In their more sophisticated form, such an exchange may be linked into the purchasing companies' own information systems (*see* the explanation of ERP in Chapter 10). Many of the large automotive, engineering and petrochemical companies, for example, have adopted such an approach. An early example of this was Dow Corning's 'Xiameter' service. Dow Corning was the global market leader in silicon, a material which has a wide range of industrial applications, from clothing and computers, to cosmetics construction. Traditionally its customers had paid top prices for pioneering technology and premium quality products, delivered with an emphasis on "solutions-based" service'. However, in the early 2000s some of its larger and more sophisticated customers wanted a different kind of supply arrangement. As one customer put it: *'I don't need these services. I know I can go and buy a tanker of this fluid at a lower price. I'll buy this but I just need low price and guaranteed delivery.'* This part of their market consisted of experienced purchasers of commonly used silicone materials who wanted the lowest price and an easy way of doing business with their supplier. Their solution was to offer a 'no-frills', limited availability service with low prices that could only be accessed on the web. This service would offer only regular products without any technical advice. They branded this service 'Xiameter' (rhymes with 'diameter'). It was limited to about 350 common silicone compounds (out of more than 7,500) ordered in high volumes. It would have a 'lean' management structure and would secure its supply from Dow Corning's manufacturing sites around the world. Customers could only place orders online (a novel approach at the time). Minimum order quantities were strictly applied. Delivery lead times were fixed by production scheduling. Standard payment terms were 30 days. All communication was via e-mail including automated order confirmation, shipping notices and invoices. E-mail enquiries had a one-day guaranteed response. Any customers deviating from these rules faced additional charges. Order cancellation fees were set at 5 per cent of the order's value. Expedited orders incurred a 10 per cent supplement and late payment carried an 18 per cent annual interest charge.

Managing ongoing supply

Managing supply relationships is not just a matter of choosing the right suppliers and then leaving them to get on with day-to-day supply. It is also about ensuring that suppliers are given the right information and encouragement to maintain smooth supply and that internal inconsistency does not negatively affect their ability to supply. A basic requirement is that some mechanism should be set up that ensures the two-way flow of information between customer and supplier. It is easy for both suppliers and customers simply to forget to inform each other of internal developments that could affect supply. Customers may see suppliers as having the responsibility for ensuring appropriate supply 'under any circumstances'. Or, suppliers themselves may be reluctant to inform customers of any potential problems with supply because they see it as risking the relationship. Yet, especially if customer and supplier see themselves as 'partners', the free flow of information, and a mutually supportive tolerance of occasional problems, is the best way to ensure smooth supply. Often day-to-day supplier relationships are damaged because of internal inconsistencies. For example, one part of a business may be asking a supplier for some special service beyond the strict nature of their agreement, while another part of the business is not paying suppliers on time.[8]

Service-level agreements

Some organisations bring a degree of formality to supplier relationships by encouraging (or requiring) all suppliers to agree service-level agreements (SLAs). SLAs are formal definitions of the dimensions of service and the relationship between suppliers and the organisation. The type of issues covered by such an agreement could include response times, the range of services, dependability of service supply, and so on. Boundaries of responsibility and appropriate performance measures could also be agreed. For example, an SLA between an information systems support unit and a research unit in the laboratories of a large pharmaceutical company could define such performance measures as:

- the types of information network services which may be provided as 'standard'
- the range of special information services which may be available at different periods of the day
- the minimum 'up time', i.e. the proportion of time the system will be available at different periods of the day
- the maximum response time and average response time to get the system fully operational should it fail
- the maximum response time to provide 'special' services, and so on.

Although SLAs are described here as mechanisms for governing the ongoing relationship between suppliers and customers, they often prove inadequate because they are seen as being useful in setting up the terms of the relationship, but then are only used to resolve disputes. For SLAs to work effectively, they must be treated as working documents that establish the details of ongoing relationships *in the light of experience*. Used properly, they are a repository of the knowledge that both sides have gathered through working together. Any SLA that stays unchanged over time is, at the very least, failing to encourage improvement in supply.

How can suppliers be developed?

In any relationship other than pure market-based transactional relationships, it is in a customer's long-term interests to take some responsibility for developing supplier capabilities. Helping a supplier to improve not only enhances the service (and hopefully price) from the supplier, it may also lead to greater supplier loyalty and long-term commitment. This is why some particularly successful businesses (including Japanese automotive manufacturers) invest in supplier development teams whose responsibility is to help suppliers to improve their own operations processes. Of course, committing the resources to help suppliers is only worthwhile if it improves the effectiveness of the supply chain as a whole. Nevertheless, the potential for such enlightened self-interest can be significant.

How customers and suppliers see each other[9]

One of the major barriers to supplier development is the mismatch between how customers and suppliers perceive both what is required and how the relationship is performing. Exploring potential mismatches is often a revealing exercise, both for customers and suppliers. Figure 7.7 illustrates this. It shows that gaps may exist between four sets of ideas. As a customer you (presumably) have an idea about what you really want from a supplier. This may, or may not, be formalised in the form of a service level agreement. But no SLA can capture everything about what is required. There may be a gap between how you as a customer interpret what is required and how the supplier interprets it. This is the *requirements perception gap*. Similarly, as a customer, you (again presumably) have a view on how your supplier is performing in terms of fulfilling your requirements. That may not coincide with how your supplier believes it is performing. This is the *fulfilment perception gap*. Both these gaps are a function of the effectiveness of the communication between supplier and customer. But there are also two other gaps. The gap between what you want from your supplier and how they are performing indicates the type of development that, as a customer, you should be giving to your supplier. Similarly, the gap between your supplier's perceptions of your needs

> **OPERATIONS PRINCIPLE**
> *Unsatisfactory supplier relationships can be caused by requirements and fulfilment perception gaps.*

Animated diagram

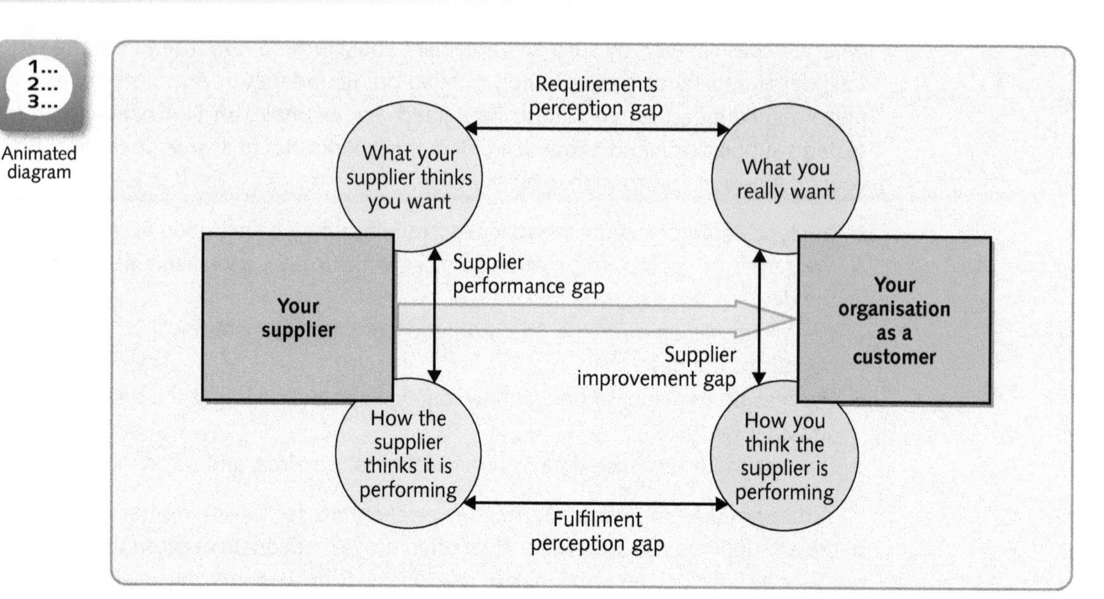

Figure 7.7 Explore the potential perception mismatches to understand supplier development needs

and its performance indicates how they should initially see themselves improving their own performance. Ultimately, of course, their responsibility for improvement should coincide with their customer's views of requirements and performance.

DIAGNOSTIC QUESTION

How should the demand side be managed?

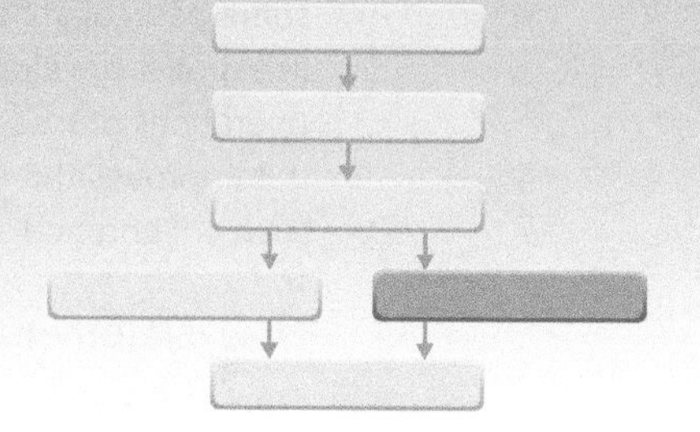

Video

The management of demand side relationships will depend partly on the nature of demand, in particular how uncertain it is. Knowing the exact demands that customers are going require allows a supplier to plan its own internal processes in a systematic manner. This type of demand is called 'dependent' demand; it is relatively predictable because it is dependent upon some factor which is itself predictable. For example, supplying tyres to an automobile factory involves examining the manufacturing schedules in the car plant and deriving the demand for tyres from these. If 200 cars are to be manufactured on a particular day, then it is simple to calculate that 1,000 tyres will be demanded by the car plant (each car has five tyres). Because of this, the tyres can be ordered from the tyre manufacturer to a delivery schedule which is closely in line with the demand for tyres from the plant. In fact, the demand for every part of the car plant will be derived from the assembly schedule for the finished cars. Manufacturing instructions and purchasing requests will all be dependent upon this figure. Managing internal process networks when external demand is dependent is largely a matter of calculating, in as precise a way as possible, the internal consequences of demand. MRP, treated in Chapter 10, is the best known dependent demand approach.

But not all operations have such predictable demand. Some operations are subject to independent demand. There is a random element in demand which is virtually independent of any obvious factors. They are required to supply demand without having any firm forward visibility of customer orders. A drive-in tyre replacement service will need to manage a stock of tyres. In that sense it is exactly the same task that faced the supplier of tyres to the car plant, but demand is very different. It cannot predict either the volume or the specific needs of customers. It must make decisions on how many and what type of tyres to stock, based on demand forecasts and in the light of the risks it is prepared to run of being out of stock. Managing internal process networks when external demand is independent involves making 'best guesses' concerning future demand, attempting to put the resources in place to satisfy this demand, and attempting to respond quickly if actual demand does not match the forecast. Inventory planning and control, treated in Chapter 9, is a typical approach.

Logistics services

Logistics means moving products to customers. Sometimes the term 'physical distribution management', or simply 'distribution', is used as being analogous to logistics. Logistics is now frequently outsourced to 'third party' logistics (or 3PL) providers, which vary in terms of the range and integration of their services. At the simplest level, the 'haulage' and 'storage' businesses either move goods around or they store them in warehouses. Clients take responsibility for all planning. Physical distribution companies bring haulage and storage together, collecting clients' products, putting them into storage facilities and delivering them to the end customer as required. 'Contract' logistics service providers tend to have more sophisticated clients with more complex operations. Total 'supply chain management' (or 4PL) providers offer to manage supply chains from end to end, often for several customers simultaneously. Doing this requires a much greater degree of analytical and modelling capability, business process reengineering and consultancy skills.

Logistics management and the internet

Internet-based communication has had a significant impact on physical distribution management. Information can be made available more readily along the distribution chain, so that transport companies, warehouses, suppliers and customers can share knowledge of where goods are in the chain (and sometimes where they are going next). This allows the operations within the chain to coordinate their activities more readily. It also gives the potential for some significant cost savings. For example, an important issue for transportation companies is back-loading. When the company is contracted to transport goods from A to B, its vehicles may have to return from B to A empty. Back-loading means finding a potential customer who wants their goods transported from B to A in the right time-frame. With the increase in information availability through the internet, the possibility of finding a back-load increases. Companies that can fill their vehicles on both the outward and return journeys will have significantly lower costs per distance travelled than those whose vehicles are empty for half the total journey. Similarly, internet-based technology that allows customers visibility of the progress of distribution can be used to enhance the perception of customer service. 'Track-and-trace' technologies, for example, allow package distribution companies to inform and reassure customers that their service is being delivered as promised.

Automatic identification technologies

Tracing the progress of items through a supply chain has involved the use of bar codes to record progress. During manufacture, bar codes are used to identify the number of products passing through a particular point in the process. In warehouses, bar codes are used to keep track of how many products are stored at particular locations. But bar codes have disadvantages. It is sometimes difficult to align the item so that the bar code can be read conveniently, items can only be scanned one by one, and the bar code only identifies the *type* of item not a specific item itself. That is, the code identifies that an item is, say, a can of one type of drink rather than one specific can. These drawbacks can be overcome through the use of 'automated identification' or Auto-ID. Usually this involves Radio Frequency Identification (RFID). Here an Electronic Product Code (ePC) that is a unique number, 96 bits long, is embedded in a memory chip or smart tag. These tags are put on individual items so that each item has its own unique identifying code. At various points during its manufacture, distribution, storage and sale each smart tag can be scanned by a wireless radio frequency 'reader'. This transmits the item's embedded identify code to a network such as the internet, describing, for example, when and where it was made, where it has been stored, etc. This information can then be fed into control systems. It is also controversial – see the Critical Commentary, later.

Customer development

Earlier in the chapter, Figure 7.7 illustrated some of the gaps in perception and performance that can occur between customers and suppliers. The purpose then was to demonstrate the nature of supplier development. The same approach can be used to analyse the nature of requirements and performance with customers. In this case the imperative is to understand customer perceptions, both of their requirements and their view of your performance, and feed these into your own performance improvement plans. What is less common, but can be equally valuable, is to use these gaps (shown in Figure 7.8) to examine the question of whether customer requirements and perceptions of performance are either accurate or reasonable. For example, customers may be placing demands on suppliers without fully considering their consequences. It may be that slight modifications in what is demanded would not inconvenience customers and yet would provide significant benefits to suppliers that could then be passed on to customers. Similarly, customers may be incompetent at measuring supplier performance, in which case the benefits of excellent supplier service will not be recognised. So, just as customers have a responsibility to help develop their own supplier's performance, in their own as well as their supplier's interests, suppliers have a responsibility to develop their customer's understanding of how supply should be managed.

> **OPERATIONS PRINCIPLE**
> *Unsatisfactory customer relationships can be caused by requirement and fulfilment perception gaps.*

Animated diagram

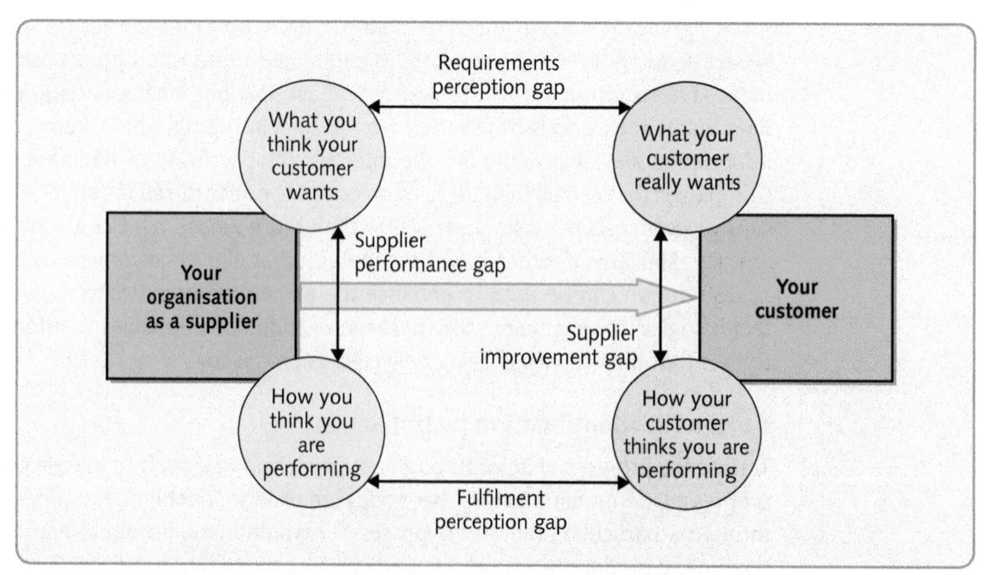

Figure 7.8 Explore the potential perception mismatches to understand customer development needs

DIAGNOSTIC QUESTION

Are supply chain dynamics under control?

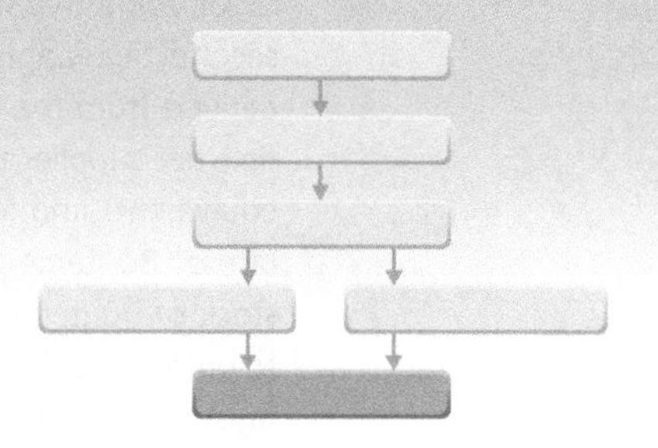

Video

There are dynamics that exist between firms in supply chains that cause errors, inaccuracies and volatility, and these increase for operations further upstream in the supply chain. This effect is known as the 'bull whip effect'[10], so called, because a small disturbance at one end of the chain causes increasingly large disturbances as it works its way towards the end. Its main cause is a perfectly understandable and rational desire by the different links in the supply chain to manage their levels of activity and inventory sensibly. To demonstrate this, examine the production rate and stock levels for the supply chain shown in Table 7.2. This is a four-stage supply chain where an original equipment manufacturer (OEM) is served by three tiers of suppliers. The demand from the OEM's market has been running at a rate of 100 items per period, but in period 2, demand reduces to 95 items per period. All stages in the supply chain work on the principle that they will keep in stock one period's demand. This is a simplification but not a gross one. Many operations gear their inventory levels to their demand rate. The column headed 'stock' for each level of supply shows the starting stock at the beginning of the period and the finish stock at the end of the period. At the beginning of period 2, the OEM has 100 units in stock (that being the rate of demand up to period 2). Demand in period 2 is 95 and so the OEM knows that it would need to produce sufficient items to finish up at the end of the period with 95 in stock (this being the new demand rate). To do this, it need only manufacture 90 items; these, together with five items taken out of the starting stock, will supply demand and leave a finished stock of 95 items. The beginning of period 3 finds the OEM with 95 items in stock. Demand is also 95 items and therefore its production rate to maintain a stock level of 95 will be 95 items per period. The original equipment manufacturer now operates at a steady rate of producing 95 items per period. Note, however, that a change in demand of only five items has produced a fluctuation of ten items in the OEM's production rate.

Table 7.2 Fluctuations of production levels along supply chain in response to small change in end customer demand

	Third-tier supplier		*Second-tier supplier*		*First-tier supplier*		*Original equipment mfr.*		
Period	*Prodn.*	*Stock*	*Prodn.*	*Stock*	*Prodn.*	*Stock*	*Prodn.*	*Stock*	*Demand*
1	100	100 100	100	100 100	100	100 100	100	100 100	100
2	20	100 60	60	100 80	80[b]	100[a] 90[c]	90[d]	100 95	95
3	180	60 120	120	80 100	100	90 95	95	95 95	95
4	60	120 90	90	100 95	95	95 95	95	95 95	95
5	100	90 95	95	95 95	95	95 95	95	95 95	95
6	95	95 95	95	95 95	95	95 95	95	95 95	95

Starting stock (a) + production (b) = finishing stock (c) + demand, that is production in previous tier down (d): see explanation in text. All stages in the supply chain keep one period's inventory: c = d.

Carrying this same logic through to the first-tier supplier, at the beginning of period 2, the second-tier supplier has 100 items in stock. The demand which it has to supply in period 2 is derived from the production rate of the OEM. This has dropped down to 90 in period 2. The first-tier supplier therefore has to produce sufficient to supply the demand of 90 items (or the equivalent) and leave one month's demand (now 90 items) as its finish stock. A production rate of 80 items per month will achieve this. It will therefore start period 3 with an opening stock of 90 items, but the demand from the OEM has now risen to 95 items. It therefore has to produce sufficient to fulfil this demand of 95 items and leave 95 items in stock. To do this, it must produce 100 items in period 3. After period 3 the first-tier supplier then resumes a steady state, producing 95 items per month. Note again, however, that the fluctuation has been even greater than that in the OEM's production rate, decreasing to 80 items a period, increasing to 100 items a period, and then achieving a steady rate of 95 items a period. Extending the logic back to the third-tier supplier, it is clear that the further back up the supply chain an operation is placed, the more drastic are the fluctuations.

OPERATIONS PRINCIPLE
Demand fluctuations become progressively amplified as their effects work back up the supply chain.

This relatively simple demonstration ignores any time lag in material and information flow between stages. In practice there will be such a lag, and this will make the fluctuations even more marked. Figure 7.9 shows the net result of all these effects in a typical supply chain. Note the increasing volatility further back in the chain.

Controlling supply chain dynamics

The first step in improving supply chain performance involves attempting to reduce the bull whip effect. This usually means coordinating the activities of the operations in the chain in several ways:[11]

Share information throughout the supply chain

OPERATIONS PRINCIPLE
The bull whip effect can be reduced by information sharing, aligning planning and control decisions, improving flow efficiency, and better forecasting.

One reason for the bull whip effect is that each operation in the chain reacts only to the orders placed by its *immediate* customer. They have little overview of what is happening throughout the chain. But if chain-wide information is shared throughout the chain, it is unlikely that such wild fluctuations will occur. With information transmitted throughout the chain, all the operations can monitor true demand, free of distortions. So, for example, information regarding supply problems, or shortages, can be transmitted

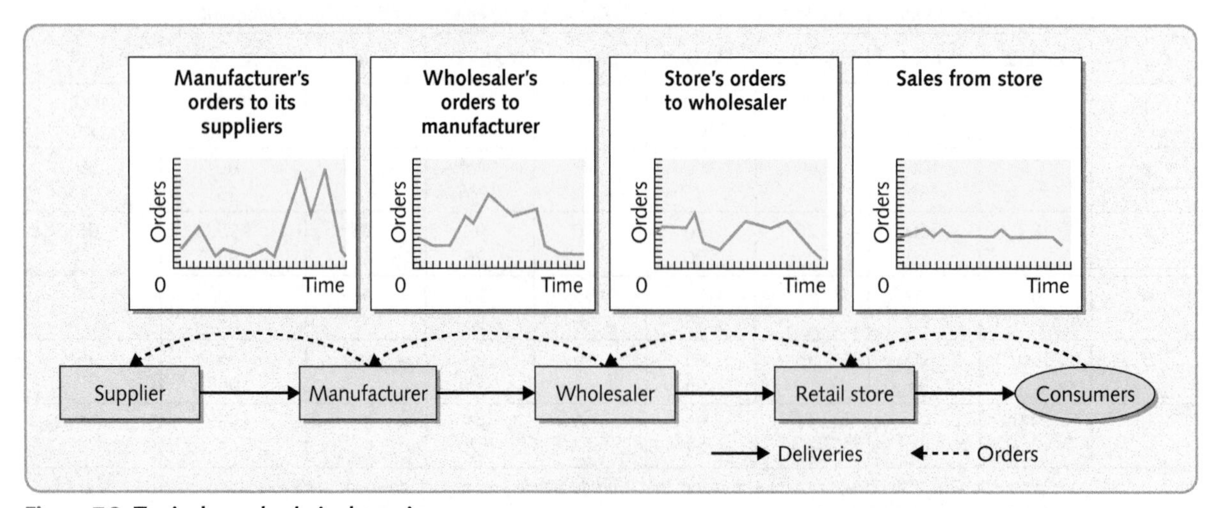

Figure 7.9 Typical supply chain dynamics

Animated diagram

down the chain so that downstream customers can modify their schedules and sales plans accordingly. For example, the electronic point-of-sale (EPOS) systems, used by many retailers, make information on current demand downstream in the supply chain available to upstream operations. Sales data from checkouts or cash registers is consolidated and transmitted to the warehouses, transportation companies and supplier operations in the supply chain. This means that suppliers can be aware of the 'real' movements in the market.

Align all the channels of information and supply

Channel alignment means the adjustment of scheduling, material movements, stock levels, pricing and other sales strategies so as to bring all the operations in the chain into line with each other. This goes beyond the provision of information. It means that the systems and methods of planning and control decision making are harmonised through the chain. For example, even when using the same information, differences in forecasting methods or purchasing practices can lead to fluctuations in orders between operations in the chain. One way of avoiding this is to allow an upstream supplier to manage the inventories of its downstream customer. This is known as vendor-managed inventory (VMI). So, for example, a packaging supplier could take responsibility for the stocks of packaging materials held by a food manufacturing customer. In turn, the food manufacturer takes responsibility for the stocks of its products that are held in its customer's (the supermarket's) warehouses.

Increase operational efficiency throughout the chain

'Operational efficiency' in this context means the efforts that each operation in the chain makes to reduce its own complexity, the cost of doing business with other operations in the chain, and its throughput time. The cumulative effect of this is to simplify throughput in the whole chain. For example, imagine a chain of operations whose performance level is relatively poor: quality defects are frequent, the lead time to order products and services is long, delivery is unreliable and so on. The behaviour of the chain would be a continual sequence of errors and effort wasted in replanning to compensate for the errors. Poor quality would mean extra and unplanned orders being placed, and unreliable delivery and slow delivery lead times would mean high safety stocks. Just as important, most operations managers' time would be spent coping with the inefficiency. By contrast, a chain whose operations had high levels of operations performance would be more predictable and have faster throughput, both of which would help to minimise supply chain fluctuations.

Improve forecasts

Improved forecast accuracy also helps to reduce the bull whip effect. Bull whip is caused by the demand pattern, lead times, forecasting mechanisms and the replenishment decisions used to order product from production facilities or suppliers. Improving the accuracy of your forecasts directly reduces the inventory holding requirements that will achieve customer service level targets. Reducing lead-times means that you need to forecast less far into the future and thus lead times have a large impact on bull whip and inventory costs. The exact nature of how bull whip propagates in a supply chain is also dependent on the nature of the demand pattern. Negatively correlated demands require less inventory in the supply chain that positively correlated demand patters, for example. But bull whip is not unavoidable. By using sophisticated replenishment policies, designed using control engineering principles, many businesses have been able to eliminate bullwhip effects. Sometimes this comes at a cost. Extra inventory may be required in parts of the chain, or customer service levels reduce. But more often bullwhip avoidance creates a 'win-win'. It reduces inventory requirements and improves customer service.

Critical commentary

Each chapter contains a short critical commentary on the main ideas covered in the chapter. Its purpose is not to undermine the issues discussed in the chapter, but to emphasise that, although we present a relatively orthodox view of operation, there are other perspectives.

- This emphasis on understanding the end customer in a supply chain has led some authorities to object to the very term *supply* chain. Rather, they say, they should be referred to as *demand* chains. Their argument is based on the idea that the concept of 'supply' implies a 'push' mentality. Any emphasis on pushing goods through a supply chain should be avoided. It implies that customers should consume what suppliers see fit to produce. On the other hand, referring to 'demand chains' puts proper emphasis on the importance of seeing customers as pulling demand through the chain. Nevertheless, 'supply chain' is still the most commonly used term.

- Although the SCOR model is increasingly adopted, it has been criticised for under-emphasising people issues. The SCOR model assumes, but does not explicitly address, the human resource base skill set, notwithstanding the model's heavy reliance on supply chain knowledge to understand the model and methodology properly. Often external expertise is needed to support the process. This, along with the nature of the SCC membership also imply that the SCOR model may be appropriate only for relatively large companies that are more likely to have the necessary business capabilities to implement the model. Many small- to medium-sized companies may find difficulty in handle full-scale model implementation. Some critics would also argue that the model lacks a link to the financial plans of a company making it very difficult to highlight the benefits obtainable, as well as inhibiting senior management support.

- The use of technology in supply chain management is not always universally welcomed. Even e-procurement is seen by some as preventing closer partnership-type relationships that, in the long run, may be more beneficial. Similarly, track-and-trace technology is seen by some as a waste of time and money. '*What we need,*' they argue, '*is to know that we can trust the delivery to arrive on time; we do not need the capability to waste our time finding out where the delivery is.*' The idea of RFID also opens up many ethical issues. People see its potential and its dangers in very different ways. Take the following two statements:[12]

> *'We are on the brink of a revolution of "smart products" that will interconnect everyday objects, consumers and manufacturers in a dynamic cycle of world commerce. . . . The vision of the Auto-ID centre is to create a universal environment in which computers understand the world without help from human beings.'*

'Supermarket cards and other retail surveillance devices are merely the opening volley of the marketers' war against consumers. If consumers fail to oppose these practices now our long-term prospects may look like something from a dystopian science fiction novel . . . though many Auto-ID proponents appear focused on inventory and supply chain efficiency, others are developing financial and consumer applications that, if adopted, will have chilling effects on consumers' ability to escape the oppressive surveillance of manufacturers, retailers, and marketers. Of course, government and law enforcement will be quick to use the technology to keep tabs on citizens as well.'

- It is this last issue which particularly scares some civil liberties activists. Keeping track of items within a supply chain is a relatively uncontentious issue. Keeping track of items when those items are identified with a particular individual going about their everyday lives, is far more problematic. So, beyond the check-out for every arguably beneficial application there is also potential for misuse. For example, smart tags could drastically reduce theft because items could automatically report when they are stolen; their tags serving as a homing device pinpoint their exact location. But, similar technology could be used to trace any citizen, honest or not.

SUMMARY CHECKLIST

This checklist comprises questions that can be usefully applied to any type of operations and reflect the major diagnostic questions used within the chapter.

- ☐ Is it understood that the performance of any one operation is partly a function of all the other operations in the supply chain?
- ☐ Are supply chain concepts applied internally as well as externally?
- ☐ Are supply chain objectives understood in the context of the whole chain rather than the single operation?
- ☐ Which product or service groups are 'functional' and which are 'innovative'?
- ☐ Which products or service groups need 'lean' and which need 'agile' supply chain management?
- ☐ Is the position on the 'transactional to partnership' spectrum understood for each customer and supplier relationship?
- ☐ Are customer and supplier relationships at an appropriate point on the transactional to partnership spectrum?
- ☐ Are 'partnership' relationships *really* partnerships or are they just called that?
- ☐ Are suppliers and potential suppliers rigorously assessed using some scoring procedure?
- ☐ Are the trade-offs inherent in supplier selection understood?
- ☐ Is the approach to single- or multi-sourcing appropriate?
- ☐ Is the purchasing activity making full use of internet-based mechanisms?
- ☐ Are service level agreements used? Do they develop over time?
- ☐ Is sufficient effort put into supplier development?
- ☐ Are actual and potential mismatches of perception in the supplier relationships explored?
- ☐ Is the difference between dependent and independent demand understood?
- ☐ Is the potential for outsourcing logistics services regularly explored?
- ☐ Could new technologies such as RFID have any benefit?
- ☐ Has the idea of customer development been explored?
- ☐ Have mechanisms for reducing the impact of the bull whip effect been explored?
- ☐ Has there been a risk assessment to assess supply chain vulnerability?

CASE STUDY

Supplying fast fashion[13]

Garment retailing has changed. No longer is there a standard look that all retailers adhere to for a whole season. Fashion is fast, complex and furious. Different trends overlap and fashion ideas that are not even on a store's radar screen can become 'must haves' within six months. Many retail businesses with their own brands, such as H&M and Zara, sell up-to-the-minute fashionability at low prices in stores that are clearly focused on one particular market. In the world of fast fashion, catwalk designs speed their way into high street stores at prices anyone can afford. The quality of the garment means that it may only last one season, but fast fashion customers don't want yesterday's trends. As *Newsweek* puts it, *'. . . being a "quicker picker-upper" is what made fashion retailers H&M and Zara successful. [They] thrive by practising the new science of "fast fashion", compressing product development cycles as much as six times.'* But the retail operations that customers see are only the end part of the supply chains that feeds them. And these have also changed.

At its simplest level, the fast fashion supply chain has four stages. First, the garments are designed, after which they are manufactured, they are then distributed to the retail outlets where they are displayed and sold in retail operations designed to reflect the business's brand values. In this short case study, we examine two fast fashion operations, Hennes and Mauritz (known as H&M) and Zara, together with United Colours of Benetton (UCB), a similar chain, but with a different market positioning.

United Colours of Benetton. Almost 50 years ago Luciano Benetton took the world of fashion by storm by selling the bright, casual sweaters designed by his sister across Europe (and later the rest of the world), promoted by controversial advertising. By 2005, the Benetton Group was present in 120 countries throughout the world. Selling casual garments, mainly under its United Colours of Benetton (UCB) and its more fashion-oriented Sisley brands, it produces 110 million garments a year, over 90 per cent of them in Europe. Its retail network of over 5,000 stores produces revenue of around €2 billion. Benetton products are seen as less 'high fashion' but of higher quality and durability, and with higher prices, than H&M and Zara.

H&M. Established in Sweden in 1947, H&M now sell clothes and cosmetics in over 1,000 stores in 20 countries around the world. The business concept is 'fashion and quality at the best price'. With more than 40,000 employees, and revenues of around SEK 60,000 million, its biggest market is Germany, followed by Sweden and the UK. H&M are seen by many as the originator of the fast fashion concept. Certainly they have years of experience at driving down the price of up-to-the-minute fashions. *'We ensure the best price,'* they say, *'by having few middlemen, buying large volumes, having extensive experience of the clothing industry, having a great knowledge of which goods should be bought from which markets, having efficient distribution systems, and being cost-conscious at every stage.'*

Source: KevinFoy/Alamy Images

Zara. The first store opened almost by accident in 1975 when Amancio Ortega Gaona, a women's pyjama manufacturer, was left with a large cancelled order. The shop he opened was intended only as an outlet for cancelled orders. Now, Inditex, the holding group that includes the Zara brand, has over 1,300 stores in 39 countries with sales of over €3 billion. The Zara brand accounts for over 75 per cent of the group's total retail sales, and is still based in north-west Spain. By 2003 it had become the world's fastest growing volume garment retailer. The Inditex group also has several other branded chains including Pull and Bear, and Massimo Dutti. In total it employs almost 40,000 people in a business that is known for a high degree of vertical integration compared with most fast fashion companies. The company believes that it is their integration along the supply chain that allows them to respond to customer demand fast and flexibly while keeping stock to a minimum.

Design

All three businesses emphasise the importance of design in this market. Although not *haute couture*, capturing design trends is vital to success. Even the boundary between high and fast fashion is starting to blur. In 2004, H&M recruited high fashion designer Karl Lagerfeld, previous noted for his work with more exclusive brands. For H&M his designs were priced for value rather than exclusivity. *'Why do I work for H&M? Because I believe in inexpensive clothes, not 'cheap' clothes,'* said Lagerfeld. Yet most of H&M's products come from over 100 designers in Stockholm who work with a team of 50 pattern designers, around 100 buyers and a number of budget controllers. The department's task is to find the optimum balance between the three components comprising H&M's business concept – fashion, price and quality. Buying volumes and delivery dates are then decided.

Zara's design functions are organised in a different way to most similar companies. Conventionally, the design input comes from three *separate* functions: the designers themselves, market specialists, and buyers who place orders on to suppliers. At Zara the design stage is split into three product areas: women's, men's and children's garments. In each area, designers, market specialists and buyers are co-located in design halls that also contain small workshops for trying out prototype designs. The market specialists in all three design halls are in regular contact with Zara retail stores, discussing customer reaction to new designs. In this way, the retail stores are not the end of the whole supply chain but the beginning of the design stage of the chain. Zara's around 300 designers, whose average age is 26, produce approximately 40,000 items per year of which about 10,000 go into production.

Benetton also has around 300 designers, who not only design for all their brands, but also are engaged in researching new materials and clothing concepts. Since 2000, the company has moved to standardise their range globally. At one time more than 20 per cent of its ranges were customised to the specific needs of each country, now only between 5 and 10 per cent of garments are customised. This reduced the number of individual designs offered globally by over 30 per cent, strengthening the global brand image and reducing production costs.

Both H&M and Zara have moved away from the traditional industry practice of offering two 'collections' a year, for Spring/Summer and Autumn/Winter. Their 'seasonless cycle' involves the continual introduction of new products on a rolling basis throughout the year. This allows designers to learn from customers' reactions to their new products and incorporate them quickly into more new products. The most extreme version of this idea is practiced by Zara. A garment will be designed; a batch manufactured and 'pulsed' through the supply chain. Often the design is never repeated, it may be modified and another batch produced, but there are no 'continuing' designs as such. Even Benetton, have increased the proportion of what they call 'flash' collections, small collections that are put into its stores during the season.

Manufacturing

At one time Benetton focused its production on its Italian plants. Then it significantly increased its production outside Italy to take advantage of lower labour costs. Non-Italian operations include factories in North Africa, Eastern Europe and Asia. Yet each location operates in a very similar manner. A central, Benetton owned, operation performs some manufacturing operations (especially those requiring expensive technology) and coordinates the more labour intensive production activities that are performed by a network of smaller contractors (often owned and managed by ex-Benetton employees). These contractors may in turn sub-contract some of their activities. The company's central facility in Italy allocates production to each of the non-Italian networks, deciding what and how much each is to produce. There is some specialisation, for example, jackets are made in Eastern Europe while T-shirts are made in Spain. Benetton also has a controlling share in its main supplier of raw materials, to ensure fast supply to its factories. Benetton are also known for the practice of dying garments after assembly rather than using died thread or fabric. This postpones decisions about colours until late in the supply process so that there is a greater chance of producing what is needed by the market.

H&M does not have any factories of its own, but instead works with around 750 suppliers. Around half of production takes place in Europe and the rest mainly in Asia. It has 21 production offices around the world that between them are responsible for coordinating the suppliers who produce over half a billion items a year for H&M. The relationship between production offices and suppliers is vital, because it allows fabrics to be bought in early. The actual dyeing and cutting of the garments can then be decided at a later stage in the production The later an order can be placed on suppliers, the less the risk of buying the wrong thing. Average supply lead times vary from three weeks up to six months, depending on the nature of the goods. *'The most important thing,'* they say, *'is to find the optimal time to order each item. Short lead times are not always best. For some high-volume fashion basics, it is to our advantage to place orders far in advance. Trendier garments require considerably shorter lead times.'*

Zara's lead times are said to be the fastest in the industry, with a 'catwalk to rack' time as little as 15 days. According to one analyst, this is because they *'owned most of the manufacturing capability used to make their products, which they use as a means of exciting and stimulating customer demand.'* About half of Zara's products are produced in its network of 20 Spanish factories, which, like at Benetton, tend to concentrate on the more capital intensive operations such as cutting and dyeing. Sub-contractors are used for most labour intensive operations like sewing. Zara buy around 40 per cent of their fabric from its own wholly-owned subsidiary, most of which is in undyed form for dyeing after assembly. Most Zara factories and their sub-contractors work on a single shift system to retain some volume flexibility.

Distribution

Both Benetton and Zara have invested in highly automated warehouses, close to their main production centres that store, pack and assemble individual orders for their retail networks. These automated warehouses represent a major investment for both companies. In 2001, Zara caused some press comment by announcing that it would open a second automated warehouse even though, by its own calculations, it was only using about half its existing warehouse capacity. More recently, Benetton caused some controversy by announcing that it was exploring the use of RFID tags to track its garments.

At H&M, while the stock management is primarily handled internally, physical distribution is subcontracted. A large part of the flow of goods is routed from production site to the retail country via H&M's transit terminal in Hamburg. Upon arrival the goods are inspected and allocated to the stores or to the centralised store stock room. The centralised store stock room, within H&M referred to as 'Call-Off Warehouse' replenishes stores on item level according to what is selling.

Retail

All H&M stores (average size 1,300 square metres) are owned and solely run by H&M. The aim is to *'create a comfortable and inspiring atmosphere in the store that makes it simple for customers to find what they want and to feel at home'*. This is similar to Zara stores, although they tend to be smaller (average size 800 square metres). Perhaps the most remarkable characteristic of Zara stores is that garments rarely stay in the store for longer than two weeks. Because product designs are often not repeated and are produced in relatively small batches, the range of garments displayed in the store can change radically every two or three week. This encourages customers both to avoid delaying a purchase and to revisit the store frequently.

Since 2000 Benetton has been reshaping its retail operations. At one time the vast majority of Benetton retail outlets were small shops run by third parties. Now these small stores have been joined by several, Benetton owned and operated, larger stores (average size 1,500 to 3,000 square metres). These mega-stores can display the whole range of Benetton products and reinforce the Benetton shopping experience.

QUESTION

Compare and contrast the approaches taken by H&M, Benetton and Zara to managing their supply chain.

APPLYING THE PRINCIPLES

Hints

Some of these exercises can be answered by reading the chapter. Others will require some general knowledge of business activity and some might require an element of investigation. Hints on how they can all be answered are to be found in the eText at **www.pearsoned.co.uk/slack**.

1. If you were the owner of a small local retail shop, what criteria would you use to select suppliers for the goods that you wish to stock in your shop? Visit three shops that are local to you and ask the owners how they select their suppliers. In what way were their answers different from what you thought they might be?

2. What is your purchasing strategy? How do you approach buying the products and services that you need (or want)? Classify the types and products and services that you buy and record the criteria you use to purchase each category. Discuss these categories and criteria with others. Why are their views different?

3. Visit a C2C (consumer-to-consumer) auction site (for example eBay) and analyse the function of the site in terms of the way it facilitates transactions. What does such a site have to get right to be successful?

4 The example of the bull whip effect shown in Table 7.2 shows how a simple 5 per cent reduction in demand at the end of supply chain causes fluctuations that increase in severity the further back an operation is placed in the chain.

(a) Using the same logic and the same rules (i.e. all operations keep one period's demand as inventory), what would the effect on the chain be if demand fluctuated period by period between 100 and 95? That is, period 1 had a demand of 100, period 2 had a demand of 95, period 3 had a demand of 100, period 4 had a demand of 95, and so on?

(b) What happens if all operations in the supply chain decide to keep only half of each period's demand as inventory?

(c) Find examples of how supply chains try to reduce this bull whip effect.

5 Visit the websites of some distribution and logistics companies. For example, you might start with some of the following: www.eddiestobart.co.uk, www.norbert-dentressangle.com, www.accenture.com (under 'services' look for supply chain management), www.logisticsonline.com.

(a) What do you think are the market promises that these companies make to their clients and potential clients?

(b) What are the operations capabilities they need to carry out these promises successfully?

Notes on chapter

1 Sources include: (2010) 'Ocado wins world e-tailer of the year: online supermarket scoops award at World Retail Awards in Berlin', Ocado press release (Jenny Davey).

2 Sonne, P. (2010) 'Off the catwalk: Burberry gets a makeover', *The Wall Street Journal*, 9 September.

3 Fisher, M.L. (1997), 'What is the right supply chain for your product', *Harvard Business Review*, March–April.

4 We are grateful to Carsten Dittrich for very significant help with this section.

5 Kapour, V. and Gupta, A. (1997), 'Aggressive sourcing: a free-market approach', *Sloan Management Review*, Fall.

6 Parkhe, A. (1993) 'Strategic alliance structuring', *Academy of Management Journal*, Vol 36, pp 794–829.

7 Source: Grad, C. (2000) 'A network of supplies to be woven into the web', *Financial Times*, 9 February.

8 Lee, L. and Dobler, D.W. (1977) *Purchasing and Materials Management*, McGraw-Hill.

9 Harland, C.M. (1996), 'Supply chain management relationships, chains and networks', *British Journal of Management*, Vol. 1, No. 7.

10 Lee, H.L., Padmanabhan, V. and Whang, S. (1997) 'The bull whip effect in supply chains', *Sloan Management Review*, Spring.

11 Thanks to Stephen Disney at Cardiff Business School, UK, for help with this section.

12 Sources: MIT Auto-ID website; Alberch, K. (2002), 'Supermarket cards: tip of the surveillance iceberg', *Denver University Law Review*, June.

13 All data from public sources and reflects period 2004–05.

TAKING IT FURTHER

Bolstorff, P. and Rosenbaum, R. (2008) *Supply Chain Excellence – A Handbook for Dramatic Improvement Using the SCOR Model* (2nd edn), American Management Association.

Bolstorff, P. (2004), 'Supply chain by the numbers', *Logistics Today*, July, 46–50.

Christopher, M. (2010) *Logistics and Supply Chain Management*, Financial Times Prentice Hall. *A comprehensive treatment on supply chain management from a distribution perspective by one of the gurus of supply chain management.*

Fisher, M.L. (1997) 'What is the right supply chain for your product?', *Harvard Business Review*, Vol. 75, No. 2. *A particularly influential article that explores the issue of how supply chains are not all the same.*

Harrison, A. and van Hoek, R. (2007) *Logistics Management and Strategy: Competing through the Supply Chain*, Financial Times Prentice Hall. *A short but readable book that explains may of the modern ideas in supply chain management including lean supply chains and agile supply chains.*

Hines, P. and Rich, N. (1997) 'The seven value stream mapping tools', *International Journal of Operations and Production Management*, Vol. 17, No. 1. *Another academic paper, but one that explores some practical techniques that can be used to understand supply chains.*

Parmigiania A., Klassenb R.D., Russoa M.V. (2011) 'Efficiency meets accountability: performance implications of supply chain configuration, control, and capabilities', *Journal of Operations Management*, Vol. 29, pp 212–223. *Excellent review and discussion of broad supply issues.*

Presutti Jr., William D. and Mawhinney, J.R. (2007), 'The supply chain finance link', *Supply Chain Management Review*, September, 32–38.

USEFUL WEBSITES

www.opsman.org *Definitions, links and opinions on operations and process management.*

http://www.cio.com/research/scm/edit/012202_scm *Site of CIO's Supply Chain Management Research Center. Topics include procurement and fulfilment, with case studies.*

http://www.stanford.edu/group/scforum/ *Stanford University's supply chain forum. Interesting debate.*

http://www.rfidc.com/ *Site of the RFID Centre that contains RFID demonstrations and articles to download.*

http://www.spychips.com/ *Vehemently anti-RFID site. If you want to understand the nature of some activist's concern over RFID, this site provides the arguments.*

http://www.cips.org/ *The Chartered Institute of Purchasing & Supply (CIPS) is an international organisation, serving the purchasing and supply profession and dedicated to promoting best practice. Some good links.*

http://www.supply-chain.org/cs/root/home *The Supply Chain Council homepage.*

Interactive quiz

For further resources including examples, animated diagrams, self-test questions, Excel spreadsheets and video materials please explore the eText on the companion website at **www.pearsoned.co.uk/slack**.

8 Capacity management

Introduction

Providing the capability to satisfy current and future demand is a fundamental responsibility of operations management. It is at the heart of trade-offs between customer service and cost. Insufficient capacity leaves customers not served and excess capacity incurs increased costs. In this chapter we deal with *medium-term* capacity management, also sometimes referred to as *aggregate* capacity management. The essence of medium-term capacity management is to reconcile, at a general level, the aggregated supply of capacity with the aggregated level of demand (see Figure 8.1).

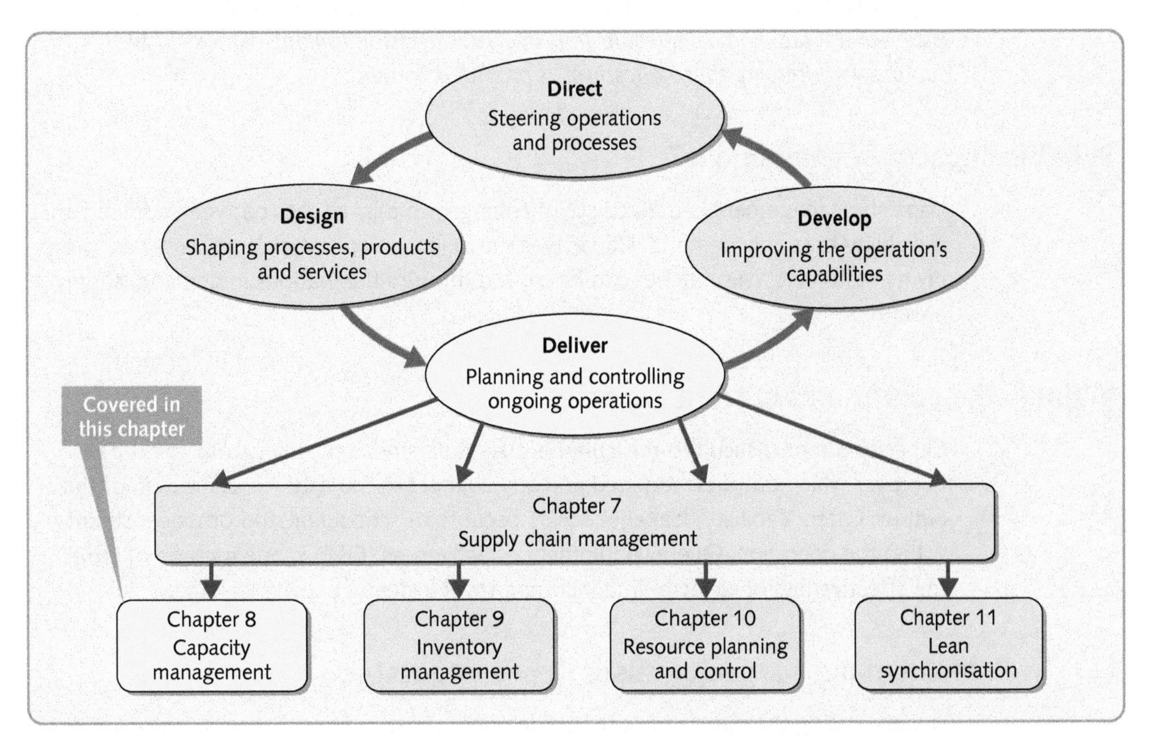

Figure 8.1 Capacity management is the activity of coping with mismatches between demand and the ability to supply demand

EXECUTIVE SUMMARY

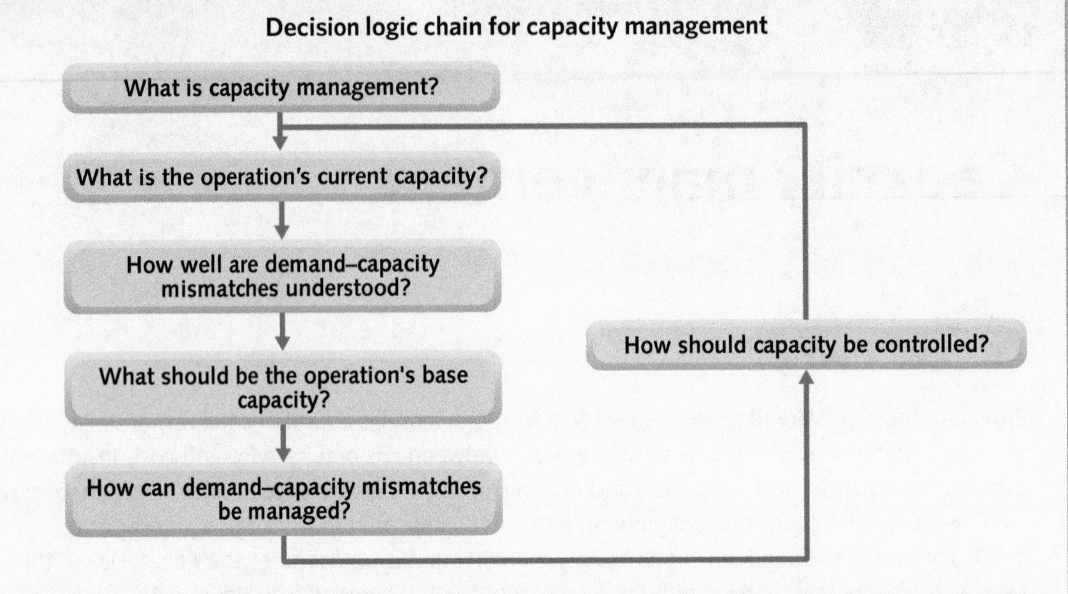

Each chapter is structured around a set of diagnostic questions. These questions suggest what you should ask in order to gain an understanding of the important issues of a topic, and, as a result, improve your decision making. An executive summary addressing these questions is provided below.

What is capacity management?

Capacity management is the activity of coping with mismatches between demand and the ability to supply demand. Capacity is the ability an operation or process has to supply its customers. Mismatches can be caused through fluctuations in demand, supply or both.

What is the operation's current capacity?

Capacity can be difficult to measure because it depends on activity mix, the duration over which output is required and any changes in the actual specification of the output. Often, capacity 'leakage' occurs because of scheduling and other constraints within the operation. Overall equipment effectiveness (OEE) is one method of judging the effectiveness of capacity that incorporates the idea of activity leakage.

How well are demand–capacity mismatches understood?

Understanding the nature of potential demand–capacity mismatches is central to capacity management. A key issue is the nature of demand and capacity fluctuations, especially the degree to which they are predictable. If fluctuations are predictable, they can be planned in advance to minimise their costs. If fluctuations are unpredictable, the main objective is to react to them quickly. Accurate simple forecasting is an advantage because it converts unpredictable variation into predictable variation. However, a broader approach to enhancing market knowledge generally can reveal more about the options for managing mismatches.

What should be the operation's base capacity?

Capacity planning often involves setting a base level of capacity and then planning capacity fluctuations around it. The level at which base capacity is set depends on three main factors: the relative importance of the operation's performance objectives, the perishability of the operation's outputs, and the degree of variability in demand or supply. High service levels, high perishability of an operation's outputs and a high degree of variability either in demand or supply, all indicate a relatively high level of base capacity.

How can demand–capacity mismatches be managed?

Demand–capacity mismatches usually call for some degree of capacity adjustment over time. There are three pure methods of achieving this, although in practice a mixture of all three may be used. A 'level capacity' plan involves no change in capacity and requires that the operation absorb demand–capacity mismatches, usually through under- or over-utilisation of its resources, or the use of inventory. The 'chase demand' plan involves the changing of capacity through such methods as overtime, varying the size of the work force, subcontracting, etc. The 'manage demand' plan involves an attempt to change demand through pricing or promotion methods, or changing product or service mix to reduce fluctuations in activity levels. Yield management is a common method of coping with mismatches when operations have relatively fixed capacities. Cumulative representations are sometimes used to plan capacity.

How should capacity be controlled?

In practice, capacity management is a dynamic process with decisions reviewed period by period. It is essential that capacity decisions made in one period reflect the knowledge accumulated from experiences in previous periods.

DIAGNOSTIC QUESTION

What is capacity management?

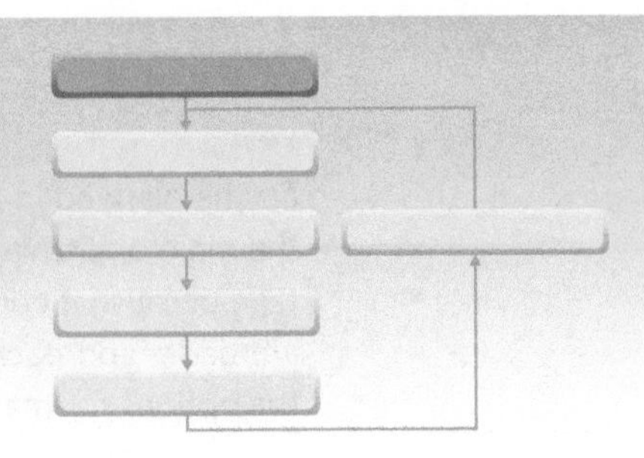

Video

Capacity is the output that an operation (or single process) can deliver in a defined unit of time. It reflects an 'ability to supply', at least in a quantitative sense. Capacity management is the activity of coping with mismatches between the demand on an operation and its ability to supply. Demand is the quantity of products or services that customers request from an operation or process at any point in time. A mismatch between demand and capacity can occur because demand fluctuates over time, or capacity fluctuates over time, or both.

Defining capacity as 'the ability to supply' is taking a broad view of the term. The 'ability to supply' depends not only on the limitations of the previous stage in a supply network, operation or process, but on all the stages up to that point. So, for example, the capacity of an ice cream manufacturer is a function not only of how much ice cream its factories can produce at any point in time, but also of how much packaging material, raw material supplies, and so on, that its suppliers can provide. It may have the factories to make 10,000 kilos of ice cream a day, but if its suppliers of dairy produce can only supply 7,000 kilos a day, then the effective capacity (in terms of 'ability to supply') is only 7,000 kilos per day. Of course, if demand remains steady any operation will attempt to make sure that its supply capacity does not limit its own ability to supply. But, capacity management is concerned with fluctuations in demand *and* supply. It involves coping with the dynamics of delivering products and services to customers. Balancing the individual capacities of each part of the network is therefore a more difficult, and ever changing, task.

OPERATIONS PRINCIPLE

Any measure of capacity should reflect the ability of an operation or process to supply demand.

It is worth noting that 'coping' with mismatches between demand and capacity may not mean that capacity should match demand. An operation could take the deliberate decision to fail to meet demand or fail to fully exploit its ability to supply. For example, a hotel may not make any effort to meet demand in peak periods because doing so would incur unwarranted capital costs. It is therefore content to leave some demand unsatisfied, although it may increase its prices to reflect this. Similarly, a flower grower may not supply the entirety of its potential supply (crop) if doing so would simply depress market prices and reduce its total revenue.

Levels of capacity management

The activity of coping with demand–capacity mismatches needs addressing over various timescales. In the long term, physical capacity needs adjusting to reflect the growth or decline in long-term demand. This task involves 'streaming on' or closing down relatively large units of physical capacity over a time period, possibly stretching into years. This activity was treated when we discussed the design of supply networks in Chapter 3. In addition, and within the physical constraints imposed by long-term capacity, most operations will need to cope with demand–capacity mismatches in the medium term, where 'medium term' may mean anything from one day to one year. It is this level that we discuss in this chapter. At an even shorter term, individual processes may need to cope with demand–capacity mismatches minute-by-minute or day-by-day. This is an issue for 'resource planning and control', examined in Chapter 10.

The following two examples illustrate the nature of capacity management.

EXAMPLE

Panettone: how Italy's bakers cope with seasonal demand[1]

Panettone has become a national symbol of the Italian Christmas. The light and fluffy, dome-shaped confection is dotted with sultanas and candied citrus peel, and is *the* Italian Christmas cake. Traditionally made in Milan, Italy, about 40 million of them are consumed throughout Italy over the holiday period. Now, they are becoming popular around the world. Over a million are exported to the US, while an endorsement from Delia Smith, a celebrity chef, caused a surge in demand in Britain with a well publicised recipe for trifle made with panettone. This boost to production is good news for the big Italian manufacturers, but although volumes are higher, the product is still seasonal, which poses a problem for even the experienced Milanese confectioners. Smaller 'artisan' producers simply squeeze a few batches of panettone into their normal baking schedules as Christmas approaches. But for the large industrial producers who need to make millions for the Christmas season, it is not possible. And no pannetone manufacturer is larger than the Bauli group. It is one of the foremost manufacturers of confectionery in Europe. Founded over 70 years ago, and in spite of its mass production approach, it has a reputation for quality and technological improvement. The company's output of pannetone accounts for 38 per cent of Italian sales. The key to its success, according to the company, is in having *'combined the skill of homemade recipes with high technology [and] quality guaranteed by high standards that are unattainable in craftsman production, but that can only be reached by selecting top quality raw materials, by thousands of tests and checks on the entire production line and the production process'*. In fact, the company says that its size is an advantage. *'High investment in research and technology allow us to manage natural fermentation and guarantee a uniform quality that artisanal bakeries find hard to achieve.'*

Source: Gabby Messina/Getty Images

In fact, although Bauli has diversified into year-round products like croissants and biscuits, it has acquired a leadership role in the production of products for festive occasions. Seasonal cakes account for over 50 per cent of its turnover of around €420 million. So successful has it been in its chosen markets, in 2009 it bought Motta and Alemagna, the two big Milanese brands that pioneered the manufacture of panettone. So how does Bauli cope with such seasonality? Partly it is by hiring large numbers of temporary seasonal workers to staff its dedicated production lines. At peak times there can be 1,200 seasonal workers in the factory, more than its permanent staff of around 800. It also starts to build up inventories before demand begins to increase for the Christmas peak. Production of panettone lasts about four months, starting in September. *'Attention to ingredients and the use of new technologies in production give a shelf life of five months without preservatives,'* says Michele Bauli, Deputy Chairman who comes from the firm's founding family. Temporary workers are also hired to bake other seasonal cakes such as the *colomba*, a dove-shaped Easter treat, which keeps them occupied for a month and a half in the spring.

EXAMPLE

The Penang Mutiara[2]

One of the vacation regions of the world, South-East Asia has many luxurious hotels. One of the best is the Penang Mutiara, a top-of-the-market hotel which nestles in the lush greenery of Malaysia's Indian Ocean coast. Owned and managed by PERNAS, Malaysia's largest hotel group, the Mutiara has to cope with fluctuating demand, notwithstanding the region's relatively constant benign climate. *'Managing a hotel of this size is an immensely complicated task,'* says the hotel's manager. *'Our customers have every right to be demanding. Quality of service has to be impeccable. Staff must be courteous and yet also friendly towards our guests. And of course they must have the knowledge to be able to answer guests' questions.*

Source: Courtesy of Mutiara Beach Resort, Penang

Most of all, though, great service means anticipating our guests' needs, thinking ahead so that you can identify what, and how much, they are likely to demand.'

The hotel tries to anticipate guests' needs in a number of ways. If guests have been to the hotel before, their likely preferences will have been noted from their previous visit. *'A guest should not be kept waiting. This is not always easy but we do our best. For example, if every guest in the hotel tonight decided to call room service and request a meal instead of going to the restaurants, our room service department would obviously be grossly overloaded, and customers would have to wait an unacceptably long time before the meals were brought up to their rooms. We can predict this to some extent, but also we keep a close watch on how demand for room service is building up. If we think it's going to get above the level where response time to customers would become unacceptably long, we will call in staff from other restaurants in the hotel. Of course, to do this we have to make sure that our staff are multi-skilled. In fact we have a policy of making sure that restaurant staff can always do more than one job. It's this kind of flexibility which allows us to maintain fast response to the customer.'*

Although the hotel needs to respond to some short-term fluctuations in demand for individual services, it can predict likely demand relatively accurately because, each day, the actual number of guests is known. Most guests book their stay well in advance, so activity levels for the hotel's restaurants and other services can be planned ahead. Demand does vary throughout the year, peaking during holiday periods, and the hotel has to cope with these seasonal fluctuations. They do this partly by using temporary part-time staff. In the back-office parts of the hotel this isn't a major problem. In the laundry, for example, it is relatively easy to put on an extra shift in busy periods by increasing staffing levels. However, this is more of a problem in the parts of the hotel that have direct contact with the customer. New or temporary staff can't be expected to have the same customer contact skills as the regular staff. The solution to this is to keep the temporary staff as far in the background as possible and make sure that only skilled, well-trained staff interact with the customer. So, for example, a waiter who would normally take orders, serve the food, and take away the dirty plates, would in peak times restrict his or her activities to taking orders and serving the food. The less skilled part of the job, taking away the plates, could be left to temporary staff.

What do these two examples have in common?

The obvious similarity between both these operations is that they have to cope with fluctuating customer demand. Also both operations have found ways to cope with these fluctuations, at least up to a point. This is good because both operations would suffer in the eyes of their customers and in terms of their own efficiency, if they could not cope. Yet in one important respect the two operations are different. Although both have to cope with variation in demand, and although the demand on both operations is a mixture of the predictable and the unpredictable, the balance between predictable variation in demand and unpredictable variation in demand is different. Demand at the Penang Mutiara is largely predictable. Seasonal fluctuations are related to known holiday periods and most customers book their stays well in advance. Pannetone producers, on the other hand, have to cope with demand, part of which is unpredictable. While the traditional Italian market is predictable, the growing international market is far less so (for example the surge caused by short-term publicity in the UK).

In both these examples, any mismatches between demand and capacity derive from predictable and unpredictable variation in demand. As well as demand fluctuations, some operations also have to cope with predictable and unpredictable variation in capacity (if it is defined as 'the

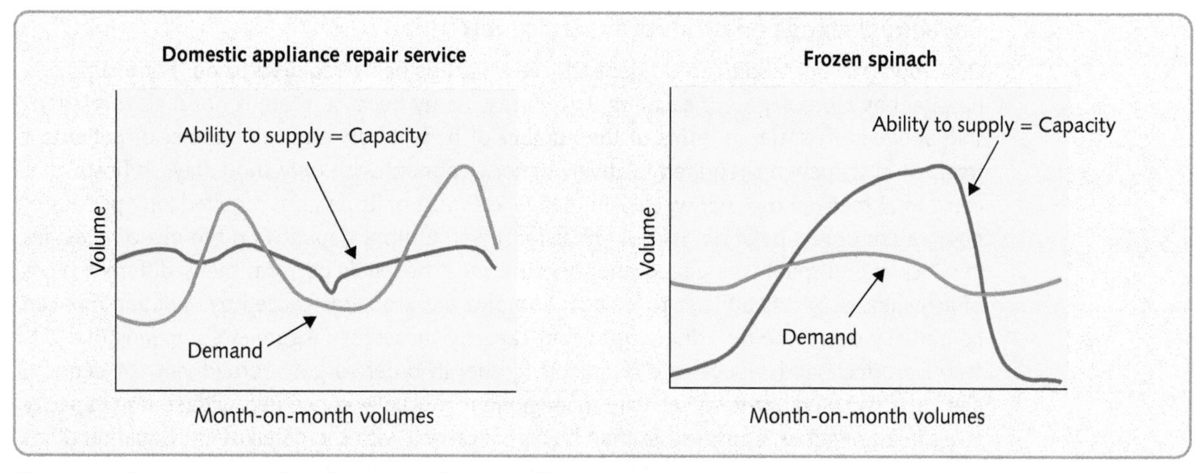

Figure 8.2 Demand-capacity mismatches for an appliance repair service and a frozen spinach business

> OPERATIONS PRINCIPLE
> *Capacity management decisions should reflect both predictable and unpredictable variations in capacity and demand.*

ability to supply'). For example, Figure 8.2 shows the demand and capacity variation of two businesses. The first is a domestic appliance repair service. Both demand and capacity vary month on month. Capacity varies because the field service operatives in the business prefer to take their vacations at particular times of the year. Nevertheless, capacity is relatively stable throughout the year. Demand, by contrast, fluctuates more significantly. It would appear that there are two peaks of demand through the year, with peak demand being approximately twice the level of the low point in demand. The second business is a food manufacturer producing frozen spinach. The demand for this product is relatively constant throughout the year but the capacity of the business (not in terms the capacity of its factories, but its ability to supply) varies significantly. During the growing and harvesting season capacity is high, but it falls off almost to zero for part of the year. Yet although the mismatch between demand and capacity is driven primarily by fluctuations in demand in the first case, and capacity in the second case, the essence of the capacity management activity is essentially similar for both.

DIAGNOSTIC QUESTION

What is the operation's current capacity?

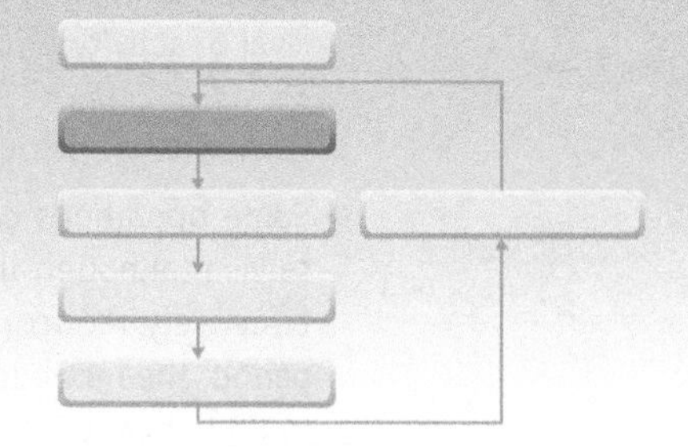

Video

Every operation and process needs to know their capacity because if they have too little they cannot meet demand (bad for revenue) and if they have too much they are paying for more capacity than they need (bad for costs). So a first step in managing capacity is being able to measure current capacity. This sounds simple, but often is not. In fact only when the operation is relatively standardised and repetitive is capacity easy to define unambiguously. Any measure of capacity will contain a number of assumptions, each of which may be necessary to give an estimate, but each of which obscures some aspect of reality. Again, taking capacity as 'the ability to supply', these assumptions relate to the mix of products or services supplied, the time over which they are supplied, and the specification of what is supplied.

> OPERATIONS PRINCIPLE
> *Capacity is a function of product-service mix, duration, and product-service specification.*

Capacity depends on product or service mix

How much an operation can do depends on what it is being required to do. For example, a hospital has a problem in measuring its capacity, partly because there is not a clear relationship between its scale (in terms of the number of beds it has) and the number of patients it treats. If all its patients required relatively minor treatment with only short stays in hospital, it could treat many people per week. Alternatively, if most of its patients required long periods of observation or recuperation, it could treat far fewer. Output depends on the mix of activities in which the hospital is engaged and, because most hospitals perform many different types of activities, output is difficult to predict. Some of the problems caused by variation mix can be partially overcome by using aggregated capacity measures. 'Aggregated' means that different products and services are bundled together in order to get a broad view of demand and capacity. Medium-term capacity management is usually concerned with setting capacity levels in aggregated terms, rather than being concerned with the detail of individual products and services. Although this may mean some degree of approximation, especially if the mix of products or services being produced varies significantly, it is usually acceptable, and is a widely used practice in medium-term capacity management. For example, a hotel might think of demand and capacity in terms of 'room nights per month'. This ignores the number of guests in each room and their individual requirements, but it is a good first approximation. A computer manufacturer might measure demand and capacity in the number of units it is capable of making per month, ignoring any variation in models.

Capacity depends on the duration over which output is required

Capacity is the output that an operation can deliver *in a defined unit of time.* The level of activity and output that may be achievable over short periods of time is not the same as the capacity that is sustainable on a regular basis. For example, a tax return processing office, during its peak periods at the end (or beginning) of the financial year, may be capable of processing 120,000 applications a week. It does this by extending the working hours of its staff, discouraging its staff from taking vacations during this period, avoiding any potential disruption to its IT systems (not allowing upgrades during this period, etc.), and maybe just by working hard and intensively. Nevertheless, staff do need vacations, they can't work long hours continually, and eventually the information system will have to be upgraded. The capacity that is possible to cope with peak times is not sustainable over long periods. Often, capacity is taken to be the level of activity or output that can be sustained over an extended period time.

Capacity depends on the specification of output

Some operations can increase their output by changing the specification of the product or service (although this is more likely to apply to a service). For example, a postal service may effectively reduce its delivery dependability at peak times. So, during the busy Christmas period, the number of letters delivered the day after being posted may drop from 95 per cent to 85 per cent. This may not always matter to the customer, who understands that the postal service is especially overloaded at this time. Similarly, accounting firms may avoid long 'relationship building' meetings with clients during busy periods. Important though these are, they can usually be deferred to less busy times. The important task is to distinguish between the 'must do' elements of the service that should not be sacrificed and the 'nice to do' parts of the service that can be omitted or delayed in order to increase capacity.

Capacity 'leakage'

Even after allowing for all the difficulties inherent in measuring capacity, the theoretical capacity of a process (the capacity that it was designed to have) is not always achieved in practice. Some reasons for this are, to some extent, predictable. Different products or services may have different requirements, so the process will need to be stopped while it is changed over. Maintenance will need to be performed. Scheduling difficulties could mean further lost time. Not all of these losses are necessarily avoidable; they may occur because of the market and technical demands on the process. However, some of the reduction in capacity can be the result of less predictable events. For example, labour shortages, quality problems, delays in the delivery of bought-in products and services, and machine, or system, breakdown, can all reduce capacity. This reduction in capacity is sometimes called 'capacity leakage'.

Overall equipment effectiveness[3]

The overall equipment effectiveness (OEE) measure is a popular method of judging the effectiveness of capacity that incorporates the concept of capacity leakage. It is based on three aspects of performance:

- *the time* that equipment is available to operate
- *the speed,* or throughput rate, of the equipment
- *the quality* of the product or service it produces.

Overall equipment effectiveness is calculated by multiplying an availability rate by a performance (or speed) rate multiplied by a quality rate. Figure 8.3 illustrates this. Some of the

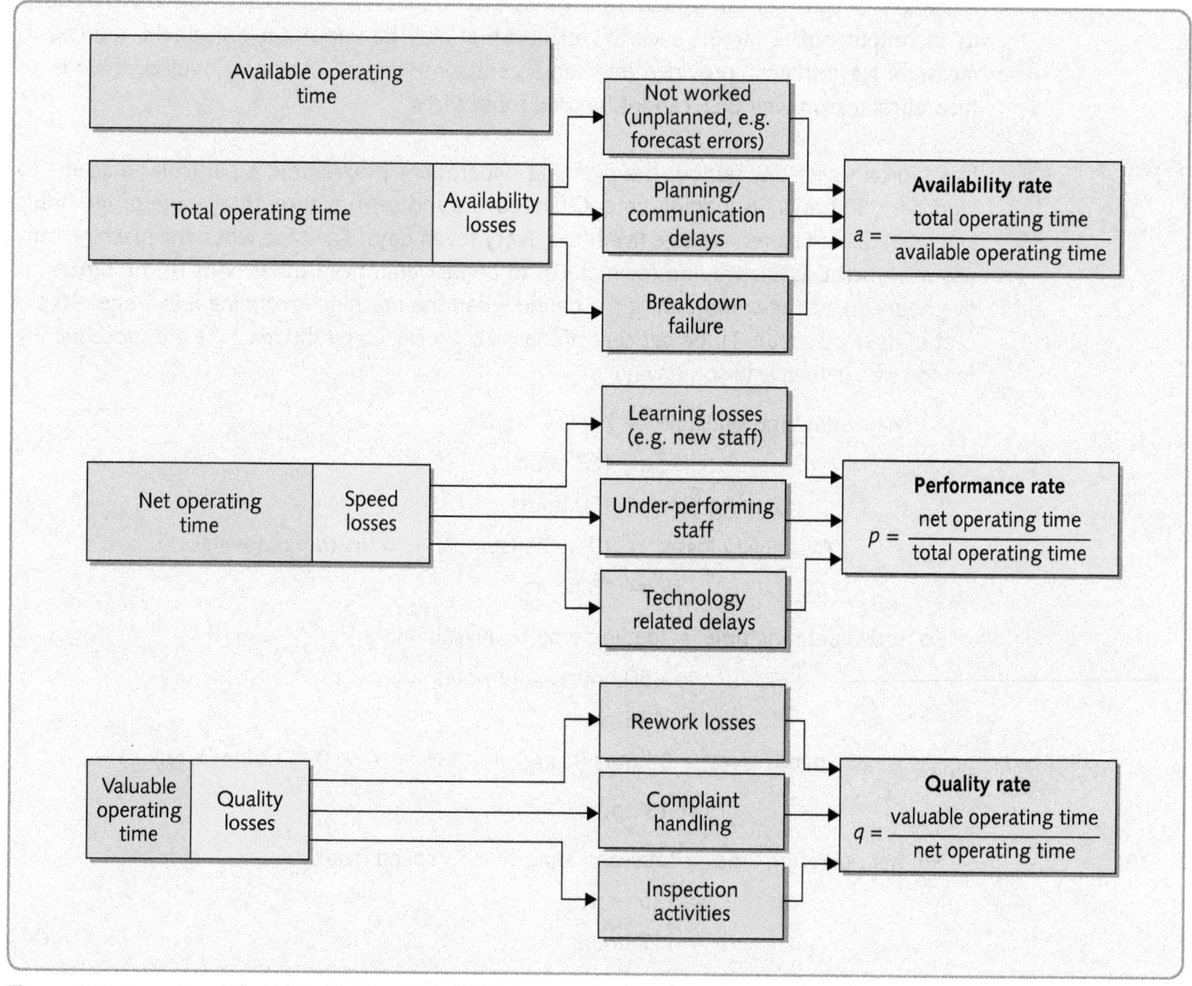

Figure 8.3 Overall equipment effectiveness (OEE)

reduction in the available capacity of a piece of equipment (or any process) is caused by time losses such as set-up and changeover losses (when the equipment or process is being prepared for its next activity), and breakdown failures (when the machine is being repaired). Some capacity is lost through speed losses such as when equipment is idling (for example when it is temporarily waiting for work from another process) and when equipment is being run below its optimum work rate. Finally, not everything processed by a piece of equipment will be error free. So some capacity is lost through quality losses.

Taking the notation in Figure 8.3.

$$\text{OEE} = a \times p \times q$$

For equipment to operate effectively, it needs to achieve high levels of performance against all three of these dimensions. Viewed in isolation, these individual metrics are important indicators of plant performance, but they do not give a complete picture of the machine's *overall* effectiveness. This can only be understood by looking at the combined effect of the three measures, calculated by multiplying the three individual metrics together. All these losses to the OEE performance can be expressed in terms of units of time – the design cycle time to produce one good part. So, a reject of one part has an equivalent time loss. In effect, this means that an OEE represents the valuable operating time as a percentage of the capacity something was designed to have.

OEE can be used for service operations and processes, but it is more difficult to do so. Of the three factors (time, speed and quality) only time is straightforward. There is no direct equivalent of speed or throughput rate that is easy to measure objectively. Similarly with quality of output, softer factors such as 'relationship' may be important but, again, difficult to measure. Nevertheless, provided one can accept some degree of approximation, there is no theoretical reason why OEE cannot be used for services.

WORKED EXAMPLE

In a typical seven-day period, the planning department programme a particular machine to work for 150 hours, its loading time. Changeovers and set-ups take an average of ten hours and breakdown failures average five hours every seven days. The time when the machine cannot work because it is waiting for material to be delivered from other parts of the process is five hours on average, and during the period when the machine is running it averages 90 per cent of its rated speed. Three per cent of the parts processed by the machine are subsequently found to be defective in some way.

$$\begin{aligned}
\text{Maximum time available} &= 7 \times 24 \text{ hours} \\
&= 168 \text{ hours} \\
\text{Loading time} &= 150 \text{ hours} \\
\text{Availability losses} &= 10 \text{ hours (setups)} + 5 \text{ hrs (breakdowns)} \\
&= 15 \text{ hours}
\end{aligned}$$

$$\begin{aligned}
\text{So, total operating time} &= \text{loading time} - \text{availability} \\
&= 150 \text{ hours} - 15 \text{ hours} \\
&= 135 \text{ hours} \\
\text{Speed losses} &= 5 \text{ hours (idling)} + ((135 - 5) \times 0.1) \text{ hours (running)} \\
&= 18 \text{ hours}
\end{aligned}$$

$$\begin{aligned}
\text{So, net operating time} &= \text{total operating time} - \text{speed losses} \\
&= 135 - 18 \\
&= 117 \text{ hours} \\
\text{Quality losses} &= 117 \text{ (net operating time)} \times 0.03 \text{ (error rate)} \\
&= 3.51 \text{ hours}
\end{aligned}$$

$$\text{So, valuable operating time} = \text{net operating time} - \text{quality losses}$$
$$= 117 - 3.51$$
$$= 113.49 \text{ hours}$$

$$\text{Therefore, availability rate} = a = \frac{\text{total operating time}}{\text{loading time}}$$
$$= \frac{135}{150} = 90\%$$

$$\text{and performance rate} = p = \frac{\text{net operating time}}{\text{total operating time}}$$
$$= \frac{117}{135} = 86.67\%$$

$$\text{and quality rate} = q = \frac{\text{valuable operating time}}{\text{net operating time}}$$
$$= \frac{113.49}{117} = 97\%$$

$$\text{OEE } (a \times p \times q) = 75.6\%$$

DIAGNOSTIC QUESTION

How well are demand-capacity mismatches understood?

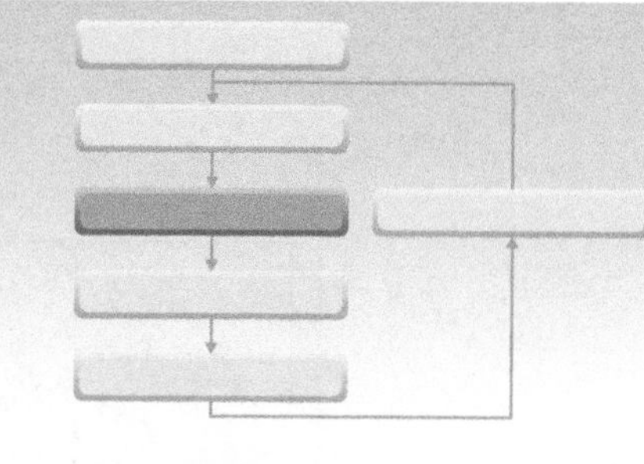

Video

A sound understanding of the nature of potential mismatches between demand and capacity is necessary for effective capacity management. For most businesses, this equates to understanding of how demand might vary (although the same logic would apply to variation in capacity). In particular, the balance between predictable and unpredictable variation in demand affects the nature of capacity management. When demand is predictable (usually under conditions of 'dependent demand', see previous chapter), capacity may need adjusting, but the adjustments can be planned in advance, preferably to minimise the costs of making the change. With unpredictable variation in demand (usually under conditions of 'independent demand'), if an operation is to react to it at all, it must do so quickly; otherwise the change in capacity will have little effect on the operation's ability to cope with the changed demand. Figure 8.4 illustrates how the objective and tasks of capacity management vary depending on the balance between predictable and unpredictable variation.

Enhanced market knowledge makes capacity planning easier

Capacity planning has to cope with mismatches between capacity and demand. Therefore, a deep understanding of the market forces that will generate demand is, if not an absolute prerequisite, nevertheless particularly important. This goes beyond the idea of forecasting as the prediction of uncontrollable events. Enhanced market knowledge is a broader concept and is illustrated in Figure 8.5. When the main characteristic of demand–supply mismatch is unpredictable variation, then forecasting in its conventional sense is important because it

		Unpredictable variation: Low	Unpredictable variation: High
Predictable variation	High	*Objective* – Adjust planned capacity as efficiently as possible *Capacity management tasks* • Evaluate optimum mix of methods for capacity fluctuation • Work on how to reduce cost of putting plan into effect	*Objective* – Adjust planned capacity as efficiently as possible and enhance capability for further fast adjustments *Capacity management tasks* • Combination of those for predictable and unpredictable variation
	Low	*Objective* – Make sure the base capacity is appropriate *Capacity management tasks* • Seek ways of providing steady capacity effectively	*Objective* – Adjust capacity as fast as possible *Capacity management tasks* • Identify sources of extra capacity and/or uses for surplus capacity • Work on how to adjust capacity and/or uses of capacity quickly

Figure 8.4 The nature of capacity management depends on the mixture of predictable and unpredictable demand and capacity variation

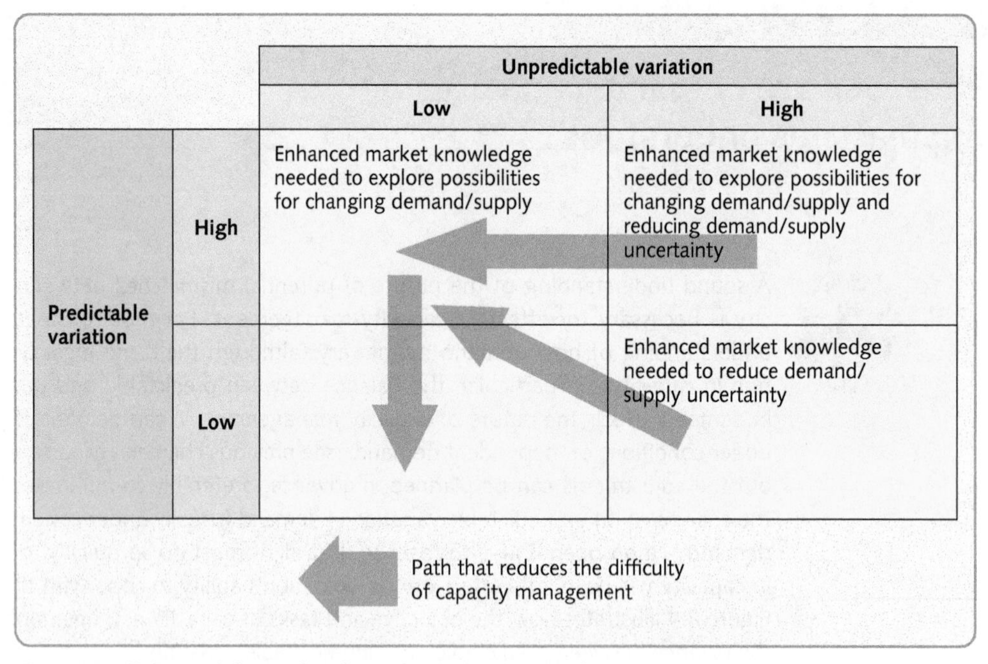

Figure 8.5 Enhanced demand and supply market knowledge can make capacity management easier

> **OPERATIONS PRINCIPLE**
> *The greater an operation's market knowledge, the more capacity management will focus on predictable demand capacity mismatches.*

converts unpredictable variation into predictable variation. But, when the main capacity management is predictable variation, then better forecasting as such is of limited value because demand–supply mismatches are, by definition, already known. What is useful under these circumstances is not so much a knowledge of what demand–supply mismatches will be, but rather how they can be changed. So, for example, can a major customer be persuaded to move their demand to a quieter period? Will increasing prices at peak periods shift demand to off-peak periods? Can new storage techniques allow the supply of food ingredients throughout the year?

If capacity management is largely a matter of coping with significant, but predictable, mismatches between demand and supply, then knowledge about how markets can be changed is important. However, when unpredictable variation is high, the first task is to transform variation from unpredictable to predictable through better forecasting. Of course, forecasting cannot eliminate predictable variation, but it is a first step towards minimising the negative effects of variation on capacity management.

Making forecasts useful for capacity management

Without some understanding of future demand and supply fluctuations, it is not possible to plan effectively for future events, only to react to them. This is why it is important to understand how forecasts are made. Forecasting was discussed in the supplement to Chapter 3, and it clearly helps the capacity management activity to have accurate forecasts. But, in addition to accuracy there are a number of other issues that make forecasts more (or less) useful as inputs to capacity planning.

Forecasts can never be perfectly accurate all the time. Yet sometimes forecast errors are more damaging that at other times. For example, if a process is operating at a level close to its maximum capacity, over-optimistic forecasts could lead the process to committing itself to unnecessary capital expenditure to increase its capacity. Inaccurate forecasts for a process operating well below its capacity limit will also result in extra cost, but probably not to the same extent. So the effort put into forecasting should reflect the varying sensitivity to forecast error. Forecasts also need to be expressed in units that are useful for capacity planning. If forecasts are expressed only in money terms and give no indication of the demands that will be placed on an operation's capacity, they will need to be translated into realistic expectations of demand, expressed in the same units as the capacity. Even worse, forecasts should not be expressed in money terms, such as sales, when those sales are themselves a consequence of capacity planning. For example, some retail operations use sales forecasts to allocate staff hours throughout the day. Yet sales will also be a function of staff allocation. Better to use forecasts of 'traffic', the number of customers who potentially could want serving if there are sufficient staff to serve them. Perhaps most importantly, forecasts should give an indication of relative uncertainty. Demand in some periods is more uncertain than others. The importance of this is that the operations managers need an understanding of when increased uncertainty makes it necessary to have reserve capacity. A probabilistic forecast allows this type of judgment between possible plans that would virtually guarantee the operation's ability to meet actual demand, and plans that minimise costs. Ideally, this judgement should be influenced by the nature of the way the business wins orders: price-sensitive markets may require a risk-avoiding cost minimisation plan that does not always satisfy peak demand, whereas markets that value responsiveness and service quality may justify a more generous provision of operational capacity. Remember though, the idea that 'better forecasting' is needed for effective capacity management is only partly true. A better approach would be to say that it is enhanced market knowledge (both of demand and supply) generally that is important.

Better forecasting or better operations responsiveness?

OPERATIONS PRINCIPLE
Attempting to increase market knowledge and attempting to increase operations flexibility present alternative approaches to capacity management, but are not mutually exclusive.

The degree of effort (and cost) to devote to forecasting is often a source of heated debate within businesses. This often comes down to two opposing arguments. One goes something like this: *'Of course it is important for forecasts to be as accurate as possible; we cannot plan operations capacity otherwise. This invariable means we finish up with too much capacity (thereby increasing costs), or too little capacity (thereby losing revenue and dissatisfying customers).'* The counter argument is very different: *'Demand will always be uncertain, that is the nature of demand. Get used to it. The only way to satisfy customers is to make the operation sufficiently responsive to cope with demand, almost irrespective of what it is.'* Both these arguments have some merit, but both are extreme positions. In practice, operations must find some balance between having better forecasts and being able to cope without perfect forecasts.

Trying to get forecasts right has particular value where the operation finds it difficult or impossible to react to unexpected demand fluctuations in the short term. Internet-based retailers at some holiday times, for example, find it difficult to flex the quantity of goods they have in stock in the short term. Customers may not be willing to wait. On the other hand, other types of operation working in intrinsically uncertain markets may develop fast and flexible processes to compensate for the difficulty in obtaining accurate forecasts. For example, fashion garment manufacturers try to overcome the uncertainty in their market by shortening their response time to new fashion ideas (cat walk to rack time) and the time taken to replenish stocks in the stores (replenishment time). Similarly, when the cost of not meeting demand is very high, processes also have to rely on their responsiveness rather than accurate forecasts. For example, accident and emergency departments in hospitals must be responsive even if it means under-utilised resources at times.

DIAGNOSTIC QUESTION

What should be the operation's base capacity?

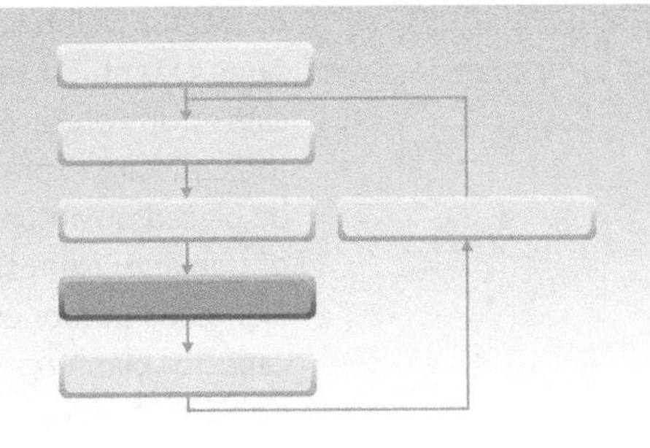

Video

The most common way of planning capacity is to decide on a 'base level' of capacity and then adjust it periodically up or down to reflect fluctuations in demand. In fact, the concept of 'base' capacity is unusual because, although nominally it is the capacity level from which increases and decreases in capacity level are planned, in very unstable markets, where fluctuations are significant, it may never occur. Also, these two decisions of 'what should the base level of capacity be?' and 'how do we adjust capacity around that base to reflect demand?' are interrelated. An operation could set its base level of capacity at such a high level compared to demand that there is no need ever to adjust capacity levels because they will never exceed the base level of capacity. However, this is clearly wasteful, which is why most operations will adjust their capacity level over time. Nevertheless, although the two decisions are interrelated it is usually worthwhile setting a nominal base level of capacity before going on to consider how it can be adjusted.

OPERATIONS PRINCIPLE

The higher the base level of capacity, the less capacity fluctuation is needed to satisfy demand.

Setting base capacity

The base level of capacity in any operation is influenced by many factors, but should be clearly related to three in particular.

- The relative importance of the operation's performance objectives.
- The perishability of the operation's outputs.
- The degree of variability in demand or supply.

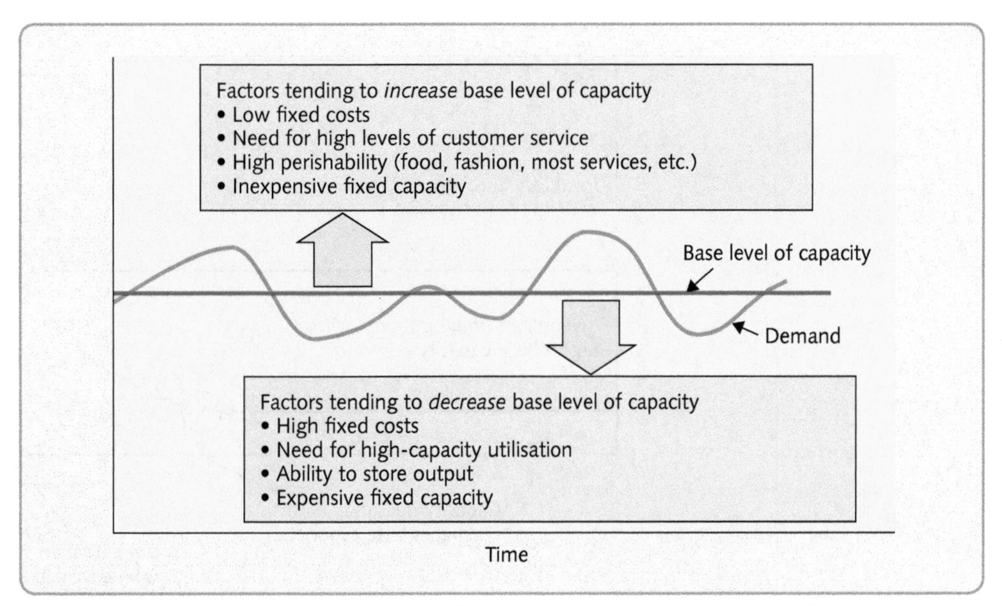

Figure 8.6 The base level of capacity should reflect the relative importance of the operation's performance objectives

The operation's performance objectives

Base levels of capacity should be set primarily to reflect an operation's performance objectives (see Figure 8.6). For example, setting the base level of capacity high compared to average demand will result in relatively high levels of under-utilisation of capacity and therefore high costs. This is especially true when an operation's fixed costs are high and therefore the consequences of under-utilisation are also high. Conversely, high base levels of capacity result in a capacity 'cushion' for much of the time, so the ability to flex output to give responsive customer service will be enhanced. When the output from the operation is capable of being stored, there may also be a trade-off between fixed capital and working capital in where the base capacity level is set. A high level of base capacity can require considerable investment (unless the cost per unit of capacity is relatively low). Reducing the base level of capacity would reduce the need for capital investment but (where possible) may require inventory to be built up to satisfy future demand and therefore increased levels of working capital. For some operations, building up inventory is either risky because products have a short shelf life (for example perishable food, high performance computers, or fashion items) or because the output cannot be stored at all (most services).

The perishability of the operation's outputs

When either supply or demand is perishable, base capacity will need to be set at a relatively high level because inputs to the operation or outputs from the operation cannot be stored for long periods. For example, a factory that produces frozen fruit will need sufficient freezing, packing and storage capacity to cope with the rate at which the fruit crop is being harvested during its harvesting season. Similarly, a hotel cannot store its accommodation services. If an individual hotel room remains unoccupied, the ability to sell for that night has 'perished'. In fact, unless a hotel is fully occupied every single night, its capacity is always going to be higher than the average demand for its services.

The degree of variability in demand or supply

Variability, either in demand or capacity (or processing rate), will reduce the ability of an operation to process its inputs. That is, it will reduce its effective capacity. This effect was explained

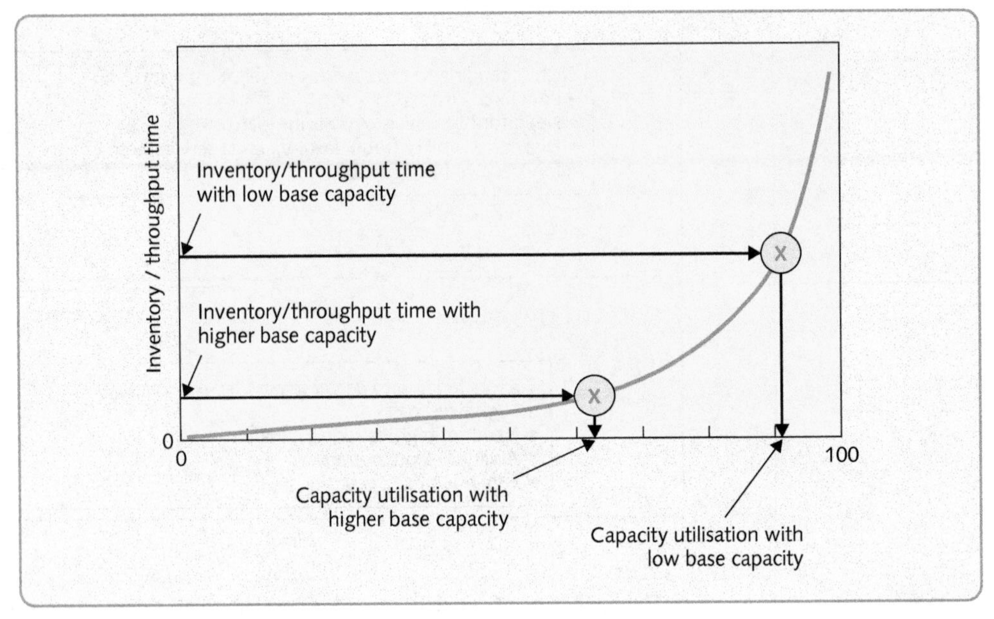

Figure 8.7 The effect of variability on the utilisation of capacity

in Chapter 5 when the consequences of variability in individual processes were discussed. As a reminder, the greater the variability in arrival time or activity time at a process the more the process will suffer both high throughput times and reduced utilisation. This principle holds true for whole operations, and because long throughout put times mean that queues will build up in the operation, high variability also affects inventory levels. This is illustrated in Figure 8.7. The implication of this is that the greater the variability, the more extra capacity will need to be provided to compensate for the reduced utilisation of available capacity. Therefore, operations with high levels of variability will tend to set their base level of capacity relatively high in order to provide this extra capacity.

DIAGNOSTIC QUESTION

How can demand–capacity mismatches be managed?

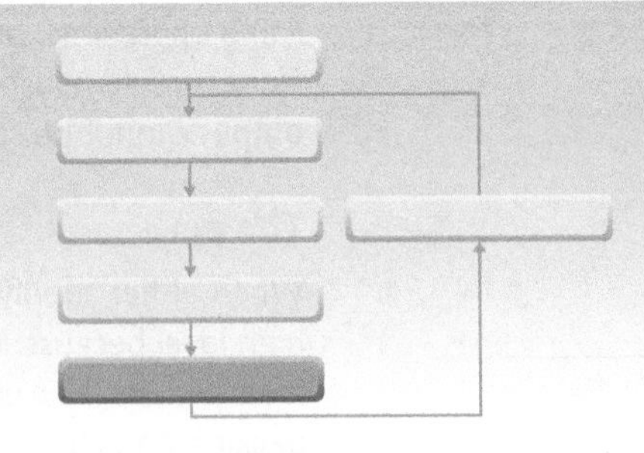

Almost all operations have to cope with varying demand or supply; therefore they will need to consider adjusting capacity around its nominal base level. There are three 'pure' plans available for treating such variation, although in practice, most organisations will use a mixture of all of them, even if one plan dominates.

- Ignore demand fluctuations and keep nominal capacity levels constant (level capacity plan).
- Adjust capacity to reflect the fluctuations in demand (chase demand plan).
- Attempt to change demand (demand management).

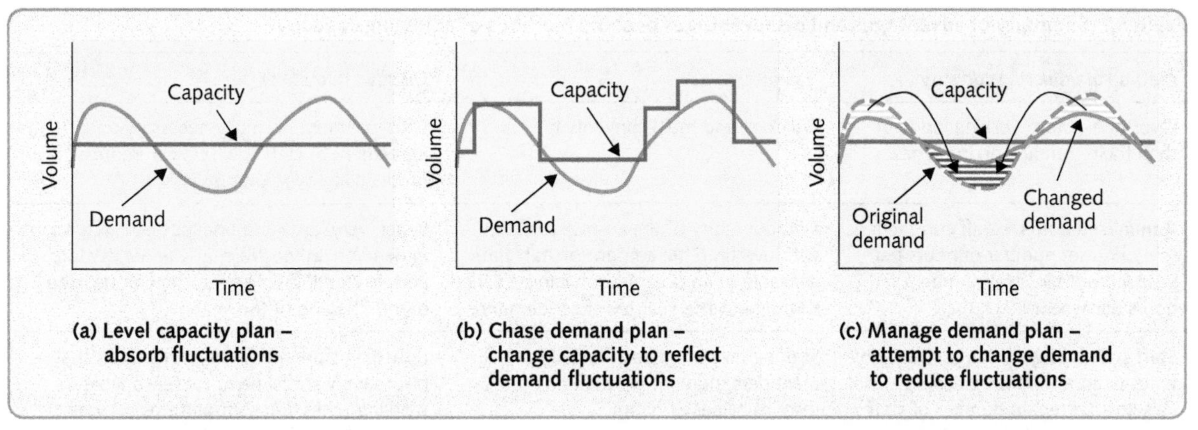

Figure 8.8 Managing demand–capacity mismatches using 'level capacity', 'chase demand' and 'manage demand' plans

Level capacity plan

Animated diagram

In a level capacity plan, the processing capacity is set at a uniform level throughout the planning period, regardless of the fluctuations in forecast demand. This means that the same number of staff operate the same processes and should therefore be capable of producing the same aggregate output in each period. Where non-perishable materials are processed, but not immediately sold, they can be transferred to finished goods inventory in anticipation of later sales. When inventory is not possible, as in most services, operations demand fluctuations are absorbed through under-utilisation of the operation's resources and/or and failure to meet demand immediately (see Figure 8.8(a)). The more demand fluctuates, the higher is either inventory or under-utilisation when using a level capacity plan. Both are expensive, but may be considered if the cost of building inventory is low compared with changing output levels, or in service operations, if the opportunity costs of individual lost sales is very high, for example, in the high-margin retailing of jewellery and in (real) estate agents. Setting capacity below the forecast peak demand level will reduce the degree of under-utilisation, but, in the periods where demand is expected to exceed capacity, customer service may deteriorate.

Chase demand plan

Animated diagram

Chase demand plans attempt to match capacity closely to the varying levels of forecast demand, as in Figure 8.8(b). This is much more difficult to achieve than a level capacity plan, as different numbers of staff, different working hours, and even different amounts of equipment may be necessary in each period. For this reason, pure chase demand plans are unlikely to appeal to operations producing standard, non-perishable products, especially where operations are capital-intensive. The chase demand policy would require a level of physical capacity (as opposed to effective capacity), all of which would only be used occasionally. A pure chase demand plan is more usually adopted by operations that cannot store their output, such as a call centre. It avoids the wasteful provision of excess staff that occurs with a level capacity plan, and yet should satisfy customer demand throughout the planned period. Where inventory is possible, a chase demand policy might be adopted in order to minimise it.

The chase demand approach requires that capacity is adjusted by some means. There are a number of different methods for achieving this, although they may not all be feasible for all types of operation. Some of these methods are shown in Table 8.1.

Table 8.1 Summary of advantages and disadvantages of some methods of adjusting capacity

Method of adjusting capacity	*Advantages*	*Disadvantages*
Overtime – staff working longer than their normal working times	Quickest and most convenient	Extra payment normally necessary and agreement of staff to work, can reduce productivity over long periods
Annualised hours – staff contracting to work a set number of hours per year rather than a set number of hours per week.	Without many of the costs associated with overtime the amount of staff time available to an organisation can be varied throughout the year to reflect demand.	When very large and unexpected fluctuations in demand are possible, all the negotiated annual working time flexibility can be used before the end of the year.
Staff scheduling – arranging working times (start and finish times) to vary the aggregate number of staff available for work at any time.	Staffing levels can be adjusted to meet demand without changing job responsibilities or hiring new staff.	Providing start and finish (shift) times that both satisfy staffs' need for reasonable working times and shift patterns, as well as providing appropriate capacity, can be difficult.
Varying the size of the workforce – hiring extra staff during periods of high demand and laying them off as demand falls, or hire and fire.	Reduces basic labour costs quickly.	Hiring costs and possible low productivity while new staff go through the learning curve. Lay-offs may result in severance payments and possible loss of morale in the operation and loss of goodwill in the local labour market.
Using part-time staff – recruit staff who work for less than the normal working day (at the busiest periods).	Good method of adjusting capacity to meet predictable short-term demand fluctuations	Expensive if the fixed costs of employment for each employee (irrespective of how long he or she works) are high.
Skills flexibility – designing flexibility in job design and job demarcation so that staff can transfer across from less busy parts of the operation.	Fast method of reacting to short-term demand fluctuations	Investment in skills training needed and may cause some internal disruption.
Sub-contracting/outsourcing – buying, renting or sharing capacity or output from other operations.	No disruption to the operation.	Can be very expensive because of sub-contractor's margin and sub-contractor may not be as motivated to give same service, or quality. Also a risk of leakage of knowledge.
Change output rate – expecting staff (and equipment) to work faster than normal.	No need to provide extra resources.	Can only be used as a temporary measure, and even then can cause staff dissatisfaction, a reduction in the quality of work, or both.

Changing capacity when variation is unpredictable

Both the mix of methods used to change capacity and how they are implemented will depend on the balance between predictable and unpredictable variation. As we discussed earlier, the objective of capacity management when demand variation is predictable is to affect the changes as efficiently as possible. Whereas when demand fluctuations are unpredictable, the objective is usually to change capacity as fast as possible. In the latter case, it is necessary to understand the flexibility of the resources that may be used to increase capacity. In this case we are using flexibility to mean both how much capacity can be changed and how fast it can be changed. In fact, the degree of change and the response time required to make the change are almost always related. The relationship can be shown in what is termed a 'range–response' curve. Figure 8.9 shows one of these for a call centre. It shows that within a few minutes of demand for the call centre's services increasing, it has the ability to switch a proportion of its calls to the company's other call centres. However, not everyone in these other call centres is

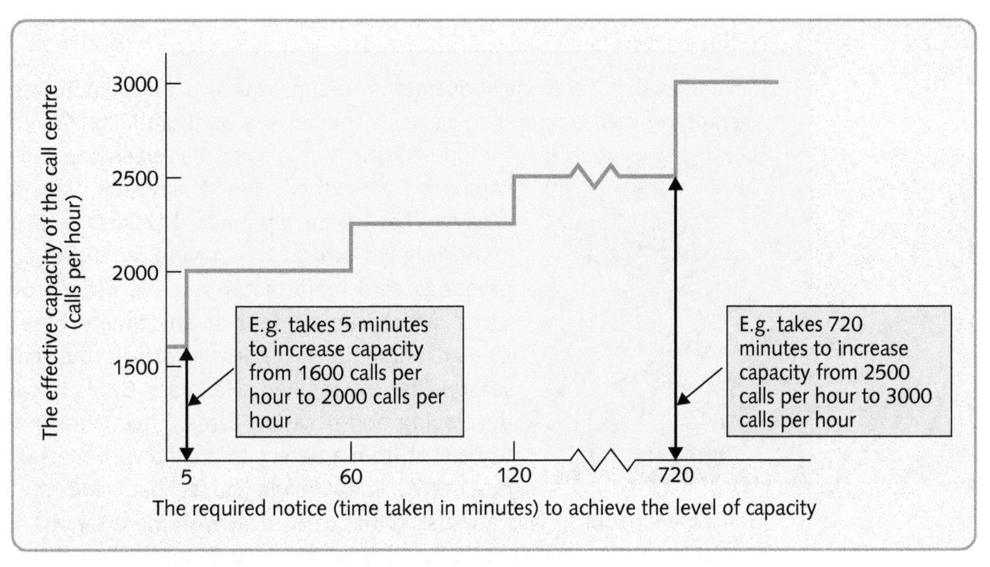

Figure 8.9 The range-response curve for increasing capacity at a call centre

trained to take such calls, therefore any further increase in capacity must come from bringing in staff currently not on shift. Eventually, the call centre will hit its limits of physical capacity (computers, telephone lines, etc.). Any further capacity increase will have to wait until more physical capacity is added.

Manage demand plan

Animated diagram

The objective of demand management is to change the pattern of demand to bring it closer to available capacity, usually by transferring customer demand from peak periods to quiet periods, as was shown in Figure 8.8(c). There are a number of methods for achieving this.

- *Constraining customer access* – customers may only be allowed access to the operation's products or services at particular times. For example, reservation and appointment systems in hospitals.
- *Price differentials* – adjusting price to reflect demand. That is, increasing prices during periods of high demand and reducing prices during periods of low demand.
- *Scheduling promotion* – varying the degree of market stimulation through promotion and advertising in order to encourage demand during normally low periods.
- *Service differentials* – allowing service levels to reflect demand (implicitly or explicitly), allowing service to deteriorate in periods of high demand and increase in periods of low demand. If this strategy is used explicitly, customers are being educated to expect varying levels of service and hopefully move to periods of lower demand.

A more radical approach attempts to create alternative products or services to fill capacity in quiet periods. It can be an effective demand management method but, ideally, new products or services should meet three criteria: (a) they can be produced on the same processes; (b) they have different demand patterns to existing offerings; (c) they are sold through similar marketing channels. For example, ski resorts may provide organised mountain activity holidays in the summer, and garden tractor companies may make snow movers in the autumn and winter. However, the apparent benefits of filling capacity in this way must be weighted against the risks of damaging the core product or service, and the operation must be fully capable of serving both markets.

EXAMPLE

Hallmark Cards[4]

Companies that traditionally operate in seasonal markets can demonstrate some considerable ingenuity in their attempts to develop counter-seasonal products. One of the most successful industries in this respect has been the greetings card industry. Mother's Day, Father's Day, Halloween, Valentine's Day and other occasions have all been promoted as times to send (and buy) appropriately designed cards. Now, having run out of occasions to promote, greetings card manufacturers have moved on to 'non-occasion' cards, which can be sent at any time. These have the considerable advantage of being less seasonal, thus making the companies' seasonality less marked. Hallmark Cards has been the pioneer in developing non-occasion cards. Their cards include those intended to be sent from a parent to a child with messages such as 'Would a hug help?', 'Sorry I made you feel bad', and 'You're perfectly wonderful – it's your room that's a mess'. Other cards deal with more serious adult themes such as friendship ('You're more than a friend, you're just like family') or even alcoholism ('This is hard to say, but I think you're a much neater person when you're not drinking'). Now Hallmark Cards have founded a 'loyalty marketing group' that *'helps companies communicate with their customers at an emotional level'*. It promotes the use of greetings cards for corporate use, to show that customers and employees are valued. But how will 'e-cards' send by e-mail or via online social networks affect seasonal demand? They can be 'sent' speedily without buying a stamp and they can include animations and songs. The card industry maintain that e-cards are not a serious threat, because people use them as a supplement to physical cards or for less important dates when they would not bother putting a card in the post. They may even help to level out demand. Hallmark, for example, offer their own 'e-cards'. They permit people to send free e-cards from their websites, but encourage them to purchase annual subscriptions for access to the most elaborate designs.

Source: Jim Holden/Alamy Images

Yield management

In operations which have relatively fixed capacities, such as airlines and hotels, it is important to use the capacity of the operation for generating revenue to its full potential. One approach used by such operations is called yield management.[5] This is really a collection of methods, some of which we have already discussed, which can be used to ensure that an operation maximises its potential to generate profit. Yield management is especially useful where:

- capacity is relatively fixed
- the market can be fairly clearly segmented
- the service cannot be stored in any way
- the services are sold in advance
- the marginal cost of making a sale is relatively low.

Airlines, for example, fit all these criteria. They adopt a collection of methods to try to maximise the yield (i.e. profit) from their capacity. Over-booking capacity may be used to compensate for passengers who do not show up for the flight. If the airline did not fill this seat it would lose the revenue from it, so airlines regularly book more passengers onto flights than the capacity of the aircraft can cope with. However, if more passengers show up than they expect, the airline will have a number of upset passengers (although they may be able to offer financial inducements for the passengers to take another flight). By studying past data on flight demand, airlines try to balance the risks of over-booking and under-booking. Operations may also use price discounting at quiet times, when demand is unlikely to fill capacity. Airlines will also sell heavily discounted tickets to agents who then themselves take the risk of finding customers for them. This type of service may also be varied. For example, the relative demand for first-, business-, and economy-class seats varies throughout the year. There is no point discounting tickets in a class for which demand will be high. Yield management tries to adjust the availability of the different classes of seat to reflect their demand. They will also vary the number of seats available in each class by upgrading or even changing the configuration of airline seats.

Using cumulative representations to plan capacity

Practice note

When an operation's output can be stored, a useful method of assessing the feasibility and consequences of adopting alternative capacity plans is the use of cumulative demand and supply curves. These plot (or calculate) both the cumulative demand on an operation, and its cumulative ability to supply, over time. For example, Figure 8.10 shows the forecast

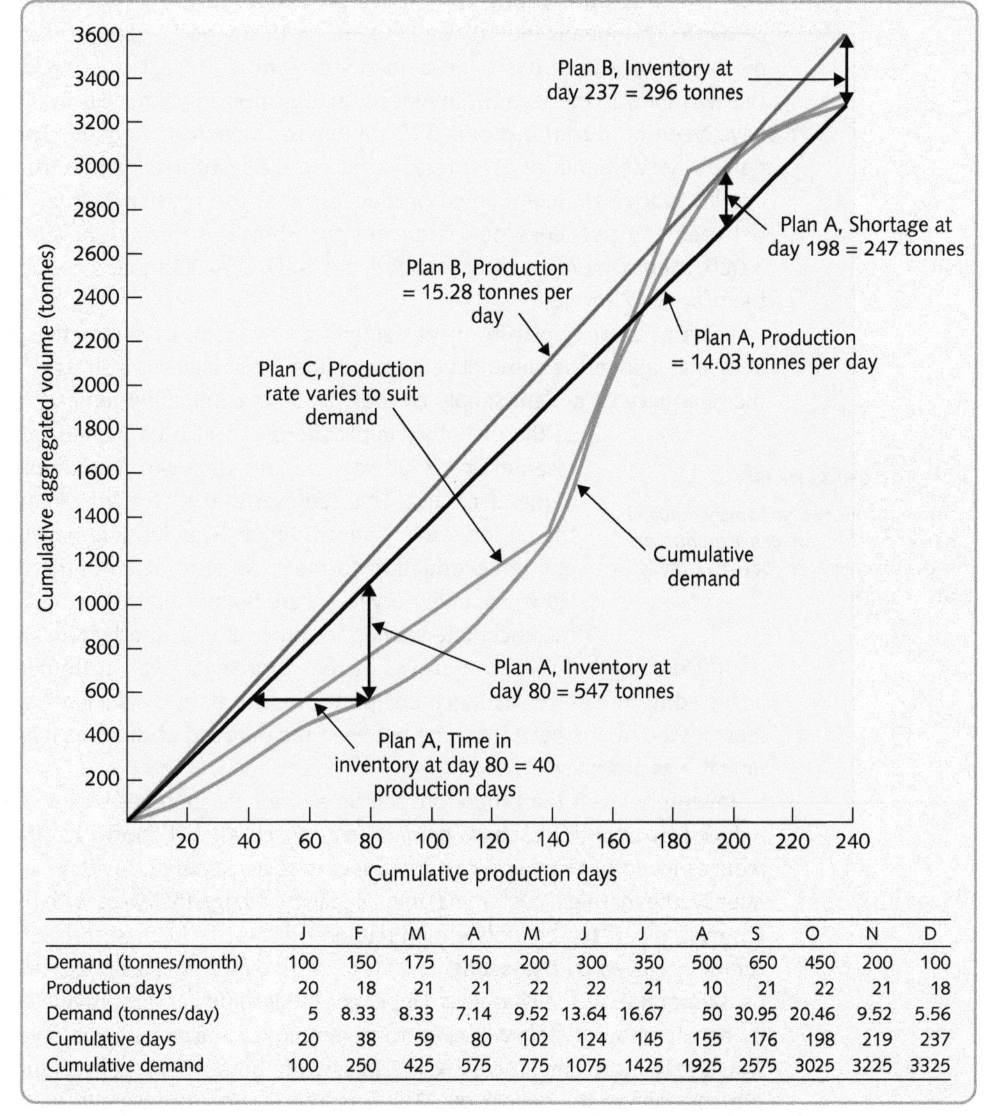

	J	F	M	A	M	J	J	A	S	O	N	D
Demand (tonnes/month)	100	150	175	150	200	300	350	500	650	450	200	100
Production days	20	18	21	21	22	22	21	10	21	22	21	18
Demand (tonnes/day)	5	8.33	8.33	7.14	9.52	13.64	16.67	50	30.95	20.46	9.52	5.56
Cumulative days	20	38	59	80	102	124	145	155	176	198	219	237
Cumulative demand	100	250	425	575	775	1075	1425	1925	2575	3025	3225	3325

Figure 8.10 Cumulative representation of demand and three capacity plans

aggregated demand for a chocolate factory which makes confectionery products. Demand for its products in the shops is greatest in December. To meet this demand and allow time for the products to work their way through the supply chain, the factory must supply a demand which peaks in September. But the cumulative representation of demand against available supply time (production days) shown in Figure 8.10 reveals that, although total demand peaks in September, because of the restricted number of available production days, the peak demand per production day occurs a month earlier in August. It also shows that the effective fluctuation in demand over the year is even greater than it seemed. The ratio of monthly peak demand to monthly lowest demand is 6.5:1, but the ratio of peak to lowest demand per production day is 10:1. Demand per production day is more relevant to operations managers, because production days represent the 'ability to supply'.

The feasibility and consequences of a capacity plan can be assessed on this basis. Figure 8.10 also shows a level capacity plan (A) that assumes production at a rate of 14.03 tonnes per productive day. This meets cumulative demand by the end of the year, so total over-capacity is equal to or greater than under-capacity. However, if one of the aims of the plan is to supply demand when it occurs, the plan is inadequate. Up to around day 168, the line representing cumulative production is above that representing cumulative demand. This means that at any time during this period, more product has been produced by the factory than has been demanded from it. In fact the vertical distance between the two lines is the level of inventory at that point in time. So by day 80, 1,122 tonnes have been produced but only 575 tonnes have been demanded. The surplus of production above demand, or inventory, is therefore 547 tonnes. When the cumulative demand line lies above the cumulative production line, the reverse is true. The vertical distance between the two lines now indicates the shortage, or lack of supply. So by day 198, 3,025 tonnes have been demanded but only 2,778 tonnes produced. The shortage is therefore 247 tonnes.

For any capacity plan to meet demand as it occurs, its cumulative production line must always lie above the cumulative demand line. This makes it a straightforward task to judge the adequacy of a plan, simply by looking at its cumulative representation. An impression of the inventory implications can also be gained from a cumulative representation by judging the area between the cumulative production and demand curves. This represents the amount of inventory carried over the period. Level capacity plan (B) is feasible because it always ensures enough production to meet demand at any time throughout the year. However, inventory levels are high using this plan. It may even mean that the chocolate spends so much time in the factory's inventory, that it has insufficient shelf life when it arrives at the company's retail customers. Assuming a 'first-in-first-out' inventory management principle, the time product stays in inventory will be represented by the horizontal line between the demand at the time it is 'demanded' and the time it was produced.

OPERATIONS PRINCIPLE

For any capacity plan to meet demand as it occurs, its cumulative production line must always lie above its cumulative demand line.

Inventory levels (and therefore the time products spend as part of the inventory) can be reduced by adopting a chase demand plan, such as that shown as (C) in Figure 8.10. This reduces inventory-carrying costs, but incurs costs associated with changing capacity levels. Usually, the marginal cost of making a capacity change increases with the size of the change. For example, if the chocolate manufacturer wishes to increase capacity by 5 per cent, this can be achieved by requesting its staff to work overtime – a simple, fast and relatively inexpensive option. If the change is 15 per cent, overtime cannot provide sufficient extra capacity and temporary staff will need to be employed – a more expensive solution which also would take more time. Increases in capacity of above 15 per cent might only be achieved by sub-contracting some work out. This would be even more expensive.

DIAGNOSTIC QUESTION

How should capacity be controlled?

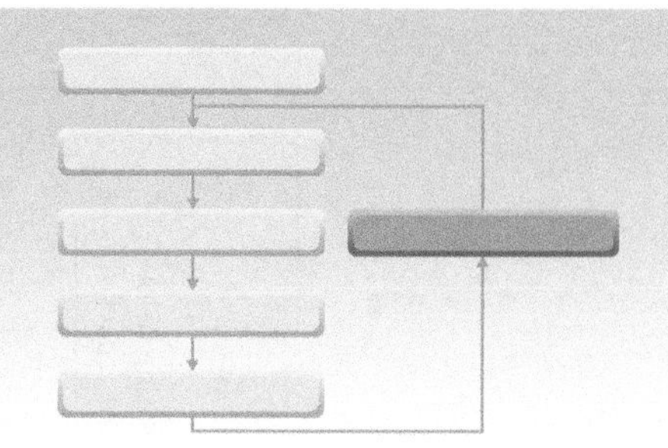

Video

Although planning capacity levels in advance, and even planning how to respond to unexpected changes in demand, is an important part of capacity management, it does not fully reflect the dynamic nature of the activity. Capacity management must react to *actual* demand and *actual* capacity as it occurs. Period by period, operations management considers its forecasts of demand, its understanding of current capacity and, if outputs can be stocked, how much inventory has been carried forward from the previous period. Based on all this information, it makes plans for the following period's capacity. During the next period, demand might or might not be as forecast and the actual capacity of the operation might or might not turn out as planned (because of the capacity leakage discussed earlier). But whatever the actual conditions during that period, at the beginning of the next period the same types of decisions must be made, in the light of the new circumstances. Figure 8.11 shows how this works in practice. It shows the overall performance of an operation's capacity management as a function of the way it manages capacity and way it manages (or forecasts) demand.

The success of capacity management is generally measured by some combination of costs, revenue, working capital and customer satisfaction (which goes on to influence revenue). This

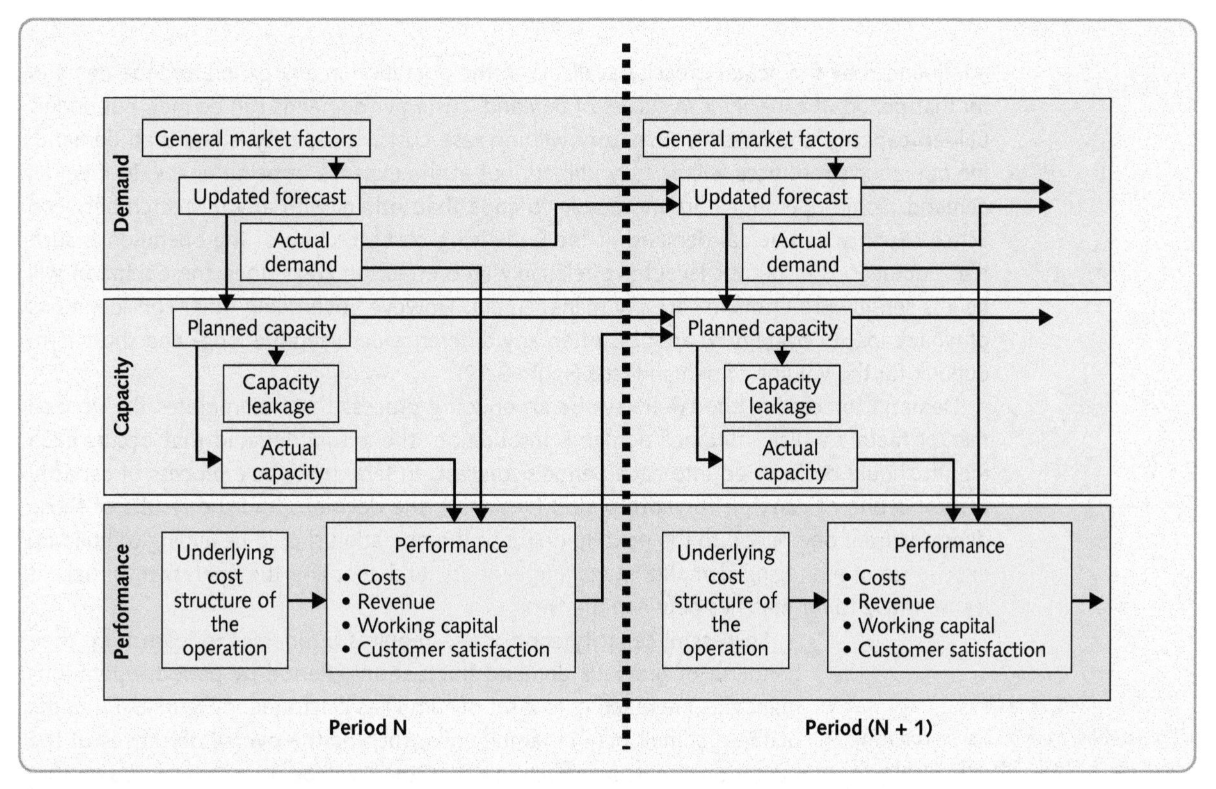

Figure 8.11 How should capacity be controlled – the dynamics of capacity management

		Short-term outlook for volume		
		Decreasing below current capacity	Level with current capacity	Increasing above current capacity
Long-term outlook for volume	Decreasing below current capacity	Reduce capacity (semi) permanently. For example; reduce staffing levels; reduce supply agreements.	Plan to reduce capacity (semi) permanently.For example, freeze recruitment; modify supply agreements.	Increase capacity temporarily. For example, increase working hours, and/or hire temporary staff; modify supply agreements.
	Level with current capacity	Reduce capacity temporarily. For example, reduce staff working hours; modify supply agreements.	Maintain capacity at current level.	Increase capacity temporarily. For example, increase working hours, and/or hire temporary staff; modify supply agreements.
	Increasing above current capacity	Reduce capacity temporarily. For example, reduce staff working hours, but plan to recruit; modify supply agreements.	Plan to increase capacity above current level; plan to increase supply agreements.	Increase capacity (semi) permanently. For example, hire staff; increase supply agreements.

Figure 8.12 Capacity management strategies are partly dependent on the long-term and short-term outlook for volumes

is influenced by the actual capacity available to the operation in any period and the demand for that period. If capacity is in excess of demand, customer demands can be met, but under-utilised capacity and possibly inventory will increase costs. If capacity is less than demand, the operation's resources will be fully utilised, but at the expense of being unable to meet all demand. Some operations are more able to cope than others with any mismatch between actual capacity and actual demand. If the underlying cost structure of the operation is such that fluctuations in output level have relatively little effect on costs, then the operation will be less sensitive to errors in capacity management. However, overriding other considerations of what capacity strategy to adopt is often any difference between the long- and short-term outlook for the volume of demand (see Figure 8.12).

Demand forecasting should always be an ongoing process that incorporates the general market factors which influence demand. In addition, the actual demand that occurs each month should be factored into each period's forecast. In fact, the whole process of capacity control is one of carrying forward, period by period, the decisions and the results of those decisions from one period to the next. In doing so the operation should be aiming to build up experience of managing demand, managing capacity, and adapting the operation to make it less sensitive to mismatches between the two.

Successful capacity control also requires businesses to learn from their handling of previous demand fluctuations. Period by period, operations managers are reacting to a set of stimuli as illustrated in Figure 8.11. Some of these stimuli may be ambiguous, such as the overall objectives of the operation and its approach to risk. Others will be uncertain, such as future demand and (to a lesser extent) future capacity. This is a complex decision-making process that depends on more than the availability and accuracy

OPERATIONS PRINCIPLE

The learning from managing capacity in practice should be captured and used to refine both demand forecasting and capacity planning.

of information (although this is important). It also depends on the ability to refine decision-making behaviour through learning from past successes and mistakes. For example, some managers may have the tendency to over-react to immediate stimuli by frequently increasing or decreasing capacity as forecasts of future demand are adjusted. If so, some mechanism will need to be put in place that smoothes both forecasts and the response to them.

Critical commentary

Each chapter contains a short critical commentary on the main ideas covered in the chapter. Its purpose is not to undermine the issues discussed in chapter, but to emphasise that, although we present a relatively orthodox view of operation, there are other perspectives.

- For such an important topic, there is surprisingly little standardisation in how capacity is measured. Not only is a reasonably accurate measure of capacity needed for operations management, it is also needed to decide whether it is worth investing in extra physical capacity such as machines. Yet not all practitioners would agree with the way in which capacity has been defined or measured in this chapter (although it does represent orthodox practice). One school of thought is that whatever capacity efficiency measures are used, they should be useful as diagnostic measures which can highlight the root causes of inefficient use of capacity. The idea of overall equipment effectiveness (OEE) described earlier is often put forward as a useful way of measuring capacity efficiencies.

- The other main point of controversy in capacity management concerns the use of varying staff levels. To many, the idea of fluctuating the workforce to match demand, either by using part-time staff or by hiring and firing, is more than just controversial. It is regarded as unethical. It is any business's responsibility, they argue, to engage in a set of activities which are capable of sustaining employment at a steady level. Hiring and firing merely for seasonal fluctuations, which can be predicted in advance, is treating human beings in a totally unacceptable manner. Even hiring people on a short-term contract, in practice, leads to them being offered poorer conditions of service and leads to a state of permanent anxiety as to whether they will keep their jobs. On a more practical note, it is pointed out that, in an increasingly global business world where companies may have sites in different countries, those countries that allow hiring and firing are more likely to have their plants 'downsized' than those where legislation makes this difficult.

SUMMARY CHECKLIST

This checklist comprises questions that can be usefully applied to any type of operations and reflect the major diagnostic questions used within the chapter.

- ☐ Is the importance of effective capacity management fully understood?
- ☐ Is the operation's current capacity measured?
- ☐ If so, are all the assumptions inherent in the measurement of capacity made fully explicit?
- ☐ What capacity 'leakage' is normal, and have options for minimising capacity leakage been explored?
- ☐ Is there scope for using the overall equipment effectiveness (OEE) measure of capacity?
- ☐ What is the balance between predictable variation and unpredictable variation in demand and capacity?
- ☐ Realistically, what potential is there for making unpredictable variability more predictable through better forecasting?
- ☐ Does an understanding of the market include the extent to which the behaviour of customers and/or suppliers can be influenced to reduce variability?
- ☐ Does the operations base capacity reflect all the factors that should be influencing its level?
- ☐ Have alternative methods of adjusting (or not) capacity been fully explored and assessed?
- ☐ If variation is unpredictable, have methods of speeding up the operation's reaction to demand–capacity mismatches been explored?
- ☐ Is there scope for using cumulative representations of demand and capacity for planning purposed?
- ☐ Is the method of deciding period-by-period capacity levels effective?
- ☐ How does the method of deciding period-by-period capacity levels reflect previous experience?

CASE STUDY

Blackberry Hill Farm

'Six years ago I had never heard of agri-tourism. As far as I was concerned, I had inherited the farm and I would be a farmer all my life.' (Jim Walker, Blackberry Hill Farm)

The 'agri-tourism' that Jim is referring to is 'a commercial enterprise at a working farm, or other agricultural centre, conducted for the enjoyment of visitors that generates supplemental income for the owner'. *'Farming has become a tough business,'* says Jim. *'Low world prices, a reduction in subsidies, and increasingly uncertain weather patterns have made it a far more risky business than when I first inherited the farm. Yet, because of our move into the tourist trade we are flourishing. Also . . . I've never had so much fun in my life'.* But, Jim warns, agri-tourism isn't for everyone. *'You have to think carefully. Do you really want to do it? What kind of life style do you want? How open-minded are you to new ideas? How business-minded are you? Are you willing to put a lot of effort into marketing your business? Above all, do you like working with people? If you had rather be around cows than people, it isn't the business for you.'*

History

Blackberry Hill Farm was a 200-hectare mixed farm in the south of England when Jim and Mandy Walker inherited it 15 years ago. It was primarily a cereal growing operation with a small dairy herd, some fruit and vegetable growing and mixed woodland that was protected by local preservation laws. Six years ago it had become evident to Jim and Mandy that they may have to rethink how the farm was being managed. *'We first started a pick-your-own (PYO) operation because our farm is close to several large centres of population. Also the quantities of fruit and vegetables that we were producing were not large enough to interest the commercial buyers. Entering the PYO market was a reasonable success and in spite of making some early mistakes, it turned our fruit and vegetable growing operation from making a small loss to making a small profit. Most importantly, it gave us some experience of how to deal with customers face-to-face and of how to cope with unpredictable demand. The biggest variable in PYO sales is weather. Most business occurs at the weekends between late spring and early autumn. If rain keeps customers away during part of those weekends, nearly all sales have to occur in just a few days.'*

Source: Fancy/Veer/Corbis

Within a year of opening up the PYO operation Jim and Mandy had decided to reduce the area devoted to cereals and increase their fruit and vegetable growing capability. At the same time they organised a Petting Zoo that allowed children to mix with, feed and touch various animals.

'We already had our own cattle and poultry but we extended the area and brought in pigs and goats. Later we also introduced some rabbits, ponies and donkeys, and even a small bee keeping operation.' At the same time the farm started building up its collection of 'farm heritage' exhibits. These were static displays of old farm implements and 'recreations' of farming processes together with information displays. This had always been a personal interest of Jim's and it allowed him to convert two existing farm outbuilding to create a 'Museum of Farming Heritage'.

The year after, they introduced tractor rides for visitors around the whole farm and extended the petting zoo and farming tradition exhibits further. But the most significant investment was in the 'Preserving Kitchen'. *'We had been looking for some way of using the surplus fruits and vegetable that we occasionally accumulated and also for some kind of products that we could sell in a farm shop. We started the Preserving Kitchen to make jams and fruit, vegetables and sauces preserved in jars. The venture was an immediate success. We started making just 50 kilograms of preserves a week, within three months that had grown 300 kilogrammes a week and we are now producing around 1,000 kilogrammes a week, all under the 'Blackberry Hill Farm' label.'* The following year the preserving kitchen was extended and

Table 8.2(a) Number of visitors last year

Month	Total visitors	Month	Total visitors
January	1,006	August	15,023
February	971	September	12,938
March	2,874	October	6,687
April	6,622	November	2,505
May	8,905	December	3,777
June	12,304	**Total**	**88,096**
July	14,484	Average	7,341.33

Table 8.2(b) Farm opening times*

January to Mid-March	Wednesday–Sunday	10.00–16.00
Mid-March to May	Tuesday–Sunday	9.00–18.00
May to September	All week	8.30–19.00
October to November	Tuesday–Sunday	10.00–16.00
December	Tuesday–Sunday	9.00–18.00

*Special evening events at Easter, summer weekends and Christmas.

a viewing area added. *'It was a great attraction from the beginning,'* says Mandy. *'We employed ladies from the local village to make the preserves. They are all extrovert characters, so when we asked them to dress up in traditional 'farmers wives' type clothing they were happy to do it. The visitors love it, especially the good natured repartee with our ladies. The ladies also enjoy giving informal history lessons when we get school parties visiting us.'*

Within the last two years, the farm has further extended its preserving kitchen, farm shop, exhibits and petting zoo. It has also introduced a small adventure playground for the children, a café serving drinks and its own produce, a picnic area and a small bakery. The bakery was also open to view by customers and staffed by bakers in traditional dress. *'It's a nice little visitor attraction,'* says Mandy, *'and it gives us another opportunity to squeeze more value out of our own products.'* Table 8.2(a) shows last year's visitor numbers; Table 8.2(b) shows the farm's opening times.

Demand

The number of visitors to the farm was extremely seasonal. From a low point in January and February, when most people just visited the farm shop, the spring and summer months could be very busy, especially on public holidays. The previous year Mandy had tracked the number of visitors arriving at the farm each day. *'It is easy to record the number of people visiting the farm attractions, because they pay the entrance charge. What we had not done before is include the people who just visited the farm shop and bakery that can be accessed both from within the farm and from the car park. We estimate that the number of people visiting the shop but not the farm ranges from 74 per cent in February down to around 15 per cent in August.'* Figure 8.13 shows the number of visitors in the previous year's August. *'What*

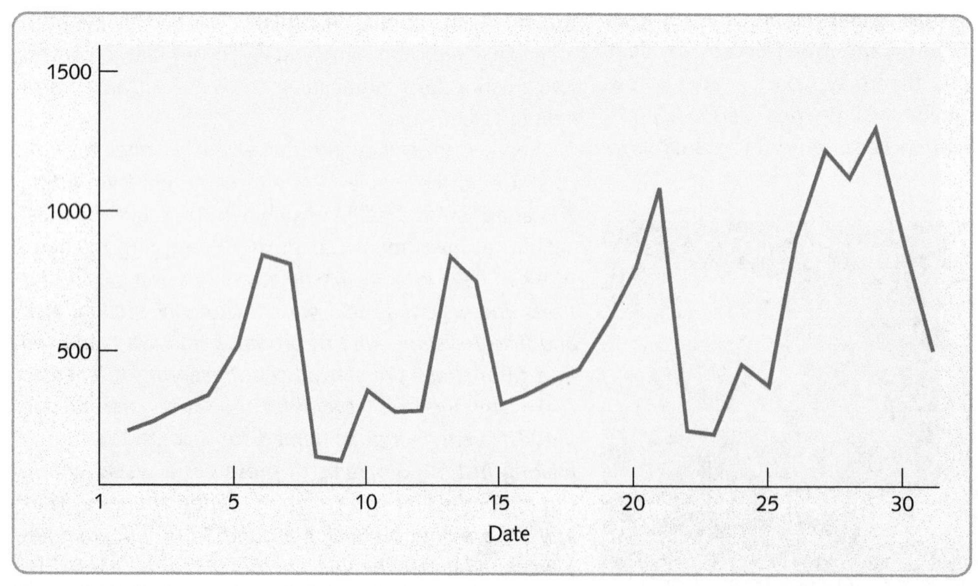

Figure 8.13 Daily number of visitors in August last year

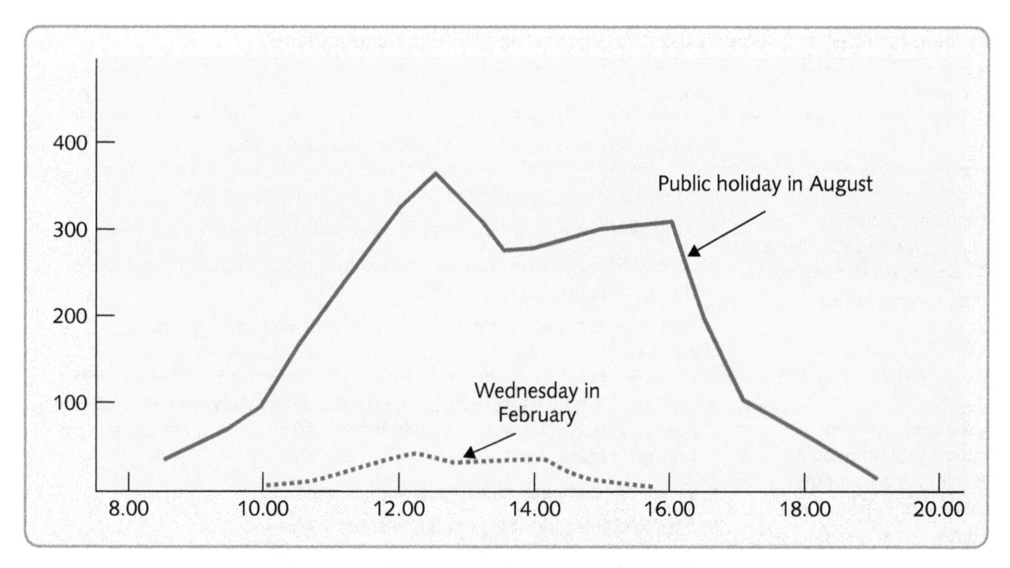

Figure 8.14 Visitor arrivals, public holiday in August and a Wednesday in February

our figures do not include are those people who visit the shop but don't buy anything. This is unlikely to be a large number.'

Mandy had also estimated the average stay at the farm and/or farm shop. She reckoned that in winter time the average stay was 45 minutes, but in August in climbed to 3.1 hours.

Current issues

Both Jim and Mandy agreed that their lives had fundamentally changed over the last few years. Income from visitors and from the Blackberry Hill brand of preserves now accounted for 70 per cent of the farm's revenue. More importantly, the whole enterprise was significantly more profitable than it had ever been. Nevertheless, the farm faced a number of issues.

The first was the balance between its different activities. Jim was particularly concerned that the business remained a genuine farm. *'When you look at the revenue per hectare, visitor and production activities bring in far more revenue than conventional agricultural activities. However, if we push the agri-tourism too far we become no better than a theme park. We represent something more than this to our visitors. They come to us partly because of what we represent as well as what we actually do. I am not sure that we would want to grow much more. Anyway, more visitors would mean that we have to extend the car park. That would be expensive, and although it would be necessary, it does not directly bring in any more revenue. There are already parking problems during peak period and we have had complaints from the police that our visitors park inappropriately on local roads.'*

'There is also the problem of complexity. Every time we introduce a new attraction, the whole business gets that little bit more complex to manage. Although we enjoy it tremendously, both Mandy and I are spreading ourselves thinly over an ever widening range of activities.' Mandy was also concerned over this. *'I'm starting to feel that my time is being taken up in managing the day-to-day problems of the business. This does not leave time either for thinking about the overall direction in which we should be going, or spending time talking with the staff. That is why we both see this coming year as a time for consolidation and for smoothing out the day-to-day problems of managing the business, particularly the queuing, which is getting excessive at busy times. That is why this year we are limiting ourselves to just one new venture for the business.'*

Staff management was also a concern for Mandy. The business had grown to over 80 (almost all part-time and seasonal) employees. *'We have become a significant employer in the area. Most of our employees are still local people working part-time for extra income but we are also now employing 20 students during the summer period and, last year, eight agricultural students from Eastern Europe. But now, labour is short in this part of the country and it is becoming more difficult to attract local people, especially to produce Blackberry Hill Farm Preserves. Half of the Preserving Kitchen staff work all year, with the other employed during the summer and autumn periods. But most of them would prefer guaranteed employment throughout the year.'*

Table 8.3 gives more details of some of the issues of managing the facilities at the farm, and Table 8.4 shows the demand and production of preserves month by month through the previous year.

Table 8.3 The farm's main facilities and some of the issues concerned with managing them

Facility	*Issues*
Car park	• 85 car parking spaces, 4 × 40-seater tour bus spaces.
Fixed exhibits, etc. Recreation of old farmhouse kitchen, recreation of barnyard, old-fashioned milking parlour, various small exhibits on farming past and present, adventure playground, ice cream and snack stands.	• Most exhibits in, or adjacent to, the farm museum. • At peak times helpers are dressed in period costume to entertain visitors. • Feedback indicates customers find exhibits more interesting than they thought they would. • Visitors free to look when they wish absorbs demand from busy facilities.
Tractor rides One tractor towing decorated covered cart with maximum capacity of 30 people. Tour takes around 20 minutes on average (including stops). Waits ten minutes between tours except at peak times when tractor circulates continuously.	• Tractor acts both as transport and entertainment. Approximately 60 per cent of visitors stay on for the whole tour; 40 per cent use it as 'hop-on hop-off' facility. • Overloaded at peak times, long queues building. • Feedback indicates it is popular, except for queuing. • Jim reluctant to invest in further cart and tractor.
Pick-your-own area Largest single facility on the farm. Use local press, dedicated telephone line (answering machine) and website to communicate availability of fruit and vegetables. Check-out and weighing area next to farm shop, also displays picked produce and preserves etc. for sale.	• Very seasonal and weather dependent, both for supply and demand. • Farm plans for a surplus over visitor demand: uses surplus in preserves. • Six weighing/paying stations at undercover checkout area. Queues develop at peak times. Feedback indicates some dissatisfaction with this. • Can move staff from farm shop to help with checkout in busy periods, but farm shop also tends to be busy at the same time. • Considering using packers at pay stations to speed up the process.
Petting Zoo Accommodation for smaller animals, including sheep and pigs. Large animals (cattle, horses) brought to viewing area daily. Visitors can view all animals and handle/stroke most animals under supervision.	• Approximately 50 per cent of visitors view Petting Zoo. • Number of staff in attendance varies between 0 (off-peak) and 5 (peak periods). • The area can get congested during peak periods. • Staff need to be skilled at managing children.
Preserving Kitchen Boiling vats, mixing vats, jar sterilising equipment, etc. Visitor viewing area can hold 15 people comfortably. Average length of stay 7 minutes in off-season, 14 minutes in peak season.	• Capacity of kitchen is theoretically 4,500 kilogrammes per month on a five-day week and 6,000 kilogrammes on a seven-day week. • In practice, capacity varies with season because of interaction with visitors. Can be as low as 5,000 kg on a seven-day week in summer, or up to 5,000 kg on a five-day week in winter. • Shelf life of products is on average 12 months. • Current storage area can hold 16,000 kilogrammes.
Bakery Contains mixing and shaping equipment, commercial oven, cooling racks, display stand, etc. Just installed doughnut-making machine. All pastries contain farm's preserved fruit.	• Starting to become a bottleneck since doughnut-making machine installed; visitors like watching it. • Products also on sale at farm shop adjacent to bakery. • Would be difficult to expand this area because of building constraints.
Farm shop and café Started by selling farm's own products exclusively. Now sells a range of products from farms in the region and wider. Started selling frozen menu dishes (lasagne, goulash, etc.) produced off-peak in the preserving kitchen.	• The most profitable part of the whole enterprise, Jim and Mandy would like to extend the retailing and café operation. • Shop includes area for cooking displays, cake decoration, fruit dipping (in chocolate), etc. • Some congestion in shop at peak times but little visitor dissatisfaction. • More significant queuing for café in peak periods. • Considering allowing customers to place orders before they tour the farm's facilities and collect their purchases later. • Retailing more profitable per square metre than café.

Table 8.4 Preserve demand and production (previous year)

Month	*Demand (kg)*	*Cumulative demand (kg)*	*Production (kg)*	*Cumulative product (kg)*	*Inventory (kg)*
January	682	682	4,900	4,900	4,218
February	794	1,476	4,620	9,520	8,044
March	1,106	2,582	4,870	14,390	11,808
April	3,444	6.026	5,590	19,980	13,954
May	4,560	10,586	5,840	25,820	15,234
June	6,014	16,600	5,730	31,550	14,950
July	9,870	26,470	5,710	37,260	10,790
August	13,616	40,086	5,910	43,170	3,084
September	5,040	45,126	5,730	48,900	3,774
October	1,993	47,119	1,570*	50,470	3,351
November	2,652	49,771	2,770*	53,240	3,467
December	6,148	55,919	4,560	57,800	1,881
Average demand	4,660			Average inventory	7,880

*Technical problems reduced production level.

Where next?

By the 'consolidation' and improvement of 'day-to-day' activities, Jim and Mandy mean that they wanted to increase their revenue, while at the same time reducing the occasional queues that they knew could irritate their visitors, preferably without any significant investment in extra capacity. They are also concerned to be able to offer more stable employment to the Preserving Kitchen 'ladies' throughout the year, who would produce at a near constant rate. However, they were not sure if this could be done without storing the products for so long that their shelf life would be seriously affected. There was no problem with the supply of produce to keep production level, less than 2 per cent of the fruit and vegetables that go into their preserves are actually grown on the farm. The remainder were bought at wholesale markets, although this was not generally understood by customers.

Of the many ideas being discussed as candidates for the 'one new venture' for next year, two are emerging as particularly attractive. Jim likes the idea of developing a Maize Maze, a type of attraction that had become increasingly popular in Europe and North America in the last five years. It involves planting a field of maize (corn) and, once grown, cutting through a complex serious of paths in the form of a maze. Evidence from other farms indicate that a maze would be extremely attractive to visitors and Jim reckons that it could account for up to an extra 10,000 visitors during the summer period. Designed as a separate activity with its own admission charge, it would require an investment of around £20,000, but generate more than twice that in admission charges as well as attracting more visitors to the farm itself.

Mandy favours the alterative idea – that of building up their business in organised school visits. *'Last year we joined the National Association of Farms for Schools. Their advice is that we could easily become one of the top school attractions in this part of England. Educating visitors about farming tradition is already a major part of what we do. And many of our staff have developed the skills to communicate to children exactly what farm life used to be like. We would need to convert and extend one of our existing underused farm outbuildings to make a 'school room' and that would cost between and £30,000 and £35,000. And although we would need to discount our admission charge substantially, I think we could break even on the investment within around two years.'*

QUESTIONS

1 How could the farm's day-to-day operations be improved?

2 What advice would you give Jim and Mandy regarding this year's 'new venture'?

APPLYING THE PRINCIPLES

Hints

Some of these exercises can be answered by reading the chapter. Others will require some general knowledge of business activity and some might require an element of investigation. Hints on how they can all be answered are to be found in the eText at **www.pearsoned.co.uk/slack.**

1 A Pizza Company has a demand forecast for the next 12 months which is shown in the table below. The current workforce of 100 staff can produce 1,000 cases of pizzas per month.

(a) Prepare a production plan which keeps the output level. How much warehouse space would the company need for this plan?

(b) Prepare a demand chase plan. What implications would this have for staffing levels, assuming that the maximum amount of overtime would result in production levels of only 10 per cent greater than normal working hours?

Pizza demand forecast

Month	*Demand (cases per month)*
January	600
February	800
March	1000
April	1500
May	2000
June	1700
July	1200
August	1100
September	900
October	2500
November	3200
December	900

2 Consider how airlines cope with balancing capacity and demand. In particular, consider the role of yield management. Do this by visiting the website of a low-cost airline, and for a number of flights price the fare that is being charged by the airline from tomorrow onwards. In other words, how much would it cost if you needed to fly tomorrow, how much if you needed to fly next week, how much if you needed to fly in two weeks etc. Plot the results for different flights and debate the findings.

3 Calculate the over equipment efficiency (OEE) of the following facilities by investigating their use:

(a) a lecture theatre

(b) a cinema

(c) a coffee machine

Discuss whether it is worth trying to increase the OEE of these facilities and, if it is, how you would go about it.

4 How should a business work out what it is prepared to pay for increasingly sophisticated weather forecasts?

5 What seem to be the advantages and disadvantages of the strategy adopted by Hallmark Cards described earlier in the chapter? What else could Hallmark do to cope with demand fluctuations?

Notes on chapter

1 Sources include: 'A piece of cake: Panettone season arrives' (2009), *The Economist* 10 December; Bauli website: http://www.bauligroup.it/en/.
2 Source: interview with company staff.
3 With special thanks to Philip Godfrey and Cormac Campbell of OEE Consulting Ltd. (www.oeeconsulting.com).
4 Sources include: 'Greeting cards: message of hope – card companies believe technology will bring a smile to special occasions' (2010), *The Economist* 6 May; Robinette, S. (2001) 'Get emotional', *Harvard Business Review*, May.
5 Kimes, S. (1989) 'Yield management: a tool for capacity-constrained service firms', *Journal of Operations Management*, Vol. 8, No. 4.

TAKING IT FURTHER

Brandimarte, P. and Villa, A. (1999) *Modelling Manufacturing Systems: From aggregate planning to real time control*. Springer, New York, NY. *Very academic although it does contain some interesting pieces if you need to get 'under the skin' of the subject.*

Buxey, G. (1993) 'Production planning and scheduling for seasonal demand', *International Journal of Operations and Production Management*, Vol. 13, No. 7. *Another academic paper but one that takes an understandable and systematic approach.*

Fisher, M.L., Hammond, J.H. and Obermeyer, W. (1994), 'Making supply meet demand in an uncertain world', *Harvard Business Review*, Vol. 72, No. 3, May–June.

USEFUL WEBSITES

www.opsman.org *Definitions, links and opinions on operations and process management.*

http://www.dti.gov.uk/er/index *Website of the Employment Relations Directorate who have developed a framework for employers and employees which promotes a skilled and flexible labour market founded on principles of partnership.*

http://www.worksmart.org.uk/index.php *This site is from the Trades Union Congress. Its aim is 'to help today's working people get the best out of the world of work'.*

http://www.eoc-law.org.uk/ *This website aims to provide a resource for legal advisers and representatives who are conducting claims on behalf of applicants in sex discrimination and equal pay cases in England and Wales. This site covers employment-related sex discrimination only.*

http://www.dol.gov/index.htm *US Department of Labor's site with information regarding using part-time employees.*

http://www.downtimecentral.com/ *Lots of information on operational equipment efficiency (OEE).*

Interactive quiz

For further resources including examples, animated diagrams, self-test questions, Excel spreadsheets and video materials please explore the eText on the companion website at **www.pearsoned.co.uk/slack**.

9

Inventory management

Introduction

Operations managers often have an ambivalent attitude towards inventories. They can be costly, tying up working capital. They are also risky because items held in stock could deteriorate, become obsolete or just get lost. They can also take up valuable space in the operation. On the other hand, they can provide some security in an uncertain environment. Knowing that you have the items in stock is a comforting insurance against unexpected demand. This is the dilemma of inventory management: in spite of the cost and the other disadvantages associated with holding stocks, they do facilitate the smoothing of supply and demand. In fact they only exist because supply and demand are not exactly in harmony with each other (see Figure 9.1).

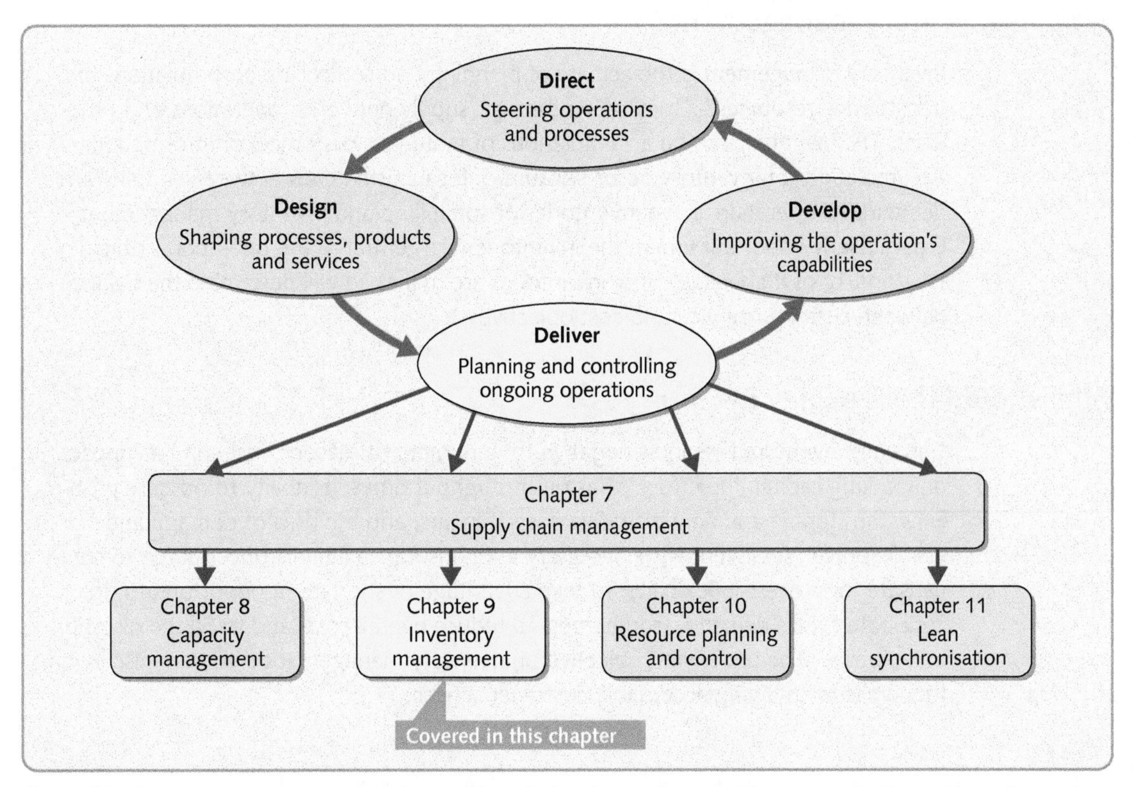

Figure 9.1 Inventory management is the activity of planning and controlling accumulations of transformed resources as they move through supply networks, operations and processes

EXECUTIVE SUMMARY

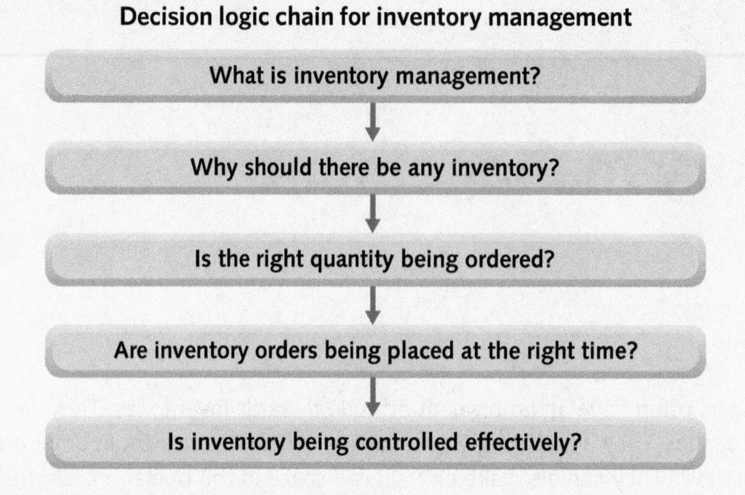

Each chapter is structured around a set of diagnostic questions. These questions suggest what you should ask in order to gain an understanding of the important issues of a topic, and, as a result, improve your decision making. An executive summary addressing these questions is provided below.

What is inventory management?

Inventory management is the activity of planning and controlling accumulations of transformed resources as they move through supply networks, operations and processes. The inventory can be accumulations of materials, customers or information. Accumulations of inventory occur because of local mismatches between supplier and demand. All operations have inventories of some kind and inventory management is particularly important where the inventories are central to the operation's objectives and/or of high value. How inventories are managed will determine the balance between customer service and cost objectives.

Why should there be any inventory?

Generally inventory is seen as negative for a number of reasons, including its impact on working capital, the effect it has on throughput times, its ability to obscure problems, the storage and administrative costs it incurs, and the risks of damage and obsolescence. Yet inventory is necessary as an insurance against uncertainty, to compensate for process inflexibility, to take advantage of short-term opportunities, to anticipate future demand, (sometimes) to reduce overall costs and to fill the distribution pipeline. The underlying objective of inventory management is to minimise inventory while maintaining acceptable customer service.

Is the right quantity being ordered?

A key inventory decision is the 'order quantity' decision. Various formulae exist that attempt to identify the order quantity that minimises total costs under different circumstances. One approach to this problem, the news vendor problem, includes the effects of probabilistic demand in determining order quantity.

Are inventory orders being placed at the right time?

Broadly, there are two approaches to this. The reorder point approach is to time reordering at the point in time where stock will fall to zero minus the order lead time. A variation of this is to reorder at the equivalent inventory level (the reorder level approach). Reordering at a fixed point or level are termed continuous review methods because they require continuous monitoring of stock levels. A different approach, called the periodic review approach, places orders at predetermined times, but varies the order depending on the level of inventory at that time. Both continuous and periodic review can be calculated on a probabilistic basis to include safety stocks.

Is inventory being controlled effectively?

The most common inventory control approach is based on the Pareto (80:20) curve. It classifies stocked items by the usage value (their usage rate multiplied by their value). High usage value items are deemed A class and controlled carefully, whereas low usage value items (B and C class) are controlled less intensely. However, this approach often has to be modified to take account of slow-moving items. Inventory information systems are generally used to keep track of inventory, forecast demand and place orders automatically.

DIAGNOSTIC QUESTION

What is inventory management?

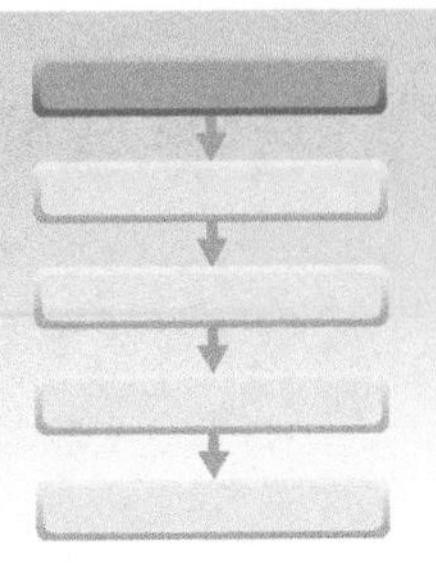

Video

Inventory is a term we use to describe the accumulations of materials, customers or information as they flow through processes or networks. Occasionally the term is also used to describe transforming resources, such as rooms in hotels or automobiles in a vehicle hire firm. Physical inventory (sometimes called 'stock') is the accumulation of physical materials such as components, parts, finished goods or physical (paper) information records. Queues are accumulations of customers, physical as in a queueing line or people in an airport departure lounge, or waiting for service at the end of phone lines. Databases are stores for accumulations of digital information, such as medical records or insurance details. Managing these accumulations is what we call 'inventory management'. And it's important. Material inventories in a factory can represent a substantial proportion of cash tied up in working capital. Minimising them can release large quantities of cash. However, reducing them too far can lead to customers' orders not being fulfilled. Customers held up in queues for too long can get irritated, angry, and possibly leave, so reducing revenue. Databases are critical for storing digital information and while storage may be inexpensive, maintaining databases may not be.

All processes, operations and supply networks have inventories (accumulations) of materials, customers and information

Most things that flow do so in an uneven way. Rivers flow faster down steep sections or where they are squeezed into a ravine. Over relatively level ground they flow slowly, and form pools or even large lakes where there are natural or man-made barriers blocking their path. It's the same in operations. Passengers in an airport flow from public transport or their vehicles, then have to queue at several points including check-in, security screening and immigration. They then have to wait again (a queue even if they are sitting) in the departure lounge as they are joined (batched) with other passengers to form a group of several hundred people who are ready to board the aircraft. They are then squeezed down the air bridge as they file in one at a time to board the plane. Likewise in a tractor assembly plant, stocks of components such as gearboxes, wheels, lighting circuits, etc. are brought into the factory in tens or hundreds and are then stored next to the assembly line ready for use. Finished tractors will also be stored until the transporter comes to take them away in ones or tens to the dealers or directly to the end customer. Similarly, a government tax department collects information about us and our finances from various sources, including our employers, our tax forms, information from banks or other investment companies, and stores this in databases until they are checked, sometimes by people, sometimes automatically, to create our tax codes and/or tax bills. In fact, because most operations involve flows of materials, customers and/or information, at some points they are likely to have material and information inventories and queues of customers waiting for goods or services (see Table 9.1).

Inventories are often the result of uneven flows. If there is a difference between the timing or the rate of supply and demand at any point in a process or network, then accumulations will occur. A common analogy is the water tank shown in Figure 9.2. If, over time, the rate of supply of water to the tank differs from the rate at which it is demanded, a tank of water (inventory) will be needed to maintain supply. When the rate of supply exceeds the rate of demand, inventory increases; when the rate of demand exceeds the rate of supply, inventory decreases.

Table 9.1 Examples of inventory held in processes, operations or supply networks

Process, operation or supply network	*'Inventories'*		
	Physical inventories	*Queues of customers*	*Information in databases*
Hotel	Food items, drinks, toilet items	At check-in and check-out	Customer details, loyalty card holders, catering suppliers
Hospital	Dressings, disposable instruments, blood	Patients on a waiting list, patients in bed waiting for surgery, patients in recovery wards	Patient medical records
Credit card application process	Blank cards, form letters	Customers waiting on the phone	Customer's credit and personal information
Computer manufacturer	Components for assembly, packaging materials, finished computers ready for sale	Customers waiting for delivery of their computer	Customers' details, supplier information

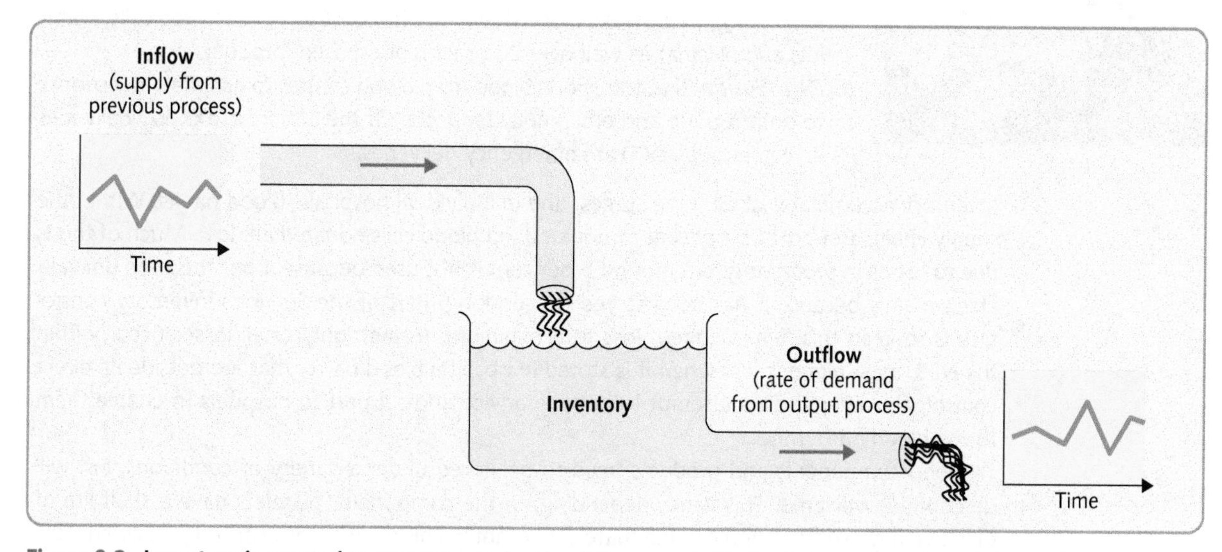

Figure 9.2 Inventory is created to compensate for the differences in timing between supply and demand

So if an operation or process can match supply and demand rates, it will also succeed in reducing its inventory levels. But most organisations must cope with unequal supply and demand, at least at some points in their supply chain. Both the following organisations depend on the ability to manage supply and demand inequality through their inventory management.

There is a complication when using this 'water flow' analogy to represent flows and accumulations (inventories) of information. Inventories of information can either be stored because of uneven flow, in the same way as materials and people, or stored because the operation needs to use the information to process something in the future. For example, an internet retail operation will process each order it receives, and inventories of information may accumulate because of uneven flows as we have described. But, in addition, during order processing customer details could be permanently stored in a database. This information will then be used, not only for future orders from the same customer, but also for other processes, such as targeting promotional activities. In this case the inventory of information has turned from a transformed resource into a transforming resource, because it is being used to transform other information rather than being transformed itself. So, whereas managing physical material concerns ordering and holding the right amounts of goods or materials to deal with the variations in flow, and managing queues is about about the level of resources to deal with demand, a database is the accumulation of information but may not cause an interruption to the flow. Managing databases is about the organisation of the data, its storage, security and retrieval (access and search).

EXAMPLE

The UK's National Blood Service[1]

For blood services, such as the UK's National Blood Service (NBS), the consequences of running out of stock can literally be a matter of life or death. Many people owe their lives to transfusions that were made possible by the efficient management of blood, stocked in a supply network that stretches from donation centres through to hospital blood banks. The NBS supply chain has three main stages.

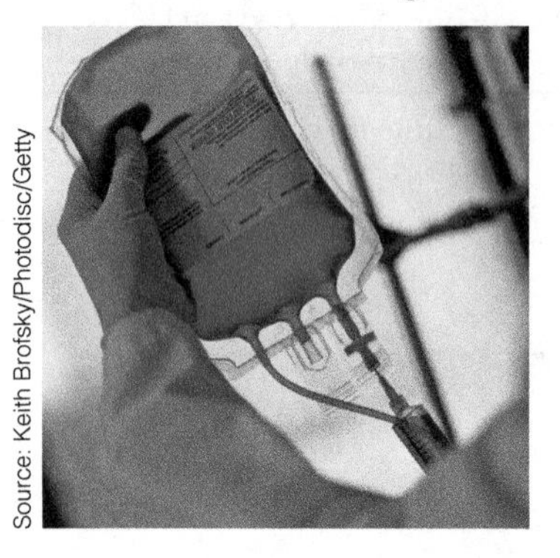

Source: Keith Brofsky/Photodisc/Getty

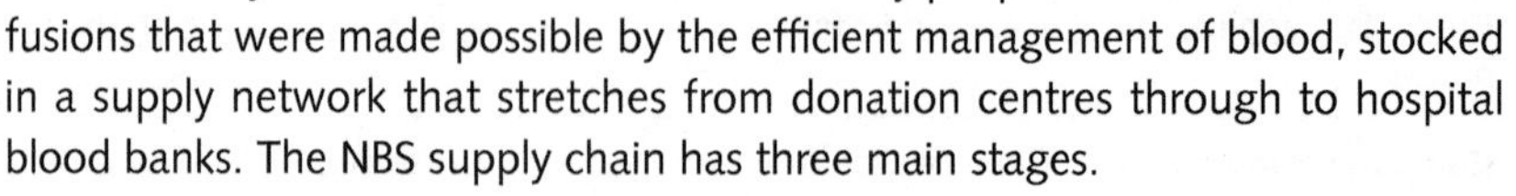

- *Collection*, that involves recruiting and retaining blood donors, encouraging them to attend donor sessions (at mobile or fixed locations) and transporting the donated blood to their local blood centre.
- *Processing*, that breaks blood down into its constituent parts (red cells, platelets and plasma) as well over 20 other blood-based 'products'.
- *Distribution*, that transports blood from blood centres to hospitals in response to both routine and emergency requests. Of the Service's 200,000 deliveries a year, about 2,500 are emergency deliveries.

Inventory accumulates at all three stages, and in individual hospitals' blood banks. Within the supply chain, around 11.5 per cent of donated red blood cells donated are lost. Much of this is due to losses in processing, but around 5 per cent is not used because it has 'become unavailable', mainly because it has been stored for too long. Part of the Service's inventory control task is to keep this 'time expired' loss to a minimum. In fact, only small losses occur within the NBS, most blood is lost when it is stored in hospital blood banks that are outside its direct control. However, it does attempt to provide advice and support to hospitals to enable them to use blood efficiently.

Blood components and products need to be stored under a variety of conditions, but will deteriorate over time. This varies depending on the component; platelets have a shelf life of only five days and demand can fluctuate significantly. This makes stock control particularly difficult. Even red blood cells that have a shelf life of 35 days may not be acceptable to hospitals if they are close to their 'use by date'. Stock accuracy is crucial. Giving a patient the wrong type of blood can be fatal.

At a local level demand can be affected significantly by accidents. One serious accident involving a cyclist used 750 units of blood, which completely exhausted the available supply (miraculously, he survived). Large-scale accidents usually generate a surge of offers from donors wishing to make immediate donations. There is also a more predictable seasonality to the donating of blood, with a low period during the summer vacation. Yet there is always an unavoidable tension between maintaining sufficient stocks to provide a very high level of supply dependability to hospitals and minimising wastage. Unless blood stocks are controlled carefully, they can easily go past the 'use by date' and be wasted. But avoiding outdated blood products is not the only inventory objective at NBS. It also measures the percentage of requests that it was able to meet in full, the percentage emergency requests delivered within two hours, the percentage of units banked to donors bled, the number of new donors enrolled, and the number of donors waiting longer than 30 minutes before they are able to donate. The traceability of donated blood is also increasingly important. Should any problems with a blood product arise, its source can be traced back to the original donor.

EXAMPLE

Mountains of grit[2]

Students of operations management from Singapore to Saudi Arabia will maybe not have a full appreciation of how important this decision is in the colder parts of the world, but, believe me, road gritting is big news every winter where snow and ice can cause huge disruption to everyday life – but not every time it snows and, more interestingly, not everywhere it snows. The local government authorities around northern Europe and America differ significantly in how well they cope with freezing weather, usually by spreading grit (actually rock salt, a mixture of salt and grit) on the roads. So how do the authorities decide how much grit to stock up in preparation for winter, and when to spread it on the roads? For example, in the UK, when snow is forecast, potential trouble spots are identified by networks of sensors embedded in the road surface to measure climatic conditions. Each sensor is connected by cable or mobile phone technology to an automatic weather station by the roadside. The siting of the sensors is important. They must be sited either on a representative stretch of road (no nearby trees, buildings or bridges, which offer some protection from the cold), or traditional cold spots. The weather stations then beam back data about air and road temperatures, wind speed and direction, and the wetness of roads. Salt levels are also measured to ensure that grit already spread has not been blown away by wind or washed away by rain. It has been known for cold weather to be forecast and the gritting trucks to be dispatched, only for the weather to change, with snow turning to rain, which washes away the grit. Then when temperatures suddenly drop again the rain freezes on the road.

Source: Paul Drabble/Corbis

Forecasting how much grit will be needed is even more difficult. Long-range weather forecasts are notoriously inaccurate, so no one knows just how bad a coming winter will be. To make matters worse, the need for road grit depends on more than just the total volume of snow. Local authorities can use the same amount of salt on one 30-cm snowfall as one 5-cm snowfall. Furthermore, the number of snowy days is important in determining how much grit will be needed. In the skiing areas of central Europe, most winter days will have snow predictably, while parts of the UK could have little or no snow one winter and many weeks of snow the next.

Supplies of road grit can also vary, as can its price. There are many reasons for this. Mainly of course, if a bad winter is forecast, all authorities in an area will want to buy the same grit, which will reduce supply and put prices up. Also salt mines can flood, especially in winter. Nor is it cheap to transport grit from one area to another – it is a low value but heavy material. As a consequence some authorities organise purchasing groups to get better prices before the season starts. Getting more salt during the season may be possible but prices are higher and supply is not guaranteed. In addition, an authority has to decide how fast to use up its inventory of grit. At the start of the winter period, authorities may be cautious about gritting because, once used, the grit cannot be used again, and who knows what the weather will be like later in the season. But in the final analysis, the decision of how large an inventory of grit to buy and how to use it is a balance between risks and consequences. Build up too big an inventory of grit and it may not all be used with the cost of carrying it over to next year being borne by local taxpayers. Build up too small an inventory and incur the wrath of local voters when the roads are difficult to negotiate. Of course a perfect weather forecast would help!

What do these two examples have in common?

Both of these organisations depend on their ability to manage inventory. In doing so, both are attempting to manage the trade-off that lies at the heart of all inventory management: balancing the costs of holding stock against the customer service that comes from having appropriate stock levels. Too high stock levels have a cost. This may be simply working capital in the case of road grit, or it could be the cost of blood becoming outdated and being wasted in the blood service. But without an appropriate level of inventory, customers suffer poor service. This means potentially disrupting local traffic flow in the case of road gritting. But a failure of supply for the blood service may have even more drastic consequences. So, for both operations at each point in the inventory system, operations managers need to manage the day-to-day tasks of running the system. Requests for grit or blood will be received from internal or external customers; they will be supplied, and demand will gradually deplete the inventory. Orders will need to be placed for replenishment of the stocks; deliveries will arrive and require storing. And at each stage of managing the inventory the competing demands of costs and service levels will need trading off. To manage this trade-off, we first need to understand the reasons for not having inventory, the reasons for having it, and then understand the tools available to help make these balancing decisions.

DIAGNOSTIC QUESTION

Why should there be any inventory?

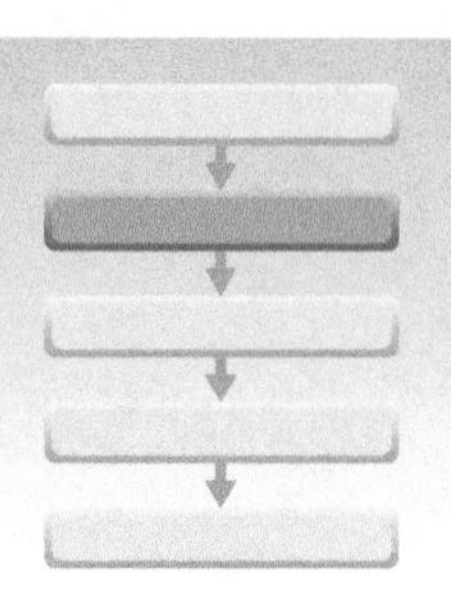

Video

There are plenty of reasons to avoid accumulating inventory where possible. Table 9.2 identifies some of these, particularly those concerned with cost, space, quality and operational/organisational issues.

So why have inventory?

On the face of it, it may seem sensible to have a smooth and even flow of materials, customers and information through operational processes and networks and thus not have any accumulations. In fact, inventories provide many advantages for both operations and their customers. If a customer has to go to a competitor because a part is out of stock or because they have had to wait too long or because the company insists on collecting all their personal details each time they call, the value of inventories seems undisputable. The task of operations

OPERATIONS PRINCIPLE
Inventory should only accumulate when the advantages of having it outweigh the disadvantages.

Table 9.2 Some reasons to avoid inventories

	'Inventories'		
	Physical inventories	***Queues of customers***	***Digital information in databases***
Cost	Ties up working capital and there could be high administrative and insurance costs	Primarily time–cost to the customer, i.e. wastes customers' time	Cost of set-up, access, update and maintenance
Space	Requires storage space	Requires areas for waiting or phone lines for held calls	Requires memory capacity, may require secure and/or special environment
Quality	May deteriorate over time, become damaged or obsolete	May upset customers if they have to wait too long, may lose customers	Data may be corrupted or lost or become obsolete
Operational/ organisational	May hide problems (see lean synchronisation, Chapter 13)	May put undue pressure on the staff and so quality is compromised for throughput	Databases need constant management, access control, updating and security

management is to allow inventory to accumulate only when its benefits outweigh its disadvantages. The following are some of the benefits of inventory.

Physical inventory is an insurance against uncertainty

Inventory can act as a buffer against unexpected fluctuations in supply and demand. For example, a retail operation can never forecast demand perfectly over the lead time. It will order goods from its suppliers such that there is always a minimum level of inventory to cover against the possibility that demand will be greater than expected during the time taken to deliver the goods. This is buffer, or safety inventory. It can also compensate for the uncertainties in the process of the supply of goods into the store. The same applies with the output inventories, which is why hospitals always have a supply of blood, sutures and bandages for immediate response to accident and emergency patients. Similarly, auto servicing services, factories and airlines may hold selected critical spare parts inventories so that maintenance staff can repair the most common faults without delay. Again, inventory is being used as an 'insurance' against unpredictable events.

Physical inventory can counteract a lack of flexibility

Where a wide range of customer options is offered, unless the operation is perfectly flexible, stock will be needed to ensure supply when it is engaged on other activities. This is sometimes called cycle inventory. For example, suppose a baker makes three types of bread. Because of the nature of the mixing and baking process, only one kind of bread can be produced at any time. The baker will have to produce each type of bread in batches large enough to satisfy the demand for each kind of bread between the times when each batch is ready for sale. So, even when demand is steady and predictable, there will always be some inventory to compensate for the intermittent supply of each type of bread.

Physical inventory allows operations to take advantage of short-term opportunities

Sometimes opportunities arise that necessitate accumulating inventory, even when there is no immediate demand for it. For example, a supplier may be offering a particularly good deal on selected items for a limited time period, perhaps because they want to reduce their own finished goods inventories. Under these circumstances a purchasing department may opportunistically take advantage of the short-term price advantage.

Physical inventory can be used to anticipate future demands

Medium-term capacity management (covered in Chapter 8) may use inventory to cope with demand–capacity. Rather than trying to make a product (such as chocolate) only when it is needed, it is produced throughout the year ahead of demand and put into inventory until it is needed. This type of inventory is called anticipation inventory and is most commonly used when demand fluctuations are large but relatively predictable.

Physical inventory can reduce overall costs

Holding relatively large inventories may bring savings that are greater than the cost of holding the inventory. This may be when bulk-buying gets the lowest possible cost of inputs, or when large order quantities reduce both the number of orders placed and the associated costs of administration and material handling. This is the basis of the 'economic order quantity' (EOQ) approach that will be treated later in this chapter.

Physical inventory can increase in value

Sometimes the items held as inventory can increase in value and so become an investment. For example, dealers in fine wines are less reluctant to hold inventory than dealers in wine that does not get better with age. (However, it can be argued that keeping fine wines until they are at their peak is really part of the overall process rather than inventory as such.) A more obvious example is inventories of money. The many financial processes within most organisations will try to maximise the inventory of cash they hold because it is earning them interest.

Physical inventory fills the processing 'pipeline'

'Pipeline' inventory exists because transformed resources cannot be moved instantaneously between the point of supply and the point of demand. When a retail store places an order, its supplier will 'allocate' the stock to the retail store in its own warehouse, pack it, load it onto its truck, transport it to its destination, and unload it into the retailer's inventory. From the time that stock is allocated (and therefore it is unavailable to any other customer) to the time it becomes available for the retail store, it is pipeline inventory. Especially in geographically dispersed supply networks, pipeline inventory can be substantial.

Queues of customers help balance capacity and demand

This is especially useful if the main service resource is expensive, for example doctors, consultants, lawyers or expensive equipment such as CAT scans. By waiting a short time after their arrival, and creating a queue of customers, the service always has customers to process. This is also helpful where arrival times are less predictable, for example where an appointment system is not used or not possible.

Queues of customers enable prioritisation

In cases where resources are fixed and customers are entering the system with different levels of priority, the formation of a queue allows the organisation to serve urgent customers while keeping other less urgent ones waiting. In the UK it is not usual to have to wait three to four hours for treatment in an accident and emergency ward, with more urgent cases 'jumping the queue' for treatment.

Queuing gives customers time to choose

Time spent in a queue gives customers time to decide what products/services they require for example, customers waiting in a fast-food restaurant have time to look at the menu so that when they get to the counter they are ready to make their order without holding up the server.

Queues enable efficient use of resources

By allowing queues to form, customers can be batched together to make efficient use of operational resources. For example, a queue for an elevator makes better use of its capacity, in an airport by calling customers to the gate, staff can load the aircraft more efficiently and quickly.

Databases provide efficient multi-level access

Databases are relatively cheap ways of storing information and providing many people with access, although there may be restrictions or different levels of access. The doctor's receptionist will be able to call up your records to check your name and address and make an appointment; the doctor will then be able to call up the appointment and the patient's records; the pharmacist will be able to call up the patient's name and prescriptions and cross-check for other prescriptions and known allergies, etc.

Databases of information allow single data capture

There is no need to capture data at every transaction with a customer or supplier, though checks may be required.

Table 9.3 Some ways in which physical inventory may be reduced

Reason for holding inventory	*Example*	*How inventory could be reduced*
As an insurance against uncertainty	Safety stocks for when demand or supply is not perfectly predictable	• Improve demand forecasting • Tighten supply, e.g. through service level penalties
To counteract a lack of flexibility	Cycle stock to maintain supply when other products are being made	• Increase flexibility of processes, e.g. by reducing changeover times (see Chapter 11) • Using parallel processes producing output simultaneously (see Chapter 5)
To take advantage of relatively short-term opportunities	Suppliers offer 'time limited' special low-cost offers	• Persuade suppliers to adopt 'everyday low prices' (see Chapter 11)
To anticipate future demands	Build up stocks in low-demand periods for use in high-demand periods	• Increase volume flexibility by moving towards a 'chase demand' plan (see Chapter 8)
To reduce overall costs	Purchasing a batch of products in order to save delivery and administration costs	• Reduce administration costs through purchasing process efficiency gains • Investigate alternative delivery channel that reduce transport costs
To fill the processing 'pipeline'	Items being delivered to customer	• Reduce process time between customer request and dispatch of items • Reduce throughput time in the downstream supply chain (see Chapter 7)

Databases of information speed up the process

Amazon, for example, stores, if you agree, your delivery address and credit card information so that purchases can be made with a single click, making it fast and easy for the customer.

Reducing physical inventory

The objective of most operations managers who manage physical inventories is to reduce the overall level (and/or cost) of inventory while maintaining an acceptable level of customer service. Table 9.3 above identifies some of the ways in which inventory may be reduced.

DIAGNOSTIC QUESTION

Is the right quantity being ordered?

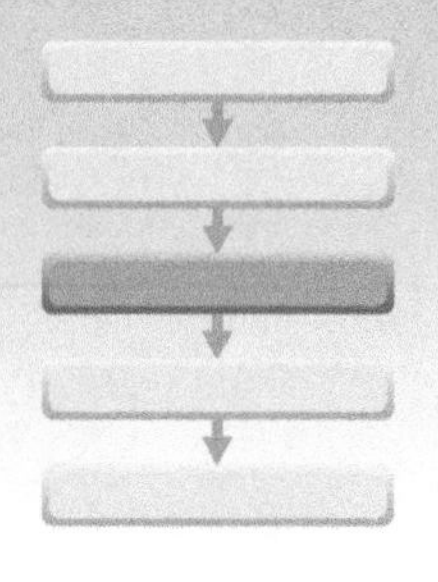

Video

To illustrate this decision, consider how we manage our domestic inventory. We implicitly make decisions on order quantity, that is, how much to purchase at one time by balancing two sets of costs: the costs associated with going out to purchase the food items and the costs associated with holding the stocks. The option of holding very little or no inventory of food and purchasing each item only when it is needed it requires little money because purchases are made only when needed, but involves buying several times a day, which is inconvenient. Conversely, making one journey to the local superstore every few months and purchasing all the provisions we would need until our next visit reduces purchasing time and costs but requires a very large amount of money each time the trip is made – money which could otherwise be in the bank and earning interest. We might also have to invest in extra cupboard units and a very large freezer. Somewhere between these extremes lies an ordering strategy that will minimise the total costs and effort involved in purchasing food.

Inventory costs

A similar range of costs apply in commercial order-quantity decisions as in the domestic situation. These are costs of placing an order, including preparing the documentation, arranging for the delivery to be made, arranging to pay the supplier for the delivery, and the general costs of keeping all the information that allows us to do this. An 'internal order' on processes within an operation has equivalent costs. Price discount costs for large orders or extra costs for small orders may also influence how much to purchase. If inventory cannot supply demand, there will be costs to us incurred by failing to supply customers. External customers may take their business elsewhere. Internal stock-outs could lead to idle time at the next process, inefficiencies and eventually, again, dissatisfied external customers. There are the working capital costs of funding the lag between paying suppliers and receiving payment from customers. Storage costs are the costs associated with physically storing goods, such as renting, heating and lighting a warehouse, as well as insuring the inventory. While stored as inventory there is a risk of obsolescence costs if the inventory is superseded (in the case of a change in fashion) or deteriorates with age (in the case of most foodstuffs).

Some of these costs will decrease as order size is increased; the first three costs (cost of placing an order, price discount costs, and stock out costs) are like this. The other costs (working capital, storage and obsolescence costs) generally increase as order size is increased, but it may not be the same organisation that incurs each cost. For example, sometimes suppliers agree to hold consignment stock. This means that they deliver large quantities of inventory to their customers to store but will only charge for the goods as and when they are used. In the meantime they remain the supplier's property so do not have to be financed by the customer, who does however provide storage facilities.

Inventory profiles

An inventory profile is a visual representation of the inventory level over time. Figure 9.3 shows a simplified inventory profile for one particular stock item in a retail operation. Every

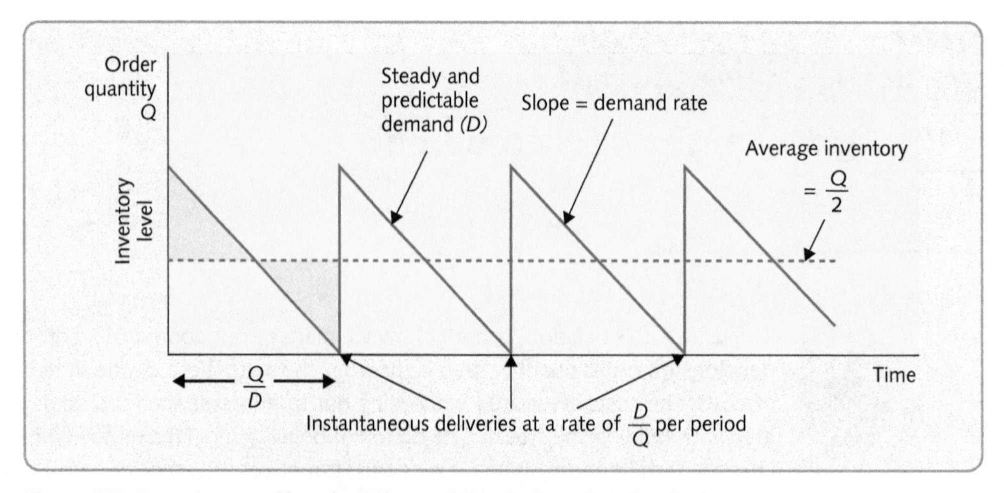

Figure 9.3 Inventory profiles chart the variation in inventory level

time an order is placed, Q items are ordered. The replenishment order arrives in one batch instantaneously. Demand for the item is then steady and perfectly predictable at a rate of D units per month. When demand has depleted the stock of the items entirely, another order of Q items instantaneously arrives, and so on. Under these circumstances:

$$\text{The average inventory} = \frac{Q}{2} \text{ (because the two shaded areas in Figure 9.3 are equal)}$$

$$\text{The time interval between deliveries} = \frac{Q}{D}$$

$$\text{The frequency of deliveries} = \text{the reciprocal of the time interval} = \frac{D}{Q}$$

The economic order quantity (EOQ) formula

The economic order quantity (EOQ) approach attempts to find the best balance between the advantages and disadvantages of holding stock. For example, Figure 9.4 shows two alternative order-quantity policies for an item. Plan A, represented by the unbroken line, involves ordering in quantities of 400 at a time. Demand in this case is running at 1,000 units per year. Plan B, represented by the dotted line, uses smaller but more frequent replenishment orders. This time only 100 are ordered at a time, with orders being placed four times as often. However, the average inventory for plan B is one-quarter of that for plan A.

To find out whether either of these plans, or some other plan, minimises the total cost of stocking the item, we need some further information, namely the total cost of holding one unit in stock for a period of time (C_h) and the total costs of placing an order (C_o).

In this case the cost of holding stocks is calculated at £1 per item per year and the cost of placing an order is calculated at £20 per order.

We can now calculate total holding costs and ordering costs for any particular ordering plan as follows:

$$\text{Holding costs} = \text{holding cost/unit} \times \text{average inventory}$$

$$= C_h \times \frac{Q}{2}$$

$$\text{Ordering costs} = \text{ordering cost} \times \text{number of orders per period}$$

$$= C_o \times \frac{D}{Q}$$

So,

$$\text{total cost, } C_t \neq \frac{C_h Q}{2} + \frac{C_o D}{Q}$$

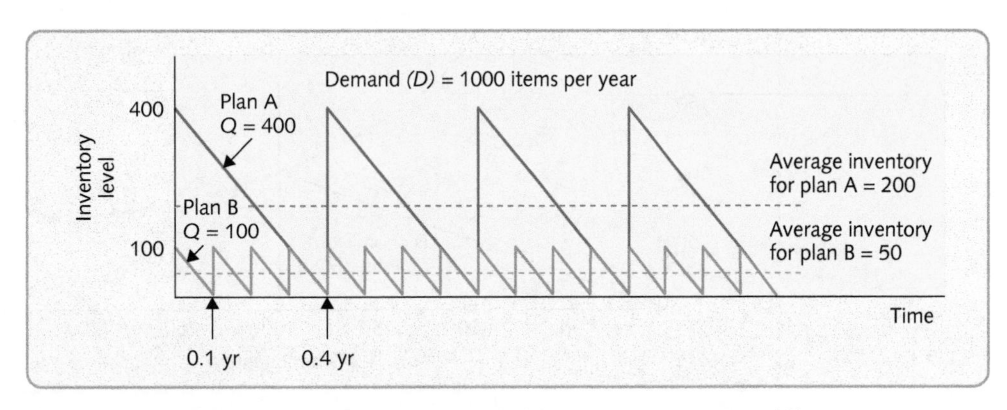

Figure 9.4 Two alternative inventory plans with different order quantities (Q)

We can now calculate the costs of adopting plans with different order quantities. These are illustrated in Table 9.4. As we would expect with low values of Q, holding costs are low but ordering costs are high, because orders have to be placed very frequently. As Q increases, the holding costs increase but the costs of placing orders decrease. In this case the order quantity, Q, which minimises the sum of holding and order costs, is 200. This 'optimum' order quantity is called the economic order quantity (EOQ). This is illustrated graphically in Figure 9.5.

Table 9.4 Costs of adoption of plans with different order quantities

Demand (D) = 1,000 units per year Holding costs (C_h) = £1 per item per year.
Order costs (C_o) = £20 per order

Order quantity (Q)	*Holding costs* $(0.5Q \times C_h)$	+	*Order costs* $((D/Q) \times C_o)$	=	*Total costs*
50	25		20 × 20 = 400		425
100	50		10 × 20 = 200		250
150	75		6.7 × 20 = 134		209
200	100		5 × 20 = 100		200*
250	125		4 × 20 = 80		205
300	150		3.3 × 20 = 66		216
350	175		2.9 × 20 = 58		233
400	200		2.5 × 20 = 50		250

*Minimum total cost.

Animated diagram

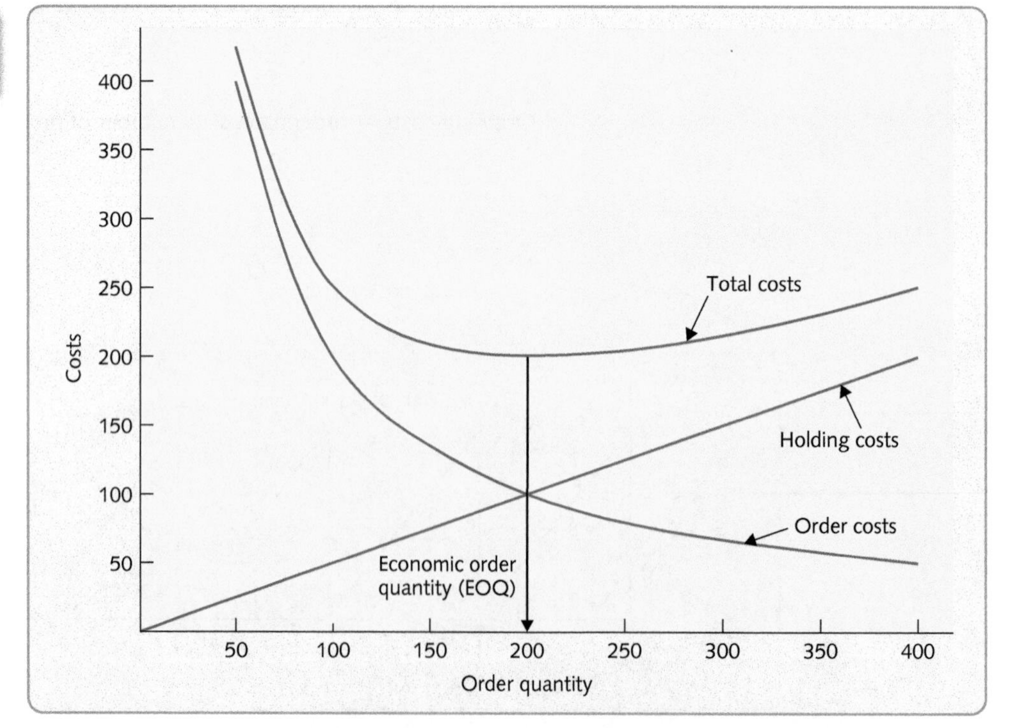

Figure 9.5 Inventory-related costs minimise at the 'economic order quantity' (EOQ)

A more elegant method of finding the EOQ is to derive its general expression. This can be done using simple differential calculus as follows. From before:

$$\text{Total cost} = \text{holding cost} + \text{order cost}$$

$$C_t = \frac{C_h Q}{2} + \frac{C_o D}{Q}$$

The rate of change of total cost is given by the first differential of C_t with respect to Q:

$$\frac{dC_t}{dQ} = \frac{C_h}{2} - \frac{C_o D}{Q^2}$$

The lowest cost will occur when $\frac{dC_t}{dQ} = 0$, that is:

$$0 = \frac{C_h}{2} - \frac{C_o D}{Q_o^2}$$

where Q_o = the EOQ. Rearranging this expression gives:

$$Q_o = \text{EOQ} = \sqrt{\frac{2C_o D}{C_h}}$$

When using the EOQ:

$$\text{Time between orders} = \frac{\text{EOQ}}{D}$$

$$\text{Order frequency} = \frac{D}{\text{EOQ}} \text{ per period}$$

Sensitivity of the EOQ

The graphical representation of the total cost curve in Figure 9.5 shows that, although there is a single value of Q which minimises total costs, any relatively small deviation from the EOQ will not increase total costs significantly. In other words, costs will be near-optimum provided a value of Q which is reasonably close to the EOQ is chosen. Put another way, small errors in estimating either holding costs or order costs will not result in a significant change in the EOQ. This is a particularly convenient phenomenon because, in practice, both holding and order costs are not easy to estimate accurately. The other implication is that, because the total cost curve is not symmetrical, it is usually better to have slightly more than slightly less inventory.

OPERATIONS PRINCIPLE

For any stock replenishment activity there is a theoretical 'optimum' order quantity that minimises total inventory-related costs.

WORKED EXAMPLE

A building materials supplier obtains its bagged cement from a single supplier. Demand is reasonably constant throughout the year, and last year the company sold 2,000 tonnes of this product. It estimates the costs of placing an order at around £25 each time an order is placed, and calculates that the annual cost of holding inventory is 20 per cent of purchase cost. The company purchases the cement at £60 per tonne. How much should the company order at a time?

$$\text{EOQ for cement} = \sqrt{\frac{2C_o D}{C_h}}$$

$$= \sqrt{\frac{2 \times 25 \times 2000}{0.2 \times 60}}$$

$$= \sqrt{\frac{100{,}000}{12}}$$

$$= 91.287 \text{ tonnes}$$

After calculating the EOQ the operations manager feels that placing an order for 91.287 tonnes exactly seems somewhat over-precise. Why not order a convenient 100 tonnes?

Total cost of ordering plan for Q = 91.287:

$$= \frac{C_h Q}{2} + \frac{C_o D}{Q}$$

$$= \frac{(0.2 \times 60) \times 91.287}{2} + \frac{25 \times 2000}{91.287}$$

$$= £1095.454$$

Total cost of ordering plan for Q = 100:

$$= \frac{(0.2 \times 60) \times 100}{2} + \frac{25 \times 2000}{100}$$

$$= £1100$$

The extra cost of ordering 100 tonnes at a time is £1100 − £1095.45 = £4.55. The operations manager therefore should feel confident in using the more convenient order quantity.

Gradual replacement – the economic batch quantity (EBQ) model

The simple inventory profile shown in Figure 9.3 assumes that each complete replacement order arrives at one point in time. However, replenishment may occur over a time period rather than in one lot, for example where an internal order is placed for a batch of parts to be produced on a machine. The machine will start to produce items and ship them in a more or less continuous stream into inventory, but at the same time demand is removing items from the inventory. Provided the rate at which items are being supplied to the inventory (P) is higher than the demand rate (D), then the inventory will increase. After the batch has been completed the machine will be reset (to produce some other part), and demand will continue to deplete the inventory level until production of the next batch begins. The resulting profile is shown in Figure 9.6. This is typical for inventories supplied by batch processes, and the minimum-cost batch quantity for this profile is called the economic batch quantity (EBQ). It is derived as follows:

$$\text{Maximum stock level} = M$$

$$\text{Slope of inventory build-up} = P - D$$

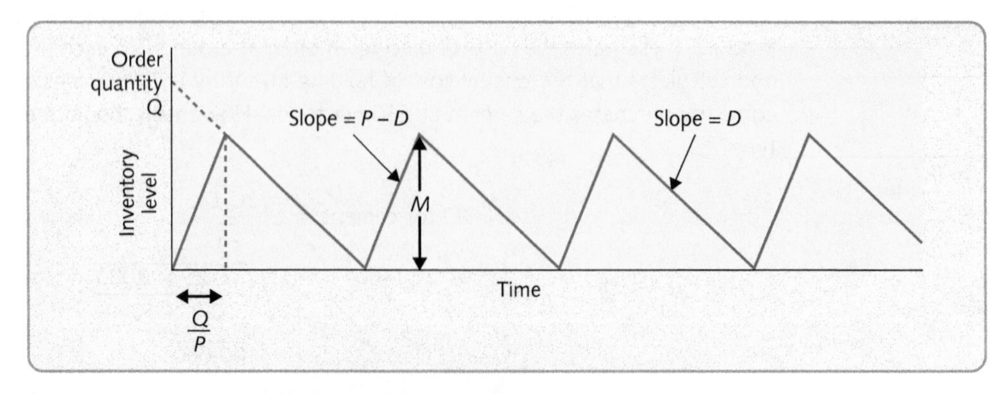

Figure 9.6 Inventory profile for gradual replacement of inventory

Also, as is clear from Figure 9.6:

$$\text{Slope of inventory build-up} = M \div \frac{Q}{P} = \frac{MP}{Q}$$

So,

$$\frac{MP}{Q} = P - D$$

$$M = \frac{Q(P - D)}{P}$$

$$\text{Average inventory level} = \frac{M}{2} = \frac{Q(P - D)}{2P}$$

As before:

$$\text{Total cost} = \text{holding cost} + \text{order cost}$$

$$C_t = \frac{C_h Q(P - D)}{2P} + \frac{C_o D}{Q}$$

$$\frac{dC_t}{dQ} = \frac{C_h(P - D)}{2P} - \frac{C_o D}{Q^2}$$

Again, equating to zero and solving Q gives the minimum-cost order quantity EBQ:

$$\text{EBQ} = \sqrt{\frac{2C_o D}{C_h(1-(D/P))}}$$

WORKED EXAMPLE

The manager of a bottle-filling plant which bottles soft drinks needs to decide how long a 'run' of each type of drink to process. Demand for each type of drink is reasonably constant at 80,000 per month (a month has 160 production hours). The bottling lines fill at a rate of 3,000 bottles per hour, but take an hour to clean and reset between different drinks. The cost (of labour and lost production capacity) of each of these changeovers has been calculated at £100 per hour. Stock-holding costs are counted at £0.1 per bottle per month.

$$D = 80{,}000 \text{ per month} = 500 \text{ per hour}$$

$$\text{EBQ} = \sqrt{\frac{2C_o D}{C_h(1-(D/P))}} = \sqrt{\frac{2 \times 100 \times 80{,}000}{0.1(1-(500/3000))}}$$

$$\text{EBQ} = 13{,}856$$

The staff who operate the lines have devised a method of reducing the changeover time from 1 hour to 30 minutes. How would that change the EBQ?

$$\text{New } C_o = £50$$

$$\text{New EBQ} = \sqrt{\frac{2 \times 50 \times 80{,}000}{0.1(1-(500/3000))}} = 9798$$

If customers won't wait – the news vendor problem

A special case of the inventory order quantity decision is when an order quantity is purchased for a specific event or time period, after which the items are unlikely to be sold. A simple example of this is the decision taken by a newspaper vendor of how many newspapers to stock for the day. If the news vendor should run out of papers, customers will either go elsewhere or decide not to buy a paper that day. Newspapers left over at the end of the day are worthless and demand for the newspapers varies day-by-day. In deciding how many newspapers to carry, the news vendor is in effect balancing the risk and consequence of running out of newspapers against that of having newspapers left over at the end of the day. Retailers and manufacturers of high-class leisure products, such as some books and popular music CDs, face the same problem. For example, a concert promoter needs to decide how many concert T-shirts to order emblazoned with the logo of the main act. The profit on each T-shirt sold at the concert is £5 and any unsold T-shirts are returned to the company that supplies them, but at a loss to the promoter of £3 per T-shirt. Demand is uncertain but is estimated to be between 200 and 1000. The probabilities of different demand are as follows:

Demand level	200	400	600	800
Probability	0.2	0.3	0.4	0.1

How many T-shirts should the promoter order? Table 9.5 shows the profit which the promoter would make for different order quantities and different levels of demand.

We can now calculate the *expected* profit which the promoter will make for each order quantity by weighting the outcomes by their probability of occurring.

If the promoter orders 200 T-shirts:

$$\begin{aligned}\text{Expected profit} &= 1000 \times 0.2 + 1000 \times 0.3 + 1000 \times 0.4 + 1000 \times 0.1 \\ &= £1000\end{aligned}$$

If the promoter orders 400 T-shirts:

$$\begin{aligned}\text{Expected profit} &= 400 \times 0.2 + 2000 \times 0.3 + 2000 \times 0.4 + 2000 \times 0.1 \\ &= £1680\end{aligned}$$

If the promoter orders 600 T-shirts:

$$\begin{aligned}\text{Expected profit} &= -200 \times 0.2 + 1400 \times 0.3 + 3000 \times 0.4 + 3000 \times 0.1 \\ &= £1880\end{aligned}$$

Table 9.5 Pay-off matrix for T-shirt order quantity (profit or loss in £s)

Demand level	*200*	*400*	*600*	*800*
Probability	0.2	0.3	0.4	0.1
Promoter orders 200	1000	1000	1000	1000
Promoter orders 400	400	2000	2000	2000
Promoter orders 600	−200	1400	3000	3000
Promoter orders 800	−800	800	2400	4000

If the promoter orders 800 T-shirts:

$$\text{Expected profit} = -800 \times 0.2 + 800 \times 0.3 + 2400 \times 0.4 + 4000 \times 0.1$$
$$= £1440$$

The order quantity which gives the maximum profit is 600 T-shirts, which results in a profit of £1880.

The importance of this approach lies in the way it takes a probabilistic view of part of the inventory calculation (demand). Something we shall use again in this chapter.

DIAGNOSTIC QUESTION

Are inventory orders being placed at the right time?

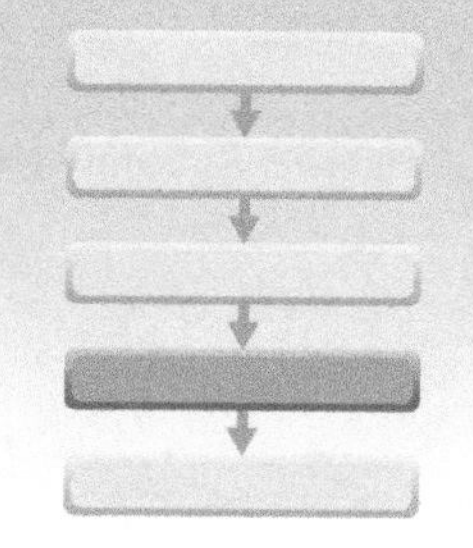

Video

When we assumed that orders arrived instantaneously and demand was steady and predictable, the decision on when to place a replenishment order was self-evident. An order would be placed as soon as the stock level reached zero, it would arrive instantaneously and prevent any stock-out occurring. When there is a lag between the order being placed and it arriving in the inventory, we can still calculate the timing of a replacement order simply, as shown in Figure 9.7. The lead time for an order to arrive is in this case two weeks, so the re-order point (ROP) is the point at which stock will fall to zero minus the order lead time. Alternatively, we can define the point in terms of the level that the inventory will have reached when a replenishment order needs to be placed. In this case this occurs at a re-order level (ROL) of 200 items.

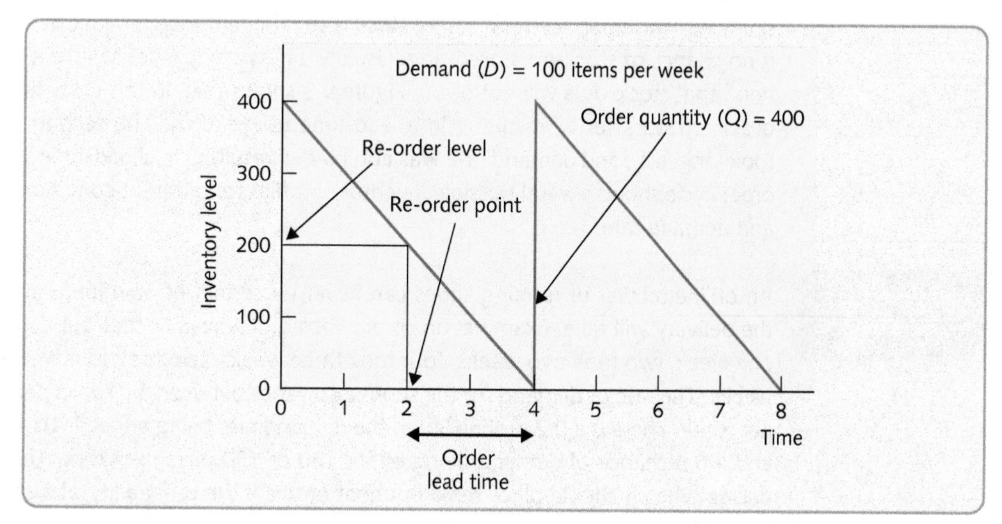

Figure 9.7 Re-order level (ROL) and re-order point (ROP) are derived from the order lead time and demand rate

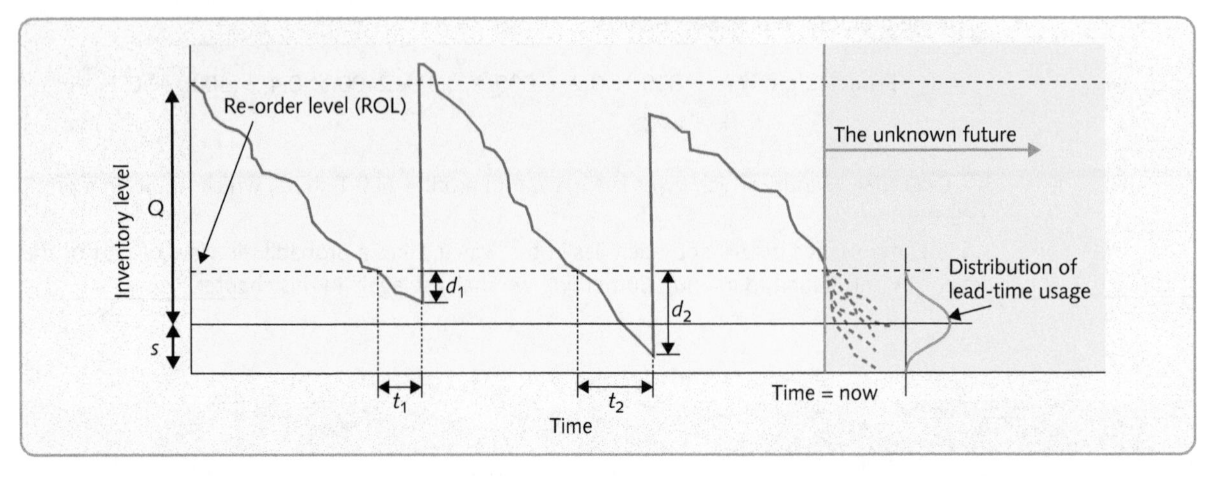

Figure 9.8 Safety stock (s) helps to avoid stock-outs when demand and/or order lead time are uncertain

However, this assumes that both the demand and the order lead time are perfectly predictable. In most cases this is not so. Both demand and the order lead time are likely to vary to produce a profile that looks something like that in Figure 9.8. In these circumstances it is necessary to make the replenishment order somewhat earlier than would be the case in a purely deterministic situation. This will result in, on average, some 'safety' stock still being in the inventory when the replenishment order arrives. The earlier the replenishment order is placed, the higher will be the expected level of safety stock when the replenishment order arrives. But because of the variability of both lead time (t) and demand rate (d), the safety stock at the time of replenishment will vary. The main consideration in setting safety stock is the probability that the stock will not have run out before the replenishment order arrives. This depends on the lead-time usage distribution. This is a combination of the distributions that describe lead-time variation and the demand rate during the lead time. If safety stock is set below the lower limit of this distribution then there will be shortages every single replenishment cycle. If safety stock is set above the upper limit of the distribution, there is no chance of stock-outs occurring. Usually, safety stock is set to give a predetermined likelihood that stock-outs will not occur. Figure 9.8 shows that, in this case, the first replenishment order arrived after t_1, resulting in a lead-time usage of d_1. The second replenishment order took longer, t_2, and demand rate was also higher, resulting in a lead-time usage of d_2. The third order cycle shows several possible inventory profiles for different conditions of lead-time usage and demand rate.

OPERATIONS PRINCIPLE

For any stock replenishment activity, the timing of replenishment should reflect the effects of uncertain lead-time and uncertain demand during that lead-time.

WORKED EXAMPLE

An online retailer of running shoes can never be certain of how long, after placing an order, the delivery will take. Examination of previous orders reveals that out of ten orders: one took one week, two took two weeks, four took three weeks, two took four weeks and one took five weeks. The rate of demand for the shoes also varies between 110 pairs per week and 140 pairs per week. There is a 0.2 probability of the demand rate being either 110 or 140 pairs per week, and a 0.3 chance of demand being either 120 or 130 pairs per week. The company needs to decide when it should place replenishment orders if the probability of a stock-out is to be less than 10 per cent.

Both lead time and the demand rate during the lead time will contribute to the lead-time usage. So the distributions that describe each will need to be combined. Figure 9.9 and Table 9.6 show how this can be done. Taking lead time to be either one, two, three, four

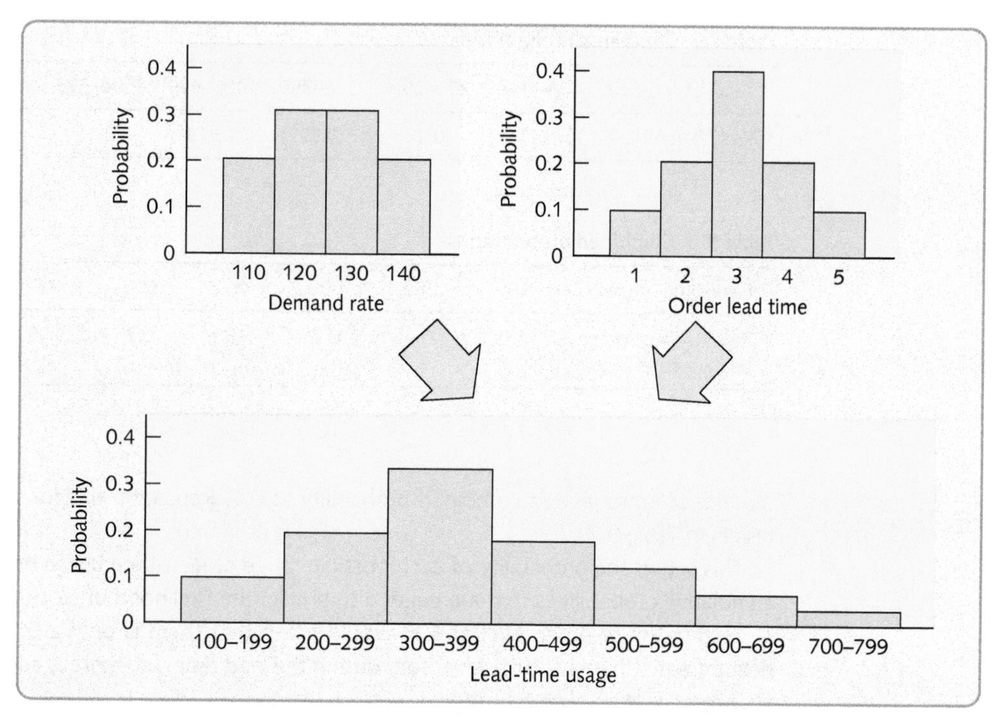

Figure 9.9 The probability distributions for order lead time and demand rate combine to give the lead-time usage distribution

Table 9.6 Matrix of lead-time and demand-rate probabilities

			Lead-time probabilities				
			1	2	3	4	5
			0.1	0.2	0.4	0.2	0.1
Demand-rate probabilities	110	0.2	110 (0.02)	220 (0.04)	330 (0.08)	440 (0.04)	550 (0.02)
	120	0.3	120 (0.03)	240 (0.06)	360 (0.12)	480 (0.06)	600 (0.03)
	130	0.3	130 (0.03)	260 (0.06)	390 (0.12)	520 (0.06)	650 (0.03)
	140	0.2	140 (0.02)	280 (0.04)	420 (0.08)	560 (0.04)	700 (0.02)

or five weeks, and demand rate to be either 110, 120, 130 or 140 pairs per week, and also assuming the two variables to be independent, the distributions can be combined as shown in Table 9.6. Each element in the matrix shows a possible lead-time usage with the probability of its occurrence. So if the lead time is one week and the demand rate is 110 pairs per week, the actual lead-time usage will be $1 \times 110 = 110$ pairs. Since there is a 0.1 chance of the lead time being one week, and a 0.2 chance of demand rate being 110 pairs per week, the probability of both these events occurring is $0.1 \times 0.2 = 0.02$.

We can now classify the possible lead-time usages into histogram form. For example, summing the probabilities of all the lead-time usages which fall within the range 100–199 (all

Table 9.7 Combined probabilities

Lead-time usage	100–199	200–299	300–399	400–499	500–599	600–699	700–799
Probability	0.1	0.2	0.32	0.18	0.12	0.06	0.02

Table 9.8 Combined probabilities

Lead-time usage (x)	100	200	300	400	500	600	700	800
Probability of usage being greater than x	1.0	0.9	0.7	0.38	0.2	0.08	0.02	0

the first column) gives a combined probability of 0.1. Repeating this for subsequent intervals results in Table 9.7.

This shows the probability of each possible range of lead-time usage occurring, but it is the cumulative probabilities that are needed to predict the likelihood of stock-out (see Table 9.8).

Setting the re-order level at 600 would mean that there is only a 0.08 chance of usage being greater than available inventory during the lead time, i.e. there is a less than 10 per cent chance of a stock-out occurring.

Continuous and periodic review

The approach we have described is often called the continuous review approach. To make the decision in this way the stock level of each item must be reviewed continuously and an order placed when the stock level reaches its re-order level. The virtue of this approach is that, although the timing of orders may be irregular (depending on the variation in demand rate), the order size (Q) is constant and can be set at the optimum economic order quantity. But continually checking inventory levels may be time-consuming. An alternative, and simpler, approach, but one which sacrifices the use of a fixed (and therefore possibly optimum) order quantity, is 'periodic review'. Here, rather than ordering at a predetermined re-order level, the periodic approach orders at a fixed and regular time interval. So the stock level of an item could be found, for example, at the end of every month and a replenishment order placed to bring the stock up to a predetermined level. This level is calculated to cover demand between the replenishment order being placed and the following replenishment order arriving. Plus safety stocks will need to be calculated, in a similar manner to before, based on the distribution of usage over this period.

Two-bin and three-bin systems

Keeping track of inventory levels is especially important in continuous review approaches to re-ordering. A simple and obvious method of indicating when the re-order point has been reached is necessary, especially if there are a large number of items to be monitored. The simple two-bin system involves storing the re-order point quantity plus the safety inventory quantity in the second bin and using parts from the first bin. When the first bin empties, it is the signal to order the next re-order quantity. Different 'bins' are not always necessary to operate this type of system. For example, a common practice in retail operations is to store the second 'bin' quantity upside-down behind or under the first 'bin' quantity. Orders are then placed when the upside-down items are reached.

DIAGNOSTIC QUESTION

Is inventory being controlled effectively?

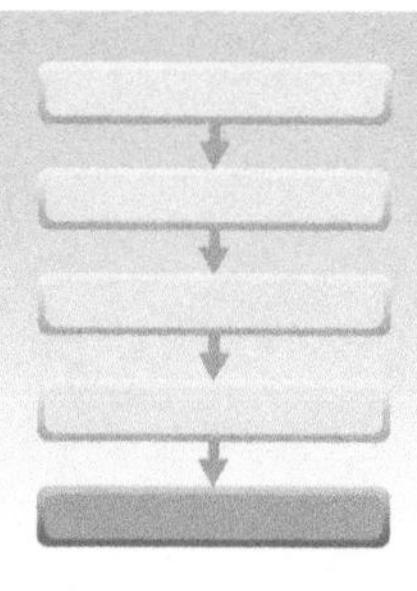

Video

Even probabilistic models are still simplified compared with the complexity of real stock management. Coping with many thousands of stocked items, supplied by many hundreds of different suppliers, with possibly tens of thousands of individual customers, makes for a complex and dynamic operations task. Controlling such complexity requires an approach that discriminates between different items so that each has a degree of control that is appropriate to its importance. It also requires an information system to keep track of inventories.

Inventory priorities – the ABC system

Some stocked items are more important than others. Some might have a high usage rate, so if they ran out many customers would be disappointed. Others might be of particularly high value, so excessively high inventory levels would be particularly expensive. One common way of discriminating between different stock items is to rank them by their usage value (usage rate multiplied by value). Items with a particularly high usage value are deemed to warrant the most careful control, whereas those with low usage values need not be controlled quite so rigorously. Generally, a relatively small proportion of the total range of items contained in an inventory will account for a large proportion of the total usage value. This phenomenon is known as the Pareto, or 80/20 rule. It is called this because, typically, 80 per cent of an operation's sales are accounted for by only 20 per cent of all stocked item types. (This idea is also used elsewhere in operations management, see for example, Chapter 13). Here the relationship is used to classify items into A, B or C categories, depending on their usage value.

OPERATIONS PRINCIPLE

Different inventory management decision rules are needed for different classes of inventory.

- Class A items are those 20 per cent or so of high usage value items that account for around 80 per cent of the total usage value.
- Class B items are those of medium usage value, usually the next 30 per cent of items which often account for around 10 per cent of the total usage value.
- Class C items are those low usage value items which, although comprising around 50 per cent of the total types of items stocked, probably only account for around 10 per cent of the total usage value of the operation.

Although annual usage and value are the two criteria most commonly used to determine a stock classification system, other criteria might also contribute towards the (higher) classification of an item. The consequence of stock-out might give higher priority to some items that would seriously delay or disrupt operations, if they were not in stock. Uncertainty of supply may also give some items priority, as might high obsolescence or deterioration risk.

WORKED EXAMPLE

Table 9.9 shows all the parts stored by an electrical wholesaler. The 20 different items stored vary in terms of both their usage per year and cost per item as shown. However, the wholesaler has ranked the stock items by their usage value per year. The total usage value per year is £5,569,000. From this it is possible to calculate the usage value per year of each item as a percentage of the total usage value, and from that a running cumulative total of the usage value as shown. The wholesaler can then plot the cumulative percentage of all stocked items against

Table 9.9 Warehouse items ranked by usage value

Stock no.	*Usage (items/year)*	*Cost (£/item)*	*Usage value (£000/year)*	*% of total value*	*Cumulative % of total value*
A/703	700	20.00	14 000	25.41	25.14
D/012	450	2.75	1238	22.23	47.37
A/135	1000	0.90	900	16.16	63.53
C/732	95	8.50	808	14.51	78.04
C/735	520	0.54	281	5.05	83.09
A/500	73	2.30	168	3.02	86.11
D/111	520	0.22	114	2.05	88.16
D/231	170	0.65	111	1.99	90.15
E/781	250	0.34	85	1.53	91.68
A/138	250	0.30	75	1.34	93.02
D/175	400	0.14	56	1.01	94.03
E/001	80	0.63	50	0.89	94.92
C/150	230	0.21	48	0.86	95.78
F/030	400	0.12	48	0.86	96.64
D/703	500	0.09	45	0.81	97.45
D/535	50	0.88	44	0.79	98.24
C/541	70	0.57	40	0.71	98.95
A/260	50	0.64	32	0.57	99.52
B/141	50	0.32	16	0.28	99.80
D/021	20	0.50	10	0.20	100.00
Total			5 569	100.00	

the cumulative percentage of their value. So, for example, the part with stock number A/703 is the highest value part and accounts for 25.14 per cent of the total inventory value. As a part, however, it is only one-twentieth, or 5 per cent, of the total number of items stocked. This item together with the next highest value item (D/012) account for only 10 per cent of the total number of items stocked, yet account for 47.37 per cent of the value of the stock, and so on.

This is shown graphically in Figure 9.10. The first four part numbers (20 per cent of the range) are considered Class A whose usage will be monitored very closely. The six next part numbers (30 per cent of the range) are to be treated as Class B items with slightly less effort devoted to their control. All other items are classed as Class C items whose stocking policy is reviewed only occasionally.

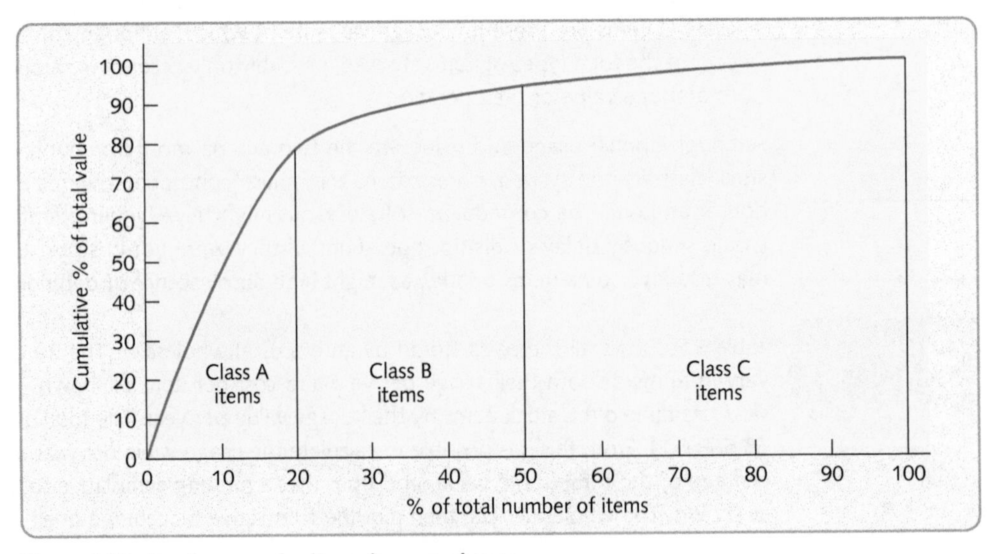

Figure 9.10 Pareto curve for items in a warehouse

Inventory information systems

Most inventories of any significant size are managed by information systems. This is especially so since data capture has been made more convenient through the use of bar-code readers, radio frequency identification (RFID) and the point-of-sale recording of sales transactions. Many commercial systems of stock control are available, although they tend to share certain common functions.

- **Updating stock records.** Every time an inventory transaction takes place the position, status and possibly value of the stock will have changed. This information must be recorded so that operations managers can determine their current inventory status at any time.
- **Generating orders.** Both the 'how much' and the 'when' to order decisions can be made by a stock control system. Originally almost all computer systems calculated order quantities by using the EOQ formulae. Now more sophisticated probabilistic algorithms are used, based on examining the marginal return on investing in stock. The system will hold all the information which goes into the ordering algorithm but might periodically check to see if demand or order lead times, or any of the other parameters, have changed significantly and recalculate accordingly. The decision on when to order, on the other hand, is a far more routine affair which computer systems make according to whatever decision rules operations managers have chosen to adopt: either continuous review or periodic review.
- **Generating inventory reports.** Inventory control systems can generate regular reports of stock value which can help management monitor its inventory control performance. Similarly, customer service performance, such as the number of stock-outs or the number of incomplete orders, can be regularly monitored. Some reports may be generated on an exception basis. That is, the report is only generated if some performance measure deviates from acceptable limits.
- **Forecasting.** Inventory replenishment decisions should ideally be made with a clear understanding of forecast future demand. Inventory control systems usually compare actual demand against forecast and adjust forecasts in the light of actual levels of demand.

Common problems with inventory systems

Our description of inventory systems has been based on the assumption that operations:

- have a reasonably accurate idea of costs such as holding cost, or order cost
- have accurate information that really does indicate the actual level of stock and sales.

In fact, data inaccuracy often poses one of the most significant problems for inventory managers. This is because most computer-based inventory management systems are based on what is called the perpetual inventory principle. This is the simple idea that stock records are (or should be) automatically updated every time that items are recorded as having been received into an inventory or taken out of the inventory. So,

$$\text{opening stock level} + \text{receipts in} - \text{dispatches out} = \text{new stock level}$$

Any errors in recording these transactions, and/or in handling the physical inventory, can lead to discrepancies between the recorded and actual inventory, and these errors are perpetuated until physical stock checks are made (usually quite infrequently). In practice there are many opportunities for errors to occur, if only because inventory transactions are numerous. This means that it is surprisingly common for the majority of inventory records to be in inaccurate. The underlying causes of errors include:

OPERATIONS PRINCIPLE

The maintenance of data accuracy is vital for the day-to-day effectiveness of inventory management systems.

- keying errors: entering the wrong product code
- quantity errors: a mis-count of items put into or taken from stock
- damaged or deteriorated inventory: not recorded as such, or not correctly deleted from the records when it is destroyed

- the wrong items being taken out of stock, but the records not being corrected when they are returned to stock
- delays between the transactions being made and the records being updated
- items stolen from inventory (common in retail environments, but also not unusual in industrial and commercial inventories).

Critical commentary

Each chapter contains a short critical commentary on the main ideas covered in the chapter. Its purpose is not to undermine the issues discussed in the chapter, but to emphasise that, although we present a relatively orthodox view of operation, there are other perspectives.

- The approach to determining order quantity that involves optimising the costs of holding stock against the costs of ordering stock, typified by the EOQ and EBQ models, has always been subject to criticisms. Originally these concerned the validity of some of the assumptions of the model; more recently they have involved the underlying rationale of the approach itself. Criticisms include: that the assumptions included in the EOQ models are simplistic; that the real costs of stock in operations are not as assumed in EOQ models; that cost minimisation is not an appropriate objective for inventory management.

- The last criticism is particularly significant. Many organisations (such as supermarkets and wholesalers) make the most of their revenue and profits simply by holding and supplying inventory. Because their main investments are in the inventory, it is critical that they make a good return on this capital by ensuring that it has the highest possible 'stock turn' and/or gross profit margin. Alternatively, they may also be concerned to maximise the use of space by seeking to maximise the profit earned per square metre. The EOQ model does not address these objectives. Similarly for products that deteriorate or go out of fashion, the EOQ model can result in excess inventory of slower-moving items. In fact the EOQ model is rarely used in such organisations, and there is more likely to be a system of periodic review for regular ordering of replenishment inventory. For example, a typical builder's supply merchant might carry around 50,000 different items of stock (SKUs). However, most of these cluster into larger families of items such as paints, sanitary ware or metal fixings. Single orders are placed at regular intervals for all the required replenishments in the supplier's range, and these are then delivered together at one time. If deliveries are made weekly then, on average, the individual item order quantities will be for only one week's usage. Less popular items, or ones with erratic demand patterns, can be individually ordered at the same time, or (when urgent) can be delivered the next day by carrier.

- The ABC approach to inventory classification is also regarded by some as misleading. Many professional inventory managers point out that it is the slow moving (C category) items that often pose the greatest challenge in inventory management. Often these slow moving items, although only accounting for 20 per cent of sales, require a large part (typically between one half and two thirds) of the total investment in stock. This is why slow-moving items are a real problem. Moreover, if errors in forecasting or ordering result in excess stock in 'A class' fast-moving items, it is relatively unimportant in the sense that excess stock can be sold quickly. However, excess stock in slow-moving C items will be there a long time. According to some inventory managers, it is the A items that can be left to look after themselves; it is the B and even more the C items that need controlling.

SUMMARY CHECKLIST

This checklist comprises questions that can be usefully applied to any type of operations and reflect the major diagnostic questions used within the chapter.

- ☐ Have all inventories been itemised and costed?
- ☐ Have all the costs and negative effects of inventory been assessed?
- ☐ What proportion of inventory is there:
  - as an insurance against uncertainty?
  - to counteract a lack of flexibility?
  - to allow operations to take advantage of short-term opportunities?
  - to anticipate future demand?
  - to reduce overall costs?
  - because it can increase in value?
  - because it is in the processing pipeline?
- ☐ Have methods of reducing inventory in these categories been explored?
- ☐ Have cost minimisation methods been used to determine order quantity?
- ☐ Do these use a probabilistic estimate of demand?
- ☐ Have the relative merits of continuous and period inventory review been assessed?
- ☐ Are probabilistic estimates of demand and lead time used to determine safety stock levels?
- ☐ Are items controlled by their usage value?
- ☐ Does the inventory information system integrate all inventory decisions?

CASE STUDY

supplies4medics.com

Founded at the height of the 'dot.com bubble' of the late 1990s, supplies4medics.com has become one of Europe's most successful direct mail suppliers of medical hardware and consumables to hospitals, doctors' and dentists' surgeries, clinics, nursing homes and other medical-related organisations. Its physical and online catalogues list just over 4,000 items, categorised by broad applications such as 'hygiene consumables' and 'surgeons' instruments'. Quoting their website:

'We are the pan-European distributors of wholesale medical and safety supplies . . . We aim to carry everything you might ever need: from nurses' scrubs to medical kits, consumables for operations, first aid kits, safety products, chemicals, fire-fighting equipment, nurse and physicians' supplies, etc. Everything is at affordable prices – and backed by our very superior customer service and support – supplies4medics is your ideal source for all medical supplies. Orders are normally despatched same-day, via our European distribution partner, the Brussels Hub of DHL. You should therefore receive your complete order within one week, but you can request next day delivery if required, for a small extra charge. You can order our printed catalogue on the link at the bottom of this page, or shop on our easy-to-use on-line store.'

Source: Courtesy of Exel plc

Last year turnover grew by over 25 per cent to about €120 million, a cause for considerable satisfaction in the company. However, profit growth was less spectacular, and market research suggested that customer satisfaction, although generally good, was slowly declining. Most worrying, inventory levels had grown faster than sales revenue, in percentage terms. This was putting a strain on cash flow, requiring the company to borrow more cash to fund the rapid growth planned for the next year. Inventory holding is estimated to be costing around 15 per cent per annum, taking account of the cost of borrowing, insurance, and all warehousing overheads.

Pierre Lamouche, the Head of Operations summarised the situation faced by his department:

'As a matter of urgency, we are reviewing our purchasing and inventory management systems! Most of our existing re-order levels (ROL) and re-order quantities (ROQ) were set several years ago, and have never been recalculated. Our focus has been on rapid growth through the introduction of new product lines. For more recently introduced items, the ROQs were based only on forecast sales, which actually can be quite misleading. We estimate that it costs us, on average, €50 to place and administer every purchase order. In the meantime, sales of some products have grown fast, while others have declined. Our average inventory (stock) cover is about ten weeks, but . . . amazingly . . . we still run out of critical items! In fact, on average, we are currently out of stock of about 500 SKUs (Stock Keeping Units) at any time. As you can imagine, our service level is not always satisfactory with this situation. We really need help to conduct a review of our system, so have employed a mature intern from the local business school to review our system. He has first asked my team to provide information on a random, representative sample of 20 items from the full catalogue range, which is copied below.'

Table 9.10 Representative sample of 20 catalogue items

Sample number	*Catalogue reference number**	*Sales unit description***	*Sales unit cost (Euro)*	*Last 12 months' sales (units)*	*Inventory as at last year end (units)*	*Re-order quantity (units)*
1	11036	Disposable aprons (10pk)	2.40	100	0	10
2	11456	Ear-loop masks (Box)	3.60	6000	120	1000
3	11563	Drill type 164	1.10	220	420	250
4	12054	Incontinence pads large	3.50	35400	8500	10000
5	12372	150ml syringe	11.30	430	120	100
6	12774	Rectal speculum 3-prong	17.40	65	20	20
7	12979	Pocket organiser blue	7.00	120	160	500
8	13063	Oxygen trauma kit	187.00	40	2	10
9	13236	Zinc oxide tape	1.50	1260	0	50
10	13454	Dual head stethoscope	6.25	10	16	25
11	13597	Disp. latex catheter	0.60	3560	12	20
12	13999	Roll-up wheelchair ramp	152.50	12	44	50
13	14068	WashClene tube	1.40	22500	10500	8000
14	14242	Cervical collar	12.00	140	24	20
15	14310	Head wedge	89.00	44	2	10
16	14405	Three-wheel scooter	755.00	14	5	5
17	14456	Neonatal trach. tube	80.40	268	6	100
18	14675	Mouldable strip paste	10.20	1250	172	100
19	14854	Sequential comp. pump	430.00	430	40	50
20	24943	Toilet safety frame	25.60	560	18	20

*Reference numbers are allocated sequentially as new items are added to catalogue.
**All quantities are in sales units (e.g. item, box, case, pack).

QUESTIONS

1 Prepare a spreadsheet-based ABC analysis of usage value. Classify as follows:

A Items: top 20 per cent of usage value
B Items: next 30 per cent of usage value
C Items: remaining 50 per cent of usage value.

2 Calculate the inventory weeks for each item, for each classification and for all the items in total. Does this suggest that the Operations Manager's estimate of inventory weeks is correct? If so, what is your estimate of the overall inventory at the end of the base year, and how much might that have increased during the year?

3 Based on the sample, analyse the underlying causes of the availability problem described in the text.

4 Calculate the EOQs for the A Items.

5 What recommendations would you give to the company?

APPLYING THE PRINCIPLES

Hints

Some of these exercises can be answered by reading the chapter. Others will require some general knowledge of business activity and some might require an element of investigation. Hints on how they can all be answered are to be found in the eText at **www.pearsoned.co.uk/slack.**

1 Read the example of the National Blood Service at the beginning of the chapter.

(a) What are the factors that constitute inventory holding costs, order costs and stock-out costs in a National Blood Service?

(b) What makes this particular inventory planning and control example so complex?

(c) How might the National Blood Service inventory management affect its ability to collect blood?

2 Estimate the annual usage value and average inventory level (or value) and space occupied by 20 representative items of food used within your household, or that of your family. Using Pareto analysis, categorise this into usage value groups (e.g. A,B,C), and calculate the average stock turn for each group. Does this analysis indicate a sensible use of capital and space, and if not, what changes might you make to the household's shopping strategy?

3 Obtain the last few years' Annual Report and Accounts (you can usually download these from the company's website) for two materials-processing operations (as opposed to customer- or information-processing operations) within one industrial sector. Calculate each operation's stock–turnover ratio and the proportion of inventory to current assets over the last few years. Try to explain what you think are the reasons for any differences and trends you can identify and discuss the likely advantages and disadvantages for the organisations concerned.

4 Visit a large petrol (gas) filling station to meet the manager. Discuss and analyse the inventory planning and control system used for fuel and other items in the shop, such as confectionery and lubricants. You should then obtain data to show how the system is working, (for example, re-order points and quantities, use of forecasts to predict patterns of demand) and if possible, prepare graphs showing fluctuations in inventory levels for the selected products.

5 Using product information obtained from web searches, compare three inventory management systems (or software packages) that could be purchased by the General Manager of a large state-run hospital who wishes to gain control of inventory throughout the organisation. What are the claimed benefits of each system, and how do they align to the theories presented in this chapter? What disadvantages might be experienced in using these approaches to inventory management. What resistance might be presented by the hospital's staff, and why?

Notes on chapter

1 Source: NBS website and discussion with association staff.

2 Sources include: 'How do they know when to grit roads?' (2010), BBC magazine website 18 December; Grimm, E. 'Prepare for winter' (2009), *Daily Chronicle*, 30 November.

TAKING IT FURTHER

Emmett, S. and Granville, D. (2007) *Excellence in Inventory Management: How to Minimise Costs and Maximise Service*, Cambridge Academic. *Practical and thorough examination of professional inventory management.*

Flores, B.E. and Whybark, D.C. (1987) 'Implementing multiple criteria ABC analysis', *Journal of Operations Management*, Vol 7, No 1. *An academic paper but one that gives some useful hints on the practicalities of ABC analysis.*

Mather, H. (1984) *How to Really Manage Inventories*, McGraw-Hill. *A practical guide by one of the more influential production management authors.*

Wild, T. (2002) *Best Practice in Inventory Management*, Butterworth-Heinemann. *A straightforward and readable practice-based approach to the subject.*

USEFUL WEBSITES

www.opsman.org *Definitions, links and opinions on operations and process management.*

http://www.inventoryops.com/dictionary.htm *A great source for information on Inventory Management and Warehouse Operations.*

http://www.mapnp.org/libary/ops_mgnt/ops_mgnt.htm *General 'private' site on operations management, but with some good content.*

http://www.apics.org *Site of APICS: a US 'educational society for resource managers'.*

http://www.inventorymanagement.com *Site of the Centre for Inventory management. Cases and links.*

Interactive quiz

For further resources including examples, animated diagrams, self-test questions, Excel spreadsheets and video materials please explore the eText on the companion website at www.pearsoned.co.uk/slack.

10

Resource planning and control

Introduction

If materials or information or customers are to flow smoothly through processes, operations and supply networks, the value-adding resources at each stage must be managed to avoid unnecessary delay, but also use operations resources efficiently. This is the activity of resource planning and control. It is a subject with many technical issues. We cover the best known of these materials requirements planning (MRP) in the supplement to this chapter (see Figure 10.1).

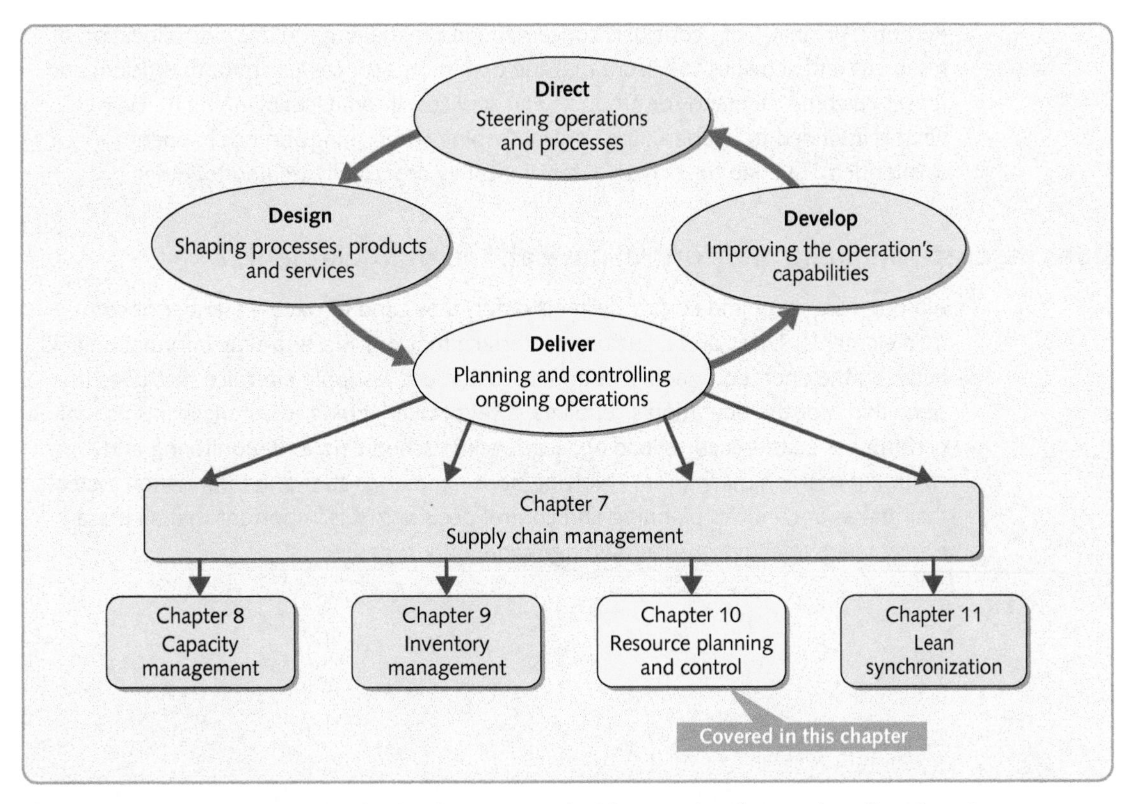

Figure 10.1 Resource planning and control is concerned with managing the ongoing allocation of resources and activities to ensure that the operation's processes are both efficient and reflect customer demand for products and services

EXECUTIVE SUMMARY

Decision logic chain for resource planning and control

What is resource planning and control?
↓
Does resource planning and control have all the right elements?
↓
Is resource planning and control information integrated?
↓
Are core planning and control activities effective?

Each chapter is structured around a set of diagnostic questions. These questions suggest what you should ask in order to gain an understanding of the important issues of a topic, and, as a result, improve your decision making. An executive summary addressing these questions is provided below.

What is resource planning and control?

Resource planning and control is concerned with managing the ongoing allocation of resources and activities to ensure that the operation's processes are both efficient and reflect customer demand for products and services. In practice, planning (deciding what is intended to happen) and control (coping when things do not happen as intended), overlap to such an extent that they are usually treated together.

Does resource planning and control have all the right elements?

Although planning and control systems differ, they tend to have a number of common elements. These are: a customer interface that forms a two-way information link between the operation's activities and its customers; a supply interface that does the same thing for the operation's suppliers; a set of overlapping 'core' mechanisms that perform basic tasks such as loading, sequencing, scheduling and monitoring and control; a decision mechanism involving both operations staff and information systems that makes or confirms planning and control decisions. It is important that all these elements are effective in their own right and work together.

Is resource planning and control information integrated?

Resource planning and control involves vast amounts of information. Unless all relevant information is integrated it is difficult to make informed planning and control decisions. The most common method of doing this is through the use of integrated 'enterprise resource planning' (ERP) systems. These are information systems that have grown out of the more specialised and detailed material requirements planning (MRP) systems that have been common in the manufacturing sector for many years. MRP is treated in the supplement to this chapter. Investment in ERP systems often involves large amounts of capital and staff time. It also may mean a significant overhaul of the way the business organises itself. Not all investments in ERP have proved successful.

Are core planning and control activities effective?

Unless the resource planning and control system makes appropriate decisions at a detailed level, it cannot be effective. These detailed decisions fall into four overlapping categories. Loading is the activity of allocating work to individual processes or stages in the operation. Sequencing is the activity of deciding the order or priority in which a number of jobs will be processed. Scheduling is the activity of producing a detailed timetable showing when activities should start and end. Monitoring and control is the activity of detecting any deviation from what has been planned and acting to cope and re-plan as necessary. The theory of constraints (TOC) is a useful concept in resource planning and control that emphasises the role of bottleneck stages or processes in planning and control.

DIAGNOSTIC QUESTION

What is resource planning and control?

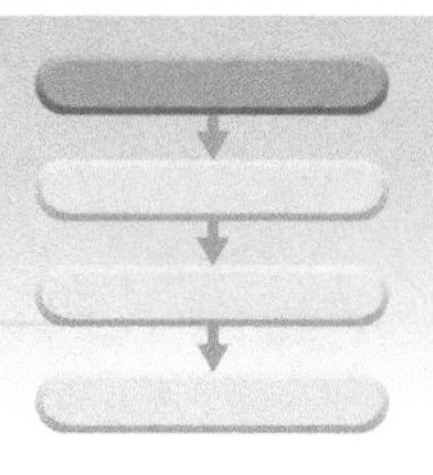

Video

Resource planning and control is concerned with managing the ongoing allocation of resources and activities to ensure that the operation's processes are both efficient and reflect customer demand for products and services. Planning and control activities are distinct but often overlap. Formally, planning determines what is *intended* to happen at some time in the future, while control is the process of *coping* when things do not happen as intended. Control makes the adjustments which help the operation to achieve the objectives that the plan has set, even when the assumptions on which the plan was based do not hold true.

Look at the resource planning and control activities in the following two organisations. One, Air France, is a very large and very complex network of operations and processes. The other, the service section of a BMW automotive dealership, is far smaller. However, although the challenges are different, the task of planning and controlling each operation's resources is surprisingly similar.

EXAMPLE

Operations control at Air France[1]

'In many ways a major airline can be viewed as one large planning problem which is usually approached as many independent, smaller (but still difficult) planning problems. The list of things which need planning seems endless: crews, reservation agents, luggage, flights, through trips, maintenance, gates, inventory, equipment purchases. Each planning problem has its own considerations, its own complexities, its own set of time horizons, its own objectives, but all are interrelated.'

Source: Courtesy of MediaLibrary/Air France

Air France has 80 flight planners working 24-hour shifts in their flight planning office at Roissy, Charles de Gaulle. Their job is to establish the optimum flight routes, anticipate any problems such as weather changes and minimise fuel consumption. Overall the goals of the flight planning activity are first, and most important, safety followed by economy and passenger comfort. Increasingly powerful computer programs process the mountain of data necessary to plan the flights, but in the end many decisions still rely on human judgement. Even the most sophisticated expert systems only serve as support for the flight planners. Planning Air France's schedule is a massive job that includes the following:

- *Frequency*. For each airport how many separate services should the airline provide?
- *Fleet assignment*. Which type of plane should be used on each leg of a flight?
- *Banks*. At any airline hub where passengers arrive and may transfer to other flights to continue their journey, airlines like to organise flights into 'banks' of several planes which arrive close together, pause to let passengers change planes, and all depart close together.
- *Block times*. A block time is the elapsed time between a plane leaving the departure gate at an airport and arriving at its gate in the arrival airport. The longer the allowed block time the more likely a plane will keep to schedule even if it suffers minor delays, but the fewer flights can be scheduled.
- *Planned maintenance*. Any schedule must allow time for planes to have time at a maintenance base.

- *Crew planning*. Pilot and cabin crew must be scheduled to allocate pilots to fly planes on which they are licensed and to keep within the maximum 'on duty' allowances.
- *Gate plotting*. If many planes are on the ground at the same time there may be problems in loading and unloading them simultaneously.
- *Recovery*. Many things can cause deviations from any plan in the airline industry. Allowances must be built in that allow for recovery.

For flights within and between Air France's 12 geographic zones, the planners construct a flight plan that will form the basis of the actual flight only a few hours later. All planning documents need to be ready for the flight crew who arrive two hours before the scheduled departure time. Being responsible for passenger safety and comfort, the captain always has the final say and, when satisfied, co-signs the flight plan together with the planning officer.

EXAMPLE

Joanne manages the schedule[2]

Joanne Cheung is the Senior Service Adviser at a premier BMW dealership. She and her team act as the interface between customers who want their cars serviced and repaired, and the 16 technicians who carry out the work in their state-of-the-art workshop. *'There are three types of work that we have to organise,'* says Joanne. *'The first is performing repairs on customers' vehicles. They usually want this doing as soon as possible. The second type of job is routine servicing. It is usually not urgent, so customers are generally willing to negotiate a time for this. The remainder of our work involves working on the pre-owned cars which our buyer has bought in to sell on to customers. Before any of these cars can be sold they have to undergo extensive checks. To some extent we treat these categories of work slightly differently. We have to give good service to our internal car buyers, but there is some flexibility in planning these jobs. At the other extreme, emergency repair work for customers has to be fitted into our schedule as quickly as possible. If someone is desperate to have their car repaired at very short notice, we sometimes ask them to drop their car in as early as they can and pick it up as late as possible. This gives us the maximum amount of time to fit it into the schedule.'*

Source: Nigel Slack

'There are a number of service options open to customers. We can book short jobs in for a fixed time and do it while they wait. Most commonly, we ask the customer to leave the car with us and collect it later. To help customers we have ten loan cars which are booked out on a first-come, first-served basis. Alternatively, the vehicle can be collected from the customer's home and delivered back there when it is ready. Our four drivers who do this are able to cope with up to 12 jobs a day.

'Most days we deal with 50 to 80 jobs, taking from half an hour up to a whole day. To enter a job into our process all Service Advisers have access to the computer-based scheduling system. On-screen it shows the total capacity we have day-by-day, all the jobs that are booked in, the amount of free capacity still available, the number of loan cars available, and so on. We use this to see when we have the capacity to book a customer in, and then enter all the customer's details. BMW have issued 'standard times' for all the major jobs. However, you have to modify these standard times a bit to take account of circumstances. That is where the Service Adviser's experience comes in.'

'We keep all the most commonly used parts in stock, but if a repair needs a part which is not in stock, we can usually get it from the BMW parts distributors within a day. Every evening our planning system prints out the jobs to be done the next day and the parts which are likely to be needed for each job. This allows the parts staff to pick out the parts for each job so that the technicians can collect them first thing the next morning without any delay.'

'Every day we have to cope with the unexpected. A technician may find that extra work is needed, customers may want extra work doing, and technicians are sometimes ill, which reduces our capacity. Occasionally parts may not be available so we have to arrange with the customer for the vehicle to be rebooked for a later time. Every day up to four or five customers just don't turn up. Usually they have just forgotten to bring their car in so we have to rebook them in at a later time. We can cope with most of these uncertainties because our technicians are flexible in terms of the skills they have and also are willing to work overtime when needed. Also, it is important to manage customers' expectations. If there is a chance that the vehicle may not be ready for them, it shouldn't come as a surprise when they try and collect it.'

What do these two examples have in common?

The system set up by both Air France and the BMW dealership has a number of common elements. First, there is some kind of acknowledgement that there should be an effective *customer interface* that translates the needs of customers into their implications for the operation. This involves setting the timetable of flights (frequency, timing etc.) and the interfaces between flights (banks) in Air France. On a more individual scale Joanne needs to judge the degree of urgency of each job and feed back to the customer, managing their expectations where appropriate. Both planning and control systems also have a *supply interface* that translates the operation's plans in terms of the supply of parts, or fuel, ground services, crew availability, etc. At the heart of each company's activities is a set of *core mechanics* that load capacity, prioritise, schedule, monitor and control the operation. The job of this decision making is to reconcile the needs of the customers and the operation's resources in some way. For Joanne this involves attempting to maximise the utilisation of her workshop resources while keeping customers satisfied. Air France also has similar objectives, with customer comfort and safety being paramount. Also each operation is attempting some *information integration* that involves both computer assisted information handling and the skills and experience of planning and control staff.

DIAGNOSTIC QUESTION

Does resource planning and control have all the right elements?

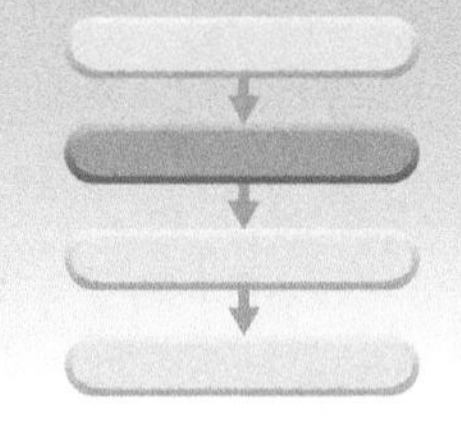

Video

Figure 10.2 illustrates the elements that should be present in all planning and control systems. In more sophisticated systems they may even be extended to include the integration of this core operations resource planning and control task with other functional areas of the firm such as finance, marketing, and personnel. We deal with this cross-functional perspective when we discuss enterprise resource planning (ERP) later.

How does the system interface with customers?

The part of the resource planning and control system that manages the way customers interact with the business on a day-to-day basis is called the 'customer interface' or sometimes 'demand management'. This is a set of activities that interface with both individual customers and the market more broadly. Depending on the business, these activities may include customer negotiation, order entry, demand forecasting, order promising, updating customers, keeping customer histories, post-delivery customer service and physical distribution.

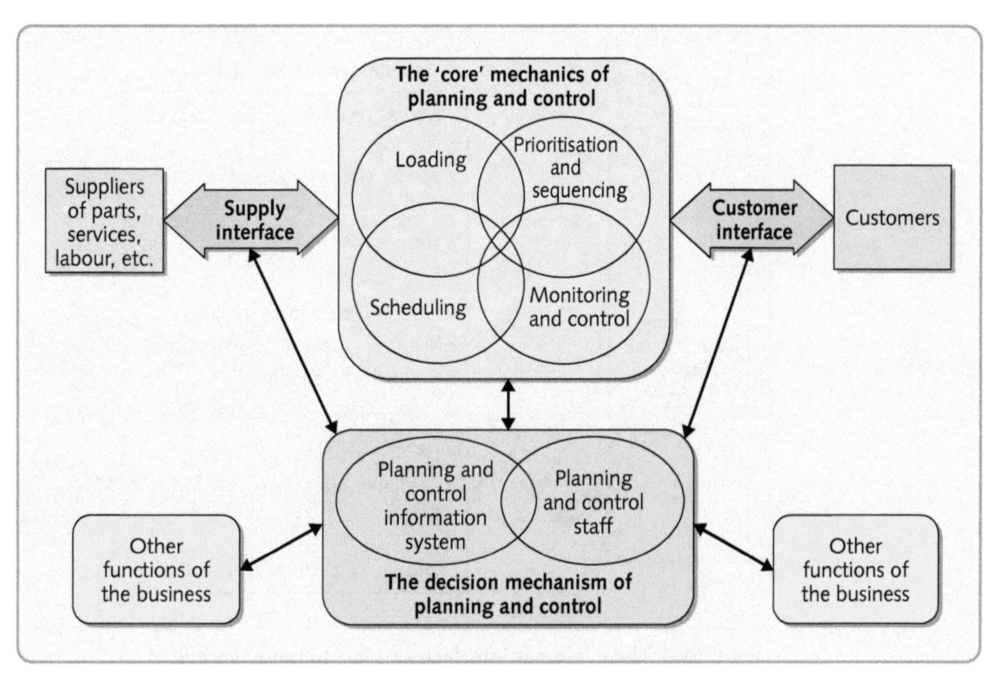

Figure 10.2 The key elements of a resource planning and control system

The customer interface defines the customer experience

Practice note

The customer interface is important because it defines the nature of the customer experience. It is the public face of the operation (the 'line of visibility' as it was called in Chapter 8). Therefore, it needs to be managed like any other 'customer processing' process, where the quality of the service, as the customer sees it, is defined by the gap between customers' expectations and their perceptions of the service they receive. Figure 10.3 illustrates a typical customer experience of interacting with a planning and control customer interface. The experience itself will start before any customer contact is initiated. Customer expectations will have been influenced by the way the business presents itself through promotional activities, the ease with which channels of communication can be used (for example, design of the website), and so on, the question being, 'Does the communication channel give any indication of the kind of service response (for example, how long will we have to wait?) that the customer can expect?' At the first point of contact when an individual customer requests services or products, their request must be understood, delivery possibly negotiated, and a delivery promise made. Prior to the delivery of the service or product, the customer may or may not change their mind, which in turn may or may not involve re-negotiating delivery promises. Similarly, customers may require or value feedback as to the progress of their request. At the point of delivery, not only are the products and services handed over to the customer, but there may also be an opportunity to explain the nature of the delivery and gauge customers' reactions. Following the completion of the delivery there may also be some sort of post-delivery action, such as a phone call to confirm that all is well.

> **OPERATIONS PRINCIPLE**
> *Customers' perceptions of an operation will partially be shaped by the customer interface of its planning and control system.*

As is usual with such customer experiences, the managing of customer expectations is particularly important in the early stages of the experience. For example, if there is a possibility that a delivery may be late (perhaps because of the nature of the service being requested) then that possibility is established as an element in the customer's expectations. As the experience continues, various interactions with the customer interface serve to build up customer perceptions of the level of support and care exhibited by the operation.

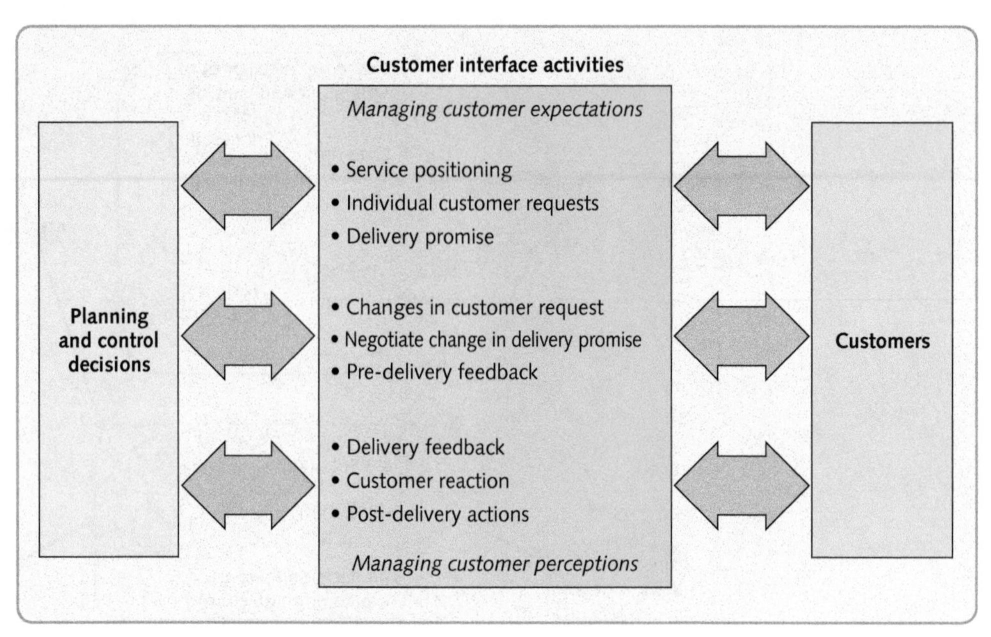

Figure 10.3 The customer interface as a 'customer experience'

The customer interface should reflect the operation's objectives

In managing a customer's experience, the customer interface element of the planning and control system is, in effect, operationalising the business' operations objectives. It may have to prioritise one type of customer over another. It may have to encourage some types of customer to transact business more than other (possibly less profitable) types of customer. It will almost certainly have to trade off elements of customer service against the efficiency and utilisation of the operations resources. No matter how sophisticated the customer interface technology, or how skilled the customer interface staff, this part of the planning and control system cannot operate effectively without clear priorities derived from the operation's strategic objectives.

The customer interface acts as a trigger function

Acceptance of an order should prompt the customer interface to trigger the operation's processes. Exactly what is triggered will depend on the nature of the business. For example, some building and construction companies, because they are willing to build almost any kind of construction, will keep relatively few of their own resources within the business, but rather hire them in when the nature of the job becomes evident. This is a 'resource-to-order' operation where the customer interface triggers the task of hiring in the relevant equipment (and possibly labour) and purchasing the appropriate materials. If the construction company confined itself to a narrower range of construction tasks, thereby making the nature of demand slightly more predictable, it would be likely to have its own equipment and labour permanently within the operation. Here, accepting a job would only need to trigger the purchase of the materials to be used in the construction, and the business is a 'produce to order' operation. Some construction companies will construct pre-designed standard houses or apartments ahead of any firm demand for them. If demand is high, customers may place requests for houses before they are started or during their construction. In this case, the customer will form a backlog of demand and must wait. However, the company is also taking the risk of holding a stock of unsold houses. Operations of this type are called 'produce-ahead-of-order'.

How does the system interface with suppliers?

Practice note

The supplier interface provides the link between the activities of the operation itself and those of its suppliers. The timing and level of activities within the operation or process will have implications for the supply of products and services to the operation. Suppliers need to be informed so that they can make products and services available when needed. In effect this is the mirror image of the customer interface. As such, the supplier interface is concerned with managing the supplier experience to ensure appropriate supply. Because the customer is not directly involved in this does not make it any less important. Ultimately, customer satisfaction will be influenced by supply effectiveness because that in turn influences delivery to customers. Using the expectations–perception gap to judge the quality of the supplier interface function may at first seem strange. After all, suppliers are not customers as such. Yet, it is important to be a 'quality customer' to suppliers because this increases the chances of receiving high quality service from them. This means that suppliers fully understand customer expectations because they have been made clear and unambiguous.

> **OPERATIONS PRINCIPLE**
> *An operation's planning and control system can enhance or inhibit the ability of its suppliers to support delivery effectiveness.*

The supplier interface has both a long- and short-term function. It must be able to cope with different types of long-term supplier relationship, and also handle individual transactions with suppliers. To do the former it must understand the requirements of all the processes within the operation and also the capabilities of the suppliers (in large operations, there could be thousands of suppliers). Figure 10.4 shows a simplified sequence of events in the management of a typical supplier–operation interaction which the supplier interface must facilitate. When the planning and control activity requests supply, the supplier interface must have identified potential suppliers and might also be able to suggest alternative materials or services if necessary. Formal requests for quotations may be sent to potential suppliers if no supply agreement exists. These requests might be sent to several suppliers or a smaller group, who may be 'preferred' suppliers. Just as it was important to manage customer expectations, it is important to manage supplier expectations, often prior to any formal supply of products or services. This issue was discussed in Chapter 7 as supplier development. To handle individual transactions, the supplier interface will need to issue formal purchase orders. These may be stand-alone

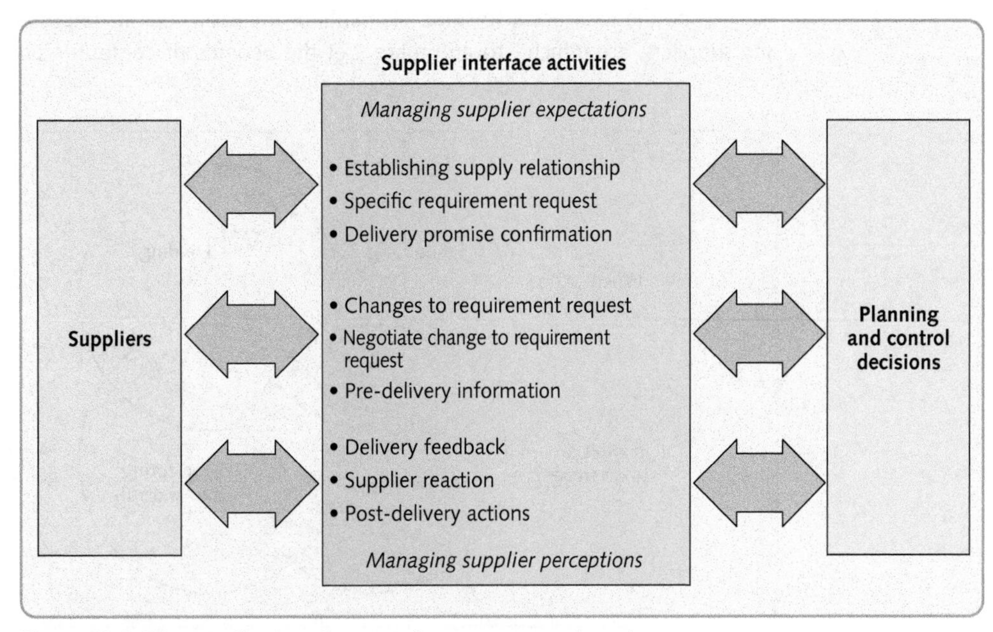

Figure 10.4 The supplier interface as a 'customer experience'

documents or, more likely, electronic orders. Whatever the mechanisms, it is an important activity because it often forms the legal basis of the contractual relationship between the operation and its supplier. Delivery promises will need to be formally confirmed. While waiting for delivery, it may be necessary to negotiate changes in supply, track progress to get early warning of potential changes to delivery. Also supplier delivery performance needs to be established and communicated with follow up as necessary.

How does the system perform basic planning and control calculations?

Resource planning and control requires the reconciliation of supply and demand in terms of the level and timing of activities within an operation or process. To do this four overlapping activities are performed. These are loading, sequencing, scheduling, and monitoring and control. However, some caution is needed when using these terms. Different organisations may use them in different ways, and even textbooks in the area may adopt different definitions. Although these four activities are very closely interrelated, they do address different aspects of the resource planning and control task. Loading allocates tasks to resources in order to assess *what* level of activity will be expected of each part of the operation. Scheduling is more concerned with *when* the operation or process will do things. Sequencing is a more detailed set of decisions that determines *in what order* jobs pass through processes. Monitoring and control involves checking if *activities are going to* plan by observing what is actually happing in practice, and making adjustments as necessary (see Figure 10.5). This part of the planning and control system can be regarded as the engine room of the whole system insomuch as it calculates the consequences of planning and control decisions. Without understanding how these basic mechanisms work it is difficult to understand how any operation is being planned and controlled. Because of their importance, we treat the four interrelated activities later in the chapter.

Does the system integrate human with 'automated' decision making?

Although computer-based resource planning and control systems are now widespread in many industries, much of the decision making is still carried out partially by people. This is always likely to be the case because some elements of the task, such as negotiating with customers and suppliers, are difficult to automate. Yet the benefits of computer-aided decision making

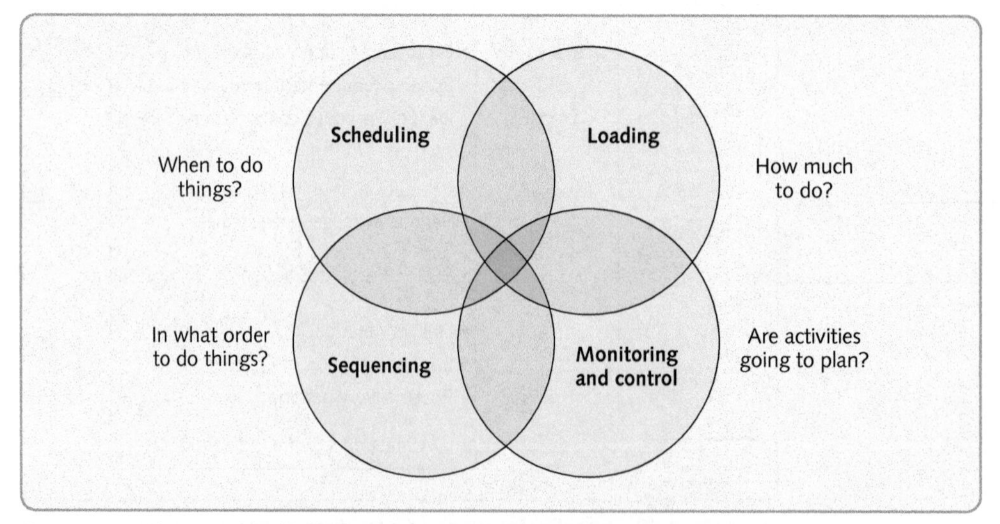

Figure 10.5 The 'core mechanisms' of planning and control

are difficult to ignore. Unlike humans, computer-based planning and control can cope with immense complexity, both in terms of being able to model the inter-relationship between decisions and in terms of being able to store large quantities of information. However, humans are generally better at many of the 'soft' qualitative tasks that can be important in planning and control. In particular, humans are good at the following.

- *Flexibility, adaptability and learning*. Humans can cope with ambiguous, incomplete, inconsistent and redundant goals and constraints. In particular they can deal with the fact that planning and control objectives and constraints may not be stable for longer than a few hours.
- *Communication and negotiation*. Humans are able to understand and sometimes influence the variability inherent in an operation. They can influence job priorities and sometimes processing times. They can negotiate between internal processes and communicate with customers and suppliers in a way that could minimise misunderstanding.
- *Intuition*. Humans can fill in the blanks of missing information that are required to plan and control. They can accumulate the tacit knowledge about what is, and what may be, really happening with the operation's processes.

These strengths of human decision making versus computer decision making provide a clue as to what should be the appropriate degree of automation built into decision making in this area. When planning and controlling stable and relatively straightforward processes that are well understood, decision making can be automated to a greater degree than processes that are complex, unstable and poorly understood.

DIAGNOSTIC QUESTION

Is resource planning and control information integrated?

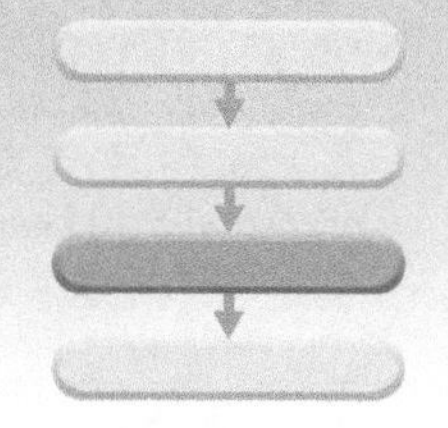

Video

One of the most important issues in resource planning and control is managing the sometimes vast amounts of information generated, not just from the operations function, but from almost every other function of the business. Unless all relevant information is brought together and integrated it is difficult to make informed planning and control decisions. This is what enterprise resource planning (ERP) is about. It has been defined as: '*a complete enterprise-wide business solution. The ERP system consists of software support modules such as: marketing and sales, field service, product design and development, production and inventory control, procurement, distribution, industrial facilities management, process design and development, manufacturing, quality, human resources, finance and accounting, and information services. Integration between the modules is stressed without the duplication of information.*'[3]

OPERATIONS PRINCIPLE

Planning and control systems should integrate information from all relevant organisational functions.

The origins of ERP

Enterprise resource planning has spawned a huge industry devoted to developing the computer systems needed to drive it. The (now) large companies which have grown almost exclusively on the basis of providing ERP systems include SAP, Oracle and Baan. Yet ERP is the one of the latest (and most important) stages in a development that started with materials requirements planning (MRP), an approach that became popular during the 1970s, although the planning and control logic that underlies it had been known for some time. It is a method (simple in principle but complex in execution); of translating a statement of required output into a plan for all the activities that must take place to achieve the required output. What popularised MRP was the availability of computer power to drive the basic planning and control mathematics in a fast, efficient, and most importantly, flexible manner. MRP is treated in the supplement to this chapter. Manufacturing resource planning (MRP II) expanded out of MRP during the 1980s. This extended concept has been described as a game plan for planning and monitoring all the resources of a manufacturing company: manufacturing, marketing, finance and engineering. Again, it was a technology innovation that allowed the development. Local Area Networks (LANs) together with increasingly powerful desktop computers allowed a much higher degree of processing power and communication between different parts of a business.

The strength of MRP and MRP II lay always in the fact that it could explore the *consequences* of any changes to what an operation was required to do. So, if demand changed, the MRP system would calculate all the 'knock-on' effects and issue instructions accordingly. The same principle applies to ERP, but on a much wider basis. ERP systems allow decisions and databases from all parts of the organisation to be integrated so that the consequences of decisions in one part of the organisation are reflected in the planning and control systems of the rest of the organisation (Figure 10.6).

ERP changes the way companies do business

Practice note

Arguably the most significant issue in many company's decision to buy an off-the-shelf ERP system is that of its compatibility with the company's current business processes and practices. Experience of ERP installation suggests that it is extremely important to make sure that one's current way of doing business will fit (or can be changed to fit) with a standard ERP package. One of the most common reasons for not installing ERP is incompatibility between the assumptions in the software and the operating practice of core business processes. If a business's current processes do not fit, they can either change their processes to fit the ERP package, or modify the software within the ERP package to fit their processes.

OPERATIONS PRINCIPLE

ERP systems are only fully effective if the way a business organises its processes is aligned with the underlying assumption of its ERP system.

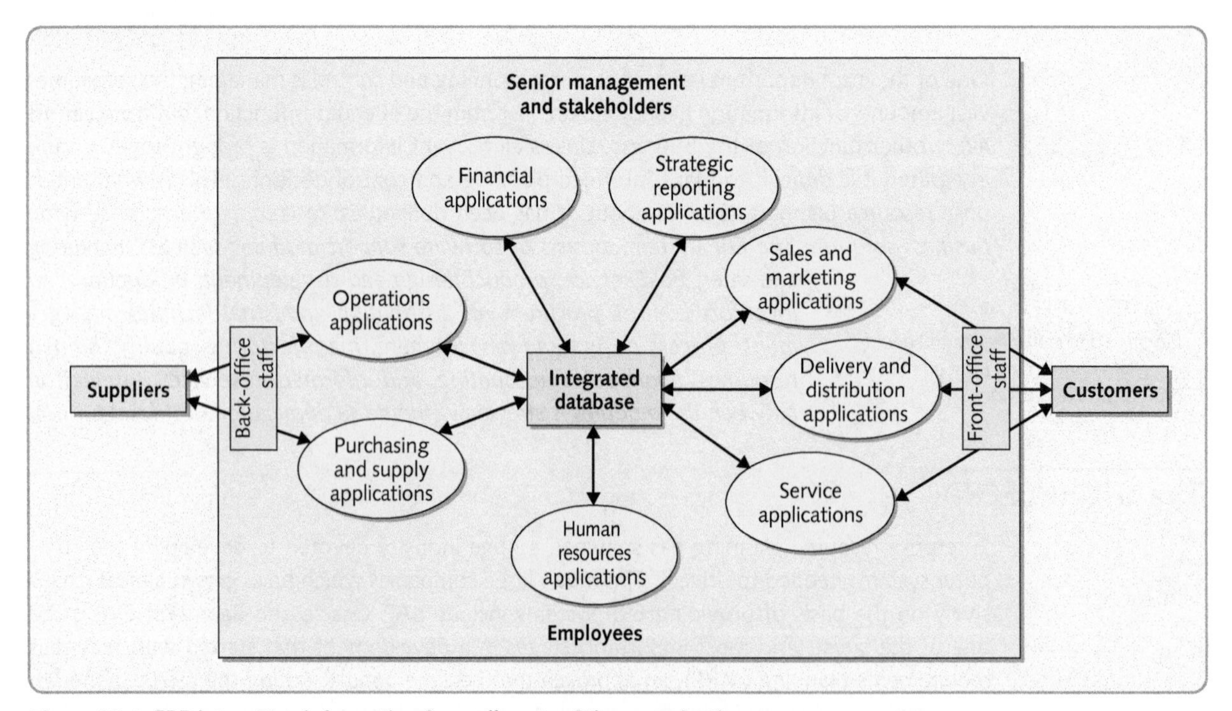

Figure 10.6 ERP integrates information from all parts of the organisation

Both of these options involve costs and risks. Changing business practices that are working well will involve re-organisation costs as well introducing the potential for errors to creep into the processes. Adapting the software will both slow down the project and introduce potentially dangerous software 'bugs' into the system. It would also make it difficult to upgrade the software later on.

ERP installation can be particularly expensive. Attempting to get new systems and databases to talk to old (sometimes called *legacy*) systems can be very problematic. Not surprisingly, many companies choose to replace most, if not all, their existing systems simultaneously. New common systems and relational databases help to ensure the smooth transfer of data between different parts of the organisation.

In addition to the integration of systems, ERP usually includes other features which make it a powerful planning and control tool:

- It is based on a client/server architecture; that is, access to the information systems is open to anyone whose computer is linked to central computers.
- It can include decision support facilities that enable operations decision-makers to include the latest company information.
- It is often linked to external extranet systems, such as the electronic data interchange (EDI) systems, which are linked to the company's supply chain partners.
- It can be interfaced with standard applications programs which are in common use by most managers, such as spreadsheets etc.
- Often, ERP systems are able to operate on most common platforms such as Windows or UNIX, or Linux.

The benefits of ERP

ERP is generally seen as having the potential to significantly improve the performance of many companies in many different sectors. This is partly because of the very much enhanced visibility that information integration gives, but it is also a function of the discipline that ERP demands. Yet this discipline is itself a 'double-edged' sword. On one hand, it 'sharpens up' the management of every process within an organisation, allowing best practice (or at least common practice) to be implemented uniformly through the business. No longer will individual idiosyncratic behaviour by one part of a company's operations cause disruption to all other processes. On the other hand, it is the rigidity of this discipline that is both difficult to achieve and (arguably) inappropriate for all parts of the business. Nevertheless, the generally accepted benefits of ERP are as follows:

- Greater visibility of what is happening in all parts of the business.
- Forcing the business process-based changes that potentially make all parts of the business more efficient.
- Improved control of operations that encourages continuous improvement (albeit within the confines of the common process structures).
- More sophisticated communication with customers, suppliers, and other business partners, often giving more accurate and timely information.
- Integrating whole supply chains including suppliers' suppliers and customers' customers.

Web-integrated ERP

An important justification for embarking on ERP is the potential it gives to link up with the outside world. For example, it is much easier for an operation to move into internet-based trading if it can integrate its external internet systems into its internal ERP systems. However, as has been pointed out by some critics of the ERP software companies, ERP vendors were not prepared for the impact of e-commerce and had not made sufficient allowance in their products for the need to interface with internet-based communication channels. The result of this has been that whereas the internal complexity of ERP systems was designed only to be intelligible to systems experts, the internet has meant that customers and suppliers (who are non-experts) are demanding access to the same information.

One problem is that different types of external company often need different types of information. Customers need to check the progress of their orders and invoicing, whereas suppliers and other partners want access to the details of operations planning and control. Not only that, but they want access all the time. The internet is always there, but ERP systems are often complex and need periodic maintenance. This can mean that every time the ERP system is taken off-line for routine maintenance or other changes, the website also goes off-line. To combat this, some companies configure their ERP and e-commerce links in such a way that they can be decoupled so that ERP can be periodically shut down without affecting the company's web presence.

Supply network ERP

The step beyond integrating internal ERP systems with immediate customers and suppliers is to integrate it with the systems of other businesses throughout the supply network. This is often exceptionally complicated. Not only do different ERP systems have to communicate together, they have to integrate with other types of system. For example, sales and marketing functions often use systems such as customer relationship management (CRM) systems that manage the complexities of customer requirements, promises and transactions. Getting ERP and CRM systems to work together is itself often difficult. Nevertheless, such web-integrated ERP, or 'c-commerce' (collaborative commerce) applications are emerging and starting to make an impact on the way companies do business. Although a formidable task, the benefits are potentially great. The costs of communicating between supply network partners could be dramatically reduced and the potential for avoiding errors as information and products move between partners in the supply chain are significant. Yet such transparency also brings risks. If the ERP system of one operation within a supply chain fails for some reason, it may block the effective operation of the whole integrated information system throughout the network.

DIAGNOSTIC QUESTION

Are core planning and control activities effective?

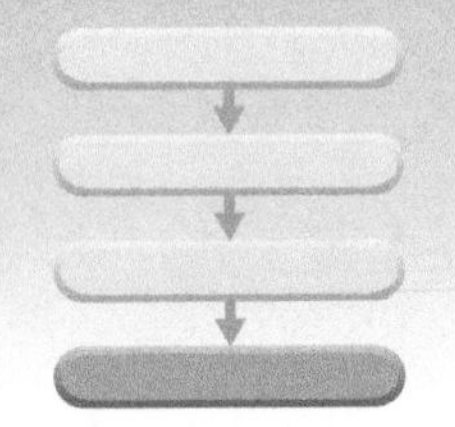

Video

All resource planning and control activity eventually relies on a set of calculations that guide how much work to load onto different parts of the operation, when different activities should be performed, in what order individual jobs should be done, and how processes can be adjusted if they have deviated from plan. These calculations can be thought of as the 'engine room' of the whole resource planning and control system. Although the algorithms that guide the calculations are often embedded within computer-based systems, it is worthwhile understanding some of the core ideas on which they are based. These fall into four overlapping categories: loading, scheduling, sequencing, and monitoring and control.

Loading

Loading is the amount of work that is allocated to a process or stage or whole process. It is a capacity-related issue that will attempt to reconcile how much the operation or the process is expected to do with how much the operation or process can do. Essentially the loading activity calculates the consequences on individual parts of the operation of the operation's overall workload. It may or may not assume realistic capacity limits on what can be loaded. If it does, it is called finite loading; if not, it is called infinite loading. Finite loading is an approach which only allocates work to a work centre (a person, a machine, or perhaps a group of people or machines) up to a set limit. This limit is the estimate of capacity for the work centre (based on the times available for loading). Work over and above this capacity is not accepted. Figure 10.7(a) shows that the load on the work centre is not allowed to exceed the capacity limit. Finite loading is particularly relevant for operations where:

> **OPERATIONS PRINCIPLE**
> *For any given level of demand a planning and control system should be able to indicate the implications for the loading on any part of the operation.*

- *it is possible to limit the load* – for example, it is possible to run an appointment system for a general medical practice or a hairdresser;
- *it is necessary to limit the load* – for example, for safety reasons only a finite number of people and weight of luggage are allowed on aircraft;
- *the cost of limiting the load is not prohibitive* – for example, the cost of maintaining a finite order book at a specialist sports car manufacturer does not adversely affect demand, and may even enhance it.

Infinite loading is an approach to loading work which does not limit accepting work, but instead tries to cope with it. Figure 10.7(b) illustrates a loading pattern where capacity constraints have not been used to limit loading. Infinite loading is relevant for operations where:

- *it is not possible to limit the load* – for example, an accident and emergency department in a hospital should not turn away arrivals needing attention;
- *it is not necessary to limit the load* – for example, fast-food outlets are designed to flex capacity up and down to cope with varying arrival rates of customers during busy periods, customers accept that they must queue for some time before being served – unless this is extreme, the customers might not go elsewhere;

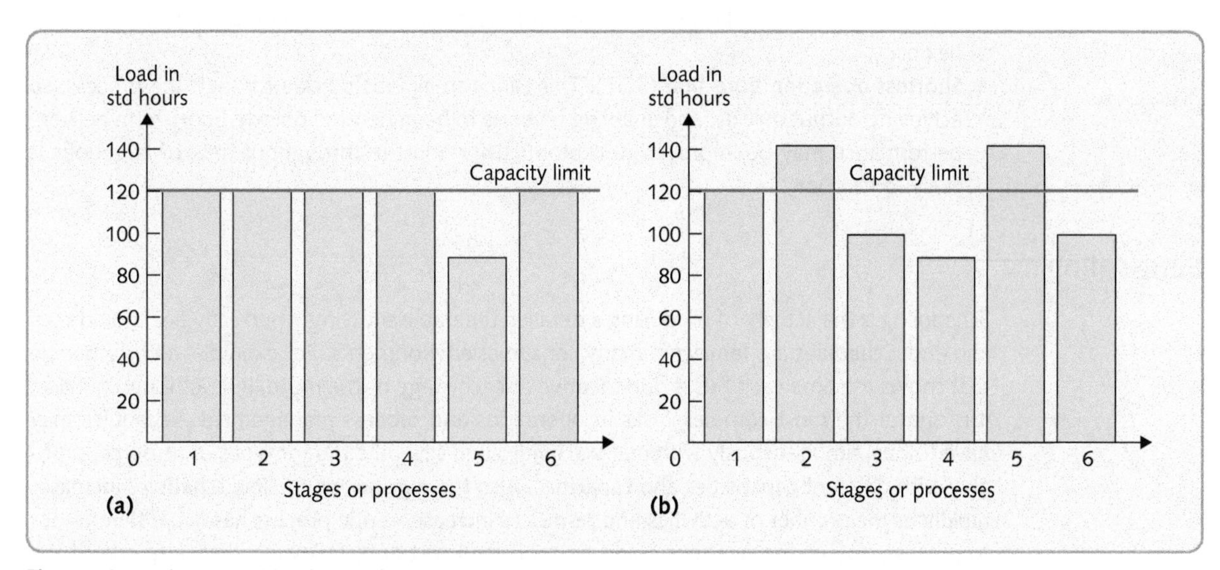

Figure 10.7 (a) Forward loading; (b) backward loading

- *the cost of limiting the load is prohibitive* – for example, if a retail bank turned away customers at the door because a set amount were inside, customers would feel less than happy with the service.

In complex planning and control activities where there are multiple stages, each with different capacities and with a varying mix arriving at the facilities, such as a machine shop in an engineering company, the constraints imposed by finite loading make loading calculations complex and not worth the considerable computational power which would be needed.

Sequencing

Practice note

After the 'loading' of work onto processes, the order or sequence in which it will be worked on needs to be determined. This task is called 'sequencing'. The priorities given to work in an operation are often determined by some pre-defined set of sequencing rules. Some of these are summarised below.

- **Customer priority.** This allows an important or aggrieved customer, or item, to be prioritised irrespective of their order of arrival. Some banks, for example, give priority to important customers. Accident and emergency departments in hospitals must rapidly devise a schedule that prioritises patients presenting symptoms of a serious illness. Hospitals have developed 'triage systems', whereby medical staff hurriedly sort through the patients to determine their relative urgency.
- **Due date (DD).** Work is sequenced according to when it is 'due' for delivery, irrespective of the size of each job or the importance of each customer. For example, a support service in an office block, such as a reprographic unit, may sequence the work according to when the job is needed. Due date sequencing usually improves delivery reliability and average delivery speed, but may not provide optimal productivity.
- **Last in first out (LIFO).** This is usually selected for practical reasons. For example, unloading an elevator is more convenient on a LIFO basis, as there is only one entrance and exit. LIFO has a very adverse effect on delivery speed and reliability.
- **First in first out (FIFO).** Also called 'first come, first served' (FCFS), this is a simple and equitable rule, used especially when queues are evident to customers, as in theme parks.
- **Longest operation time first (LOT).** Executing the longest job first has the advantage of utilising work centres for long periods, but, although utilisation may be high (therefore cost relatively low), this rule does not take into account delivery speed, delivery reliability or flexibility.
- **Shortest operation time first (SOT).** This sends small jobs quickly through the process, so achieving output quickly, and enabling revenue to be generated quickly. Short-term delivery performance may be improved, but productivity and the throughput time of large jobs is likely to be poor.

Scheduling

Scheduling is the activity of producing a detailed timetable showing when activities should start and end. Schedules are familiar in many consumer environments. For example, a bus schedule that shows the time each bus is due to arrive at each stage of the route. But, although familiar, it is one of the most complex tasks in operations and process management. Schedules may have to deal simultaneously with many activities and several different types of resource, probably with different capabilities and capacities. Also the number of possible schedules increases rapidly as the number of activities and resources increase. If one process has five different jobs to process, any of the five jobs could be processed first and, following that, any one of the remaining four jobs, and so on. This means that there are: $5 \times 4 \times 3 \times 2 = 120$ different

OPERATIONS PRINCIPLE
An operation's planning and control system should allow for the effects of alternative schedules to be assessed.

schedules possible. More generally, for n jobs there are $n!$ (factorial n, or $n \times (n - 1) \times (n - 2) \ldots \times 1$)) different ways of scheduling the jobs through a single process or stage. If there is more than one process or stage, there are $(n!)^m$ possible schedules. Where n is the number of jobs and m is the number of processes or stages. In practical terms, this means that there are often many millions of feasible schedules, even for relatively small operations. This is why scheduling rarely attempts to provide an 'optimal' solution but rather satisfies itself with an 'acceptable' feasible one.

Gantt charts

Animated diagram

The most common method of scheduling is by use of the Gantt chart. This is a simple device which represents time as a bar, or channel, on a chart. The start and finish times for activities can be indicated on the chart and sometimes the actual progress of the job is also indicated. The advantages of Gantt charts are that they provide a simple visual representation both of what should be happening and of what actually is happening in the operation. Furthermore, they can be used to 'test out' alternative schedules. It is a relatively simple task to represent alternative schedules (even if it is a far from simple task to find a schedule which fits all the resources satisfactorily). Figure 10.8 illustrates a Gantt chart for a specialist software developer. It indicates the progress of several jobs as they are expected to progress through five stages of the process. Gantt charts are not an optimising tool, they merely facilitate the development of alternative schedules by communicating them effectively.

Scheduling work patterns

Where the dominant resource in an operation is its staff, then the schedule of work times effectively determines the capacity of the operation itself. Scheduling needs to make sure enough people are working at any time to provide a capacity appropriate for the level of demand. Operations such as call centres and hospitals, which must respond directly to customer demand, will need to schedule the working hours of their staff with demand in mind. For example, Figure 10.9 shows the scheduling of shifts for a small technical 'hot line'

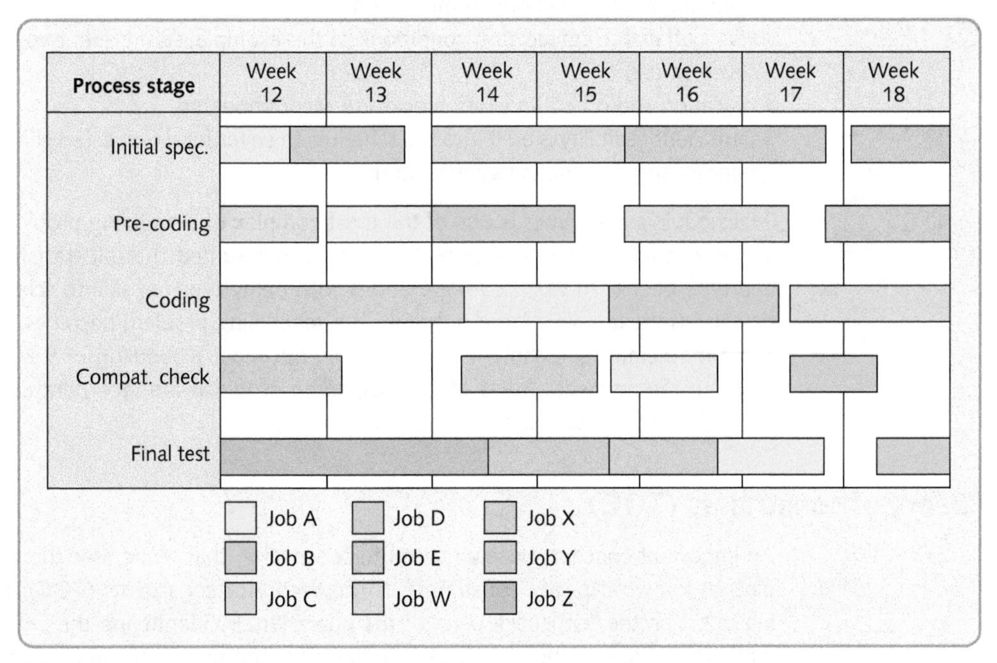

Figure 10.8 Gantt chart showing the schedule for jobs at each process stage

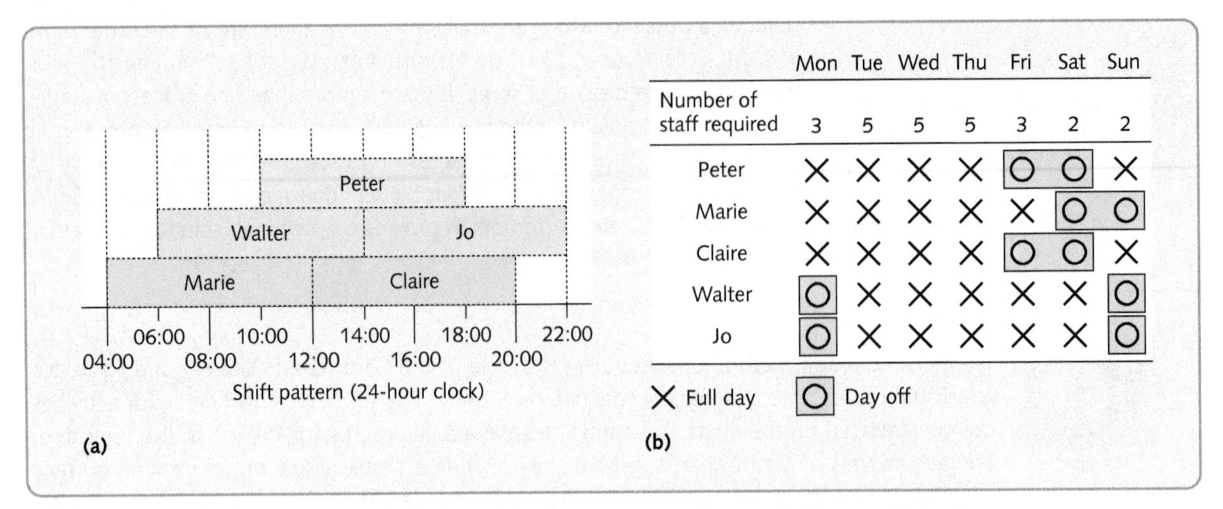

Figure 10.9 Shift allocation for the technical 'hot line' on (a) a daily basis and (b) a weekly basis

support service for the software company. Its service times are 04.00 hrs to 20.00 hrs on Monday, 04.00 hrs to 22.00 hrs Tuesday to Friday, 06.00 hrs to 22.00 hrs on Saturday, and 10.00 hrs to 20.00 hrs on Sunday. Demand is heaviest Tuesday to Thursday, starts to decrease on Friday, is low over the weekend and starts to increase again on Monday. The scheduling task for this kind of problem can be considered over different time scales, two of which are shown in Figure 10.9. During the day, working hours need to be agreed with individual staff members. During the week, days off need to be agreed. During the year, vacations, training periods, and other blocks of time where staff are unavailable need to be agreed. All this has to be scheduled such that:

- capacity matches demand
- the length of each shift is neither excessively long nor too short to be attractive to staff
- working at unsocial hours is minimised
- days off match agreed staff conditions (in this example, staff prefer two consecutive days off every week)
- vacation and other 'time-off' blocks are accommodated
- sufficient flexibility is built into the schedule to cover for unexpected changes in supply (staff illness) and demand (surge in customer calls).

Scheduling staff times is one of the most complex of scheduling problems. In the relatively simple example shown in Figure 10.9 we have assumed that all staff have the same level and type of skill. In very large operations with many types of skill to schedule and uncertain demand (for example a large hospital), the scheduling problem becomes extremely complex. Some mathematical techniques are available but most scheduling of this type is, in practice, solved using heuristics (rules of thumb), some of which are incorporated into commercially available software package.

Theory of constraints (TOC)

An important concept, closely related to scheduling, that recognises the importance of planning to known capacity constraints, is the theory of constraints (TOC). It focuses scheduling effort on the bottleneck parts of the operation. By identifying the location of constraints, working to remove them, and then looking for the next constraint, an operation is always focusing on the part that critically determines the pace of output. The approach which uses this idea is called optimised production technology (OPT). Its development and the marketing of it as a proprietary software product were originated by Eliyahu Goldratt. It helps to schedule production systems to the pace dictated by the most heavily loaded resources, that is, bottlenecks. If the rate of activity in any part of the system exceeds that of the bottleneck, then items are being produced that cannot be used. If the rate of working falls below the pace at the bottleneck, then the entire system is under-utilised. The 'principles' of underlying OPT demonstrate this focus on bottlenecks.

OPT principles

1. Balance flow, not capacity. It is more important to reduce throughput time rather than achieving a notional capacity balance between stages or processes.
2. The level of utilisation of a non-bottleneck is determined by some other constraint in the system, not by its own capacity. This applies to stages in a process, processes in an operation, and operations in a supply network.
3. Utilisation and activation of a resource are not the same. According to the TOC a resource is being *utilised* only if it contributes to the entire process or operation creating more output. A process or stage can be *activated* in the sense that it is working, but it may only be creating stock or performing other non-value added activity.
4. An hour lost (not used) at a bottleneck is an hour lost for ever out of the entire system. The bottleneck limits the output from the entire process or operation, therefore the under-utilisation of a bottleneck affects the entire process or operation.
5. An hour saved at a non-bottleneck is a mirage. Non-bottlenecks have spare capacity anyway. Why bother making them even less utilised?
6. Bottlenecks govern both throughput and inventory in the system. If bottlenecks govern flow, then they govern throughput time, which in turn governs inventory.
7. You do not have to transfer batches in the same quantities as you produce them. Flow will probably be improved by dividing large production batches into smaller ones for moving through a process.
8. The size of the process batch should be variable, not fixed. Again, from the EBQ model, the circumstances that control batch size may vary between different products.
9. Fluctuations in connected and sequence-dependent processes add to each other rather than averaging out. So, if two parallel processes or stages are capable of a particular average output rate, in series they will never be able to achieve the same average output rate.
10. Schedules should be established by looking at all constraints simultaneously. Because of bottlenecks and constraints within complex systems, it is difficult to work out schedules according to a simple system of rules. Rather, all constraints need to be considered together.

Monitoring and control

Having created a plan for the operation through loading, sequencing and scheduling, each part of the operation has to be monitored to ensure that activities are happening as planned. Any deviation can be rectified through some kind of intervention in the operation, which itself will probably involve some re-planning. Figure 10.10 illustrates a simple view of control. The output from a work centre is monitored and compared with the plan which indicates what the work centre is supposed to be doing. Deviations from this plan are taken into account through a re-planning activity and the necessary

OPERATIONS PRINCIPLE

A planning and control system should be able to detect deviations from plans within a timescale that allows an appropriate response.

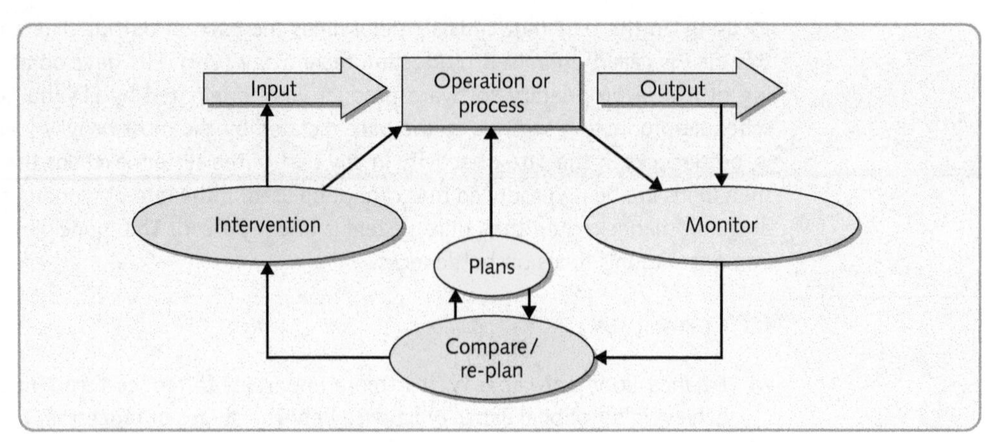

Figure 10.10 A simple model of control

interventions made (on a timely basis) to the work centre which will ensure that the new plan is carried out. Eventually, however, some further deviation from planned activity will be detected and the cycle is repeated.

Push and pull control

An element of control is periodic intervention into processes and operations. A key distinction is between intervention signals that push work through processes and operations and those that pull work only when it is required. In a *push* system of control, activities are scheduled by means of a central system and completed in line with central instructions, such as an MRP system (see the supplement to this chapter). Each work centre pushes out work without considering whether the succeeding work centre can make use of it. Deviations from plan are noted by the central operations planning and control system, and plans adjusted as required. In a *pull* system of control, the pace and specification of what is done are set by the succeeding 'customer' workstation, which 'pulls' work from the preceding (supplier) workstation. The customer acts as the only 'trigger' for movement. If a request is not passed back from the customer to the supplier, the supplier cannot produce anything or move any materials. A request from a customer not only triggers production at the supplying stage, but also prompts the supplying stage to request a further delivery from its own suppliers. In this way, demand is transmitted back through the stages from the original point of demand by the original customer.

Push systems of control are more formal and require significant decision-making or computing power when it is necessary to re-plan in the light of events. But push control can cope with very significant changes in circumstances, such as major shifts in output level or product mix. By contrast, pull control is more self-adjusting in the sense that the natural rules that govern relationships between stages or processes cope with deviations from plan without reference to any higher decision-making authority. But there are limits to the extent that this can cope with major fluctuations in demand. Pull control works best when conditions are relatively stable. Understanding the differences between push and pull is also important because they have different effects in terms of their propensities to accumulate inventory. Pull systems are far less likely to result in inventory build-up and therefore have advantages in terms of the lean synchronisation of flow (covered in Chapter 13).

OPERATIONS PRINCIPLE

Pull control reduces the build-up on inventory between processes or stages.

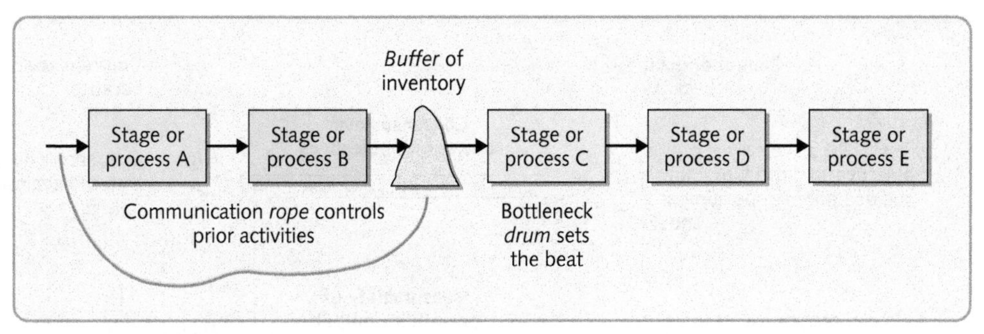

Figure 10.11 The drum, buffer, rope concept

Drum, buffer, rope control

The drum, buffer, rope concept comes from the theory of constraints (TOC) described earlier. It is an idea that helps to decide exactly *where* control should occur. Again, the TOC emphasises the role of the bottleneck on work flow. If the bottleneck is the chief constraint, it should be the control point of the whole process. The bottleneck should be the *drum* because it sets the 'beat' for the rest of the process to follow. Because it does not have sufficient capacity, a bottleneck is (or should be) working all the time. Therefore, it is sensible to keep a *buffer* of inventory in front of it to make sure that it always has something to work on. Also, because it constrains the output of the whole process, any time lost at the bottleneck will affect the output from the whole process. So it is not worthwhile for the parts of the process before the bottleneck to work to their full capacity. All they would do is produce work that would accumulate further along in the process up to the point where the bottleneck is constraining flow. Therefore, some form of communication between the bottleneck and the input to the process is needed to make sure that activities before the bottleneck do not overproduce. This is called the *rope* (see Figure 10.11).

OPERATIONS PRINCIPLE

The constraints of bottleneck processes and activities should be a major input to the planning and control activity.

The degree of difficulty in controlling operations

The simple monitoring control model in Figure 10.10 helps in understanding the basic functions of the monitoring and control activity. But it is a simplification. Some simple technology-dominated processes may approximate to it, but many other operations do not. In fact, the specific criticisms cited in the Critical Commentary at the end of this chapter provide a useful set of questions which can be used to assess the degree of difficulty associated with control of any operation:

- Is there consensus over what the operation's objectives should be?
- How well can the output from the operation be measured?
- Are the effects of interventions into the operation predictable?
- Are the operation's activities largely repetitive?

Figure 10.12 illustrates how these four questions can form dimensions of 'controllability'. It shows three different operations. The food-processing operation is relatively straightforward to control, while the child care service is particularly difficult. The tax advice service is somewhere in between.

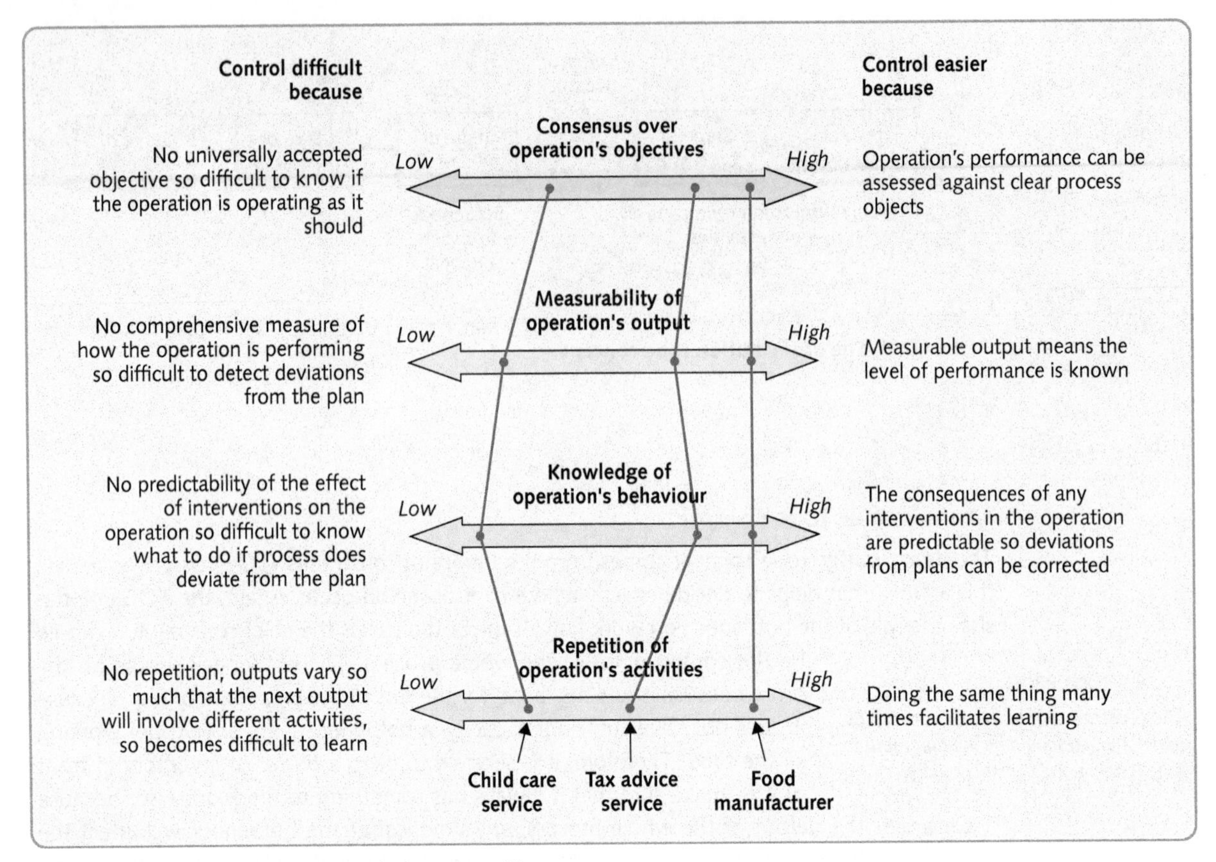

Figure 10.12 How easy is an operation to control?

Critical commentary

Each chapter contains a short critical commentary on the main ideas covered in the chapter. Its purpose is not to undermine the issues discussed in the chapter, but to emphasise that, although we present a relatively orthodox view of operation, there are other perspectives.

- Far from being the magic ingredient which allows operations to fully integrate all their information, ERP is regarded by some as one of the most expensive ways of getting zero or even negative return on investment. For example, the American chemicals giants, Dow Chemical, spent almost half a billion dollars and seven years implementing an ERP system which became outdated almost as soon as it was implemented. One company, FoxMeyer Drug, claimed that the expense and problems which it encountered in implementing ERP eventually drove it into bankruptcy. One problem is that ERP implementation is expensive. This is partly because of the need to customise the system, understand its implications on the organisation, and train staff to use it. Spending on what some call the *ERP ecosystem* (consulting, hardware, networking and complementary applications) has been estimated as being twice the spending on the software itself. But it is not only the expense which has disillusioned many companies, it is also the returns they have had for their investment. Some studies show that the vast majority of companies implementing ERP are disappointed with the effect it has had on their businesses. Certainly many companies find that they have to (sometimes fundamentally) change the way they organise their operations in order to fit in with ERP systems. This organisational impact of ERP (which has been described as the corporate equivalent of root-canal work) can have a significantly disruptive effect on the organisation's operations.

- If one accepts only some of the criticisms of ERP, it does pose the question as to why companies have invested such large amounts of money in it. Partly it was the attraction of turning the company's information systems into a 'smooth running and integrated machine'. The prospect of such organisational efficiency is attractive to most managers, even if it does presuppose a very simplistic model of how organisations work in practice. After a while, although organisations could now see the formidable problems in ERP implementation, the investments were justified on the basis that, 'even if we gain no significant advantage by investing in ERP, we will be placed at a disadvantage by *not* investing in it because all our competitors are doing so'. There is probably some truth in this; sometimes businesses have to invest just to stand still.

- Most of the perspectives on control taken in this chapter are simplifications of a far more messy reality. They are based on models used to understand mechanical systems such as car engines. But anyone who has worked in real organisations knows that organisations are not machines. They are social systems, full of complex and ambiguous interactions. Simple models such as these assume that operations objectives are always clear and agreed, yet organisations are political entities where different and often conflicting objectives compete. Local government operations, for example, are overtly political. Furthermore, the outputs from operations are not always easily measured. A university may be able to measure the number and qualifications of its students, for example, but it cannot measure the full impact of its education on their future happiness. Also, even if it is possible to work out an appropriate intervention to bring an operation back into 'control', most operations cannot perfectly predict what effect the intervention will have. Even the largest of burger bar chains does not know *exactly* how a new shift allocation system will affect performance. Also, some operations never do the same thing more than once anyway. Most of the work done by construction operations are one-offs. If every output is different, how can 'controllers' ever know what is supposed to happen? Their plans themselves are mere speculation.

SUMMARY CHECKLIST

This checklist comprises questions that can be usefully applied to any type of operations and reflect the major diagnostic questions used within the chapter.

- ☐ Is appropriate effort devoted to planning and controlling the operation's resources and activities?
- ☐ Have any recent failures in planning and control been used to reconsider how the planning and control system operates?
- ☐ Does the system interface with customers so as to encourage a positive customer experience?
- ☐ Does the planning and control system interface with suppliers so as to promote a supplier experience that is in your long-term interests?
- ☐ Does the system perform basic planning and control calculations in an appropriate and realistic manner?
- ☐ Is the balance between human and automated decision making understood and appropriate for the circumstances?
- ☐ How well is resource planning and control information integrated?
- ☐ Have the advantages and disadvantages of moving to a sophisticated (but expensive!) ERP system been investigated?
- ☐ If so, have the possibilities of web integration and supply chain scope been investigated?
- ☐ Are bottlenecks accounted for in the way planning and control decisions are made?
- ☐ If not, have bottlenecks been identified and their effect on the smooth flow of items through the operation been evaluated?

CASE STUDY

subText Studios, Singapore

'C.K. One' was clearly upset. Since he had founded *subText* in the fast growing South East Asian computer-generated imaging (CGI) market, three years ago, this was the first time that he had needed to apologise to his clients. In fact, it had been more than an apology; he had agreed to reduce his fee, though he knew that didn't make up for the delay. He admitted that, up to that point, he hadn't fully realised just how much risk there was, both reputational and financial, in failing to meet schedule dates. It wasn't that either he or his team was unaware of the importance of reliability. On the contrary. 'Imagination', 'expertise' and 'reliability' all figured prominently in their promotional literature, mission statements, and so on. It was just that the 'imagination' and 'expertise' parts had seemed to be the things that had been responsible for their success so far. Of course, it had been bad luck that, after more than a year of perfect reliability (not one late job), the two that had been late in the first quarter of 2004 had been particularly critical. *'They were both for new clients,'* said CK, *'and neither of them indicated just how important the agreed delivery date was to them. We should have known, or found out, I admit. But it's always more difficult with new clients, because without a track record with them, you don't really like even to admit the possibility of being late.'*

Source: Rob Melnychuk/Digital Vision/Getty Images

The company

After studying computer science up to Masters level at the National University of Singapore, C.K. Ong had worked for four years in CGI workshops in and around the Los Angeles area of California and then taken his MBA at Stanford. It was there that his fellow students had named him C.K. One, partly because of his fondness for the fragrance and partly because his outgoing leadership talents usually left him as the leader of whatever group he was working in. After that, *'. . . The name just kinda stuck,'* even when he returned to Singapore to start *subText* Studios. While in California, CK had observed that a small but growing part of the market for computer generated imaging services was in the advertising industry. *'Most CGI work is still connected with the movie industry,'* admitted CK, *'However, two important factors have emerged over the last four or five years. First, the ad agencies have realised that, with one or two notable exceptions, the majority of their output is visually less arresting than most of the public are used to seeing at the movies. Second, the cost of sophisticated CGI, once something of a barrier to most advertising budgets, is starting to fall rapidly. Partly this is because of cheaper computing power and partly because the scarcity of skilled CGI experts who also have creative talent is starting to rectify itself.'* CK had decided to return to Singapore both for family reasons and because the market in the area was growing quickly and, unlike Hong Kong that had a large movie industry with its ancillary service industries, Singapore had few competitors.

The company was set up on a similar but slightly simpler basis to the companies CK had worked for in California. At the heart of the company were the three 'core' departments that dealt sequentially with each job taken on. These three departments were 'Pre-production', 'Production' and 'Post-production'.

- **Pre-production** was concerned with taking and refining the brief as specified by the client, checking with and liaising with the client to iron out any ambiguities in the brief, story-boarding the sequences, and obtaining outline approval of the brief from the client. In addition, pre-production also acted as account liaison with the client and were also responsible for estimating the resources and timing for each job. They also had nominal responsibility for monitoring the job through the remaining two stages, but generally they only did this if the client needed to be consulted during the production and post-production processes. The Supervising Artists in each department were responsible for the control of the jobs in their departments.

- **Production** involved the creation of the imagery itself. This could be a complex and time-consuming process involving the use of state-of-the-art workstations and CGI software. Around 80 per cent of all production work was carried out in-house, but for some jobs other specialist workshops were contracted. This was only done for work that *subText* either could not do, or would find difficult to do. Contracting was hardly ever used simply to increase capacity because the costs of doing so could drastically reduce margins.
- **Post-production** had two functions; the first was to integrate the visual image sequences produced by Production with other effects such as sound effects, music, voice overs, etc.; the second was to cut, edit and generally produce the finished 'product' in the format required by the client.

Each of the three departments employed teams of two people. *'It's a trick I learnt working for a workshop in L.A.,'* said CK. *'Two people working together both enhance the creative process (if you get the right two people!), and provide a discipline for each other. Also, it allows for some flexibility, both in the mixing of different talents and in making sure that there is always at least one person from the team present at any time who knows the progress and status of any job.'* Pre-production had two teams, Production three teams, and Post-production two teams. In addition, CK himself would sometimes work in the core departments, particularly in Pre-production, when he had the time, although this was becoming less common. His main role was in marketing the company's services and business development generally. *'I am the external face of the company and partly my job is to act as a cut-out, particularly for the Production and Post-production people. The last thing I want is them being disturbed by the clients all the time. I also try and help out around the place when I can, especially with the creative and storyboarding work. The problem with doing this is that, particularly for Pre-production and Post-production, a single extra person assisting does not always help to get the job done faster. In fact, it can sometimes confuse things and slow things down. This is why, for Pre-production and Post-production work, one team is always exclusively devoted to one job. We never allow either one team to be working on two jobs at the same time, or have both teams working on one job. It just doesn't work because of the confusion it creates. That doesn't apply to Production. Usually (but not always) the Production work can be parcelled up so that two or even all three of the teams could be working on different parts of it at the same time. Provided there is close coordination between the teams and provided that they are all committed to pulling it together at the end, there should be a more or less inverse relationship between the number of bodies working on the job and the length of time it takes. In fact, with the infamous 'fifty three slash F' job that's exactly what we had to do. However, not withstanding what I just said about shortening the time, we probably did lose some efficiency there by having all three teams working on it.'*

'We pay our teams in the three core departments, a salary based on their experience and a yearly bonus. For that they are expected to, within reason, work until the job is finished. It varies, but most of us work at least ten-hour days relatively frequently. That level of work is factored in to the time estimates we make for each stage of the process. And, although we can be a little inaccurate sometimes, I don't think it's anything to do with a lack of motivation or pace of work. It's just that this type of thing is sometimes difficult to estimate.'

The 'fifty three slash F' job

The 'fifty three slash F' job, recently finished (late) and delivered to the client (dissatisfied) had been the source of much chaos, confusion and recrimination over the last two or three weeks. Although the job was only three days late, it had caused the client (the Singapore office of a US Advertising agency) to postpone a presentation to its own client. Worse, *subText* had given only five days notice of late delivery, trying until the last to pull back onto schedule.

The full name of the job that had given them so much trouble was 04/53/F. The 04 related to the year in which the job was started, the 53 was the client's reference number and the F the job identifier (at the start of the year the first job would be labelled A, then B and so forth with AA, BB etc. being used subsequently). Table 10.1 shows the data for all the jobs started in 2004 up to the current time (day 58, every working day was numbered throughout the year). Figure 10.13 shows the schedule for this period. The job had been accepted on day 18 and had seemed relatively straightforward, although it was always clear that it would be a long production job. It was also clear that time was always going to be tight. There were 32 days in which to finish a job that was estimated to take 30 days.

'We had been negotiating for this job for two or three weeks and we were delighted to get it. It was important to us because the client, although small in Singapore, had interests all over the world. We saw it as a way into potentially some major business in the future. In hindsight we underestimated how much having three teams working on the production stage of this job at one point or other would increase its complexity. OK, it was not an easy piece of CGI to carry off, but we probably would have been OK if we had organised the CGI stage better. It was also real bad luck that, in our efforts to deliver the 'fifty three slash F' job on time, we also disrupted the

Table 10.1 ***subText*** **Studios Singapore – planning data for day 02 to day 58 2004**

Job (04)	Day in	Estimated total time	Actual total time	Due date	Actual delivery	Pre-prod Est.	Pre-prod Actual	Prod Est.	Prod Actual	Post-prod Est.	Post-prod Actual
06/A	−4	29	30	40	34	6	8	11	10	12	12
11/B	−4	22	24	42	31	4	5.5	7	7.5	11	11
04/C	2	31	30.5	43	40	9	9.5	12	13	10	9
54/D	5	28	34	55	58	10	12	12	17	6	5
31/E	15	34	25	68	57	10	11	12	14	12	–
53/F	18	32	49	50	53	6	10	18	28	8	11
24/G	25	26	20	70	–	9	11	9	9	8	–
22/H	29	32	26	70	–	10	12	14	14	8	–
22/I	33	30	11	75	–	10	11	12	–	8	–
09/J	41	36	14	81	–	12	14	14	–	10	–
20/K	49	40	–	89	–	12	–	14	–	14	–

'fifty four slash D' job that turned out to be the only other new client we have had this year.' (CK Ong)

The job had proved difficult from the start. The pre-production stage took longer than estimated, mainly because the client's creative team changed just before the start of *subText* beginning the work. But it was the actual CGI itself that proved to be the major problem. Not only was the task intrinsically difficult, it was difficult to parcel it up into separate packages that could be coordinated for working on by the two teams allocated to the job.

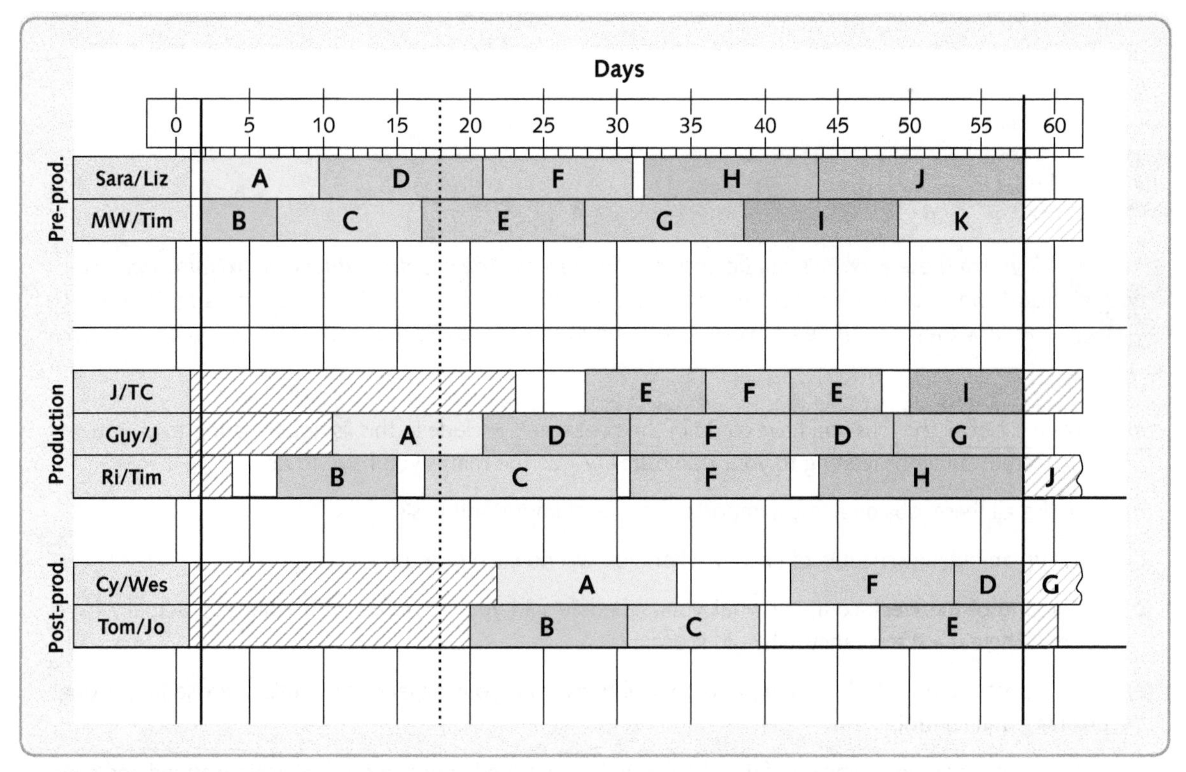

Figure 10.13 ***subText*** **Studios Singapore – actual schedule for day 02 to day 58 2004**

More seriously, it became apparent within two or three days of starting the production work, that they would need the help of another studio for some of the effects. Although the other studio was a regular supplier at short notice, this time they were too busy with their own work to help out. Help eventually came from a specialist studio in Hong Kong: *'The subcontracting delay was clearly a problem, but it was only half way through the production phase that we first realised just how much difficulty the 'fifty three slash F' job was in. It was at that stage that we devoted all our production resources to finishing it. Unfortunately, even then, the job was late. The decision eventually to put all three teams on to the 'fifty three slash F' job was not easy because we knew that it would both disrupt other jobs and potentially cause more coordination problems. However, when I accept jobs into production I am accepting that I will try and do whatever it takes to pass it over to post-production on the date agreed. We did miss out that time, but you can't say we didn't give it everything we've got and technically the job was brilliant, even the client admits that.'* ('TC' Ashwan, Supervising CGI Artist, Production Department)

'No way will we be doing that again'

'No way will we be doing that again', said CK to the core teams when they met to pick over what had gone wrong. *'We are desperately in need of a more professional approach to keeping track of our activities. There is no point in me telling everyone how good we are if we then let them down. The problem is that I don't want to encourage a 'command and control' culture in the studio. We depend on all staff feeling that they have the freedom to explore seemingly crazy options that may just lead to something real special. We aren't a factory. But we do need to get a grip on our estimating so that we have a better idea of how long each job really will take. After that each of the core departments can be responsible for their own planning.'*

QUESTIONS

1. What went wrong with the 'fifty three slash F' job and how could the company avoid making the same mistakes again?
2. What would you suggest that *subText* do to tighten up their planning and control procedures?

APPLYING THE PRINCIPLES

Hints

Some of these exercises can be answered by reading the chapter. Others will require some general knowledge of business activity and some might require an element of investigation. Hints on how they can all be answered are to be found in the eText at **www.pearsoned.co.uk/slack.**

1 (a) Make a list of all the jobs you have to do in the next week. Include in this list jobs relating to your work and/or study and jobs relating to your domestic life – all the things you have to do.

(b) Prioritise all these jobs on a 'most important' to 'least important' basis.

(c) Draw up an outline schedule of exactly when you will do each of these jobs.

(d) At the end of the week compare what your schedule said you *would* do with what you actually *have* done. If there is a discrepancy, why did it occur?

(e) Draw up your-own list of planning and control rules from your experience in this exercise in personal planning and control.

2 Revisit the example at the beginning of the chapter which explained how one car dealership planned and controlled its workshop activities. Try and visit a local car servicing operation and talk to them about their approach

to planning and control. Note – it may be less sophisticated than the operation described in this example, or if it is a franchised operation, it may have a different approach to the one described in the example.

(a) What do you think the planning and control 'system' (it may be computer-based or not) has to do well in order to ensure efficient and effective scheduling?

(b) What do you think are the ideal qualities for people with similar jobs to Joanne Cheung's?

3 From your own experience of making appointments at your General Practitioner's surgery, or by visiting whoever provides you with primary medical care, reflect on how patients are scheduled to see a doctor or nurse.

(a) What do you think planning and control objectives are for a General Practitioners surgery?

(b) How could your own medical practice be improved?

4 Read the following descriptions of two cinemas.

Kinepolis in Brussels is one of the largest cinema complexes in the world, with 28 screens, a total of 8,000 seats, and four showings of each film every day. It is equipped with the latest projection technology. All the film performances are scheduled to start at the same times every day: 4 pm, 6 pm, 8 pm and 10.30 pm. Most customers arrive in the 30 minutes before the start of the film. Each of the 18 ticket desks has a networked terminal and a ticket printer. For each customer, a screen code is entered to identify and confirm seat availability for the requested film. Then the number of seats required is entered and the tickets are printed, though these do not allocate specific seat positions. The operator then takes payment by cash or credit card and issues the tickets. This takes an average of 19.5 seconds, a further five seconds is needed for the next customer to move forward. An average transaction involves the sale of approximately 1.7 tickets.

The UCI cinema in Birmingham has eight screens. The cinema incorporates many 'state-of-the-art' features, including the high-quality THX sound system, fully computerised ticketing and a video games arcade off the main hall. In total the eight screens can seat 1,840 people; the capacity (seating) of each screen varies, so the cinema management can allocate the more popular films to the larger screens and use the smaller screens for the less popular films. The starting times of the eight films at UCI are usually staggered by ten minutes, with the most popular film in each category (children's, drama, comedy, etc.) being scheduled to run first. Because the films are of different durations, and since the manager must try to maximise the utilisation of the seating, the scheduling task is complex. Ticket staff are continually aware of the remaining capacity of each 'screen' through their terminals. There are up to four ticket desks open at any one time. The target time per overall transaction is 20 seconds. The average number of ticket sales per transaction is 1.8. All tickets indicate specific seat positions, and these are allocated on a first-come-first-served basis.

(a) Reflect on the main differences between the two cinemas from the perspectives of their operations managers. What are the advantages and disadvantages of the two different methods of scheduling the films onto the screens?

(b) Find out the running times and classification of eight popular films. Try to schedule these onto the UCI screens, taking account of what popularity you might expect at different times. Allow at 20 minutes for emptying, cleaning and admitting the next audience, and 15 minutes for advertising, before the start of the film.

(c) Visit your local cinema (meet the manager if you can). Compare the operations with those at Kinepolis and UCI, particularly in terms of scheduling.

Notes on chapter

1 Sources: Farman, J. (1999) 'Les Coulisses du Vol', Air France – talk presented by Richard E. Stone, NorthWest Airlines at the IMA Industrial Problems Seminar, 1998.

2 Source: Interview with Joanne Cheung, Steve Deeley and other staff at Godfrey Hall, BMW Dealership, Coventry.

3 Goldratt, E.Y. and Cox, J. (1984) *The Goal*, North River Press.

TAKING IT FURTHER

Dickersbach J. and Keller G. (2010) *Production Planning and Control with SAP ERP* (2nd edn), SAP PRESS/GALILEO PRESS. *Technical but thorough.*

Goldratt, E.Y. and Cox, J. (2004) *The Goal: A Process of Ongoing Improvement*, Gower Publishing Ltd. *Don't read this if you like good novels but do read this if you want an enjoyable way of understanding some of the complexities of scheduling. It particularly applies to the drum, buffer, rope concept described in this chapter and it also sets the scene for the discussion of Optimised Production Technology.*

Magal, S.R. and Word J. (2010) *Integrated Business Processes with ERP Systems*, John Wiley & Sons. *It's written in partnership with SAP, the leading seller of ERP systems, but it does cover all of the key processes supported by modern ERP systems.*

Vollmann, T.E., Berry, W.L. and Whybark, D.C. (2010) *Manufacturing Planning and Control System for Supply Chain Management*, (6th edn), McGraw-Hill Higher Education. *This is the bible of production planning and control. It deals with all the issues in this part of this textbook.*

USEFUL WEBSITES

www.opsman.org *Definitions, links and opinions on operations and process management.*

http://www.bpic.co.uk/ *Some useful information on general planning and control topics.*

http://www.erpfans.com/ *Yes, even ERP has its own fan club! Debates and links for the enthusiast.*

http://www.sap.com/index.epx 'Helping to build better businesses for more than three decades', *SAP has been the leading worldwide supplier of ERP systems for ages. They should know how to do it by now!*

http://www.sapfans.com/ *Another fan club, this one is for SAP enthusiasts.*

http://www.apics.org *The American professional and education body that has its roots in planning and control activities.*

Interactive quiz

For further resources including examples, animated diagrams, self-test questions, Excel spreadsheets and video materials please explore the eText on the companion website at www.pearsoned.co.uk/slack.

Materials requirements planning (MRP)

Introduction

Materials requirements planning (MRP) is an approach to calculating how many parts or materials of particular types are required and what times they are required. This requires data files which, when the MRP program is run, can be checked and updated. Figure 10.14 shows how these files relate to each other. The first inputs to materials requirements planning are customer orders and forecast demand. MRP performs its calculations based on the combination of these two parts of future demand. All other requirements are derived from, and dependent on, this demand information.

Master production schedule

The master production schedule (MPS) forms the main input to materials requirements planning and contains a statement of the volume and timing of the end products to be made. It drives all the production and supply activities that eventually will come together to form the end products. It is the basis for the planning and utilisation of labour and equipment, and it determines the provisioning of materials and cash. The MPS should include all sources of demand, such as spare parts, internal production promises, etc. For example, if a manufacturer of earth excavators plans an exhibition of its products and allows a project team to raid the stores so that it can build two pristine examples to be exhibited, this is likely to leave the factory short of parts. MPS can also be used in service organisations. For example, in a hospital theatre there is a master schedule that contains a statement of which operations are planned and when. This can be used to provision materials for the operations, such as the sterile instruments, blood and dressings. It may also govern the scheduling of staff for operations.

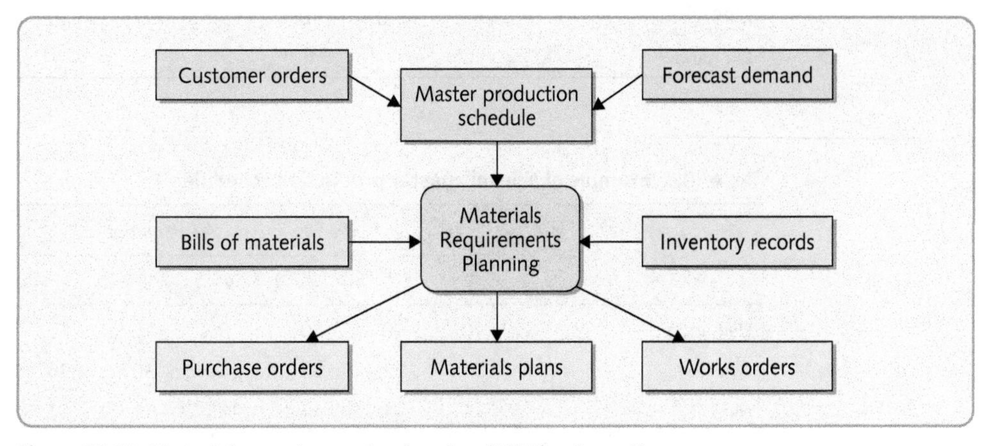

Figure 10.14 Materials requirements planning (MRP) schematic

The master production schedule record

Master production schedules are time-phased records of each end product, which contain a statement of demand and currently available stock of each finished item. Using this information, the available inventory is projected ahead in time. When there is insufficient inventory to satisfy forward demand, order quantities are entered on the master schedule line. Table 10.2 is a simplified example of part of a master production schedule for one item. In the first row the known sales orders and any forecast are combined to form 'Demand'. The second row, 'Available', shows how much inventory of this item is expected to be in stock at the end of each weekly period. The opening inventory balance, 'On hand', is shown separately at the bottom of the record. The third row is the master production schedule, or MPS; this shows how many finished items need to be completed and available in each week to satisfy demand.

'Chase demand' or level master production schedules

In the example in Table 10.2, the MPS increases as demand increases and aims to keep available inventory at 0. The master production schedule is 'chasing' demand (see Chapter 10) and so adjusting the provision of resources. An alternative 'levelled' MPS for this situation is shown in Table 10.3. Level scheduling involves averaging the amount required to be completed to smooth out peaks and troughs; it generates more inventory than the previous MPS.

'Available to promise' (ATP)

The master production schedule provides the information to the sales function on what can be promised to customers and when delivery can be promised. The sales function can load known sales orders against the master production schedule and keep track of what is available to promise (ATP) (see Table 10.4). The ATP line in the master production schedule shows the maximum that is still available in any one week, against which sales orders can be loaded.

Table 10.2 Example of a master production schedule

		Week number								
		1	*2*	*3*	*4*	*5*	*6*	*7*	*8*	*9*
Demand		10	10	10	10	15	15	15	20	20
Available		20	10	0	0	0	0	0	0	0
MPS		0	0	10	10	15	15	15	20	20
On hand	30									

Table 10.3 Example of a 'level' master production schedule

		Week number								
		1	*2*	*3*	*4*	*5*	*6*	*7*	*8*	*9*
Demand		10	10	10	10	15	15	15	20	20
Available		31	32	33	34	30	26	22	13	4
MPS		11	11	11	11	11	11	11	11	11
On hand	30									

Table 10.4 Example of a level master production schedule including available to promise

		Week number								
		1	*2*	*3*	*4*	*5*	*6*	*7*	*8*	*9*
Demand		10	10	10	10	15	15	15	20	20
Sales orders		10	10	10	8	4				
Available		31	32	33	34	30	26	22	13	4
ATP		31	1	1	3	7	11	11	11	11
MPS		11	11	11	11	11	11	11	11	11
On hand	30									

The bill of materials (BOM)

From the master schedule, MRP calculates the required volume and timing of assemblies, sub-assemblies and materials. To do this it needs information on what parts are required for each product. This is called the 'bill of materials'. Initially it is simplest to think about these as a product structure. The product structure in Figure 10.15 is a simplified structure showing the parts required to make a simple board game. Different 'levels of assembly' are shown with the finished product (the boxed game) at level 0, the parts and sub-assemblies that go into the boxed game at level 1, the parts that go into the sub-assemblies at level 2, and so on.

A more convenient form of the product structure is the 'indented bill of materials'. Table 10.5 shows the whole indented bill of materials for the board game. The term 'indented'

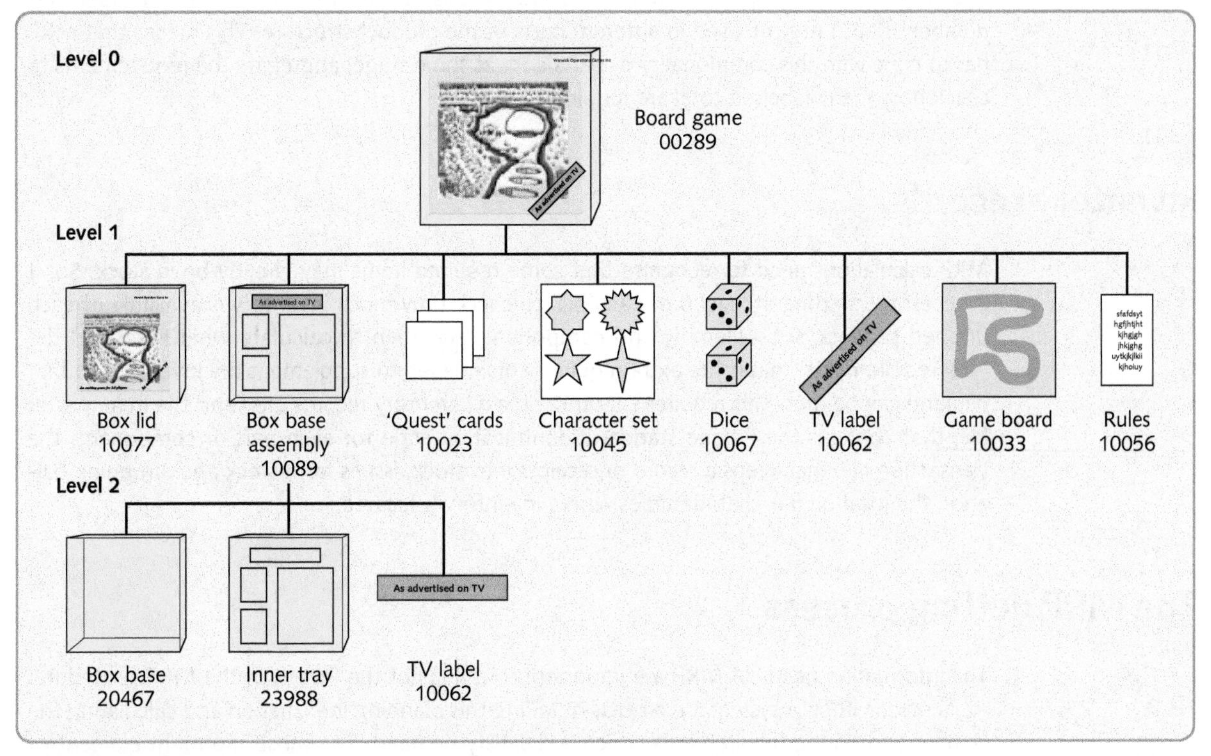

Figure 10.15 Product structure for a simple board game

Table 10.5 Indented bill of materials for board game

Part number: 00289
Description: Board game
Level: 0

Level	*Part number*	*Description*	*Quantity*
0	00289	Board game	1
.1	10077	Box lid	1
.1	10089	Box base assembly	1
..2	20467	Box base	1
..2	10062	TV label	1
..2	23988	Inner tray	1
.1	10023	Quest cards set	1
.1	10045	Character set	1
.1	10067	Die	2
.1	10062	TV label	1
.1	10033	Game board	1
.1	10056	Rules booklet	1

refers to the indentation of the level of assembly, shown in the left-hand column. Multiples of some parts are required; this means that MRP has to know the required number of each part to be able to multiply up the requirements. Also, the same part (for example, the TV label, part number 10062) may be used in different parts of the product structure. This means that MRP has to cope with this commonality of parts and, at some stage, aggregate the requirements to check how many labels in total are required.

Inventory records

MRP calculations need to recognise that some required items may already be in stock. So, it is necessary, starting at level 0 of each bill, to check how much inventory is available of each finished product, sub-assembly and component, and then to calculate what is termed the 'net' requirements, that is the extra requirements needed to supplement the inventory so that demand can be met. This requires that three main inventory records are kept: the item master file, that contains the unique standard identification code for each part or component; the transaction file, that keeps a record of receipts into stock, issues from stock and a running balance; the location file, that identifies where inventory is located.

The MRP netting process

The information needs of MRP are important, but it is not the 'heart' of the MRP procedure. At its core, MRP is a systematic process of taking this planning information and calculating the volume and timing requirements which will satisfy demand. The most important element of this is the MRP netting process.

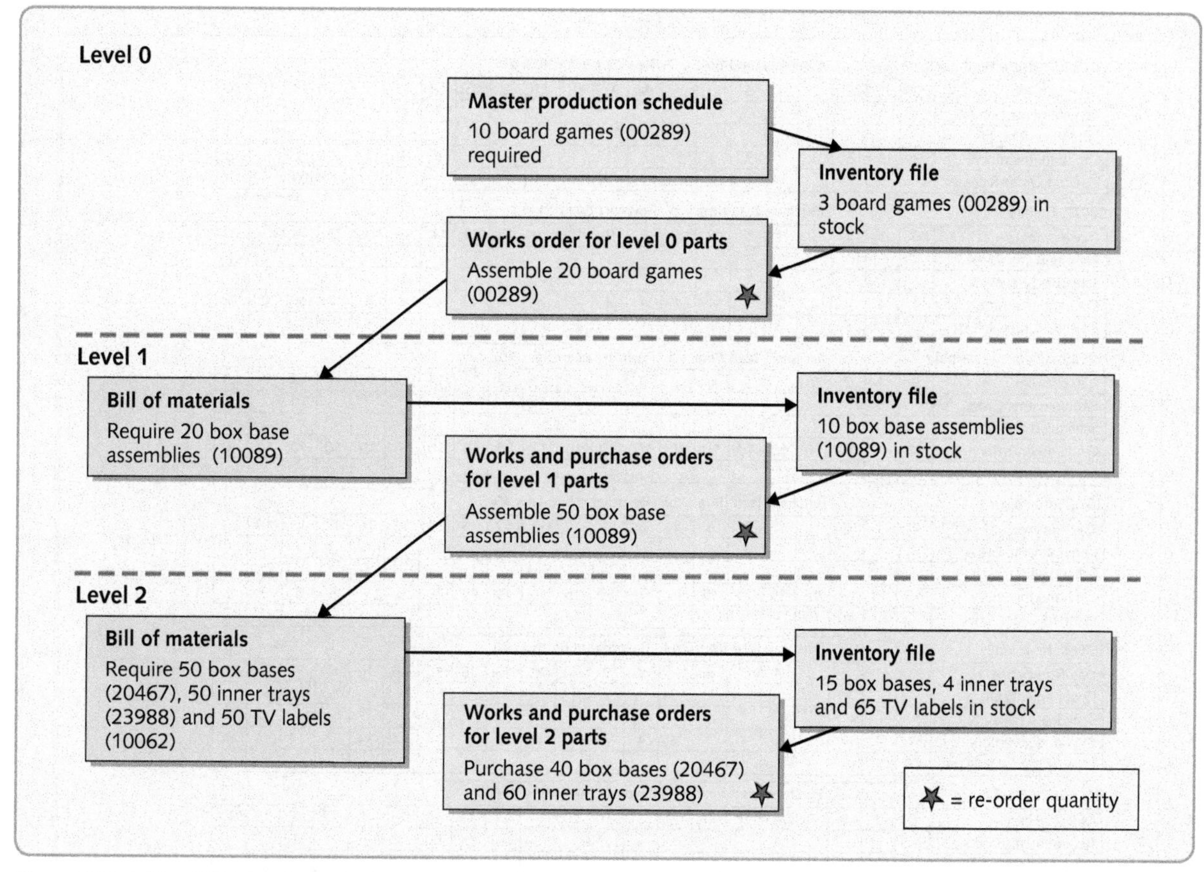

Figure 10.16 The MRP netting calculations for the simple board game

Figure 10.16 illustrates the process that MRP performs to calculate the volumes of materials required. The master production schedule is 'exploded', examining the implications of the schedule through the bill of materials, checking how many sub-assemblies and parts are required. Before moving down the bill of materials to the next level, MRP checks how many of the required parts are already available in stock. It then generates 'works orders', or requests, for the net requirements of items. These form the schedule which is again exploded through the bill of materials at the next level down. This process continues until the bottom level of the bill of materials is reached.

Back-scheduling

In addition to calculating the volume of materials required, MRP also considers when each of these parts is required, that is, the timing and scheduling of materials. It does this by a process called back-scheduling which takes into account the lead time (the time allowed for completion of each stage of the process) at every level of assembly. Again using the example of the board game, assume that ten board games are required to be finished by a notional planning day which we will term day 20. To determine when we need to start work on all the parts that make up the game, we need to know all the lead times that are stored in MRP files for each part (see Table 10.6).

Using the lead time information, the program is worked backwards to determine the tasks that have to be performed and the purchase orders that have to be placed. Given the lead times and inventory levels shown in Table 10.6, the MRP records shown in Figure 10.17 can be derived.

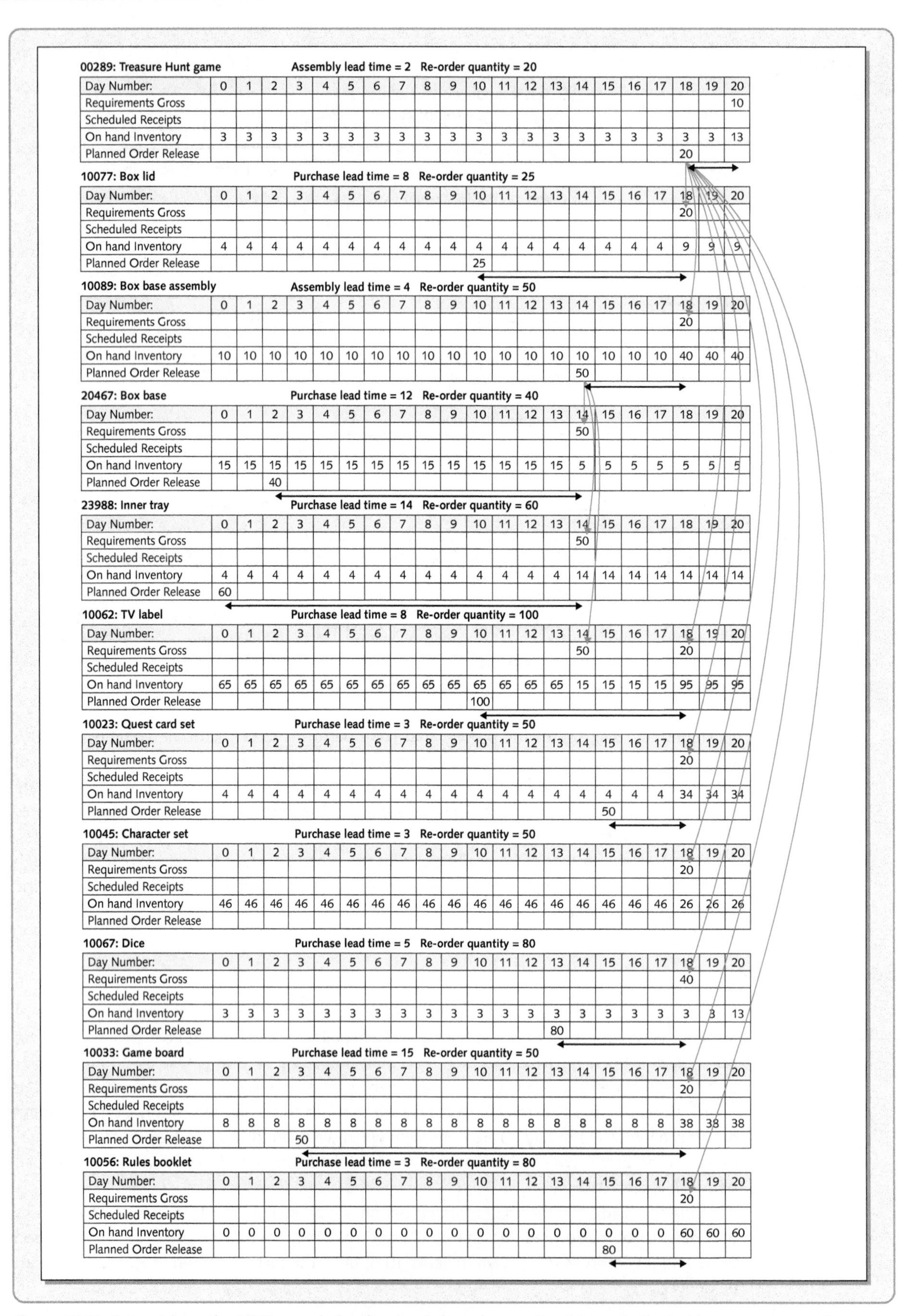

00289: Treasure Hunt game — Assembly lead time = 2 Re-order quantity = 20

Day Number:	0	1	2	3	4	5	6	7	8	9	10	11	12	13	14	15	16	17	18	19	20
Requirements Gross																					10
Scheduled Receipts																					
On hand Inventory	3	3	3	3	3	3	3	3	3	3	3	3	3	3	3	3	3	3	3	3	13
Planned Order Release																			20		

10077: Box lid — Purchase lead time = 8 Re-order quantity = 25

Day Number:	0	1	2	3	4	5	6	7	8	9	10	11	12	13	14	15	16	17	18	19	20
Requirements Gross																			20		
Scheduled Receipts																					
On hand Inventory	4	4	4	4	4	4	4	4	4	4	4	4	4	4	4	4	4	4	9	9	9
Planned Order Release											25										

10089: Box base assembly — Assembly lead time = 4 Re-order quantity = 50

Day Number:	0	1	2	3	4	5	6	7	8	9	10	11	12	13	14	15	16	17	18	19	20
Requirements Gross																			20		
Scheduled Receipts																					
On hand Inventory	10	10	10	10	10	10	10	10	10	10	10	10	10	10	10	10	10	10	40	40	40
Planned Order Release															50						

20467: Box base — Purchase lead time = 12 Re-order quantity = 40

Day Number:	0	1	2	3	4	5	6	7	8	9	10	11	12	13	14	15	16	17	18	19	20
Requirements Gross															50						
Scheduled Receipts																					
On hand Inventory	15	15	15	15	15	15	15	15	15	15	15	15	15	15	5	5	5	5	5	5	5
Planned Order Release			40																		

23988: Inner tray — Purchase lead time = 14 Re-order quantity = 60

Day Number:	0	1	2	3	4	5	6	7	8	9	10	11	12	13	14	15	16	17	18	19	20
Requirements Gross															50						
Scheduled Receipts																					
On hand Inventory	4	4	4	4	4	4	4	4	4	4	4	4	4	4	14	14	14	14	14	14	14
Planned Order Release	60																				

10062: TV label — Purchase lead time = 8 Re-order quantity = 100

Day Number:	0	1	2	3	4	5	6	7	8	9	10	11	12	13	14	15	16	17	18	19	20
Requirements Gross															50				20		
Scheduled Receipts																					
On hand Inventory	65	65	65	65	65	65	65	65	65	65	65	65	65	65	15	15	15	15	95	95	95
Planned Order Release											100										

10023: Quest card set — Purchase lead time = 3 Re-order quantity = 50

Day Number:	0	1	2	3	4	5	6	7	8	9	10	11	12	13	14	15	16	17	18	19	20
Requirements Gross																			20		
Scheduled Receipts																					
On hand Inventory	4	4	4	4	4	4	4	4	4	4	4	4	4	4	4	4	4	4	34	34	34
Planned Order Release																50					

10045: Character set — Purchase lead time = 3 Re-order quantity = 50

Day Number:	0	1	2	3	4	5	6	7	8	9	10	11	12	13	14	15	16	17	18	19	20
Requirements Gross																			20		
Scheduled Receipts																					
On hand Inventory	46	46	46	46	46	46	46	46	46	46	46	46	46	46	46	46	46	46	26	26	26
Planned Order Release																					

10067: Dice — Purchase lead time = 5 Re-order quantity = 80

Day Number:	0	1	2	3	4	5	6	7	8	9	10	11	12	13	14	15	16	17	18	19	20
Requirements Gross																			40		
Scheduled Receipts																					
On hand Inventory	3	3	3	3	3	3	3	3	3	3	3	3	3	3	3	3	3	3	3	3	13
Planned Order Release														80							

10033: Game board — Purchase lead time = 15 Re-order quantity = 50

Day Number:	0	1	2	3	4	5	6	7	8	9	10	11	12	13	14	15	16	17	18	19	20
Requirements Gross																			20		
Scheduled Receipts																					
On hand Inventory	8	8	8	8	8	8	8	8	8	8	8	8	8	8	8	8	8	8	38	38	38
Planned Order Release				50																	

10056: Rules booklet — Purchase lead time = 3 Re-order quantity = 80

Day Number:	0	1	2	3	4	5	6	7	8	9	10	11	12	13	14	15	16	17	18	19	20
Requirements Gross																			20		
Scheduled Receipts																					
On hand Inventory	0	0	0	0	0	0	0	0	0	0	0	0	0	0	0	0	0	0	60	60	60
Planned Order Release																80					

Figure 10.17 Extract from the MRP records for the simple board game (lead times indicated by arrows ←→)

Table 10.6 Back-scheduling of requirements in MRP

Part no.	*Description*	*Inventory on-hand day 0*	*Lead time (days)*	*Re-order quantity*
00289	Board game	3	2	20
10077	Box lid	4	8	25
10089	Box base assy	10	4	50
20467	Box base	15	12	40
23988	Inner tray	4	14	60
10062	TV label	65	8	100
10023	Quest cards set	4	3	50
10045	Character set	46	3	50
10067	Die	22	5	80
10033	Game board	8	15	50
10056	Rules booklet	0	3	80

MRP capacity checks

The MRP process needs a feedback loop to check whether a plan was achievable and whether it has actually been achieved. Closing this planning loop in MRP systems involves checking production plans against available capacity and, if the proposed plans are not achievable at any level, revising them. All but the simplest MRP systems are now closed-loop systems. They use three planning routines to check production plans against the operation's resources at three levels.

- *Resource requirements plans (RRPs)* involve looking forward in the long term to predict the requirements for large structural parts of the operation, such as the numbers, locations and sizes of new plants.
- *Rough-cut capacity plans (RCCPs)* are used in the medium to short term to check the master production schedules against known capacity bottlenecks, in case capacity constraints are broken. The feedback loop at this level checks the MPS and key resources only.
- *Capacity requirements plans (CRPs)* look at the day-to-day effects of the works orders issued from the MRP on the loading individual process stages.

11

Lean synchronisation

Introduction

Lean synchronisation aims to meet demand instantaneously, with perfect quality and no waste. This involves supplying products and services in perfect synchronisation with the demand for them, using 'lean' or 'just-in-time' (JIT) principles. These principles were once a radical departure from traditional operations practice, but have now become orthodox in promoting the synchronisation of flow through processes, operations and supply networks (see Figure 11.1).

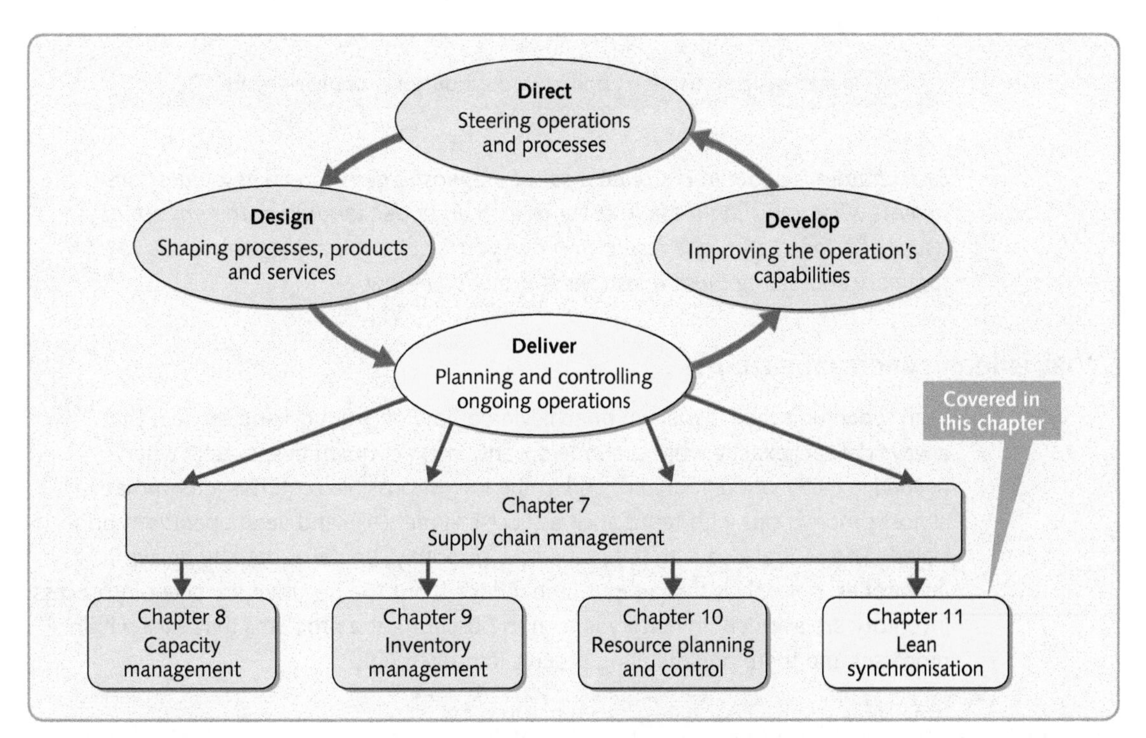

Figure 11.1 Lean synchronisation is the aim of achieving a flow of products and services that always delivers exactly what customers want, in exact quantities, exactly when needed, exactly where required, and at the lowest possible cost

EXECUTIVE SUMMARY

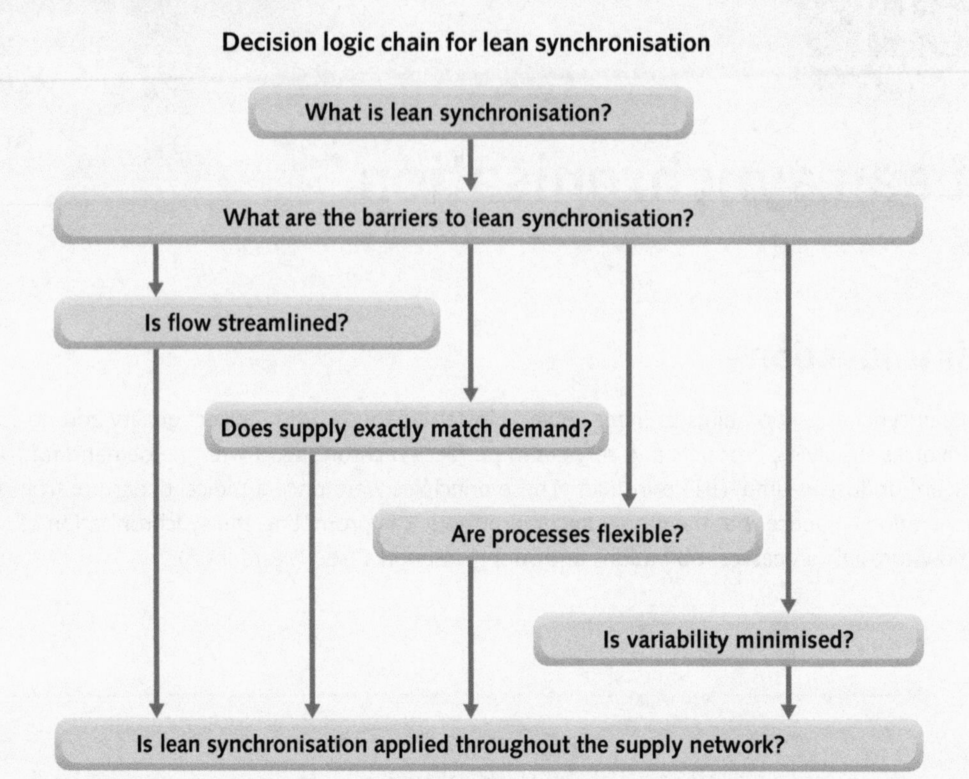

Each chapter is structured around a set of diagnostic questions. These questions suggest what you should ask in order to gain an understanding of the important issues of a topic, and, as a result, improve your decision making. An executive summary addressing these questions is provided below.

What is lean synchronisation?

Lean synchronisation is the aim of achieving a flow of products and services that always delivers exactly what customers want, in exact quantities, exactly when needed, exactly where required, and at the lowest possible cost. It is a term that is almost synonymous with terms such as 'just-in-time' (JIT) and 'lean operations principles'. The central idea is that if items flow smoothly, uninterrupted by delays in inventories, not only is throughput time reduced, but the negative effects of in-process inventory are avoided. Inventory is seen as obscuring the problems that exist within processes and therefore inhibiting process improvement.

What are the barriers to lean synchronisation?

The aim of lean synchronisation can be inhibited in three ways. First is the failure to eliminate waste in all parts of the operation. The causes of waste are more extensive than is generally understood. The second is a failure to involve all the people within the operation in the shared task of smoothing flow and eliminating waste. Japanese proponents of lean synchronisation often use a set of 'basic working practices' to ensure involvement. Third is the failure to adopt continuous improvement principles. Because pure lean synchronisation is an aim rather than something that can be implemented quickly, it requires the continual application of incremental improvement steps to reach it.

Is flow streamlined?

Long process routes are wasteful and cause delay and inventory build up. Physically reconfiguring processes to reduce distance travelled and aid cooperation between staff can help to streamline flow. Similarly, ensuring flow visibility helps to make improvement to flow easier. Sometimes this can involve small-scale technologies that can reduce fluctuations in flow volume.

Does supply exactly match demand?

The aim of lean synchronisation is to meet demand exactly; neither too much nor too little and only when it is needed. Achieving this often means pull control principles. The most common method of doing this is the use of kanbans, simple signalling devices that prevent the accumulation of excess inventory.

Are processes flexible?

Responding exactly to demand only when it is need often requires a degree of flexibility in processes, both to cope with unexpected demand and to allow processes to change between different activities without excessive delay. This often means reducing changeover times in technologies.

Is variability minimised?

Variability in processes disrupts flow and prevents lean synchronisation. Variability includes quality variability and schedule variability. Statistical process control (SPC) principles are useful in reducing quality variability. The use of levelled scheduling and mixed modelling can be used to reduce flow variability and total productive maintenance (TPM) can reduce variability caused by breakdowns.

Is lean synchronisation applied throughout the network?

The same benefits of lean synchronisation that apply within operations can also apply between operations. Furthermore, the same principles that can be used to achieve lean synchronisation within operations can be used to achieve it between operations. This is more difficult, partly because of the complexity of flow and partly because supply networks are prone to the type of unexpected fluctuations that are easier to control within operations.

DIAGNOSTIC QUESTION

What is lean synchronisation?

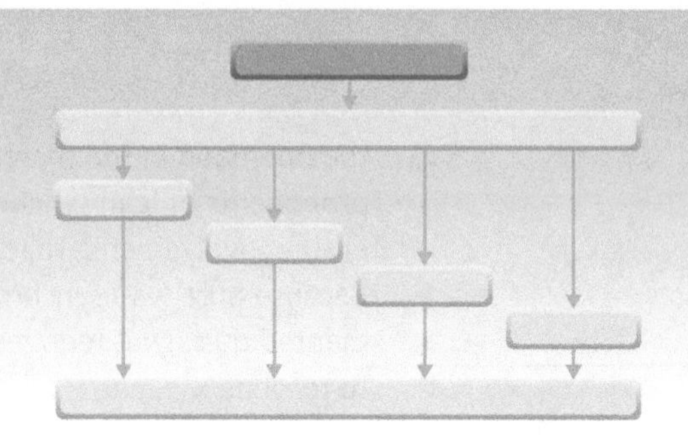

Synchronisation means that the flow of products and services always delivers exactly what customers want (perfect quality), in exact quantities (neither too much nor too little), exactly when needed (not too early or too late), and exactly where required (not to the wrong location). *Lean* synchronisation is to do all this at the lowest possible cost. It results in items flowing rapidly and smoothly through processes, operations and supply networks.

The benefits of synchronised flow

The best way to understand how lean synchronsation differs from more traditional approaches to managing flow is to contrast the two simple processes in Figure 11.2. The traditional approach assumes that each stage in the process will place its output in an inventory that 'buffers' that stage from the next one downstream in the process. The next stage down will then (eventually) take outputs from the inventory, process them, and pass them through to the next buffer inventory. These buffers are there to 'insulate each stage from its neighbours, making each stage relatively independent so that if, for example, stage A stops operating for some reason, stage B can continue, at least for a time. The larger the buffer inventory, the greater the degree of insulation between the stages. This insulation has to be paid for in terms

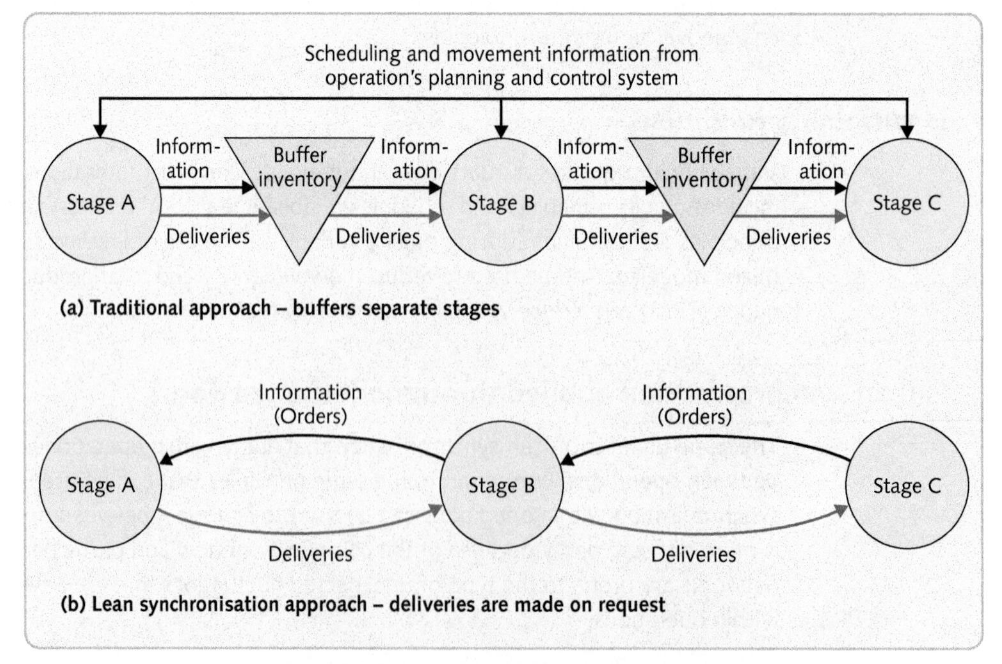

Figure 11.2 (a) Traditional and (b) lean synchronised flow between stages

of inventory and slow throughput times because items will spend time waiting in the buffer inventories.

But, the main 'learning' argument against this traditional approach lies in the very conditions it seeks to promote, namely the insulation of the stages from one another. When a problem occurs at one stage, the problem will not immediately be apparent elsewhere in the system. The responsibility for solving the problem will be centred largely on the people within that stage, and the consequences of the problem will be prevented from spreading to the whole system. However, contrast this with the pure lean synchronised process illustrated in Figure 11.2. Here items are processed and then passed directly to the next stage 'just-in-time' for them to be processed further. Problems at any stage have a very different effect in such a system. Now if stage A stops processing, stage B will notice immediately and stage C very soon after. Stage A's problem is now quickly exposed to the whole process, which is immediately affected by the problem. This means that the responsibility for solving the problem is no longer confined to the staff at stage A. It is now shared by everyone, considerably improving the chances of the problem being solved, if only because it is now too important to be ignored. In other words, by preventing items accumulating between stages, the operation has increased the chances of the intrinsic efficiency of the plant being improved.

Non-synchronised approaches seek to encourage efficiency by protecting each part of the process from disruption. The lean synchronised approach takes the opposite view. Exposure of the system (although not suddenly, as in our simplified example) to problems can both make them more evident and change the 'motivation structure' of the whole system towards solving the problems. Lean synchronisation sees accumulations of inventory as a 'blanket of obscurity' that lies over the production system and prevents problems being noticed. This same argument can be applied when, instead of queues of material, or information (inventory), an operation has to deal with queues of customers. Table 11.1 shows how certain aspects of inventory are analogous to certain aspects of queues.

OPERATIONS PRINCIPLE

Buffer inventory used to insulate stages or processes localises the motivation to improve.

Table 11.1 Inventories of materials – information or customers have similar characteristics

	Inventory		
	Of material (queue of material)	***Of information (queue of information)***	***Of customers (queue of people)***
Cost	Ties up working capital	Less current information and so worth less	Wastes customers' time
Space	Needs storage space	Needs memory capacity	Needs waiting area
Quality	Defects hidden, possible damage	Defects hidden, possible data corruption	Gives negative perception
Decoupling	Makes stages independent	Makes stages independent	Promotes job specialisation/fragmentation
Utilisation	Stages kept busy by work in progress	Stages kept busy by work in data queues	Servers kept busy by waiting customers
Coordination	Avoids need for synchronisation	Avoids need for straight-through processing	Avoids having to match supply and demand

Source: Adapted from Fitzsimmons, J.A. (1990) 'Making Continual Improvement: A Competitive Strategy for Service Firms' *in* Bowen, D.E., Chase, R.B., Cummings, T.G. and Associates (eds) *Service Management Effectiveness*, Jossey-Bass.

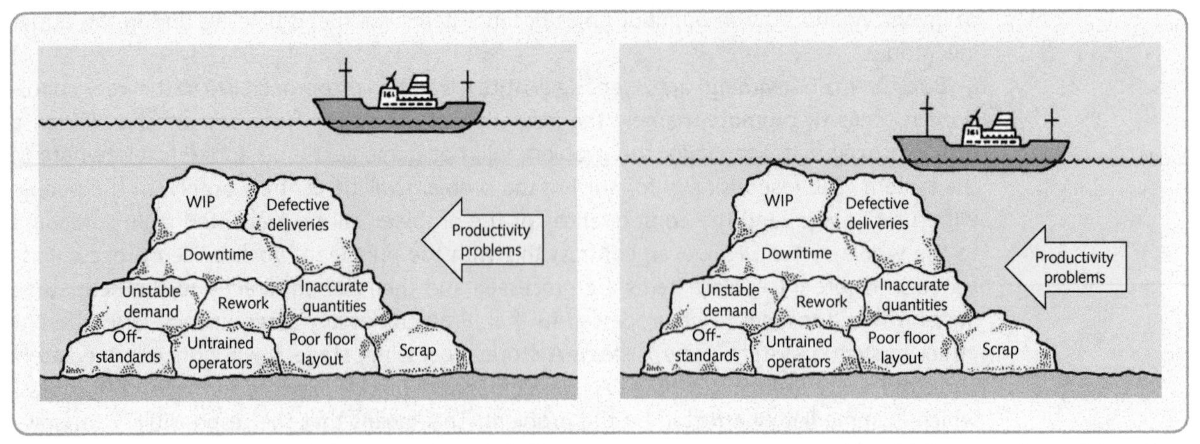

Figure 11.3 Reducing the level of inventory (water) allows operations management (the ship) to see the problems in the operation (the rocks) and work to reduce them

The river and rocks analogy

The idea of obscuring effects of inventory is often illustrated diagrammatically, as in Figure 11.3. The many problems of the operation are shown as rocks in a river bed that cannot be seen because of the depth of the water. The water in this analogy represents the inventory in the operation. Yet, even though the rocks cannot be seen, they slow the progress of the river's flow and cause turbulence. Gradually reducing the depth of the water (inventory) exposes the worst of the problems which can be resolved, after which the water is lowered further, exposing more problems, and so on. The same argument will also apply for the flow between whole processes, or whole operations. For example, stages A, B and C in Figure 11.2 could be a supplier operation, a manufacturer and a customer's operation, respectively.

Synchronisation, 'lean' and 'just-in-time'

Different terms are used to describe what here we call lean synchronisation. Our shortened definition – *'lean synchronisation aims to meet demand instantaneously, with perfect quality and no waste'* – could also be used to describe the general concept of 'lean', or 'just-in-time' (JIT). The concept of 'lean' stresses the elimination of waste, while 'just-in-time' emphasises the idea of producing items only when they are needed. But all three concepts overlap to a large degree, and no definition fully conveys the full implications for operations practice. Here we use the term lean synchronisation because it best describes the impact of these ideas on flow and delivery.

Two operations that have implemented lean synchronisation are briefly described below. One is the company that is generally credited with doing most to develop the whole concept; the other is a not-for-profit hospital that nevertheless has derived benefits from adopting some of the principles.

EXAMPLE

Toyota[1]

Seen as the leading practitioner and the main originator of the lean approach, the Toyota Motor Company has progressively synchronised all its processes simultaneously to give high quality, fast throughput and exceptional productivity. It has done this by developing a set of practices that has largely shaped what we now call 'lean' or 'just-in-time' but which Toyota calls the Toyota Production System (TPS). The TPS has two themes: 'just-in-time' and *jidoka*. Just-in-time is defined as the rapid and coordinated movement of parts throughout the production system and supply network to meet customer demand. It is operationalised by means of *heijunka* (levelling and smoothing the flow of items), *kanban* (signalling to the preceding process that more parts are needed) and *nagare* (laying out processes to achieve smoother flow of parts throughout the production process). *Jidoka* is described as 'humanising the interface

Source: Courtesy of Toyota Motor Manufacturing (UK) Ltd

between operator and machine'. Toyota's philosophy is that the machine is there to serve the operator's purpose. The operator should be left free to exercise his/her judgement. *Jidoka* is operationalised by means of fail-safeing (or machine *jidoka*), line-stop authority (or human *jidoka*), and visual control (at-a-glance status of production processes and visibility of process standards).

Toyota believe that both just-in-time and *jidoka* should be applied ruthlessly to the elimination of waste, where waste is defined as 'anything other than the minimum amount of equipment, items, parts and workers that are absolutely essential to production'. Fujio Cho of Toyota identified seven types of waste that must be eliminated from all operations processes. They are: waste from over production; waste from waiting time; transportation waste; inventory waste; processing waste; waste of motion; waste from product defects. Beyond this, authorities on Toyota claim that its strength lies in understanding the differences between the tools and practices used with Toyota operations and the overall philosophy of their approach to lean synchronisation. This is what some have called the apparent paradox of the Toyota production system, 'namely, that activities, connections and production flows in a Toyota factory are rigidly scripted, yet at the same time Toyota's operations are enormously flexible and adaptable. Activities and processes are constantly being challenged and pushed to a higher level of performance, enabling the company to continually innovate and improve'.

One influential study of Toyota identified four rules that guide the design, delivery, and development activities within the company.

1. All work shall be highly specified as to content, sequence, timing and outcome.
2. Every customer–supplier connection must be direct and there must be an unambiguous yes or no method of sending requests and receiving responses.
3. The route for every product and service must be simple and direct.
4 Any improvement must be made in accordance with the scientific method, under the guidance of a teacher, and at the lowest possible level in the organisation.

EXAMPLE

Lean hospitals[2]

In one of the increasing number of healthcare services to adopt lean principles, the Bolton Hospitals National Health Service Trust in the north of the UK, has reduced one of its hospital's mortality rate from one injury by more than a third. David Fillingham, Chief Executive of Bolton Hospitals NHS Trust said, *'We had far more people dying from fractured hips than should have been dying.'* Then the trust greatly reduced its mortality rate for fractured neck of femur by redesigning the patient's stay in hospital to reduce or remove the waits between 'useful activity'. The mortality rate fell from 22.9 per cent to 14.6 per cent, which is the equivalent of 14 more patients surviving every six months. At the same time, the average length of stay fell by a third from 34.6 days to 23.5 days.

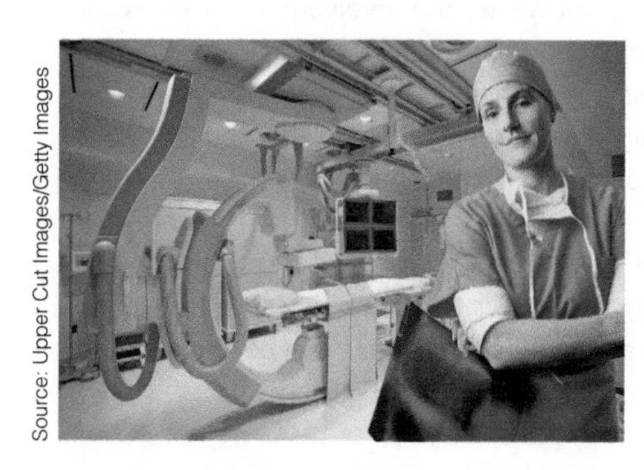
Source: Upper Cut Images/Getty Images

The trust held five 'rapid improvement events', involving employees from across the organisation who spent several days examining processes and identifying alternative ways how to improve them. Some management consultants were also used but strictly in an advisory role. In addition third-party experts were brought in. These included staff from the Royal Air Force, who has been applying lean principles to running aircraft carriers. The value of these outsiders was not only their expertise. *'They asked all sorts of innocent, naïve questions,'* said Mr Fillingham, *'to which, often, no member of staff has an answer'*. Other lean-based improvement initiatives included examining the patient's whole experience from start to finish so that delays (some of which could prove fatal) could be removed on their journey to the operating theatre, radiology processes were speeded up and unnecessary paperwork was eliminated. Cutting the length of stay and reducing process complications should also start to reduce costs, although Mr Fillingham says that it could take several years for the savings to become substantial. Not only that, but staff are also said to be helped by the changes because they can spend more time helping patients rather than doing non-value added activities.

Meanwhile at Salisbury district hospital in the south of the UK, lean principles have reduced delays in waiting for the results of tests from the ultrasound department. Waiting lists have been reduced from 12 weeks to between two weeks and zero after an investigation showed that 67 per cent of demand was coming from just 5 per cent of possible ultrasound tests: abdominal, gynaecological and urological. So all work was streamed into routine 'green' streams and complex 'red' ones. This is like having different traffic lanes on a motorway dedicated to different types of traffic, with fast cars in one lane and slow trucks in another. Mixing both types of work is like mixing fast cars and slow-moving trucks in all lanes. The department then concentrated on doing the routine 'green' work more efficiently. For example, the initial date scan used to check the age of a foetus took only two minutes, so a series of five-minute slots were allocated just for these. *'The secret is to get the steady stream of high-volume, low-variety chugging down the ultrasound motorway,'* says Kate Hobson, who runs the department. Streaming routine work in this way has left more time to deal with the more complex jobs, yet staff are not overloaded. They are more likely to leave work on time and also believe that the department is doing a better job, all of which has improved morale says Kate Hobson. *'I think people feel their day is more structured now. It's not that madness, opening the doors and people coming at you.'* Nor has this more disciplined approach impaired the department's ability to treat really urgent jobs. In fact it has stopped leaving space in its schedule for emergencies – the, now standard, short waiting time is usually sufficient for urgent jobs.

What do these two examples have in common?

Here are two types of operation separated by product, culture, size, location, and their route to adopting lean synchronisation principles. Toyota took decades to develop a fully integrated and coherent philosophy to managing their operations and have become one of the world's leading and most profitable automotive companies as a result. The hospitals are far earlier in their path to lean synchronisation, yet they have adopted and adapted several ideas from the lean synchronisation philosophy and gained benefits. The exact interpretation of what 'lean' means in practice will differ (and 'lean' is something of a fashion in healthcare) but it still has the potential to improve their service delivery. That is because, notwithstanding the differences between the two operations, the basic principles of lean synchronisation remain the same, namely aiming to achieve perfect synchronisation through smooth and even flow. But, lean synchronisation is an *aim*. It is not something that can simply be implemented overnight. Both these organisations have worked hard at overcoming the barriers to lean synchronisation. These can be summarised as: the elimination of all waste; the involvement of everyone in the business; the adoption of a continuous improvement philosophy. The focus on eliminating waste uses four important methods: streamlining flow; making sure that supply matches demand exactly; increasing process flexibility; reducing the effects of variability. And, although rooted in manufacturing, the techniques of lean or just-in-time philosophies are now being extended to service operations.

Before further discussion, it is important to be clear on the distinction between the *aim* (lean synchronisation), the *approach to overcoming the barriers* to achieving lean synchronisation, the *methods of eliminating waste*, and the various *techniques* that can be used to help eliminate waste. The relationship between these elements is shown in Figure 11.4.

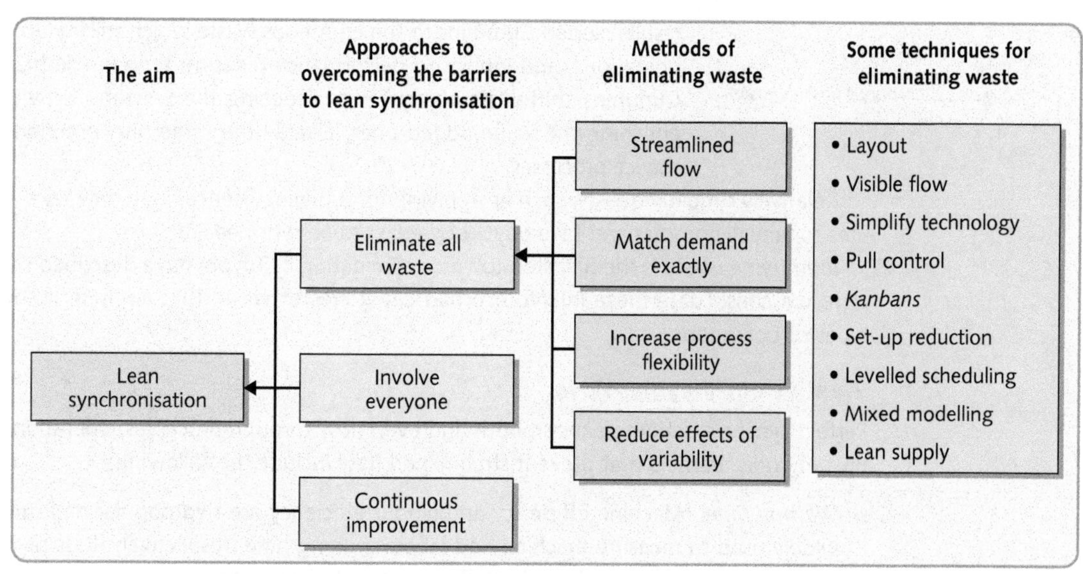

Figure 11.4 Schematic of the issues covered in this chapter

DIAGNOSTIC QUESTION

What are the barriers to lean synchronisation?

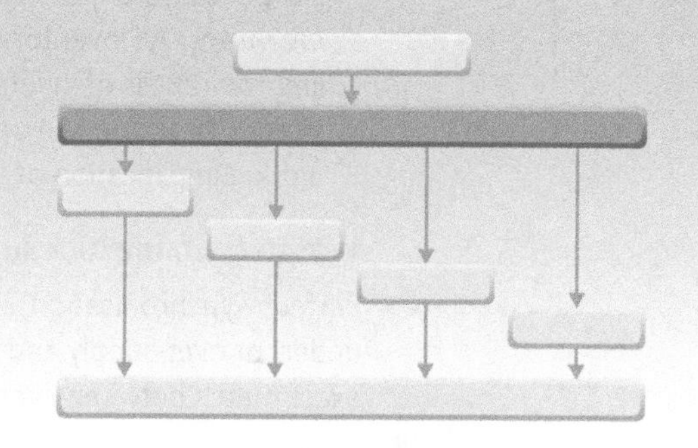

Video

The aim of pure lean synchronisation represents an ideal of smooth, uninterrupted flow without delay, waste or imperfection of any kind. The supply and demand between stages in each process, between processes in each operation, and between operations in each supply network, are all perfectly synchronised. It represents the ultimate in what customers are looking from an operation. But first one must identify the barriers to achieving this idea state. We group these under three headings:

- a failure to eliminate waste in all parts of the operation
- a failure to harness the contribution of all the people within the operation
- a failure to establish improvement as a continuous activity.

The waste elimination barrier

Practice note

Arguably the most significant part of the lean philosophy is its focus on the elimination of all forms of waste. Waste can be defined as any activity that does not add value. For example, a study by Cummins Worldwide Fortune 500, the engine company, showed that, at best, an engine was only being worked on for 15 per cent of the time it was in the factory.[3] At worst, this fell to 9 per cent, which meant that for 91 per cent of its time, the operation was adding cost to the engine, not adding value. Although already a relatively efficient manufacturer, the

> **OPERATIONS PRINCIPLE**
> *Focusing on synchronous flow exposes sources of waste.*

results alerted Cummins to the enormous waste which still lay dormant in its operations, and which no performance measure then in use had exposed. Cummins shifted its objectives to reducing the wasteful activities and to enriching the value-added ones. Exactly the same phenomenon applies in service processes.

Relatively simple requests, such as applying for a driving licence, may only take a few minutes to actually process, yet take days (or weeks) to be returned.

Identifying waste is the first step towards eliminating it. Toyota have described seven types. Here we consolidate these into four broad categories of waste that apply in many different types of operation.

Waste from irregular flow

Perfect synchronisation means smooth and even flow through processes, operations and supply networks. Barriers that prevent streamlined flow include the following:

- *Waiting time.* Machine efficiency and labour efficiency are two popular measures that are widely used to measure machine and labour waiting time, respectively. Less obvious is the time when items wait as inventory, there simply to keep operators busy.
- *Transport.* Moving items around the plant, together with double and triple handling, does not add value. Layout changes that bring processes closer together, improvements in transport methods and workplace organisation can all reduce waste.
- *Process inefficiencies.* The process itself may be a source of waste. Some operations may only exist because of poor component design, or poor maintenance, and so could be eliminated.
- *Inventory.* All inventory should become a target for elimination. However, it is only by tackling the causes of inventory, such as irregular flow, that it can be reduced.
- *Wasted motions.* An operator may look busy but sometimes no value is being added by the work. Simplification of work is a rich source of reduction in the waste of motion.

Waste from inexact supply

Perfect synchronisation supplies exactly what is wanted, exactly when it is needed. Any under- or over-supply and any early or late delivery will result in waste. Barriers to achieving an exact match between supply and demand include the following:

- *Over-production or under-production.* Supplying more than, or less than, is immediately needed by the next stage, process or operation. (This is the greatest source of waste according to Toyota).
- *Early or late delivery.* Items should only arrive exactly when they are needed. Early delivery is as wasteful as late delivery.
- *Inventory.* Again, all inventories should become a target for elimination. However, it is only by tackling the causes of inventory, such as inexact supply, that it can be reduced.

Waste from inflexible response

Customer needs can vary, in terms of what they want, how much they want, and when they want it. However, processes usually find it more convenient to change what they do relatively infrequently, because every change implies some kind of cost. That is why hospitals schedule specialist clinics only at particular times, and why machines often make a batch of similar products together. Yet responding to customer demands exactly and instantaneously requires a high degree of process flexibility. Symptoms of inadequate flexibility include the following:

- *Large batches.* Sending batch of items through a process inevitably increases inventory as the batch moves through the whole process.
- *Delays between activities.* The longer the time (and the cost) of changing over from one activity to another, the more difficult it is to synchronise flow to match customer demand instantaneously.

- *More variation in activity mix than in customer demand.* If the mix of activities in different time periods varies more than customer demand varies, then some 'batching' of activities must be taking place.

Waste from variability

Synchronisation implies exact levels of quality. If there is variability in quality levels then customers will not consider themselves as being adequately supplied. Variability therefore is an important barrier to achieving synchronised supply. Symptoms of poor variability include the following:

- *Poor reliability of equipment.* Unreliable equipment usually indicates a lack of conformance in quality levels. It also means that there will be irregularity in supplying customers. Either way, it prevents synchronisation of supply.
- *Defective products or services.* Waste caused by poor quality is significant in most operations. Service or product errors cause both customers and processes to waste time until they are corrected.

But capacity utilisation may be sacrificed in the short term

A paradox in the lean synchronisation concept is that adoption may mean some sacrifice of capacity utilisation. In organisations that place a high value on the utilisation of capacity this can prove particularly difficult to accept. But it is necessary. Return to the process shown in Figure 11.2. When stoppages occur in the traditional system, the buffers allow each stage to continue working and thus achieve high capacity utilisation. The high utilisation does not necessarily make the system as a whole produce more parts. Often the extra production goes into the large buffer inventories. In a synchronised lean process, any stoppage will affect the rest of the system, causing stoppages throughout the operation. This will necessarily lead to lower capacity utilisation, at least in the short term. However, there is no point in producing output just for its own sake. Unless the output is useful and enables the operation as a whole to produce saleable output, there is no point in producing it anyway. In fact, producing just to keep utilisation high is not only pointless, it is counter-productive, because the extra inventory produced merely serves to make improvements less likely. Figure 11.5 illustrates the two approaches to capacity utilisation.

OPERATIONS PRINCIPLE
Focusing on lean synchronisation can initially reduce resource utilisation.

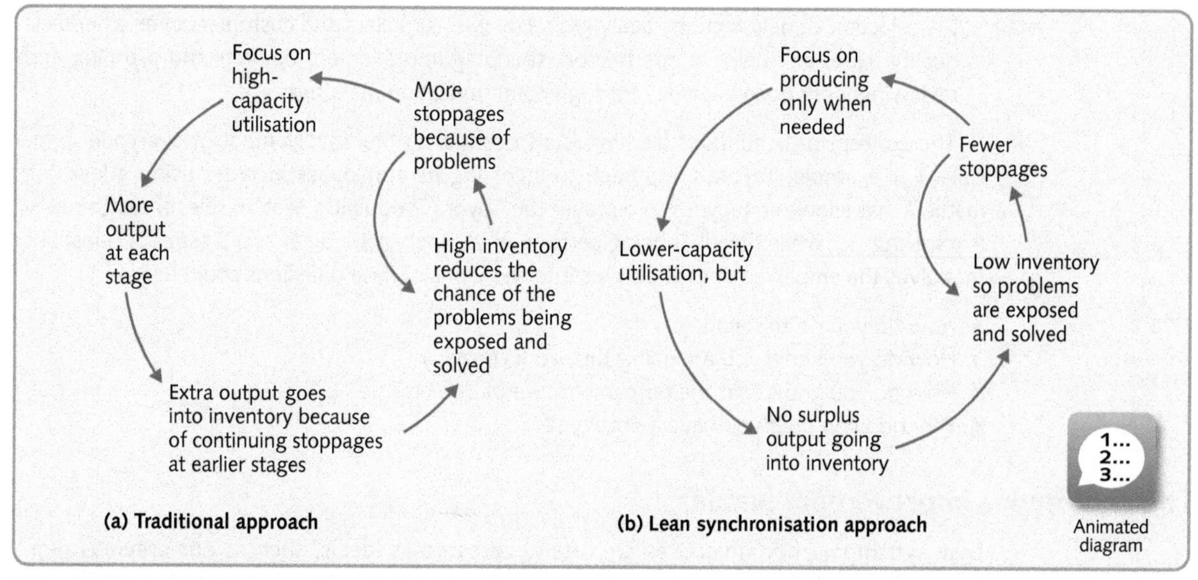

Figure 11.5 The different views of capacity utilisation in (a) traditional and (b) lean synchronisation approaches to planning and controlling flow

The involvement barrier

An organisational culture that supports lean synchronisation must place a very significant emphasis on involving everyone in the organisation. This approach to people management (sometimes called the 'respect-for-people' system, after a rough translation from the Japanese) is seen by some as the most controversial aspect of the lean philosophy. It encourages (and often requires) team-based problem-solving, job enrichment, job rotation and multi-skilling. The intention is to encourage a high degree of personal responsibility, engagement and 'ownership' of the job. Some Japanese companies refer to the operationalising of the 'involvement of everyone' principle by adopting 'basic working practices'. They are held to be the basic preparation of the operation and its employees for implementing lean synchronisation. They include the following:

- **Discipline.** Work standards that are critical for the safety of staff, the environment and quality must be followed by everyone all the time.
- **Flexibility.** It should be possible to expand responsibilities to the extent of people's capabilities. This applies as much to managers as it does to shop-floor personnel. Barriers to flexibility, such as grading structures and restrictive practices, should be removed.
- **Equality.** Unfair and divisive personnel policies should be discarded. Many companies implement the egalitarian message through to company uniforms, consistent pay structures which do not differentiate between full-time staff and hourly-rated staff, and open-plan offices.
- **Autonomy.** Delegate responsibility to people involved in direct activities so that management's task becomes one of supporting processes. Delegation includes giving staff the responsibility for stopping processes in the event of problems, scheduling work, gathering performance monitoring data, and general problem solving.
- **Development of personnel.** Over time, the aim is to create more company members who can support the rigours of being competitive.
- **Quality of working life (QWL).** This may include, for example, involvement in decision-making, security of employment, enjoyment and working area facilities.
- **Creativity.** This is one of the indispensable elements of motivation. Creativity in this context means not just doing a job, but also improving how it is done, and building the improvement into the process.
- **Total people involvement.** Staff take on more responsibility to use their abilities to the benefit of the company as a whole. They are expected to participate in activities such as the selection of new recruits, dealing directly with suppliers and customers over schedules, quality issues and delivery information, spending improvement budgets and planning and reviewing work done each day through communication meetings.

The concept of continuous learning is also central to the 'involvement of everyone' principle. For example, Toyota's approach to involving its employees includes using a learning method that allows employees to discover the Toyota Production System rules through problem solving. So, while the job is being performed, a supervisor/trainer asks a series of questions that gives the employee deeper insights into the work[4]. These questions could be:

- How do you do this work?
- How do you know you are doing this work correctly?
- How do you know that the outcome is free of defects?
- What do you do if you have a problem?

The continuous improvement barrier

Lean synchronisation objectives are often expressed as ideals, such as our previous definition: 'to meet demand instantaneously with perfect quality and no waste'. While any operation's current performance may be far removed from such ideals, a fundamental lean belief is that it is possible to get closer to them over time. Without such beliefs to drive progress, lean proponents claim improvement is more likely to be transitory than continuous. This is why the concept of continuous improvement is such an important part of the lean philosophy. If its aims are set in terms of ideals which individual organisations may never fully achieve, then the emphasis must be on the way in which an organisation moves closer to the ideal state. The Japanese word that incorporates the idea of continuous improvement is *kaizen*. It is one of the main pillars of process improvement and is explained fully in Chapter 13.

Techniques to address the four sources of waste

Of the three barriers to achieving lean synchronisation (reduce waste, involve everyone and adopt continuous improvement), the last two are addressed further in Chapter 13. Therefore the rest of this chapter is devoted to what could be called the 'core' of lean synchronisation. These are a collection of 'just-in-time' tools and techniques that are the means of cutting out waste. Although many of these techniques are used to reduce waste generally within processes, operations, and supply networks, we will group the approaches to reducing waste under the four main headings: streamlining flow; matching demand exactly; increasing process flexibility; reducing the effects of variability.

DIAGNOSTIC QUESTION

Is flow streamlined?

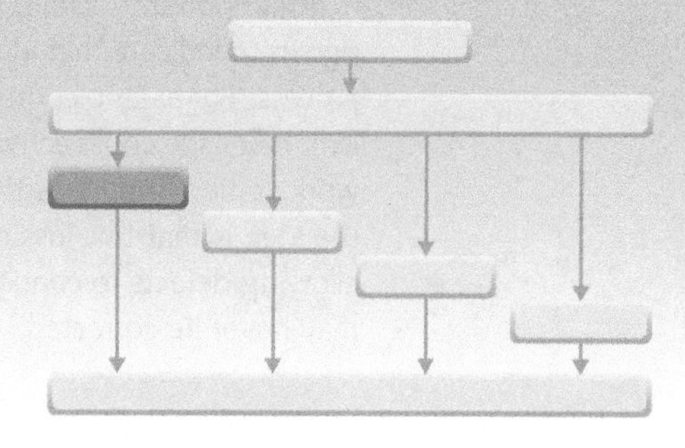

Video

The smooth flow of materials, information and people in the operation is a central idea of lean synchronisation. Long process routes provide opportunities for delay and inventory build-up, add no value, and slow down throughput time. So, the first contribution any operation can make to streamlining flow is to reconsider the basic layout of its processes. Primarily, reconfiguring the layout of a process to aid lean synchronisation involves moving it down the 'natural diagonal' of process design that was discussed in Chapter 4. Broadly speaking, this means moving from functional layouts towards cell-based layouts, or from cell-based layouts towards product layouts. Either way, it is necessary to move towards a layout that brings more systematisation and control to the process flow. At a more detailed level, typical layout techniques include: placing workstations close together so that inventory physically just cannot build up because there is no space for it to do so, and arranging workstations in such a way that all those who contribute to a common activity are in sight of each other and can provide mutual help, for example by facilitating movement between workstations to balance capacity.

OPERATIONS PRINCIPLE

Simple, transparent flow exposes sources of waste.

Examine the shape of process flow

The pattern that flow makes within or between processes is not a trivial issue. Processes that have adopted the practice of curving line arrangements into U-shaped or 'serpentine' arrangements can have a number of advantages (U shapes are usually used for shorter lines and serpentines for longer lines). One authority[5] sees the advantages of this type of flow patterns as *staffing flexibility and balance*, because the U shape enables one person to tend several jobs, *rework*, because it is easy to return faulty work to an earlier station, *free flow*, because long straight lines interfere with cross travel in the rest of the operation, and *teamwork*, because the shape encourages a team feeling.

Ensure visibility

Appropriate layout also includes the extent to which all movement is transparent to everyone within the process. High visibility of flow makes it easier to recognise potential improvements to flow. It also promotes quality within in a process because the more transparent the operation or process, the easier it is for all staff to share in its management and improvement. Problems are more easily detectable and information becomes simple, fast and visual. Visibility measures include the following:

- Clearly indicated process routes using signage.
- Performance measures clearly displayed in the workplace.
- Coloured lights used to indicate stoppages.
- An area is devoted to displaying samples of one's own and competitors' process outputs, together with samples of good and defective output.
- Visual control systems (e.g. *kanbans*, discussed later).

An important technique used to ensure flow visibility is the use of simple, but highly visual signals to indicate that a problem has occurred, together with operational authority to stop the process. For example, on an assembly line, if an employee detects some kind of quality problem, he or she could activate a signal that illuminates a light (called an 'andon' light) above the work station and stops the line. Although this may seem to reduce the efficiency of the line, the idea is that this loss of efficiency in the short term is less than the accumulated losses of allowing defects to continue on in the process. Unless problems are tackled immediately, they may never be corrected.

Use small-scale simple process technology

There may also be possibilities to encourage smooth streamlined flow through the use of small-scale technologies, that is, using several small units of process technology (for example, machines), rather than one large unit. Small machines have several advantages over large ones. First, they can process different products and services simultaneously. For example, in Figure 11.6 one large machine produces a batch of A, followed by a batch of B, and followed by a batch of C. However, if three smaller machines are used they can each produce A, B or C simultaneously. The system is also more robust. If one large machine breaks down, the whole system ceases to operate. If one of the three smaller machines breaks down, it is still operating at two-thirds effectiveness. Small machines are also easily moved, so that layout flexibility is enhanced, and the risks of making errors in investment decisions are reduced. However, investment in capacity may increase in total because parallel facilities are needed, so utilisation may be lower (see the earlier arguments).

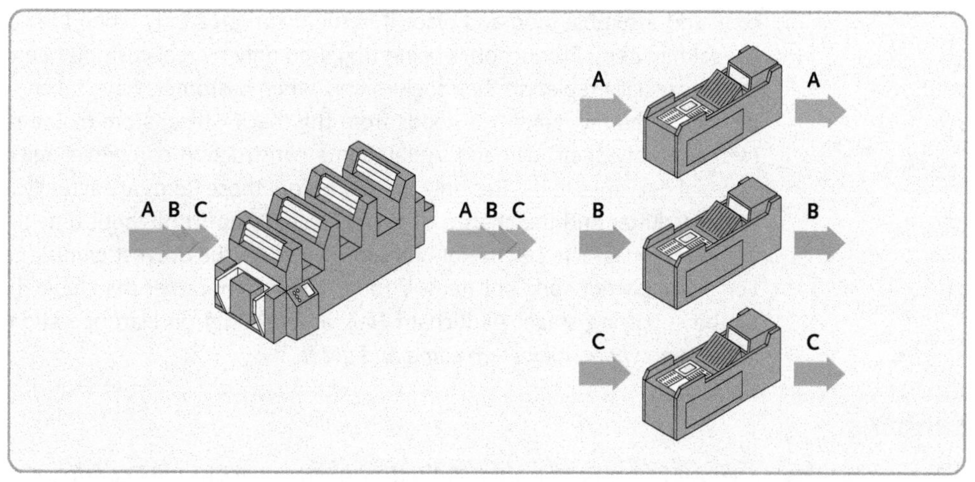

Figure 11.6 Using several small machines rather than one large one allows simultaneous processing, is more robust, and is more flexible

DIAGNOSTIC QUESTION

Does supply exactly match demand?

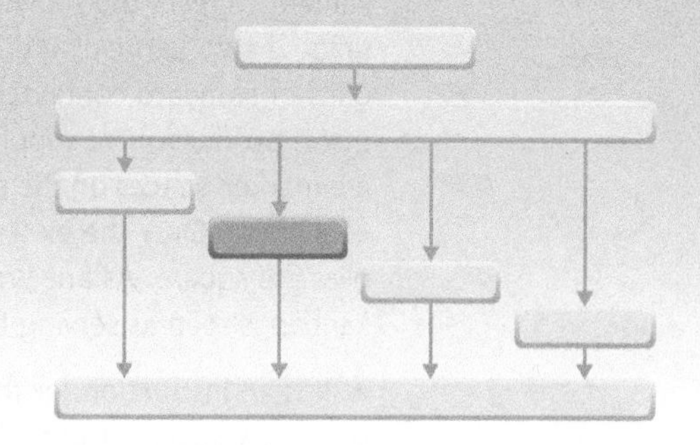

Video

The value of the supply of products or services is always time dependent. Something that is delivered early or late often has less value than something that is delivered exactly when it is needed. We can see many everyday examples of this. For example, parcel delivery companies charge more for guaranteed faster delivery. This is because our real need for the delivery is often for it to be as fast as possible. The closer to instantaneous delivery we can get the more value the delivery has for us and the more we are willing to pay for it. In fact delivery of information earlier than it is required can be even more harmful than late delivery because it results in information inventories that serve to confuse flow through the process. For example, an Australian tax office used to receive applications by mail, open the mail and send it through to the relevant department who, after processing it, sent it to the next department. This led to piles of unprocessed applications building up within its processes, causing problems in tracing applications, and losing them, sorting through and prioritising applications, and worst of all, long throughput times. Now they only open mail when the stages in front can process it. Each department requests more work only when they have processed previous work.

OPERATIONS PRINCIPLE

Delivering only and exactly what is needed and when it is needed, smoothes flow and exposes waste.

Pull control

The exact matching of supply and demand is often best served by using 'pull control' wherever possible (discussed in Chapter 10). At its simplest, consider how some fast-food restaurants cook and assemble food and place it in the warm area only when the customer-facing server has sold an item. Production is being triggered only by real customer demand. Similarly supermarkets usually replenish their shelves only when customers have taken sufficient products off the shelf. The movement of goods from the 'back office' store to the shelf is triggered only by the 'empty-shelf' demand signal. Some construction companies make it a rule to call for material deliveries to its sites only the day before those items are actually needed. This not only reduces clutter and the chance of theft, it speeds up throughput time and reduces confusion and inventories. The essence of pull control is to let the downstream stage in a process, operation or supply network, pull items through the system rather than have them 'pushed' to them by the supplying stage. As Richard Hall, an authority on lean operations put it, *'Don't send nothing nowhere, make 'em come and get it.'*[6]

Kanbans

Practice note

The use of kanbans is one method of operationalising pull control. *Kanban* is the Japanese for card or signal. It is sometimes called the 'invisible conveyor' that controls the transfer of items between the stages of an operation. In its simplest form, it is a card used by a customer stage to instruct its supplier stage to send more items. Kanbans can also take other forms. In some Japanese companies, they are solid plastic markers or even coloured ping-pong balls. Whichever kind of kanban is being used, the principle is always the same; the receipt of a kanban triggers the movement, production or supply of one unit or a standard container of units. If two kanbans are received, this triggers the movement, production or supply of two units or standard containers of units, and so on. Kanbans are the only means by which movement, production or supply can be authorised. Some companies use 'kanban squares'. These are marked spaces on the shop floor or bench that are drawn to fit one or more work pieces or containers. Only the existence of an empty square triggers production at the stage that supplies the square. As one would expect, at Toyota the key control tool is its kanban system. The kanban is seen as serving three purposes:

- It is an instruction for the preceding process to send more.
- It is a visual control tool to show up areas of over-production and lack of synchronisation.
- It is a tool for *kaizen* ('continuous improvement'). Toyota's rules state that 'the number of kanbans should be reduced over time'.

DIAGNOSTIC QUESTION

Are processes flexible?

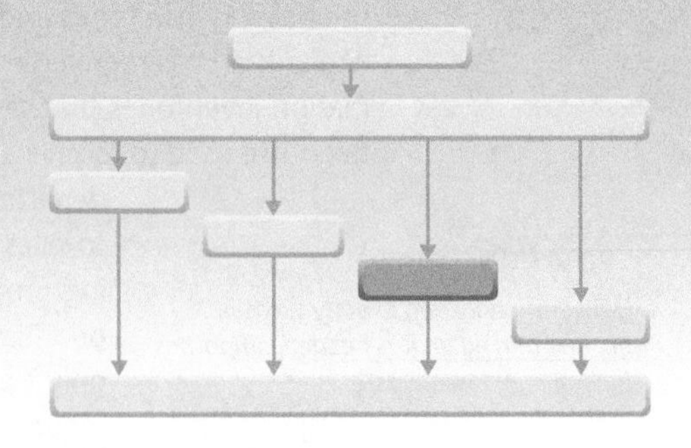

Video

Responding exactly and instantaneously to customer demand implies that operations resources need to be sufficiently flexible to change both what they do and how much they do of it without incurring high cost or long delays. In fact, flexible processes (often with flexible technologies) can significantly enhance smooth and synchronised flow. For example, new publishing technologies allow professors to assemble printed and e-learning course material customised to the needs of individual courses or even individual students. In this case flexibility is allowing customised, small batches to be delivered 'to order'. In another example, a firm of lawyers used to take ten days to prepare its bills for customers. This meant that customers were not asked to pay until ten days after the work had been done. Now they use a system that, everyday, updates each customer's account. So, when a bill is sent it includes all work up to the day before the billing date. The principle here is that process inflexibility also delays cash flow.

OPERATIONS PRINCIPLE

Change-over flexibility reduces waste and smoothes flow.

Reduce setup times

For many technologies, increasing process flexibility means reducing set-up times, defined as the time taken to change over the process from one activity to the next. Compare the time it takes you to change the tyre on your car with the time taken by a Formula 1 team. Set-up reduction can be achieved by a variety of methods such as cutting out time taken to search for tools and equipment, the pre-preparation of tasks which delay changeovers, and the constant practice of set-up routines. Set-up time reduction is also called single minute exchange of dies, (SMED), because this was the objective in some manufacturing operations. The other common approach to set-up time reduction is to convert work which was previously performed while the machine was stopped (called *internal* work) to work that is performed while the machine is running (called *external* work). There are three major methods of achieving the transfer of internal setup work to external work:[7]

- Pre-prepare equipment instead of having to do it while the process is stopped. Preferably, all adjustment should be carried out externally.
- Make equipment capable of performing all required tasks so that changeovers become a simple adjustment.
- Facilitate the change of equipment, for example by using simple devices such as roller conveyors.

Fast changeovers are particularly important for airlines because they can't make money from aircraft that are sitting idle on the ground. It is called 'running the aircraft hot' in the industry. For many smaller airlines, the biggest barrier to running hot is that their markets are not large enough to justify passenger flights during the day *and* night. So, in order to avoid aircraft being idle over night, they must be used in some other way. That was the motive behind Boeing's 737 'Quick Change' (QC) aircraft. With it, airlines have the flexibility to use it for passenger flights during the day and, with less than a one-hour changeover (set-up) time, use it as a cargo airplane throughout the night. Boeing engineers designed frames that hold entire rows of seats that could smoothly glide on and off the aircraft allowing 12 seats to be rolled into place at once. When used for cargo, the seats are simply rolled out and replaced by special cargo containers designed to fit the curve of the fuselage and prevent damage to the interior. Before reinstalling the seats the sidewalls are thoroughly cleaned so that, once the seats are in place, passengers cannot tell the difference between a QC aircraft and a normal 737. Airlines, like Aloha Airlines that serves Hawaii, particularly value the aircraft's flexibility. It allows them to provide frequent reliable services in both passenger and cargo markets. So the aircraft that has been carrying passengers around the islands during the day can be used to ship fresh supplies over night to the hotels that underpin the tourist industry.

DIAGNOSTIC QUESTION

Is variability minimised?

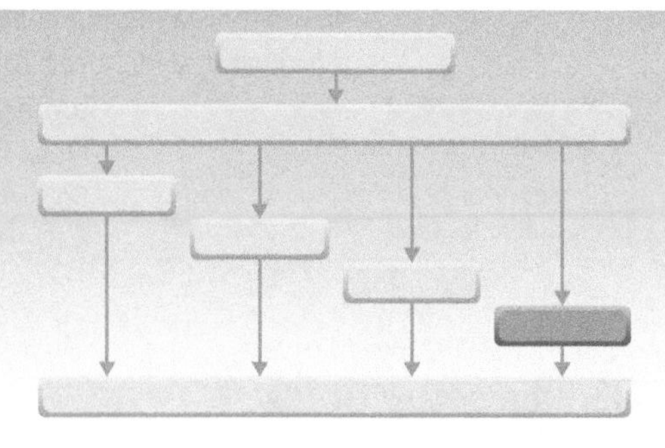

Video

One of the biggest causes of the variability that will disrupt flow and prevent lean synchronisation is variation in the quality of items. This is why a discussion of lean synchronisation should always include an evaluation of how quality conformance is ensured within processes. In particular, the principles of statistical process control (SPC) can be used to understand quality variability. Chapter 12 and its supplement on SPC examine this subject, so in this section we shall focus on other causes of variability. The first of these is variability in the mix of products and services moving through processes, operations or supply networks.

OPERATIONS PRINCIPLE
Variability, in product/service quality, quantity or timing, acts against smooth flow and waste elimination.

Level schedules as much as possible

Levelled scheduling (or *heijunka*) means keeping the mix and volume of flow between stages even over time. For example, instead of producing 500 parts in one batch, that would cover the needs for the next three months; levelled scheduling would require the process to make only one piece per hour regularly. Thus, the principle of levelled scheduling is very straightforward – however, the requirements to put it into practice are quite severe, although the benefits resulting from it can be substantial. The move from conventional to levelled scheduling is illustrated in Figure 11.7. Conventionally, if a mix of products were required in a time period (usually a month), a batch size would be calculated for each product and the batches produced in some sequence. Figure 11.7(a) shows three products that are produced in a 20-day time period in a production unit.

Quantity of product A required = 3000
Quantity of product B required = 1000
Quantity of product C required = 1000
Batch size of product A = 600
Batch size of product B = 200
Batch size of product C = 200

Starting at day 1, the unit commences producing product A. During day 3, the batch of 600 As is finished and dispatched to the next stage. The batch of Bs is started but is not finished until day 4. The remainder of day 4 is spent making the batch of Cs and both batches are dispatched at the end of that day. The cycle then repeats itself. The consequence of using large batches is, first, that relatively large amounts of inventory accumulate within and between the units, and second, that most days are different from one another in terms of what they are expected to produce (in more complex circumstances, no two days would be the same).

Now suppose that the flexibility of the unit could be increased to the point where the batch sizes for the products were reduced to a quarter of their previous levels without loss of capacity (see Figure 11.7b):

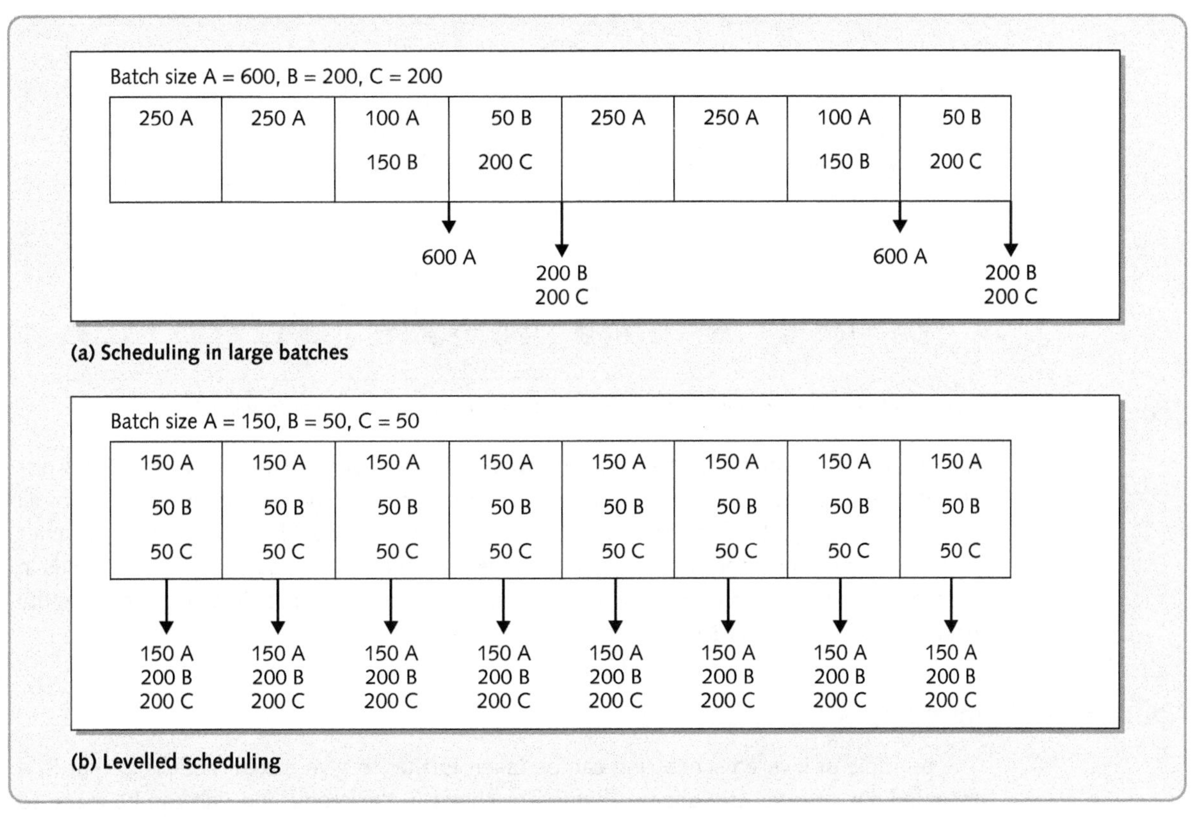

Figure 11.7 Levelled scheduling equalises the mix of products made each day

Batch size of product A = 150

Batch size of product B = 50

Batch size of product C = 50

A batch of each product can now be completed in a single day, at the end of which the three batches are dispatched to their next stage. Smaller batches of inventory are moving between each stage, which will reduce the overall level of work-in-progress in the operation. Just as significant, however, is the effect on the regularity and rhythm of production at the unit. Now every day in the month is the same in terms of what needs to be produced. This makes planning and control of each stage in the operation much easier. For example, if on day 1 of the month the daily batch of As was finished by 11.00 am, and all the batches were successfully completed in the day, then the following day the unit will know that, if it again completes all the As by 11.00 am, it is on schedule. When every day is different, the simple question, 'Are we on schedule to complete our production today?' requires some investigation before it can be answered. However, when every day is the same, everyone in the unit can tell whether production is on target by looking at the clock. Control becomes visible and transparent to all, and the advantages of regular, daily schedules can be passed to upstream suppliers.

Level delivery schedules

A similar concept to levelled scheduling can be applied to many transportation processes. For example, a chain of convenience stores may need to make deliveries of all the different types of products it sells every week. Traditionally it may have dispatched a truck loaded with one particular product around all its stores so that each store received the appropriate amount of

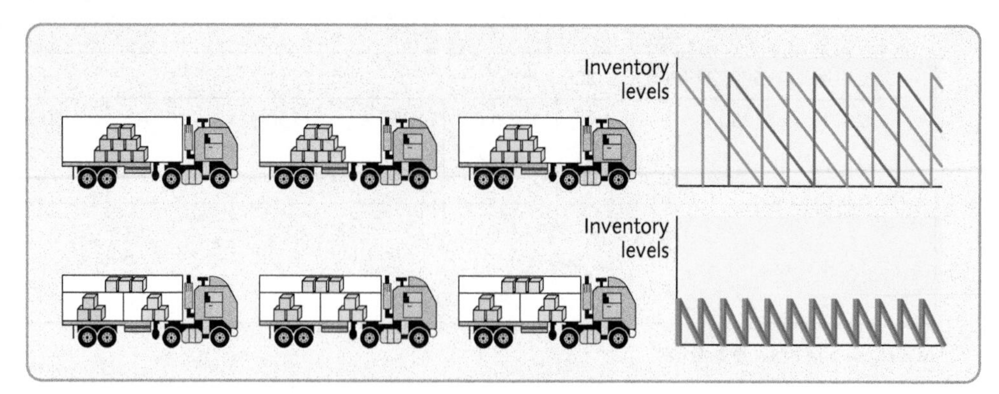

Figure 11.8 Delivering smaller quantities more often can reduce inventory levels

the product that would last them for one week. This is equivalent to the large batches discussed in the previous example. An alternative would be to dispatch smaller quantities of all products in a single truck more frequently. Then, each store would receive smaller deliveries more frequently, inventory levels would be lower and the system could respond to trends in demand more readily because more deliveries means more opportunity to change the quantity delivered to a store. This is illustrated in Figure 11.8.

Adopt mixed modelling where possible

The principle of levelled scheduling can be taken further to give mixed modelling, that is, a repeated mix of outputs. Suppose that the machines in the production unit can be made so flexible that they achieve the JIT ideal of a batch size of one. The sequence of individual products emerging from the unit could be reduced progressively as illustrated in Figure 11.9. This would produce a steady stream of each product flowing continuously from the unit. However, the sequence of products does not always fall as conveniently as in Figure 11.9. The unit production times for each product are not usually identical and the ratios of required volumes are less convenient. For example, if a process is required to produce products A, B and C in the ratio 8:5:4. It could produce 800 of A, followed by 500 of B, followed by 400 of A; or 80A, 50B, and 40C. But ideally, to sequence the products as smoothly as possible, it would produce in the order . . . BACABACABACABACAB . . . repeated . . . repeated . . . etc. Doing this achieves relatively smooth flow (but does rely on significant process flexibility).

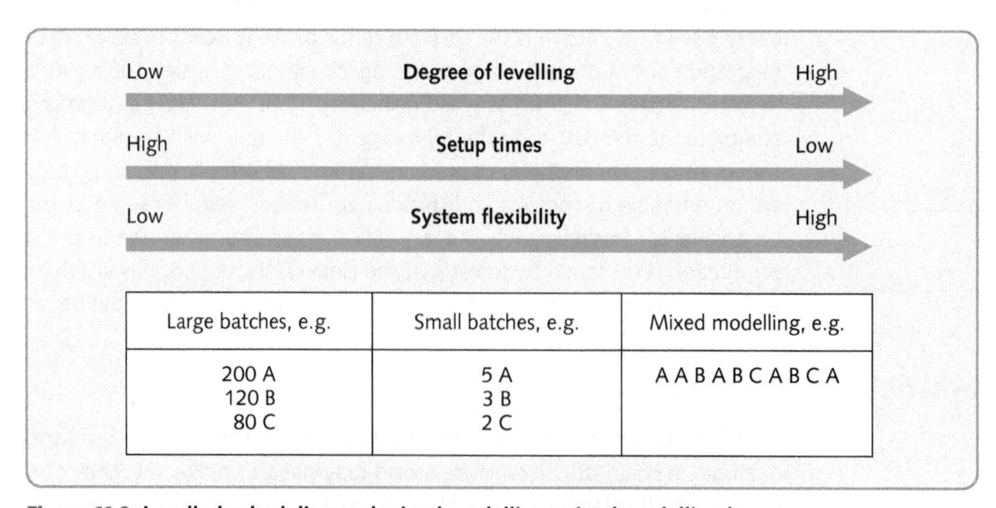

Large batches, e.g.	Small batches, e.g.	Mixed modelling, e.g.
200 A 120 B 80 C	5 A 3 B 2 C	A A B A B C A B C A

Figure 11.9 Levelled scheduling and mixed modelling: mixed modelling becomes possible as the batch size approaches one

Adopt total productive maintenance (TPM)

Total productive maintenance aims to eliminate the variability in operations processes caused by the effect of breakdowns. This is achieved by involving everyone in the search for maintenance improvements. Process owners are encouraged to assume ownership of their machines and to undertake routine maintenance and simple repair tasks. By so doing, maintenance specialists can then be freed to develop higher-order skills for improved maintenance systems. TPM is treated in more detail in Chapter 14.

DIAGNOSTIC QUESTION

Is lean synchronisation applied throughout the supply network?

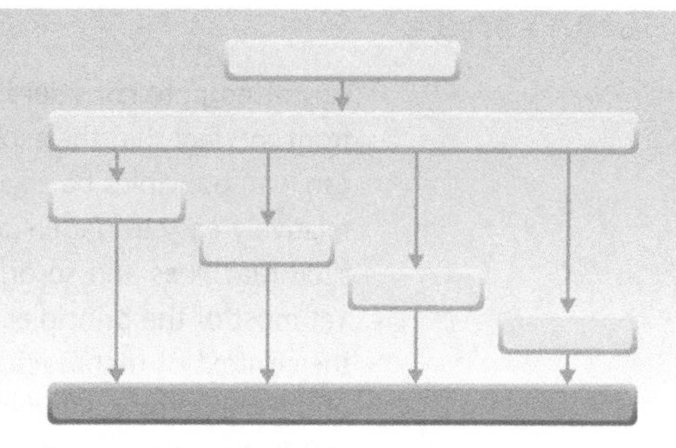

Video

Although most of the concepts and techniques discussed in this chapter are devoted to the management of stages *within* processes and processes *within* an operation, the same principles can apply to the whole supply chain. In this context, the stages in a process are the whole businesses, operations or processes between which products flow. And as any business starts to approach lean synchronisation it will eventually come up against the constraints imposed by the lack of lean synchronisation of the other operations in its supply chain. So, achieving further gains must involve trying to spread lean synchronisation practice outward to its partners in the chain. Ensuring lean synchronisation throughout an entire supply network is clearly a far more demanding task than doing the same within a single process. It is a complex task. And it becomes more complex as more of the supply chain embraces the lean philosophy. The nature of the interaction between whole operations is far more complex than between individual stages within a process. A far more complex mix of products and services is likely to be being provided and the whole network is likely to be subject to a less predictable set of potentially disruptive events. Making a supply chain adopt lean synchronisation means more than making each operation in the chain lean. A collection of localised lean operations rarely leads to an overall lean chain. Rather one needs to apply the lean synchronisation philosophy to the supply chain as a whole. Yet the advantages from truly lean chains can be significant.

> OPERATIONS PRINCIPLE
> *The advantages of lean synchronisation apply at the level of the process, the operation and the supply network.*

Essentially the principles of lean synchronisation are the same for a supply chain as they are for a process. Fast throughput throughout the whole supply network is still valuable and will save cost throughout the supply network. Lower levels of inventory will still make it easier to achieve lean synchronisation. Waste is just as evident (and even larger) at the level of the supply network and reducing waste is still a worthwhile task. Streamline flow, exact matching of supply and demand, enhanced flexibility, and minimising variability are all still tasks that will benefit the whole network. The principles of pull control can work between whole operations in the same way as they can between stages within a single process. In fact, the principles and the techniques of lean synchronisation are essentially the same no matter what level of analysis is being used. And because lean synchronisation is being implemented on a larger scale, the benefits will also be proportionally greater.

One of the weaknesses of lean synchronisation principles is that they are difficult to achieve when conditions are subject to unexpected disturbance (see the Critical Commentary at the end of this chapter). This is especially a problem with applying lean synchronisation principles in the context of the whole supply network. Whereas unexpected fluctuations and disturbances do occur within operations, local management has a reasonable degree of control that it can exert in order to reduce them. Outside the operation, within the supply network, fluctuations can also be controlled to some extent (see Chapter 7), but it is far more difficult to do so. Nevertheless, it is generally held that, although the task is more difficult and although it may take longer to achieve, the aim of lean synchronisation is just as valuable for the supply network as a whole as it is for an individual operation.

Lean service

Any attempt to consider how lean ideas apply throughout a whole supply chain must also confront the fact that these chains include service operations, often dealing in intangibles. So how can lean principles be applied in these parts of the chain? The idea of lean factory operations is relatively easy to understand. Waste is evident in over-stocked inventories, excess scrap, badly sited machines and so on. In services it is less obvious; inefficiencies are more difficult to see. Yet most of the principles and techniques of lean synchronisation, although often described in the context of manufacturing operations, are also applicable to service settings. In fact, some of the philosophical underpinning to lean synchronisation can also be seen as having its equivalent in the service sector. Take, for example, the role of inventory. The comparison between manufacturing systems that hold large stocks of inventory between stages and those that do not centres on the effect which inventory has on improvement and problem-solving. Exactly the same argument can be applied when, instead of queues of material (inventory), an operation has to deal with queues of information, or even customers.

With its customer focus, standardisation, continuous quality improvement, smooth flow, and efficiency, lean thinking has direct application in all operations, manufacturing or service. Bradley Staats and David Upton of Harvard Business School have studied how lean ideas can be applied in service operations[8]. They make three main points:

- In terms of operations and improvements, the service industries in general are a long way behind manufacturing.
- Not all lean manufacturing ideas translate from factory floor to office cubicle. For example, tools such as those that empower manufacturing workers to 'stop the line' when they encounter a problem is not directly replicable when there is no line to stop.
- Adopting lean operations principles alters the way a company learns through changes in problem-solving, coordination through connections, and pathways and standardisation.

Examples of lean service

Many of the examples of lean philosophy and lean techniques in service industries are directly analogous to those found in manufacturing industries because physical items are being moved or processed in some way. Consider the following examples:

- Supermarkets usually replenish their shelves only when customers have taken sufficient products off the shelf. The movement of goods from the 'back office' store to the shelf is triggered only by the 'empty-shelf' demand signal. *Principle – pull control*.
- An Australian tax office used to receive applications by mail, open the mail and send it through to the relevant department who, after processing it, sent it to the next department. Now they only open mail when the stages in front can process it. Each department requests

more work only when they have processed previous work. *Principle – don't let inventories build up, use pull control.*

- One construction company makes a rule of only calling for material deliveries to its sites the day before materials are needed. This reduces clutter and the chance of theft. *Principle – pull control reduces confusion.*
- Many fast-food restaurants cook and assemble food and place it in the warm area only when the customer-facing server has sold an item. *Principle – pull control reduces throughput time.*

Other examples of lean concepts and methods apply even when most of the service elements are intangible.

- Some websites allow customers to register for a reminder service that automatically e-mails reminders for action to be taken. For example, the day before a partner's birthday, in time to prepare for a meeting, etc. *Principle – the value of delivered information, like delivered items, can be time dependent. Too early and it deteriorates (you forget it), too late and it's useless (because it's too late).*
- A firm of lawyers used to take ten days to prepare its bills for customers. This meant that customers were not asked to pay until ten days after the work had been done. Now they use a system that, everyday, updates each customer's account. So, when a bill is sent it includes all work up to the day before the billing date. *Principle – process delays also delay cash flow, fast throughput improves cash flow.*
- New publishing technologies allow professors to assemble printed and e-learning course material customised to the needs of individual courses or even individual students. *Principle – flexibility allows customisation and small batch sizes delivered 'to order'.*

EXAMPLE

Pixar adopts lean[9]

It seems that lean principles (or some lean principles) can be applied even to the most unlikely of processes. None less likely than Pixar Animation Studios, the Academy Award-winning computer animation studio and makers of feature films that have resulted in an unprecedented streak of both critical and box office success including *Toy Story* (1, 2 and 3), *A Bug's Life*, *Monsters, Inc.*, *Finding Nemo*, *The Incredibles*, *Ratatouille*, *WALL-E* and *Up*. Since its incorporation, Pixar has been responsible for many important breakthroughs in the application of computer graphics (CG) for filmmaking. So, the company has attracted some of the world's finest technical, creative and production talent in the area. And such 'knowledge-based' talent is notoriously difficult to manage – certainly not the type of processes that are generally seen as being appropriate for lean synchronisation. Managing creativity involves a difficult trade-off, between encouraging the freedom to produce novel ideas, yet making sure that they work within an effective overall structure.

Source: Sipa Press/Rex Studios

Nevertheless, Pixar did get the inspiration from Toyota and the way it uses lean production. In particular the way Toyota has encouraged continuous advice and criticism from its production line workers to improve its performance. Pixar realised that it could do the same with producing cartoon characters. Adopting constant feedback surfaces problems before they become crises, and provides creative teams with inspiration and challenge. Pixar also devotes a great deal of effort to persuading its creative staff to work together. In similar companies, people may collaborate on specific projects, but are less good at focusing on what's going on elsewhere in the business. Pixar, however, tries to cultivate a sense of collective responsibility. Staff even show unfinished work to one another in daily meetings, so get used to giving and receiving constructive criticism.

Lean supply chains are like an air traffic control systems[10]

The concept of the lean supply chain has been likened to an air traffic control system, in that it attempts to provide continuous 'real-time visibility and control' to all elements in the chain. This is the secret of how the world's busiest airports handle thousands of departures and arrivals daily. All aircraft are given an identification number that shows up on a radar map. Aircraft approaching an airport are detected by the radar and contacted using radio. The control tower precisely positions the aircraft in an approach pattern which it coordinates. The radar detects any small adjustments that are necessary, which are communicated to the aircraft. This real-time visibility and control can optimise airport throughput while maintaining extremely high safety and reliability.

Contrast this to how most supply chains are coordinated. Information is captured only periodically, probably once a day, and any adjustments to logistics, output levels at the various operations in the supply chain are adjusted, and plans rearranged. But imagine what would happen if this was how the airport operated, with only a 'radar snapshot' once a day. Coordinating aircraft with sufficient tolerance to arrange take-offs and landings every two minutes would be out of the question. Aircraft would be jeopardised, or alternatively, if aircraft were spaced further apart to maintain safety, throughput would be drastically reduced. Yet this is how most supply chains have traditionally operated. They use a daily 'snapshot' from their ERP systems (see Chapter 10 for an explanation of ERP). This limited visibility means operations must either space their work out to avoid 'collisions' (i.e. missed customer orders) thereby reducing output, or they must 'fly blind' thereby jeopardising reliability.

Lean and agile

One continuing debate on how lean principles can be applied across the supply chain concerns whether supply network should be lean or 'agile'. Professor Martin Christopher of Cranfield University defines agility as '*rapid strategic and operational adaptation to large scale, unpredictable changes in the business environment. Agility implies responsiveness from one end of the supply chain to the other. It focuses upon eliminating the barriers to quick response, be they organizational or technical.*' Other definitions stress that agility is the capability of operating profitably in a competitive environment of continually changing customer opportunities. The clue lies in how the word 'agile' is often defined; it implies being responsive, quick moving, flexible, nimble, active and constantly ready to change. But some proponents of operational agility go further than this. They see agility as also implying a rejection of a planning paradigm that makes any assumption of a predictable future. Like lean, it is more of a philosophy than an approach. Agile encourages a better match to what customers want by placing an emphasis on producing 'emergent' demand as opposed to rigid plans or schedules. Furthermore, rather than uncertainty and change being seen as things to be 'coped with' or preferably avoided, it should be embraced so that agility becomes changing faster than one's customer. Even less ambitious approaches to agility see it as more than simply organisational flexibility. It involves an organisational mastery of uncertainty and change, where people within the organisation, and their capacity to learn from change and their collective knowledge, are regarded as the organisation's greatest assets because they allow the operation to respond effectively to uncertainty and change. Continually inventing innovative business processes solutions to new market demands becomes a key operations objective.

All this seems very different to the underlying assumptions of the lean philosophy. Again, look at the word: lean means 'thin, having no superfluous fat, skinny, gaunt, undernourished'. Lean attempts to eliminate waste and provide value to the customer throughout the entire supply chain. It thrives on standardisation, stability, defined processes and repeatability – not at all the way agility has been described. Lean is also a well defined (although frequently misunderstood) concept. Agility, on the other hand, is a far newer and less 'operationalised' set of relatively strategic objectives. But some operational level distinctions can be inferred.

The type of principles needed to support a lean philosophy include such things as: simple processes; waste elimination; simple (if any) IT; the use of manual and robust planning and control as well as pull control and kanbans with overall MRP. Agile philosophies, by contrast, require: effective demand management to keep close to market needs; a focus on customer relationship management; responsive supply coordination, visibility across the extended supply chain; continuous rescheduling and quick response to changing demand; short planning cycles; integrated knowledge management; fully exploited e-commerce solutions.

So are lean and agile philosophies fundamentally opposed? Well, yes and no. Certainly they have differing emphases. Saying that lean equals synchronised, regular flow and low inventory, and agile equals responsiveness, flexibility and fast delivery, may be something of a simplification, but it more or less captures the distinction between the two. But because they have different objectives and approaches does not mean that they cannot co-exist. Nor does it mean that there is a 'lean versus agile' argument to be resolved. The two approaches may not be complimentary, as some consultants claim, but both do belong to the general collection of methodologies that are available to help companies meet the requirements of their markets. In the same way as it was wrong to think that JIT would replace MRP, so 'agile' is not a substitute for lean[11].

However, agile and lean are each more appropriate for differing market and product/service conditions. Put simply, if product/service variety or complexity is high, and demand predictability low then you have the conditions in which agile principles keep an operation ready to cope with instability in the business environment. Conversely, if product/service variety is low, and demand predictability high, then a lean approach can exploit the stable environment to achieve cost efficiency and dependability. So the two factors of product/service variety or complexity and demand uncertainty influence whether agile or lean principles should dominate. But what of the conditions where complexity and uncertainty are not related in this manner? Figure 11.10 illustrates how complexity and uncertainty affect the adoption of lean, agile, and other approaches to organising the flow in a supply chain.

- When complexity is low, and demand uncertainty is also low (operations that produce commodities), lean planning and control is appropriate.
- When is complexity is low, and demand uncertainty is high (operations that produce fashion-based products/services), agile planning and control is appropriate.
- When complexity is high, and demand uncertainty is also high (operations that produce 'super value' products/services) project or requirements planning and control (for example MRPII, see Chapter 10) is appropriate.
- When product/service complexity is high, and demand uncertainty is low (operations that produce 'consumer durable type products/services), a combination of agile and lean planning and control is appropriate.

This last category, shown as the bottom left quadrant in Figure 11.10 has been rather clumsily called 'leagile'. Leagile is based on the idea that both lean and agile practices can be employed within supply chains. It envisages an inventory decoupling point that is the separation between the responsive (and therefore agile) 'front end' of the supply chain that reacts fast and flexibly to customer demand, and the efficient (and therefore lean). This is not a new idea and in product-based supply chains involves 'making to *forecast*' before the decoupling point and 'making (or assembling, adapting or finishing) to *order* after it. The idea has many similarities with the idea of 'mass customisation'. However, it is difficult to transpose the idea directly into supply chains that deal exclusively in non-tangible services.

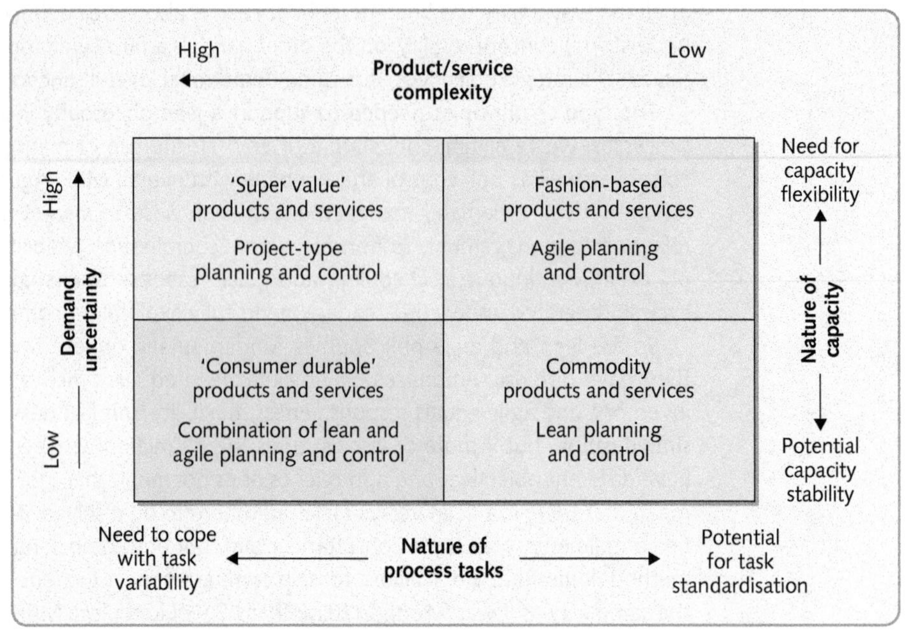

Figure 11.10 The degree of complexity of products/services and the uncertainty of demand influence the relative emphasis of lean or agile supply-chain principles

Critical commentary

Each chapter contains a short critical commentary on the main ideas covered in the chapter. Its purpose is not to undermine the issues discussed in the chapter, but to emphasise that, although we present a relatively orthodox view of operation, there are other perspectives.

- Lean synchronisation principles can be taken to an extreme. When just-in-time ideas first started to have an impact on operations practice in the West, some authorities advocated the reduction of between-process inventories to zero. While in the long term this provides the ultimate in motivation for operations managers to ensure the efficiency and reliability of each process stage, it does not admit the possibility of some processes always being intrinsically less than totally reliable. An alternative view is to allow inventories (albeit small ones) around process stages with higher than average uncertainty. This at least allows some protection for the rest of the system. The same ideas apply to just-in-time delivery between factories. The Toyota Motor Corp., often seen as the epitome of modern JIT, has suffered from its low inter-plant inventory policies. Both the Kobe earthquake and fires in supplier plants have caused production at Toyota's main factories to close down for several days because of a shortage of key parts. Even in the best-regulated manufacturing networks, one cannot always account for such events.

- One of the most counter-intuitive issues in lean synchronisation is the way it appears to downplay the idea of capacity under-utilisation. And it is true that, when moving towards lean synchronisation, fast throughput time and smooth flow *is* more important than the high utilisation which can result in inventory build-up. However, this criticism is not really valid in the long term. Consider the relationship between capacity utilisation and process throughput time (or inventory), shown in Figure 11.11. The improvement path envisaged by adopting lean synchronisation is shown as moving from the state that most businesses find themselves in (high utilisation but long throughput times) towards the lean synchronisation ideal (short throughput times). Although, inevitably, this means moving towards a position of lower capacity utilisation, lean synchronisation also stresses a reduction in all types of process variability. As this begins to become reality, the improvement path moves towards the point where throughput time is short and capacity utilisation high. It manages to do this because of the reduction in process variability.

- Not all commentators see lean synchronisation-influenced people-management practices as entirely positive. The JIT approach to people management can be viewed as patronising. It may be, to some extent, less autocratic than some Japanese management practice dating from earlier times. However, it is certainly not in line with some of the job design philosophies which place a high emphasis on contribution and commitment. Even in Japan, the JIT approach is not without its critics. Kamata wrote an autobiographical description of life as an employee at a Toyota plant called *Japan in the Passing Lane*.[12] His account speaks of 'the inhumanity and the unquestioning adherence' of working under such a system. Similar criticisms have been voiced by some trade union representatives.

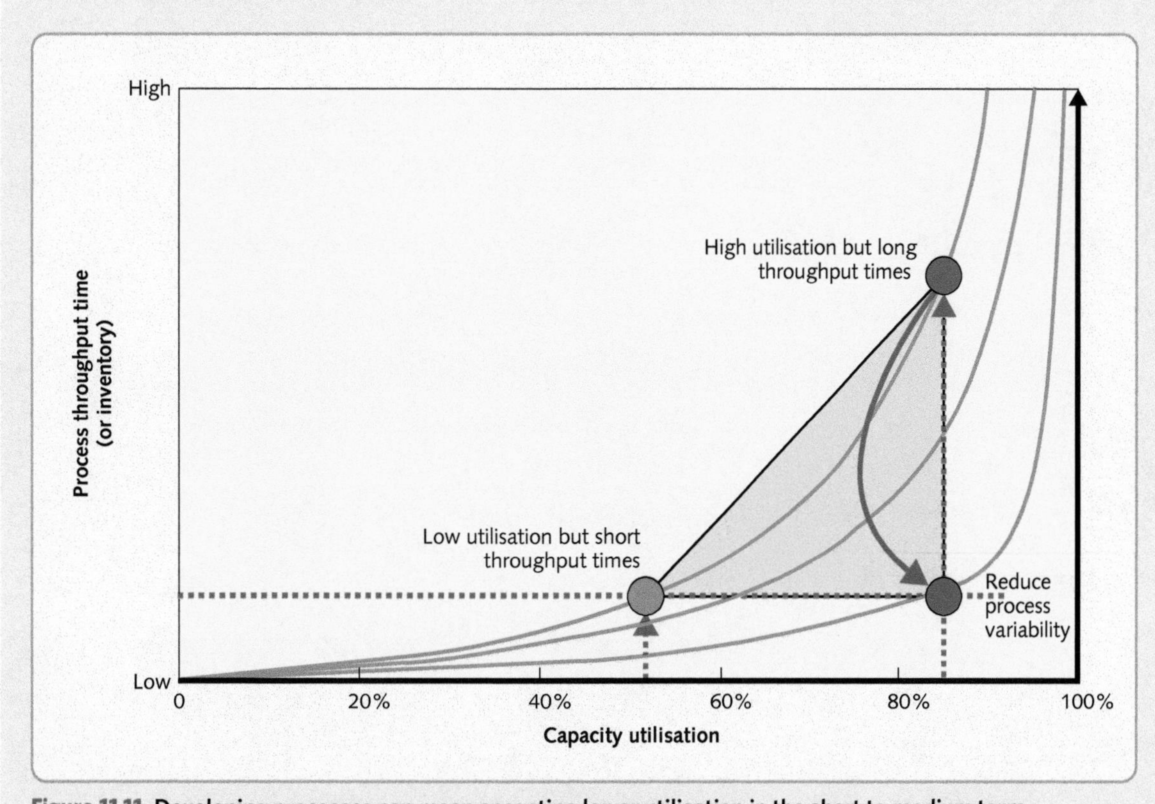

Figure 11.11 Developing processes can mean accepting lower utilisation in the short to medium term

Any textbook of this type has to segment the ideas and knowledge contained within its subject so as to treat them in such a way as to explain it, to communicate each set of ideas as clearly as possible. Yet doing this inevitably means imposing artificial boundaries between the various topics. No more so than in the case of lean synchronisation. There are some particularly evangelical proponents of the lean philosophy who object strongly to separating out the whole concept of lean into a separate chapter. The underlying ideas of lean, they say, have now comprehensively replaced those ideas described as 'traditional' at the beginning of this chapter. Rather, lean principles should be foundation for the whole of operations and process management. Lean principles have something to tell us about everything in the subject from quality management to inventory management, from job design to product design. And they are right of course. Nevertheless, the ideas behind lean synchronisation are both counter-intuitive and important enough to warrant separate treatment. Also lean in its pure form is not necessarily equally applicable to every situation (refer to the discussion about lean and agile). Hence the inclusion of this chapter that focuses on this topic. Remember though, lean synchronisation is one of those topics (like operations strategy, quality and improvement) that has a particularly strong influence over the whole subject.

SUMMARY CHECKLIST

This checklist comprises questions that can be usefully applied to any type of operations and reflect the major diagnostic questions used within the chapter.

- ☐ Are the benefits of attempting to achieve lean synchronisation well understood within the business?
- ☐ Notwithstanding that the idea derives from manufacturing operations, have the principles been considered for non-manufacturing processes within the business?
- ☐ Is the extent of waste within operations and processes fully understood?
- ☐ Can the flow of items through processes be made more regular?
- ☐ How much inventory of items is building up because of inexact supply?
- ☐ How much waste is caused because of inflexibility in the operation's processes?
- ☐ How much waste is caused because of variability (especially of quality) within the operation's processes?
- ☐ Are capacity utilisation performance measures likely to prove a barrier to achieving lean synchronisation?
- ☐ Does the culture of the organisation encourage the involvement in the improvement process of all people in the organisation?
- ☐ Are the ideas of continuous improvement understood?
- ☐ Are the ideas of continuous improvement used in practice?
- ☐ Are the various techniques used to promote lean synchronisation understood and practised?
- ☐ Is the concept of lean synchronisation applied throughout the supply network?
- ☐ Has the possibility of blending push (such as MRP) and pull (such as lean synchronisation) been considered?

CASE STUDY

Implementing lean at CWHT

by Nicola Burgess, Warwick University

In July 2011, the acting Chief Executive of Chaswick and Wallasey NHS Hospital Trust (CWHT) took the decision to implement lean thinking across the organisation. The idea was to bring the hospital into line with other hospitals in the UK where lean improvement initiatives were becoming increasingly widespread. In fact, the trust had started to experiment with lean principles two years earlier. The external consultants used for this initial work were not asked to bid for the new initiative. Despite their global reputation for world-class engineering it was felt that the previous lean work had been disjointed, fragmented and too focused on the need to optimise departments around targets. This time the trust wanted a much more coherent and joined up approach to lean implementation. This time they were thinking about 'transformation' into becoming a total *lean organisation*. All the consultants who were bidding for the contract emphasised building 'soft skills' training and 'project facilitation' that would equip the organisation with an internal change team capable of rolling out lean throughout the hospital. The internal change team was led by a 'Head of Lean' and several Lean Leaders, many of whom joined the team as part of a secondment from their clinical roles. The 'Lean team' comprised 11 staff employed on two-year contracts.

During the period September–December 2011, preparations for implementing lean throughout the trust took place led by the external consultants who had won the contract: 'Change M'. Eighteen projects were designed to take place across three streams of work to reflect a number of patient pathways throughout the organisation. The aim was to move staff out of their functional 'silos' and to help them see their role within the whole patient pathway rather than within a single function. Meanwhile, the lean facilitators underwent training in project facilitation and change management skills. The project roll-out was set to begin January 2012.

A new Chief Executive

In October 2011 a new Chief Executive Sir William Oberon was appointed to begin work in January. He had an impressive record, with Chief Executive roles at a number of hospitals, including a world-leading heart specialist hospital, and had overseen successful turnarounds in two of the worst performing hospitals in England. Following his appointment, Sir William immediately instructed

Source: Alex Segre/Alamy Images

consultants TOC-Health (with whom he had worked in the past) to enter CWHT in December 2011 and begin work straight away in the Accident and Emergency (A&E) department. A new Director of Operations also began in post in December 2011. This made the assignment difficult for TOC-Health. Adam Smith, Client Director of TOC-Health explained: *'It was a funny situation really. We arrived in the Trust before William had taken up his appointment and before the new Director of Operations. We had to introduce ourselves to the Lean Team. It was rather embarrassing and awkward, but William had said: "I want it done very quickly Adam". It's not usually the way we work.'*

TOC-Health had been employed with very clear responsibility to sort out problems in A&E. The Lean Team had been asked to steer clear of the work. At the time UK healthcare targets specified that 98 per cent of patients must be seen within four hours of arrival and so CWHT had implemented a new Clinical Decision Unit (CDU) to help expedite people out of A&E so as not to breach the target. Unfortunately, at the end of 2011, the Trust was still operating at around 95 per cent which meant that the trust would not obtain a good performance rating. In addition, the trust had a large number of patient outliers (patients in the wrong beds, on the wrong wards) and some financial overruns. In addition they were also struggling on other important targets. Speaking about Sir William's decision to take on the Chief Executive role, the Director of TOC-Health quoted Sir William's words: *'CWHT has this fantastic new building, it's just ridiculous it's not meeting it targets. The hospital is punching well below its weight – the size of the prize is huge!'*

Approaches to improvement

Talking about the approach of TOC-Health, their Director explained: *'The whole point about our approach is fast, focused breakthroughs in performance. You must identify the one true bottleneck and focus on fixing that. In our opinion, if you improve process by process you are chasing your tail, you're just never going to get there; it will take you so long that by the time you've improved, it will have changed anyway.'* It soon became clear that the two consultancy firms had very different approaches to the number of improvement projects that should run concurrently. TOC-Health was focused around the idea that an organisation should not have many disparate projects on the go simultaneously, rather they should focus on just one (the bottleneck). Change M, on the other hand were happy to let many projects take place in various parts of the organisation using what they called the Rapid Improvement Event (RIE) approach.

Meanwhile, although the Head of Lean had begun her projects on schedule, the instruction to keep away from A&E where TOC-Health was working meant her planned activities had to be rescheduled. Nor was she happy with the changes in responsibilities. *'I think we had a reasonably clear understanding of how lean would be implemented until we had a change of Chief Executive. I now feel we don't have a clear way forward to becoming a lean organisation. The emphasis has shifted to get some events done and get some money out; that isn't what lean is about.'* Similar concerns had been expressed about how to measure the success and benefits of the Lean Team. *'Again the emphasis has shifted. Originally it was about having a positive impact, getting people involved in lean, engaging and empowering them towards continuous improvement and following a set of key principles, but now it's changed to "save some money", and people are forgetting the cultural side of it.'*

The 'principles' that the Head of Lean was referring to had been adapted from the lessons learned from lean practitioners in healthcare.* The main principles were as follows:

1 Focus on the patient (not the organisation and its employees, suppliers, etc.) and design care around them in order to determine what real value represents.
2 Identify what represents value for the patient (along the whole value stream or patient pathway) and get rid of everything else.
3 Reduce the time required to go from start to finish along every pathway (which creates more value at less cost).
4 Pursue principles 1, 2, and 3 endlessly through continuous improvement that engages everyone (doctors, nurses, technicians, managers, suppliers, and patients and their families) who 'touch' the patient pathways.

*Particularly a book called *On the Mend* by John Toussaint and Roger Gerard (www.onthemendbook.org)

A new arrival

In February, and much to the Head of Lean's surprise, a third set of consultants was appointed to focus on the application of Work Study Method to operating theatres. *'I think the timescales have changed. Before, there was a recognition that we're in it for the long haul, it wasn't going to be a quick fix. I think now the driver is that "you will become a high performing trust come hell or high water and if what we need to do to get there is to bring a hundred management consultants in who've all got a different approach then that's what we'll do". My worry is that in the longer term we'll fall over again because actually all we've done is stick sticking plaster over again which is what we were doing before.'*

The impact of lean

Consultants and nurses in the trust were divided on the impact of lean. Those who had experienced Change M's rapid improvement events (RIEs) in their area tended to be enthusiastic about the benefits and the changes they had made. Small, but significant, changes could produce benefits including reduced confusion, increased staff morale and better patient flow. For example, improved prominence and clarity of signage stopped patients getting lost, and leaving clinicians to wait for them. A reduction in stock levels produced cost and space savings as well as reducing the amount of time spent looking for the correct items. In one store cupboard 25,000 pairs of surgical gloves were identified from 500 different suppliers. Another RIE blew the myth on the effectiveness of the Medical Records Department: *'It was amazing. We just exploded the myth that when you didn't get case notes in a clinical area it was medical records fault, but it hardly ever was. Consultants had notes in their cars, they had them at home, we had a thousand notes in the secretary's offices, and we wondered why we couldn't get case notes! Two people walked seven miles a day looking for them – they were all over the place. Now that was a good RIE because we did manage to sort out medical records and create some semblance of order in their lives.'*

Yet those who had no direct involvement in the lean activity are sceptical: *'We're not making cars, people are different and the processes that we put people through repeatedly are more complicated than the processes that you go through to make a car. These ideas may be OK in manufacturing, but all it has resulted in here are teams of expensive consultants crawling all over the hospital'* (Consultant Surgeon). But there were some converts according to the Head of Lean. *'A consultant (medical) came to me at the beginning of the week saying, "This is all a load of rubbish. There's no point in mapping the process, we all know what happens: the patient goes from there to there and this is the solution and this is what we need to do". During the middle of the improvement week, the Consultant said: "I never realised what actually does happen in reality." By the end of the week the Consultant's mindset has changed to: "Actually this has been great because I never understood, I only saw my bit of it".'*

Although frustrated by the confusion caused by using multiple consultants, the Head of Lean was optimistic. *'We are starting to see some quite significant, if limited, results. The real issue is getting everyone to change the way they behave. It is tackling doctors who are used to doing their own thing and having no performance measures. It is negotiating with suppliers familiar with a culture that allows them to offer new apparatus with little attention to cost or clinical benefits. It is gradually persuading nurses that constantly working around problems in the care delivery process will not make deep-seated problems go away. It is slowly educating administrators to accept that that you cannot simply run broken processes harder. Ultimately we have seen that lean can potentially work in healthcare. What we have yet to discover is a method for communicating the benefits and value of lean to others, and quantifying this value in a manner that is significant at an executive level of the organisation.'*

QUESTIONS

1 What complexities and barriers to lean implementation are demonstrated in the case study?

2 How do the complexities and barriers identified above relate to your own organisation?

3 How might an organisation overcome these barriers?

APPLYING THE PRINCIPLES

Hints

Some of these exercises can be answered by reading the chapter. Others will require some general knowledge of business activity and some might require an element of investigation. Hints on how they can all be answered are to be found in the eText at **www.pearsoned.co.uk/slack**.

1 Re-examine the description of the Toyota production system at the beginning of the chapter.

(a) List all the different techniques and practices which Toyota adopts. Which of these would you call just-in-time philosophies and which are just-in-time techniques?

(b) How are operations objectives (quality, speed, dependability, flexibility, cost) influenced by the practices which Toyota adopts?

2 Consider this record of an ordinary flight.

'Breakfast was a little rushed but left the house at 6.15. Had to return a few minutes later, forgot my passport. Managed to find it and leave (again) by 6.30. Arrived at the airport 7.00, dropped Angela off with bags at terminal and went to the long-term car park. Eventually found a parking space after ten minutes. Waited eight minutes for the courtesy bus. Six-minute journey back to the terminal, we start queuing at the check-in counters by 7.24. Twenty-minute wait. Eventually get to check-in and find that we have been allocated seat at different ends of the plane. Staff helpful but takes eight minutes to sort it out. Wait in queue for security checks for ten minutes. Security decide I look suspicious and search bags for three minutes. Waiting in lounge by 8.05. Spend one hour five minutes in lounge reading computer magazine and looking at small plastic souvenirs. Hurrah, flight is called 9.10! Takes two minutes to rush to the gate and queue for further five minutes at gate. Through the gate and on to air bridge that is continuous queue going onto plane, takes four minutes but finally in seats by 9.21. Wait for plane to fill up with other passengers for 14 minutes. Plane starts to taxi to runway at 9.35. Plane queues to take-off for ten minutes. Plane takes off 9.45. Smooth flight to Amsterdam: 55 minutes. Stacked in queue of planes waiting to land for ten minutes. Touch down at Schiphol Airport 10.50. Taxi to terminal and wait 15 minutes to disembark. Disembark at 11.05 and walk to luggage collection (calling at lavatory on way); arrive luggage collection 11.15. Wait for luggage eight minutes. Through customs (not searched by Netherlands security who decide I look

trustworthy) and to taxi rank by 11.26. Wait for taxi four minutes. Into taxi by 11.30; 30-minute ride into Amsterdam. Arrive at hotel 12.00.'

(a) Analyse the journey in terms of value added time (actually going somewhere) and non-value added time (the time spent queueing, etc.).

(b) Visit the websites of two or three airlines and examine their business class and first class services to look for ideas that reduce the non-value-added time for customers who are willing to pay the premium.

(c) Next time you go on a journey, time each part of the journey and perform a similar analysis.

3 Examine the value-added versus non-value-added times for some other services. For example:

(a) Handing an assignment in for marking if you are currently studying for a qualification. What is the typical elapsed time between handing the assignment in and receiving it back with comments? How much of this elapsed time do you think is value-added time?

(b) Posting a letter (the elapsed time is between posting the letter in the box and it being delivered to the recipient).

(c) Taking a garment to be professionally dry cleaned.

4 Using an internet search engine, enter 'kanban', and capture who uses such devices for planning and control. Contrast the ways in which they are used.

5 Consider how set-up reduction principles can be used on the following:

(a) Changing a tyre at the side of the road (following a puncture).

(b) Cleaning out an aircraft and preparing it for the next flight between its inbound flight landing and disembarking its passengers, and the same aircraft being ready to take-off on its outbound flight.

(c) The time between the finish of one surgical procedure in a hospital's operating theatre, and the start of the next one.

(d) The 'pitstop' activities during a Formula One race (how does this compare to (a) above?).

6 In the chapter the example of Boeing's success in enabling aircraft to convert between passenger and cargo operations was described.

(a) If the changeover between 'passengers' and 'cargo' took two hours instead of one hour, how much impact do you think it would have on the usefulness of the aircraft?

(b) For an aircraft that carries passengers all the time, what is the equivalent of set-up reduction? And why might it be important?

Notes on chapter

1 Spears, S. and Bowen, H.K. (1999), 'Decoding the DNA of the Toyota production system', *Harvard Business Review*, October, pp 96–106.

2 Source: Mathieson, S.A. (2006), 'NHS should embrace lean times', *Guardian*, 8 June.

3 Lee, D.C. (1987) 'Set-up time reduction: making JIT work' in Voss, C.A. (ed), *Just-in-time Manufacture*, IFS/Springer-Verlag.

4 Spears, S. and Bowen, H.K. *op. cit.*

5 Harrison, A. (1992) *Just-in-time Manufacturing in Perspective*, Prentice Hall.

6 Hall R. (1983) *Zero Inventories*, McGraw Hill, New York.

7 Yamashina, H. 'Reducing set-up times makes your company flexible and more competitive', unpublished, quoted in Harrison A., *op. cit.*

8 Reported in Hanna, J. (2007), 'Bringing lean principles to service industries', *Harvard* Business Review, October.

9 Source: 'Planning for the sequel: how Pixar's leaders want to make their creative powerhouse outlast them' (2010), *The Economist* 17 June.

10 This great metaphor seems to have originated from the consultancy '2think': http://www.2think.biz/index.htm.

11 Kruse, G. (2002) 'IT enabled lean agility', *Control*, November.

12 Kamata, S. (1983) *Japan in the Passing Lane*, Allen and Unwin.

TAKING IT FURTHER

Bicheno, J. and Holweg, M. (2010) *The Lean Toolbox: The Essential Guide to Lean Transformation* (4th edn), PICSIE Books. *A practical guide from two of the European authorities on all matters lean.*

Holweg, M. (2007) 'The genealogy of lean production', *Journal of Operations Management Vol. 25, 420–437.*

Mann, D. (2010) *Creating a Lean Culture* (2nd edn), Productivity Press. *Treats the soft side of lean.*

Schonberger, R.J. (1996) *World Class Manufacturing: The Next Decade*, The Free Press. *One of the really influential authors who established JIT in the West. Now seen as over-simplistic but worth looking at to understand pure JIT.*

Spear, S. and Bowen, H.K. (1999) 'Decoding the DNA of the Toyota production system', *Harvard Business Review*, September–October. *Revisits the leading company as regards JIT practice and re-valuates the underlying philosophy behind the way it manages its operations. Recommended.*

Womack, J.P., Jones, D.T. and Roos, D. (1990) *The Machine that Changed the World*, Rawson Associates. *Arguably the most influential book on operations management practice of the last fifty years. Firmly rooted in the automotive sector but did much to establish JIT.*

Womack, J. P. and Jones, D. T. (2003) *Lean Thinking: Banish Waste and Create Wealth in Your Corporation*, Free Press. *Some of the lessons from 'The Machine that Changed the World' but applied in a broader context.*

USEFUL WEBSITES

www.opsman.org *Definitions, links and opinions on operations and process management.*

http://www.lean.org/ *Site of the Lean Enterprise Unit, set up by one of the founders of the lean thinking movement.*

http://www.iet.org/index.cfm *The site of the Institution Electrical Engineers (that includes Manufacturing engineers surprisingly) has material on this and related topics as well as other issues covered in this book.*

http://www.mfgeng.com *The manufacturing engineering site.*

Interactive quiz

For further resources including examples, animated diagrams, self-test questions, Excel spreadsheets and video materials please explore the eText on the companion website at **www.pearsoned.co.uk/slack**

12 Quality management

Introduction

All businesses are concerned with quality, usually because they have come to understand that high quality can be a significant competitive advantage. But 'quality management' has come to mean more than avoiding errors. It is also seen as an approach to the way processes should be managed and, more significantly, improved, generally. This is because quality management focuses on the very fundamental of operations and process management – the ability to produce and deliver the products and services that the market requires, both in the short and long terms. A grasp of quality management principles is the foundation of any improvement activity (see Figure 12.1).

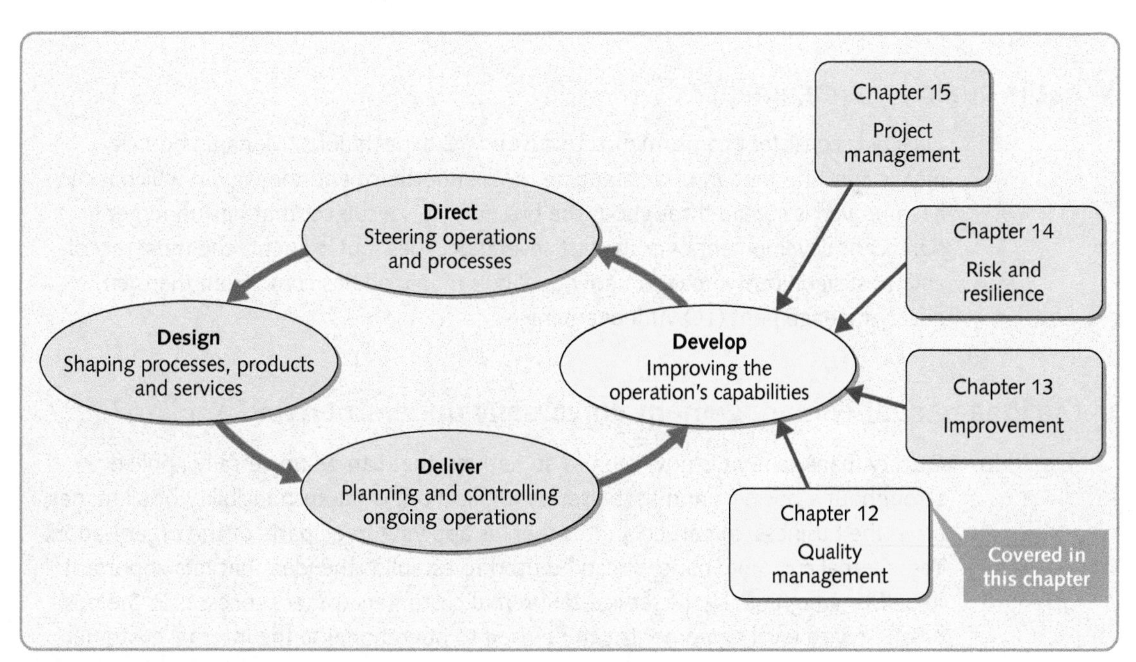

Figure 12.1 Quality management is the activity of ensuring consistent conformance to customers' expectations

EXECUTIVE SUMMARY

Each chapter is structured around a set of diagnostic questions. These questions suggest what you should ask in order to gain an understanding of the important issues of a topic, and, as a result, improve your decision making. An executive summary addressing these questions is provided below.

What is quality management?

Quality is consistent conformance to customers' expectations. Managing quality means ensuring that an understanding of its importance and the way in which it can be improved is spread throughout the business. It is a subject that has undergone significant development over the last several decades, but arguably the most recent and most significant impact on how quality is managed has come from the total quality management (TQM) movement.

Is the idea of quality management universally understood and applied?

Quality management is now seen as something that can be universally applied throughout a business and that also, by implication, is the responsibility of all managers in the business. In particular, it is seen as applying to all parts of the organisation. The internal customer concept can be used to establish the idea that it is important to deliver a high quality of service to internal customers (other processes in the business). Service level agreements can be used to operationalise the internal customer concept. Just as important is the idea that quality also applies to every individual in the business. Everyone has the ability to impair quality, so everyone also has the ability to improve it.

Is quality adequately defined?

Quality needs to be understood from the customer's point of view because it is defined by the customer's perceptions and expectations. One way of doing this is to use a quality gap model. This starts from the fundamental potential gap between customers' expectations and perceptions and deconstructs the various influences on perceptions and expectations. Gaps between these factors can then be used to diagnose possible root causes of quality problems. A further development is to define the quality characteristics of products or services in terms of their functionality, appearance, reliability, durability, recovery and contact.

Is quality adequately measured?

Without measuring quality it is difficult to control it, and the various attributes of quality can be measured either as a variable (measured on a continuously variable scale) or attribute (a binary acceptable or not acceptable judgement). One approach to measuring quality is to express all quality-related issues in cost terms. Quality costs are usually categorised as prevention costs (incurred in trying to prevent errors), appraisal costs (associated with checking for errors), internal failure costs (errors that are corrected with the operation) and external failure costs (errors that experienced by customers). Generally, it is held that increasing expenditure on prevention will bring a more than equivalent reduction in other quality-related costs.

Is quality adequately controlled?

Control means monitoring and responding to any deviations from acceptable levels of quality. One of the most common ways of doing this is through statistical process control (SPC). This technique not only attempts to reduce the variation in quality performance, so as to enhance process knowledge, but also is used to detect deviations outside the 'normal' range of quality variation.

Does quality management always lead to improvement?

Very often quality improvements are not sustained because there is no set of systems and procedures to support and embed them within the operation's day-to-day routines. The best known system for doing this is the ISO 9000 approach adopted now throughout the world. Of the other systems, one of the most widely known is the EFQM excellence model. Once only known as the basis of the European Quality Award, it is now extensively used as a self-assessment tool that allows organisations to assess their own quality systems.

DIAGNOSTIC QUESTION

What is quality management?

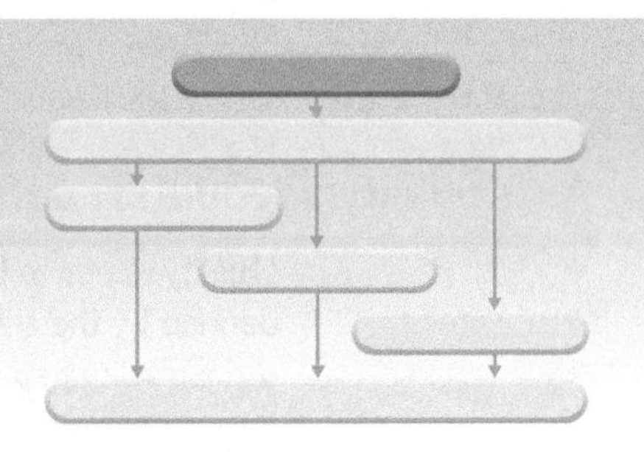

Video

There are many definitions of 'quality': 'conformance to specification'; being 'fit for purpose'; 'achieving appropriate specification'; and so on. The one we use here is *'quality is consistent conformance to customers' expectations'* because it includes both the idea of quality as *specification* (what the product or service can do), and quality as *conformance* (there are no errors, so it always does what it is supposed to do). Not surprisingly, for such an important topic, it has a history. Approaches to quality management have always been of interest to any business that aspired to satisfy its customers. Arguably, the most significant of the approaches to quality management was total quality management (TQM) that became popular with all types of business in the late 1970s and 1980s, although it was based on earlier work by several management thinkers. Feigenbaum popularised the term 'total quality management' in 1957. After that it was developed through the work of several 'quality gurus' including Deming, Juran, Ishikawa, Taguchi and Crosby (see Taking it further at the end of the chapter).

It can be viewed as a logical extension of the way in which quality-related practice has progressed. Originally quality was achieved by inspection – screening out defects before customers noticed them. Then the 'quality control' (QC) concept developed a more systematic approach to not only detecting, but also solving quality problems. 'Quality assurance' (QA) widened the responsibility for quality to include functions other than direct operations, such as human resources, accounting and marketing. It also made increasing use of more sophisticated statistical quality techniques. TQM included much of what went before but developed its own distinctive themes, especially in its adoption of a more 'all embracing' approach. Since the fashionable peak of TQM, there has been some decline in the status of TQM, yet its ideas, many of which are included in this chapter, have become accepted quality practice. The two businesses described in the following examples both incorporate TQM ideas in their approach to quality, especially their inclusion of every employee.

EXAMPLE

The Four Seasons Hotel Canary Wharf[1]

The first Four Seasons Hotel opened over 45 years ago. Since then the company has grown to 81 properties in 34 countries. Famed for its quality of service, the hotel group has won countless awards including the prestigious Zagat survey, numerous AAA Five Diamond Awards and it is also one of only 14 organisations that have been on the *FORTUNE* magazine's list of '100 Best Companies to Work For' every year since it was launched in 1998, thus ranking it as 'top hotel chain' internationally. From its inception the group has had the same guiding principle: 'To make the quality of our service our competitive advantage'. The company has what it calls its Golden Rule: 'Do to others (guests and staff) as you would wish others to do to you'. It is a simple rule, but it guides the whole organisation's approach to quality.

Source: Photographers Direct/OttmarBeirwagen

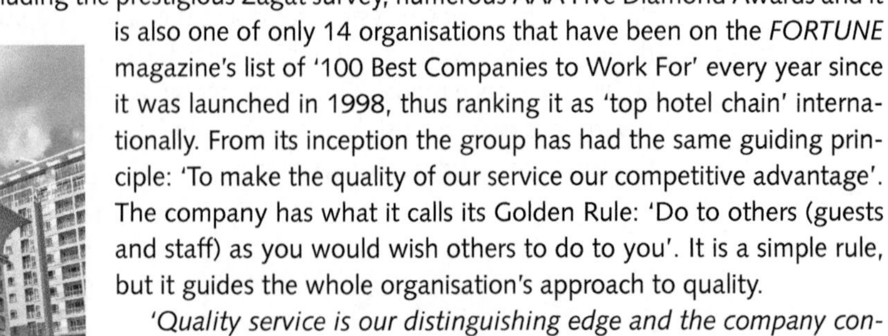

'Quality service is our distinguishing edge and the company continues to evolve in that direction. We are always looking for better, more creative and innovative ways of serving our guests,' says Michael Purtill, the General Manager of the Four Seasons Hotel,

Canary Wharf, London. *'We have recently refined all of our operating standards across the company enabling us to further enhance the personalised, intuitive service that all our guests receive. All employees are empowered to use their creativity and judgement in delivering exceptional service and making their own decisions to enhance our guests' stay. For example, one morning an employee noticed that a guest had a flat tyre on their car and decided on his own accord to change it for them, which was very much appreciated by the guest.'*

'The Golden Rule means that we treat our employees with dignity, respect and appreciation. This approach encourages them to be equally sensitive to our guests' needs and offer sincere and genuine service that exceeds expectations. Just recently one of our employees accompanied a guest to the hospital and stayed there with him for entire afternoon. He wanted to ensure that the guest wasn't alone and was given the medical attention he needed. The following day that same employee took the initiative to return to the hospital (even though it was his day off) to visit and made sure that that guest's family in America was kept informed about his progress. We ensure that we have an ongoing focus on recognising these successes and publicly praise and celebrate all individuals who deliver these warm, spontaneous, thoughtful touches.

'At Four Seasons we believe that our greatest asset and strength are our people. We pay a great deal of attention to selecting the right people with an attitude that takes great pride in delivering exceptional service. We know that motivated and happy employees are essential to our service culture and are committed to developing our employees to their highest potential. Our extensive training programmes and career development plans are designed with care and attention to support the individual needs of our employees as well as operational and business demands. In conjunction with traditional classroom based learning, we offer tailor-made internet-based learning featuring exceptional quality courses for all levels of employee. Such importance is given to learning and development that the hotel has created two specialised rooms designated to learning and development. One is intended for group learning and the other is equipped with private computer stations for internet based individual learning. There is also a library equipped with a broad variety of hospitality-related books, CDs and DVDs that can be taken home at any time. This encourages our employees to learn and develop at an individual pace. This is very motivating for our employees and in the same instance their development is invaluable to the growth of our company. Career-wise, the sky is the limit and our goal is to build lifelong, international careers with Four Seasons.

'Our objective is to exceed guest expectations and feedback from our guests and our employees are an invaluable barometer of our performance. We have created an in-house database that is used to record all guest feedback (whether positive or negative). We also use an online guest survey and guest comment cards which are all personally responded to and analysed to identify any potential service gaps. We continue to focus on delivering individual personalised experiences and our Guest History database remains vital in helping us to achieve this. All preferences and specific comments about service experience are logged on the database. Every comment and every preference is discussed and planned for, for every guest, for every visit. It is our culture that sets Four Seasons apart; the drive to deliver the best service in the industry that keeps their guests returning again and again.'

EXAMPLE

Ryanair[2]

Ryanair was Europe's original and is still Europe's largest low-cost airline (LCA) and whatever else can be said about its strategy, it does not suffer from any lack of clarity. It has grown by offering low-cost basic services and has devised an operations strategy which is in line with its market position. The efficiency of the airlines' operations supports its low-cost market position. Turn-around time at airports is kept to a minimum. This is achieved partly because there are no meals to be loaded onto the aircraft and partly through improved employee productivity. All the aircraft in the fleet are identical, giving savings through standardisation of parts, maintenance

Source: Peter Titmuss/Alamy Images

and servicing. It also means large orders to a single aircraft supplier and therefore the opportunity to negotiate prices down. Also, because the company often uses secondary airports, landing and service fees are much lower. Finally, the cost of selling its services is reduced where possible. Ryanair has developed its own low-cost internet booking service. In addition, the day-to-day experiences of the company's operations managers can also modify and refine these strategic decisions. For example, Ryanair changed its baggage handling contractors at Stanstead airport in the UK after problems with misdirecting customers' luggage. The company's policy on customer service is also clear. *'We patterned Ryanair after Southwest Airlines, the most consistently profitable airline in the US,'* says Michael O'Leary, Ryanair's Chief Executive. *'Southwest founder Herb Kelleher created a formula for success that works by flying only one type of airplane – the 737 – using smaller airports, providing no-frills service on-board, selling tickets directly to customers and offering passengers the lowest fares in the market. We have adapted his model for our marketplace and are now setting the low-fare standard for Europe. Our customer service is about the most well defined in the world. We guarantee to give you the lowest air fare. You get a safe flight. You get a normally on-time flight. That's the package. We don't, and won't, give you anything more. Are we going to say sorry for our lack of customer service? Absolutely not. If a plane is cancelled, will we put you up in a hotel overnight? Absolutely not. If a plane is delayed, will we give you a voucher for a restaurant? Absolutely not.'*

What do these two examples have in common?

Seemingly, not much. The guests at the Four Seasons are paying for exceptional service at a top-range hotel and have high expectations. Ryanair, on the other hand, does not even pretend to offer anything close to a 'luxurious' quality of service. Indeed, some see Ryanair as not competing through their service at all. But they are wrong. Ryanair may not be interested in giving luxurious service, but it does try to conform to what it promises its customers (a reliable, basic service). In fact both Four Seasons and Ryanair try to conform to what they promise, even though exactly what they promise is very different. Both businesses define quality in terms of the expectations and perceptions of customers. This means seeing things *from a customer's point of view*. Customers are seen not regarded as being *external* to the organisation but as the most important *part* of it. Both see quality, not as a single attribute, but a combination of many different things, some of which are difficult to define. There is also an emphasis on every part of the business and every individual having responsibility for ensuring quality.

> **OPERATIONS PRINCIPLE**
> *Quality is multi-faceted, its individual elements differ for different operations.*

DIAGNOSTIC QUESTION

Is the idea of quality management universally understood and applied?

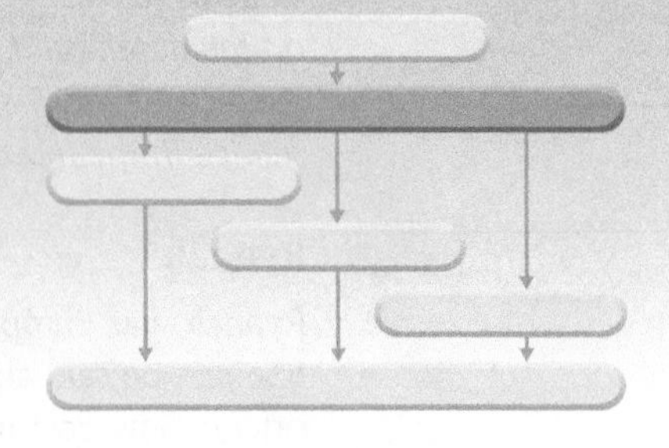

Video

If an operation is to fully understand customers' expectations and to match or exceed them in a consistent manner, it needs to take a universal or *total* approach to quality. Adopting a universal approach means that an understanding of *why* quality is important and *how* quality can be improved permeates the entire organisation. This idea was popularised by proponents of total quality management (TQM), who saw TQM as the ideal unifying philosophy that could unite the whole business behind customer-focused improvement. In particular, two questions are worth asking: first, does quality apply to all parts of the organisation; second, does every person in the organisation contribute to quality.

Does quality apply to all parts of the organisation?

If quality management is to be effective, every process must work properly together. This is because every process affects and in turn is affected by others. Called the **internal customer concept**, it is recognition that every part of an organisation is both an internal customer and, at the same time, an internal supplier for other parts of the organisation. This means that errors in the service provided within an organisation will eventually affect the product or service that reaches the external customer. So, one of the best ways of satisfying external customers is to satisfy internal customers. This means that each process has a responsibility to manage its own internal customer–supplier relationships by clearly defining their own and their customers' exact requirements. In fact the exercise replicates what should be going on for the whole operation and its external customers.

OPERATIONS PRINCIPLE

An appreciation of, involvement in, and commitment to quality should permeate the entire organisation.

Service-level agreements

Some operations bring a degree of formality to the internal customer concept by requiring processes to agree service-level agreements (SLAs) with each other. SLAs are formal definitions of the service and the relationship between two processes. The type of issues that would be covered by such an agreement could include response times, the range of services, dependability of service supply, and so on. Boundaries of responsibility and appropriate performance measures could also be agreed. For example, an SLA between an information systems help desk and the processes that are its internal customers could define such performance measures as:

- the types of information network services that may be provided as 'standard'
- the range of special information services that may be available at different periods of the day
- the minimum 'up time', i.e. the proportion of time the system will be available at different periods of the day
- the maximum response time and average response time to get the system fully operational should it fail
- the maximum response time to provide 'special' services, and so on.

SLAs are best thought of as an approach to deciding service priorities between processes, and as a basis for improving process performance from the internal customers' perspective. At their best they can be the mechanism for clarifying exactly how processes can contribute to the operations as a whole. (See the Critical Commentary at the end of the chapter for a more cynical view.)

Does every person in the organisation contribute to quality?

A total approach to quality should include every individual in the business. People are the source of both good and bad quality and it is everyone's personal responsibility to get quality right. This applies not only to those people who can affect quality directly and have the capability to make mistakes that are immediately obvious to customers, for example those who serve customers face-to-face or physically make products. It also applies to those who are less directly involved in producing products and services. The keyboard operator who mis-keys data, or the product designer who fails to investigate thoroughly the conditions under which products will be used in practice, could also set in motion a chain of events that customers eventually see as poor quality.

It follows that, if everyone has the ability to impair quality, they also have the ability to improve it – if only by 'not making mistakes'. But their contribution is expected to go beyond a commitment not to make mistakes, they are expected to bring something positive to the way they perform their jobs. Everyone is capable of improving the way in which they do their own jobs and practically everyone is capable of helping others in the organisation to improve theirs. Neglecting the potential inherent in all people is neglecting a powerful source of improvement.

EXAMPLE

Even Google suffers 'human error'[3]

For a 40-minute period one busy Saturday in January (2009), all search results on Google, the most popular search engine in the world, were flagged as potentially harmful, with users warned that the site 'may harm your computer'. Users were being warned that all search results were dangerous and were advised to pick another search engine. Google's search service has been hit by technical problems, with users unable to access search results. So what had happened? '*It was human error*,' wrote Marissa Mayer, Vice President of search products and user experience, on the official Google blog. Google work with an organisation called stopbadware.org who find out, using customer complaints, which sites install malicious software on people's computers and therefore should carry a warning. The list of dangerous sites is regularly updated and handed to Google, but when Google updated the list on this Saturday, it mistakenly flagged all sites as potentially dangerous. '*We will carefully investigate this incident and put more robust file checks in place to prevent it from happening again*,' said Marissa Mayer.

DIAGNOSTIC QUESTION

Is 'quality' adequately defined?

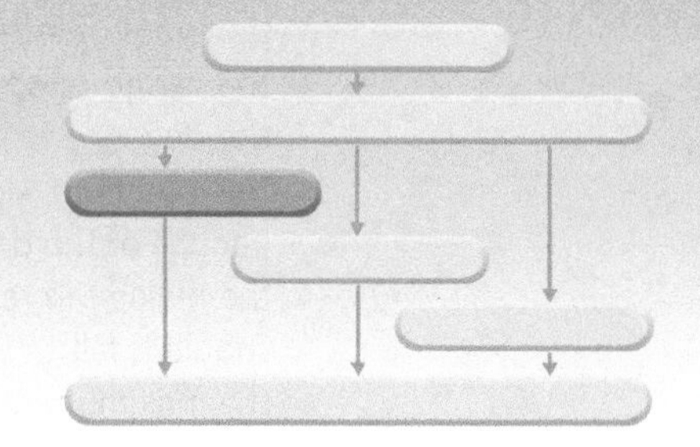

Video

Quality is consistent conformance to customers' expectations. It needs to be understood from a customer's point of view because, to the customer, the quality of a particular product or service is whatever he or she perceives it to be. However, individual customer's expectations may be different. Past experiences, individual knowledge and history will all shape customers' individual expectations. Perceptions are not absolute. Exactly the same product or service may be perceived in different ways by different customers. Also, in some situations, customers may be unable to judge the 'technical' specification of the service or product. They may then use surrogate measures as a basis for their perception of quality. For example, after seeking financial advice from an adviser it might be difficult immediately to evaluate the technical quality of the advice, especially if no better solution presents itself. In reality a judgement of the quality of the advice may be based on perceptions of trustworthiness, relationship, the information that was provided, or the way in which it was provided.

OPERATIONS PRINCIPLE

Perceived quality is governed by the magnitude and direction of the gap between customers' expectations and their perceptions of a product or service.

Closing the gaps–alignment in quality

If the product or service experience was better than expected, then the customer is satisfied and quality is perceived to be high. If the product or service was less than his or her expectations, then quality is low and the customer may be dissatisfied. If the product or service matches expectations then the perceived quality of the product or service is seen to be acceptable. These relationships are summarised in Figure 12.2.

Both customers' expectations and perceptions are influenced by a number of factors, some of which cannot be controlled by the operation and some of which can, at least to a certain extent. Figure 12.3 shows some of the factors that will influence the gap between expectations and perceptions and the potential gaps between some of these factors. This approach to

Video

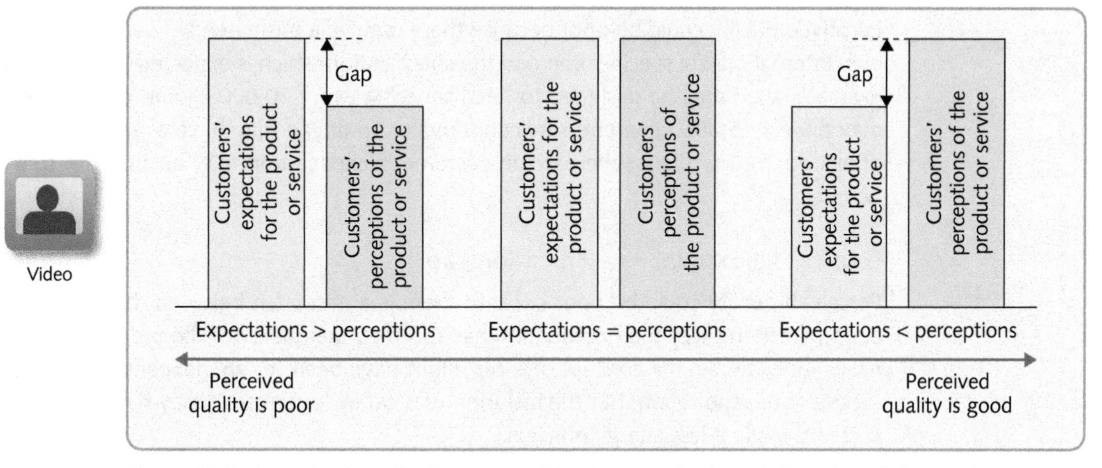

Figure 12.2 Perceived quality is governed by the magnitude and direction of the gap between customers' expectations and their perceptions of the product or service

Practice note

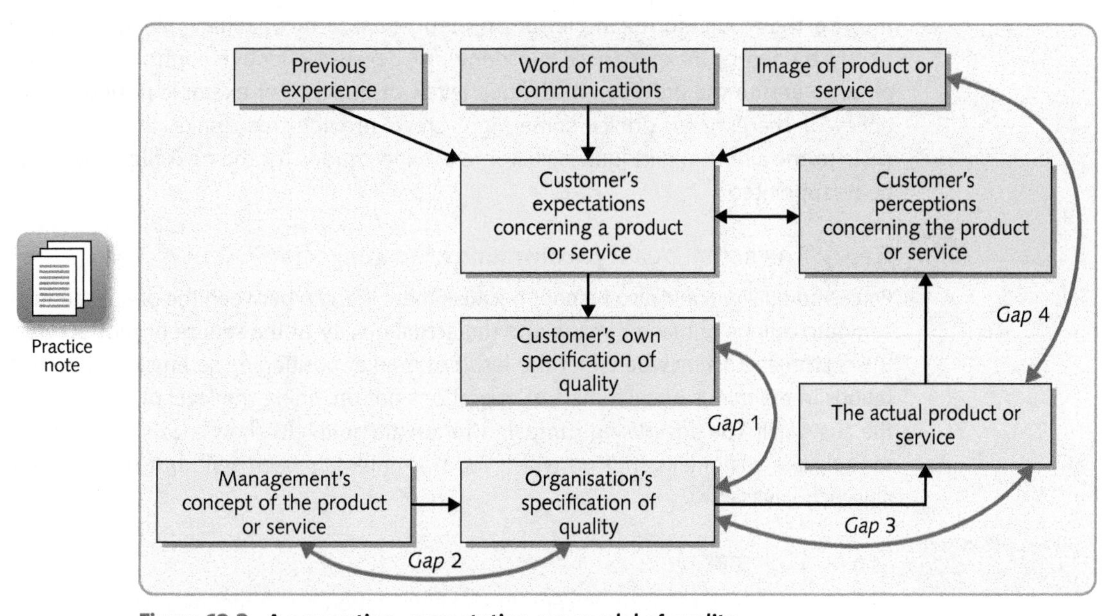

Figure 12.3 A perception–expectation gap model of quality

defining quality is called a 'gap model', of quality. The model shown in Figure 12.3 is adapted from one developed by Zeithaml, Berry and Parasuraman[4], primarily to understand how quality in service operations can be managed and identify some of the problems in so doing. However, this approach is now also used in all types of operation.

Diagnosing quality problems

Describing perceived quality in this way allows a diagnosis of quality problems. If the perceived quality gap is such that customers' perceptions of the product or service fail to match their expectations of it, then the reason (or reasons) must lie in other gaps elsewhere in the model. Four other gaps could explain a perceived quality gap between customers' perceptions and expectations.

Gap 1: The customer's specification–operation's specification gap

Perceived quality could be poor because there may be a mismatch between the organisation's own internal quality specification and the specification which is expected by the customer. For example, a car may be designed to need servicing every 10,000 kilometres but the customer may expect 15,000 kilometre service intervals. An airline may have a policy of charging for drinks during the flight whereas the customer's expectation may be that the drinks would be free.

Gap 2: The concept–specification gap

Perceived quality could be poor because there is a mismatch between the product or service concept and the way the organisation has specified the quality of the product or service internally. For example, the concept of a car might have been for an inexpensive, energy-efficient means of transportation, but the inclusion of a catalytic converter may have both added to its cost and made it less energy efficient.

Gap 3: The quality specification–actual quality gap

Perceived quality could be poor because there is a mismatch between the actual quality of the service or product provided by the operation and its internal quality specification. This may be the result, for example, of an inappropriate or unachievable specification, or of poorly trained or inexperienced personnel, or because effective control systems are not in place to ensure the provision of defined levels of quality. For example if, despite an airline's policy of charging for drinks, some flight crews provide free drinks, they add unexpected costs to the airline and influence customers' expectations for the next flight, when they may be disappointed.

Gap 4: The actual quality–communicated image gap

Perceived quality could also be poor because there is a gap between the organisation's external communications or market image and the actual quality of the service or product delivered to the customer. This may be either the result of market positioning setting unachievable expectations in the minds of customers or operations not providing the level of quality expected by the customer. The advertising campaign for an airline might show a cabin attendant offering to replace a customer's shirt on which food or drink has been spilt, but this service may not always be available.

EXAMPLE

Tea and Sympathy[5]

Defining quality in terms of perception and expectation can sometimes reveal some surprising results. For example, Tea and Sympathy is a British restaurant and café in the heart of New York's West Village. Over the last ten years it has become a fashionable landmark in a city with

Source: Eleanor Bentall/Corbis

one of the broadest range of restaurants in the world. Yet it is tiny, around a dozen tables packed into an area little bigger than the average British sitting room. Not only expatriate Brits but also native New Yorkers and celebrities queue to get in. As the only British restaurant in New York, it has a novelty factor, but also it has become famous for the unusual nature of its service. '*Everyone is treated in the same way*,' says Nicky Perry one of the two ex-Londoner's who run it. '*We have a firm policy that we don't take any shit.*' This robust attitude to the treatment of customers is reinforced by 'Nicky's Rules' that are printed on the menu:

1 Be pleasant to the waitresses – remember Tea and Sympathy girls are always right.
2 You will have to wait outside the restaurant until your entire party is present: no exceptions.
3 Occasionally, you may be asked to change tables so that we can accommodate all of you.
4 If we don't need the table you may stay all day, but if people are waiting it's time to naff off.
5 These rules are strictly enforced. Any argument will incur Nicky's wrath. You have been warned.

Most of the waitresses are also British and enforce Nicky's Rules strictly. If customers object they are thrown out. Nicky says that she has had to train 'her girls' to toughen up. '*I've taught them that when people cross the line they can tear their throats out as far as I'm concerned. What we've discovered over the years is that if you are really sweet, people see it as a weakness.*' People get thrown out of the restaurant about twice a week and yet customers still queue for the genuine shepherd's pie, a real cup of tea, and of course, the service.

Quality characteristics

Much of the 'quality' of a product or service will have been specified in its design, but not all the design details are useful in defining quality. Rather it is the *consequences* of the design that are perceived by customers. These consequences of the design are called *quality characteristics*. Table 12.1 shows a list of quality characteristics that are generally useful applied to a service (flight) and a product (car).

Table 12.1 Quality characteristics for a motor car and an air journey

Quality characteristics	*Car*	*Flight*
Functionality – how well the product or service does its job, including its performance and features	Speed, acceleration, fuel consumption, ride quality, road-holding, etc.	Safety and duration of journey, on-board meals and drinks, car and hotel booking services
Appearance – the sensory characteristics of the product or service: its aesthetic appeal, look, feel, sound and smell	Aesthetics, shape, finish, door gaps, etc.	Decor and cleanliness of aircraft, lounges and crew
Reliability – the consistency of the product's or service's performance over time, or the average time for which it performs within its tolerated band of performance	Mean time to failure	Keeping to the published flight times
Durability – the total useful life of the product or service assuming occasional repair or modification	Useful life (with repair)	Keeping up with trends in the industry
Recovery – the ease with which problems with the product or service can be rectified or resolved	Ease of repair	Resolution of service failures
Contact – the nature of the person-to-person contact which might take place; it could include the courtesy, empathy, sensitivity and knowledge of contact staff	Knowledge and courtesy of sales and service staff	Knowledge, courtesy and sensitivity of airline staff

DIAGNOSTIC QUESTION

Is 'quality' adequately measured?

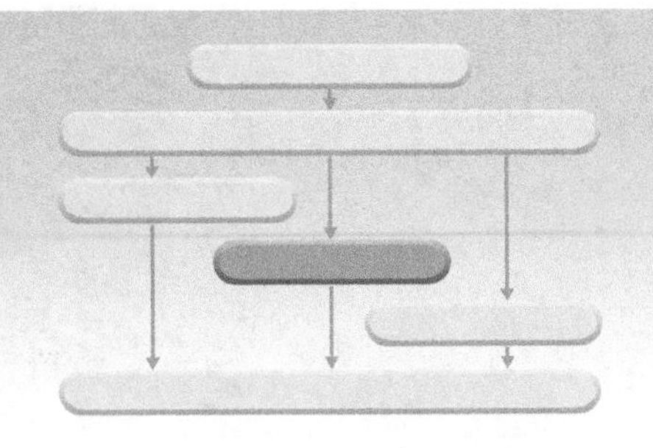

Video

Some quality characteristics are relatively easy to measure. For example, is the gap between a car door and pillar less than 5mm? Other more difficult-to-measure quality characteristics, such as 'appearance', need to be decomposed into their constituent elements, such as 'colour match', 'surface finish' and 'the number of visible scratches', all of which are capable of being measured in a relatively objective manner. They may even be quantifiable. However, decomposing quality characteristics into their measurable sub-components can result in some loss of meaning. A quantified list of 'colour match', the 'smoothness' of the surface finish and the 'number of visible scratches' does not cover factors such as 'aesthetics', a characteristic that is difficult to measure, but nonetheless important. Some quality characteristics cannot themselves be measured at all. The 'courtesy' of airline staff, for example, has no objective quantified measure, yet airlines place a great deal of importance on the need to ensure courtesy in their staff. In cases like this, the operation will have to attempt to measure customer *perceptions* of courtesy.

Variables and attributes

The measures used to describe quality characteristics are of two types: variables and attributes. Variable measures are those that can be measured on a continuously variable scale (for example, length, diameter, weight or time). Attributes are those that are assessed by judgement and have two states (for example, right or wrong, works or does not work, looks OK or not OK). Table 12.2 categorises some of the measures that might be used for the quality characteristics of the car and the flight.

Measuring the 'costs of quality'

One approach to measuring aggregated quality is to express all quality-related issues in cost terms. This is the 'cost of quality approach (usually taken to refer to both costs and benefits of quality). These costs of quality are usually categorised as *prevention costs, appraisal costs, internal failure costs* and *external failure costs*. Table 12.3 illustrates the type of factors that are included in these categories.

Understand the relationship between quality costs[6]

At one time it was assumed that failure costs reduce as the money spent on appraisal and prevention increases. There must be a point beyond which the cost of improving quality gets larger than the benefits that it brings. Therefore there must be an optimum amount of quality effort to be applied in any situation that minimises the total costs of quality. Figure 12.4(a) sums up this idea.

More recently the 'optimum quality effort' approach has been challenged. First, why should any operation accept the *inevitability* of errors? Some occupations seem to be able to accept a zero-defect standard (even if they do not always achieve it). No one accepts the inevitability of pilots crashing a certain proportion of their aircraft, or nurses dropping a certain number of babies. Second, failure costs are generally underestimated. They are usually taken to include the cost of 'reworking' defective products, 're-serving' customers, scrapping parts and materials,

Table 12.2 Variable and attribute measures for quality characteristics

	Car		Flight	
Characteristic	*Variable*	*Attribute*	*Variable*	*Attribute*
Functionality	Acceleration and braking characteristics from test bed	Is the ridge quality satisfactory?	Number of journeys which actually arrived at the destination (i.e. didn't crash!)	Was the food acceptable?
Appearance	Number of blemishes visible on car	Is the colour to specification?	Number of seats not cleaned satisfactorily	Is the crew dressed smartly?
Reliability	Average time between faults	Is the reliability satisfactory?	Proportion of journeys that arrived on time	Were there any complaints?
Durability	Life of the car	Is the useful life as predicted?	Number of times service innovations lagged competitors	Generally, is the airline updating its services in a satisfactory manner?
Recovery	Time from fault discovered to fault repaired	Is the serviceability of the car acceptable?	Proportion of service failures resolved satisfactorily	Do customers feel that staff deal satisfactorily with complaints?
Contact	Level of help provided by sales staff (1 to 5 scale)	Did customers feel well served (yes or no)?	The extent to which customers feel well treated by staff (1 to 5 scale)	Did customers feel that the staff were helpful (yes or no)?

Table 12.3 The categories of quality cost

Category of quality related cost	*Includes such things as:*
Prevention costs – those costs incurred in trying to prevent problems, failures and errors from occurring in the first place	• Identifying potential problems and putting the process right before poor quality occurs • Designing and improving the design of products and services and processes to reduce quality problems • Training and development of personnel in the best way to perform their jobs • Process control
Appraisal costs – those costs associated with controlling quality to check to see if problems or errors have occurred during and after the creation of the product or service	• The setting up of statistical acceptance sampling plans • The time and effort required to inspect inputs, processes and outputs • Obtaining processing inspection and test data • Investigating quality problems and providing quality reports • Conducting customer surveys and quality audits
Internal failure costs – failure costs that are associated with errors dealt with inside the operation	• The cost of scrapped parts and materials • Reworked parts and materials • The lost production time as a result of coping with errors • Lack of concentration due to time spent troubleshooting rather than improvement
External failure costs – failure costs that are associated with errors being experienced by customers	• Loss of customer goodwill affecting future business • Aggrieved customers who may take up time • Litigation (or payments to avoid litigation) • Guarantee and warranty costs • The cost to the company of providing excessive capability (too much coffee in the pack and too much information to a client)

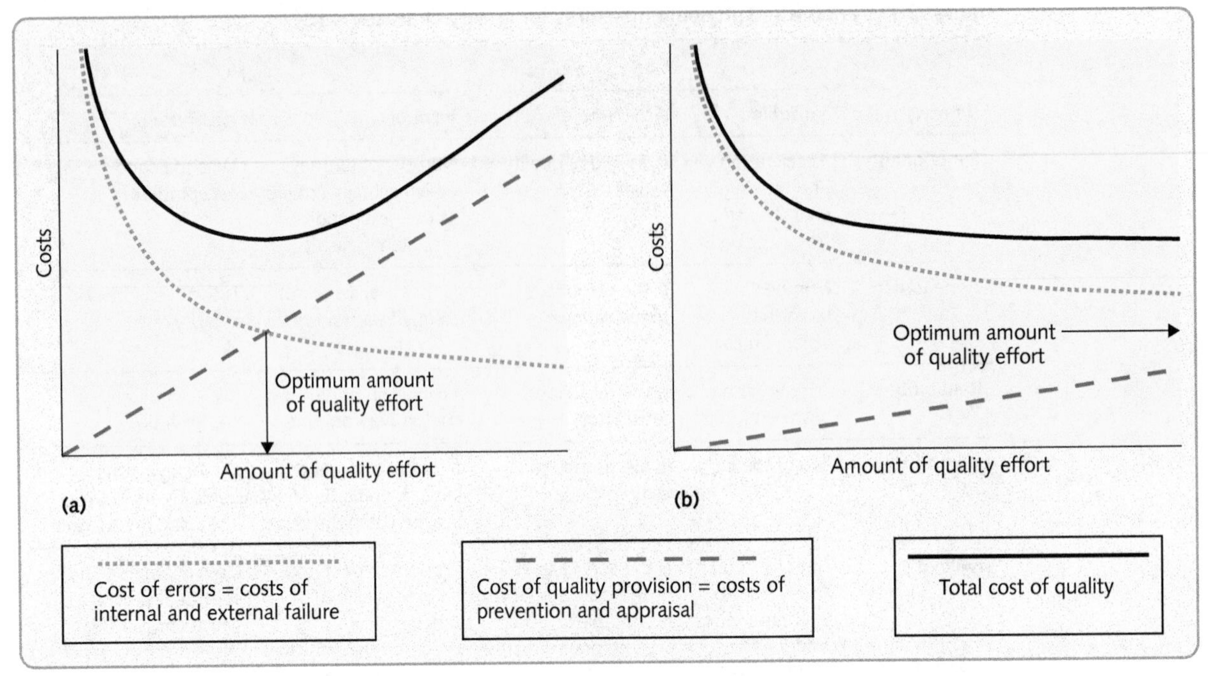

Figure 12.4 (a) The traditional cost of quality model; (b) a more modern view

the loss of goodwill, warranty costs, etc. These are important, but in practice, the real cost of poor quality should include all the management time wasted in organising rework and rectification and, more important, the loss of concentration and the erosion of confidence between processes within the operation. Third, it implies that prevention costs are inevitably high. But by stressing the importance of quality to every individual, preventing errors becomes an integral part of everyone's work. More quality is not only achieved by using more inspectors, we all have a responsibility for our own quality and all should be capable of 'doing things right first time'. This may incur some costs – training, automatic checks, anything which helps to prevent errors occurring in the first place – but not such a steeply inclined cost curve as in the 'optimum-quality' theory. Finally the 'optimum-quality level' approach, by accepting compromise, does little to challenge operations managers and staff to find ways of improving quality.

Put these corrections into the optimum-quality effort calculation and the picture looks very different (see Figure 12.4b). If there is an 'optimum', it is a lot further to the right, in the direction of putting more effort (but not necessarily cost) into quality.

The TQM-influenced quality cost model

TQM rejected the optimum-quality level concept. Rather, it concentrated on how to reduce all known and unknown failure costs. So, rather than placing most emphasis on appraisal (so that 'bad products and service don't get through to the customer'), it emphasised prevention (to stop errors happening in the first place). This has a significant, positive effect on internal failure costs, followed by reductions both in external failure costs and, once confidence has been firmly established, also in appraisal costs. Eventually even prevention costs can be stepped down in absolute terms, though prevention remains a significant cost in relative terms. Figure 12.5 illustrates this idea, showing how initially total quality costs may rise as investment in some aspects of prevention is increased. Once this relationship between categories of quality cost is accepted it shifts the emphasis from a reactive, approach to quality (waiting for errors to happen, then screening them out), to a more proactive, 'getting it right first time' approach (doing something before errors happen).

> OPERATIONS PRINCIPLE
> *Effective investment in preventing quality errors can significantly reduce appraisal and failure costs.*

Animated diagram

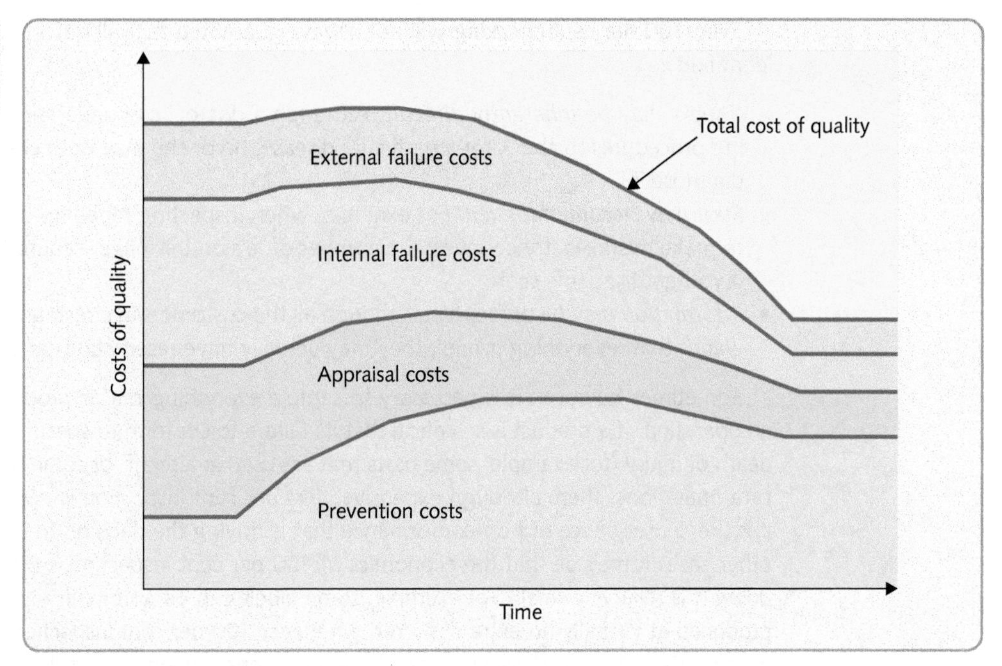

Figure 12.5 Increasing the effort spent on preventing errors occurring in the first place brings a more than equivalent reduction in other cost categories

DIAGNOSTIC QUESTION

Is 'quality' adequately controlled?

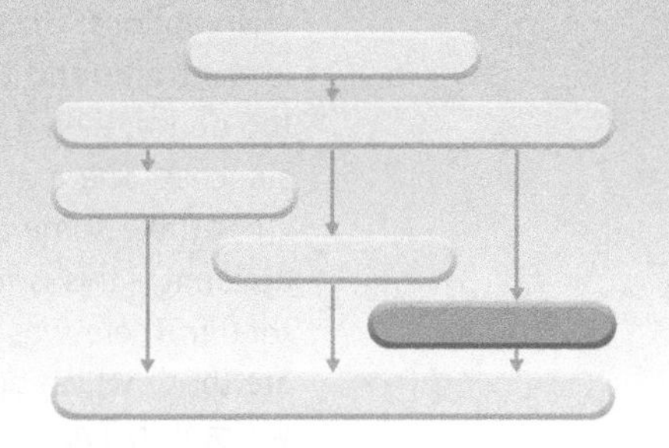

Video

After quality has been defined and measured, processes will need to check that their quality conforms to whatever quality standards are deemed appropriate. This does not necessarily mean checking everything, sampling may be more appropriate.

Check every product and service or take a sample?

There are several reasons why checking everything may not be sensible:

- *It could be dangerous to check everything.* A doctor, for example, checks just a small sample of blood rather than it all. The characteristics of this sample are taken to represent those of the rest of the patient's blood.
- *Checking of everything might destroy the product or interfere with the service.* A lamp manufacturer cannot check the life of every single light bulb leaving the factory; they would all be destroyed. Nor would it be appropriate for a head waiter to check whether customers are enjoying the meal every 30 seconds.
- *Checking everything may be too costly.* For example, it just might not be feasible to check every single item from a high-volume plastic moulding machine or to check the feelings of every single bus passenger every day.

Even 100 per cent checking will not always guarantee that all defects or problems will be identified.

- *Checks may be inherently difficult.* Although a doctor may undertake all the correct testing procedures to check for a particular disease, he or she may not necessarily be certain to diagnose it.
- *Staff may become fatigued.* For example, when inspecting repetitive items where it is easy to make mistakes (try counting the number of 'e's on this page – count them again and see if you get the same score).
- *Information may be unreliable.* Although all the customers in a restaurant may tell the head waiter that 'everything is fine', they may actually have reservations about their experience.

Sometimes however, it is necessary to sample everything that is produced by a process or an operation. If a product is so critical that its failure to conform to specification would result in death or injury (for example, some parts that are used in aircraft, or some services within health care operations) then, although expensive, 100 per cent inspection is necessary. In such cases it is the consequence of non-conformance that is driving the decision to inspect everything. In other cases it may be that the economics of 100 per cent inspection are such that the cost of doing it is relatively small. For example, some labels can be automatically scanned as they are produced at virtually no extra cost. Yet, whenever 100 per cent inspection is adopted, there is another risk – that of classifying something as an error when, in fact, it conforms to specification. This distinction is summarised in what is often referred to as type I and type II errors.

Type I and type II errors

Checking quality by sampling, although requiring less time than checking everything, does have its own problems. Take the example of someone waiting to cross a street. There are two main options: cross (take some action), or continue waiting (take no action). If there is a break in the traffic and the person crosses, or if that person continues to wait because the traffic is too dense, then a correct decision has been made (the action was appropriate for the circumstances). There are two types of incorrect decisions or errors. One would be a decision to cross (take some action) when there is not an adequate break in the traffic, resulting in an accident – this is referred to as a type I error. Another would be a decision not to cross even though there was an adequate gap in the traffic – this is called a type II error. Type I errors are those which occur when a decision was made to do something and the situation did not warrant it. Type II errors are those which occur when nothing was done, yet a decision to do something should have been taken as the situation did indeed warrant it. So, there are four outcomes, summarised in Table 12.4.

Statistical process control (SPC)

The most common method of checking the quality of a sampled product or service so as to make inferences about all the output from a process is called statistical process control (SPC). SPC is concerned with sampling the process during the production of the goods or the delivery

Table 12.4 Type I and type II errors for a pedestrian crossing the road

		Road conditions	
		Safe (action was appropriate)	**Unsafe** (action was not appropriate)
Decision	**Cross** (take some action)	Correct decision	**Type I error**
	Wait (take no action)	**Type II error**	Correct decision

of service. Based on this sample, decisions are made as to whether the process is 'in control', that is, operating as it should be. If there seems to be a problem with the process, then it can be stopped (if possible and appropriate) and the problem identified and rectified. For example, an international airport may regularly ask a sample of customers if the cleanliness of its restaurants is satisfactory. If an unacceptable number of customers in one sample are found to be unhappy, airport managers may have to consider improving the procedures in place for cleaning tables.

Control charts

The value of SPC is not just to make checks of a single sample but to monitor the results of many samples over a period of time. It does this by using control charts. Control charts record some aspect of quality (or performance generally) over time to see if the process seems to be performing as it should (called *in control*), or not (called *out of control*). If the process does seem to be going out of control, then steps can be taken *before* there is a problem.

Figure 12.6 shows a typical control chart. Charts that look something like these, can be found in almost any operation. They could, for example, represent the percentage of customers in a sample of 1,000 who, each week, were dissatisfied with the service they received from two call centres. In chart (a), measured customer dissatisfaction has been steadily increasing over time. There is evidence of a clear (negative) trend that management may wish to investigate. In chart (b), although there is little evidence of any trend in average dissatisfaction, the variability in performance seems to be increasing. Again the operation may want to investigate the causes.

Looking for *trends* is an important use of control charts. If the trend suggests the process is getting steadily worse, then it will be worth investigating the process. If the trend is steadily improving, it may still be worthy of investigation to try to identify what is happening that is making the process better. An even more important use of control charts is to investigate the *variation* in performance.

Why variation is a bad thing

OPERATIONS PRINCIPLE
High levels of variation reduce the ability to detect changes in process performance.

Although a trend, such as that shown in Figure 12.6(a), clearly indicates deteriorating performance, the variation shown in Figure 12.6(b) can be just as serious. Variation is a problem because it masks any changes in process behaviour. Figure 12.7 shows the performance of two processes both of which change their behaviour at the same time. The process on

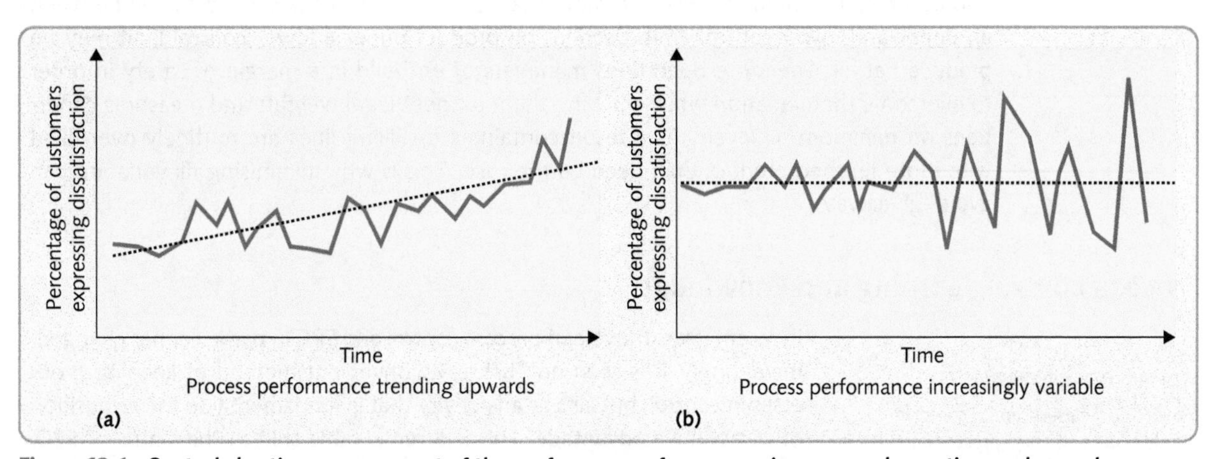

Figure 12.6 Control charting – any aspect of the performance of a process is measured over time and may show trends in average performance, and/or changes in the variation of performance over time

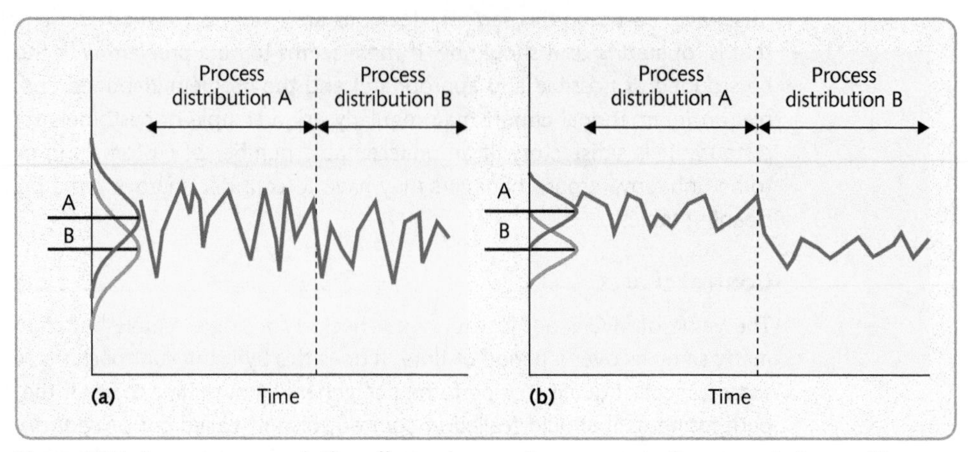

Figure 12.7 Low process variation allows changes in process performance to be readily detected

the left has such a wide natural variation that it is not immediately apparent that any change has taken place. Eventually it will become apparent, but it may take some time. By contrast, the performance of process represented by the chart on the right has a far narrower band of variation, so the same change in average performance is more easily noticed. The narrower the variation of a process, the more obvious are any changes that might occur, and the easier it is to make a decision to intervene. SPC is discussed much further in the supplement to this chapter. It is also one of the core ideas in the Six Sigma improvement approach that is discussed in the next chapter.

EXAMPLE

What a Giveaway[7]

Another negative effect of wide-process variability is particularly evident in any process that fills any kind of container with weighed product. Even small reductions in the variability in filling levels can translate into major savings. This is because of what is known as 'giveaway' or 'over-fill', caused by the necessity to ensure that containers are not legally underweight. Although slightly different regulations may apply in various parts of the world, any process that produces products that have an 'e' after the stated weight on the container must produce products with an average weight greater than the declared weight on the container, with the average weight being determined by sampling. In addition, there are two other legal conditions. First, no more than 2.5 per cent of the sample can lie between an upper and lower control limit. Second, no product under a lower control limit may be produced at all. Therefore operations managers often build in a margin of safety in order to overcome that variation while allowing them to meet legal weights and measures conditions on minimum fill levels. As a result containers on filling lines are routinely over-filled with more finished product than need be the case. This is why minimising fill variation can avoid 'giveaway'.

Process control, learning and knowledge

OPERATIONS PRINCIPLE

Statistical-based control gives the potential to enhance process knowledge.

In recent years the role of process control and SPC in particular has changed. Increasingly, it is seen not just as a convenient method of keeping processes in control, but also as an activity that is fundamental to the acquisition of competitive advantage. This is a remarkable shift in the status of SPC. Traditionally it was seen as one of the most *operational*, immediate and 'hands-on' operations management techniques. Yet it is now seen as contributing to an operation's *strategic* capabilities. This is how the logic of the argument goes:

1 SPC is based on the idea that process variability indicates whether a process is in control or not.
2 Processes are brought into *control* and improved by progressively reducing process variability. This involves eliminating the assignable causes of variation.
3 One cannot eliminate assignable causes of variation without gaining a better understanding of how the process operates. This involves *learning* about the process, where its nature is revealed at an increasingly detailed level.
4 This learning means that *process knowledge* is enhanced, which in turn means that operations managers are able to predict how the process will perform under different circumstances. It also means that the process has a greater capability to carry out its tasks at a higher level of performance.
5 This increased *process capability* is particularly difficult for competitors to copy. It cannot be bought 'off-the-shelf'. It only comes from time and effort being invested in controlling operations processes. Therefore, process capability leads to strategic advantage.

In this way, process control leads to learning which enhances process knowledge and builds difficult-to-imitate process capability.

DIAGNOSTIC QUESTION

Does quality management always lead to improvement?

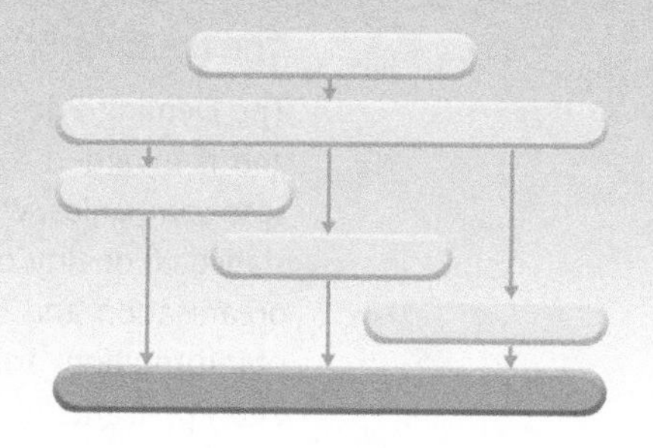

Video

No amount of effort put into quality initiatives can guarantee improvement in process performance. In fact some surveys show up to half of quality programmes provide only disappointing, if any, permanent improvement. Improving quality is not something that happens simply by getting everyone in an organisation to 'think quality'. Very often improvements do not stick because there is no set of systems and procedures to support and embed them into the operation's day-to-day routines. 'Quality systems' are needed.

A quality system is *'the organisational structure, responsibilities, procedures, processes and resources for implementing quality management'*.[8] It should cover all facets of a business's operations and processes, and define the responsibilities, procedures and processes that ensure the implementation of quality improvement. The best known quality system is ISO 9000.

The ISO 9000 approach

The ISO 9000 series is a set of worldwide standards that establishes requirements for companies' quality management systems. It is being used worldwide to provide a framework for quality assurance. By 2000, ISO 9000 had been adopted by more than a quarter of a million organisations in 143 countries. Originally its purpose was to provide an assurance to the purchasers of products or services by defining the procedures, standards and characteristics of the control system that governed the process that produced them. In 2000, ISO 9000 was substantially revised. Rather than using different standards for different functions within a business, it took a 'process' approach and focused on the outputs from any operation's process, rather than the detailed procedures that had dominated the previous version. This process orientation requires operations to define and record core processes and sub-processes. ISO 9000 (2000) also stresses four other principles.

- Quality management should be customer focused and customer satisfaction measured using surveys and focus groups. Improvement against customer standards should be documented.
- Quality performance should be measured, and relate to products and services themselves, the processes that created them, and customer satisfaction. Furthermore, measured data should always be analysed.
- Quality management should be improvement driven. Improvement must be demonstrated in both process performance and customer satisfaction.
- Top management must demonstrate their commitment to maintaining and continually improving management systems. This commitment should include communicating the importance of meeting customer and other requirements, establishing a quality policy and quality objectives, conducting management reviews to ensure the adherence to quality policies, and ensuring the availability of the necessary resources to maintain quality systems.

The Deming Prize

The Deming Prize was instituted by the Union of Japanese Scientists and Engineers in 1951 and is awarded to those companies, initially in Japan, but more recently opened to overseas companies, which have successfully applied 'company-wide quality control' based upon statistical quality control. There are ten major assessment categories: policy and objectives, organisation and its operation, education and its extension, assembling and disseminating of information, analysis, standardisation, control, quality assurance, effects and future plans. The applicants are required to submit a detailed description of quality practices. This is a significant activity in itself and some companies claim a great deal of benefit from having done so.

The Malcolm Baldrige National Quality Award

In the early 1980s the American Productivity and Quality Center recommended that an annual prize, similar to the Deming Prize, should be awarded in America. The purpose of the awards was to stimulate American companies to improve quality and productivity, to recognise achievements, to establish criteria for a wider quality effort and to provide guidance on quality improvement. The main examination categories are: leadership, information and analysis, strategic quality planning, human resource utilisation, quality assurance of products and services, quality results and customer satisfaction. The process, like that of the Deming Prize, includes a detailed application and site visits.

The EFQM Excellence Model[9]

Over 20 years ago Western European companies formed the European Foundation for Quality Management (EFQM). An important objective of the EFQM is to recognise quality

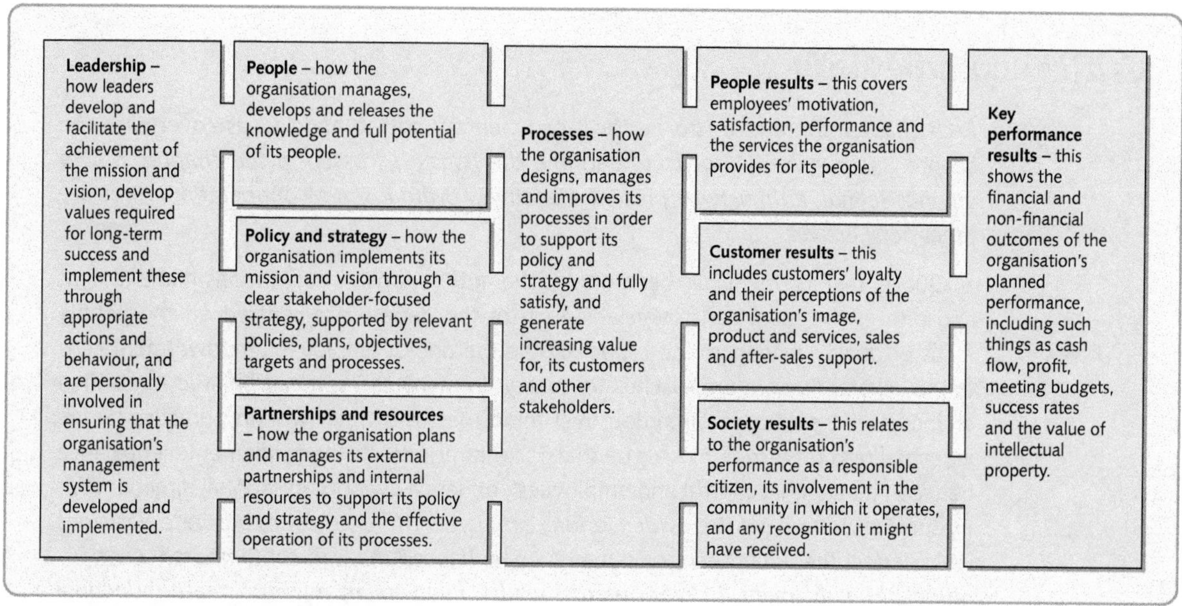

Figure 12.8 The EFQM Excellence Model

achievement. Because of this, it launched the European Quality Award (EQA), awarded to the most successful exponent of total quality management in Europe each year. To receive a prize, companies must demonstrate that their approach to total quality management has contributed significantly to satisfying the expectations of customers, employees and others with an interest in the company for the past few years. In 1999, the model on which the European Quality Award was based was modified and renamed 'The EFQM Excellence Model'. The changes made were not fundamental but did attempt to reflect some new areas of management and quality thinking (for example, partnerships and innovation) and placed more emphasis on customer and market focus. The model is based on the idea that the outcomes of quality management in terms of what it calls 'people results', 'customer results', 'society results' and 'key performance results' are achieved through a number of 'enablers'. These enablers are leadership and constancy of purpose, policy and strategy, how the organisation develops its people, partnerships and resources, and the way it organises its processes. These ideas are incorporated as shown in Figure 12.8. The five enablers are concerned with how results are being achieved, while the four 'results' are concerned with what the company has achieved and is achieving.

Self-assessment

The EFQM defines *self-assessment* as 'a comprehensive, systematic, and regular review of an organisation's activities and results referenced against a model of business excellence'. The main advantage of using such models for self-assessment seems to be that companies find it easier to understand some of the more philosophical concepts of quality management when they are translated into specific areas, questions and percentages. Self-assessment also allows organisations to measure their progress in achieving the benefits of quality management.

Critical commentary

Each chapter contains a short critical commentary on the main ideas covered in the chapter. Its purpose is not to undermine the issues discussed in the chapter, but to emphasise that, although we present a relatively orthodox view of operation, there are other perspectives.

- Quality management has been one of the hottest topics in operations management and one of the most controversial. Much of the debate has centred on the people focus of quality management, especially the rhetoric of employee empowerment central to several modern approaches to quality. In many cases, it can be little more than an increase in employee discretion over minor details of their working practice. Some industrial relations academics argue that TQM rarely affects the fundamental imbalance between managerial control and employees' influence over organisational direction. For example, '*... there is little evidence that employee influence over corporate decisions which affect them has been, or can ever be, enhanced through contemporary configuration of involvement. In other words, whilst involvement might increase individual task discretion, or open up channels for communication, the involvement programme is not designed to offer opportunities for employees to gain or consolidate control over the broader environment in which their work is located.*'[10]

- Other criticisms concern the appropriateness of some mechanisms, such as service level agreements (SLAs). Some see the strength of SLAs as the degree of formality they bring to customer–supplier relationships, but there also drawbacks. The first is that the 'pseudo-contractual' nature of the formal relationship can work against building partnerships. This is especially true if the SLA includes penalties for deviation from service standards. The effect can sometimes be to inhibit rather than encourage joint improvement. The second is that SLAs tend to emphasise the 'hard' and measurable aspects of performance rather than the 'softer' but often more important aspects. So a telephone may be answered within four rings, but how the caller is treated in terms of 'friendliness' may be far more important.

- Similarly, and notwithstanding its widespread adoption (and its revision to take into account some of its perceived failing), ISO 9000 is not seen as beneficial by all authorities. Criticisms include the following:
  - The whole process of documenting processes, writing procedures, training staff and conducting internal audits is expensive and time-consuming.
  - Similarly, the time and cost of achieving and maintaining ISO 9000 registration are excessive.
  - It is too formulaic. It encourages operations to 'manage by manual', substituting a 'recipe' for a more customised and creative approach to managing operations improvement.

SUMMARY CHECKLIST

This checklist comprises questions that can be usefully applied to any type of operations and reflect the major diagnostic questions used within the chapter.

- ☐ Does everyone in the business really believe in the importance of quality, or is it just one of those things that people say without really believing it?
- ☐ Is there an accepted definition of quality used within the business?
- ☐ Do people understand that there are many different definitions and approaches to quality, and do they understand why the business has chosen its own particular approach?
- ☐ Do all parts of the organisation understand their contribution to maintaining and improving quality?
- ☐ Are service level agreements used to establish concepts of internal customer service?
- ☐ Is some form of gap model used to diagnose quality problems?
- ☐ Is quality defined in terms of a series of quality characteristics?
- ☐ Is quality measured using all relevant quality characteristics?
- ☐ Is the cost of quality measured?
- ☐ Are quality costs categorised as prevention, appraisal, internal failure and external failure costs?
- ☐ Is quality adequately controlled?
- ☐ Has the idea of statistical process control (SPC) been explored as a mechanism for controlling quality?
- ☐ Do individual processes have any idea of their own variability of quality performance?
- ☐ Have quality systems been explored, such as ISO 9000 and the EFQM excellence model?

CASE STUDY

Turnround at the Preston plant

'Before the crisis, the quality department was just for looks. We certainly weren't used much for problem solving; the most we did was inspection. Data from the quality department was brought to the production meeting and they would all look at it, but no one was looking behind it.' (Quality Manager, Preston Plant)

The Preston plant of Rendall Graphics was located in Preston, Vancouver, across the continent from their headquarters in Massachusetts. The plant had been bought from the Georgetown Corporation by Rendall in March 2000. Precision coated papers for ink-jet printers accounted for the majority of the plant's output, especially paper for specialist uses. The plant used coating machines that allowed precise coatings to be applied. After coating, the conversion department cut the coated rolls to the final size and packed the sheets in small cartons.

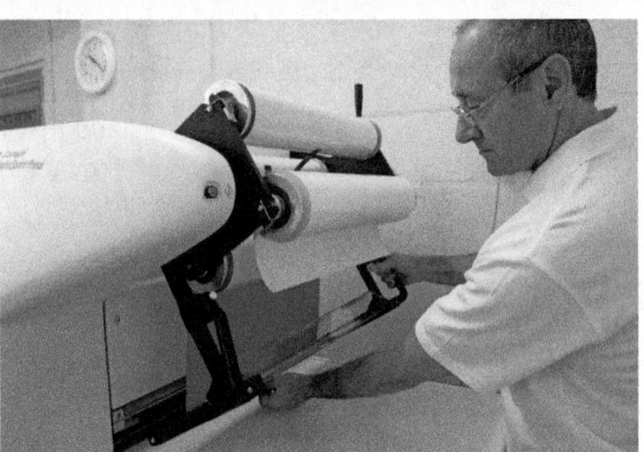

Source: MikeBooth/Alamy Images

The curl problem

In late 1998 Hewlett Packard (HP), the plant's main customer for ink-jet paper, informed the plant of some problems it had encountered with paper curling under conditions of low humidity. There had been no customer complaints to HP, but their own personnel had noticed the problem, and they wanted it fixed. Over the next seven or eight months a team at the plant tried to solve the problem. Finally, in October of 1999 the team made recommendations for a revised and considerably improved coating formulation. By January 2000 the process was producing acceptably. However, 1999 had not been a good year for the plant. Although sales were reasonably buoyant the plant was making a loss of around $2 million for the year. In October 1999, Tom Branton, previously accountant for the business, was appointed as Managing Director.

Slipping out of control

In the spring of 2000, productivity, scrap and re-work levels continued to be poor. In response to this the operations management team increased the speed of the line and made a number of changes to operating practice in order to raise productivity.

'Looking back, changes were made without any proper discipline, and there was no real concept of control. We were always meeting specification, yet we didn't fully understand how close we really were to not being able to make it. The culture here said, "If it's within specification then it's OK" and we were very diligent in making sure that the product which was shipped ***was*** *in specification. However, Hewlett Packard gets "process charts" that enables them to see more or less exactly what is happening right inside your operation. We were also getting all the reports but none of them were being internalised; we were using them just to satisfy the customer. By contrast, HP have a statistically-based analytical mentality that says to itself, "You might be capable of making this product but we are thinking two or three product generations forward and asking ourselves, will you have the capability then, and do we want to invest in this relationship for the future?'* (Tom Branton)

The spring of 2000 also saw two significant events. First, Hewlett Packard asked the plant to bid for the contract to supply a new ink-jet platform, known as the Vector project, a contract that would secure healthy orders for several years. The second event was that the plant was acquired by Rendall. *'What did Rendall see when they bought us? They saw a small plant on the Pacific coast losing lots of money.'* (Finance Manager, Preston Plant)

Rendall were not impressed by what they found at the Preston plant. It was making a loss and had only just escaped from incurring a major customer's disapproval over the curl issue. If the plant did not get the

Vector contract, its future looked bleak. Meanwhile the chief concern continued to be productivity. But also, once again, there were occasional complaints about quality levels. However, HP's attitude caused some bewilderment to the operations management team. *'When HP asked questions about our process the operations guys would say, "Look we're making roll after roll of paper, it's within specification. What's the problem?"'* (Quality Manager, Preston Plant)

But it was not until summer that the full extent of HP's disquiet was made. *'I will never forget June 2000. I was at a meeting with HP in Chicago. It was not even about quality. But during the meeting one of their engineers handed me a control chart, one that we supplied with every batch of product. He said, "Here's your latest control chart. We think you're out of control and you don't know that you're out of control and we think that we are looking at this data more than you are." He was absolutely right, and I fully understood how serious the position was. We had our most important customer telling us we couldn't run our processes just at the time we were trying to persuade them to give us the Vector contract.'* (Tom Branton)

The crisis

Tom immediately set about the task of bringing the plant back under control. They first of all decided to go back to the conditions which prevailed in the January, when the curl team's recommendations had been implemented. This was the state before productivity pressures had caused the process to be adjusted. At the same time the team worked on ways of implementing unambiguous 'shut-down rules' that would allow operators to decide under what conditions a line should be halted if they were in doubt about the quality of the product they were making.

'At one point in May of 2000 we had to throw away 64 jumbo rolls of out-of-specification product. That's over $100,000 of product scrapped in one run. Basically that was because they had been afraid to shut the line down. Either that or they had tried to tweak the line while it was running to get rid of the defect. The shut-down guidelines in effect say, "We are not going to operate when we are not in a state of control". Until then our operators just couldn't win. If they failed to keep the machines running we would say, "You've got to keep productivity up". If they kept the machines running but had quality problems as a result, we criticised them for making garbage. Now you get into far more trouble for violating process procedures than you do for not meeting productivity targets.' (Engineer, Preston Plant)

This new approach needed to be matched by changes in the way the communications were managed in the plant. *'We did two things that we had never done before. First each production team started holding daily reviews of control chart data. Second, one day a month we took people away from production and debated the control chart data. Several people got nervous because we were not producing anything. But it was necessary. For the first time you got operators from the three shifts meeting together and talking about the control chart data and other quality issues. Just as significantly we invited HP up to attend these meetings. Remember these weren't staged meetings, it was the first time these guys had met together and there was plenty of heated discussion, all of which the Hewlett Packard representatives witnessed.'* (Engineer, Preston Plant)

At last something positive was happening in the plant and morale on the shop floor was buoyant. By September 2000 the results of the plant's teams efforts were starting to show. Processes were coming under control, quality levels were improving and, most importantly, personnel both on the shop floor and in the management team were beginning to get into the 'quality mode' of thinking. Paradoxically, in spite of stopping the line periodically, the efficiency of the plant was also improving.

Yet the Preston team did not have time to enjoy their emerging success. In September of 2000 the plant learned that it would not get the Vector project because of their recent quality problems. Then Rendall decided to close the plant. *'We were losing millions, we had lost the Vector project, and it was really no surprise. I told the senior management team and said that we would announce it probably in April of 2001. The real irony was that we knew that we had actually already turned the corner.'* (Tom Branton)

Notwithstanding the closure decision, the management team in Preston set about the task of convincing Rendall that the plant could be viable. They figured it would take three things. First, it was vital that they continue to improve quality. Progressing with their quality initiative involved establishing full statistical process control (SPC). Second, costs had to be brought down. Working on cost reduction was inevitably going to be painful. The first task was to get an understanding of what should be an appropriate level of operating costs. *'We went through a zero-based assessment to decide what an ideal plant would look like, and the minimum number of people needed to run it.'* (Tom Branton)

By December of 2000 there were 40 per cent fewer people in the plant than two months earlier. All departments were affected. The quality department shrank more than most, moving from 22 people down to 6. *'When the plant was considering down-sizing they asked me, "How can we run a lab with six technicians?" I said, "Easy. We just make good paper in the first place, and then we don't have to inspect all the garbage. That alone would save an immense amount of time.'* (Quality Manager, Preston Plant)

Third, the plant had to create a portfolio of new product ideas which could establish a greater confidence in future sales. Several new ideas were under active investigation. The most important of which was 'Protowrap', a wrap for newsprint that could be repulped. It was a product that was technically difficult. However the plant's newly acquired capabilities allowed the product to be made economically.

Out of the crisis

In spite of their trauma, the plant's management team faced Christmas of 2000 with increasing optimism. They had just made a profit for the first time for over two years. By spring of 2001 even HP, at a corporate level, were starting to take notice. It was becoming obvious that the Preston plant really had made a major change. More significantly, HP had asked the plant to bid for a new product. April 2001 was a good month for the plant. It had chalked up three months of profitability and HP formally gave the new contract to Preston. Also in April, Rendall reversed their decision to close the plant.

QUESTIONS

1 What are the most significant events in the story of how the plant survived because of its adoption of quality-based principles?

2 The plant's processes eventually were brought under control. What were the main benefits of this?

3 SPC is an operational level technique of ensuring quality conformance. How many of the benefits of bringing the plant under control would you class as strategic?

APPLYING THE PRINCIPLES

Hints

Some of these exercises can be answered by reading the chapter. Others will require some general knowledge of business activity and some might require an element of investigation. Hints on how they can all be answered are to be found in the eText at **www.pearsoned.co.uk/slack**.

1 Using the four categories of quality-related costs, make a list of the costs that fall within each category for the following operations:

(a) a university library

(b) a washing machine manufacturer

(c) a nuclear electricity generating station

(d) a church.

2 Consider how a service level agreement could be devised for the following:

(a) the service between a library and its general customers

(b) the service given by a motor vehicle rescue service to its customers

(c) the service given by a university audio visual aids department to both academic staff and students.

3 Using an internet search engine (such as Google.com) look at the consultancy organisations that are selling help and advice on quality management. How do they try to sell quality improvement approaches to prospective customers?

4 Visit the website of the European Foundation for Quality Management (www.efqm.org). Look at the companies that have won or been finalists in the European Quality Awards and try to identify the characteristics

which make them 'excellent' in the opinion of the EFQM. Investigate how the EFQM promotes its model for self-assessment purposes.

5 Find two products, one a manufactured food item (for example, a packet of breakfast cereal, a packet of biscuits, etc.) and the other a domestic electrical item (for example, electric toaster, coffee maker, etc.).

(a) Identify the important quality characteristics for these two products.

(b) How could each of these quality characteristics be specified?

(c) How could each of these quality characteristics be measured?

6 Many organisations check up on their own level of quality by using 'mystery shoppers'. This involves an employee of the company acting the role of a customer and recording how they are treated by the operation. Choose two or three high-visibility operations (for example, a cinema, a department store, the branch of a retail bank, etc.) and discuss how you would put together a mystery shopper approach to testing their quality. This may involve you determining the types of characteristics you would wish to observe, the way in which you would measure these characteristics, an appropriate sampling rate, and so on. Try out your mystery shopper plan by visiting these operations.

Notes on chapter

1 Source: Interview with Michael Purtill, General Manager, Four Seasons Canary Wharf Hotel.

2 Sources include: Ryanair website; Keenan, S. (2002), 'How Ryanair puts passengers in their place', *Times*, 19 June.

3 'Human error' hits Google search' (2009), BBC news website: http://news.bbc.co.uk, 31 January.

4 Parasuraman, A. *et al.* (1985) 'A Conceptual Model of Service Quality and Implications for Future Research', *Journal of Marketing*, Vol 49, Fall.

5 Mechling, L. (2002) 'Get ready for a storm in a tea shop', *Independent*, 8 March; company website (2011).

6 Source: Plunkett, J.J. and Dale, B.S. (1987) 'A Review of the literature in quality-related costs', *The International Journal of Quality and Reliability Management*, Vol 4, No 1.

7 Source: Wheatley, M. (2010), 'Filling time on the production line', *Engineering and Technology* magazine, 8 November.

8 Dale, B.G. (ed) (1999) *Managing Quality*, Blackwell.

9 Source: The EFQM website: www.efqm.org.

10 Hyman, J. and Mason, B. (1995) *Management Employees Involvement and Participation*, Sage.

TAKING IT FURTHER

Bounds, G., Yorks, L., Adams, M. and Ranney, G. (1994) *Beyond Total Quality Management: Towards the Emerging Paradigm*, McGraw-Hill. *A useful summary of the state of play in total quality management at about the time it was starting to lose its status as the only approach to managing quality.*

Crosby, P.B. (1979) *Quality is Free*, McGraw-Hill. *One of the gurus. It had a huge impact in its day. Read it if you want to know what all the fuss was about.*

Dale B.G., Ton van der Wiele and Jos van Iwaarden (2007) *Managing Quality*, (5th edn), Wiley-Blackwell. *This is the fifth edition of a book that has long been one of the best respected texts in the area. A comprehensive and balanced guide to the area.*

Goetsch D.L. and Davis S. (2009) *Quality Management for Organisational Excellence: Introduction to Total Quality* (6th edn), Pearson Education. *Up-to-date account of the topic.*

Feigenbaum, A.V. (1986) *Total Quality Control*, McGraw-Hill. *A more comprehensive book than those by some of the other quality gurus.*

Hoyle D. (2006) *Quality Management Essentials*, Butterworth-Heinemann. *Practical.*

Pande, P.S., Neuman, R.P. and Kavanagh, R.R. (2000) *The Six Sigma Way*, McGraw-Hill, New York. *There are many books written by consultants for practising managers on the now fashionable Six Sigma Approach (see supplement to chapter). This one is readable and informative.*

USEFUL WEBSITES

www.opsman.org *Definitions, links and opinions on operations and process management.*

http://www.bfq.co.uk/ *The British Quality Foundation is a not-for-profit organisation promoting business excellence.*

http://www.juran.com *The Juran Institute's mission statement is to provide clients with the concepts, methods and guidance for attaining leadership in quality.*

http://www.asq.org/ *The American Society for Quality site. Good professional insights.*

http://www.quality.nist.gov/ *American Quality assurance Institute. Well established institution for all types of business quality assurance.*

http://www.gslis.utexas.edu/~rpollock/tqm.html *Non-commercial site on Total Quality Management with some good links.*

http://www.iso.org/iso/en/ISOOnline.frontpage *Site of the International Standards Organisation that runs the ISO 9000 and ISO 14000 families of standards. ISO 9000 has become an international reference for quality management requirements.*

Interactive quiz

For further resources including examples, animated diagrams, self-test questions, Excel spreadsheets and video materials please explore the eText on the companion website at www.pearsoned.co.uk/slack.

Statistical process control (SPC)

Introduction

The purpose of statistical process control (SPC) is to both control process performance, keeping it within acceptable limits, and to improve process performance by reducing the variation in performance from its target level. It does this by applying statistical techniques to understand the nature of performance variation over time. For those who are anxious about the 'statistical' part of SPC, don't be. Essentially SPC is based on principles that are both practical and intuitive. The statistical element is there to help rather than complicate quality decisions.

Variation in process performance

The core instrument in SPC is the control chart. These were explained earlier in the chapter. They are an illustration of the dynamic performance of a process, measuring how some aspect of process performance varies over time. All processes vary to some extent. No machine will give precisely the same result each time it is used. All materials vary a little. People in the process differ marginally in how they do things each time they perform a task. Given this, it is not surprising that any measure of performance quality (whether attribute or variable) will also vary. Variations that derive from these *normal* or *common causes* of variation can never be entirely eliminated (although they can be reduced).

For example, at a call centre for a utility company, customer care operatives answer queries over accounts, service visits, and so on. The length of time of each call will vary depending on the nature of the enquiry and the customer's needs. There will be some variation around an average call time. When the process of answering and responding to customer enquiries is stable, the computer system that intercepts and allocates calls to customer care operatives could be asked to randomly sample the length of each call. As this data builds up, the histogram showing call times could develop as is shown in Figure 12.9. The first calls could lie anywhere within the natural variation of the process but are more likely to be close to the average call length (Figure 12.9(a)). As more calls are measured they would clearly show a tendency to be close to the process average (see Figure 12.9(b and c)). Eventually the data will show a smooth histogram that can be drawn into a smoother distribution that will indicate the underlying process variation (the distribution shown in Figure 12.9(f)).

Often this type of variation can be described by a *normal distribution*. (Even if this raw data does not conform to a normal distribution, it can be manipulated to approximate to one by using sampling – see later.) It is a characteristic of normal distributions that 99.7 per cent of the measures will lie within ± 3 standard deviations of the distribution (standard deviation is a measure of how widely the distribution is spread or *dispersed*).

The central limit theorem

Not all processes will vary in their performance according to a normal distribution. However, if a sample is taken from any type of distribution, the distribution of the average of the sample

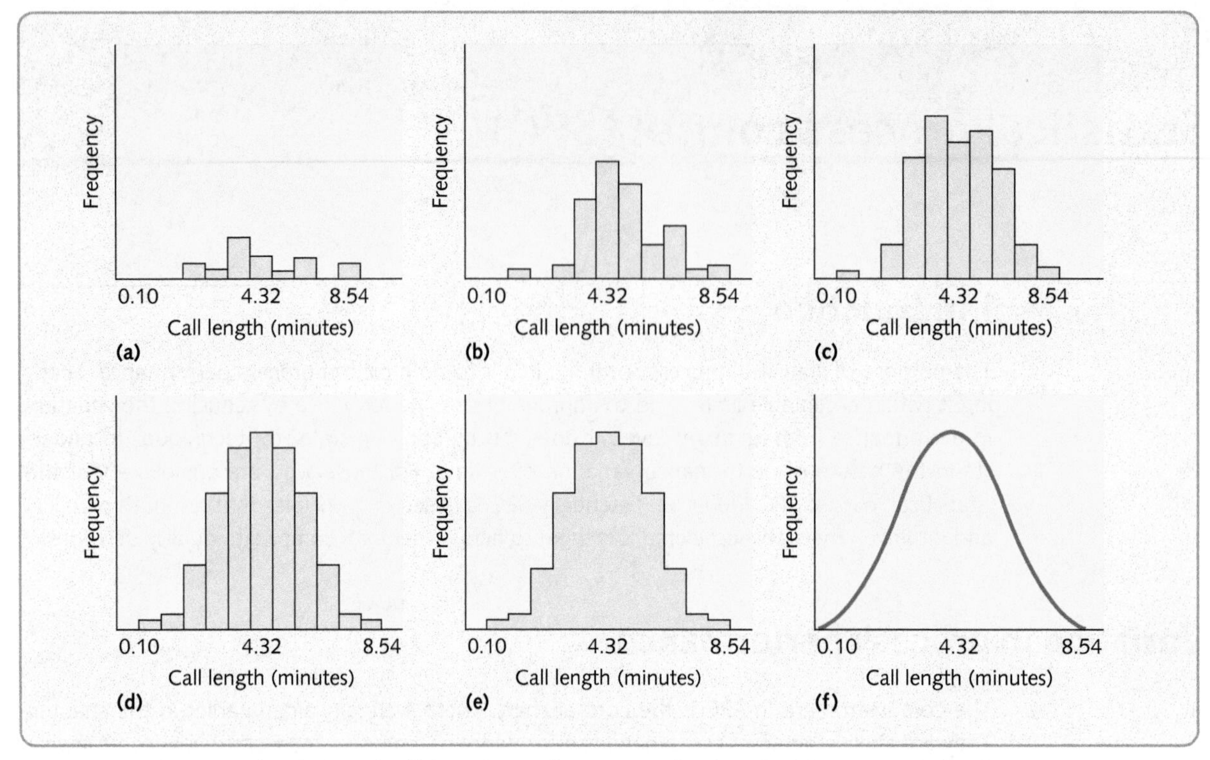

Figure 12.9 The natural variation of call times in a call centre can be described by a normal distribution

(sample mean) *will* approximate to a normal distribution. For example, there is an equal probability of any number between one and six being thrown on a six-sided, unweighted die. The distribution is rectangular with an average of 3.5 as shown in Figure 12.10(a). But if a die is thrown (say) six times repeatedly and the average of the six throws calculated, the sample

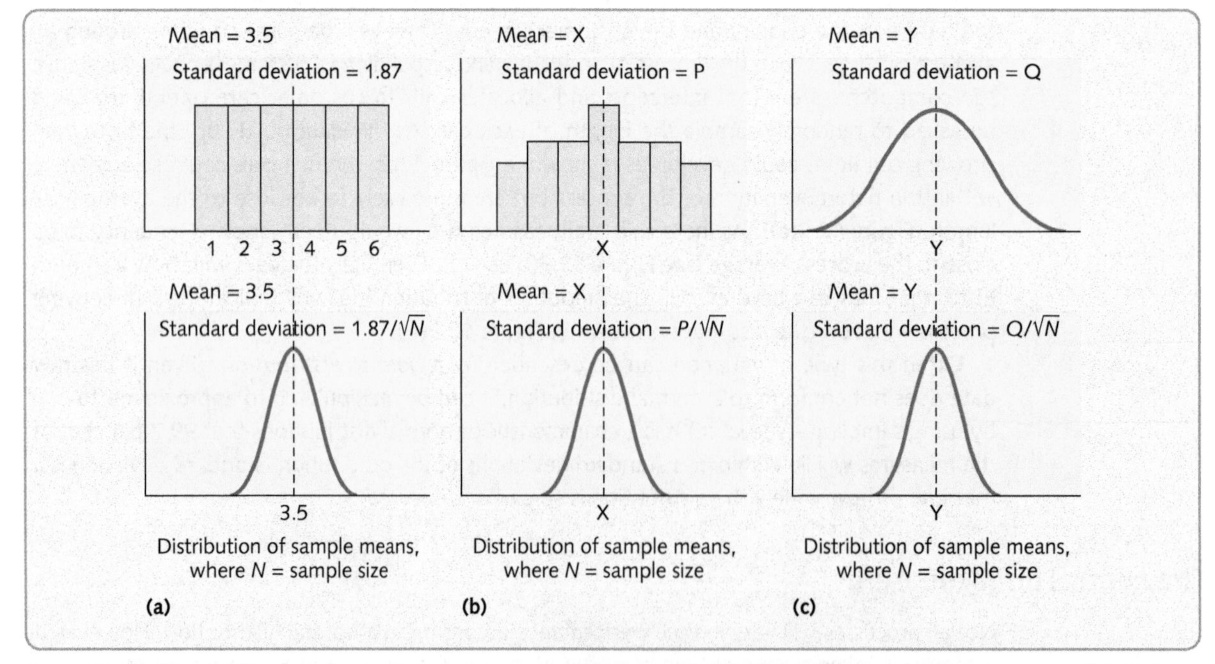

Figure 12.10 The distribution of sample means (averages) from any distribution will approximate to a normal distribution

average will also be 3.5, but the standard deviation of the distribution will be the standard deviation of the original rectangular distribution divided by the square of the sample size. More significantly, the shape of the distribution will be close to normal and so can be treated the same way as a normal distribution. This becomes important when control limits are calculated, see later.

Is the process 'in control'?

Not all variation in process performance is the result of common causes. There may be something wrong with the process that is assignable to an abnormal and preventable cause. Machinery may have worn or been set up badly. An untrained person may not be following the prescribed procedure for the process. The causes of such variation are called *assignable or abnormal causes*. The question for operations management is whether the results from any particular sample, when plotted on the control chart, simply represent the variation due to *common* causes or due to some specific and correctable, *assignable* cause. Figure 12.11(a), for example, shows the control chart for the average call length of samples of customer calls in a utility's call centre. Like any process the results vary, but the last three points seem to be lower than usual. The question is whether this is natural variation or the symptom of some more serious cause. Is the variation the result of common causes, or does it indicate assignable causes (something abnormal) occurring in the process?

To help make this decision, control limits can be added to the control charts that indicate the expected extent of 'common-cause' variation. If any points lie outside these control limits then the process can be deemed *out of control* in the sense that variation is likely to be due to assignable causes. These can be set in a statistically revealing manner based on the probability that the mean of a particular sample will differ by more than a set amount from the mean of the population from which it is taken. Figure 12.11(b) shows the same control chart as Figure 12.11(a) with the addition of control limits put at ±3 standard deviations (of the population of sample means) away from the mean of sample averages. It shows that the probability of the final point on the chart being influenced by an assignable cause is very high indeed. When the process is exhibiting behaviour which is outside its normal 'common-cause' range, it is said to be 'out of control'.

However, we cannot be absolutely certain that the process is out of control. There is a small but finite chance that the point is a rare but natural result at the tail of its distribution. Stopping the process under these circumstances would represent a type I error because the process is

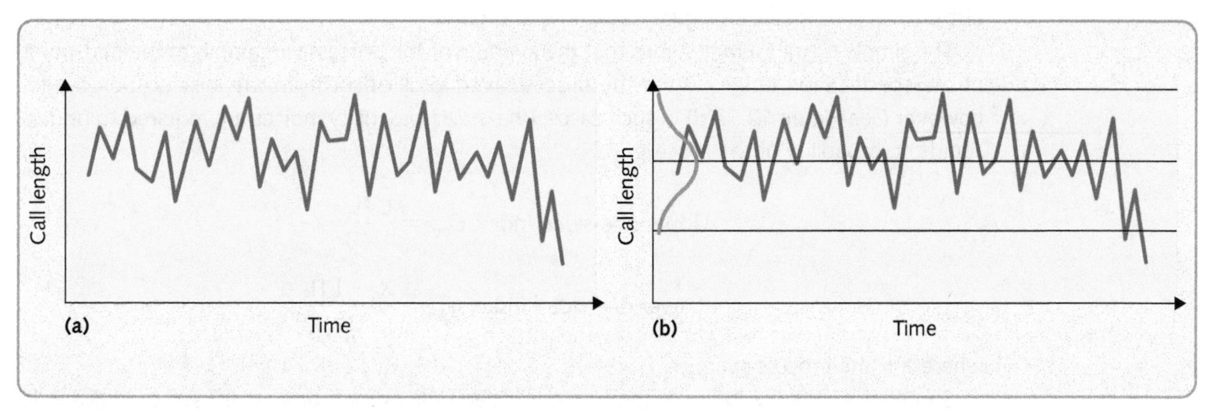

Figure 12.11 Control chart for the average call length in a call centre: (a) without control limit; (b) with control limits derived from the natural variation of the process

Table 12.5 Type I and type II errors in SPC

		Actual process state	
		In control	**Out of control**
Decision	**Stop process**	**Type I error**	Correct decision
	Leave alone	Correct decision	**Type II error**

actually in control. Alternatively, ignoring a result which in reality is due to an assignable cause is a type II error (see Table 12.5). Control limits that are set at three standard deviations either side of the population mean are called the upper control limit (UCL) and lower control limit (LCL). There is only a 0.3 per cent chance of any sample mean falling outside these limits by chance causes (that is, a chance of a type I error of 0.3 per cent).

Process capability

Using control charts to assess whether the process is in control is an important internal benefit of SPC. An equally important question for any operations manager would be: 'Is the variation in the process performance acceptable to external customers?' The answer will depend on the acceptable range of performance that will be tolerated by the customers. This range is called the *specification range*. Returning to the call time example, if the call time is too small then the organisation might offend customers, if it is too large, the organisation is 'giving away' too much of its time.

Process capability is a measure of the acceptability of the variation of the process. The simplest measure of capability (C_p) is given by the ratio of the specification range to the 'natural' variation of the process (i.e. ± 3 standard deviations):

$$C_p = \frac{\text{UTL} - \text{LTL}}{6s}$$

where UTL = the upper tolerance limit
LTL = the lower tolerance limit
s = the standard deviation of the process variability.

Generally, if the C_p of a process is greater than 1, it is taken to indicate that the process is 'capable', and a C_p of less than 1 indicates that the process is not 'capable', assuming that the distribution is normal (see Figure 12.12 a, b and c).

The simple C_p measure assumes that the average of the process variation is at the mid-point of the specification range. Often the process average is offset from the specification range, however (see Figure 12.12 d) in such cases, *one-sided* capability indices are required to understand the capability of the process:

$$\text{Upper one-sided index } C_{pu} = \frac{\text{UTL} - X}{3s}$$

$$\text{Lower one-sided index } C_{pl} = \frac{X - \text{LTL}}{3s}$$

where X = the process average.

Sometimes only the lower of the two one-sided indices for a process is used to indicate its capability (C_{pk}):

$$C_{pk} = \min(C_{pu}, C_{pl})$$

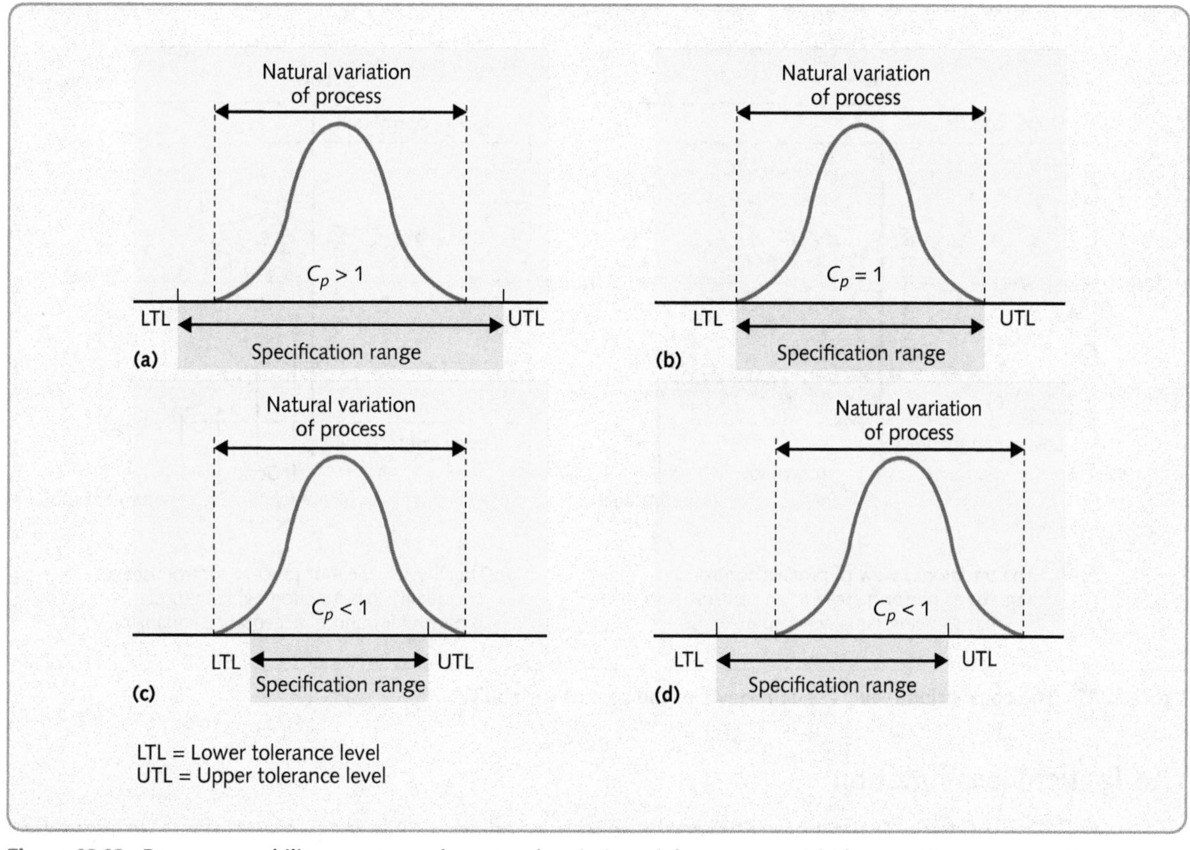

Figure 12.12 Process capability compares the natural variation of the process with the specification range that is required

WORKED EXAMPLE

In the case of the call centre process, described previously, process capability can be calculated as follows:

$$\text{Specification range} = 16 \text{ minutes} - 1 \text{ minute} = 15 \text{ minutes}$$

$$\text{Natural variation of process} = 6 \times \text{standard deviation}$$

$$= 6 \times 2 = 12 \text{ minutes}$$

$$C_p = \text{process capability}$$

$$= \frac{\text{UTL} - \text{LTL}}{6s}$$

$$= \frac{16 - 1}{6 \times 2} = \frac{15}{12}$$

$$= 1.25$$

If the natural variation of the process changed to have a process average of 7 minutes but the standard deviation of the process remained at 2 minutes:

$$C_{pu} = \frac{16 - 7}{3 \times 2} = \frac{9}{6} = 1.5$$

$$C_{pl} = \frac{7 - 1}{3 \times 2} = \frac{6}{6} = 1.0$$

$$C_{pl} = \min(1.5, 1.0)$$

$$= 1.00$$

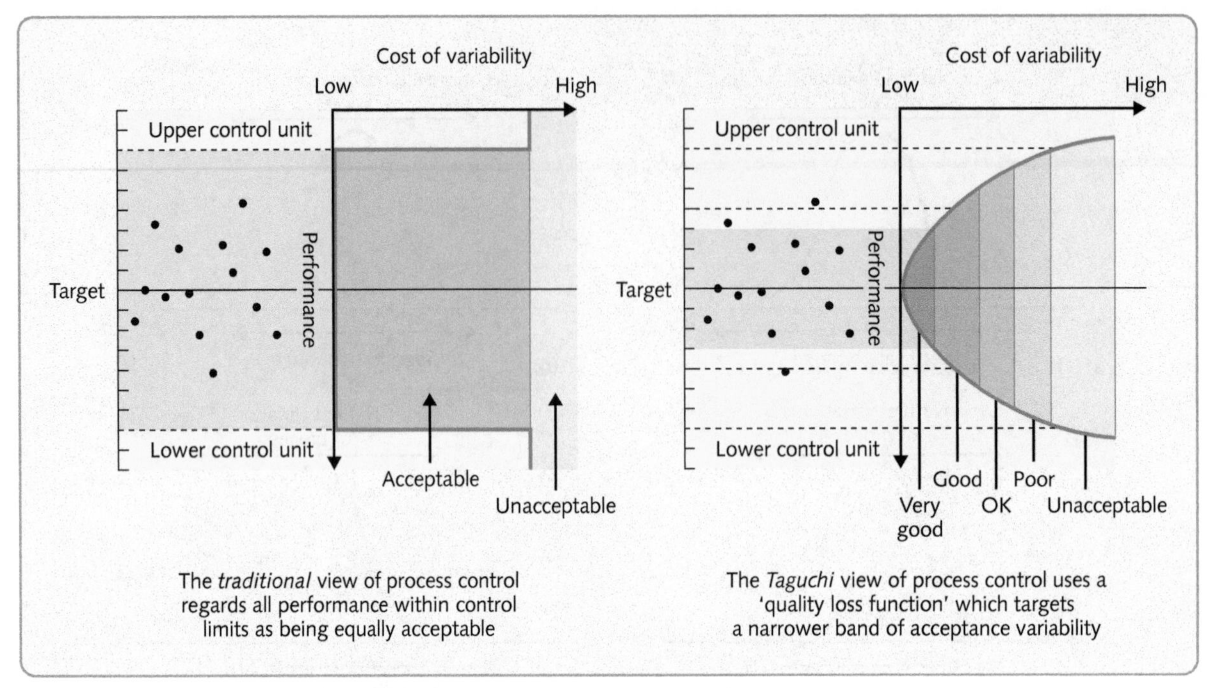

Figure 12.13 The conventional and Taguchi views of the cost of variability

The Taguchi loss function

Genichi Taguchi has criticised the concept of an acceptable range of variation.[1] He suggested that the consequences of being 'off-target' (that is, deviating from the required process average performance) were inadequately described by simple control limits. Instead, he proposed a quality loss function (QLF) – a mathematical function which includes all the costs of poor quality. These include wastage, repair, inspection, service, warranty and generally, what he termed, 'loss to society' costs. This loss function is expressed as follows:

$$L = D^2C$$

where L = total loss to society costs
D = deviation from target performance
C = a constant

Figure 12.13 illustrates the difference between the conventional and Taguchi approaches to interpreting process variability. The more graduated approach of the QLF shows losses increasing quadratically as performance deviates from target. Because of this there will be an appropriate motivation to progressively reduce process variability. This is sometimes called the *target-oriented* quality philosophy.

Control charts for variables

The most commonly used type of control chart employed to control variables is the *$\bar{X}$–R chart*. In fact this is really two charts in one. One chart is used to control the sample average or mean ($\bar{X}$). The other is used to control the variation within the sample by measuring the range (*R*). The range is used because it is simpler to calculate than the standard deviation of the sample.

The means ($\bar{X}$) chart can pick up changes in the average output from the process being charted. Changes in the means chart would suggest that the process is drifting generally away from its supposed process average, although the variability inherent in the process may not have changed. The range (R) chart plots the range of each sample, which is the difference between the largest and the smallest measurement in the samples. Monitoring sample range gives an indication of whether the variability of the process is changing, even when the process average remains constant.

Control limits for variables control chart

As with attributes control charts, a statistical description of how the process operates under normal conditions (when there are no assignable causes) can be used to calculate control limits. The first task in calculating the control limits is to estimate the grand average or population mean ($\overline{\overline{X}}$) and average range ($\overline{R}$) using m samples, each of sample size n. The population mean is estimated from the average of a large number (m) of sample means:

$$\overline{\overline{X}} = \frac{\overline{X}_1 + \overline{X}_2 + \ldots \overline{X}_m}{m}$$

The average range is estimated from the ranges of the large number of samples:

$$\overline{R} = \frac{R_1 + R_2 + \ldots R_m}{m}$$

The control limits for the sample means chart are:

$$\text{Upper control limit (UCL)} = \overline{\overline{X}} + A_2\overline{R}$$
$$\text{Lower control limit (LCL)} = \overline{\overline{X}} - A_2\overline{R}$$

The control limits for the range charts are:

$$\text{Upper control limit (UCL)} = D_4\overline{R}$$
$$\text{Lower control limit (LCL)} = D_3\overline{R}$$

The factors A_2, D_3 and D_4 vary with. sample size and are shown in Table 12.6.

Table 12.6 Factors for the calculation of control limits

Sample size n	A_2	D_3	D_4
2	1.880	0	3.267
3	1.023	0	2.575
4	0.729	0	2.282
5	0.577	0	2.115
6	0.483	0	2.004
7	0.419	0.076	1.924
8	0.373	0.136	1.864
9	0.337	0.184	1.816
10	0.308	0.223	1.777
12	0.266	0.284	1.716
14	0.235	0.329	1.671
16	0.212	0.364	1.636
18	0.194	0.392	1.608
20	0.180	0.414	1.586
22	0.167	0.434	1.566
24	0.157	0.452	1.548

The LCL for the means chart may be negative (for example, temperature or profit may be less than zero) but it may not be negative for a range chart (or the smallest measurement in the sample would be larger than the largest). If the calculation indicates a negative LCL for a range chart then the LCL should be set to zero.

WORKED EXAMPLE

GAM (Groupe A Maquillage) is a contract cosmetics company that manufactures and packs cosmetics and perfumes for other companies. One of its plants operates a filling line that automatically fills plastic bottles with skin cream and seals the bottles with a screw-top cap. The tightness with which the screw-top cap is fixed is an important aspect of quality. If the cap is screwed on too tightly, there is a danger that it will crack; if screwed on too loosely it might come loose. Either outcome could cause leakage. The plant had received some complaints of product leakage, possibly caused by inconsistent fixing of the screw-tops. Tightness can be measured by the amount of turning force (torque) that is required to unfasten the tops. The company decided to take samples of the bottles coming out of the filling-line process, test them for their unfastening torque and plot the results on a control chart. Several samples of four bottles are taken during a period when the process is regarded as being in control.

The following data are calculated from this exercise:

$$\text{The grand average of all samples } \overline{\overline{X}} = 812 \text{ g/cm}^3$$
$$\text{The average range of the sample } \overline{R} = 6 \text{ g/cm}^3$$

Control limits for the means ($\overline{X}$) chart were calculated as follows:

$$\begin{aligned} \text{UCL} &= \overline{\overline{X}} + A_2\overline{R} \\ &= 812 + (A_2 \times 6) \end{aligned}$$

From Table 12.6, we know for a sample size of four $A_2 = 0.729$. Thus:

$$\begin{aligned} \text{UCL} &= 812 + (0.729 \times 6) \\ &= 816.37 \\ \text{LCL} &= \overline{\overline{X}} - (A_2\overline{R}) \\ &= 812 - (0.729 \times 6) \\ &= 807.63 \end{aligned}$$

Control limits for the range chart (R) were calculated as follows:

$$\begin{aligned} \text{UCL} &= D_4 \times \overline{R} \\ &= 2.282 \times 6 \\ &= 13.69 \\ \text{LCL} &= D_3\overline{R} \\ &= 0 \times 6 \\ &= 0 \end{aligned}$$

After calculating these averages and limits for the control chart, the company regularly took samples of four bottles during production, recorded the measurements and plotted them as shown in Figure 12.14. This control chart reveals that only with difficulty can the process average be kept in control. Occasional operator interventions are required. Also the process range is moving towards (and once exceeding) the upper control limit. The process also seems to be becoming more variable. (After investigation it was discovered that, because of faulty

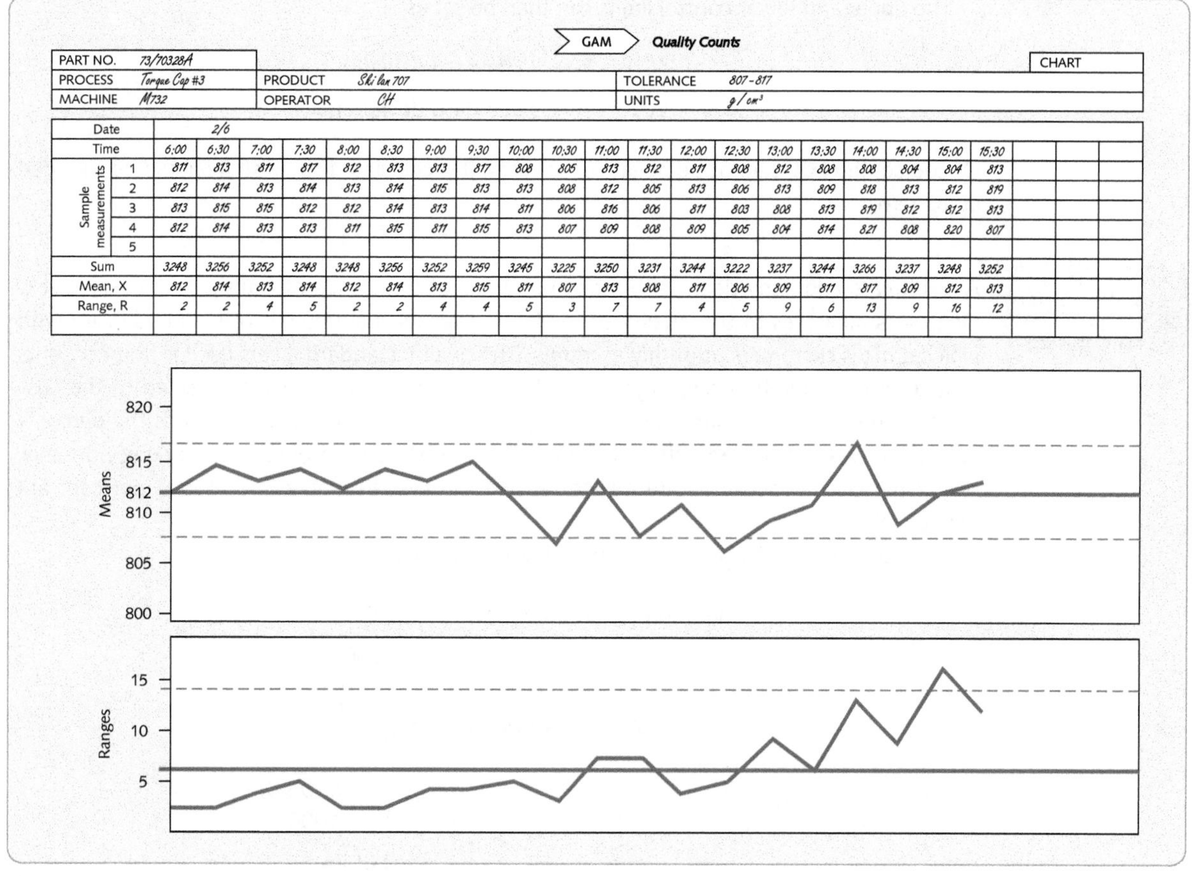

GAM Quality Counts

PART NO. 73/70328A			CHART
PROCESS Torque Cap #3	PRODUCT Ski lan 707	TOLERANCE 807-817	
MACHINE M732	OPERATOR CH	UNITS g/cm³	

Date	2/6																			
Time	6:00	6:30	7:00	7:30	8:00	8:30	9:00	9:30	10:00	10:30	11:00	11:30	12:00	12:30	13:00	13:30	14:00	14:30	15:00	15:30
Sample measurements 1	811	813	811	817	812	813	813	817	808	805	813	812	811	808	812	808	808	804	804	813
2	812	814	813	814	813	814	815	813	813	808	812	805	813	806	813	809	818	813	812	819
3	813	815	815	812	812	814	813	814	811	806	816	806	811	803	808	813	819	812	812	813
4	812	814	813	813	811	815	811	815	813	807	809	808	809	805	804	814	821	808	820	807
5																				
Sum	3248	3256	3252	3248	3248	3256	3252	3259	3245	3225	3250	3231	3244	3222	3237	3244	3266	3237	3248	3252
Mean, X	812	814	813	814	812	814	813	815	811	807	813	808	811	806	809	811	817	809	812	813
Range, R	2	2	4	5	2	2	4	4	5	3	7	7	4	5	9	6	13	9	16	12

Figure 12.14 The completed control form for GAM's torque machine showing the mean ($\overline{X}$) and range ($\overline{R}$) charts

maintenance of the line, skin cream was occasionally contaminating the part of the line that fitted the cap resulting in erratic tightening of the caps.)

Control charts for attributes

Attributes have only two states – 'right' or 'wrong', for example – so the statistic calculated is the proportion of wrongs (p) in a sample. (This statistic follows a binomial distribution.) Control charts using p are called 'p-charts'. When calculating control limits, the population mean ($\overline{p}$) (the actual, normal or expected proportion of 'defectives') may not be known. Who knows, for example, the actual number of city commuters who are dissatisfied with their journey time? In such cases the population mean can be estimated from the average of the proportion of 'defectives' ($\overline{p}$) from m samples each of n items, where m should be at least 30 and n should be at least 100:

$$\overline{p} = \frac{p^1 + p^2 + p^3 \dots p^n}{m}$$

One standard deviation can then be estimated from:

$$\sqrt{\frac{\overline{p}(1 - \overline{p})}{n}}$$

The upper and lower control limits can then be set as:

$$\text{UCL} = \overline{p} + 3 \text{ standard deviations}$$
$$\text{LCL} = \overline{p} - 3 \text{ standard deviations}$$

Of course, the LCL cannot be negative, so when it is calculated to be so it should be rounded up to zero.

WORKED EXAMPLE

A credit card company deals with many hundreds of thousands of transactions every week. One of its measures of the quality of service it gives its customers is the dependability with which it mails customers' monthly accounts. The quality standard it sets itself is that accounts should be mailed within two days of the 'nominal post date' which is specified to the customer. Every week the company samples 1,000 customer accounts and records the percentage which was not mailed within the standard time. When the process is working normally, only 2 per cent of accounts are mailed outside the specified period, that is, 2 per cent are 'defective'.

Control limits for the process can be calculated as follows:

$$\text{Mean proportion defective, } \overline{p} = 0.02$$
$$\text{Sample size } n = 1000$$
$$\text{Standard deviation } s = \sqrt{\frac{\overline{p}(1-\overline{p})}{n}} = \sqrt{\frac{0.02(0.98)}{1000}} = 0.0044$$

With the control limits at $\overline{p} \pm 3s$:

$$\text{Upper control limit (UCL)} = 0.02 + 3(0.0044) = 0.0332 = 3.32\%$$
$$\text{and lower control limit (LCL)} = 0.02 - 3(0.0044) = 0.0068 = 0.68\%$$

Figure 12.15 shows the company's control chart for this measure of quality over the last few weeks, together with the calculated control limits. It also shows that the process is in control.

Sometimes it is more convenient to plot the actual number of defects (c) rather than the proportion (or percentage) of defectives, on what is known as a c-chart. This is very similar to the p-chart but the sample size must be constant and the process mean and control limits are calculated using the following formulae:

$$\text{Process mean } \overline{c} = \frac{c_1 + c_2 + c_3 \ldots c_m}{m}$$

$$\text{Control limits} = \overline{c} \pm 3\sqrt{c}$$

where c = number of defects
m = number of samples.

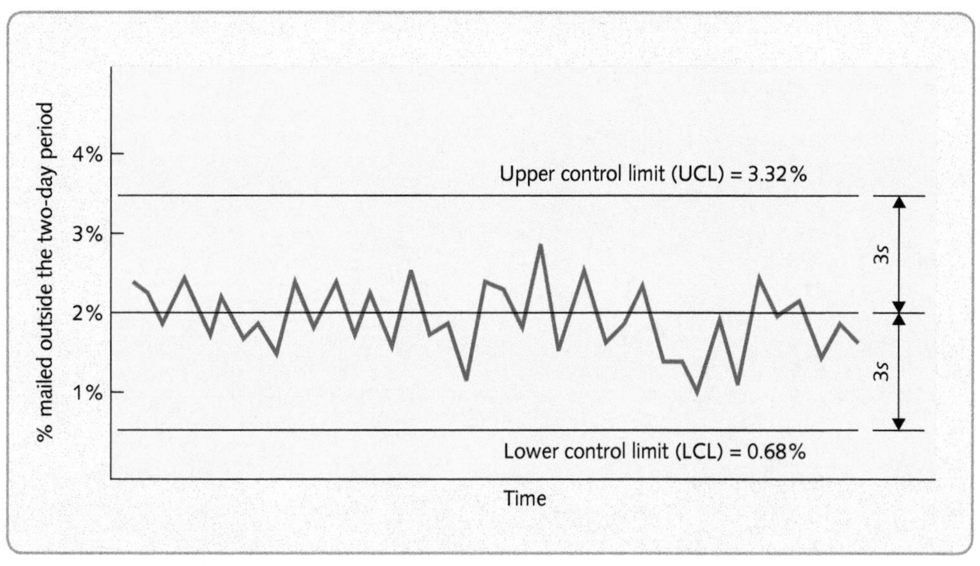

Figure 12.15 Control chart for the percentage of customer accounts which are mailed outside their two-day period

Note on supplement

1 Taguchi, G. and Clausing, D. (1990) 'Robust Quality', *Harvard Business Review*, Vol 68, No 1, pp 65–75. For more details of the Taguchi approach see Stuart, G. (1993) *Taguchi Methods: A Hands-on Approach*, Addison-Wesley.

13 Improvement

Introduction

All operations, no matter how well managed, are capable of improvement. In fact in recent years the emphasis has shifted markedly towards making improvement one of the main responsibilities of operations managers. And although the whole of this book is focused on improving the performance of individual processes, operations and whole supply networks, there are some issues that relate to the activity of improving itself. In any operation, whatever is improved, and however it is done, the overall direction and approach to improvement needs to be addressed (see Figure 13.1).

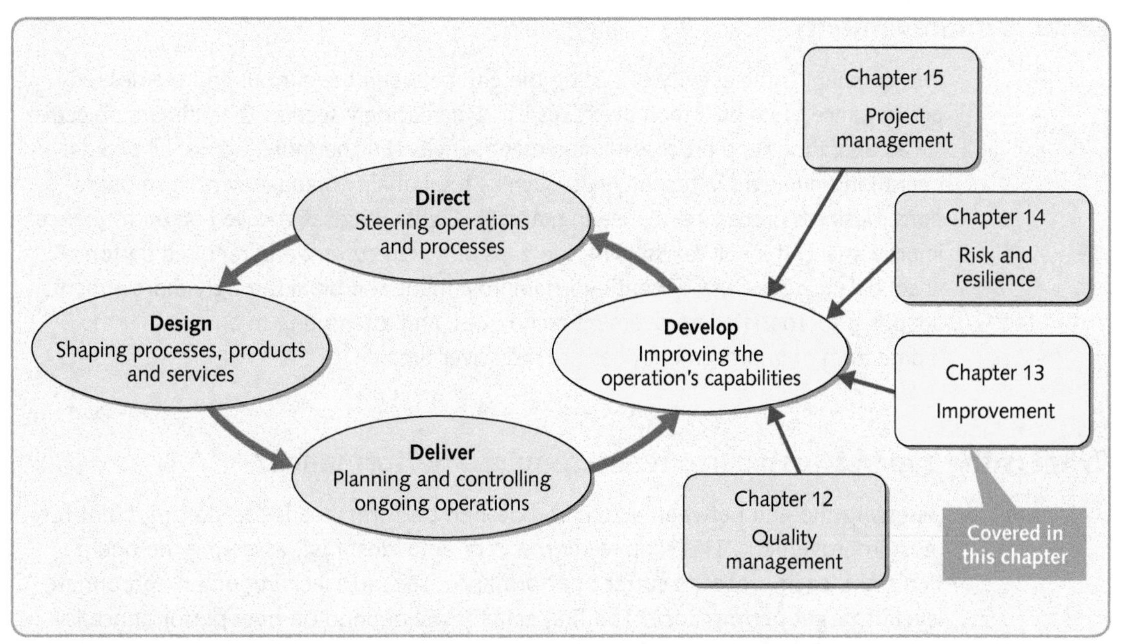

Figure 13.1 Improvement is the activity of closing the gap between the current and the desired performance of an operation or process

EXECUTIVE SUMMARY

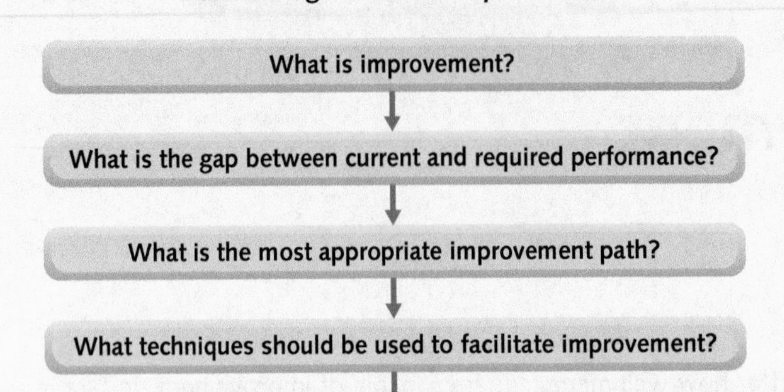

Each chapter is structured around a set of diagnostic questions. These questions suggest what you should ask in order to gain an understanding of the important issues of a topic, and, as a result, improve your decision making. An executive summary addressing these questions is provided below.

What is improvement?

Improvement is the activity of closing the gap between the current and the desired performance of an operation or process. It is increasingly seen as the ultimate objective for all operations and process management activity. Furthermore, almost all popular operations initiatives in recent years, such as total quality management, lean operations, business process reengineering, and Six Sigma, have all focused on performance improvement. It involves assessing the gaps between current and required performance, balancing the use of continuous improvement and breakthrough improvement, adopting appropriate improvement techniques, and attempting to ensure that the momentum of improvement does not fade over time.

What is the gap between current and required performance?

Assessing the gap between actual and desired performance is the starting point for most improvement. This requires two sets of activities: first, assessing the operation's and each process's current performance; second, deciding on an appropriate level of target performance. The first activity will depend on how performance is measured within the operation. This involves deciding what aspects of performance to measure, which are the most important aspects of performance, and what detailed measures should be used to assess each factor. The balanced score card approach is an approach to performance measurement that is currently influential in many organisations. Setting targets for performance can be done in a number of ways. These include historically based targets, strategic targets that reflect strategic objectives, external performance targets that relate to external and/or competitor operations, and absolute performance targets based on the theoretical upper limit of performance. Benchmarking is an important input to this part of performance improvement.

What is the most appropriate improvement path?

Two improvement paths represent different philosophies of improvement, although both may be appropriate at different times. They are breakthrough improvement and continuous improvement. Breakthrough improvement focuses on major and dramatic changes that are intended to result in dramatic increases in performance. The business process reengineering approach is typical of breakthrough improvement. Continuous improvement focuses on small but never-ending improvements that become part of normal operations life. Its objective is to make improvement part of the culture of the organisation. Often continuous improvement involves the use of multi-stage improvement cycles for regular problem solving. The Six Sigma approach to improvement brings many existing ideas together and can be seen as a combination of continuous and breakthrough improvement.

What techniques should be used to facilitate improvement?

Almost all techniques in operations management contribute directly or indirectly to the performance improvement. However, some more general techniques have become popularly associated with improvement. These include scatter diagrams (correlation), cause–effect diagrams, Pareto analysis, and why–why analysis.

How can improvement be made to stick?

One of the biggest problems in improvement is to preserve improvement momentum over time. One factor that inhibits improvement becoming accepted as a regular part of operations activity is the emphasis on the fashionability of each new improvement approach. Most new improvement ideas contain some worthwhile elements but none will provide the ultimate answer. There must be some overall management of the improvement process that can absorb the best of each new idea. And, although authorities differ to some extent, most emphasise the importance of an improvement strategy, top management support and training.

DIAGNOSTIC QUESTION

What is improvement?

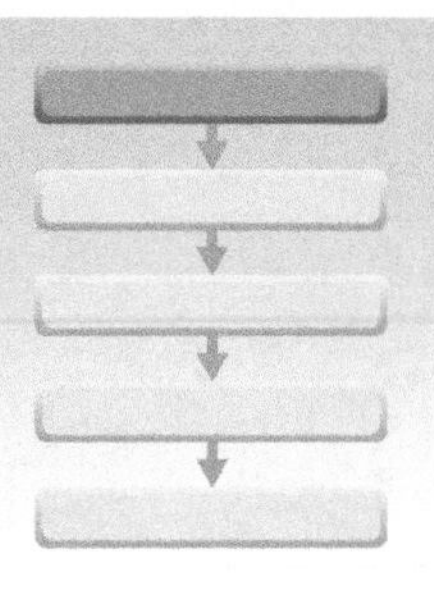

Video

Improvement comes from closing the gap between what you are and what you want to be. Or, in a specifically operations context, it comes from closing the gap between current and desired performance. Performance improvement is the ultimate objective of operations and process management. It has also become the subject of innumerable ideas that have been put forward as particularly effective methods of ensuring improvement. These include many that are described in this book, for example, total quality management (TQM), 'lean' operations, business process reengineering (BPR), Six Sigma, and so on. All of these, and other, ideas have something to contribute. What is important is that all managers develop an understanding of the underlying elements of improvement. The following two examples illustrate many of these.

OPERATIONS PRINCIPLE

Performance improvement is the ultimate objective of operations and process management.

EXAMPLE

Reinventing Singapore's libraries[1]

Is the traditional library still relevant in a world where information and entertainment is dominated by digital media? Borrowing from local libraries in most parts of the world has declined in recent years. Why? Libraries are often shut when people want to use them and when they are open they can be a dull with shelves full of old books, queues to borrow or return books, unhelpful or unapproachable staff who insist that users are quiet. Furthermore, libraries are increasingly under threat from the internet, an easy-to-use and efficient source of information. Yet some libraries are working hard to change this image. The National Library Board of Singapore (NLB) created and implemented a radical improvement plan to transform the library system in Singapore, designed to move its libraries from being merely a functional space to welcoming social spaces. Libraries have cafés, are located in shopping centres and are integrated into the community who see them as inviting havens to relax, meet people and share in the company of other book-lovers. So successful was the improvement initiative that there has been a significant increase in usage, and many libraries in other parts of the world have copied the Singapore model in their own countries.

Source: World of Asia/Alamy Images

One of the first steps in their improvement journey was to develop a set of 'service standards' covering five areas: customer service, printed collections, housekeeping, programmes, and staffing and administration. They cover such things as the running of programmes, opening times, shelving of books, the answering of telephones, and response time to e-mails. Audit teams monitor libraries against these service standards with awards for the best-performing libraries and individuals. Importantly, processes are continually being reviewed and improved. The NLB changed the librarys' performance evaluation process from one relying solely on supervisory discretion to a more objective system that also required input from peers and customers. Metrics and standards for operations and front-line customer service were established. Performance measurements, such as a 'complaints versus compliments' ratio, were introduced. Training was also seen as vital to improvement, with staff expected to undertake around 60 hours of training each year. For front-line staff this might include customer service courses on dealing with difficult customers or handling complaints.

After a while staff began to see little successes based on changes they had proposed. This improved their confidence and led to greater service improvement. NLB facilitated this with sessions they called 'ask stupid questions' (ASQ), where staff were free to ask 'stupid' questions. However, as one senior manager put it, *'In my view there are no stupid questions, there are only stupid answers'*. These initiatives were also supported by staff suggestion schemes, a performance bonus linking success to remuneration, training needs analysis and personal development based on individual training plans.

At a more structured level, some of the innovative improvements were the result of business process re-engineering (BPR). Their BPR methodology included several key phases:

- *Envisioning* – to reaffirm NLB's vision for its future operations and specific initiatives.
- *Core process identification* – to identify focus areas for BPR, set stretch performance targets and establish working groups for each focus area.
- *Process analysis* – to map out current business processes in full detail, collect data, perform model-based simulations, and diagnose critical issues within these processes.
- *Process redesign* – to develop new process designs with the intent to attain quantum improvements as defined by the stretch targets.
- *Blueprinting* – to consolidate all redesign recommendations and to discern organisation-wide issues such as technology infrastructure, human resources, organisation structure and customer services.
- *Implementation planning* – to develop a complete masterplan of implementation activities necessary to bring NLB from the current state towards the desired future state.

The BPR team, reporting to the Chief Executive, carried out a review of the core internal processes. The main objective was to develop a set of integrated designs and strategies for the NLB to manage its fast expanding network of libraries for the 21st century. Participants were required to identify, analyse and redesign the NLB's core business processes to remove redundancies and inneficiencies. The core processes included:

- *Time-to-market process* – the activities needed to bring book titles to the libraries, including title selection, acquisition, cataloguing, processing, transportation, packaging and shelf display.
- *Time-to-checkout process* – the activities between the time when a library user has an interest in a title and the time when the user leaves the library with the title checked out.
- *Time-to-shelf process* – the activities between the time when a library user turns up at a book return point and the time when the returned books are made available on the shelf for the next user.
- *Time-to-information process* – the activities between the time when a customer poses his information enquiry to the library and the time when the customer has received and accepted the information product.
- *Library planning, set-up and renewal process* – the activities in the planning, construction/renovation, moving-in and other logistics for setting up new libraries and upgrading existing ones.

Along with using a BPR approach, the NLB were keen to exploit technology fully. It was the first public library in the world to prototype radio-frequency identification (RFID) to create its Electronic Library Management System. RFID tags, or transponders contained in 'smart labels', receive and respond to radio-frequency queries from an RFID transceiver, which enables the remote and automatic retrieval, storing and sharing of information. Unlike barcodes, which need to be manually scanned, RFID simply broadcasts its presence and automatically sends data about the item to electronic readers. They have installed RFID tags in 10 million books making NLB one of the largest users of RFID tags in the world. Customers now have to spend little time queuing, as book issuing is automatic, as are book returns. Indeed, books can be returned to any book-drop at any library where RFID enables fast and easy sorting. Yet the NLB had to be very careful that they did not alienate or intimidate customers with their use of technology. Younger people are comfortable with the sort of technology that uses computers for returning, borrowing and searching for books, creating and checking accounts, and even paying fines. But some older customers needed help, so they enlisted a group of volunteers (including senior citizens) to get them familiar with the new technology.

But even with the stress on BPR methods and technology, NLB still stress the paramount importance of their staff's contribution. *'When there are challenges our staff respond and achieve wonderful results. That's because the organisation is very open to change. And that is the key thing; it's an attitude. Innovations are possible because everybody has the mindset that they would like to see things done differently. Improving all aspects of the library is an integral part of all my colleagues' jobs. Managers and librarians are expected to be involved in improvement projects, many of which are cross-functional, involving people from various functions working on specific issues raised by either staff or customers. We also often form small improvement teams. They will brainstorm for ideas, test out their suggestions and when they are successful we will have sessions where the teams from across the libraries will discuss and share what they have done. Where appropriate we will adopt the proven ideas or schemes for implementation in all our libraries. While this is seen as a normal and natural part of the job, we do reward staff for their ideas.'*

EXAMPLE

Improvement at Heineken[2]

Heineken International brews beer that is sold around the world and operates in over 170 countries with brands such as Heineken and Amstel. Its Zoeterwoude facility, a packaging plant that fills bottles and cans in The Netherlands faced two challenges. First, it needed to improve its operations processes to reduce its costs. Second, it needed to improve the efficiency of its existing lines in order to increase their capacity, without which it would have to invest in a new packaging line. The goal of a 20 per cent improvement in operating efficiency was set because it was seen as challenging yet achievable. It was also decided to focus the improvement project around two themes: (a) obtaining accurate operational data on which improvement decisions could be based; (b) changing the culture of the operation to promote fast and effective decision-making. Before the improvement, project staff at the Zoeterwoude plant had approached problem-solving as an ad hoc activity, only to be done when circumstances made it unavoidable. By contrast, the improvement initiative taught the staff to use various problem-solving techniques such as cause–effect and Pareto diagrams (discussed later in this chapter).

Source: Bloomberg/Alamy Images

'Until we started using these techniques,' says Wilbert Raaijmakers, Heineken Netherlands Brewery Director, *'there was little consent regarding what was causing any problems. There was poor communication between the various departments and job grades. For example, maintenance staff believed that production stops were caused by operating errors, while operators were of the opinion that poor maintenance was the cause.'* The use of better information, analysis and improvement techniques helped the staff to identify and treat the root causes of problems. With many potential improvements to make, staff teams were encouraged to set priorities that would reflect the overall improvement target. There was also widespread use of benchmarking performance against targets to gauge progress.

Improvement teams had been 'empowered, organised and motivated' before the improvement initiative, through the company's 'cultural change' programme. *'Its aim,'* according to Wilbert Raaijmakers, *'was to move away from a command-and-control situation and evolve towards a more team-oriented organisation.'* Fundamental to this was a programme to improve the skills and knowledge of individual operators through special training programmes. Nevertheless, the improvement initiative exposed a number of further challenges. For example, the improvement team discovered that people were more motivated to improve when the demand was high, but it was more difficult to motivate them when production pressures were lower. To overcome this, communication was improved so that staff were kept fully informed of future production levels and the upcoming schedule of training and maintenance activities that were planned during slumps in demand. The lesson being that it is difficult to convince people to change if they are not aware of the underlying reason for it. Even so, some staff much preferred to stick with their traditional methods, and some team leaders were more skilled at encouraging change than others. Many staff needed coaching, reassurance and formal training on how to take ownership of problems. But, at the end of 12 months, the improvement project had achieved its 20 per cent goal allowing the plant to increase the volume of its exports and cut its costs significantly. Yet Wilbert Raaijmakers still sees room for improvement: *'The optimisation of an organisation is a never-ending process. If you sit back and do the same thing tomorrow as you did today, you'll never make it. We must remain alert to the latest developments and stress the resulting information to its full potential.'*

What do these two examples have in common?

The improvement initiatives at these two operations, and the way they managed them, is typical of improvement projects. Both measured performance and placed information gathering at the centre of their improvement initiative. Both had a view of improvement targets that related directly to strategic objectives. Both made efforts to collect information that would allow decisions based on evidence rather than opinion. Heineken also made extensive use of simple improvement techniques that would both analyse problems and help to channel its staff's creativity at all levels. NLB used business process reengineering (BPR), exploited technology and focused on their customers' perceptions, needs and requirements. This included how customers, both old and young, could use the newer technologies. They both had to foster an environment that allowed all staff to contribute to its improvement, and both came to view improvement not as a 'one off', but rather as the start of a never-ending cycle of improvement. Most importantly, both had to decide how to organise the whole improvement initiative. Different organisations with different objectives may choose to implement improvement initiatives in a different way, but all will face a similar set of issues to these two operations, even if they choose to make different decisions.

DIAGNOSTIC QUESTION

What is the gap between current and required performance?

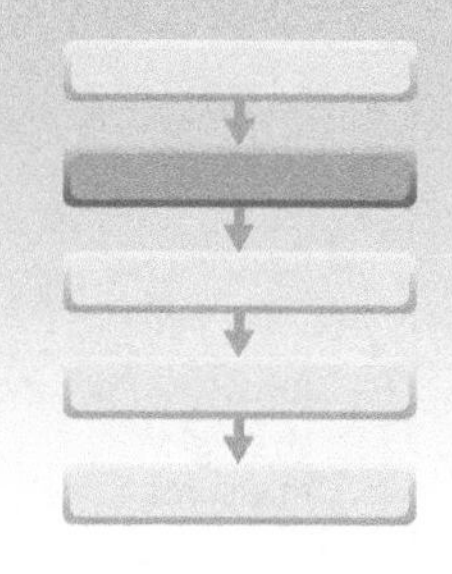

Video

The gap between how an operation or process is currently performing, and how it wishes to perform, is the key driver of any improvement initiative. The wider the gap, the more importance is likely to be given to improvement. But, in order to harness the gap as a driver of improvement, it must be addressed in some detail, both in terms of exactly what is failing to meet targets, and by how much. Answering these questions depends on the operation's ability to do three things: assess its current performance, derive a set of target levels of performance that the organisation can subscribe to, and compare current against target performance in a systematic and graphic manner that demonstrates to everyone the need for improvement.

Assessing current performance – performance measurement

OPERATIONS PRINCIPLE

Performance measurement is a prerequisite for the assessment of operations performance.

Some kind of *performance measurement* is a prerequisite for judging whether an operation is good, bad or indifferent, although this is not the only reason for investing in effective performance measurement. Without one, it would be impossible to exert any control over an operation on an ongoing basis. However, a performance measurement system that gives no help to ongoing improvement is only partially effective. Performance measurement, as we are treating it here, concerns three generic issues.

- What factors to include as performance measures?
- Which are the most important performance measures?
- What detailed measures to use?

What factors to include as performance measures?

An obvious starting point for deciding which performance measures to adopt is to use the five generic performance objectives: quality, speed, dependability, flexibility and cost. These can be broken down into more detailed measures, or they can be aggregated into 'composite' measures, such as 'customer satisfaction', 'overall service level' or 'operations agility'. These composite measures may be further aggregated by using measures such as 'achieve market objectives', 'achieve financial objectives', 'achieve operations objectives' or even 'achieve overall strategic objectives'. The more aggregated performance measures have greater strategic relevance insomuch as they help to draw a picture of the overall performance of the business, although by doing so they necessarily include many influences outside those that operations performance improvement would normally address. The more detailed performance measures are usually monitored more closely and more often, and although they provide a limited view of an operation's performance, they do provide a more descriptive and complete picture of what should be and what is happening within the operation. In practice, most organisations will choose to use performance targets from throughout the range. This idea is illustrated in Figure 13.2.

Choosing the important performance measures

One of the problems of devising a useful performance measurement system is trying to achieve some balance between having a few key measures on one hand (straightforward and simple, but may not reflect the full range of organisational objectives), or, on the other hand, having many detailed measures (complex and difficult to manage, but capable of conveying many nuances of performance). Broadly, a compromise is reached by making sure that there

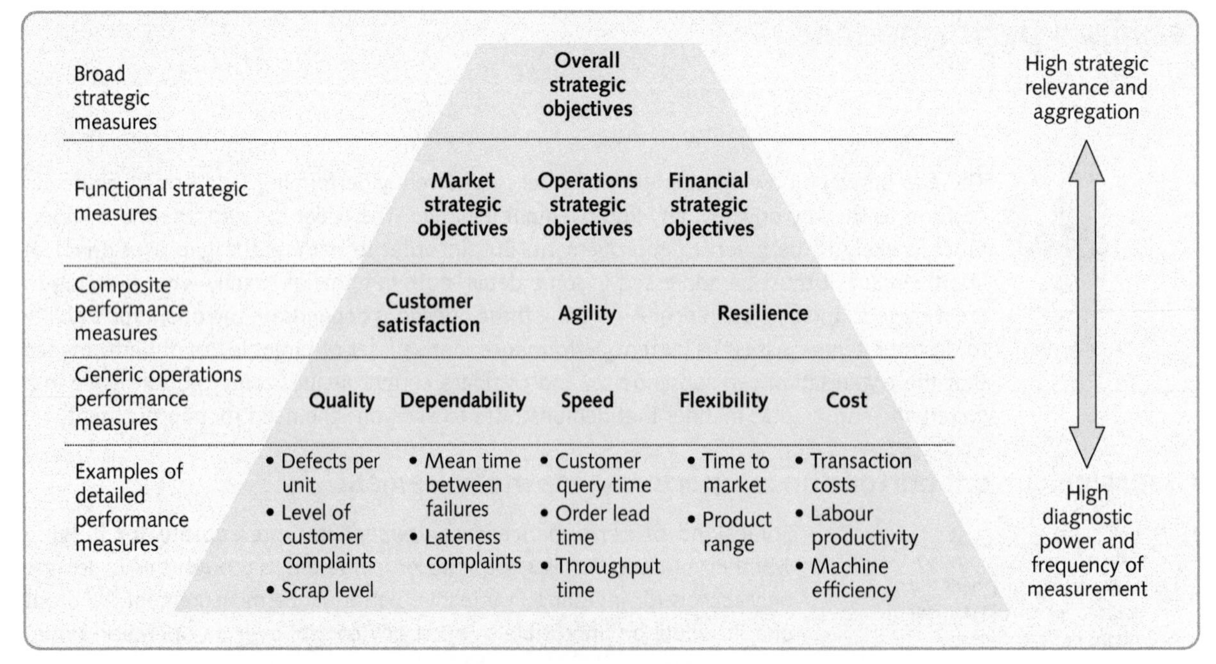

Figure 13.2 Performance measures can involve different levels of aggregation

OPERATIONS PRINCIPLE

Without strategic clarity, key performance indicators cannot be appropriately targeted.

is a clear link between the operation's overall strategy, the most important (or 'key') performance indicators (KPI's) that reflect strategic objectives, and the bundle of detailed measures that are used to 'flesh out' each key performance indicator. Obviously, unless strategy is well defined then it is difficult to 'target' a narrow range of key performance indicators.

What detailed measures to use?

The five performance objectives – quality, speed, dependability, flexibility and cost – are really composites of many smaller measures. For example, an operation's cost is derived from many factors which could include the purchasing efficiency of the operation, the efficiency with which it converts materials, the productivity of its staff, the ratio of direct to indirect staff, and so on. All of these measures individually give a partial view of the operation's cost performance, and many of them overlap in terms of the information they include. However, each of them does give a perspective on the cost performance of an operation that could be useful either to identify areas for improvement or to monitor the extent of improvement. If an organisation regards its 'cost' performance as unsatisfactory, disaggregating it into 'purchasing efficiency', 'operations efficiency', 'staff productivity', etc. might explain the root cause of the poor performance. Table 13.1 shows some of the partial measures which can be used to judge an operation's performance.

Table 13.1 Some typical partial measures of performance

Performance objective	*Some typical measures*
Quality	Number of defects per unit Level of customer complaints Scrap level Warranty claims Mean time between failures Customer satisfaction score
Speed	Customer query time Order lead time Frequency of delivery Actual *versus* theoretical throughput time Cycle time
Dependability	Percentage of orders delivered late Average lateness of orders Proportion of products in stock Mean deviation from promised arrival Schedule adherence
Flexibility	Time needed to develop new products/services Range of products/services Machine change-over time Average batch size Time to increase activity rate Average capacity/maximum capacity Time to change schedules
Cost	Minimum delivery time/average delivery time Variance against budget Utilisation of resources Labour productivity Added value Efficiency Cost per operation hour

The balanced scorecard approach

'The balanced scorecard retains traditional financial measures. But financial measures tell the story of past events, an adequate story for industrial age companies for which investments in long-term capabilities [and] customer relationships were not critical for success. These financial measures are inadequate, however, for guiding and evaluating the journey that information age companies must make to create future value through investment in customers, suppliers, employees, processes, technology, and innovation.'[3]

Generally, operations performance measures have been broadening in their scope. It is now generally accepted that the scope of measurement should, at some level, include external as well as internal, long-term as well as short-term, and 'soft' as well as 'hard' measures. The best-known manifestation of this trend is the 'balanced scorecard' approach taken by Kaplan and Norton[4]. As well as including financial measures of performance, in the same way as traditional performance measurement systems, the balanced scorecard approach also attempts to provide the important information that is required to allow the overall strategy of an organisation to be reflected adequately in specific performance measures. In addition to financial measures of performance, it also includes more operational measures of customer satisfaction, internal processes, innovation and other improvement activities. In doing so it measures the factors behind financial performance which are seen as the key drivers of future financial success. In particular, it is argued that a balanced range of measures enables managers to address the following questions (see Figure 13.3):

- How do we look to our shareholders (financial perspective)?
- What must we Excel at (internal process perspective)?
- How do our customers see us (the customer perspective)?
- How can we continue to improve and build capabilities (the learning and growth perspective)?

The balanced scorecard attempts to bring together the elements that reflect a business's strategic position, including product or service quality measures, product and service development times, customer complaints, labour productivity, and so on. At the same time it attempts to avoid performance reporting becoming unwieldy by restricting the number of measures and focusing especially on those seen to be essential. The advantages of the approach are that it presents an overall picture of the organisation's performance in a single report, and by

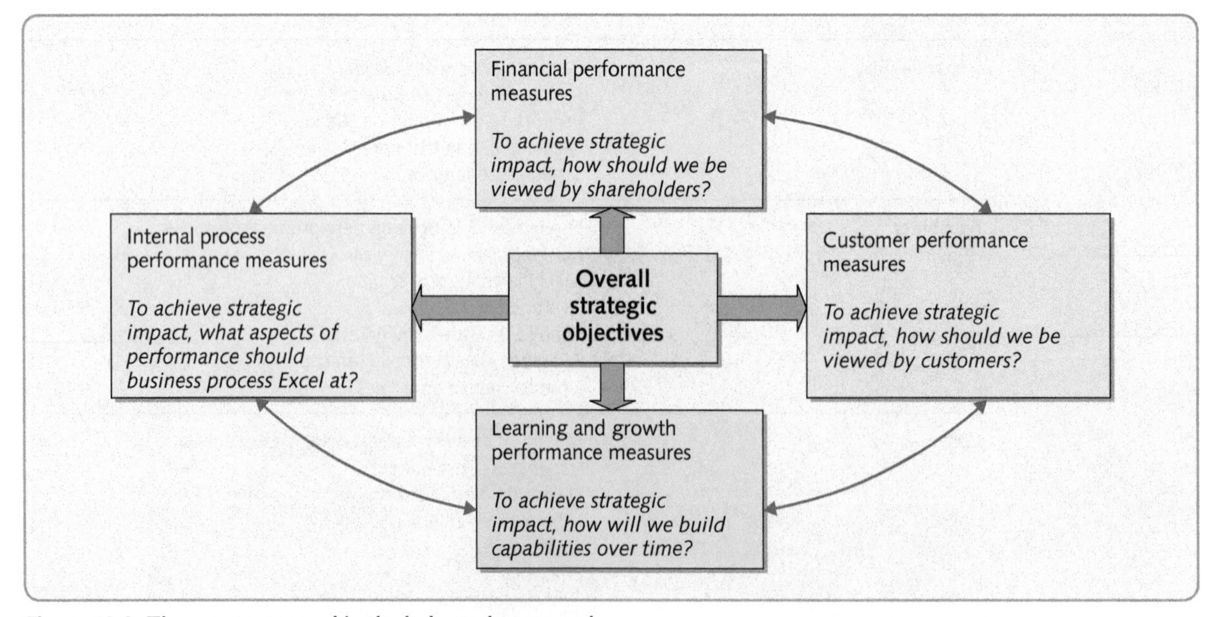

Figure 13.3 The measures used in the balanced scorecard

being comprehensive in the measures of performance it uses, encourages companies to take decisions in the interests of the whole organisation rather than sub-optimising around narrow measures. Developing a balanced scorecard is a complex process and is now the subject of considerable debate. One of the key questions that have to be considered is how specific measures of performance should be designed? Inadequately designed performance measures can result in dysfunctional behaviour, so teams of managers are often used to develop a scorecard which reflects their organisation's specific needs.

Setting target performance

A performance measure means relatively little until it is compared against some kind of target. Knowing that only one document in 500 is sent out to customers containing an error tells us relatively little unless we know whether this is better or worse than what we were achieving previously, and whether it is better or worse than what other similar operations (especially competitors) are achieving. Setting performance targets transforms performance measures into performance 'judgements'. Several approaches to setting targets can be used, including the following.

- *Historically-based targets* – targets that compare current against previous performance.
- *Strategic targets* – targets set to reflect the level of performance that is regarded as appropriate to achieve strategic objectives.
- *External performance-based targets* targets set to reflect the performance that is achieved by similar, or competitor, external operations.
- *Absolute performance targets* – targets based on the theoretical upper limit of performance.

One of the problems in setting targets is that different targets can give very different messages regarding the improvement being achieved. So, for example, in Figure 13.4, one of an operation's performance measures is 'delivery' (in this case defined as the proportion of orders delivered on time). The performance for one month has been measured at 83 per cent, but any judgement regarding performance will be dependent on the performance targets. Using a *historical* target, when compared to last year's performance of 60 per cent, this months' performance of 83 per cent is good. But, if the operation's *strategy* calls for a 95 per cent delivery performance, the actual performance of 83 per cent looks decidedly poor. The company may also be concerned with how they perform against *competitors'* performances. If competitors are currently averaging delivery performances of around 80 per cent, the company's

OPERATIONS PRINCIPLE

Performance measures only have meaning when compared against targets.

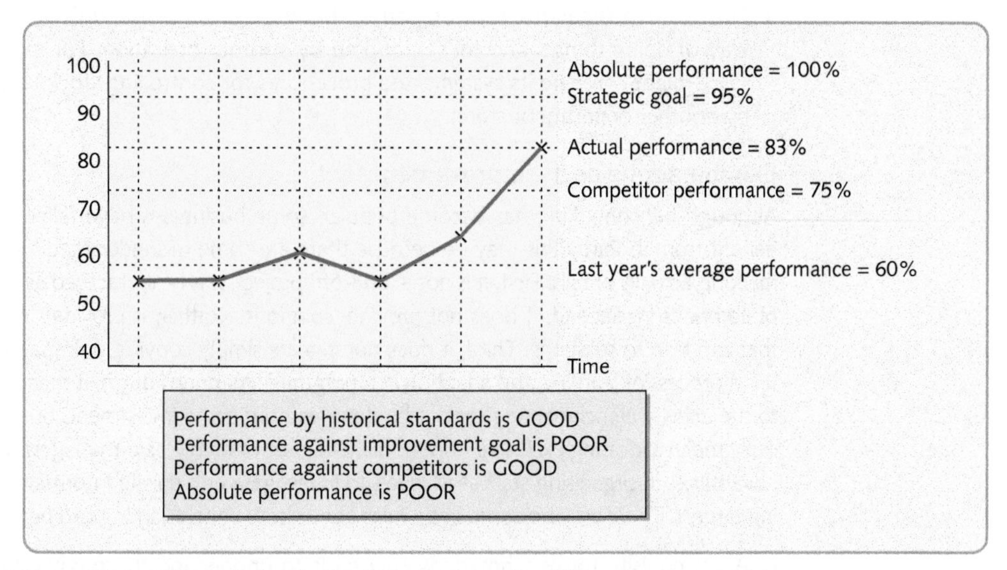

Figure 13.4 Different standards of comparison give different messages

performance looks rather good. Finally, the more ambitious managers within the company may wish to at least try and seek perfection. Why not, they argue, use an *absolute* performance standard of 100 per cent delivery on time? Against this standard the company's actual 83 per cent again looks disappointing.

Benchmarking

Benchmarking, is 'the process of learning from others' and involves comparing one's own performance or methods against other comparable operations. It is a broader issue than setting performance targets, and includes investigating other organisations' operations practice in order to derive ideas that could contribute to performance improvement. Its rationale is based on the idea that (a) problems in managing processes are almost certainly shared by processes elsewhere, and (b) that there is probably another operation somewhere that has developed a better way of doing things. For example, a bank might learn some things from a supermarket about how it could cope with demand fluctuations during the day. Benchmarking is essentially about stimulating creativity in improvement practice.

> **OPERATIONS PRINCIPLE**
> *Improvement is aided by contextualising processes and operations.*

Types of benchmarking

Practice note

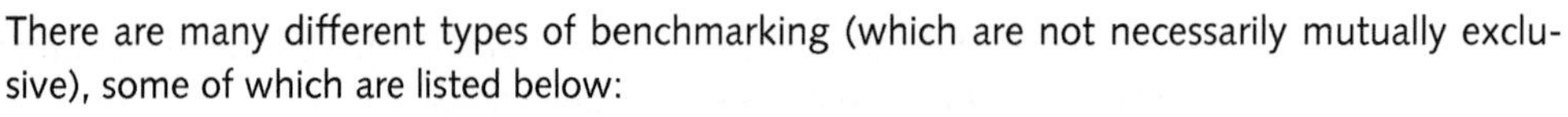

There are many different types of benchmarking (which are not necessarily mutually exclusive), some of which are listed below:

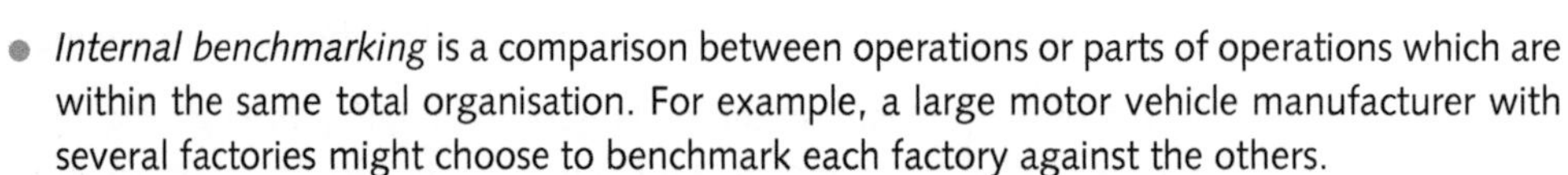

- *Internal benchmarking* is a comparison between operations or parts of operations which are within the same total organisation. For example, a large motor vehicle manufacturer with several factories might choose to benchmark each factory against the others.
- *External benchmarking* is a comparison between an operation and other operations which are part of a different organisation.
- *Non-competitive benchmarking* is benchmarking against external organisations which do not compete directly in the same markets.
- *Competitive benchmarking* is a comparison directly between competitors in the same, or similar, markets.
- *Performance benchmarking* is a comparison between the levels of achieved performance in different operations. For example, an operation might compare its own performance in terms of some or all of our performance objectives – quality, speed, dependability, flexibility and cost – against other organisations' performance in the same dimensions.
- *Practice benchmarking* is a comparison between an organisation's operations practices, or way of doing things, and those adopted by another operation. For example, a large retail store might compare its systems and procedures for controlling stock levels with those used by another department store.

Benchmarking as an improvement tool

Although benchmarking has become popular, some businesses have failed to derive maximum benefit from it. Partly this may be because there are some misunderstandings as to what benchmarking actually entails. First, it is not a 'one-off' project. It is best practised as a continuous process of comparison. Second, it does not provide 'solutions'. Rather, it provides ideas and information that can lead to solutions. Third, it does not involve simply copying or imitating other operations. It is a process of learning and adapting in a pragmatic manner. Fourth, it means devoting resources to the activity. Benchmarking cannot be done without some investment, but this does not necessarily mean allocating exclusive responsibility to a set of highly paid managers. In fact, there can be advantages in organising staff at all levels to investigate and collate information from benchmarking targets. There are also some basic rules about how benchmarking can be organised.

- A prerequisite for benchmarking success is to understand thoroughly your own processes. Without this it is difficult to compare your processes against those of other companies.

- Look at the information that is available in the public domain. Published accounts, journals, conferences and professional associations can all provide information which is useful for benchmarking purposes.
- Do not discard information because it seems irrelevant. Small pieces of information only make sense in the context of other pieces of information that may emerge subsequently.
- Be sensitive in asking for information from other companies. Don't ask any questions that you would not like to be asked yourself.

Assess the gap between actual and target performance

A comparison of actual and target performance should guide the relative priorities for improvement. A significant aspect of performance is the relative importance of the various performance measures. Because some factor of performance is relatively poor does not mean that it should be improved immediately if current performance as a whole exceeds target performance. In fact, both the relative importance of the various performance measures, and their performance against targets, need to be brought together in order to prioritise for improvement. One way of doing this is through the importance–performance matrix.

The importance-performance matrix

As its name implies, the importance–performance matrix positions each aspect of performance on a matrix according to its scores or ratings on how important each aspect of relative performance is, and what performance it is currently achieving. Figure 13.5 shows an importance–performance matrix divided into zones of improvement priority. The first zone boundary is the 'lower bound of acceptability' shown as line AB in Figure 13.5. This is the boundary between acceptable and unacceptable current performance. When some aspect of performance is rated as relatively unimportant, this boundary will be low. Most operations are prepared to tolerate lower performance for relatively unimportant performance factors. However, for performance factors that are rated more important, they will be markedly less sanguine at poor or mediocre levels of current performance. Below this minimum bound of acceptability (AB) there is clearly a need for improvement; above this line there is no immediate urgency for any improvement.

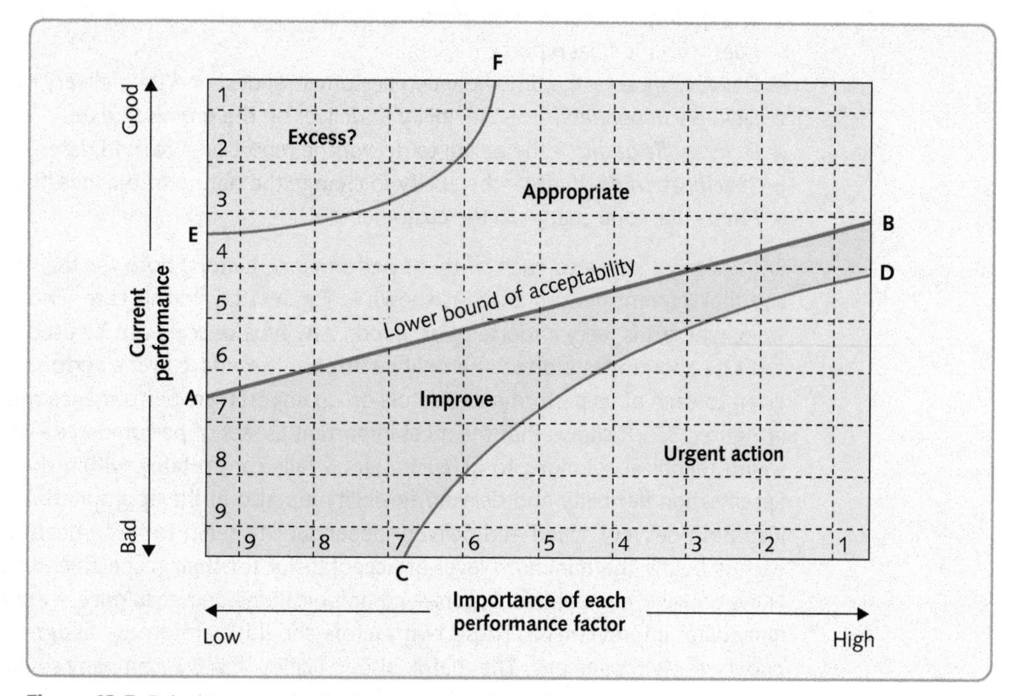

Figure 13.5 Priority zones in the importance–performance matrix

However, not all factors of performance that fall below the minimum line will be seen as having the same degree of improvement priority. A boundary approximately represented by line CD represents a distinction between an urgent priority zone and a less urgent improvement zone. Similarly, above the line AB, not all competitive factors are regarded as having the same priority. The line EF can be seen as the approximate boundary between performance levels which are regarded as 'good' or 'appropriate' on one hand and those regarded as 'too good' or 'excess' on the other. Segregating the matrix in this way results in four zones which imply very different priorities:

- *The 'appropriate' zone.* Performance factors in this area lie above the lower bound of acceptability and so should be considered satisfactory.
- *The 'improve' zone.* Lying below the lower bound of acceptability, any performance factors in this zone must be candidates for improvement.
- *The 'urgent-action' zone.* These performance factors are important to customers but current performance is unacceptable. They must be considered as candidates for immediate improvement.
- *The 'excess?' zone.* Performance factors in this area are 'high performing', but are not particularly important. The question must be asked, therefore, whether the resources devoted to achieving such a performance could be used better elsewhere.

EXAMPLE

EXL Laboratories

EXL Laboratories is a subsidiary of an electronics company. It carries out research and development as well as technical problem-solving work for a wide range of companies. It is particularly keen to improve the level of service that it gives to its customers. However, it needs to decide which aspect of its performance to improve first. It has devised a list of the most important aspects of its service:

- *The quality of its technical solutions* – the perceived appropriateness by customers.
- *The quality of its communications with customers* – the frequency and usefulness of information.
- *The quality of post-project documentation* – the usefulness of the documentation which goes with the final report.
- *Delivery speed* – the time between customer request and the delivery of the final report.
- *Delivery dependability* – the ability to deliver on the promised date.
- *Delivery flexibility* – the ability to deliver the report on a revised date.
- *Specification flexibility* – the ability to change the nature of the investigation.
- *Price* – the total charge to the customer.

EXL assigns a rating to each of these performance factors, both for their relative importance and their current performance, as shown in Figure 13.6. In this case, EXL have used a 1 to 9 scale, where 1 is 'very important' or 'good'. Any type of scale can be used.

EXL Laboratories plotted the relative importance and current performance ratings it had given to each of its performance factors on an importance–performance matrix. This is shown in Figure 13.7. It shows that the most important aspect of performance – the ability to deliver sound technical solutions to its customers – falls comfortably within the appropriate zone. Specification flexibility and delivery flexibility are also in the appropriate zone, although only just. Both delivery speed and delivery dependability seem to be in need of improvement as each is below the minimum level of acceptability for their respective importance positions. However, two competitive factors – communications and cost/price – are clearly in need of immediate improvement. These two factors should therefore be assigned the most urgent priority for improvement. The matrix also indicates that the company's documentation could almost be regarded as 'too good'.

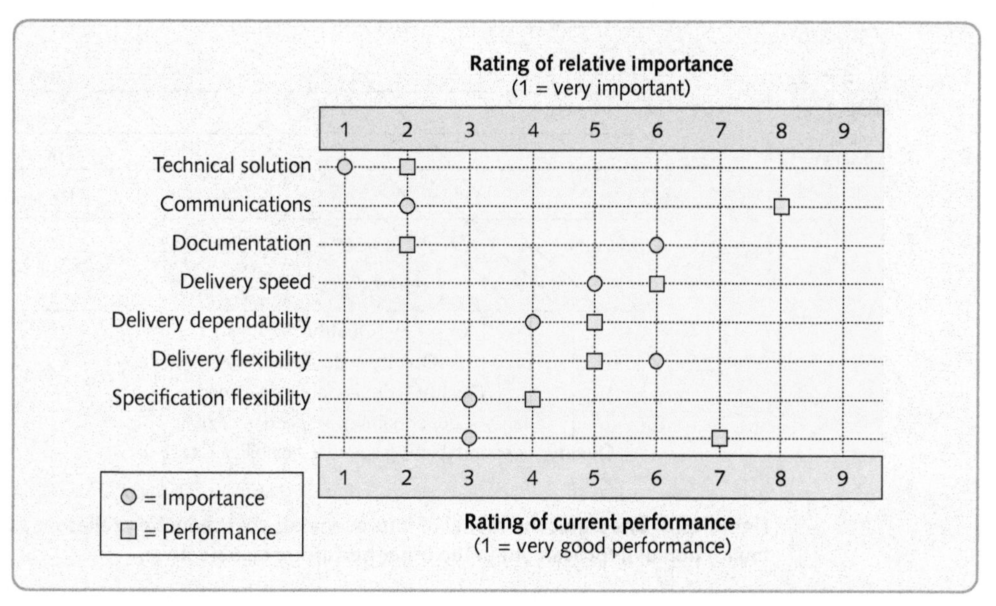

Figure 13.6 Ratings of relative 'importance' and 'current performance' at EXL Laboratories

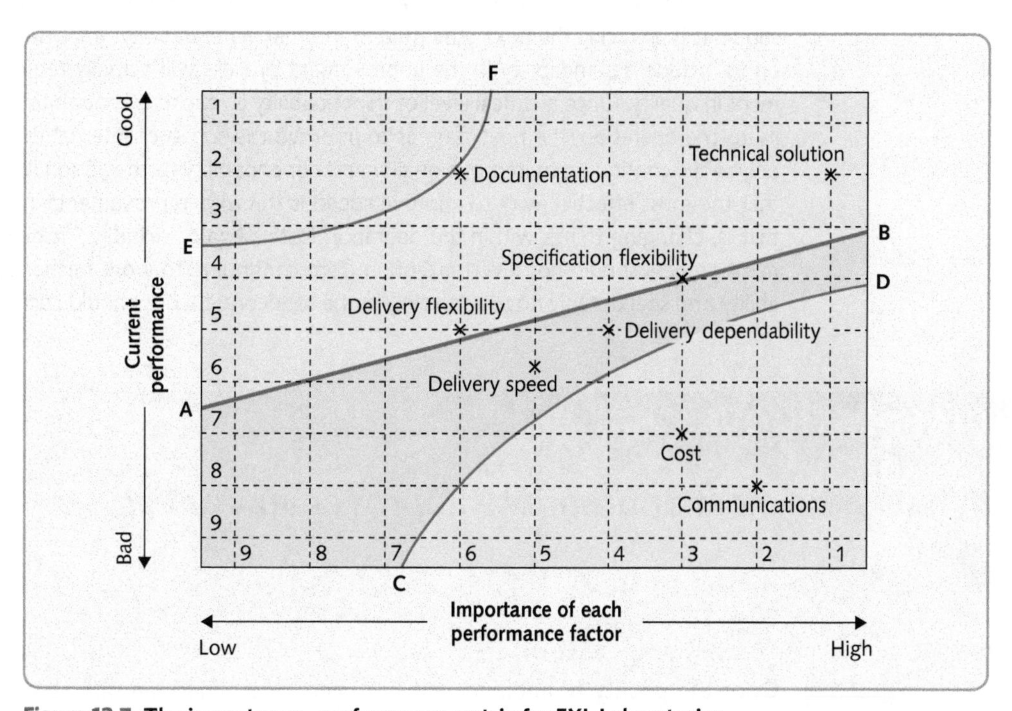

Figure 13.7 The importance–performance matrix for EXL Laboratories

The sandcone theory

As well as approaches that base improvement priority on an operation's specific circumstances, some authorities believe that there is also a generic 'best' sequence of improvement. The best-known theory is called *the sandcone theory*[5], so called because the sand is analogous to management effort and resources. Building a stable sandcone needs a stable foundation of quality upon which one can build layers of dependability, speed, flexibility and cost (see Figure 13.8). Building up improvement is thus a cumulative process, not a sequential one. Moving on to the

1... 2... 3... Animated diagram

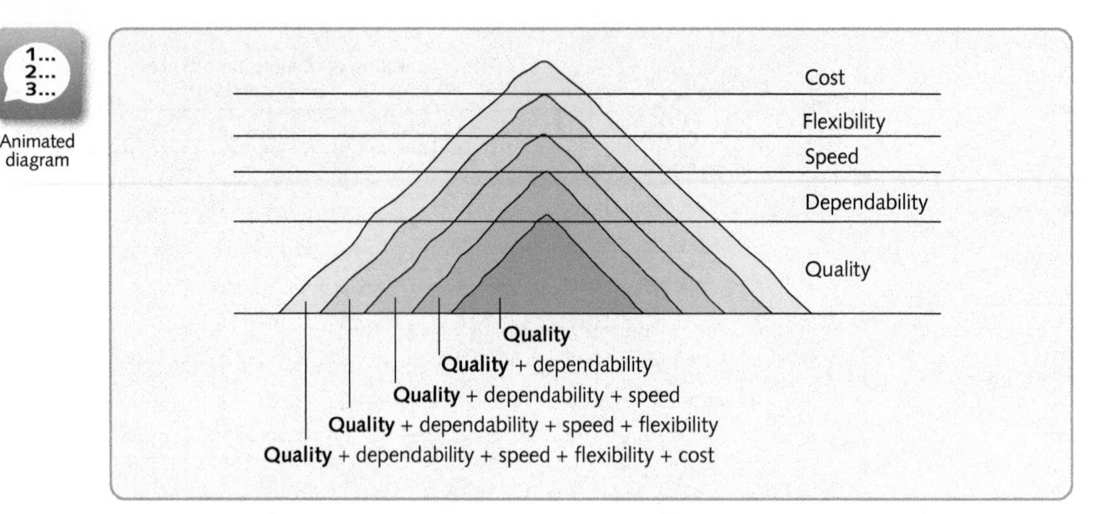

Figure 13.8 The sandcone model of improvement: cost reduction relies on a cumulative foundation of improvement in the other performance objectives

second priority for improvement does not mean dropping the first, and so on. According to the sandcone theory, the first priority should be *quality*, since this is a precondition to all lasting improvement. Only when the operation has reached a minimally acceptable level in quality should it then tackle the next issue, that of internal *dependability*. Importantly though, moving on to include dependability in the improvement process will actually require further improvement in quality. Once a critical level of dependability is reached, enough to provide some stability to the operation, the next stage is to improve the *speed* of internal throughput, but again only while continuing to improve quality and dependability further. Soon it will become evident that the most effective way to improve speed is through improvements in response *flexibility*, that is, changing things within the operation faster. Again, including flexibility in the improvement process should not divert attention from continuing to work further on quality, dependability and speed. Only now, according to the sandcone theory, should *cost* be tackled head on.

DIAGNOSTIC QUESTION

What is the most appropriate improvement path?

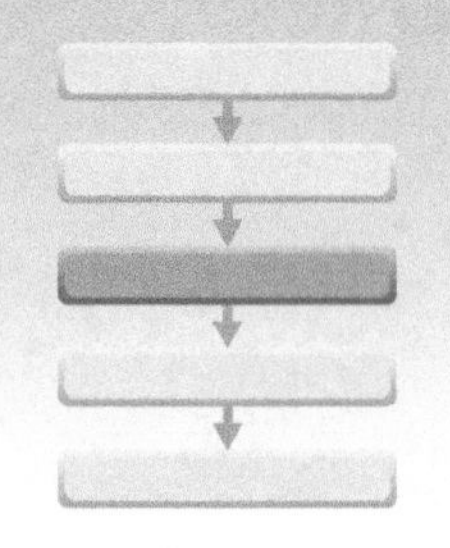

Video

Once the priority of improvement has been determined, an operation must consider the approach or path it wishes to take to reaching its improvement goals. Two paths represent different, and to some extent opposing, philosophies: *breakthrough improvement* and *continuous improvement*. Although they represent different philosophies of improvement, they are not mutually exclusive. Few operations cannot benefit from improving their operations performance on a continuous basis, and few operations would reject investing in a major improvement breakthrough leap in performance if it represented good value. For most operations, both approaches are relevant to some extent, although possibly at different points in time. But to understand how and when each approach is appropriate one must understand their underlying philosophies.

OPERATIONS PRINCIPLE

Breakthrough and continuous improvement are not mutually exclusive.

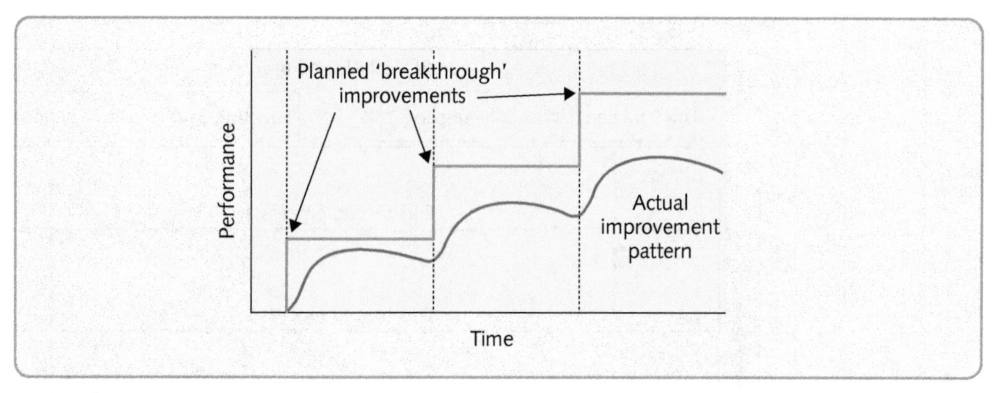

Figure 13.9 'Breakthrough' improvement may not provide the dramatic leaps in performance hoped for

Breakthrough improvement

Breakthrough (or 'innovation'-based) improvement assumes that the main vehicle of improvement is major and dramatic change in the way the operation works, for example, the total reorganisation of an operation's process structure, or the introduction of a fully integrated information system. The impact of these improvements represents a step change in practice (and hopefully performance). Such improvements can be expensive, often disrupting the ongoing workings of the operation, and frequently involving changes in the product/service or process technology. The bold line in Figure 13.9 illustrates the intended pattern of performance with several breakthrough improvements. The improvement pattern illustrated by the dotted line in Figure 13.9 is regarded by some as being more representative of what really occurs when operations rely on pure breakthrough improvement.

The business process re-engineering approach

Typical of the radical breakthrough way of tackling improvement is the business process re-engineering (BPR) approach. It is a blend of a number of ideas such as fast throughput, waste elimination through process flow charting, customer-focused operations, and so on. But it was the potential of information technologies to enable the fundamental redesign of processes that acted as the catalyst in bringing these ideas together. BPR has been defined as[6] *'the fundamental rethinking and radical redesign of business processes to achieve dramatic improvements in critical, contemporary measures of performance, such as cost, quality, service and speed.'*

Underlying the BPR approach is the belief that operations should be organised around the total process which adds value for customers, rather than the functions or activities which perform the various stages of the value-adding activity. The core of BPR is a redefinition of the processes within a total operation, to reflect the business processes that satisfy customer needs. Figure 13.10 illustrates this idea. The main principles of BPR have been summarised as follows:[7]

- Rethink business processes in a cross-functional manner which organises work around the natural flow of information (or materials or customers). This means organising around outcomes of a process rather than the tasks which go into it.
- Strive for dramatic improvements in the performance by radically rethinking and redesigning the process.
- Have those who use the output from a process perform the process. Check to see if all internal customers can be their own supplier rather than depending on another function in the business to supply them (which takes longer and separates out the stages in the process).

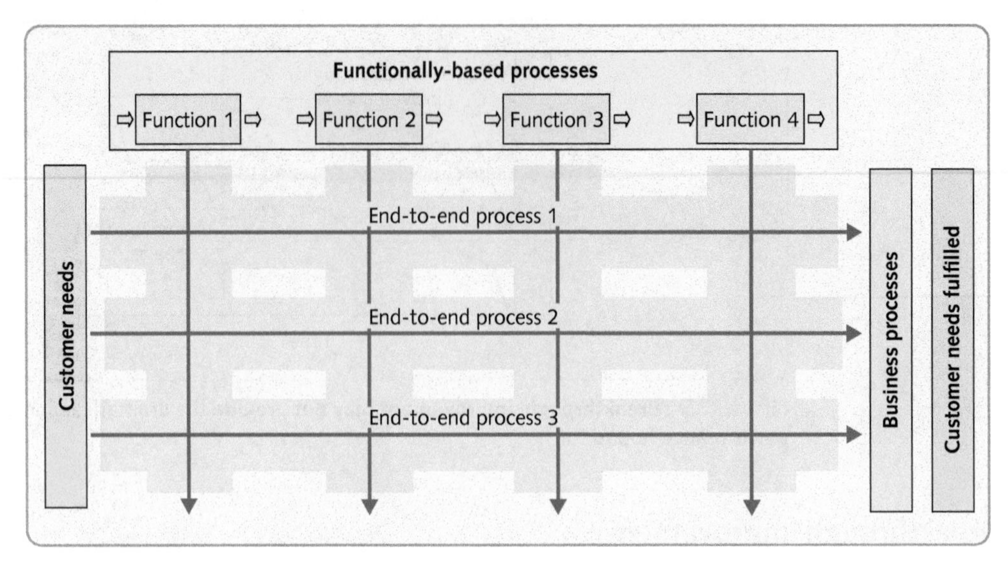

Figure 13.10 BPR advocates reorganising (re-engineering) processes to reflect the natural processes that fulfil customer needs

- Put decision points where the work is performed. Do not separate those who do the work from those who control and manage the work. Control and action are just one more type of supplier–customer relationship which can be merged.

Continuous improvement

Continuous improvement as the name implies, adopts an approach to improving performance which assumes a never-ending series of small incremental improvement steps. For example, modifying the way a product is fixed to a machine to reduce changeover time, or simplifying the question sequence when taking a hotel reservation. While there is no guarantee that such small steps towards better performance will be followed by other steps, the whole philosophy of continuous improvement attempts to ensure that they will be. It is also known as *kaizen*, defined by Masaaki Imai[8] (who has been one of the main proponents of continuous improvement) as follows: *'Kaizen means improvement. Moreover, it means improvement in personal life, home life, social life and work life. When applied to the work place, kaizen means continuing improvement involving everyone – managers and workers alike.'*

Continuous improvement is not concerned with promoting small improvements *per se*, but it does view small improvements as having one significant advantage over large ones – they can be followed relatively painlessly by others. It is not the *rate* of improvement which is important; it is the *momentum* of improvement. It does not matter if successive improvements are small; what does matter is that every month (or week, or quarter, or whatever period is appropriate) some kind of improvement has actually taken place. Continuous improvement does not always come naturally. There are specific abilities, behaviours and actions which need to be consciously developed if continuous improvement is to be sustained over the long term. Bessant and Caffyn[9] distinguish between what they call 'organisational abilities' (the ability to adopt a particular approach to continuous improvement), 'constituent behaviours' (the behaviour that staff adopt) and 'enablers' (the techniques used to progress the continuous improvement effort). They identify six generic organisational abilities, each with its own set of constituent behaviours. These are identified in Table 13.2. Examples of enablers are the improvement techniques described later in this chapter.

Table 13.2 Continuous improvement (CI) abilities and some associated behaviours[10]

Organisational ability	*Constituent behaviours*
Getting the CI habit Developing the ability to generate sustained involvement in CI	• People use formal problem-finding and solving cycle • People use simple tools and techniques • People use simple measurement to shape the improvement process • Individuals and/or groups initiate and carry through CI activities – they participate in the process • Ideas are responded to in a timely fashion – either implemented or otherwise dealt with • Managers support the CI process through allocation of resources • Managers recognise in formal ways the contribution of employees to CI • Managers lead by example, becoming actively involved in design and implementation of CI • Managers support experiment by not punishing mistakes, but instead encouraging learning from them
Focusing on CI Generating and sustaining the ability to link CI activities to the strategic goals of the company	• Individuals and groups use the organisation's strategic objectives to prioritise improvements • Everyone is able to explain what the operation's strategy and objectives are • Individuals and groups assess their proposed changes against the operation's objectives • Individuals and groups monitor/measure the results of their improvement activity • CI activities are an integral part of the individual's or group's work, not a parallel activity
Spreading the word Generating the ability to move CI activity across organisational boundaries	• People cooperate in cross-functional groups • People understand and share an holistic view (process understanding and ownership) • People are oriented towards internal and external customers in their CI activity • Specific CI projects with outside agencies (customers, suppliers, etc.) take place • Relevant CI activities involve representatives from different organisational levels
CI on the CI system Generating the ability to manage strategically the development of CI	• The CI system is continually monitored and developed • There is a cyclical planning process whereby the CI system is regularly reviewed and amended • There is periodic review of the CI system in relation to the organisation as a whole • Senior management make available sufficient resources (time, money, personnel) to support the development of the CI system • The CI system itself is designed to fit within the current structure and infrastructure • When a major organisational change is planned, its potential impact on the CI system is assessed
Walking the talk Generating the ability to articulate and demonstrate CI's values	• The 'management style' reflects commitment to CI values • When something goes wrong, people at all levels look for reasons why, rather than blame individuals • People at all levels demonstrate a shared belief in the value of small steps and that everyone can contribute, by themselves being actively involved in making and recognising incremental improvements

continued overleaf

Table 13.2 *(continued)*

Organisational ability	*Constituent behaviours*
Building the learning organisation Generating the ability to learn through CI activity	• Everyone learns from their experiences, both good and bad • Individuals seeks out opportunities for learning/personal development • Individuals and groups at all levels share their learning • The organisation captures and shares the learning of individuals and groups • Managers accept and act on all the learning that takes place • Organisational mechanisms are used to deploy what has been learned across the organisation

Improvement cycle models

An important element of continuous improvement is the idea that improvement can be represented by a never-ending process of repeatedly questioning and re-questioning the detailed working of a process. This is usually summarised by the idea of the *improvement cycle*, of which there are many, including some proprietary models owned by consultancy companies. Two of the more generally used models are: the PDCA cycle (sometimes called the Deming Cycle, named after the famous quality 'guru', W.E. Deming); the DMAIC cycle (made popular by the Six Sigma approach to improvement – see later).

OPERATIONS PRINCIPLE
Continuous improvement necessarily implies a never-ending cycle of analysis and action.

The PDCA cycle

The PDCA cycle model is shown in Figure 13.11(a). It starts with the P (for plan) stage, which involves an examination of the current method or the problem area being studied. This involves collecting and analysing data so as to formulate a plan of action which is intended to improve performance. (Some of the techniques used to collect and analyse data are explained later.) The next step is the D (for do) stage. This is the implementation stage during which the plan is tried out in the operation. This stage may itself involve a mini-PDCA cycle as the problems of implementation are resolved. Next comes the C (for check) stage where the new implemented solution is evaluated to see whether it has resulted in the expected improvement. Finally, at least for this cycle, comes the A (for act) stage. During this stage the change is consolidated or standardised if it has been successful. Alternatively, if the change has not been successful, the lessons learned from the 'trial' are formalised before the cycle starts again.

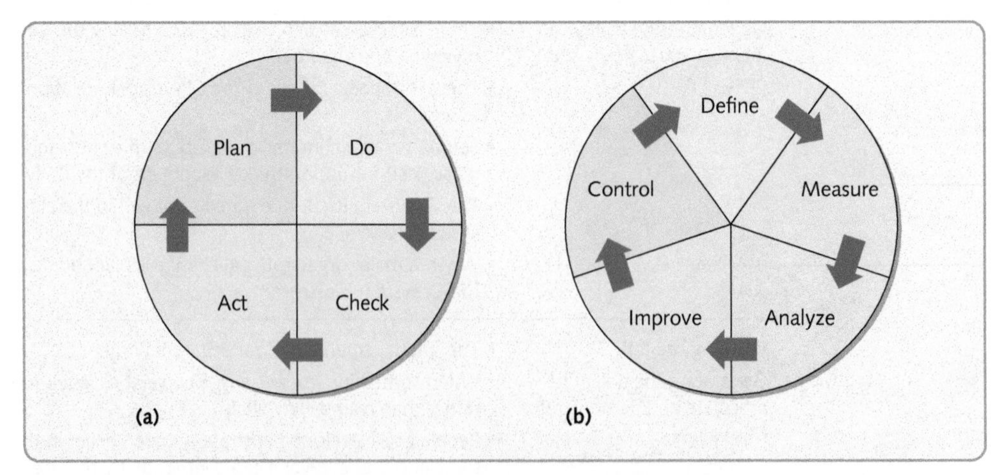

Figure 13.11 (a) the plan–do–check–act, or 'Deming' improvement cycle; (b) the define–measure–analyse–improve–control, or DMAIC Six Sigma improvement cycle

The DMAIC cycle

In some ways this cycle is more intuitively obvious than the PDCA cycle insomuch as it follows a more 'experimental' approach. The DMAIC cycle starts with defining the problem or problems, partly to understand the scope of what needs to be done and partly to define exactly the requirements of the process improvement. Often at this stage a formal goal or target for the improvement is set. After definition comes the measurement stage, important because the Six Sigma approach emphasises the importance of working with hard evidence rather than opinion. It involves validating the problem (to make sure it is really worth solving), using data to refine the problem and measuring exactly what is happening. The analysis stage can be seen as an opportunity to develop hypotheses as to what the root causes of the problem really are. Such hypotheses are validated (or not) by the analysis and the main root causes of the problem identified. Once the causes of the problem are identified, work can begin on improving the process. Ideas are developed to remove the root causes of problems, solutions are tested and those solutions that seem to work are implemented, formalised and results measured. The improved process needs then to be continually monitored and controlled to check that the improved level of performance is being sustained. Then the cycle starts again, defining the problems that are preventing further improvement.

The last point in both cycles is the most important – *'the cycle starts again'*. It is only by accepting that in a continuous improvement philosophy these cycles quite literally never stop that improvement becomes part of every person's job.

The differences between breakthrough and continuous improvement

Breakthrough improvement places a high value on creative solutions, and encourages free thinking and individualism. It is a radical philosophy insomuch as it fosters an approach to improvement which does not accept many constraints on what is possible. 'Starting with a clean sheet of paper', 'going back to first principles' and 'completely rethinking the system' are all typical breakthrough improvement principles. Continuous improvement, on the other hand, is less ambitious, at least in the short term. It stresses adaptability, teamwork and attention to detail. It is not radical; rather it builds upon the wealth of accumulated experience within the operation itself, often relying primarily on the people who operate the system to improve it. One analogy used to explain this difference is the sprint versus the marathon. Breakthrough improvement is a series of explosive and impressive sprints. Continuous improvement, like marathon running, does not require the expertise and prowess required for sprinting, but it does require that the runner (or operations manager) keeps on going. Yet notwithstanding these differences, it is possible to use both approaches. Large and dramatic improvements can be implemented as and when they seem to promise significant improvement steps, but between such occasions the operation can continue making its quiet and less spectacular kaizen improvements. Table 13.3 lists some of the differences between the two approaches.

OPERATIONS PRINCIPLE
Breakthrough improvement necessarily implies radical and/or extensive change.

The Six Sigma approach to organising improvement

One approach to improvement that combines breakthrough and continuous philosophies is *Six Sigma*. Although technically the 'Six Sigma' name derives from statistical process control (SPC), and more specifically the concept of process capability, it has now come to mean a much broader approach to improvement. The following definition gives a sense of its modern usage: *'Six Sigma is a comprehensive and flexible system for achieving, sustaining and maximising business success. Six Sigma is uniquely driven by close understanding of customer needs, disciplined use of facts, data, and statistical analysis, and diligent attention to managing, improving, and reinventing business processes.'*[12]

Table 13.3 Some features of breakthrough and continuous improvement (based on Imai)[11]

	Breakthrough improvement	*Continuous improvement*
Effect	Short-term but dramatic	Long-term and long-lasting but undramatic
Pace	Big steps	Small steps
Time-frame	Intermittent and non-incremental	Continuous and incremental
Change	Abrupt and volatile	Gradual and constant
Involvement	Select a few 'champions'	Everybody
Approach	Individualism, individual ideas and efforts	Collectivism, group efforts, systems approach
Stimulus	Technological breakthroughs, new inventions, new theories	Conventional know-how and state of the art
Risks	Concentrated – 'all eggs in one basket'	Spread – many projects simultaneously
Practical requirements	Requires large investment but little effort to maintain	Requires little investment but great effort to maintain it
Effort orientations	Technology	People
Evaluation criteria	Results for profit	Process and effects for better results

The Six Sigma concept, therefore, includes many of the issues covered in this and other chapters of this book. For example, process design and redesign, balanced scorecard measures, continuous improvement, statistical process control, ongoing process planning and control, and so on. However, at the heart of Six Sigma lies an understanding of the negative effects of variation in all types of business process. This aversion to variation was first popularised by Motorola, the electronics company, who set its objective as 'total customer satisfaction' in the 1980s, then decided that true customer satisfaction would only be achieved when its products were delivered when promised, with no defects, with no early-life failures and no excessive failure in service. To achieve this, they initially focused on removing manufacturing defects, but soon realised that many problems were caused by latent defects, hidden within the design of its products. The only way to eliminate these defects was to make sure that design specifications were tight (i.e. narrow tolerances) and its processes very capable.

Motorola's Six Sigma quality concept was so named because it required that the natural variation of processes (± 3 standard deviations) should be half their specification range. In other words, the specification range of any part of a product or service should be ± 6 the standard deviation of the process. The Greek letter sigma (σ) is often used to indicate the standard deviation of a process, hence the Six Sigma label. The Six Sigma approach also used the measure of 'defects per million *opportunities*' (DPMO). This is the number of defects that the process will produce if there were one million opportunities to do so. So difficult processes with many opportunities for defects can be compared with simple processes with few opportunities for defects.

The Six Sigma approach also holds that improvement initiatives can only be successful if significant resources and training are devoted to their management. It recommends a specially trained cadre of practitioners, many of whom should be dedicated full time to improving processes as internal consultants. The terms that have become associated with this group of experts (and denote their level of expertise) are Master Black Belt, Black Belt and Green Belt.

- *Master Black Belts* are experts in the use of Six Sigma tools and techniques as well as how such techniques can be used and implemented. They are seen as teachers who can not only guide improvement projects, but also coach and mentor Black Belts and Green Belts. Given their responsibilities, it is expected that Master Black Belts are employed full time on their improvement activities.

- *Black Belts* take a direct hand in organising improvement teams, and will usually have undertaken a minimum of 20 to 25 days training and carried out at least one major improvement project. Black Belts are expected to develop their quantitative analytical skills and also act as coaches for Green Belt. Like Master Black Belts, they are dedicated full time to improvement, and although opinions vary, some organisations recommend one Black Belt for every 100 employees.
- *Green Belts* work within improvement teams, possibly as team leaders. They have less training than Black Belts – typically around 10 to 15 days. Green Belts are not full time positions. They have normal day-to-day process responsibilities but are expected to spend at least twenty per cent of their time on improvement projects.

Devoting such a large amount of training and time to improvement is a significant investment, especially for small companies. Nevertheless, Six Sigma proponents argue that the improvement activity is generally neglected in most operations and if it is to be taken seriously, it deserves the significant investment implied by the Six Sigma approach. Furthermore, they argue, if operated well, Six Sigma improvement projects run by experienced practitioners can save far more than their cost.

The Work-Out approach[13]

The idea of including all staff in the process of improvement has formed the core of many improvement approaches. One of the best known ways of this is the 'Work-Out' approach that originated in the US conglomerate GE. Jack Welch, the then boss of GE, reputedly developed the approach to recognise that employees were an important source of brainpower for new and creative ideas, and as a mechanism for *'creating an environment that pushes towards a relentless, endless companywide search for a better way to do everything we do'*. The Work-Out programme was seen as a way to reduce the bureaucracy often associated with improvement and of *'giving every employee, from managers to factory workers, an opportunity to influence and improve GE's day-to-day operations'*. According to Welch, Work-Out was meant to help people stop *'wrestling with the boundaries, the absurdities that grow in large organisations. We're all familiar with those absurdities: too many approvals, duplication, pomposity, waste. Work-Out in essence turned the company upside* down, *so that the workers told the bosses what to do. That forever changed the way people behaved at the company. Work-Out is also designed to reduce, and ultimately eliminate all of the waste hours and energy that organisations like GE typically expend in performing day-to-day operations.'* GE also used, what it called 'town meetings' of employees. And although proponents of Work-Out emphasise the need to modify the specifics of the approach to fit the context in which it is applied, there is a broad sequence of activities implied within the approach.

- Staff, other key stakeholders and their manager hold a meeting away from the operation (a so called 'off-siter').
- At this meeting the manager gives the group the responsibility to solve a problem or set of problems shared by the group but which are ultimately the manager's responsibility.
- The manager then leaves and the group spend time (maybe two or three days) working on developing solutions to the problems, sometimes using outside facilitators.
- At the end of the meeting, the responsible manager (and sometimes the manager's boss) rejoins the group to be presented with its recommendations.
- The manager can respond in three ways to each recommendation: 'yes', 'no', or 'I have to consider it more'. If it is the last response the manager must clarify what further issues must be considered and how and when the decision will be made.

Work-Out programmes are also expensive: outside facilitators, off-site facilities and the payroll costs of a sizeable group of people meeting away from work can be substantial, even without considering the potential disruption to everyday activities. But arguably the most important implications of adopting Work-Out are cultural. In its purest form Work-Out reinforces an underlying culture of fast (and some would claim, superficial) problem-solving. It also relies on full and near universal employee involvement and empowerment together with direct dialogue between managers and their subordinates. What distinguishes the Work-Out approach from the many other types of group-based problem solving is fast decision-making and the idea that managers must respond immediately and decisively to team suggestions. But some claim that it is intolerant of staff and managers who are not committed to its values. In fact, it is acknowledged in GE that resistance to the process or outcome is not tolerated and that obstructing the efforts of the workout process is 'a career-limiting move'.

DIAGNOSTIC QUESTION

What techniques should be used to facilitate improvement?

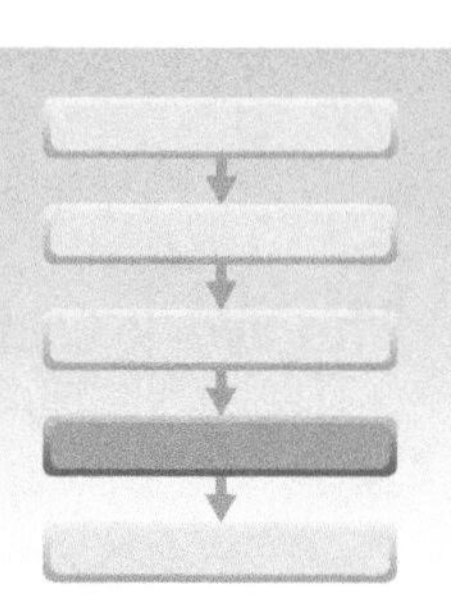

Video

All the techniques described in this book and its supplements can be regarded as 'improvement' techniques. However, some techniques are particularly useful for improving operations and processes generally. Here we select some techniques which either have not been described elsewhere or need to be re-introduced in their role of helping operations improvement particularly.

Scatter diagrams

Scatter diagrams provide a quick and simple method of identifying whether there is evidence of a connection between two sets of data: for example, the time at which you set off for work every morning and how long the journey to work takes. Plotting each journey on a graph which has departure time on one axis and journey time on the other could give an indication of whether departure time and journey time are related, and if so, how. Scatter diagrams can be treated in a far more sophisticated manner by quantifying how strong the relationship between the sets of data is. But, however sophisticated the approach, this type of graph only identifies the existence of a relationship, not necessarily the existence of a cause–effect relationship. If the scatter diagram shows a very strong connection between the sets of data, it is important evidence of a cause–effect relationship, but not proof positive. It could be coincidence!

OPERATIONS PRINCIPLE

Improvement is facilitated by relatively simple analytical techniques.

EXAMPLE

Kaston Pyral Services Ltd (1)

Kaston Pyral Services Ltd (KPS) installs and maintains environmental control, heating and air conditioning systems. It has set up an improvement team to suggest ways in which it might improve its levels of customer service. The improvement team had completed its first customer satisfaction survey. The survey asked customers to score the service they received from KPS in several ways. For example, it asked customers to score services on a scale of one to ten on promptness, friendliness, level of advice, etc. Scores were then summed to give a 'total satisfaction score' for each customer – the higher the score, the greater the satisfaction. The spread of satisfaction scores puzzled the team and they considered what factors might be causing

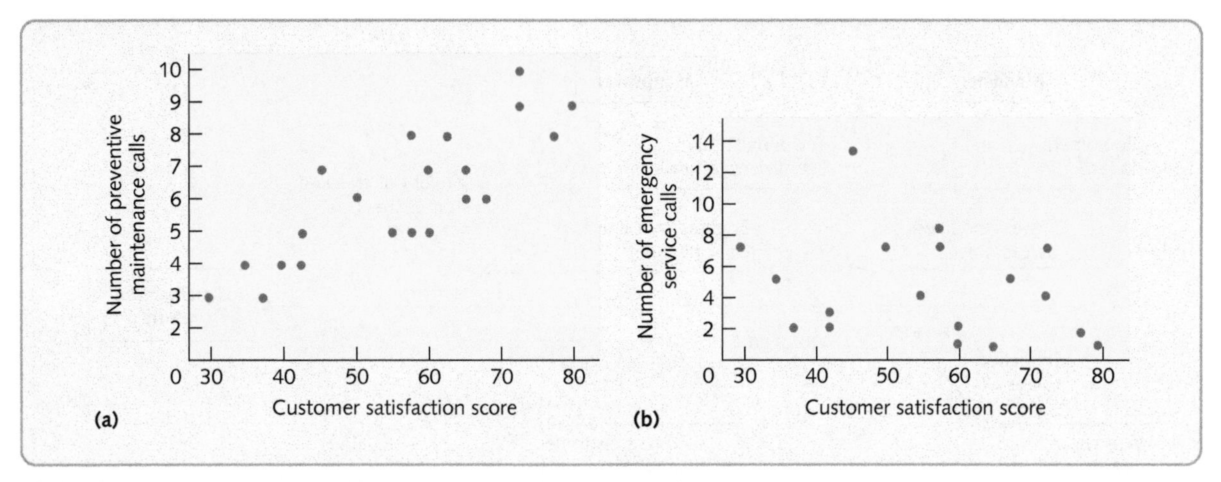

Figure 13.12 Scatter diagrams for customer satisfaction versus (a) number of preventive maintenance calls and (b) number of emergency service calls

such differences in the way their customers viewed them. Two factors were put forward to explain the differences:

1 the number of times in the past year the customer had received a preventive maintenance visit
2 the number of times the customer had called for emergency service.

All this data was collected and plotted on scatter diagrams as shown in Figure 13.12. Figure 13.12(a) shows that there seems to be a clear relationship between a customer's satisfaction score and the number of times the customer was visited for regular servicing. The scatter diagram in Figure 13.12(b) is less clear. Although all customers who had very high satisfaction scores had made very few emergency calls, so had some customers with low satisfaction scores. As a result of this analysis, the team decided to survey customers' views on its emergency service.

Cause–effect diagrams

Cause–effect diagrams are a particularly effective method of helping to search for the root causes of problems. They do this by asking what, when, where, how and why questions, but also add some possible 'answers' in an explicit way. They can also be used to identify areas where further data is needed. Cause–effect diagrams (which are also known as Ishikawa diagrams) have become extensively used in improvement programmes. This is because they provide a way of structuring group brainstorming sessions. Often the structure involves identifying possible causes under the (rather old fashioned) headings of: machinery, manpower, materials, methods and money. Yet in practice, any categorisation that comprehensively covers all relevant possible causes could be used.

EXAMPLE

Kaston Pyral Services Ltd (2)

The improvement team at KPS was working on a particular area which was proving a problem. Whenever service engineers were called out to perform emergency servicing for a customer, they took with them the spares and equipment which they thought would be necessary to repair the system. Although engineers could never be sure exactly what materials and equipment they would need for a job, they could guess what was likely to be needed and take a range of spares and equipment which would cover most eventualities. Too often, however, the engineers would find that they needed a spare that they had not brought with them. The cause–effect diagram for this particular problem, as drawn by the team, is shown in Figure 13.13.

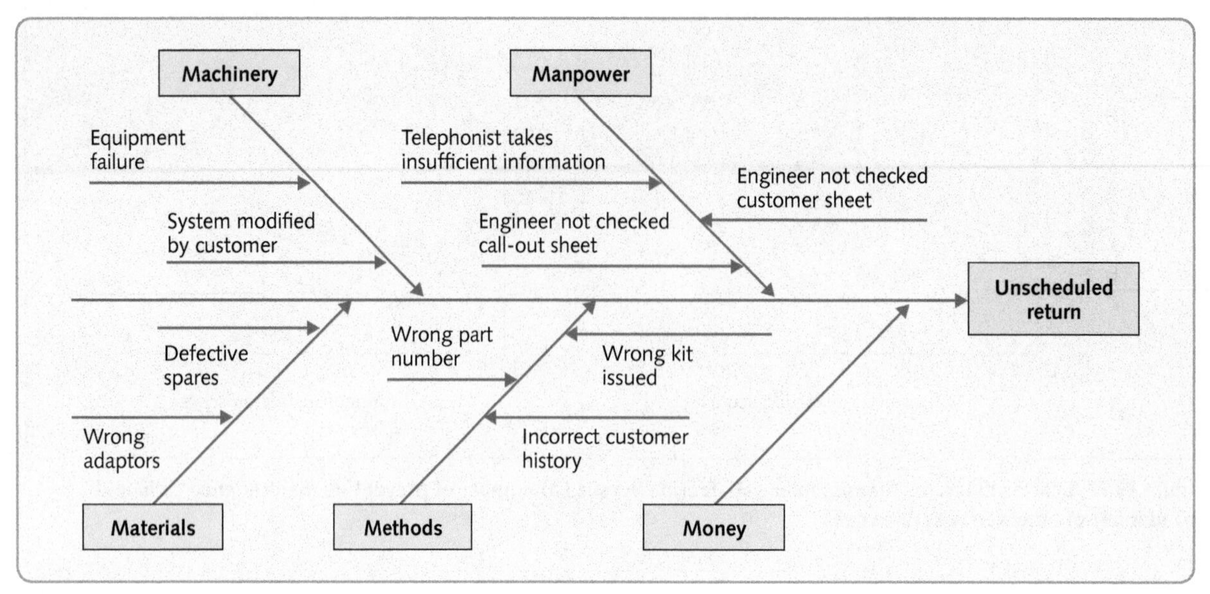

Figure 13.13 Cause–effect diagram of unscheduled returns at KPS

Pareto diagrams

In any improvement process, it is worthwhile distinguishing what is important and what is less so. The purpose of the Pareto diagram (that was first introduced in Chapter 9) is to distinguish between the 'vital few' issues and the 'trivial many'. It is a relatively straightforward technique which involves arranging items of information on the types of problem or causes of problem into their order of importance (usually measured by 'frequency of occurrence). This can be used to highlight areas where further decision-making will be useful. Pareto analysis is based on the phenomenon of relatively few causes explaining the majority of effects. For example, most revenue for any company is likely to come from relatively few of the company's customers. Similarly, relatively few of a doctor's patients will probably occupy most of his or her time.

EXAMPLE

Kaston Pyral Services Ltd (3)

The KPS improvement team which was investigating unscheduled returns from emergency servicing (the issue which was described in the cause–effect diagram in Figure 13.13) examined all occasions over the previous 12 months on which an unscheduled return had been made. They categorised the reasons for unscheduled returns as follows:

1 The wrong part had been taken to a job because, although the information which the engineer received was sound, he or she had incorrectly predicted the nature of the fault.
2 The wrong part had been taken to the job because there was insufficient information given when the call was taken.
3 The wrong part had been taken to the job because the system had been modified in some way and not recorded on KPS's records.
4 The wrong part had been taken to the job because the part had been incorrectly issued to the engineer by stores.
5 No part had been taken because the relevant part was out of stock.
6 The wrong equipment had been taken for whatever reason.
7 Any other reason.

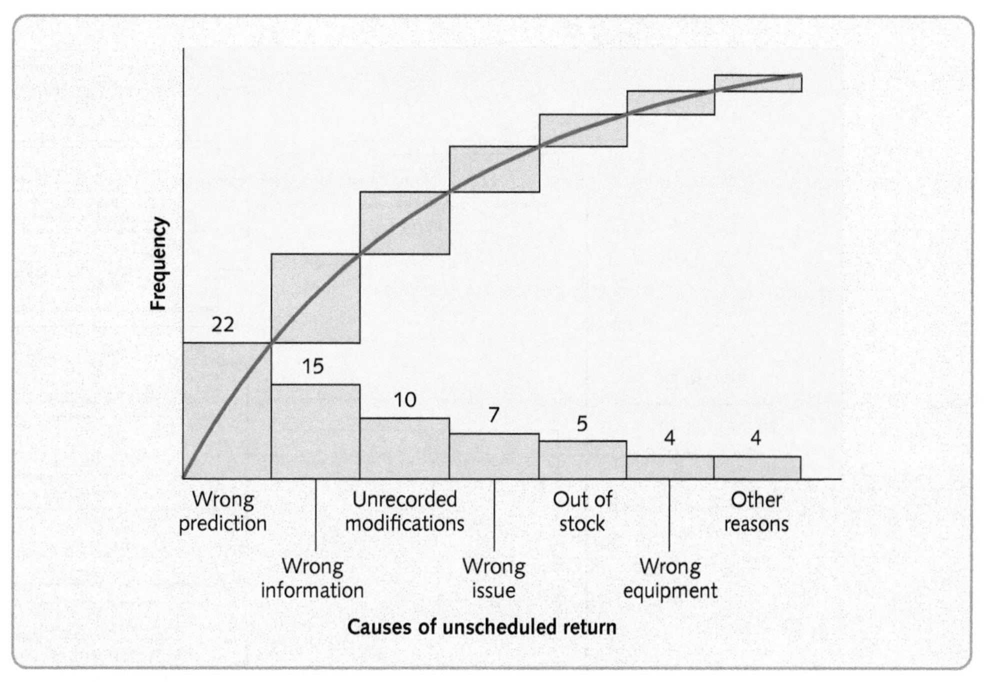

Figure 13.14 Pareto diagram for causes of unscheduled returns

The relative frequency of occurrence of these causes is shown in Figure 13.14. About a third of all unscheduled returns were due to the first category, and more than half the returns were accounted for by the first and second categories together. It was decided that the problem could best be tackled by concentrating on how to get more information to the engineers which would enable them to predict the causes of failure accurately.

Why–why analysis

Why–why analysis starts by stating the problem and asking *why* that problem has occurred. Once the major reasons for the problem occurring have been identified, each of the major reasons is taken in turn and again the question is asked *why* those reasons have occurred, and so on. This procedure is continued until either a cause seems sufficiently self-contained to be addressed by itself or no more answers to the question 'Why?' can be generated.

EXAMPLE

Kaston Pyral Services Ltd (4)

The major cause of unscheduled returns at KPS was the incorrect prediction of reasons for the customer's system failure. This is stated as the 'problem' in the why–why analysis in Figure 13.15. The question is then asked, 'Why was the failure wrongly predicted?' Three answers are proposed: first, that the engineers were not trained correctly; second, that they had insufficient knowledge of the particular product installed in the customer's location; third, that they had insufficient knowledge of the customer's particular system with its modifications. Each of these three reasons is taken in turn, and the questions are asked. 'Why is there a lack of training?', 'Why is there a lack of product knowledge?' and 'Why is there a lack of customer knowledge?' And so on.

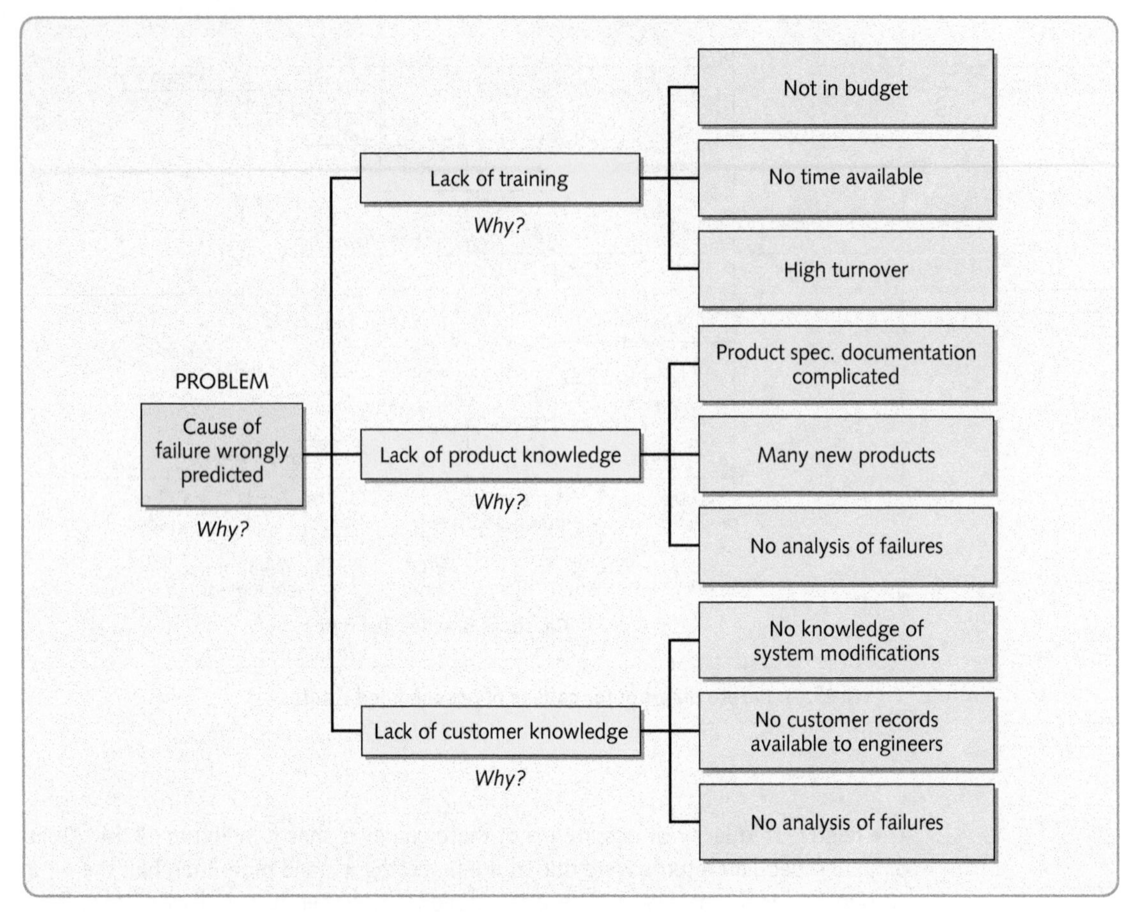

Figure 13.15 Why–why analysis for 'failure wrongly predicted'

DIAGNOSTIC QUESTION

How can improvement be made to stick?

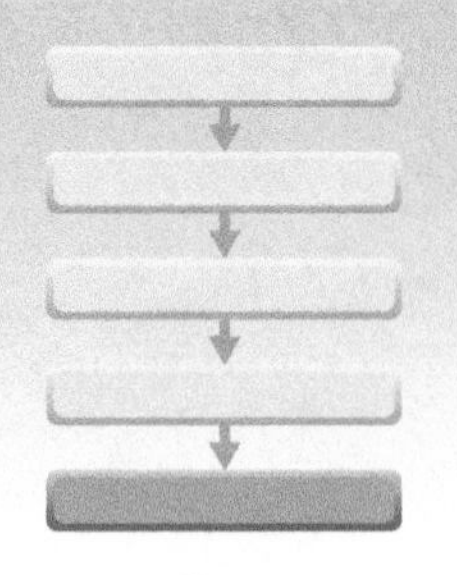

Not all of the improvement initiatives, (often launched with high expectations), will go on to fulfil their potential. Even those improvement initiatives that are successfully implemented may lose impetus over time. Sometimes this is because of managers' view of the nature of improvement, at other times it is because managers fail to manage the improvement process adequately.

Avoid becoming a victim of improvement 'fashion'

OPERATIONS PRINCIPLE

The popularity of an improvement approach is not necessarily an indicator of its effectiveness.

Improvement has, to some extent, become a fashion industry with new ideas and concepts continually being introduced as offering a novel way to improve business performance. There is nothing intrinsically wrong with this. Fashion stimulates and refreshes through introducing novel ideas. Without it, things would stagnate. The problem lies not with new improvement ideas, but rather with some managers becoming a victim of the process, where some new idea will entirely displace whatever went before. Most new ideas have something to say, but jumping from one fad to another will not only generate a backlash against any new idea, but also destroy the ability to accumulate the experience that comes from experimenting with each one.

Avoiding becoming an improvement fashion victim is not easy. It requires that those directing the improvement process take responsibility for a number of issues.

- They must take responsibility for improvement as an ongoing activity, rather than becoming champions for only one specific improvement initiative.
- They must take responsibility for understanding the underlying ideas behind each new concept. Improvement is not 'following a recipe' or 'painting by numbers'. Unless one understands *why* improvement ideas are supposed to work, it is difficult to understand *how* they can be made to work properly.
- They must take responsibility for understanding the antecedents to a 'new' improvement idea, because it helps to understand it better and to judge how appropriate it may be for one's own operation.
- They must be prepared to adapt new idea so that they make sense within the context of their own operation. 'One size' rarely fits all.
- They must take responsibility for the (often significant) education and learning effort that will be needed if new ideas are to be intelligently exploited.
- Above all they must avoid the over-exaggeration and hype that many new ideas attract. Although it is sometimes tempting to exploit the motivational 'pull' of new ideas through slogans, posters and exhortations, carefully thought-out plans will always be superior in the long run, and will help avoid the inevitable backlash that follows 'over-selling' a single approach.

Managing the improvement process

There is no absolute prescription for the way improvement should be managed. Any improvement process should reflect the uniqueness of each operation's characteristics. What appear to be almost a guarantee of difficulty in managing improvement processes, are attempts to squeeze improvement into a standard mould. Nevertheless, there are some aspects of any improvement process that appear to influence its eventual success, and should at least be debated.

OPERATIONS PRINCIPLE

There is no one universal approach to improvement.

Should an improvement *strategy* be defined?

Without thinking through the overall purpose and long-term goals of the improvement process it is difficult for any operation to know where it is going. Specifically, an improvement strategy should have something to say about:

- the competitive priorities of the organisation, and how the improvement process is expected to contribute to achieving increased strategic impact
- the roles and responsibilities of the various parts of the organisation in the improvement process
- the resources that will be available for the improvement process
- the general approach to, and philosophy of, improvement in the organisation.

Yet, too rigid a strategy can become inappropriate if the business's competitive circumstances change, or as the operation learns through experience. But, the careful modification of improvement strategy in the light of experience is not the same as making dramatic changes in improvement strategy as new improvement fashions appear.

What degree of top-management support is required?

For most authorities, the answer is unambiguous – a significant amount. Without top-management support, improvement cannot succeed. It is the most crucial factor in almost all the studies of improvement process implementation. It also goes far beyond merely allocating senior resources to the process. 'Top-management support' usually means that senior personnel must:

- understand and believe in the link between improvement and the business overall strategic impact
- understand the practicalities of the improvement process and be able to communicate its principles and techniques to the rest of the organisation
- be able to participate in the total problem-solving process to improve performance
- formulate and maintain a clear idea of the operation's improvement philosophy.

Should the improvement process be formally supervised?

Some improvement processes fail because they develop an unwieldy 'bureaucracy' to run them. But any process needs to be managed, so all improvement processes will need some kind of group to design, plan and control its efforts. However, a worthwhile goal for many improvement processes is to make themselves 'self-governing' over time. In fact there are significant advantages in terms of people's commitment in giving them responsibility for managing the improvement process. However, even when improvement is driven primarily by self-managing improvement groups, there is a need for some sort of 'repository of knowledge' to ensure that the learning and experience accumulated from the improvement process is not lost.

To what extent should improvement be group-based?

No one can really know a process quite like the people who operate it. They have access to the informal as well as the formal information networks that contain the way processes really work. But, working alone, individuals cannot pool their experience or learn from one another. So improvement processes are almost always based on teams. The issue is how these teams should be formulated, which will depend on the circumstances of the operation, its context and its objectives. For example, *quality circles*, much used in Japan, encountered mixed success in the West. A very different type of team is the *task force*, or what some US companies call a 'tiger team'. Compared with quality circles, this type of group is far more management directed and focused. Most improvement teams are between these two extremes (see Figure 13.16).

How should success be recognised?

If improvement is so important, it should be recognised, with success, effort and initiative being formally rewarded. The paradox is that, if improvement is to become part of everyday operational life, then why should improvement effort be especially rewarded? One compromise is to devise a recognition and rewards system that responds to improvement initiatives early in the improvement process, but then merges into the operation's normal reward procedures. In this way people are rewarded not just for the efficient and effective running of their processes on an ongoing basis, but also for improving their processes. Then improvement will become an everyday responsibility of all people in the operation.

How much training is required?

Training has two purposes in the development of improvement processes. The first is to provide the necessary skills that will allow staff to solve process problems and implement improvements. The second is to provide an understanding of the appropriate interpersonal, group,

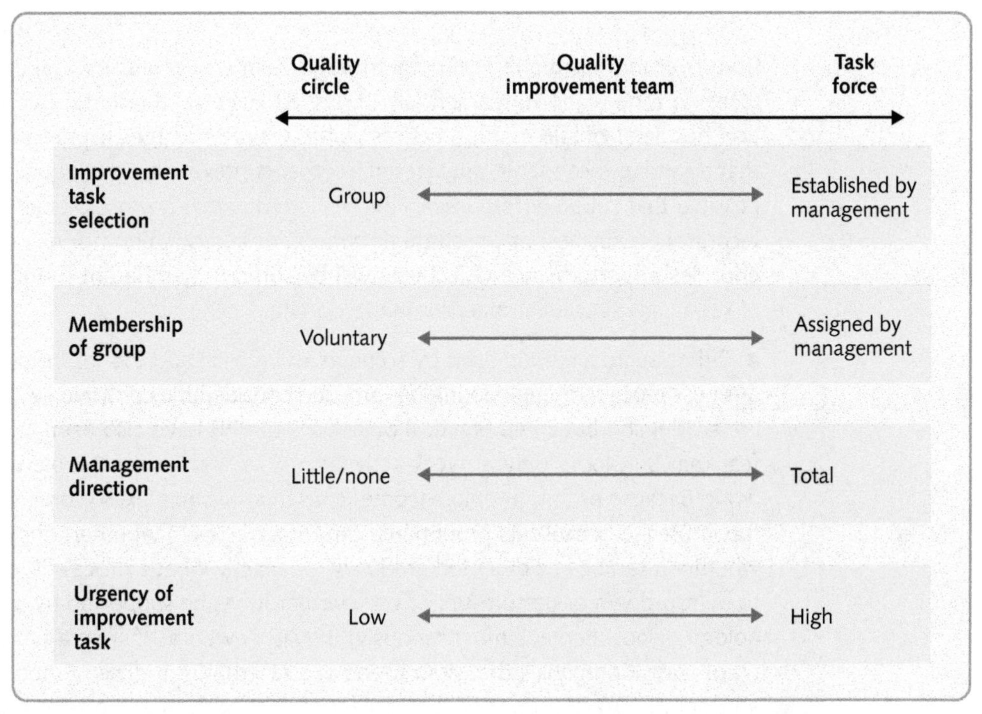

Figure 13.16 Different types of improvement groups have different characteristics

and organisational skills that are needed to 'lubricate' the improvement process. This second objective is more difficult than the first. Training and improvement techniques may take up significant time and effort, but none of this knowledge will be of much use if the organisational context for improvement mitigates against the techniques being used effectively. Although the nature of appropriate organisational development is beyond the scope of this book, it is worth noting both technique-based skills and organisational skills are enhanced if staff have a basic understanding of the core ideas and principles of operations and process management.

Critical commentary

Each chapter contains a short critical commentary on the main ideas covered in the chapter. Its purpose is not to undermine the issues discussed in the chapter, but to emphasise that, although we present a relatively orthodox view of operation, there are other perspectives.

- Many of the issues covered in this chapter are controversial, for different reasons. Some criticism concerns the effectiveness of improvement methods. For example, it can be argued that there is a fundamental flaw in the concept of benchmarking. Operations that rely on others to stimulate their creativity, especially those that are in search of 'best practice', are always limiting themselves to currently accepted methods of operating or currently accepted limits to performance. 'Best practice' is not 'best' in the sense that it cannot be bettered, it is only 'best' in the sense that it is the best one can currently find. And accepting what is currently defined as 'best' may prevent operations from ever making the radical breakthrough or improvement that takes the concept of 'best' to a new and fundamentally improved level. Furthermore, because one operation has a set of successful practices in the way it manages it process does not mean that adopting those same practices in another context will prove equally successful. It is possible that subtle differences in the resources within a process (such as staff skills or technical capabilities) or the strategic context of an operation (for example, the relative priorities of performance objectives) will be sufficiently different to make the adoption of seemingly successful practices inappropriate.

- Other approaches are seen by some as too radical and too insensitive. For example, business process re-engineering has aroused considerable controversy. Most of its critics are academics, but some practical objections to BPR have also been raised, such as the fear that BPR looks only at work activities rather than at the people who perform the work. Because of this, people become 'cogs in a machine'. Also some see BPR as being too imprecise because its proponents cannot agree as to whether it has to be radical or whether it can be implemented gradually, or exactly what a process is, or whether it has to be top-down or bottom-up, or on whether it has be supported by information technology or not. Perhaps most seriously, BPR is viewed as merely an excuse for getting rid of staff. Companies that wish to 'downsize' (that is, reduce numbers of staff within an operation) are using BPR as an excuse. This puts the short-term interests of the shareholders of the company above either their longer-term interests or the interests of the company's employees. Moreover, a combination of radical redesign together with downsizing can mean that the essential core of experience is lost from the operation. This leaves it vulnerable to any marked turbulence since it no longer has the knowledge and experience of how to cope with unexpected changes.

- Even the more gentle approach of continuous improvement is not universally welcomed. Notwithstanding its implications of empowerment and liberal attitude toward shop-floor staff, it is regarded by some worker representatives as merely a further example of management exploiting workers. Relatively established ideas such as TQM have been defined by its critics as 'management by stress'. Or, even more radically, 'TQM is like putting a vacuum cleaner next to a worker's brain and sucking out ideas. They don't want to rent your knowledge anymore, they want to own it – in the end that makes you totally replaceable.'

SUMMARY CHECKLIST

This checklist comprises questions that can be usefully applied to any type of operations and reflect the major diagnostic questions used within the chapter.

- [ ] Is the importance of performance improvement fully recognised within the operation?
- [ ] Do all operations and process managers see performance improvement as an integral part of their job?
- [ ] Is the gap between current and desired performance clearly articulated in all areas?
- [ ] Is the current performance measurement system seen as forming a basis for improvement?
- [ ] Does performance measurement focus on factors that reflect the operation's strategic objectives?
- [ ] Do performance measures allow likely problem areas to be diagnosed?
- [ ] Is some kind of balanced score card approach used that includes financial, internal, customer and learning perspectives?
- [ ] Is target performance set using an appropriate balance between historical, strategic, external and absolute performance targets?
- [ ] Are both performance and process methods benchmarked against similar operations and/or processes externally?
- [ ] Is benchmarking done on a regular basis and seen as an important contribution to improvement?
- [ ] Is some formal method of comparing actual and desired performance (such as the importance–performance matrix) used?
- [ ] To what extent does the operation have a predisposition towards breakthrough or continuous improvement?
- [ ] Have breakthrough improvement approaches such as business process re-engineering been evaluated?
- [ ] Are continuous improvement methods and problem-solving cycles used within the operation?
- [ ] If they are, has continuous improvement become a part of everyone's' job?
- [ ] Which 'abilities' and 'associated behaviours' (see Table 13.2) are evident within the operation?
- [ ] Has the Six Sigma approach to improvement been evaluated?
- [ ] Are the more common improvement techniques used to facilitate improvement within the operations?
- [ ] Does the operation show any signs of becoming a fashion victim of the latest improvement approach?
- [ ] Does the operation have a well thought-through approach to managing improvement?

CASE STUDY

Geneva Construction and Risk

'This is not going to be like last time. Then, we were adopting an improvement programme because we were told to. This time it's our idea and, if it's successful, it will be us that are telling the rest of the group how to do it.' (Tyko Mattson, Six Sigma Champion, GCR)

Tyko Mattson was speaking as the newly appointed 'Champion' at Geneva Construction and Risk Insurance, who had been charged with *'steering the Six Sigma programme until it is firmly established as part of our ongoing practice'*. The previous improvement initiative that he was referring to dated back many years to when GCR's parent company, Wichita Mutual Insurance, had insisted on the adoption of totally quality management (TQM) in all its businesses. The TQM initiative had never been pronounced a failure and had managed to make some improvements, especially in customers' perception of the company's levels of service. However, the initiative had 'faded out' during the 1990s and, even though all departments still had to formally report on their improvement projects, their number and impact was now relatively minor.

Source: Digital Vision/Getty Images

History

The Geneva Construction Insurance Company was founded in 1922 to provide insurance for building contractors and construction companies, initially in German-speaking Europe and then, because of the emigration of some family members to the USA, in North America. The company had remained relatively small and had specialised in housing construction projects until the early 1950s when it had started to grow, partly because of geographical expansion and partly because it has moved into larger (sometimes very large) construction insurance in the industrial, oil, petrochemical and power plant construction areas. In 1983 it had been bought by the Wichita Mutual Group and had absorbed the group's existing construction insurance businesses.

By 2000 it had established itself as one of the leading providers of insurance for construction projects, especially complex, high-risk projects, where contractual and other legal issues, physical exposures and design uncertainty needed 'customised' insurance responses. Providing such insurance needed particular knowledge and skills from specialists, including construction underwriters, loss adjusters, engineers, international lawyers, and specialist risk consultants. Typically, the company would insure losses resulting from contractor failure, related public liability issues, delays in project completion, associated litigation, other litigation (such as ongoing asbestos risks) and negligence issues.

The company's headquarters were in Geneva and housed all major departments including sales and marketing, underwriting, risk analysis, claims and settlement, financial control, general admin, specialist and general legal advice, and business research. There were also 37 local offices around the world, organised into four regional areas: North America; South America; Europe, Middle East and Africa; and Asia. These regional offices provided localised help and advice directly to clients and also to the 890 agents that GCR used worldwide.

The previous improvement initiative

When Wichita Mutual had insisted that CGR adopt a TQM initiative, it had gone as far as to specify exactly how it should do it and which consultants should be used to help establish the programme. Tyko Mattson shakes his head as he describes it. *'I was not with the company at that time but, looking back, it's amazing that it ever managed to do any good. You can't impose the structure of an improvement initiative from the top. It has to, at least partially, be shaped by the people who are going to be involved in it. But everything had to be done according to the handbook. The cost of quality was measured for different departments according to the handbook. Everyone had to learn the improvement techniques that were described in the handbook. Everyone had be part of a quality circle that was organised according to the handbook. We even had to have annual award ceremonies where we gave out special 'certificates of merit' to those quality circles that had achieved the type of improvement that the handbook said they should.'* The TQM initiative had been run by the 'Quality Committee', a group of eight people with representatives from all the major departments at head office. Initially, it had spent much of its time setting up the improvement groups and organising training in quality techniques. However, soon it had become swamped by the work needed to evaluate which improvement suggestions should be implemented. Soon the work load associated with assessing improvement ideas had become so great that the company decided to allocate small improvement budgets to each department on a quarterly basis that they could spend without reference to the Quality Committee. Projects requiring larger investment or that had a significant impact on other parts of the business still needed to be approved by the committee before they were implemented.

Department improvement budgets were still used within the business and improvement plans were still required from each department on an annual basis. However, the Quality Committee had stopped meeting by 1994 and the annual award ceremony had become a general communications meeting for all staff at the headquarters. *'Looking back,'* said Tyko, *'the TQM initiative faded away for three reasons. First, people just got tired of it. It was always seen as something extra rather than part of normal business life, so it was always seen as taking time away from doing your normal job. Second, many of the supervisory and middle management levels never really bought into it, I guess because they felt threatened. Third, only a very few of the local offices around the world ever adopted the TQM philosophy. Sometimes this was because they did not want the extra effort. Sometimes, however, they would argue that improvement initiatives of this type may be OK for head office processes, but not for the more dynamic world of supporting clients in the field.'*

The Six Sigma initiative

Early in 2005, Tyko Mattson, who for the last two years had been overseeing the outsourcing of some of GCR's claims processing to India, had attended a conference on 'Operations Excellence in Financial Services', and had heard several speakers detail the success they had achieved through using a Six Sigma approach to operations improvement. He had persuaded his immediate boss, Marie-Dominique Tomas, the Head of Claims for the company, to allow him to investigate its applicability to GCR. He had interviewed a number of other financial services who had implemented Six Sigma as well as a number of consultants and in September 2005 had submitted a report entitled, *'What is Six Sigma and how might it be applied in GRC?'* Extracts from this are included in the Appendix (see below). Marie-Dominique Tomas was particularly concerned that they should avoid the mistakes of the TQM initiative. *'Looking back, it is almost embarrassing to see how naive we were. We really did think that it would change the whole way that we did business. And although it did produce some benefits, it absorbed a large amount of time at all levels in the organisation. This time we want something that will deliver results without costing too much or distracting us from focusing on business performance. That is why I like Six Sigma. It starts with clarifying business objectives and works from there.'*

By late 2005, Tyko's report had been approved both by GCR and by Wichita Mutual's main board. Tyko had been given the challenge of carrying out the recommendations in his report, reporting directly to GCR's executive board. Marie-Dominique Tomas, was cautiously optimistic: *'It is quite a challenge for Tyko. Most of us on the executive board remember the TQM initiative and some are still sceptical concerning the value of such initiatives. However, Tyko's gradualist approach and his emphasis on the 'three pronged' attack on revenue, costs, and risk, impressed the board. We now have to see whether he can make it work.'*

Appendix – Extract from *What is Six Sigma and how might it be applied in GCR?*

Six Sigma – pitfalls and benefits

Some pitfalls of Six Sigma

It is not simple to implement, and is resource hungry. The focus on measurement implies that the process data is available and reasonably robust. If this is not the case, it is possible to waste a lot of effort in obtaining process performance data. It may also over-complicate things if advanced techniques are used on simple problems.

It is easier to apply Six Sigma to repetitive processes – characterised by high volume, low variety and low visibility to customers. It is more difficult to apply Six Sigma to low-volume, higher-variety and high-visibility processes where standardisation is harder to achieve and the focus is on managing the variety.

Six Sigma is not a 'quick fix'. Companies that have implemented Six Sigma effectively have not treated it as just another new initiative but as an approach that requires the long-term systematic reduction of waste. Equally, it is not a panacea and should not be implemented as one.

Some benefits of Six Sigma

Companies have achieved significant benefits in reducing cost and improving customer service through implementing Six Sigma.

Six Sigma can reduce process variation, which will have a significant impact on operational risk. It is a tried-and-tested methodology, which combines the strongest parts of existing improvement methodologies. It lends itself to being customised to fit individual company's circumstances. For example, Mestech Assurance has extended their Six Sigma initiative to examine operational risk processes.

Six Sigma could leverage a number of current initiatives. The risk self-assessment methodology, Sarbanes Oxley, the process library, and our performance metrics work are all laying the foundations for better knowledge and measurement of process data.

Six Sigma – key conclusions for GCR

Six Sigma is a powerful improvement methodology. It is not all new but what it does do successfully is to combine some of the best parts of existing improvement methodologies, tools and techniques. Six Sigma has helped many companies achieve significant benefits. It could help GCR significantly improve risk management because it focuses on driving errors and exceptions out of processes.

Six Sigma has significant advantages over other process improvement methodologies. It engages senior management actively by establishing process ownership and linkage to strategic objectives. This is seen as integral to successful implementation in the literature and by all companies interviewed who had implemented it. It forces a rigorous approach to driving out variance in processes by analysing the root cause of defects and errors and measuring improvement. It is an 'umbrella' approach, combining all the best parts of other improvement approaches.

Implementing Six Sigma across GCR is not the right approach

Companies who are widely quoted as having achieved the most significant headline benefits from Six Sigma were already relatively mature in terms of process management. Those companies, who understood their process capability, typically had achieved a degree of process standardisation and had an established process improvement culture.

Six Sigma requires significant investment in performance metrics and process knowledge. GCR is probably not yet sufficiently advanced. However, we are working towards a position where key process data are measured and known and this will provide a foundation for Six Sigma.

Why is targeted implementation recommended?

Full implementation is resource hungry. Dedicated resource and budget for implementation of improvements is required. Even if the approach is modified, resource and budget will still be needed, just to a lesser extent. However, the evidence is that the investment is well worth it and pays back relatively quickly.

There was strong evidence from companies interviewed that the best implementation approach was to pilot Six Sigma, and select failing processes for the pilot. In addition, previous internal piloting of implementations has been successful in GCR – we know this approach works within our culture.

Six Sigma would provide a platform for GSR to build on and evolve over time. It is a way of leveraging the ongoing work on processes and the risk methodology (being developed by the Operational Risk Group). This diagnostic tool could be blended into Six Sigma, giving GCR a powerful model to drive reduction in process variation and improve operational risk management.

Recommendations

It is recommended that GCR management implement a Six Sigma pilot. The characteristics of the pilot would be as follows:

- A tailored approach to Six Sigma that would fit GCR's objectives and operating environment. Implementing Six Sigma in its entirety would not be appropriate.
- The use of an external partner: GCR does not have sufficient internal Six Sigma, and external experience will be critical to tailoring the approach, and providing training.
- Establishing where GCR's sigma performance is now. Different tools and approaches will be required to advance from 2 to 3 Sigma than those required to move from 3 to 4 Sigma.
- Quantifying the potential benefits. Is the investment worth making? What would a 1 Sigma increase in performance vs. risk be worth to us?
- Keeping the methods simple, if simple will achieve our objectives. As a minimum for us that means Team Based Problem Solving and basic statistical techniques.

Next steps

1 Decide priority and confirm budget and resourcing for initial analysis to develop a Six Sigma risk improvement programme in 2006.
2 Select an external partner experienced in improvement and Six Sigma methodologies.
3 Assess GCR current state to confirm where to start in implementing Six Sigma.
4 Establish how much GCR is prepared to invest in Six Sigma and quantify the potential benefits.
5 Tailor Six Sigma to focus on risk management.
6 Identify potential pilot area(s) and criteria for assessing its suitability.
7 Develop a Six Sigma pilot plan.
8 Conduct and review the pilot programme.

QUESTIONS

1 How does the Six Sigma approach seem to differ from the TQM approach adopted by the company almost 20 years ago?
2 Is Six Sigma a better approach for this type of company?
3 Do you think Tyko can avoid the Six Sigma initiative suffering the same fate as the TQM initiative?

APPLYING THE PRINCIPLES

Hints

Some of these exercises can be answered by reading the chapter. Others will require some general knowledge of business activity and some might require an element of investigation. Hints on how they can all be answered are to be found in the eText at www.pearsoned.co.uk/slack.

1 Visit a library (for example, a university library) and consider how they could start a performance measurement programme which would enable it to judge the effectiveness with which it organises its operations. The library probably loans (if a university library) books to students on both a long-term and short-term basis, keeps an extensive stock of journals, will send off for specialist publications to specialist libraries and has an extensive online database facility. What measures of performance do you think it would be appropriate to use in this kind of operation and what type of performance standards should the library adopt?

2 (a) Devise a benchmarking programme that will benefit the course or programme that you are currently taking. In doing so, decide whether you are going to benchmark against other courses at the same institution, competitor courses at other institutions, or some other point of comparison. Also decide whether you are more interested in the performance of these other courses or the way they organise their processes, or both.

(b) Identify the institutions and courses against which you are going to benchmark your own course.

(c) Collect data on these other courses (visit them, send off for literature, or visit their website).

(d) Compare your own course against these others and draw up a list of implications for the way your course could be improved.

3 Think back to the last product or service failure that caused you some degree of inconvenience. Draw a cause–effect diagram that identifies all the main causes of why the failure could have occurred. Try and identify the frequency with which such causes happen. This could be done by talking with the staff of the operation that provided the service. Draw a Pareto diagram that indicates the relatively frequency of each cause of failure. Suggest ways in which the operation could reduce the chances of failure.

4 (a) As a group, identify a 'high visibility' operation that you all are familiar with. This could be a type of quick service restaurant, record stores, public transport systems, libraries, etc.

(b) Once you have identified the broad class of operation, visit a number of them and use your experience as customers to identify the main performance factors that are of importance to you as customers, and how each store rates against each other in terms of their performance on these same factors.

(c) Draw an importance–performance diagram for one of the operations that indicates the priority they should be giving to improving their performance.

(d) Discuss the ways in which such an operation might improve its performance and try to discuss your findings with the staff of the operation.

Notes on chapter

1 Based on a case study written by Robert Johnston, Chai Kah Hin and Jochen Wirtz, National University of Singapore, and Christopher Lovelock, Yale University. Adapted by permission.

2 Source: The EFQM website: www.efqm.org.

3 See Kaplan, R.S. and Norton, D.P. (1996), *The Balanced Scorecard*, Harvard Business School Press, Boston.

4 Kaplan, R.S. and Norton, D.P. (1996) *op. cit.*

5 Ferdows, K. and de Meyer, A. (1990) 'Lasting Improvement in Manufacturing', *Journal of Operations Management*, Vol 9, No 2. However, research for this model is mixed. For example, Patricia Nemetz questions the validity of the mode, finding more support for the idea that the sequence of improvement is generally dictated by technological (operations resource) or market (requirements) pressures: Nemetz, P. (2002) 'A longitudinal study of strategic choice, multiple advantage, cumulative model and order winner/qualifier view of manufacturing strategy', *Journal of Business and Management*, January.

6 Hammer, M. and Champy, J. (1993) *Re-engineering the Corporation*, Nicholas Brealey Publishing.

7 Hammer, M. (1990) 'Re-engineering Work: Don't Automate, Obliterate', *Harvard Business Review*, Vol 68, No 4.

8 Imai, M. (1986) *Kaizen – The Key to Japan's Competitive Success*, McGraw-Hill.

9 Bessant, J. and Caffyn, S. (1997) 'High Involvement Innovation', *International Journal of Technology Management*, Vol 14, No 1.

10 Bessant, J. and Caffyn, S. (1997) *op. cit.*

11 Imai, M. (1986), *op. cit.*

12 Pande, P.S., Neuman, R.P., and Cavanagh, R.R. (2000) *The Six Sigma Way*, McGraw-Hill, New York.

13 For further details of this approach see Schaninger, W.S., Harris, S.G. and Niebuhr, R.L. (2000) 'Adapting General Electric's Workout for use in other organizations: a template', http://www.isixsigma.com; Quinn, J. (1994) 'What a workout!' *Sales & Marketing Management, Performance Supplement* November, 58–63; Stewart, T. (1991). 'GE keeps those ideas coming', *Fortune*, 124 (4), 40–45.

TAKING IT FURTHER

Chang, R.Y. (1995) *Continuous Process Improvement: A Practical Guide to Improving Processes for Measurable Results*, Cogan Page.

Leibfried, K.H.J. and McNair, C.J. (1992) *Benchmarking: A Tool for Continuous Improvement*, HarperCollins. *There are many books on benchmarking; this is a comprehensive and practical guide to the subject.*

Pande, P.S., Neuman, R.P. and Cavanagh, R. (2002) *Six Sigma Way Team Field Book: An Implementation Guide for Project Improvement Teams*, McGraw Hill. *Obviously based on the Six Sigma principle and related to the book by the same author team recommended in Chapter 12, this is an unashamedly practical guide to the Six Sigma approach.*

USEFUL WEBSITES

www.opsman.org *Definitions, links and opinions on operations and process management.*

http://www.processimprovement.com/ *Commercial site but some content that could be useful.*

http://www.kaizen.com/ *Professional institute for kaizen. Gives some insight into practitioner views.*

http://www.ebenchmarking.com *Benchmarking information.*

http://www.quality.nist.gov/ *American Quality Assurance Institute. Well established institution for all types of business quality assurance.*

http://www.balancedscorecard.org/ *Site of an American organisation with plenty of useful links.*

Interactive quiz

For further resources including examples, animated diagrams, self-test questions, Excel spreadsheets and video materials please explore the eText on the companion website at **www.pearsoned.co.uk/slack**.

14

Risk and resilience

Introduction

One obvious way of improving operations performance is by reducing the risk of failure (or of failure causing disruption) within the operation. All operations are subject to the risk of failure of many types, for example, technology failure, supplier failure, natural and man-made disasters, and many other causes. A 'resilient' operation or process is one that can prevent failures occurring, minimise their effects, and learn how to recover from them. In an increasingly risky economic, political and social environment, resilience has become an important part of operations and process management, and in some operations, airplanes in flight or electricity supplies to hospitals, or the emergency services where failure can be literally fatal, it is vital (see Figure 14.1).

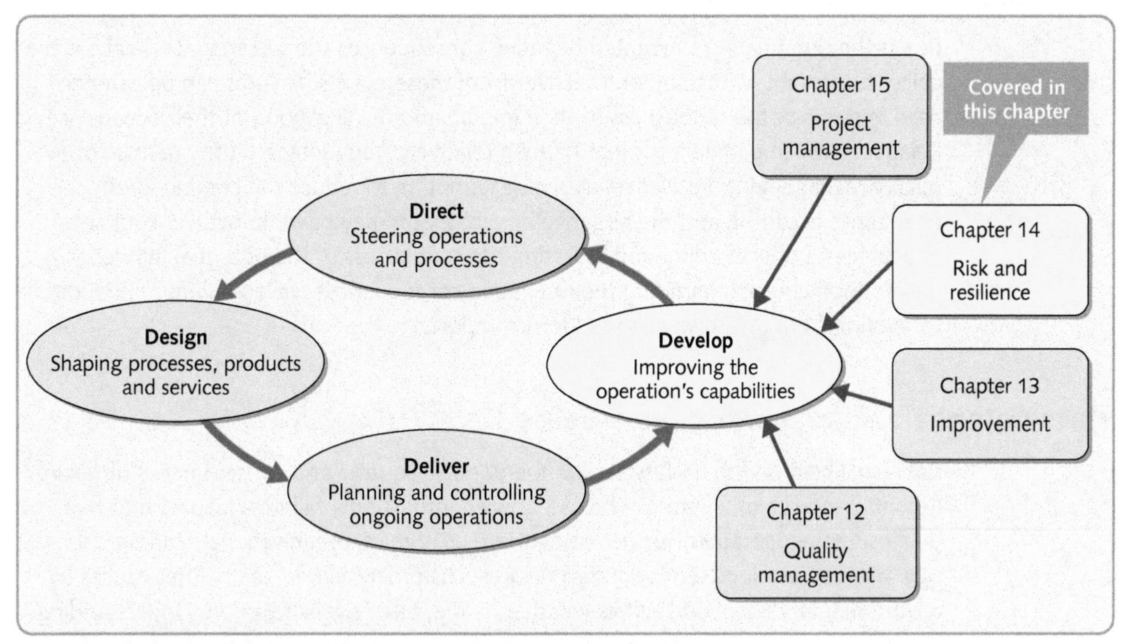

Figure 14.1 Risk is the potential for unwanted negative consequences from events; resilience is the ability to prevent, mitigate and recover from these events

EXECUTIVE SUMMARY

Each chapter is structured around a set of diagnostic questions. These questions suggest what you should ask in order to gain an understanding of the important issues of a topic, and, as a result, improve your decision making. An executive summary addressing these questions is provided below.

What are risk and resilience?

Risk is the potential for unwanted negative consequences from events. Resilience is the ability to prevent, withstand and recover from those events. Failures can be categorised in terms of the seriousness of their impact and the likelihood of their occurrence. Relatively low-impact failures that happen relatively frequently are the province of quality management. Resilience involves attempting to reduce the combined effects of a failure occurring and the negative impact that it may have. It involves four sets of activities: understanding and assessing the seriousness of the potential failures; preventing failures; minimising their negative consequences (called failure mitigation); recovering from failure so as to reduce its impact.

Have potential failure points been assessed?

Resilience begins with understanding the possible sources and consequences of failure. Potential sources of failure can be categorised into: supply failures; failures that happen inside the operation (further categorised as human, organisational, and technology failures); product/service design failures; customer failures; the failures causes by environmental disruption such as weather, crime, terrorism, and so on. Understanding why such failures occur can be aided by post-failure analysis using accident investigation, traceability, complaint analysis, fault tree analysis, and other similar techniques.

Judging the likelihood of failure may be relatively straightforward for some well-understood processes, but often has to be carried out on a subjective basis, which is rarely straightforward.

Have failure prevention measures been implemented?

Failure prevention is based on the assumption that it is generally better to avoid failures than suffer their consequences. The main approaches to failure prevention involve designing out the possibility of failure at key points in a process, providing extra but redundant resources that can provide back-up in the case of failure, installing fail-safeing mechanisms that prevent the mistakes which can cause failure, and maintaining processes so as to reduce the likelihood of failure. This last category is particularly important for many operations but can be approached in different ways such as preventive maintenance, condition-based maintenance, and total product maintenance.

Have failure mitigation measures been implemented?

Failure mitigation means isolating a failure from its negative consequences. It can involve a number of 'mitigation actions'. These include: ensuring that planning procedures are installed that provide an overarching guide to mitigation as well as demonstrating that failure is taken seriously; economic mitigation using insurance; risk sharing and hedging; spatial or temporal containment that prevents the failure spreading geographically or over time; loss reduction that removes whatever might be harmed by a failure; substitution that involves providing substitute resources to work on a failure before it becomes serious.

Have failure recovery measures been implemented?

Failure recovery is the set of actions that are taken after the negative effects of failure have occurred and that reduce the impact of the negative effects. Sometimes recovering well from a public failure can even enhance a business's reputation. However, recovery does need to be planned and procedures put in place that can discover when failures have occurred, guide appropriate action to keep everyone informed, capture the lessons learnt from the failure, and plan to absorb the lessons into any future recovery.

DIAGNOSTIC QUESTION

What are risk and resilience?

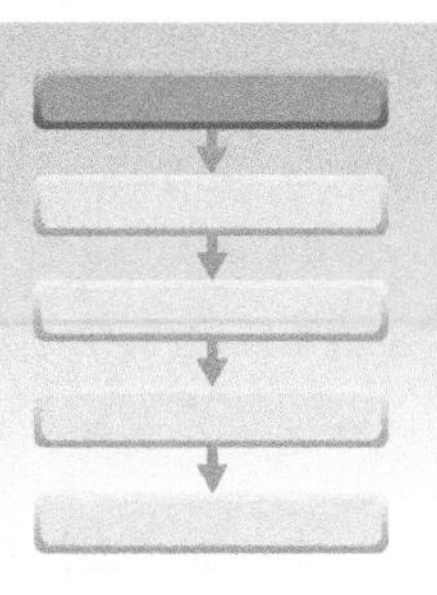

Video

Risk is the potential for unwanted negative consequences from some event. Resilience is the ability to prevent, withstand and recover from those events. Things happen in operations, or to operations, that have negative consequences, this is failure. But accepting that failure occurs is not the same thing as accepting or ignoring it. Operations do generally attempt to minimise both the likelihood of failure and the effect it will have, although the method of coping with failure will depend on how serious its negative consequences are, and how likely it is to occur. At a minor level, every small error in the delivered product or service from the operation could be considered a failure. The whole area of quality management is concerned with reducing this type of 'failure'. Other failures will have more impact on the operation, even if they do not occur very frequently. A server failure can seriously affect service and therefore customers, which is why system reliability is such an important measure of performance for IT service providers. And, if we class a failure as something that has negative consequences, some are so serious we class them as disasters, such as freak weather conditions, air crashes, and acts of terrorism. These 'failures' are treated increasingly seriously by businesses, not necessarily because their likelihood of occurrence is high (although it may be at certain times and in certain places), but because their impact is so negative.

> **OPERATIONS PRINCIPLE**
> *Failure will always occur in operations; recognising this does not imply accepting or ignoring it.*

This chapter is concerned with all types of failure other than those with relatively minor consequences. This is illustrated in Figure 14.2. Some of these failures are irritating, but relatively

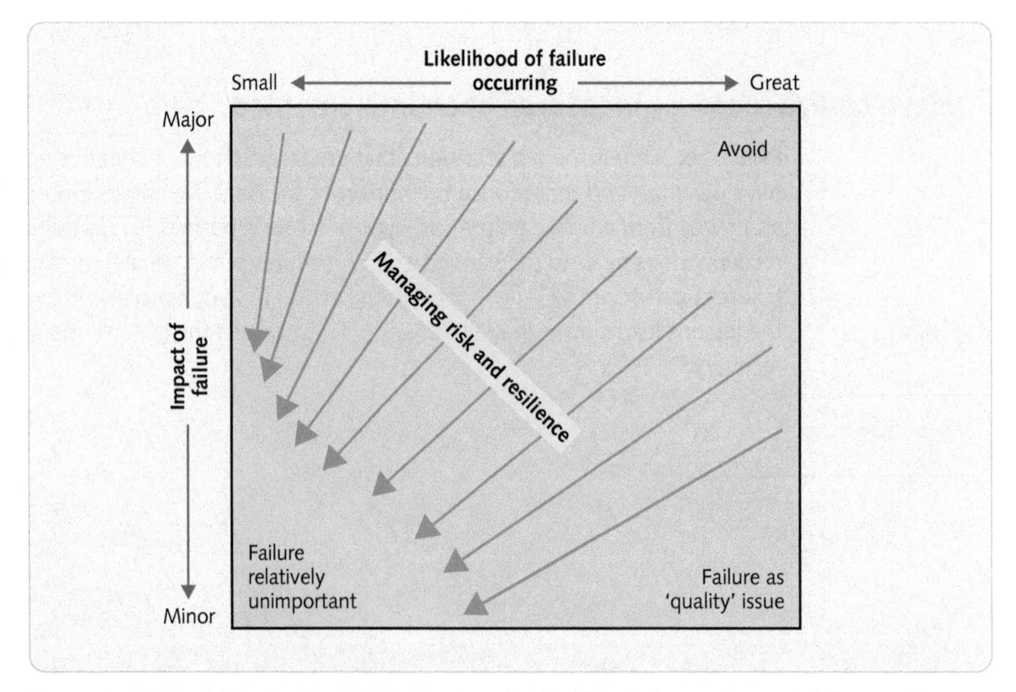

Figure 14.2 How failure is managed depends on its likelihood of occurrence and the negative consequence of failure

unimportant, especially those close to the bottom left-hand corner of the matrix in Figure 14.2. Other failures, especially those close to the top right-hand corner of matrix, are normally avoided by all business because embracing such risks would be clearly foolish. In between these two extremes is where most operations-related risks occur. In this chapter we shall be treating various aspects of these types of failure, and in particular how they can be moved in the direction of arrows in Figure 14.2.

EXAMPLE

Supply risk at Gap Inc.[1]

Gap Inc. is a $15.9 billion leading international retailer selling clothing, accessories and personal care products. Its brands include Banana Republic, Old Navy and Piperlime, but it is best known for its international chain of Gap stores throughout the United States, United Kingdom, Canada, France, Ireland and Japan, as well as Asia and the Middle East. The countries from which it sources its products are similarly international, from Sri Lanka to Lesotho, the United States to El Salvador. But an international supply base carries some significant risks. In October 2007, evidence appeared in the *Observer*, a British newspaper, that an unauthorised subcontractor had used child workers to make blouses for GapKids at a factory in Delhi. In response, Gap immediately issued a statement:

Source: Ian Dagnall/Alamy Images

'Earlier this week, the company was informed about an allegation of child labour at a facility in India that was working on one product for GapKids. An investigation was immediately launched. The company noted that a very small portion of a particular order placed with one of its vendors was apparently subcontracted to an unauthorised subcontractor without the company's knowledge or approval. This is in direct violation of the company's agreement with the vendor under its Code of Vendor Conduct. Marka Hansen, president of Gap North America, made the following statement today:

"We strictly prohibit the use of child labour. This is a non-negotiable for us – and we are deeply concerned and upset by this allegation. As we've demonstrated in the past, Gap has a history of addressing challenges like this head on, and our approach to this situation will be no exception. In 2006, Gap Inc. ceased business with 23 factories due to code violations. We have 90 people located around the world whose job is to ensure compliance with our Code of Vendor Conduct. As soon as we were alerted to this situation, we stopped the work order and prevented the product from being sold in stores. While violations of our strict prohibition on child labour in factories that produce products for the company are extremely rare, we have called an urgent meeting with our suppliers in the region to reinforce our policies. Gap Inc. has one of the industry's most comprehensive programs in place to fight for workers' rights overseas. We will continue to work with the government, NGOs, trade unions, and other stakeholder organisations in an effort to end the use of child labour."'

Gap's new policy on violations of its child-labour rules may help to limit the damage to its reputation as one of the most ethical retailers. Rather than immediately closing supplier factories that employ child workers, it now stops suppliers using children, continues to pay them, but insists that suppliers provide them with an education and guarantees them a job once they reach the legal age.

For any company sourcing its products around the world, maintaining a rigorous monitoring regime on faraway subcontractors is never going to be easy. Many of Gap's clothes are made in India, where an estimated 50 million children are employed. Yet the International Labour Organisation of the United Nations believes all companies could do more. *'If companies are capable of supervising the quality of their products, they should also be able to police their production,'* says Geir Myrstad, head of the ILO's programme to eliminate child labour.

EXAMPLE

Cadbury's salmonella outbreak[2]

In June 2007, Cadbury, founded by a Quaker family in 1824, and part of Cadbury Schweppes, one of the world's biggest confectionary companies, was fined £1 million plus costs of £152,000 for breaching food safety laws. In a national salmonella outbreak, 42 people, including children aged under ten, became ill with a rare strain of Salmonella montevideo. *'I regard this as a serious case of negligence,'* the judge said. *'It therefore needs to be marked as such to emphasise the responsibility and care which the law requires of a company in Cadbury's position.'* One prominent lawyer announced that *'Despite Cadbury's attempts to play down this significant fine, make no mistake it was intended to hurt and is one of the largest of its kind to date. This reflects no doubt the company's high profile and the length of time over which the admitted breach took place, but will also send out a blunt warning to smaller businesses of the government's intentions regarding enforcement of food safety laws.'*

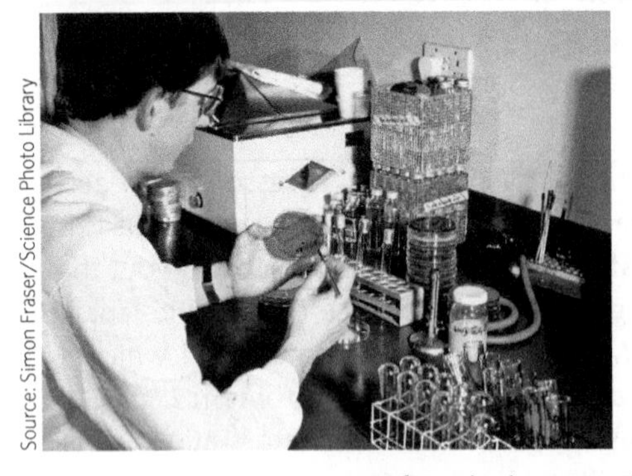
Source: Simon Fraser/Science Photo Library

Before the hearing, the company had, in fact, apologised offering its 'sincere regrets' to those affected, and pleaded guilty to nine food safety offences. But at the beginning of the incident it had not been so open. Although Cadbury said it had co-operated fully with the investigation, it admitted that it had failed to notify the authorities of positive tests for salmonella as soon as they were known within the company. While admitting its mistakes, a spokesman for the confectioner emphasised that the company had acted in good faith, a point supported by the judge when he dismissed prosecution suggestions that Cadbury had introduced the procedural changes that led to the outbreak simply as a cost-cutting measure. Cadbury, through its lawyers, said: *'Negligence we admit, but we certainly do not admit that this was done deliberately to save money and nor is there any evidence to support that conclusion.'* The judge said Cadbury had accepted that a new testing system, originally introduced to improve safety, was a *'distinct departure from previous practice'*, and was *'badly flawed and wrong'*. In a statement Cadbury said: *'Mistakenly, we did not believe that there was a threat to health and thus any requirement to report the incident to the authorities – we accept that this approach was incorrect. The processes that led to this failure ceased from June last year and will never be reinstated.'*

The company was not only hit by the fine and court costs, it had to bear the costs of recalling one million bars that may have been contaminated from sale, and face private litigation claims brought by its consumers who were affected. Cadbury said it lost around £30 million because of the recall and subsequent safety modifications, not including any private litigation claims. *The Times* reported on the case of Shaun Garratty, one of the people affected. A senior staff nurse, from Rotherham, he spent seven weeks in hospital critically ill and now he fears that his nursing career might be in jeopardy. *The Times* reported him as being 'pleased that Cadbury's had admitted guilt but now wants to know what the firm is going to do for him.' Before the incident, it said, he was a fitness fanatic and went hiking, cycling, mountain biking or swimming twice a week. He always took two bars of chocolate on the trips, usually a Cadbury's Dairy Milk and a Cadbury's Caramel bar. He also ate one as a snack each day at work. *'My gastroenterologist told me if I had not been so fit I would have died,'* said Mr. Garratty. *'Six weeks after being in hospital they thought my bowel had perforated and I had to have a laparoscopy. I was told my intestines were inflamed and swollen.'* Even since returning to work he has not fully recovered. According to one medical consultant, the illness had left him with a form of irritable bowel syndrome that he could take 18 months to recover from.

What do these two examples have in common?

Both of these operations suffered failure, in the sense that failure means a disruption to normal operation. But both companies knew that they risked the type of failure that actually occurred and both companies had procedures in place that attempted to prevent failure occurring.

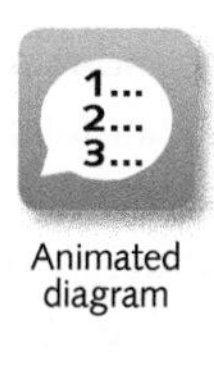

Animated diagram

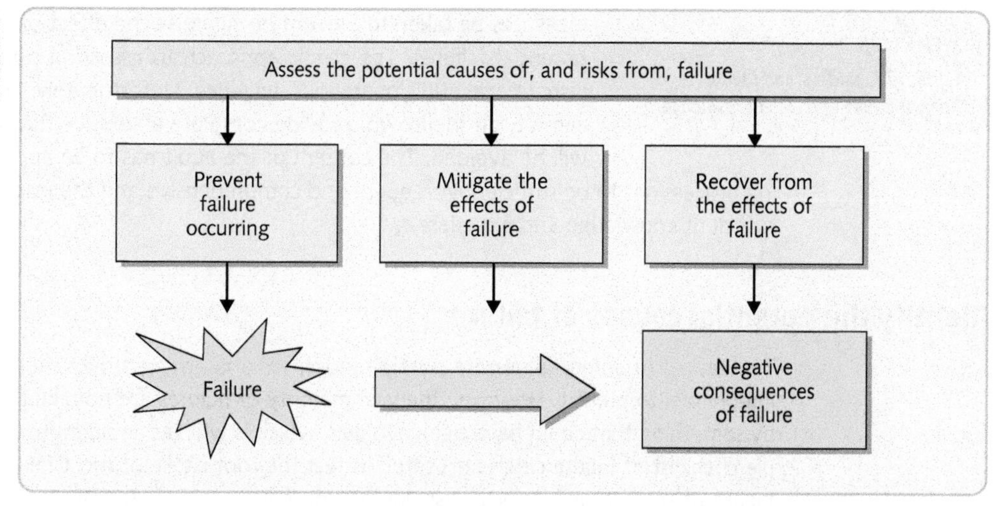

Figure 14.3 Operations and process resilience involves failure prevention, mitigating the negative consequences of failure, and failure recovery

However, notwithstanding their efforts, both companies suffered financial and reputational damage because of the failures. Also both companies were accused of, in effect, causing the failure because they were trying to save money. The difference between the two examples is that Gap's failure was external (in its suppliers' operations), whereas Cadbury's was internal (in its own processes). Yet, whether internal or external, no business can afford to ignore the risks inherent in their activities. Nor can they ignore the potential of cost cutting to affect the risks they are subjecting themselves to. Of course all business (perfectly legitimately) want to cut costs. But doing so without a good understanding of how risk is affected is potentially extremely damaging.

Obviously, some operations operate in a more risky environment than others. And, those operations with a high likelihood of failure and/or serious consequences deriving from that failure will need to give it more attention, but operations and process resilience is relevant to all businesses. They must all give attention to the four sets of activities which, in practical terms, resilience includes. The first is concerned with understanding what failures could potentially occur in the operation and assessing their seriousness. The second task is to examine ways of preventing failures occurring. The third is to minimise the negative consequences of failure (called failure or risk 'mitigation'). The final task is to devise plans and procedures that will help the operation to recover from failures when they do occur. The remainder of this chapter deals with these four tasks (see Figure 14.3).

> OPERATIONS PRINCIPLE
>
> *Resilience is governed by the effectiveness of failure prevention, mitigation and recovery.*

DIAGNOSTIC QUESTION

Have potential failure points been assessed?

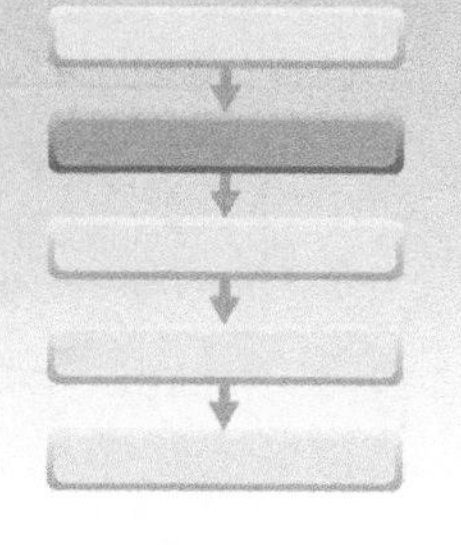

Video

A prerequisite to achieving operations and process resilience is to understand where failure might occur and what the consequences of failure might be, by reviewing all possible causes of failure. Often it is the 'failure to understand failure' that leads to excessive disruption. Each cause of failure also needs to be assessed in terms of the impact it may have. Only then can

OPERATIONS PRINCIPLE
A 'failure to understand failure' is the root cause of a lack of resilience.

measures be taken to prevent or minimise the effect of the more important potential failures. The classic approach to assessing potential failures is to inspect and audit operations activities. Unfortunately, inspection and audit cannot, on their own, provide complete assurance that undesirable events will be avoided. The content of the audit has to be appropriate, the checking process has to be sufficiently frequent and comprehensive and the inspectors have to have sufficient knowledge and experience.

Identify the potential causes of failure

The causes of some failure are purely random, like lightning strikes, and are difficult, if not impossible, to predict. However, the vast majority of failures are not like this. They are caused by something that could have been avoided, which is why, as a minimum starting point, a simple checklist of failure causes is useful. In fact the root cause of most failure is usually human failure of some type; nevertheless, identifying failure sources usually requires a more evident set, such as that illustrated in Figure 14.4. Here, failure sources are classified as: failures of supply; internal failures such as those deriving from human organisational and technological sources; failures deriving from the design of products and services; failures deriving from customer failures; general environmental failures.

Supply failure

Supply failure means any failure in the timing or quality of goods and services delivered into an operation. For example, suppliers delivering the wrong or faulty components, outsourced call centres suffering a telecoms failure, disruption to power supplies, and so on. The more an operation relies on suppliers for materials or services, the more it is at risk from failure caused by missing or sub-standard inputs. It is an important source of failure because of increasing dependence on outsourced activities in most industries, and the emphasis on keeping supply chains 'lean' in order to cut costs. For example, in early 2002 Land Rover then a division of the Ford Motor Company) had to cope with a threat to the supply of chassis for its Discovery model when the single company to which they had subcontracted its manufacture became insolvent. The receivers were demanding an up-front payment of around €60 million to continue supply, arguing that they were legally obliged to recover as much money as possible on behalf of creditors. And a single supplier agreement was a valuable asset.

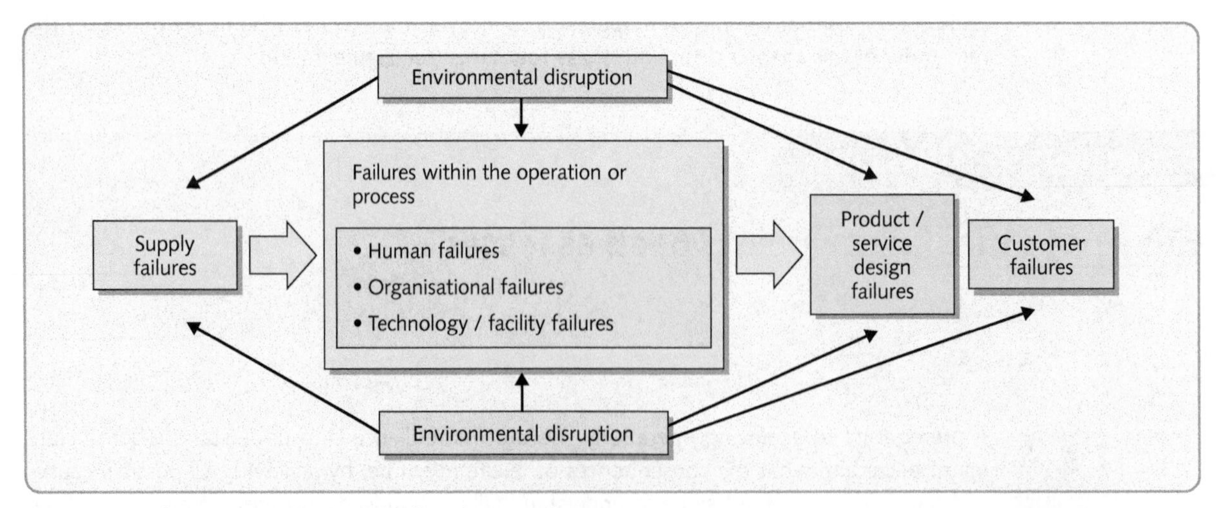

Figure 14.4 The sources of potential failure in operations

In this case, the outsourcing of a component had made the supply chain more vulnerable. But there are also other factors which have, in recent years, increased the vulnerability of supply. For example, global sourcing usually means that parts are shipped around the world on their journey through the supply chain. Micro chips manufactured in Taiwan could be assembled to printed circuit boards in Shanghai which are then finally assembled into a computer in Ireland. At the same time, many industries are suffering increased volatility in demand. Perhaps most significantly, there tends to be far less inventory in supply chains that could buffer interruptions to supply. According to one authority on supply chain management, *'Potentially the risk of disruption has increased dramatically as the result of a too-narrow focus on supply chain efficiency at the expense of effectiveness.'*[3]

Human failures

There are two broad types of human failure. The first is where key personnel leave, become ill, die, or in some way cannot fulfil their role. The second is where people are actively doing their job but are making mistakes. Understanding risk in the first type of failure involves identifying the key people without whom operations would struggle to operate effectively. These are not always the most senior individuals, but rather those fulfilling crucial roles that require special skills or tacit knowledge. Human failure through 'mistakes' also comes in two types: errors and violations. 'Errors' are mistakes in judgement; with hindsight, a person should have done something different. For example, if the manager of a sports stadium fails to anticipate dangerous crowding during a championship event. This is an error of judgement. 'Violations' are acts which are clearly contrary to defined operating procedure. For example, if a maintenance engineer fails to clean a filter in the prescribed manner, it is eventually likely to cause failure. Catastrophic failures are often caused by a combination of errors and violations, For example, one kind of accident, where an aircraft appears to be under control and yet still flies into the ground, is very rare (once in two million flights).[4] For this type of failure to occur, first, the pilot has to be flying at the wrong altitude (error). Second, the co-pilot would have to fail to cross-check the altitude (violation). Third, air traffic controllers would have to miss the fact that the plane was at the wrong altitude (error). Finally, the pilot would have to ignore the ground proximity warning alarm in the aircraft, which can be prone to give false alarms (violation).

Organisational failure

Organisational failure is usually taken to mean failures of procedures and processes and failures that derive from a business's organisational structure and culture. This is a huge potential source of failure and includes almost all operations and process management. In particular, failure in the design of processes (such as bottlenecks causing system overloading) and failures in the resourcing of processes (such as insufficient capacity being provided at peak times) need to be investigated. But there are also many other procedures and processes within an organisation that can make failure more likely. For example, remuneration policy may motivate staff to work in a way that, although increasing the financial performance of the organisation, also increases its susceptibility to failure. Examples of this can range from sales people being so incentivised that they make promises to customers that cannot be fulfilled, through to investment bankers being more concerned with profit than the risks of financial overexposure. This type of risk can derive from an organisational culture that minimises consideration of risk, or it may come from a lack of clarity in reporting relationships.

Technology/facilities failures

By 'technology and facilities' we mean all the IT systems, machines, equipment and buildings of an operation. All are liable to failure, or breakdown. The failure may be only partial, for example a machine that has an intermittent fault. Alternatively, it can be what we normally regard as a 'breakdown – a total and sudden cessation of operation. Either way, its effects could bring a large part of the operation to a halt. For example, a computer failure in a supermarket chain could paralyse several large stores until it is fixed.

Product/service design failures

In its design stage, a product or service might look fine on paper; only when it has to cope with real circumstances might inadequacies become evident. Of course, during the design process, potential risk of failure should have been identified and 'designed out'. But one only has to look at the number of 'product recalls' or service failures to understand that design failures are far from uncommon. Sometimes this is the result of a trade-off between fast time-to-market performance and the risk of the product or service failing in operation. And, while no reputable business would deliberately market flawed products or services, equally most businesses can not delay a product or service launch indefinitely to eliminate every single small risk of failure.

Customer failures

Not all failures are (directly) caused by the operation or its suppliers. Customers may 'fail' in that they misuse products and services. For example, an IT system might have been well designed, yet the user could treat it in a way that causes it to fail. Customers are not 'always right'; they can be inattentive and incompetent. However, merely complaining about customers is unlikely to reduce the chances of this type of failure occurring. Most organisations will accept that they have a responsibility to educate and train customers, and to design their products and services so as to minimise the chances of failure.

Environmental disruption

Environmental disruption includes all the causes of failure that lie outside of an operation's direct influence. This source of potential failure has risen to near the top of many firms' agenda since 11 September 2001. As operations become increasingly integrated (and increasingly dependent on integrated technologies such as information technologies), businesses are more aware of the critical events and malfunctions that have the potential to interrupt normal business activity and even stop the entire company. Typically, such disasters include the following:

- hurricanes, floods, lightning, temperature extremes
- fire
- corporate crime, theft, fraud, sabotage
- terrorism, bomb blast, bomb scare or other security attacks
- contamination of product or processes.

EXAMPLE

Slamming the door[5]

On 25 January 2003 the 'SQL Slammer' worm, a rogue program spread at frightening speed throughout the internet. It disrupted computers around the world and, at the height of the attack, its affect was such that half the traffic over the internet was being lost (see Figure 14.5). Thousands of cash dispensers in North America ceased operating and one police force was driven back to using pencils and paper when their dispatching system crashed. Yet security experts believe that the SQL Slammer did more good than harm because it highlighted weaknesses in internet security processes. Like most rogue software, it exploited a flaw in a commonly used piece of software. Much commonly used software has security flaws that can be exploited in this way. Software producers issue 'patch' software to fix flaws but this can actually direct internet terrorists to vulnerable areas in the software, and not all systems managers get around to implementing all patches. Nevertheless, every rogue program that penetrates internet security systems teaches those working to prevent security failures valuable lessons.

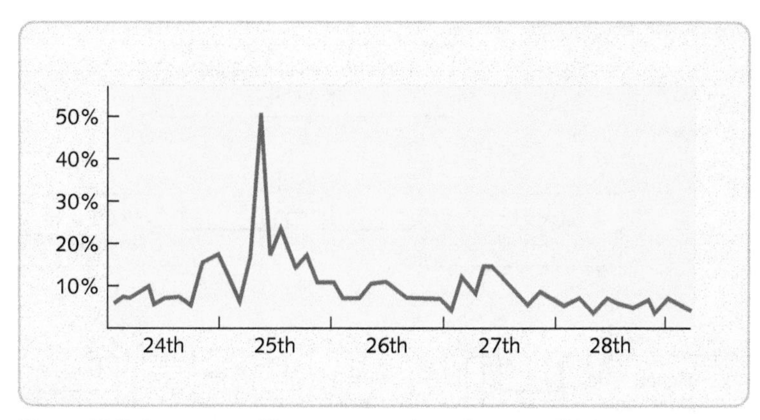

Figure 14.5 Internet traffic percentage loss, January 2003

Detecting non-evident failure

Not all failures are immediately evident. Small failures may be accumulating for a while before they become evident. Purchasing officers encountering difficulties in using a website may cause them to abandon the purchase. Within an automated materials handling line, debris may periodically accumulate that, in itself, will not cause immediate failure, but could eventually lead to sudden and dramatic failure. Even when such failures are detected, they may not always receive the appropriate attention because there are inadequate failure identification systems, or a lack of managerial support or interest in making improvements. The mechanisms available to seek out failures in a proactive way include: machine diagnostic checks; in-process checks; point-of-departure and phone interviews; customer focus groups.

Post-failure analysis

One of the critical activities of operations and process resilience is to understand why a failure has occurred. This activity is called post-failure analysis. It is used to uncover the root cause of failures. Some techniques for this were described as 'improvement techniques' in Chapter 13. Others include the following:

- **Accident investigation.** Large-scale national disasters like oil tanker spillages and aeroplane accidents are usually investigated using accident investigation, where specifically trained staff analyse the causes of the accident.
- **Failure traceability.** Some businesses (either by choice or because of a legal requirement) adopt traceability procedures to ensure that all their failures (such as contaminated food products) are traceable. Any failures can be traced back to the process which produced them, the components from which they were produced, or the suppliers who provided them.
- **Complaint analysis.** Complaints (and compliments) are a potentially valuable source for detecting the root causes of failures of customer service. Two key advantages of complaints are that they come unsolicited and also they are often very timely pieces of information that can pinpoint problems quickly. Complaint analysis also involves tracking the actual number of complaints over time, which can in itself be indicative of developing problems. The prime function of complaint analysis involves analysing the 'content' of the complaints to understand better the nature of the failure as it is perceived by the customer.
- **Fault-tree analysis.** This is a logical procedure that starts with a failure or a potential failure and works backwards to identify all the possible causes and therefore the origins of that failure. Fault-tree analysis is made up of branches connected by two types of nodes: AND nodes and OR nodes. The branches below an AND node all need to occur for the

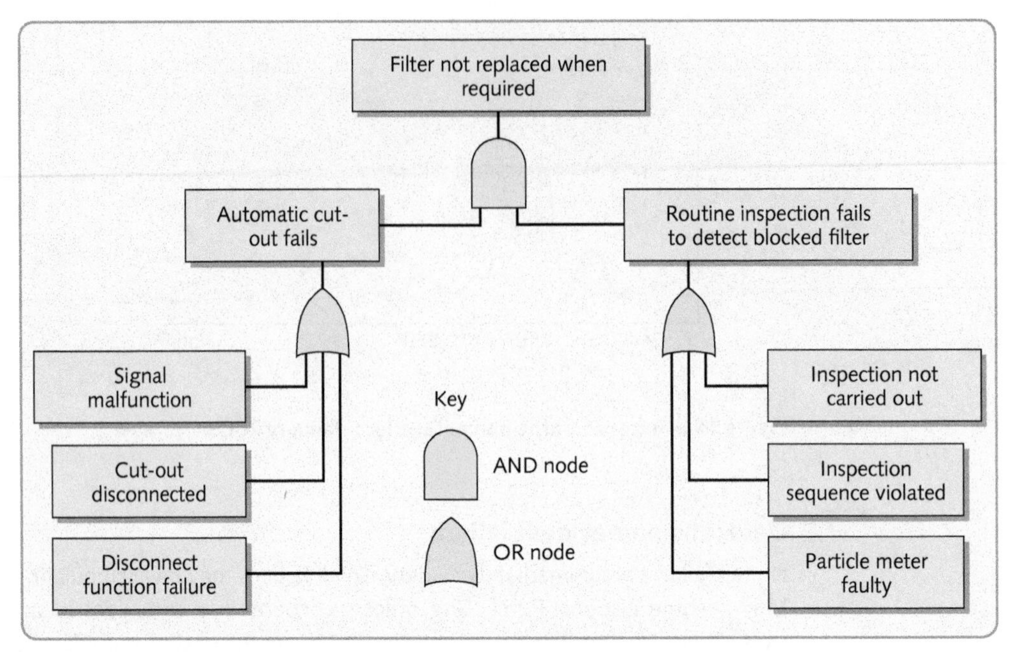

Figure 14.6 Fault-tree analysis for failure to replace filter when required

event above the node to occur. Only one of the branches below an OR node needs to occur for the event above the node to occur. Figure 14.6 shows a simple tree identifying the possible reasons for a filter in a heating system not being replaced when it should have been.

Likelihood of failure

The difficulty of estimating the chance of a failure occurring varies greatly. Some failures are the result of well-understood phenomena. A combination of rational causal analysis and historical performance data can lead to a relatively accurate estimate of failure occurring. For example, a mechanical component may fail between 10 and 17 months of its installation in 99 per cent of cases. Other types of failure are far more difficult to predict. The chances of a fire in a supplier's plant are (hopefully) low, but how low? There will be some data concerning fire hazards in this type of plant, and one may insist on regular hazard inspection reports from the supplier's insurance providers, but the estimated probability of failure will be both low and subjective.

'Objective' estimates

Estimates of failure based on historical performance can be measured in several ways, including:

- *failure rates* – how often a failure occurs
- *reliability* – the chances of a failure occurring
- *availability* – the amount of available useful operating time.

'Failure rate' and 'reliability' are different ways of measuring the same thing – the propensity of an operation, or part of an operation, to fail. Availability is one measure of the consequences of failure in the operation.

Sometimes failure is a function of time. For example, the probability of an electric lamp failing is relatively high when it is first used, but if it survives this initial stage, it could still fail at

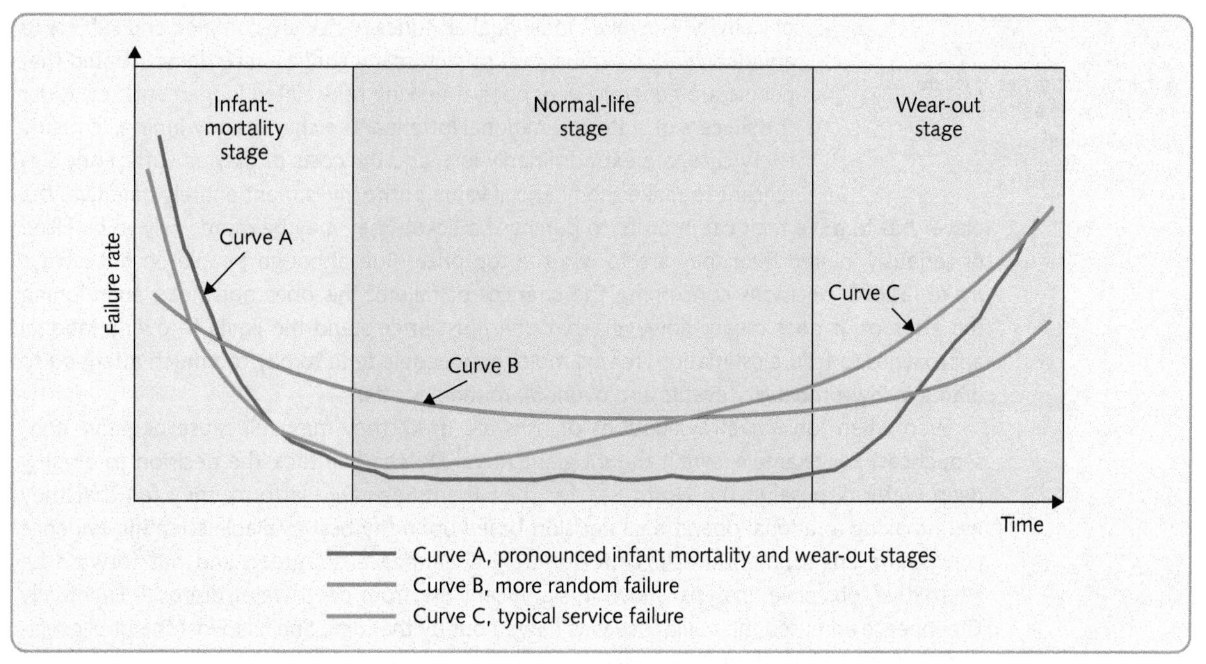

Figure 14.7 Bath-tub curves for three types of process

any point and the longer it survives, the more likely its failure becomes. Most physical parts of an operation behave in a similar manner. The curve which describes failure probability of this type is called the bath-tub curve. It comprises three distinct stages:

- the 'infant-mortality' or 'early-life' stage where early failures occur caused by defective parts or improper use
- the 'normal-life' stage when the failure rate is usually low and reasonably constant, and caused by normal random factors
- the 'wear-out' stage when the failure rate increases as the part approaches the end of its working life and failure is caused by the ageing and deterioration of parts.

Figure 14.7 illustrates three bath-tub curves with slightly different characteristics. Curve A shows a part of the operation which has a high initial infant mortality failure but then a long, low-failure, normal life followed by the gradually increasing likelihood of failure as it approaches wear-out. Curve B, while having the same stages, is far less predictable. The distinction between the three stages is less clear, with infant mortality failure subsiding only slowly and a gradually increasing chance of wear-out failure. Failure of the type shown in curve B is far more difficult to manage in a planned manner. The failure of operations which rely more on human resources than on technology, such as some services, can be closer to curve C of Figure 14.7. They may be less susceptible to component wear-out but more so to staff complacency. Without review and regeneration, the service may become tedious and repetitive, and after an initial stage of failure reduction, as problems in the service are ironed out, there can be a long period of increasing failure.

'Subjective' estimates

Failure assessment, even for subjective risks, is increasingly a formal exercise that is carried out using standard frameworks, often prompted by health and safety concerns, environmental regulations, and so on. These frameworks are similar to the formal quality inspection methods associated with quality standards like ISO 9000 that often implicitly assume unbiased

OPERATIONS PRINCIPLE
Subjective estimates of failure probability are better than no estimates at all.

objectivity. However, individual attitudes to risk are complex and subject to a wide variety of influences. In fact, many studies have demonstrated that people are generally very poor at making risk-related judgements. Consider the success of state and national lotteries. The chances of winning, in nearly every case, are extraordinarily low, and the costs of playing sufficiently significant to make the financial value of the investment entirely negative. If a player has to drive their car in order to purchase a ticket, they may be more likely to be killed or seriously injured than they are to win the top prize. But, although people do not always make rational decisions concerning the chances of failure, this does not mean abandoning the attempt. It does mean, however, that one must understand the limits to overly rational approaches to failure estimation, for example, how people tend to pay too much attention to dramatic low-probability events and overlook routine events.[6]

Even when 'objective' evaluations of risks are used, they may still cause negative consequences. For example, when the oil giant Royal-Dutch Shell took the decision to employ deep-water disposal in the North Sea for their Brent Spar Oil Platform, they felt that they were making a rational operational decision based upon the best available scientific evidence concerning environmental risk. Unfortunately Greenpeace disagreed and put forward an alternative 'objective analysis' showing significant risk from deep-water disposal. Eventually Greenpeace admitted their evidence was flawed but by that time Shell had lost the public relations battle and had altered their plans.

Failure mode and effect analysis

Practice note

One of the best known approaches to assessing the relative significance of failure is failure mode and effect analysis (FMEA). Its objective is to identify the factors that are critical to various types of failure as a means of identifying failures before they happen. It does this by providing a 'checklist' procedure built around three key questions for each possible cause of failure:

- What is the likelihood that failure will occur?
- What would the consequence of the failure be?
- How likely is such a failure to be detected before it affects the customer?

Based on a quantitative evaluation of these three questions, a risk priority number (RPN) is calculated for each potential cause of failure. Corrective actions, aimed at preventing failure, are then applied to those causes whose RPN indicates that they warrant priority (see Figure 14.8).

Animated diagram

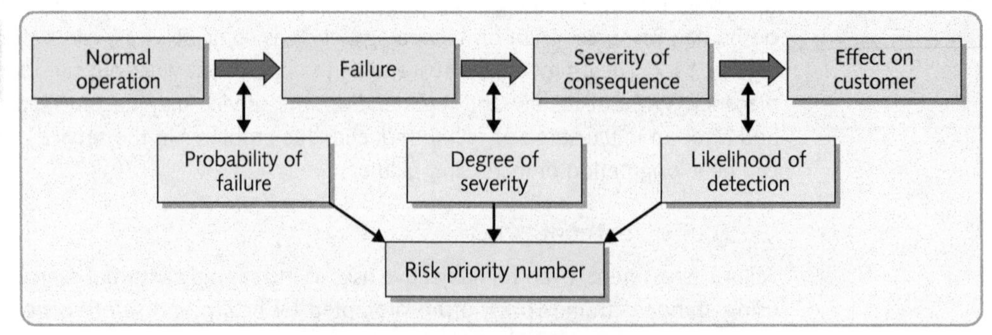

Figure 14.8 Procedure for failure modes effects analysis (FMEA)

DIAGNOSTIC QUESTION

Have failure prevention measures been implemented?

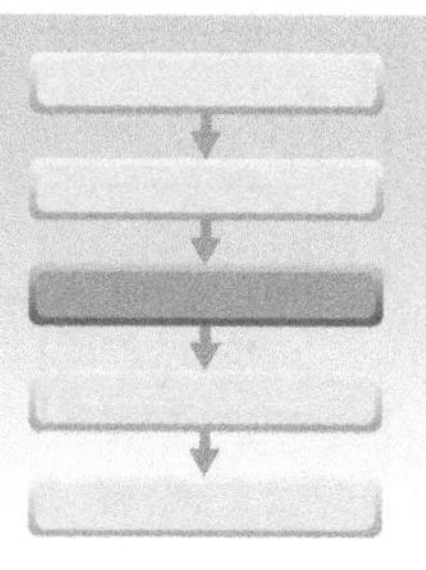

Video

It is almost always better to avoid failures and negative consequences than have to recover from them, which is why failure prevention is an important part of operations and process resilience. There are a number of approaches to this, including designing out failure points, deploying redundant resources, fail-safing and maintenance.

Designing out fail points

Process mapping, described in Chapter 5, can be used to 'engineer out' the potential fail points in operations. For example, Figure 14.9 shows a process map for an automobile repair process. The stages in the process that are particularly prone to failure and the stages which are critical to the success of the service have been marked. This will have been done by the staff of this operation metaphorically 'walking themselves through' the process and discussing each stage in turn.

Redundancy

Building in redundancy to an operation means having back-up processes or resources in case of failure. It can be an expensive solution to reduce the likelihood of the failure and is generally used when the breakdown could have a critical impact. Redundancy means doubling or even tripling some of the elements in a process so that these 'redundant' elements can come into action when one component fails. Nuclear power stations, hospitals and other public buildings have auxiliary or back-up electricity generators ready to operate in case the main electricity supply should fail. Some organisations also have 'back-up' staff held in reserve in case someone does not turn up for work or is held up on one job and is unable to move on to the next. Spacecraft have several back-up computers on board that will not only monitor the main computer but also act as a back-up in case of failure. Human bodies contain two of some organs – kidneys and eyes, for example – both of which are used in 'normal operation' but the body can cope with a failure in one of them. One response to the threat of large failures, such as terrorist activity, has been a rise in the number of companies offering 'replacement office' operations, fully equipped with normal internet and telephone communications links, and often with access to a company's current management information. Should a customer's main operation be affected by a disaster, business can continue in the replacement facility within days or even hours.

The effect of redundancy can be calculated by the sum of the reliability of the original process component and the likelihood that the back-up component will both be needed and be working.

$$R_{\mathrm{a+b}} = R_{\mathrm{a}} + (R_{\mathrm{b}} \times P_{\mathrm{(failure)}})$$

where R_{a+b} = reliability of component *a* with its back-up component *b*

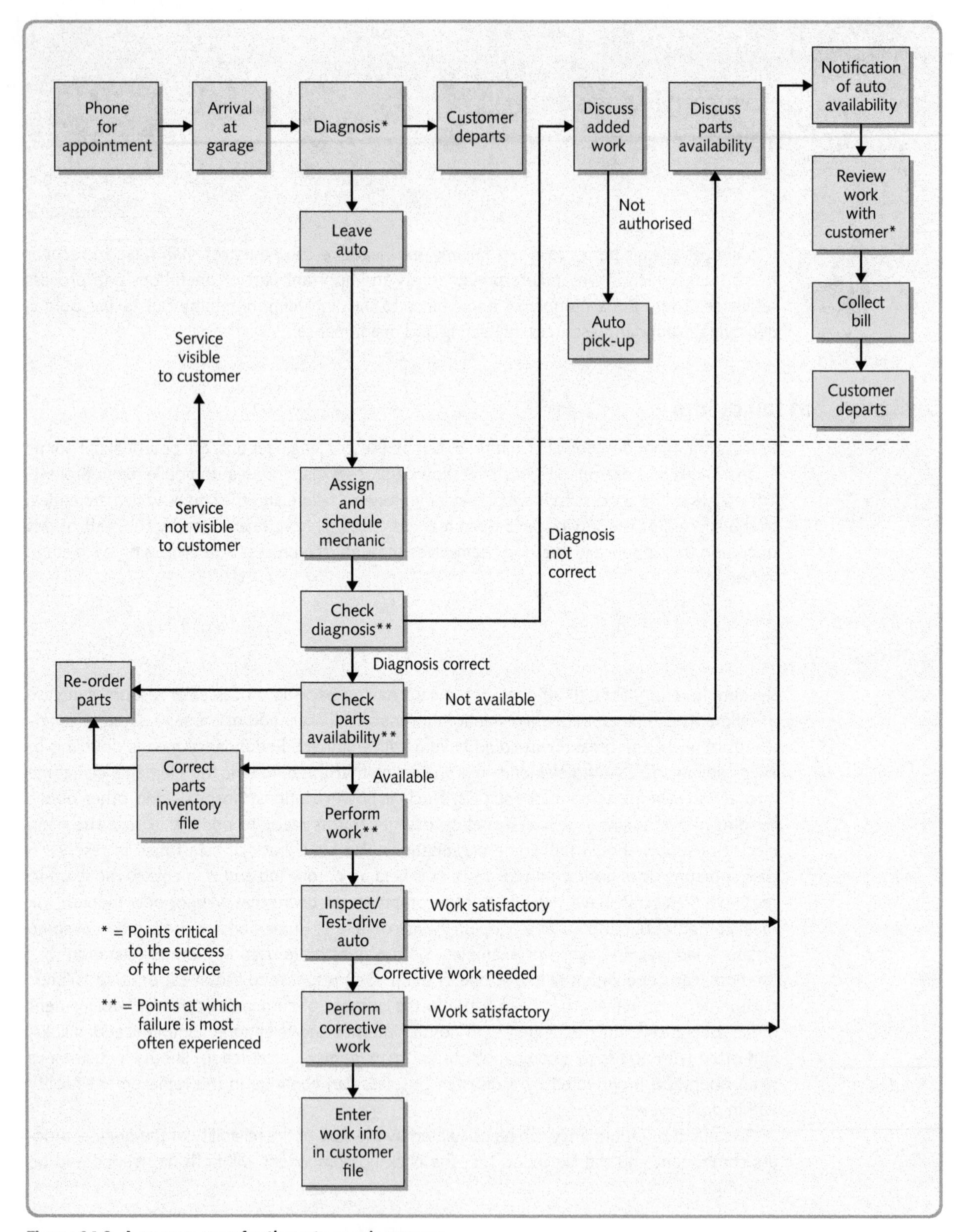

Figure 14.9 A process map for the auto repair process

R_a = reliability of *a* alone

R_b = reliability of back-up component *b*

$P_{(failure)}$ = the probability that component *a* will fail and therefore component *b* will be needed.

So, for example, a food manufacturer has two packing lines, one of which will come into action only if the first line fails. If each line has a reliability of 0.9, the lines working together (each with reliability = 0.9) will have a reliability of $0.9 + [0.9 \times (1 - 0.9)] = 0.99$.

Fail-safeing

The concept of fail-safeing has emerged since the introduction of Japanese methods of operations improvement. Called *poka-yoke* in Japan (from *yokeru* ('to prevent') and *poka* ('inadvertent errors'), the idea is based on the principle that human mistakes are to some extent inevitable. What is important is to prevent them becoming defects. Poka-yokes are simple (preferably inexpensive) devices or systems which are incorporated into a process to prevent inadvertent operator mistakes resulting in a defect.

OPERATIONS PRINCIPLE

Simple methods of fail-safeing can often be the most cost effective.

Typical poka-yokes are such devices as:

- limit switches on machines which allow the machine to operate only if the part is positioned correctly
- gauges placed on machines through which a part has to pass in order to be loaded onto, or taken off, the machine – an incorrect size or orientation stops the process
- digital counters on machines to ensure that the correct number of cuts, passes or holes have been machined
- checklists which have to be filled in, either in preparation for, or on completion of, an activity
- light beams which activate an alarm if a part is positioned incorrectly.

The same principle can also be applied to service operations, for example:

- colour-coding cash register keys to prevent incorrect entry in retail operations
- the McDonald's french-fry scoop which picks up the right quantity of fries in the right orientation to be placed in the pack
- trays used in hospitals with indentations shaped to each item needed for a surgical procedure – any item not back in place at the end of the procedure might have been left in the patient
- the paper strips placed round clean towels in hotels, the removal of which helps housekeepers to tell whether a towel has been used and therefore needs replacing
- the locks on aircraft lavatory doors, which must be turned to switch the light on
- beepers on ATMs to ensure that customers remove their cards
- height bars on amusement rides to ensure that customers do not exceed size limitations.

Maintenance

Maintenance is the term used to cover the way operations and processes try to avoid failure by taking care of their physical facilities. It is particularly important when physical facilities play a central role in the operation, such as power stations, airlines and petrochemical refineries. There are a number of approaches to maintenance, including the following.

Preventive maintenance (PM)

Attempts to eliminate or reduce the chances of failure by regularly servicing (cleaning, lubricating, replacing and checking) facilities. For example, the engines of passenger aircraft are checked, cleaned and calibrated according to a regular schedule after a set number of flying hours. Taking aircraft away from their regular duties for preventive maintenance is clearly an expensive option for any airline, but the consequences of failure while in service are considerably more serious.

Condition-based maintenance (CBM)

Attempts to perform maintenance only when the facilities require it. For example, continuous process equipment, such as that used in coating photographic paper, is run for long periods in order to achieve the high utilisation necessary for cost-effective production. Stopping the machine when it is not strictly necessary to do so would take it out of action for long periods and reduce its utilisation. Here condition-based maintenance might involve continuously monitoring the vibrations, or some other characteristic of the line. The results of this monitoring would then be used to decide whether the line should be stopped and the bearings replaced.

Total productive maintenance (TPM)

This is defined as *'the productive maintenance carried out by all employees through small group activities,'* where productive maintenance is *'maintenance management which recognises the importance of reliability, maintenance and economic efficiency in plant design'.*[7] TPM adopts team-working and empowerment principles, as well as a continuous improvement approach to failure prevention. It aims to establish good maintenance practice in operations through the pursuit of 'the five goals of TPM':[8]

1. Examine how the facilities are contributing to the effectiveness of the operation by examining all the losses which occur.
2. Achieve autonomous maintenance by allowing people to take responsibility for at least some of the maintenance tasks.
3. Plan maintenance by having a fully worked out approach to all maintenance activities, including the level of preventive maintenance which is required for each piece of equipment, the standards for condition-based maintenance and the respective responsibilities of operating staff and maintenance staff.
4. Train all staff in relevant maintenance skills so that staff have all the skills to carry out their roles.
5. Avoid maintenance altogether by 'maintenance prevention' (MP), that is, considering failure causes and the maintainability of equipment during its design stage, its manufacture, its installation and its commissioning.

How much maintenance?

Most operations plan their maintenance to include a level of regular preventive maintenance which gives a reasonably low but finite chance of breakdown. Usually the more frequent the preventive maintenance episodes, the fewer the chances of a breakdown. Infrequent preventive maintenance will cost little to provide but will result in a high likelihood (and therefore cost) of breakdown. Conversely, very frequent preventive maintenance will be expensive to provide but will reduce the cost of having to provide breakdown maintenance as in Figure 14.10(a). The total cost of maintenance appears to minimise at an 'optimum' level of preventive maintenance. However, this may not reflect reality. The cost of providing preventive maintenance in Figure 14.10(a) assumes that it is carried out by a separate set of

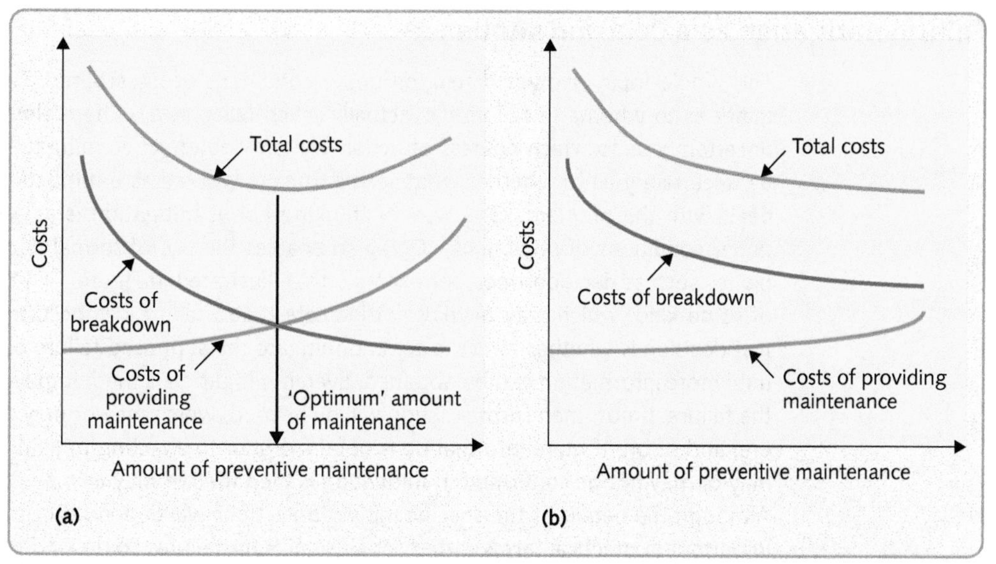

Figure 14.10 Two views of maintenance costs: (a) one model of the costs associated with preventive maintenance shows an optimum level of maintenance effort; (b) if routine preventive maintenance tasks are carried out by operators and if the real cost of breakdowns is considered, the 'optimum' level of preventive maintenance shifts toward higher levels

people (skilled maintenance staff) whose time is scheduled and accounted for separately from the 'operators' of the facilities. In many operations, however, at least some preventive maintenance can be performed by the operators themselves (which reduces the cost of providing it) and at times which are convenient for the operation (which minimises the disruption to the operation). Furthermore, the cost of breakdowns could also be higher than is indicated in Figure 14.10(a) because unplanned downtime can take away stability from the operation, preventing it being able to improve itself. Put these two ideas together and the minimising total curve and maintenance cost curve look more like Figure 14.10(b). The emphasis is shifted towards using more preventive maintenance than is generally thought appropriate.

DIAGNOSTIC QUESTION

Have failure mitigation measures been implemented?

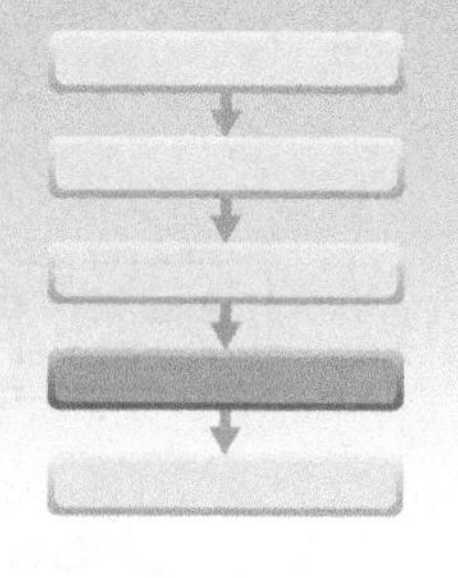

Failure mitigation means isolating a failure from its negative consequences. It is an admission that not all failures can be avoided. However, in some areas of operations management, relying on mitigation, rather than prevention, is unfashionable. For example, 'inspection' practices in quality management were based on the assumption that failures were inevitable and needed to be detected before they could cause harm. Modern total quality management places much more emphasis on prevention. Yet, in operations and process resilience, mitigation can be vital when used in conjunction with prevention in reducing overall risk.

Failure mitigation as a decision sequence

This whole topic involves managing under conditions of uncertainty. There may be uncertainty as to whether a failure has actually taken place at all. There almost certainly will be uncertainty as to which courses of action will provide effective mitigation. There may even be uncertainty as to whether what seems to have worked as a mitigation action, really has dealt with the problem. One way of thinking about mitigation is as a series of decisions under conditions of uncertainty. Doing so enables the use of formal decision analysis techniques such as decision trees, for example that illustrated in Figure 14.11. Here, an anomaly of some kind, which may or may not indicate that a failure has occurred, is detected. The first decision is whether to act to try and mitigate the supposed failure or, alternatively, wait until more information can be obtained. Even if mitigation is tried, it may or may not contain the failure. If not, then further action will be needed, which may or may not contain the failure, and so on. If more information is obtained prior to enacting mitigation, then the failure may or may not be confirmed. If mitigation is then tried, it may or may not work, and so on. Although the details of the specific mitigation actions will depend on circumstances, what is important in practical terms is that for all significant failures some kind of decision rules and mitigation planning has been established.

Failure mitigation actions

The nature of the action taken to mitigate failure will obviously depend on the nature of the failure. In most industries technical experts have established a classification of failure mitigation actions that are appropriate for the types of risk likely to be suffered. So, for example, in agriculture, government agencies and industry bodies have published mitigation strategies for such 'failures' as the outbreak of crop disease, contagious animal infections, and so on. Such documents will outline the various mitigation actions that can be taken under different circumstances and detail exactly who are responsible for each action. Although these classifications tend to be industry specific, the following generic categorisation gives a flavour of the types of mitigation actions that may be generally applicable.

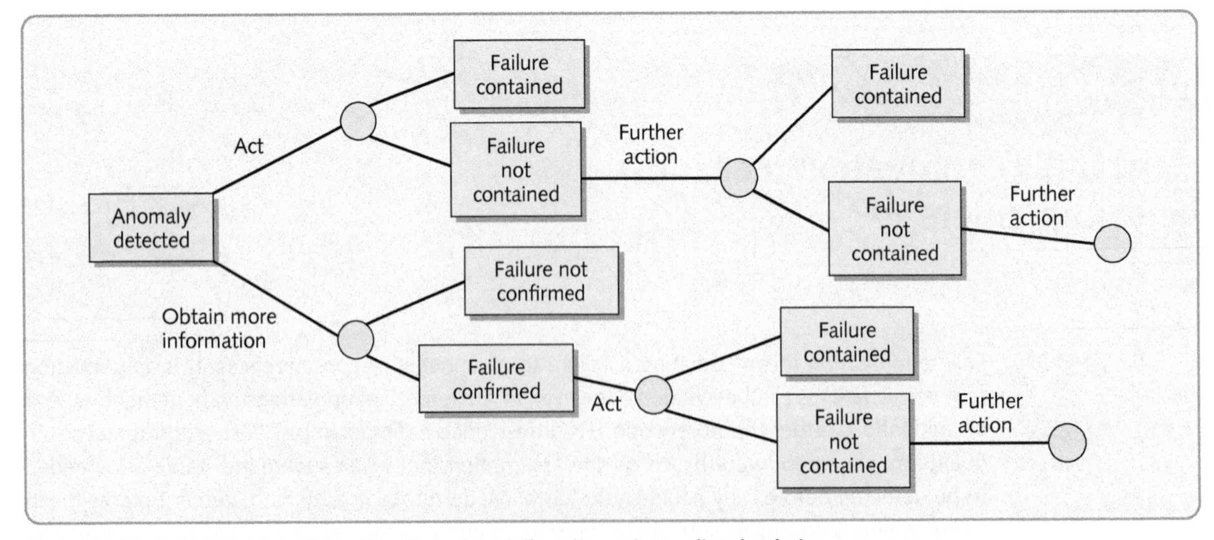

Figure 14.11 A decision tree for mitigation when failure is not immediately obvious

- **Mitigation planning** is the activity of ensuring that all possible failure circumstances have been identified and the appropriate mitigation actions identified. It is the overarching activity that encompasses all subsequent mitigation actions, and may be described in the form of a decision tree or guide rules. Almost certainly there will be some form of escalation that will guide the extra mitigation effort should early actions not prove successful. It is worth noting that mitigation planning, as well an overarching action, also provides mitigation action in its own right. For example, if mitigation planning has identified appropriate training, job design, emergency procedures, and so on, then the financial liability of a business for any losses should a failure occur will be reduced. Certainly businesses that have not planned adequately for failures will be more liable in law for any subsequent losses.
- **Economic mitigation** includes actions such as insurance against losses from failure, spreading the financial consequences of failure, and 'hedging' against failure. Insurance is the best known of these actions and is widely adopted, although ensuring appropriate insurance and effective claims management is a specialised skill in itself. Spreading the financial consequences of failure could involve, for example, spreading the equity holding in supply companies to reduce the financial consequences of such companies failing. Hedging involves creating a portfolio of ventures whose outcomes happen to be correlated so as to reduce total variability. This often takes the form of financial instruments, for example, a business may purchase a financial 'hedge' against the price risk of a vital raw material deviating significantly from a set price.
- **Containment (spatial)** means stopping the failure physically spreading to affect other parts of an internal or external supply network. Preventing contaminated food from spreading through the supply chain, for example, will depend on real time information systems that provide traceability data.
- **Containment (temporal)** means containing the spread of a failure over time. It particularly applies when information about a failure or potential failure needs to be transmitted without undue delay. For example, systems that give advanced warning of hazardous weather such as snow storms must transmit such information to local agencies such as the police and road-clearing organisations in time for them to stop the problem causing excessive disruption.
- **Loss reduction** covers any action that reduces the catastrophic consequences of failure by removing the resources that are likely to suffer those consequences. For example, the road signs that indicate evacuation routes in the event of severe weather, or the fire drills that train employees in how to escape in the event of an emergency, may not reduce the consequences of failure on buildings or physical facilities, but can dramatically help in reducing loss of life or injury.
- **Substitution** means compensating for failure by providing other resources that can substitute for those rendered less effective by the failure. It is a little like the concept of redundancy that was described earlier, but does not always imply excess resources if a failure has not occurred. For example, in a construction project, the risk of encountering unexpected geological problems may be mitigated by the existence of a separate work plan that is invoked only if such problems are found. The resources may come from other parts of the construction project, which will in turn have plans to compensate for their loss.

Table 14.1 gives some examples of each type of failure mitigation actions for three failures: the theft of money from one of a company's bank accounts, the failure of a new product technology to work adequately during the new product development process, and the outbreak of fire at a business premises.

Table 14.1 Failure mitigation actions for three failures

	Type of failure		
Failure mitigation actions	***Financial failure – theft from company account***	***Development failure – new technology does not work***	***Emergency failure – fire at premises***
Mitigation planning	Identify different types of theft that have been reported and devise mitigation actions including software to identify anomalous account behaviour.	Identify possible types of technology failure and identify contingency technologies together with plans for accessing contingency technologies.	Identify fire hazards and methods of detecting, limiting and extinguishing fires.
Economic mitigation	Insure against theft and possibly use several different accounts.	Invest in, or form partnership with, supplier of alternative technology.	Insure against fire and have more, smaller, premises.
Containment (spatial)	'Ring fence' accounts so a deficit in one account cannot be made good from another account.	Develop alternative technological solutions for different parts of the development project so that failure in one part does not affect the whole project.	Install localised sprinkler systems and fire door barriers.
Containment (temporal)	Invest in software that detects signs of possible unusual account behaviour.	Build in project milestones that indicate the possibility of eventual development failure.	Install alarm systems that indicate the occurrence of fire to everyone who may be affected (including in other premises).
Loss reduction	Build in transfer delays until approval for major withdrawals has been given; also institute plans for recovering stolen money.	Ensure the development project can use old technology if new one does not work.	Ensure means of egress and employee training are adequate.
Substitution	Ensure that reserve funds and staff to manage the transfer can be speedily brought into play.	Have fall-back work package for devoting extra resources to overcome the new technology failure.	Ensure backup team that can take over from premises rendered inoperative by fire.

EXAMPLE

Mitigating currency risk[9]

A multi-national consumer goods firm was concerned at the way its operations in certain parts of the world were exposed to currency fluctuations. The company's Russian subsidiary sourced nearly all products from its parent factories in France and Germany, while its main rivals had manufacturing facilities in Russia. Conscious of the potential volatility of the rouble, the firm needed to minimise its operating exposure to a devaluation of the currency that would leave the firm's cost structure at a serious disadvantage compared to its rivals, and without any real option but to increase their prices. In seeking to mitigate against the risk of devaluation, the firm could choose from among various financial and operations-based options. For example, financial tools were available to minimise currency exposure. Most of these allow the operation to reduce the risk of currency fluctuations but involve an 'up front' cost. Usually, the higher the risk the greater the up front cost. Alternatively the company may restructure its operations strategy in order to mitigate its currency risk. One option would be to develop its own production facilities within Russia. This may reduce, or even eliminate, the currency risk, although it may introduce other risks. A further option may be to form supply partnerships with other Russian companies. Again, this does not eliminate risks but can shift them to ones which the company feels more able to control.

More generally the company may consider creating a portfolio of operations-based strategies, such as developing alternative suppliers in different currency zones, building up excess/flexible capacity in a global production network, and creating more differentiated products that are less price-sensitive.

DIAGNOSTIC QUESTION

Have failure recovery measures been implemented?

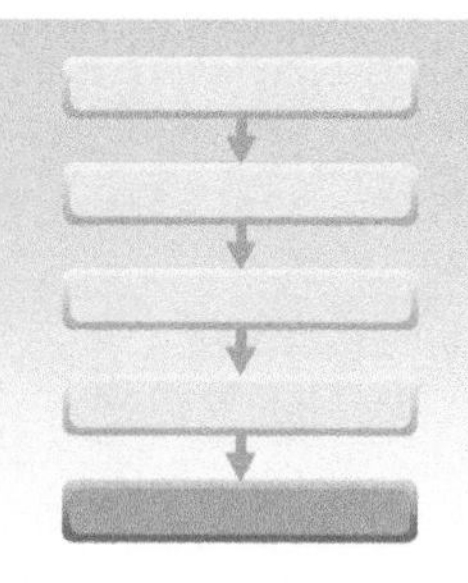

Video

Failure recovery is the set of actions that are taken after the negative effects of failure have occurred, and that reduce the impact of the negative effects. All types of operation can benefit from well-planned recovery. For example, a construction company whose mechanical digger breaks down can have plans in place to arrange a replacement from a hire company. The breakdown might be disruptive, but not as much as it might have been if the operations manager had not worked out what to do. Recovery procedures will also shape customers' perceptions of failure. Even where the customer sees a failure, it may not necessarily lead to dissatisfaction, customers may even accept that things occasionally do go wrong. If there is a metre of snow on the train lines, or if the restaurant is particularly popular, we may accept that the product or service does not work. It is not necessarily the failure itself that leads to dissatisfaction but often the organisation's response to the breakdown. Mistakes may be inevitable, but dissatisfied customers are not.

A failure may even be turned into a positive experience. If a flight is delayed by five hours, there is considerable potential for dissatisfaction. But if the airline informs passengers that the aircraft has been delayed by a cyclone at its previous destination, and that arrangements have been made for accommodation at a local hotel with a complimentary meal, passengers might then feel that they have been well treated and even recommend that airline to others. A good recovery can turn angry, frustrated customers into loyal ones. In fact one investigation[10] into customer satisfaction and customer loyalty used four scenarios to test the willingness of customers to use an operation's services again. The four scenarios were:

OPERATIONS PRINCIPLE

Successful failure recovery can yield more benefits than if the failure had not occurred.

1. The service is delivered to meet the customers' expectations and there is full satisfaction.
2. There are faults in the service delivery but the customer does not complain about them.
3. There are faults in the service delivery and the customer complains, but he/she has been fobbed off or mollified. There is no real satisfaction with the service provider.
4. There are faults in the service delivery and the customer complains and feels fully satisfied with the resulting action taken by the service providers.

Customers who are fully satisfied and do not experience any problems (1) are the most loyal, followed by complaining customers whose complaints are resolved successfully (4). Customers who experience problems but don't complain (2) are in third place and last of all come customers who do complain but are left with their problems unresolved and feelings of dissatisfaction (3).

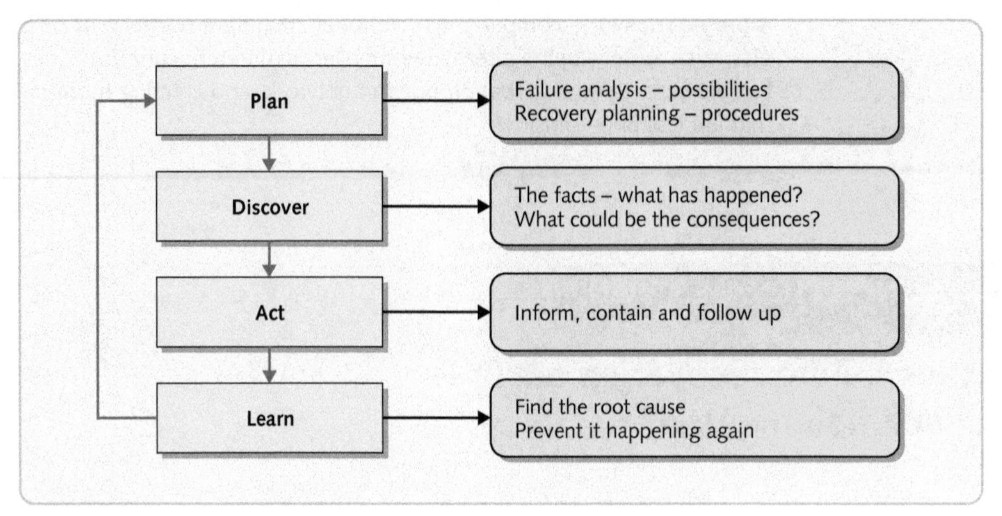

Figure 14.12 Recovery sequence for minimising the impact from failure

The recovery process

Recovery needs to be a planned process. Organisations therefore need to design appropriate responses to failure, linked to the cost and the inconvenience caused by the failure to their customers. These must first meet the needs and expectations of customers. Such recovery processes need to be carried out either by empowered front-line staff or by trained personnel who are available to deal with recovery in a way which does not interfere with day-to-day service activities. Figure 14.12 illustrates a typical recovery sequence.

Discover

The first thing any manager needs to do when faced with a failure is to discover its exact nature. Three important pieces of information are needed: first of all, what exactly has happened; second, who will be affected by the failure? third, why did the failure occur? This last point is not intended to be a detailed inquest into the causes of failure (that comes later) but it is often necessary to know something of the causes of failure in case it is necessary to determine what action to take.

Act

The discover stage could only take minutes or even seconds, depending on the severity of the failure. If the failure is a severe one with important consequences, we need to move on to doing something about it quickly. This means carrying out three actions, the first two of which could be carried out in reverse order, depending on the urgency of the situation. First, tell the significant people involved what you are proposing to do about the failure. In service operations this is especially important where the customers need to be kept informed, both for their peace of mind and to demonstrate that something is being done. In all operations, however, it is important to communicate what action is going to happen so that everyone can set their own recovery plans in motion. Second, the effects of the failure need to be contained in order to stop the consequences spreading and causing further failures. The precise containment actions will depend on the nature of the failure. Third, there needs to be some kind of follow-up to make sure that the containment actions really have contained the failure.

Learn

As discussed earlier in this chapter, the benefits of failure in providing learning opportunities should not be underestimated. In failure planning, learning involves revisiting the failure to find out its root cause and then engineering out the causes of the failure so that it will not happen again.

Plan

Learning the lessons from a failure is not the end of the procedure. Operations managers need formally to incorporate the lessons into their future reactions to failures. This is often done by working through 'in theory' how they would react to failures in the future. Specifically, this involves first identifying all the possible failures which might occur (in a similar way to the FMEA approach). Second, it means formally defining the procedures which the organisation should follow in the case of each type of identified failure.

Critical commentary

Each chapter contains a short critical commentary on the main ideas covered in the chapter. Its purpose is not to undermine the issues discussed in the chapter, but to emphasise that, although we present a relatively orthodox view of operation, there are other perspectives.

- The idea that failure can be detected through in-process inspection is increasingly seen as only partially true. Although inspecting for failures is an obvious first step in detecting them, it is not even close to being 100 per cent reliable. Accumulated evidence from research and practical examples consistently indicates that people, even when assisted by technology, are not good at detecting failure and errors. This applies even when special attention is being given to inspection. For example, airport security was significantly strengthened after 11 September 2001, yet one in ten lethal weapons that entered into airports' security systems (in order to test them) were not detected. *'There is no such thing as 100 per cent security, we are all human beings,'* says Ian Hutcheson, the Director of Security at Airport Operator BAA. No one is advocating abandoning inspection as a failure detection mechanism. Rather it is seen as one of a range of methods of preventing failure.

- Much of the previous discussion surrounding the prevention of failure has assumed a 'rational' approach. In other words, it is assumed that operations managers and customers alike will put more effort into preventing failures that are either more likely to occur or more serious in their consequences. Yet this assumption is based on a rational response to risk. In fact, being human, managers often respond to the perception of risk rather than its reality. For example, Table 14.2 shows the cost of each life saved by investment in various road and rail transportation safety (in other words, failure prevention) investments. The table shows that investing in improving road safety is very much more effective than investing in rail safety. And while no one is arguing for abandoning efforts on rail safety, it is noted by some transportation authorities that actual investment reflects more the public perception of rail deaths (low) compared with road deaths (very high).

Table 14.2 The cost per life saved of various safety (failure prevention) investments

Safety investment	*Cost per life (€M)*
Advanced train protection system	30
Train protection warning systems	7.5
Implementing recommended guidelines on rail safety	4.7
Implementing recommended guidelines on road safety	1.6
Local authority spending on road safety	0.15

SUMMARY CHECKLIST

This checklist comprises questions that can be usefully applied to any type of operations and reflect the major diagnostic questions used within the chapter.

- [ ] Does the business have an operations and process resilience policy?
- [ ] Have any possible changes in the business's vulnerability to failure been discussed and accommodated within its failure policy?
- [ ] Have all potential sources of failure been identified?
- [ ] Have any future changes in the sources of failure been identified?
- [ ] Have the impact of all potential sources of failure been assessed?
- [ ] Has the likelihood of each potential failure been assessed?
- [ ] Has the possibility of non-evident failures been addressed?
- [ ] Is post-failure analysis carried out when failure does occur?
- [ ] Are techniques such as failure mode and effect analysis (FMEA) used?
- [ ] Has due attention been paid to the possibility of designing out failure points?
- [ ] Is the concept of redundancy economically viable for any potential failures?
- [ ] Has the idea of fail-safeing (poka-yoke) been considered as a means of reducing the likelihood of failure?
- [ ] Have all approaches to process and technology maintenance been explored?
- [ ] Has the possibility that insufficient maintenance effort is being applied been investigated?
- [ ] Does the operation have a failure mitigation plan?
- [ ] Have the whole range of mitigation actions been thoroughly evaluated?
- [ ] Are specific plans in place for the use of each type of mitigation action?
- [ ] Is a well-planned recovery procedure in place?
- [ ] Does the recovery procedure cover all the steps of discover, act, learn, and plan?

CASE STUDY

Slagelse Industrial Services (SIS)[11]

Slagelse Industrial Services (SIS) had become one of Europe's most respected die casters of zinc, aluminum and magnesium parts suppliers for hundreds of companies in many industries, especially automotive and defence. The company cast and engineered precision components by combining the most modern production technologies with precise tooling and craftsmanship. Slagelse Industrial Services (SIS) began life as a classic family firm run by Erik Paulsen, who opened a small manufacturing and die-casting business in his hometown of Slagelse, a town in east Denmark, about 100 km southwest of Copenhagen. He had successfully leveraged his skills and passion for craftsmanship over many years while serving a variety of different industrial and agricultural customers. His son, Anders had spent nearly ten years working as a production engineer for a large automotive parts supplier in the UK, but eventually returned to Slagelse to take over the family firm. Exploiting his experience in mass manufacturing, Anders spent years building the firm into a larger scale industrial component manufacturer but retained his father's commitment to quality and customer service. After 20 years he sold the firm to a UK-owned industrial conglomerate and within ten years it had doubled in size again and now employed in the region of 600 people and had a turnover approaching £200 million. Throughout this period the firm had continued to target their products into niche industrial markets where their emphasis upon product quality and dependability meant they were less vulnerable to price and cost pressures. However in 2009, in the midst of difficult economic times and widespread industrial restructuring, they had been encouraged to bid for higher volume, lower margin work. This process was not very successful but eventually culminated in a tender for the design and production of a core metallic element of a child's toy (a 'transforming' robot).

Interestingly, the client firm, Alden Toys, was also a major customer for other businesses owned by SIS's corporate parent. They were adopting a preferred supplier policy and intended to have only one or two purchase points for specific elements in their global toy business. They had a high degree of trust in the parent organisation and on visiting the SIS site were impressed by the firm's depth of experience and commitment to quality. In 2010, they selected SIS to complete the design and begin trial production.

Source: EM Clements Photography

'Some of us were really excited by the prospect . . . but you have to be a little worried when volumes are much greater than anything you've done before. I guess the risk seemed okay because in the basic process steps, in the type of product if you like, we were making something that felt very similar to what we'd been doing for many years.' (SIS Operations Manager)

'Well obviously we didn't know anything about the toy market but then again we didn't really know all that much about the auto industry or the defence sector or any of our traditional customers before we started serving them. Our key competitive advantage, our capabilities, call it what you will, they are all about keeping the customer happy, about meeting and sometimes exceeding specification.' (SIS Marketing Director)

The designers had received an outline product specification from Alden Toys during the bid process and some further technical detail afterwards. Upon receipt of this final brief, a team of engineers and managers confirmed that the product could and would be manufactured using a scaled-up version of current production processes. The key operational challenge appeared to be accessing sufficient (but not too much) capacity. Fortunately, for a variety of reasons, the parent company was very supportive of the project and promised to underwrite any sensible capital expenditure plans. Although this opinion of the production challenge was widely accepted throughout the firm (and shared by Alden Toys and SIS's parent group) it was left to one specific senior engineer to actually sign both the final bid and technical completion documentation. By early 2011, the firm had begun a trial period of full volume production. Unfortunately, as would become clear later, during this design validation process SIS had effectively sanctioned a production method that would prove to be entirely inappropriate for the toy market, but it was not until 12 months later that any indication of problems began to emerge.

Throughout both North America and Europe, individual customers began to claim that their children had been 'poisoned' while playing with the end product. The threat of litigation was quickly levelled at Alden Toys and the whole issue rapidly became a 'full-blown' child health scare. A range of pressure groups and legal damage specialists supported and acted to aggregate the individual claims. Although similar accusations had been made before, the litigants and their supporters focused on the recent changes made to the production process at SIS and in particular the role of Alden Toys in managing their suppliers.

'. . . it's all very well claiming that you trust your suppliers but you simply cannot have the same level of control over another firm in another country. I am afraid that this all comes down to simple economics, that Alden Toys put its profits before children's health. Talk about trust . . . parents trusted this firm to look out for them and their families and have every right to be angry that boardroom greed was more important!' (Legal spokesperson for US litigants when being interviewed on UK TV consumer rights show)

Under intense media pressure, Alden Toys rapidly convened a high-profile investigation into the source of the contamination. It quickly revealed that an 'unauthorised' chemical had been employed in an apparently trivial metal cleaning and preparation element of the SIS production process. Although when interviewed by the US media, the parent firm's legal director emphasised there was *'no causal link established or any admission of liability by either party'*, Alden Toys immediately withdrew their order and began to signal an intent to bring legal action against SIS and its parent. This action brought an immediate end to production in this part of the operation and the inspection (and subsequent official and legal visits) had a crippling impact upon the productivity of the whole site. The competitive impact of the failure was extremely significant. After over a year of production, the new product accounted for more than a third (39 per cent) of the factory's output. In addition to major cash-flow implications, the various investigations took up lots of managerial time and the reputation of the firm was seriously affected. As the site operations manager explained, even their traditional customers expressed concerns.

'It's amazing but people we had been supplying for thirty or forty years were calling me up and asking '[Manager's name] what's going on?' and that they were worried about what all this might mean for them . . . these are completely different markets!'

QUESTIONS

1 What operational risks did SIS face when deciding to become a strategic supplier for Alden Toys?

2 What control problems did they encounter in implementing this strategy (pre- and post-investigation)?

APPLYING THE PRINCIPLES

Hints

Some of these exercises can be answered by reading the chapter. Others will require some general knowledge of business activity and some might require an element of investigation. Hints on how they can all be answered are to be found in the eText at www.pearsoned.co.uk/slack.

1 One cause of aircraft accidents is 'controlled flight into ground'. Predominantly, the reason for this is not mechanical failure but human failure such as pilot fatigue. Boeing, which dominates the commercial airline business, has calculated that over 60 per cent of all the accidents which have occurred in the past ten years had flight crew behaviour as their 'dominant cause'. For this type of failure to occur, a whole chain of minor failures must happen. First, the pilot at the controls has to be flying at the wrong altitude – there is only one chance in a thousand of this. Second, the co-pilot would have to fail to cross-check the altitude – only one chance in a hundred of this. The air traffic controllers would have to miss the fact that the plane was at

the wrong altitude (which is not strictly part of their job) – a one-in-ten chance. Finally, the pilot would have to ignore the ground proximity warning alarm in the aircraft (which can be prone to give false alarms) – a one-in-two chance.

(a) What are your views on the quoted probabilities of each failure described above occurring?

(b) How would you try to prevent these failures occurring?

(c) If the probability of each failure occurring could be reduced by a half, what would be the effect on the likelihood of this type of crash occurring?

2 Conduct a survey among colleagues, friends and acquaintances of how they cope with the possibility that their computers might 'fail', either in terms of ceasing to operate effectively, or in losing data. Discuss how the concept of redundancy applies in such failure.

3 Survey a range of people who own and/or are responsible for the performance of the following pieces of equipment. What is their approach to maintaining them and how is this influenced by the perceived seriousness of any failure?

- motor cars
- central heating systems or air conditioning systems
- domestic appliances such as dishwashers and vacuum cleaners
- furniture
- lighting or lighting systems.

4 Visit the websites of some of the many companies that offer advice and consultancy to companies wishing to review their 'business continuity' plans. Based on your investigation of these sites, identify the key issues in any business continuity plan for the following types of operation.

(a) a university

(b) an airport

(c) a container port

(d) a chemicals manufacturing plant.

In terms of its effectiveness at managing the learning process, how does a university detect failures? What could it do to improve its failure detection processes?

Notes on chapter

1 Sources include: 'Gap issues statement on media reports on child labor, San Francisco' (2011), Gap Inc. corporate website; 'Clean, wholesome and American? A storm over the use of child labour clouds Gap's pristine image (2007), *The Economist*, 1 November.

2 Source: Herman, M. and Dearball, J. (2007), 'Cadbury fined £1 million over salmonella outbreak', *Times online*, 16 July.

3 Christopher, M. (2002) 'Business is failing to manage supply chain vulnerability', *Odessey*, Issue 16, June.

4 Source: 'Air crashes, but surely . . .' (1994), *The Economist*, 4 June.

5 Norton, J. (2008) 'The typing error that gave us 30 years of spam', *Observer* 4 May.

6 Examples taken from Slack, N. and Lewis, M.A. (2011), *Operations Strategy*, (3rd Edition) Financial Times Prentice Hall.

7 Nakajima, S. (1988) *Total Productive Maintenance*, Productivity Press.

8 Nakajima, S., *ibid*.

9 Example taken from Slack, N. and M.A. Lewis (2011) *op. cit.*

10 Armistead, C.G. and Clark, G. (1992) *Customer Service and Support*, FT/Pitman Publishing.

11 Based on a case originally in Slack, N. and M.A. Lewis (2002) *Operations Strategy* (1st Edition), Financial Times Prentice Hall.

TAKING IT FURTHER

Crouhy, M., Galai, D. and Mark, R. (2006) *The Essentials of Risk Management: The Definitive Guide for the Non-risk Professional*, McGraw-Hill Professional. *Very much a book for managers, but coherent and interesting.*

Dhillon, B.S. (2002) *Engineering Maintenance: A Modern Approach*, Technomic Publishing Company. *A comprehensive book for the enthusiastic that stresses the 'cradle-to-grave' aspects of maintenance.*

Hubbard, D.W. (2009) *The Failure of Risk Management: Why it's Broken and How to fix it*, John Wiley & Sons. *Another good introduction; a polemic, but one that is clearly written.*

Löfsten, H. (1999) 'Management of industrial maintenance – economic evaluation of maintenance policies', *International Journal of Operations and Production Management*, Vol. 19, No 7. *An academic paper, but provides a useful economic rationale for choosing alternative maintenance policies.*

Narayan, V. (2004) *Effective Maintenance Management: Risk and Reliability Strategies for Optimizing Performance*, Industrial Press Inc. *This is an engineering approach that gives the subject some background theory.*

Smith, D.J. (2000) *Reliability, Maintainability and Risk*, Butterworth-Heinemann. *A comprehensive and excellent guide to all aspects of maintenance and reliability.*

USEFUL WEBSITES

www.opsman.org *Definitions, links and opinions on operations and process management.*

http://www.smrp.org/ *Site of the Society for Maintenance and Reliability Professionals, Gives an insight into practical issues.*

http://www.sre.org/ *American Society of Reliability Engineers. The newsletters give insights into reliability practice.*

http://www.rspa.com/spi/SQA.html *Lots of resources, involving reliability and poka yoke.*

http://sra.org/ *Site of the Society for Risk Analysis. Very wide scope, but interesting.*

Interactive quiz

For further resources including examples, animated diagrams, self-test questions, Excel spreadsheets and video materials please explore the eText on the companion website at **www.pearsoned.co.uk/slack**.

15

Project management

Introduction

This chapter is concerned with managing 'projects'. Some projects are complex, large-scale, have activities involving diverse resources, and last for years. Other 'projects' are far smaller affairs, possibly limited to one part of a business, and may last only a few days. Yet, although the complexity and degree of difficulty involved in managing different types of projects will vary, the essential approach to the task does not. Whether projects are large or small, internal or external, long or short; they all require defining, planning, and controlling (see Figure 15.1).

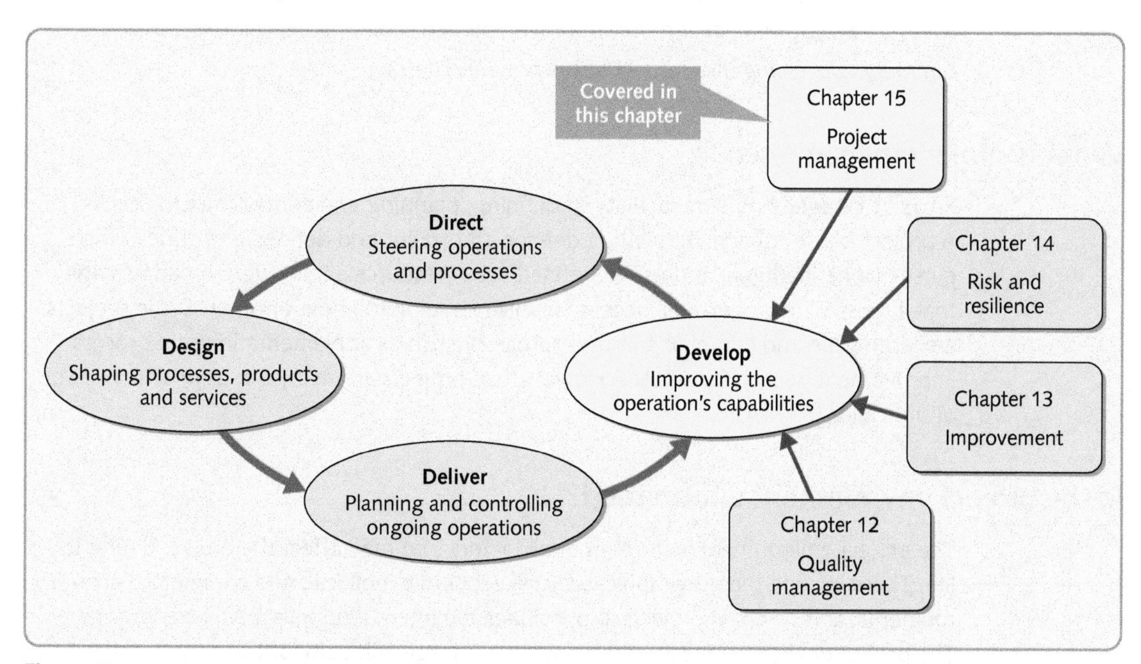

Figure 15.1 Project management is the activity of defining, planning and controlling projects

EXECUTIVE SUMMARY

Each chapter is structured around a set of diagnostic questions. These questions suggest what you should ask in order to gain an understanding of the important issues of a topic, and, as a result, improve your decision making. An executive summary addressing these questions is provided below.

What is project management?

Project management is the activity of defining, planning and controlling projects. A project is a set of activities with a defined start point and defined end state, which pursues a defined goal and uses a defined set of resources. It is a very broad activity that almost all managers will become involved in at some time or other. Some projects are large scale and complex, but most projects, such as implementation of a process improvement, will be far smaller. However, all projects are managed using a similar set of principles.

Is the project environment understood?

The project environment is the sum of all factors that may affect the project during its life. These include the geographical, social, economic, political, and commercial environments, and on small projects also includes the internal organisational environment. The project environment can also include the intrinsic difficulty of the project defined by its scale, degree of uncertainty, and complexity. Also included in a project's environment are the project stakeholders, those individuals or groups who have some kind of interest in the project. Stakeholder management can be particularly important both to avoid difficulties in the project and to maximise its chances of success. Stakeholders can be classified by their degree of interest in the project and their power to influence it.

Is the project well defined?

A project is defined by three elements: its objectives, its scope and its overall strategy. Most projects can be defined by the relative importance of three objectives. These are cost (keeping the overall project to its original budget), time (finishing the project by the scheduled finish time), and quality (ensuring that the project outcome is as was originally specified). The project scope defines it's work content and outcomes. More importantly, it should define what is not included in the project. The project strategy describes the general way in which the project is going to meet its objectives, including significant project milestones and stagegates.

Is project management adequate?

Because of their complexity and the involvement of many different parties, projects need particularly careful managing. In fact, project management is seen as a particularly demanding role with a very diverse set of skills including technical project management knowledge, interpersonal skills and leadership ability. Very often project managers need the ability to motivate staff who not only report to a manager other than themselves, but also divide their time between several different projects.

Has the project been adequately planned?

Project planning involves determining the cost and duration of the project and the level of resources that it will need. In more detail, it involves identifying the start and finish times of individual activities within the project. Generally, the five stages of project planning include identifying activities, estimating times and resources, identifying relationships and dependencies between activities, identifying time and resource schedule constraints, and fixing the final schedule. However, no amount of planning can prevent the need for replanning as circumstances dictate during the life of the project. Network planning techniques such as critical path analysis (CPA) are often used to aid the project planning process.

Is the project adequately controlled?

Project control involves monitoring the project in order to check its progress, assessing the performance of the project against the project plan, and, if necessary, intervening in order to bring the project back to plan. The process often involves continually assessing the progress of the project in terms of budgeted expenditure and progress towards meeting the project's final goal. It may also involve deciding when to devote extra resources to accelerating (also know as crashing) individual activities within the project. There are a number of proprietary computer-assisted project management packages on the market that range from relatively simple network planning programs through to complex and integrated 'enterprise project management' (EPRM) systems.

DIAGNOSTIC QUESTION

What is project management?

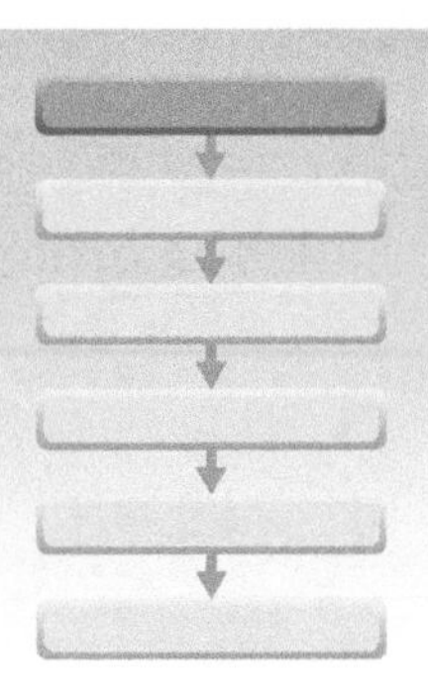

Video

A project is a set of activities with a defined start point and a defined end state, which pursues a defined goal and uses a defined set of resources. What can be defined as a 'project' can vary significantly from relatively small and local activities, through to very large 'put a man on the moon' enterprises. Project management is the activity of defining, planning and controlling projects of any type.

The activity of project management is very broad insomuch as it could encompass almost all the operations and process management tasks described in this book. Partly because of this, it could have been treated almost anywhere within the 'direct, design, delivery, develop' structure of this book. We have chosen to place it in the context of operations and process development because most of the projects that managers will be involved in are essentially improvement projects. Of course, many projects are vast enterprises with very high levels of resourcing, complexity and uncertainty that will extend over many years. Look around at the civil engineering, social, political and environmental successes (and failures) to see the evidence of major projects. Such projects require professional project management involving high level technical expertise and management skills. But so do the smaller, yet important, projects that implement the many and continuous improvements that will determine the strategic impact of operations development. This is why it is equally important to take a rigorous and systematic approach to managing improvement projects as it is to managing major projects.

At this point it is worth pointing out the distinction between 'projects' and 'programmes'. A programme, such as a continuous improvement programme, has no defined end point. Rather it is an ongoing process of change. Individual projects, such as the development of training processes, may be individual sub-sections of an overall programme, such as an integrated skills development programme. Programme management will overlay and integrate the individual projects. Generally, it is a more difficult task in the sense that it requires resource coordination, particularly when multiple projects share common resources, as emphasised in the following quotation: *'Managing projects is, it is said, like juggling three balls – cost, quality, and time. Programme management . . . is like organising a troupe of jugglers all juggling three balls and swapping balls from time to time.'*[1]

The following two projects illustrate some of the issues in project management.

EXAMPLE

Popping the Millau cork[2]

For decades, locals and French motorists had called the little bridge at Millau that was one of the few crossings on the river Tarn 'the Millau cork'. It held up all the traffic on what should have been one of the busiest north–south routes through France. No longer. In place of the little bridge is one of the most impressive and beautiful civil engineering successes of the last century. Lord Foster, the British architect who designed the bridge, described it as an attempt to enhance the natural beauty of the valley through a structure that had the 'delicacy of a butterfly', with the environment dominating the scene rather than the bridge. And

Source: Jean-Philippe Arles/Corbis

although the bridge appears to float on the clouds, it has seven pillars and a roadway of 2.5 km in length. It is also a remarkable technical achievement. At 300 metres it is the highest road bridge in the world, weighing 36,000 tonnes. The central pillar is higher than the Eiffel Tower, and took only three years to complete, notwithstanding the new engineering techniques that were needed.

Outline plans for the bridge were produced back in 1987, but, because of planning, funding and design considerations, construction did not begin until December 2001. It was completed in December 2004, on time and budget, having proved the effectiveness of its new construction technique. The traditional method of building this type of bridge (called a cable stay bridge) involves building sections of the roadway on the ground and using cranes to put them in position. Because of its height, 300 metres above the valley floor, a new technique had to be developed. First, the towers were built in the usual way, with steel reinforced concrete. The roadway was built on the high ground at either side of the valley and then pushed forward into space as further sections were added, until it met with precision (to the nearest centimetre) in the centre. This technique had never been tried before and it carried engineering risks, which added to the complexity of the project management task.

It all began with a massive recruitment drive. *'People came from all over France for employment. We knew it would be a long job. We housed them in apartments and houses in and around Millau. Eiffel gave guarantees to all the tenants and a unit was set up to help everyone with the paperwork involved in this. It was not unusual for a worker to be recruited in the morning and have his apartment available the same evening with electricity and a telephone available.'* (Jean-Pierre Martin, Chief Engineer of Groupe Eiffage and Director of Building.) Over 3,000 workers – technicians, engineers, crane drivers, carpenters, welders, winchers, metal workers, painters, concrete specialists and experts in the use of the stays and pylons that would support the bridge – contributed to the project. On the project site, 500 of them, positioned somewhere between the sky and the earth, worked in all weather to complete the project on time. *'Everyday I would ask myself what was the intense force that united these men,'* said Jean-Pierre Martin. *'They had a very strong sense of pride and they belonged to a community that was to build the most beautiful construction in the world. It was never necessary to shout at them to get them to work. Life on a construction site has many ups and downs. Some days we were frozen. Other days we were subjected to a heat wave. But even on days of bad weather, one had to force them to stay indoors. Yet often they would leave their lodgings to return to work.'*

Many different businesses were involved in building the bridge – Arcelor, Eiffel, Lafarge, Freyssinet, Potain. All of them needed coordinating in such a way that they would cooperate towards the common goal, but yet avoid any loss of overall responsibility. Jean-Pierre Martin came up with the idea of nine autonomous work groups. One group was placed at the foot of each of the seven piles that would support the bridge and two others at either end. The motto adopted by the teams was *rigueur et convivialité*, 'rigorous quality and friendly cooperation'. *'The difficulty with this type of project is keeping everyone enthusiastic throughout its duration. To make this easier we created these small groups. Each of the nine teams shifts were organised in relays between 7 and 14 hours, and 14 and 21 hours.'* So, to maintain the good atmosphere, no expense was spared to celebrate important events in the construction of the viaduct, for example, when a pile or another piece of road was completed. Sometimes, to boost the morale of the teams, and to celebrate these important events Jean-Pierre would organise a *méchouis* – a spit roast of lamb – especially popular with the many workers who were of North Africa origin.

EXAMPLE

Access HK[3]

Access HK is an independent non-profit organisation that works to fight inequality and to provide underprivileged children with educational opportunities that are otherwise unaffordable by them. Every summer, Access HK's volunteers, mostly students studying at overseas universities, return to Hong Kong to give a free Summer School to children in need. It was set up in the summer of 2001 by a group of Hong Kong students at leading UK and US universities. Since then it has organised several large-scale events to help underprivileged children, including free four-week Summer Schools during which children are taught in interactive formats on subjects such as the English language, current affairs, speech and drama. Oxford student Ng Kwan-hung, Access HK's external secretary, said: *'We share a common belief that what distinguishes one child from another is not ability but access – access to opportunities, access to education, access to love. All of us realise the importance of a good learning environment for a child's development.'* Chung Tin-Wong, a law student at Oxford and sub-committee member, added: *'We are all dedicated to providing the best education for underprivileged children.'*

Source: China Photos/Getty Images

Project managing the Summer Schools is particularly important to Access HK because the opportunities to make a difference to the underprivileged are limited largely to the holiday periods when their student volunteers are available. Project failure would mean waiting until the next year to get another chance. Also, like many charities, the budget is limited with every dollar having to count. Because of this, the student volunteers soon learn some of the arts of project management, including how to break the project down into four phases for ease of planning and control.

- **Conceptual phase** – during which the Access HK central committee agrees with the Summer School Committee its direction, aim and goal.
- **Planning phase** – when the Summer School Committee sets the time and cost parameters for the project. The time frame for the Summer School is always tight. Students volunteers only become available after they have completed their summer exams, and the Summer School must be ready to run when the primary school students have their summer break.
- **Definition and design phase** – when the detailed implementation plans for the Summer School are finalised. Communication between the team of volunteers is particularly important to ensure smooth implementation later. Many of them, although enthusiastic, have little project management experience, and therefore need the support of detailed instructions as to how to carry out their part of the project.
- **Implementation phase** – again, it is the relative inexperience of the volunteer force that dictates how the Summer School project is implemented. It is important to ensure that control mechanisms are in place that can detect any problems or deviations from the plan quickly, and help to bring it back on target.

'The success of these Summer School projects depends very much on including all our stakeholders in the process,' says one Summer School coordinator. *'All our stakeholders are important but they have different interests. The students on the Summer Schools, even if they don't articulate their objectives, need to feel that they are benefiting from the experience. Our volunteers are all bright and enthusiastic and are interested in helping to manage the process as well as taking part in it. Access HK wants to be sure that we are doing our best to fulfil their objectives and uphold their reputation. The Hong Kong government have an obvious interest in the success and integrity of the Summer Schools, and the sponsors need to be assured that their donations are being used wisely. In addition, the schools who lend us their*

buildings and many other interested parties all need to be included, in different ways and to different extents, in our project management process.'

What do these two examples have in common?

Not all projects are as large or as complex as the Millau Bridge, but the same issues will occur. Because no single project exists in isolation, the social, political and operational environment must be taken into account. For the Access HK Summer School this means including many different community groups and organisations as well as sponsors. Each project's objectives and scope also need to be clarified with stakeholders. Detailed planning, however, will be the responsibility of those who act as project managers. They are the ones who determine the time and resource commitments that the project will need, as well as identifying the long list of things that could go wrong. Back-up plans will need to be made to minimise the impact of uncertainties. The project management teams for the two examples are very different – professional and experienced in one case, volunteer and inexperienced in the other. But both have elements in common: both have an objective, a definable end result; both are temporary in that they have a defined beginning and need a temporary concentration of resources that will be redeployed once their contribution has been completed; both need to motivate the people involved in the project; both need planning; both need controlling. In other words, both need project managing.

DIAGNOSTIC QUESTION

Is the project environment understood?

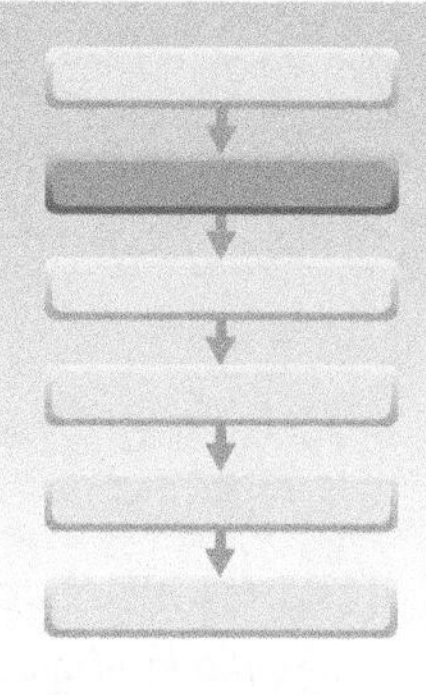

Video

The project environment comprises all the factors which may affect the project during its life. It is the context and circumstances in which the project takes place. Understanding the project environment is important because the environment affects the way in which a project will need to be managed and (just as important) the possible dangers that may cause the project to fail. Environmental factors can be considered under the following four headings.

- *Geo-social environment* – geographical, climatic and cultural factors that may affect the project.
- *Econo-political environment* – the economic, governmental and regulatory factors in which the project takes place.
- *The business environment* – industrial, competitive, supply network and customer expectation factors that shape the likely objectives of the project.
- *The internal environment* – the individual company's or group's strategy and culture, the resources available, and the interaction with other projects that will influence the project.

Project difficulty

An important element in the project management environment is the degree of difficulty of the project itself. Three factors have been proposed as determining project difficulty. These

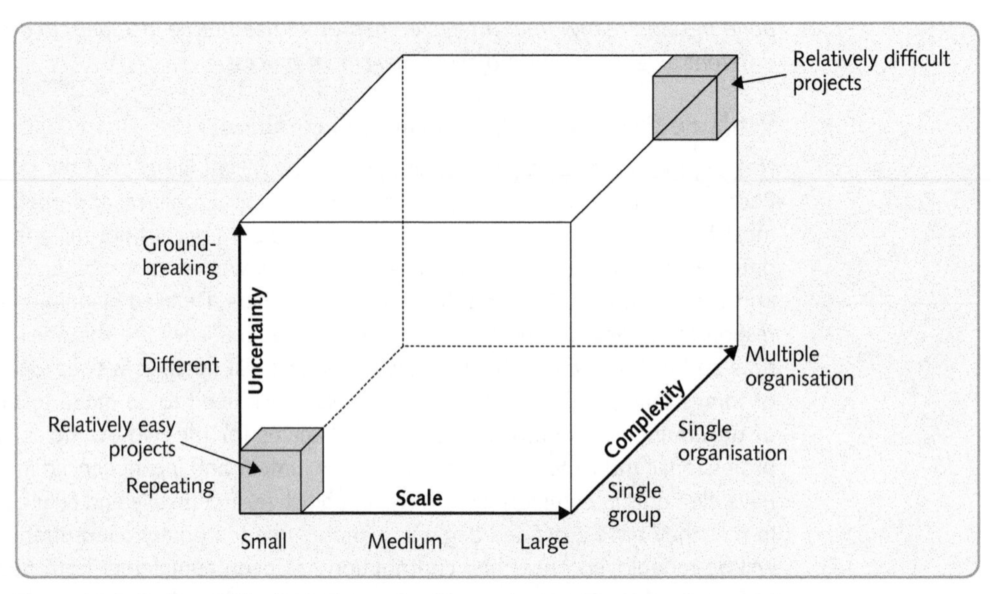

Figure 15.2 Project difficulty is determined by scale, complexity and uncertainty

are scale, uncertainty and complexity. This is illustrated in Figure 15.2. Large-scale projects involving many different types of resources with durations of many years will be more difficult to manage, both because the resources will need a high level of management effort, and because project management objectives must be maintained over a long time period. Uncertainty particularly affects project planning. Ground-breaking projects are likely to be especially uncertain with ever-changing objectives leading to planning difficulties. When uncertainty is high, the whole project planning process needs to be sufficiently flexible to cope with the consequences of change. Projects with high levels of complexity, such as multi-organisational projects, often require considerable control effort. The many separate activities, resources and groups of people involved increase the scope for things to go wrong. Furthermore, as the number of separate activities in a project increases, the ways in which they can impact on each other increases exponentially. This increases the effort involved in monitoring each activity. It also increases the chances of overlooking some part of the project which is deviating from the plan. Most significantly, it increases the 'knock-on' effect of any problem.

OPERATIONS PRINCIPLE

The difficulty of managing a project is a function of its scale, complexity and uncertainty.

Stakeholders

One way of operationalising the importance of understanding a project's environment is to consider the various 'stakeholders' who have some kind of interest in the project. The stakeholders in any project are the individuals and groups who have an interest in the project process or outcome. All projects will have stakeholders, complex projects will have many. They are likely to have different views on a project's objectives that may conflict with other stakeholders. At the very least, different stakeholders are likely to stress different aspects of a project. So, as well an ethical imperative to include as many people as possible in a project from an early stage, it is often useful in preventing objections and problems later in the project. Moreover, there can be significant direct benefits from using a stakeholder-based approach. Project managers can use the opinions of powerful stakeholders to shape the project at an early stage. This makes it more likely that they will support the project, and also can improve its quality. Communicating with stakeholders early and frequently can ensure that they fully understand the project and understand potential benefits. Stakeholder support may even help to win more

OPERATIONS PRINCIPLE

All projects have stakeholders with different interests and priorities.

Table 15.1 The rights and responsibilities of stakeholders in one IT company

The rights of stakeholders	*The responsibilities of project stakeholders*
1 To expect developers to learn and speak their language	1 Provide resources (time, money, etc.) to the project team
2 To expect developers to identify and understand their requirements	2 Educate developers about their business
3 To receive explanations of artifacts that developers use as part of working with project stakeholders, such as models they create with them (e.g. user stories or essential UI prototypes), or artifacts that they present to them (e.g. UML deployment diagrams)	3 Spend the time to provide and clarify requirements
4 To expect developers to treat them with respect	4 Be specific and precise about requirements
5 To hear ideas and alternatives for requirements	5 Make timely decisions
6 To describe characteristics that make the product easy to use	6 Respect a developer's assessment of cost and feasibility
7 To be presented with opportunities to adjust requirements to permit reuse, reduce development time, or to reduce development costs	7 Set requirement priorities
8 To be given good-faith estimates	8 Review and provide timely feedback regarding relevant work artifacts of developers
9 To receive a system that meets their functional and quality needs	9 Promptly communicate changes to requirements
	10 Own your organisation's software processes: to both follow them and actively help to fix them when needed

resources, making it more likely that projects will be successful. Perhaps most important, one can anticipate stakeholder reaction to various aspects of the project, and plan the actions that could prevent opposition, or build support.

Some (even relatively experienced) project managers are reluctant to include stakeholders in the project management process. Preferring to 'manage them at a distance' rather than allow them to interfere with the project. Others argue that the benefits of stakeholder management are too great to ignore and many of the risks can be moderated by emphasising the responsibilities as well as the rights of project stakeholders. For example, one information technology company formally identifies the rights and responsibilities of project stakeholders as shown in Table 15.1.

OPERATIONS PRINCIPLE
Project stakeholders have responsibilities as well as rights.

Managing stakeholders

Managing stakeholders can be a subtle and delicate task, requiring significant social and, sometimes, political skills. But it is based on three basic activities: identifying, prioritising, and understanding the stakeholder group.

- **Identify stakeholders.** Think of all the people who are affected by your work, who have influence or power over it, or have an interest in its successful or unsuccessful conclusion. Although stakeholders may be both organisations and people, ultimately you must communicate with people. Make sure that you identify the correct individual stakeholders within a stakeholder organisation.
- **Prioritise stakeholders.** Many people and organisations will be affected by a project. Some of these may have the power either to block or advance the project. Some may be interested in what you are doing, others may not care. Map out stakeholders using the Power–Interest Grid (see below), and classify them by their power and by their interest in the project.
- **Understand key stakeholders.** It is important to know about key stakeholders. One needs to know how they are likely to feel about and react to the project. One also needs to know how best to engage them in the project and how best to communicate with them.

Animated diagram

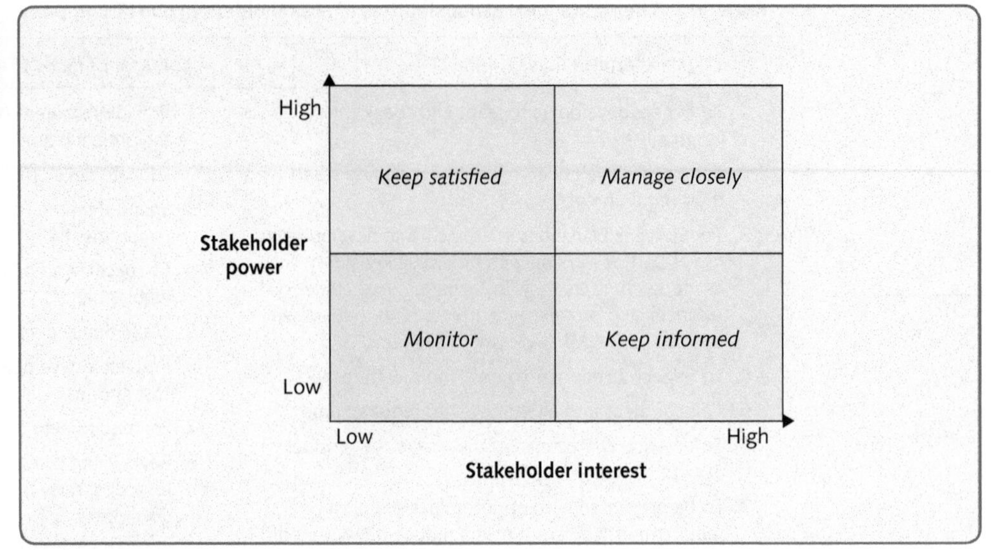

Figure 15.3 The stakeholder power–interest grid

The power-interest grid

One approach to discriminating between different stakeholders, and more importantly, how they should be managed, is to distinguish between their power to influence the project and their interest in doing so. Stakeholders who have the power to exercise a major influence over the project should never be ignored. At the very least, the nature of their interest, and their motivation, should be well understood. But not all stakeholders who have the power to exercise influence over a project will be interested in doing so, and not everyone who is interested in the project has the power to influence it. The power–interest grid, shown in Figure 15.3, classifies stakeholders simply in terms of these two dimensions. Although there will be graduations between them, the two dimensions are useful in providing an indication of how stakeholders can be managed in terms of four categories.

Stakeholders' positions on the grid give an indication of how they might be managed. High-power, interested groups must be fully engaged, with the greatest efforts made to satisfy them. High-power, less-interested groups require enough effort to keep them satisfied, but not so much that they become bored or irritated with the message. Low-power, interested groups need to be kept adequately informed, with checks to ensure that no major issues are arising. These groups may be very helpful with the detail of the project. Low-power, less-interested groups need monitoring, but without excessive communication. Some key questions that can help to understand high priority stakeholders include the following:

OPERATIONS PRINCIPLE

Different stakeholder groups will need managing differently.

- What financial or emotional interest do they have in the outcome of the project? Is it positive or negative?
- What motivates them most of all?
- What information do they need?
- What is the best way of communicating with them?
- What is their current opinion of the project?
- Who influences their opinions? Do some of these influencers therefore become important stakeholders in their own right?
- If they are not likely to be positive, what will win them around to support the project?
- If you don't think you will be able to win them around, how will you manage their opposition?

DIAGNOSTIC QUESTION

Is the project well defined?

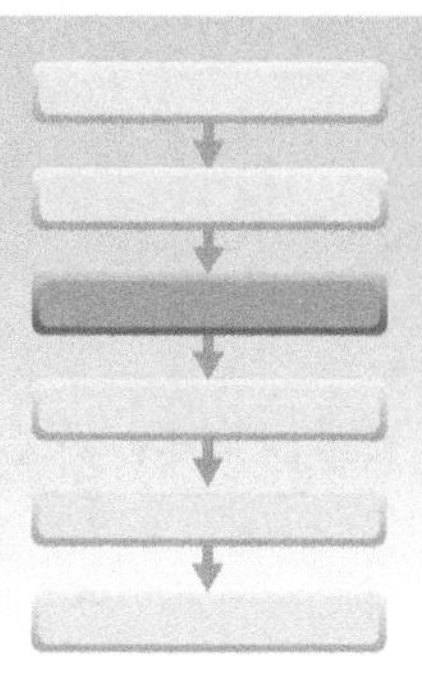

Video

Before starting the complex task of planning and executing a project, it is necessary to be clear about exactly what the project is – its definition. This is not always straightforward, especially in projects with many stakeholders. Three different elements define a project:

- its *objectives*: the end state that project management is trying to achieve
- its *scope*: the exact range of the responsibilities taken on by project management
- its *strategy*: how project management is going to meet its objectives.

Project objectives

Objectives help to provide a definition of the end point which can be used to monitor progress and identify when success has been achieved. They can be judged in terms of the five performance objectives – quality, speed, dependability, flexibility and cost. However, flexibility is regarded as a 'given' in most projects which, by definition, are to some extent one-offs, and speed and dependability are compressed into one composite objective – 'time'. This results in what are known as the 'three objectives of project management' – cost, time and quality.

The relative importance of each objective will differ for different projects. Some aerospace projects, such as the development of a new aircraft, which impact on passenger safety, will place a very high emphasis on quality objectives. With other projects, for example a research project that is being funded by a fixed government grant, cost might predominate. Other projects emphasise time: for example, the organisation of an open-air music festival has to happen on a particular date if the project is to meet its objectives. In each of these projects, although one objective might be particularly important, the other objectives can never be totally forgotten.

OPERATIONS PRINCIPLE
Different projects will place different levels of emphasis on cost, time and quality objectives.

Good objectives are those which are clear, measurable and, preferably, quantifiable. Clarifying objectives involves breaking down project objectives into three categories – the purpose, the end results and the success criteria. For example, a project that is expressed in general terms as 'improve the budgeting process' could be broken down into:

- *Purpose* – to allow budgets to be agreed and confirmed prior to the annual financial meeting.
- *End result* – a report that identifies the causes of budget delay, and which recommends new budgeting processes and systems.
- *Success criteria* – the report should be completed by 30 June, meet all departments' needs and enable integrated and dependable delivery of agreed budget statements. Cost of the recommendations should not exceed $200,000.

Project scope

The scope of a project identifies its work content and its products or outcomes. It is a boundary-setting exercise which attempts to define the dividing line between what each part of the project will do and what it won't do. Defining scope is particularly important when part of a project is being outsourced. A supplier's scope of supply will identify the legal boundaries within which the work must be done. Sometimes the scope of the project is articulated in a formal 'project specification'. This is the written, pictorial and graphical information used to define the output, and the accompanying terms and conditions.

Project strategy

The third part of a project's definition is the project strategy, which defines, in a general rather than a specific way, how the project is going to meets its objectives. It does this in two ways: by defining the phases of the project, and by setting milestones, and/or 'stagegates'. Milestones are important events during the project's life. Stagegates are the decision points that allow the project to move onto its next phase. A stagegate often launches further activities and therefore commits the project to additional costs etc. Milestone is a more passive term, which may herald the review of a part-complete project or mark the completion of a stage, but does not necessarily have more significance than a measure of achievement or completeness. At this stage the actual dates for each milestone are not necessarily determined. It is useful, however, to at least identify the significant milestones and stagegates, either to define the boundary between phases or to help in discussions with the project's customer.

DIAGNOSTIC QUESTION

Is project management adequate?

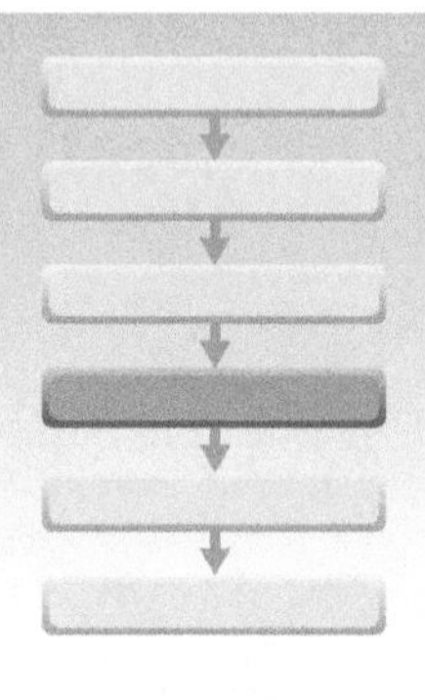

Video

In order to coordinate the efforts of many people in different parts of the organisation (and often outside it as well), all projects need a project manager. Many of a project manager's activities are concerned with managing human resources. The people working in the project team need a clear understanding of their roles in the (usually temporary) organisation. Controlling an uncertain project environment requires the rapid exchange of relevant information with the project stakeholders, both within and outside the organisation. People, equipment and other resources must be identified and allocated to the various tasks. Undertaking these tasks successfully makes the management of a project a particularly challenging operations activity.

Project management skills

The project manager is the person responsible for delivering a project. He or she leads and manages the project team, with the responsibility, if not always the authority, to run the project on a day-to-day basis. They are special people. They must posses seemingly opposing skills. They must be able to influence without necessarily having authority, pay attention to details without losing sight of the big picture, establish an open, communicative environment while remaining wedded to project objectives, and have an ability to hope for the best but plan for the worst. It is a formidable role. Ideally it involves leading, communicating, organising, negotiating, managing conflict, motivating, supporting, team building, planning, directing, problem solving, coaching and delegating.

OPERATIONS PRINCIPLE

The activity of project management requires interpersonal as well as technical skills.

In more formal terms, typical project manager responsibilities include the following:

- devising and applying an appropriate project management framework for the project
- managing the production of the required deliverables
- planning and monitoring the project
- delegating project roles within agreed reporting structures
- preparing and maintaining a project plan
- managing project risks, including the development of contingency plans
- liaison with programme management (if the project is part of a programme)
- overall progress and use of resources, initiating corrective action where necessary
- managing changes to project objectives or details
- reporting through agreed reporting lines on project progress and stage assessments
- liaison with senior management to assure the overall direction and integrity of the project
- adopting technical and quality strategy
- identifying and obtaining support and advice required for the management of the project
- managing ongoing project administration
- conducting post-project evaluation to assess how well the project was executed
- preparing any follow-on action recommendations as required.

Five characteristics in particular are seen as important in an effective project manager:[4]

- background and experience which are consistent with the needs of the project
- leadership and strategic expertise, in order to maintain an understanding of the overall project and its environment, while at the same time working on the details of the project
- technical expertise in the area of the project in order to make sound technical decisions
- interpersonal competence and the people skills to take on such roles as project champion, motivator, communicator, facilitator and politician
- proven managerial ability, in terms of a track record of getting things done.

Managing matrix tensions

In all but the simplest project, project managers usually need to reconcile the interests of both the project itself and the departments contributing resources to the project. When calling on a variety of resources from various departments, projects are operating in a 'matrix management' environment, where projects cut across organisational boundaries and involve staff who are required to report to their own line manager as well as to the project manager. Figure 15.4 illustrates the type of reporting relationship that usually occurs in matrix management structures running multiple projects. A person in department 1, assigned part-time to projects A and

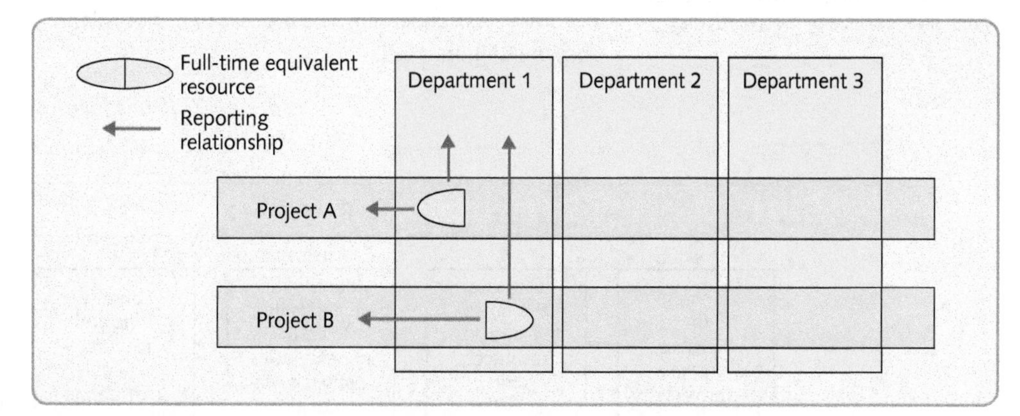

Figure 15.4 Matrix management structures often result in staff reporting to more than one project manager as well as their own department

B, will be reporting to three different managers all of whom will have some degree of authority over their activities. This is why matrix management requires a high degree of cooperation and communication between all individuals and departments. Although decision-making authority will formally rest with either the project, or departmental manager, most major decisions will need some degree of consensus. Arrangements need to be made that reconcile potential differences between project managers and departmental managers. To function effectively matrix management structures should have the following characteristics:

- There should be effective channels of communication between all managers involved, with relevant departmental managers contributing to project planning and resourcing decisions.
- There should be formal procedures in place for resolving the management conflicts that do arise.
- Project staff should be encouraged to feel committed to their projects as well as to their own department.
- Project management should be seen as the central coordinating role, with sufficient time devoted to planning the project, securing the agreement of the line managers to deliver on time and within budget.

DIAGNOSTIC QUESTION

Has the project been adequately planned?

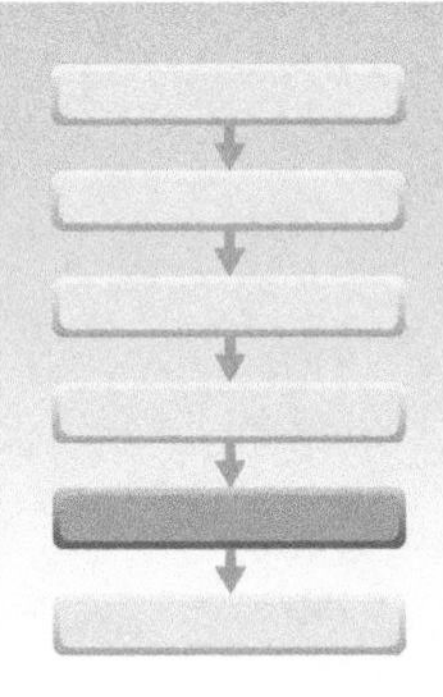

Video

All projects, even the smallest, need some degree of planning. The planning process fulfils four distinct purposes: it determines the cost and duration of the project; it determines the level of resources that will be needed; it helps to allocate work and to monitor progress; it helps to assess the impact of any changes to the project. It is a vital step at the start of the project, but it could be repeated several times during the project's life as circumstances change. This is not a sign of project failure or mismanagement. In uncertain projects, in particular, it is a normal occurrence. In fact, later stage plans typically mean that more information is available, and that the project is becoming less uncertain. The process of project planning involves five steps, shown in Figure 15.5.

OPERATIONS PRINCIPLE
A pre-requisite for project planning is some knowledge of times, resources and relationships between activities.

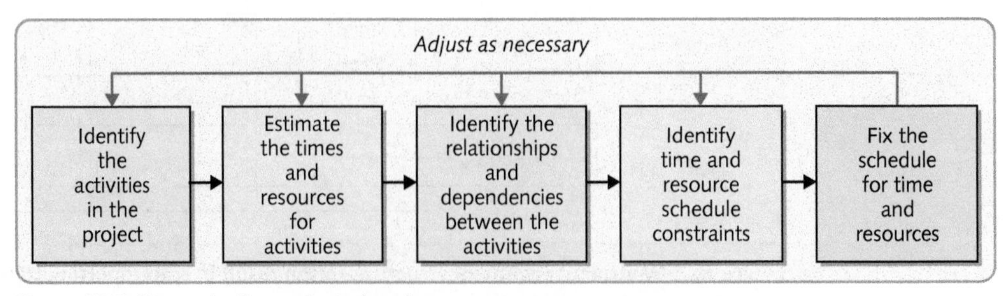

Figure 15.5 Stages in the project planning process

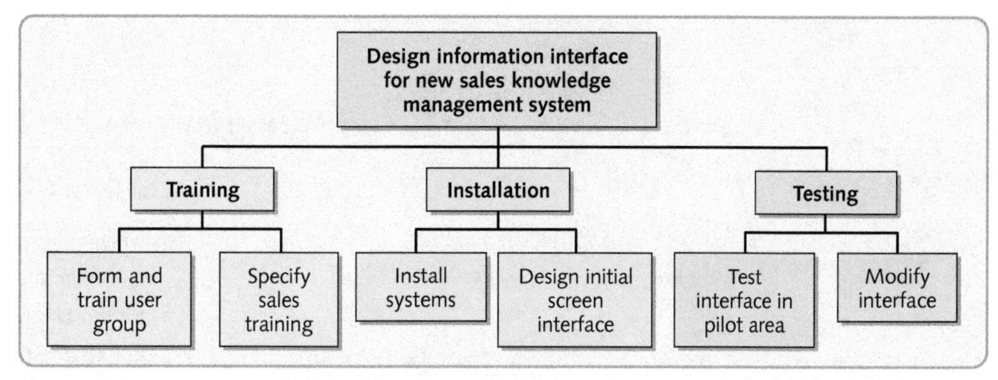

Figure 15.6 Work breakdown structure for a project to design an information interface for a new sales knowledge management system in an insurance company

Identify activities – the work breakdown structure

Some projects are too complex to be planned and controlled effectively unless they are first broken down into manageable portions. This is achieved by structuring the project into a 'family tree' that specifies the major tasks or sub-projects. These in turn are divided up into smaller tasks until a defined, manageable series of tasks, called a work package, is arrived at. Each work package can be allocated its own objectives in terms of time, cost and quality. The output from this is called the work breakdown structure (WBS). The WBS brings clarity and definition to the project planning process. It shows 'how the jigsaw fits together'. It also provides a framework for building up information for reporting purposes.

For example, Figure 15.6 shows the work breakdown structure for a project to design a new information interface (a website screen) for a new sales knowledge management system that is being installed in an insurance company. The project requires cooperation between the company's IT systems department and its sales organisation. Three types of activity will be necessary to complete the project: training, installation and testing. Each of these categories is further broken down into specific activities as shown in Figure 15.6.

Estimate times and resources

The next stage in planning is to identify the time and resource requirements of the work packages. Without some idea of how long each part of a project will take and how many resources it will need, it is impossible to define what should be happening at any time during the execution of the project. Estimates are just that, however – a systematic best guess, not a perfect forecast of reality. Estimates may never be perfect but they can be made with some idea of how accurate they might be. Table 15.2 includes time (in days) and resource (in terms of the number of IT developers needed) estimates for the sales system interface design project.

Table 15.2 Time, resource and relationships for the sales system interface design project

Code	*Activity*	*Immediate predecessor(s)*	*Duration (days)*	*Resources (developers)*
a	Form and train user group	none	10	3
b	Install systems	none	17	5
c	Specify sales training	a	5	2
d	Design initial screen interface	a	5	3
e	Test interface in pilot area	b, d	25	2
f	Modify interface	c, e	15	3

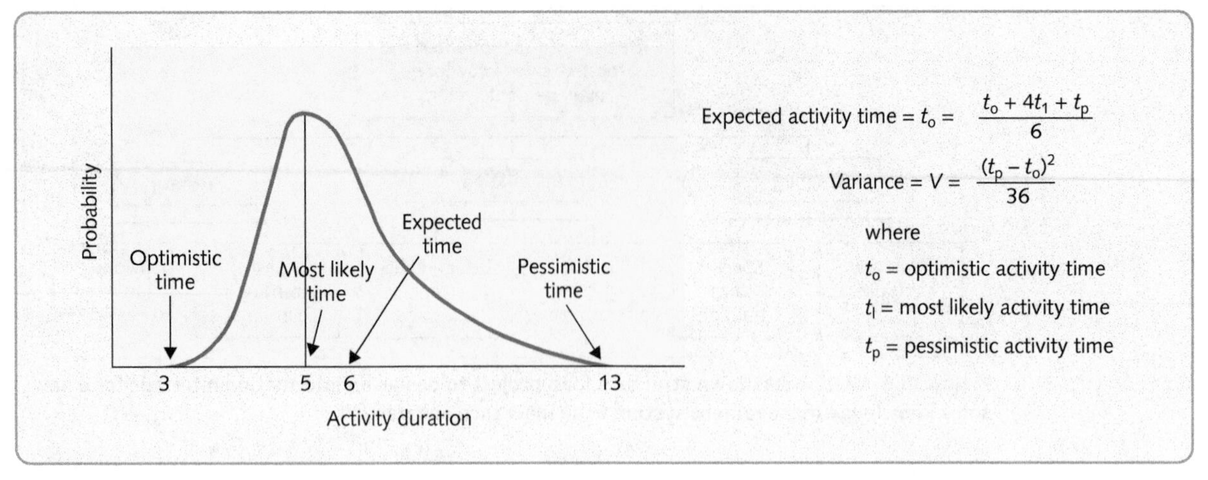

Figure 15.7 Using probabilistic time estimates

Probabilistic estimates

The amount of uncertainty in a project has a major bearing on the level of confidence which can be placed on an estimate. The impact of uncertainty on estimating times leads some project managers to use a probability curve to describe the estimate. In practice, this is usually a positively skewed distribution, as in Figure 15.7. More uncertainty increases the range of the distribution. The natural tendency of some people is to produce optimistic estimates, but these will have a relatively low probability of being correct because they represent the time which would be taken if everything went well. Most likely estimates have the highest probability of proving correct. Finally, pessimistic estimates assume that almost everything which could go wrong does go wrong. Because of the skewed nature of the distribution, the expected time for the activity will not be the same as the most likely time.

Identify the relationships and dependencies between the activities

All the activities which are identified as comprising a project will have some relationship with one another that will depend on the logic of the project. Some activities will, by necessity, need to be executed in a particular order. For example, in the construction of a house, the foundations must be prepared before the walls are built, which in turn must be completed before the roof is put in place. These activities have a dependent or series relationship. Other activities do not have any such dependence on each other. The rear garden of the house could probably be prepared totally independently of the garage being built. These two activities have an independent or parallel relationship.

In the case of the sales system interface design, Table 15.2 provided the basic information that enables the relationships between activities in the project to be established. It did this by identifying the immediate predecessor (or predecessors) for each activity. So, for example, activities **a** and **b** can be started without any of the other activities being completed. Activity **c** cannot begin until activity **a** has been completed, nor can activity **d**. Activity **e** can only start when both activities **b** and **d** have been completed, and activity **f** can only start when activities **c** and **e** have been completed.

Planning tools

Project planning is greatly aided by the use of techniques that help to handle time, resource and relationships complexity. The simplest of these techniques is the *Gantt chart* (or bar chart) which we introduced in Chapter 10. Figure 15.8 shows a Gantt chart for the activities that

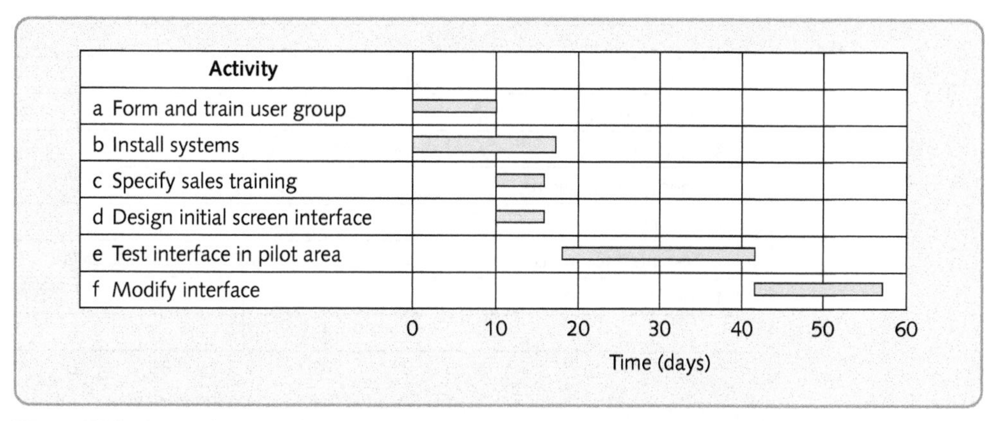

Figure 15.8 Gantt chart for the project to design an information interface for a new sales knowledge management system in an insurance company

form the sales system interface project. The bars indicate the start, duration and finish time for each activity. The length of the bar for each activity on a Gantt chart is directly proportional to the calendar time, and so indicates the relative duration of each activity. Gantt charts are the simplest way to exhibit an overall project plan, because they have excellent visual impact and are easy to understand. They are also useful for communicating project plans and status to senior managers as well as for day-to-day project control.

As project complexity increases, it becomes more necessary to identify clearly the relationships between activities, and show the logical sequence in which activities must take place. This is most commonly done by using the *critical path method* (CPM) to clarify the relationships between activities diagrammatically. The first way we can illustrate this is by using arrows to represent each activity in a project. Figure 15.9 shows this for the sales system interface design project. Each activity is represented by an arrow, in this case with the activity code and the duration in days shown next to it. The circles in Figure 15.9 represent 'events'. These are single points in time that have no duration but mark, for example, the start and finish of activities. So, in this case, event 2 represents the event 'activity **a** finishes'. Event 3 represents '*both* activities **b** and **d** finished', and so on.

The diagram shows that there are a number of chains of events that must be completed before the project can be considered as finished (event 5). In this case, activity chains **a–c–f**,

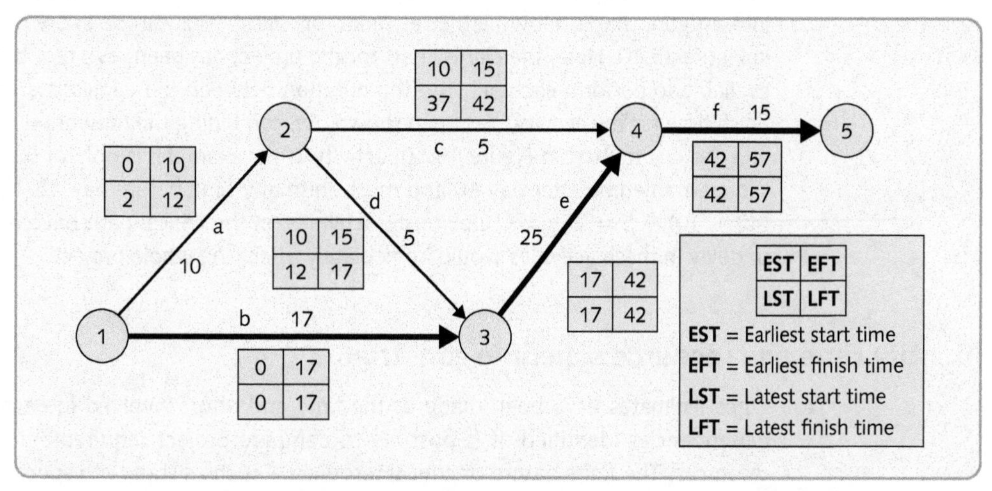

Figure 15.9 The activities, relationships, durations and arrow diagram for the new sales knowledge management system

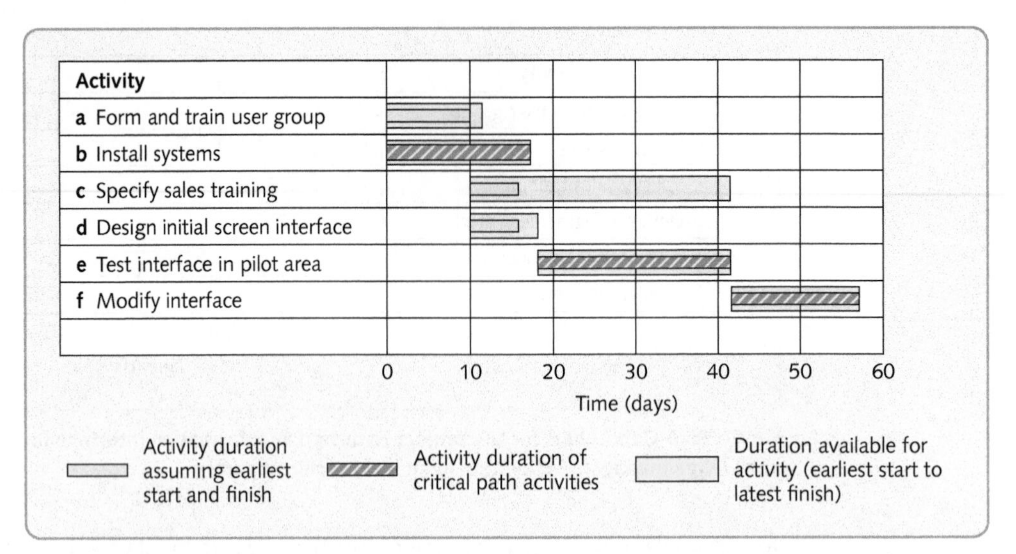

Figure 15.10 Gantt chart for the project to design an information interface for a new sales knowledge management system in an insurance company, with latest and earliest start and finish times indicated

and **a–d–e–f**, and **b–e–f**, must all be completed before the project can be considered as finished. The longest of these chains of activities is called the 'critical path' because it represents the shortest time in which the project can be finished, and therefore dictates the project timing. In this case **b–e–f** is the longest path and the earliest the project can finish is after 57 days.

Figure 15.9 also includes information concerning the earliest and latest start and finish times for each activity. This can be derived by following the logic of the arrow diagram forwards by calculating the earliest times an event can take place, and backwards by calculating the latest time an event can take place. So, for example, the earliest the project can finish (event 5) is the sum of all the activities on the critical path, **b–e–f**: 57 days. If we then take 57 days as the latest that we wish the project to finish, the latest activity **f** can start is $57 - 15 = 42$, and so on. Activities that lie on the critical path will have the same earliest and latest start times and earliest and latest finish times. That is why these activities are critical. Non-critical activities, however, have some flexibility as to when they start and finish. This flexibility is quantified into a figure that is known either as 'float' or 'slack'. This can be shown diagrammatically, as in Figure 15.10. Here, the Gantt chart for the project has been revisited, but this time the time available to perform each activity (the duration between the earliest start time and the latest finish time for the activity) has been shown. So, combining the network diagram in Figure 15.9 and the Gantt chart in Figure 15.10, activity **c**, for example, is only of 5 days duration and it can start any time after day 10 and must finish any time before day 42. Its 'float' is therefore $(42 - 10) - 5 = 27$ days. Obviously, activities on the critical path have no float; any change or delay in these activities would immediately affect the whole project.

Identify time and resource schedule constraints

Once estimates have been made of the time and effort involved in each activity, and their dependencies identified, it is possible to compare project requirements with the available resources. The finite nature of critical resources – such as staff with special skills – means that they should be taken into account in the planning process. This often has the effect of highlighting the need for more detailed re-planning.

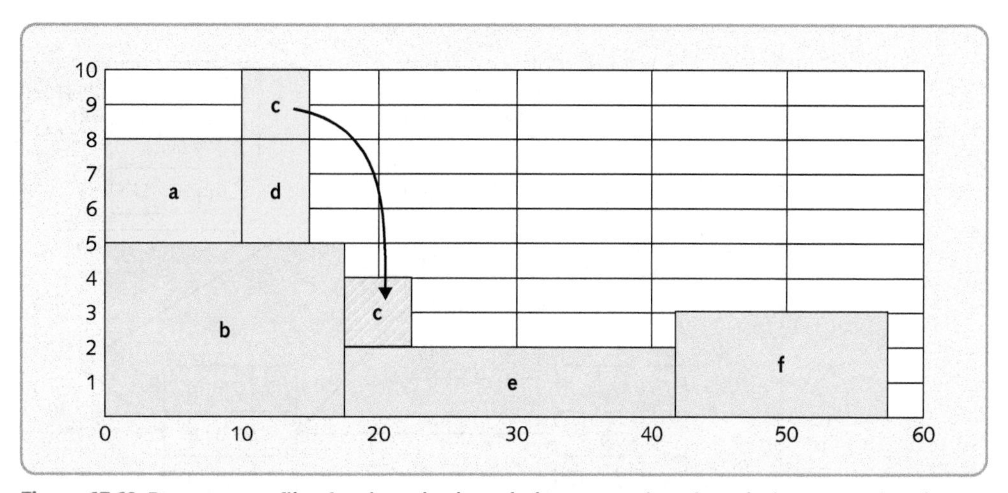

Figure 15.11 Resource profiles for the sales knowledge system interface design, assuming that all activities are started as soon as possible, and assuming that the float in activity c is used to smooth the resource profile

The logic that governs project relationships, as shown in the network diagram, is primarily derived from the technical details, but the availability of resources may also impose its own constraints, which can materially affect the relationships between activities. Return to the sales system interface design project. Figure 15.11 shows the resource profile under two different assumptions. The critical path activities (**b–e–f**) form the initial basis of the project's resource profile. These activities have no float and can only take place as shown. However, activities **a**, **c**, and **d** are not on the critical path, so project managers have some flexibility as to when these activities occur, and therefore when the resources associated with these activities will be required. From Figure 15.11, if one schedules all activities to start as soon as possible, the resource profile peaks between days 10 and 15 when 10 IT development staff are required. However, if the project manager exploits the float that activity **c** possesses and delays its start until after activity **b** has been completed (day 17), the number of IT developers required by the project does not exceed eight. In this way, float can be used to smooth out resource requirements or make the project fit resource constraints. However, it does impose further resource constrained logic on the relationship between the activities. So, for example, in this project moving activity **c** (as shown in Figure 15.11) results in a further constraint of not starting activity **c** until activity **b** has been completed.

Fix the schedule

Project planners should ideally have a number of alternatives to choose from. The one which best fits project objectives can then be chosen or developed. However, it is not always possible to examine several alternative schedules, especially in very large or very uncertain projects, as the computation could be prohibitive. However, modern computer-based project management software is making the search for the best schedule more feasible.

Variations on simple network planning

There are several variations on the simple critical path method type of network that we have used to illustrate project planning thus far. While it is beyond the scope of this book to enter into much more detail on the various ways that simple network planning can be made more sophisticated, two variants are worth mentioning. The first, activity on node networks, is simply a different approach to drawing the network diagram. The second, programme evaluation and review technique (PERT), does represent an enrichment of the basic network approach.

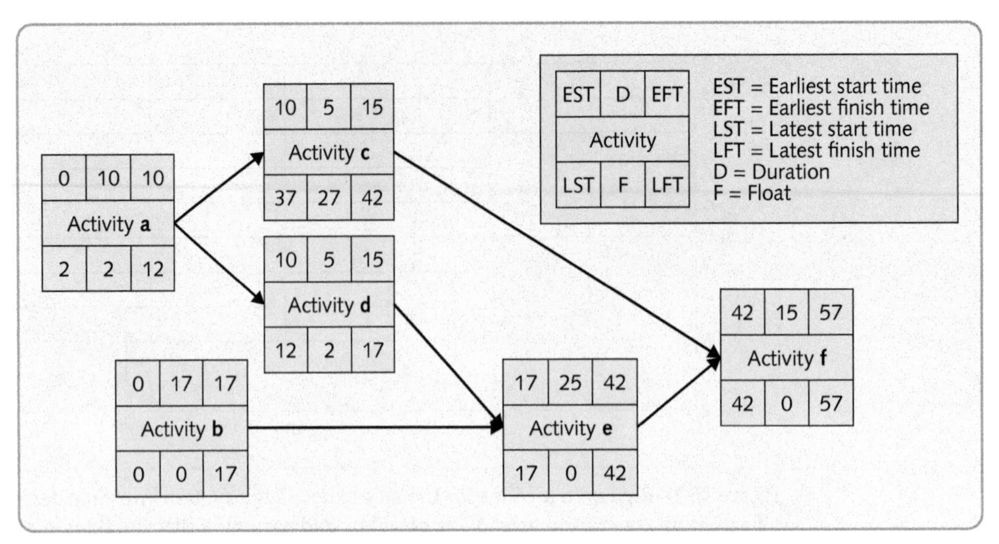

Figure 15.12 Activity on node network diagram for the sales system design project

Activity on node networks

The network we have described so far uses arrows to represent activities and circles at the junctions or nodes of the arrows to represent events. This method is called the activity on arrow (AoA) method. An alternative method of drawing networks is the activity on node (AoN) method. In the AoN representation, activities are drawn as boxes, and arrows are used to define the relationships between them. There are three advantages to the AoN method:

- It is often easier to move from the basic logic of a project's relationships to a network diagram using AoN rather than using the AoA method.
- AoN diagrams do not need dummy activities to maintain the logic of relationships.
- Most of the computer packages which are used in project planning and control use an AoN format.

Figure 15.12 shows the sales system interface design project drawn as an AoN diagram. In this case, we have kept a similar notation to the one used in the original AoA diagram. In addition, each activity box contains information on the description of the activity, its duration, and its total float.

Programme evaluation and review technique (PERT)

The programme evaluation and review technique, or PERT as it is universally known, had its origins in planning and controlling major defence projects in the US Navy. PERT had its most spectacular gains in the highly uncertain environment of space and defence projects. The technique recognises that activity durations and costs in project management are not deterministic (fixed), and that probability theory can be applied to estimates, as was shown in Figure 15.7.

OPERATIONS PRINCIPLE

Probabilistic activity time estimates facilitate the assessment of a project being completed on time.

In this type of network each activity duration is estimated on an optimistic, a most likely and a pessimistic basis, and the mean and variance of the distribution that describes each activity can be estimated as was shown in Figure 15.7. The results are shown in Table 15.3.

In this case the sum of the expected times for each of the activities on the critical path (**b–e–f**) is 58.17 days and the sum of the variances of these three activities is 6.07 days. From this, one could calculate the probability of the project overrunning by different amounts of time.

Table 15.3 PERT parameters for the sales system design project

Code	*Activity*	*Optimistic estimate*	*Most likely estimate*	*Pessimistic estimate*	*Expected time*	*Variance*
a	Form and train user group	8	10	14	10.33	1
b	Install systems	10	17	25	17.17	0.69
c	Specify sales training	4	5	6	5	0.11
d	Design initial screen interface	5	5	5	5	0
e	Test interface in pilot area	22	25	27	24.83	0.69
f	Modify interface	12	15	25	16.17	4.69

DIAGNOSTIC QUESTION

Is the project adequately controlled?

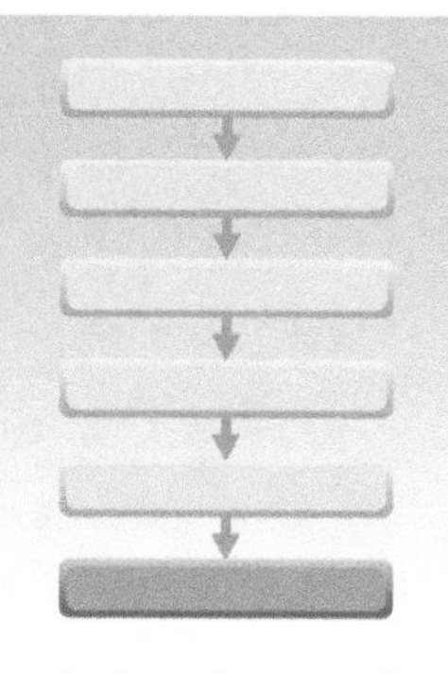

Video

All the stages in project management described so far have all taken place before the actual project takes place. Project control deals with activities during the execution of the project. Project control is the essential link between planning and doing.

The process of project control involves three sets of decisions:

- how to monitor the project in order to check on its progress
- how to assess the performance of the project by comparing monitored observations of the project with the project plan
- how to intervene in the project in order to make the changes that will bring it back to plan.

Project monitoring

Project managers have first to decide what they should be looking for as the project progresses. Usually a variety of measures are monitored. To some extent, the measures used will depend on the nature of the project. However, common measures include current expenditure to date, supplier price changes, amount of overtime authorised, technical changes to project, inspection failures, number and length of delays, activities not started on time, missed milestone, etc. Some of these monitored measures affect mainly cost, some mainly time. However, when something affects the quality of the project, there are also time and cost implications. This is because quality problems in project planning and control usually have to be solved in a limited amount of time.

Assessing project performance

A typical planned cost profile of a project through its life is shown in Figure 15.13. At the beginning of a project, some activities can be started, but most activities will be dependent on finishing. Eventually, only a few activities will remain to be completed. This pattern of a slow start followed by a faster pace with an eventual tail-off of activity holds true for almost all

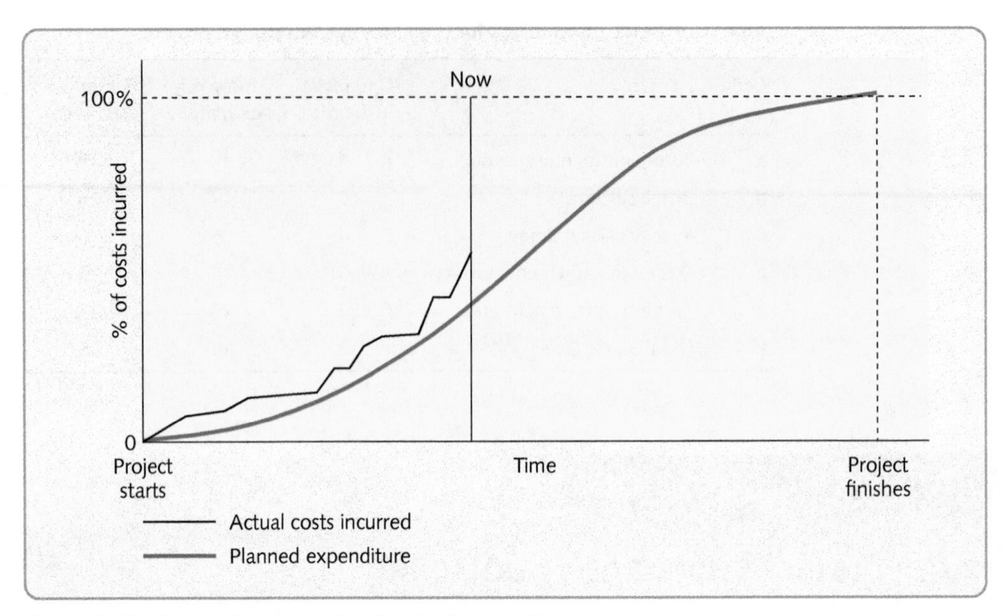

Figure 15.13 Comparing planned and actual expenditure

projects, which is why the rate of total expenditure follows an S-shaped pattern as shown in Figure 15.13, even when the cost curves for the individual activities are linear. It is against this curve that actual costs can be compared in order to check whether the project's costs are being incurred to plan. Figure 15.13 shows the planned and actual cost figures compared in this way. It shows that the project is incurring costs, on a cumulative basis, ahead of what was planned.

Intervening to change the project

If the project is obviously out of control in the sense that its costs, quality levels or times are significantly different from those planned, then some kind of intervention is almost certainly likely to be required. The exact nature of the intervention will depend on the technical characteristics of the project, but it is likely to need the advice of all the people who would be affected. Given the interconnected nature of projects – a change to one part of the project will have knock-on effects elsewhere – this means that interventions often require wide consultation. Sometimes intervention is needed even if the project looks to be proceeding according to plan. For example, the schedule and cost for a project may seem to be 'to plan', but when the project managers project activities and cost into the future, they see that problems are very likely to arise. In this case it is the trend of performance which is being used to trigger intervention.

Crashing, or accelerating, activities

Crashing activities is the process of reducing time spans on critical path activities so that the project is completed in less time. Usually, crashing activities incurs extra cost. This can be as a result of:

- overtime working
- additional resources, such as manpower
- sub-contracting.

Figure 15.14 shows an example of crashing a simple network. For each activity the duration and normal cost are specified, together with the (reduced) duration and (increased)

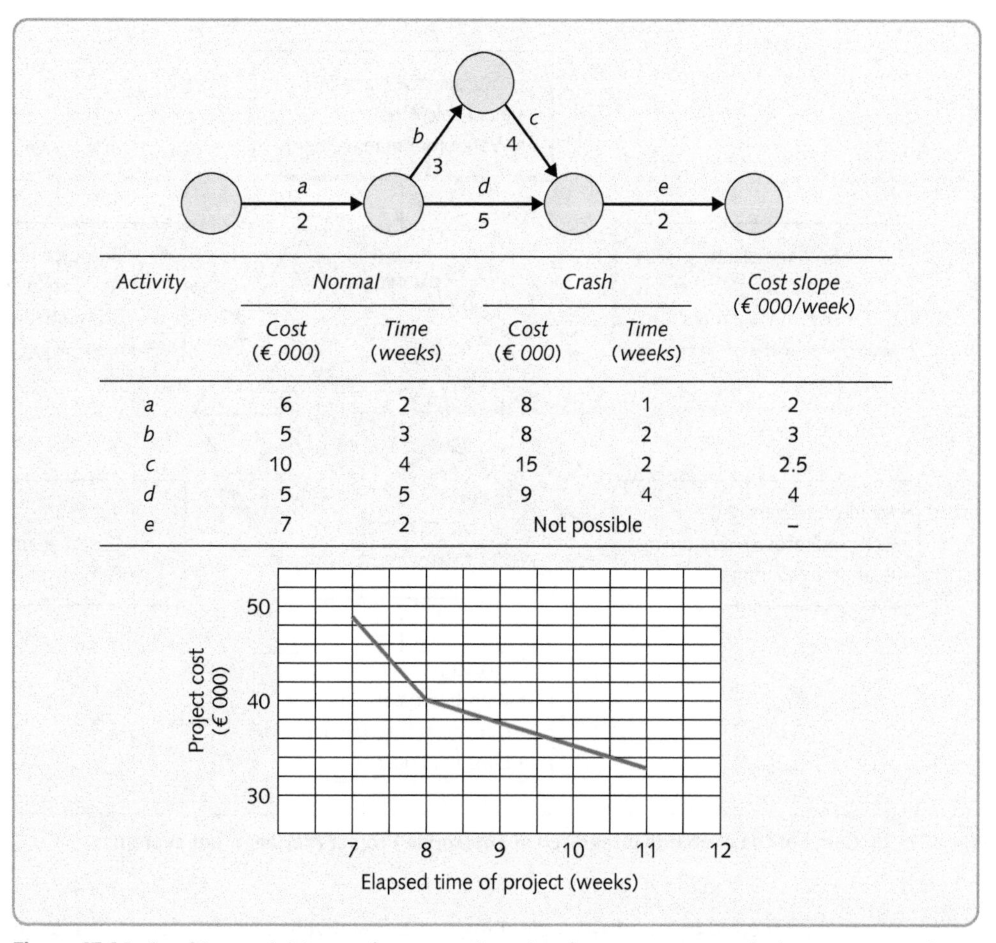

Activity	*Normal*		*Crash*		*Cost slope (€ 000/week)*
	Cost (€ 000)	*Time (weeks)*	*Cost (€ 000)*	*Time (weeks)*	
a	6	2	8	1	2
b	5	3	8	2	3
c	10	4	15	2	2.5
d	5	5	9	4	4
e	7	2	Not possible		–

Figure 15.14 Crashing activities to shorten project time becomes progressively more expensive

OPERATIONS PRINCIPLE
Only accelerating activities on the critical path(s) will accelerate the whole project.

cost of crashing them. Not all activities are capable of being crashed; here activity **e** cannot be crashed. The critical path is the sequence of activities **a**, **b**, **c**, **e**. If the total project time is to be reduced, one of the activities on the critical path must be crashed. In order to decide which activity to crash, the 'cost slope' of each is calculated. This is the cost per time period of reducing durations. The most cost-effective way of shortening the whole project then is to crash the activity on the critical path which has the lowest cost slope. This is activity **a**, the crashing of which will cost an extra €2,000 and will shorten the project by one week. After this, activity **c** can be crashed, saving a further two weeks and costing an extra €5,000. At this point all the activities have become critical and further time savings can only be achieved by crashing two activities in parallel.

The shape of the time–cost curve in Figure 15.14 is entirely typical. Initial savings come relatively inexpensively if the activities with the lowest cost slope are chosen. Later in the crashing sequence the more expensive activities need to be crashed and eventually two or more paths become jointly critical. Inevitably by that point, savings in time can only come from crashing two or more activities on parallel paths.

Computer-assisted project management

For many years, since the emergence of computer-based modelling, increasingly sophisticated software for project planning and control has become available. The rather tedious computation

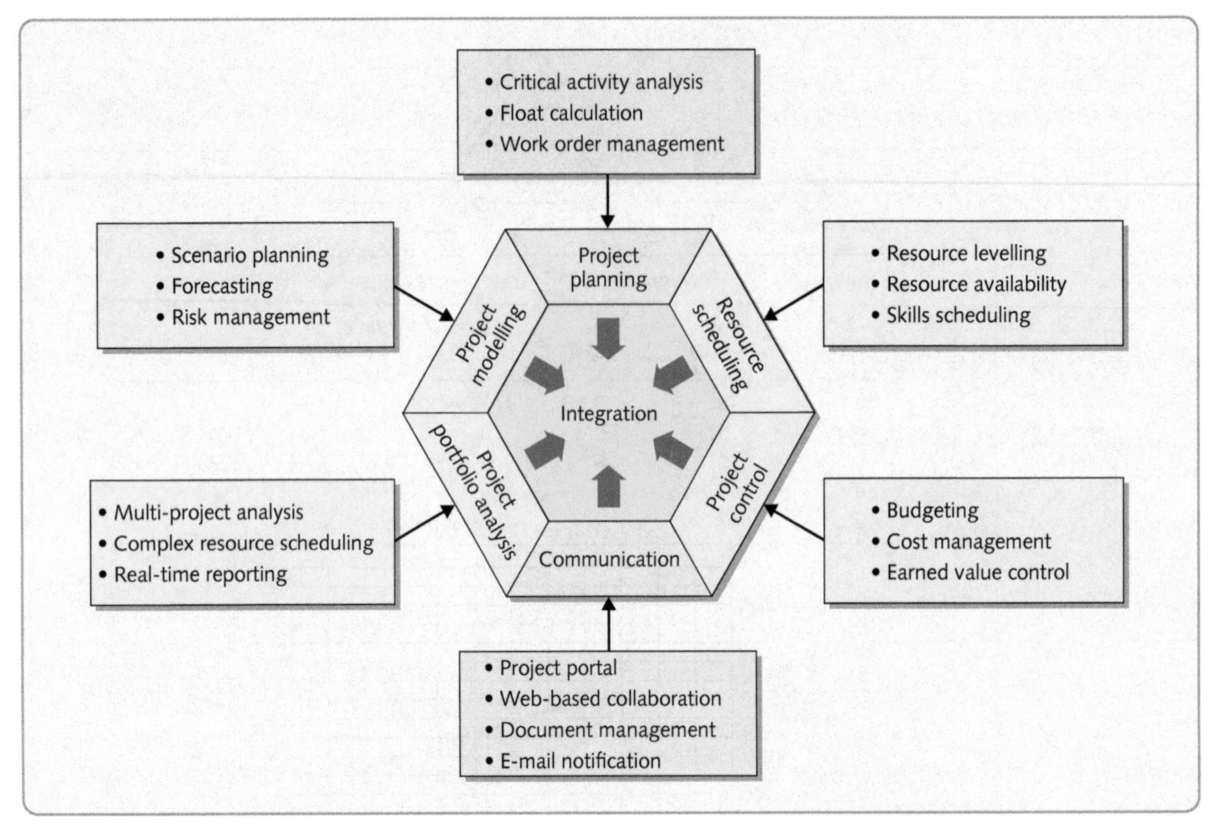

Figure 15.15 Some of the elements integrated in Enterprise Project Management systems

necessary in network planning can be relatively easily performed by project planning models. All they need are the basic relationships between activities together with timing and resource requirements for each activity. Earliest and latest event times, float and other characteristics of a network can be presented, often in the form of a Gantt chart. More significantly, the speed of computation allows for frequent updates to project plans. Similarly, if updated information is both accurate and frequent, such computer-based system can also provide effective project control data. More recently, the potential for using computer-based project management systems for communication within large and complex projects has been developed in so called Enterprise Project Management (EPM) systems.

Figure 15.15 illustrates just some of the elements that are integrated within EPM systems. Most of these activities have been treated in this chapter. Project planning involves critical path analysis and scheduling, an understanding of float, and the sending of instructions on when to start activities. Resource scheduling looks at the resource implications of planning decisions and the way project may have to be changed to accommodate resource constraints. Project control includes simple budgeting and cost management together with more sophisticated earned value control. However, EPM also includes other elements. Project modelling involves the use of project planning methods to explore alternative approaches to a project, identifying where failure might occur and exploring the changes to the project which may have to be made under alternative future scenarios. Project portfolio analysis acknowledges that, for many organisations, several projects have to be managed simultaneously. Usually these share common resources. Therefore, not only do delays in one activity within a project affect other activities in that project, they may also have an impact on completely different projects which are relying on the same resource. Finally, integrated EPM systems can help to communicate, both within a project and to outside organisations which may be contributing to the project.

Much of this communication facility is web-based. Project portals can allow all stakeholders to transact activities and gain a clear view of the current status of a project. Automatic notification of significant milestones can be made by e-mail. At a very basic level, the various documents that specify parts of the project can be stored in an online library. Some people argue that it is this last element of communication capabilities that is the most useful part of EPM systems.

Critical commentary

Each chapter contains a short critical commentary on the main ideas covered in the chapter. Its purpose is not to undermine the issues discussed in the chapter, but to emphasise that, although we present a relatively orthodox view of operation, there are other perspectives.

- When project managers talk of 'time estimates', they are really talking about guessing. By definition, planning a project happens in advance of the project itself. Therefore, no one really knows how long each activity will take. Of course, some kind of guess is needed for planning purposes. However, some project managers believe that too much faith is put in time estimates. The really important question, they claim, is not how long will something take, but how long could something take without delaying the whole project. Also, if a single most likely time estimate is difficult to estimate, then using three, as one does for probabilistic estimates, is merely over analysing what is highly dubious data in the first place.
- The idea that all project activities can be identified as entities with a clear beginning and a clear end point and that these entities can be described in terms of their relationship with each other is an obvious simplification. Some activities are more or less continuous and evolve over time. For example, take a simple project such as digging a trench and laying a communications cable in it. The activity 'dig trench' does not have to be completed before the activity 'lay cable' is started. Only two or three metres of the trench need to be dug before cable laying can commence. A simple relationship, but one that is difficult to illustrate on a network diagram. Also, if the trench is being dug in difficult terrain, the time taken to complete the activity, or even the activity itself may change, to include rock drilling activities for example. However, if the trench cannot be dug because of rock formations, it may be possible to dig more of the trench elsewhere. A contingency not allowed for in the original plan. So, even for this simple project, the original network diagram may reflect neither what will happen or could happen.

SUMMARY CHECKLIST

This checklist comprises questions that can be usefully applied to any type of operations and reflect the major diagnostic questions used within the chapter.

- ☐ Have all the factors that could influence the project been identified?
- ☐ Do these factors include both external and internal influences?
- ☐ Has the project been assessed for its intrinsic difficulty by considering its relative scale, uncertainty and complexity, when compared to other projects?
- ☐ Is the importance of stakeholder management fully understood?
- ☐ Have the rights and responsibilities of project stakeholders been defined?
- ☐ Have all stakeholders been identified?
- ☐ Have all stakeholders been prioritised in terms of their relative power and interest?
- ☐ Has sufficient attention been paid to understanding the needs and motivation of key stakeholders?
- ☐ Has the project been well-defined?
- ☐ Have the objectives of the project been defined, particularly in terms of the relative importance of cost, time, and quality?
- ☐ Has the scope of the project been defined, including the areas that the project will not include?
- ☐ Has the overall strategy of the project been defined in terms of its overall approach, its significant milestones, and any decision gateways that may occur in the project?
- ☐ Have overall project management skills within the business been generally assessed?
- ☐ For this particular project, does the project manager have skills appropriate for the project's intrinsic degree of difficulty?
- ☐ Is sufficient effort being put into the project planning process?
- ☐ Have all activities been identified and expressed in the form of a work breakdown structure?
- ☐ Have all activity times and resources been estimated using the best possible information within the organisation?
- ☐ Is there sufficient confidence in the time and resource estimates to make planning meaningful?
- ☐ Have the relationships and dependencies between activities been identified and summarised in the form of a simple network diagram?
- ☐ Have project planning tools, such as critical path analysis, been considered for the project?
- ☐ Have potential resource and time schedule constraints been built into the project plan?
- ☐ Are there mechanisms in place to monitor the progress of the project?
- ☐ Is there a formal mechanism to assess progress against project plans?
- ☐ Have mechanisms for intervening in the project to bring it back to plan been put in place?
- ☐ Is the level of computer-based project management support appropriate for the degree of difficulty of the project?

CASE STUDY

United Photonics Malaysia Sdn Bhd

Anuar Kamaruddin, Chief Operating Officer of United Photonics Malaysia (UPM), was conscious that the project in front of him was one of the most important he had handled for many years. The number and variety of the development projects underway within the company had risen sharply in the last few years, and although they had all seemed important at the time, this one – the 'Laz-skan' project – clearly justified the description given it by the President of United Photonics Corporation, the US parent of UPM: '*. . . the make-or-break opportunity to ensure the division's long-term position in the global instrumentation industry.*'

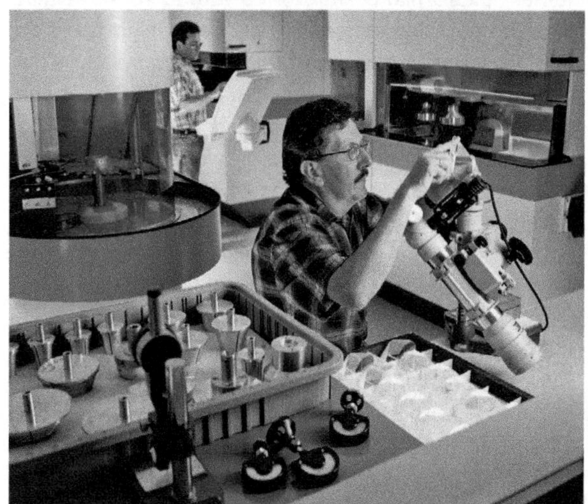
Source: William Taufic/Corbis

The United Photonics Group

United Photonics Corporation had been founded in the 1920s (as the Detroit Gauge Company), a general instrument and gauge manufacturer for the engineering industry. By expanding its range into optical instruments in the early 1930s, it eventually moved also into the manufacture of high-precision and speciality lenses, mainly for the photographic industry. Its reputation as a specialist lens manufacturer led to such a growth in sales that by 1969 the optical side of the company accounted for about 60 per cent of total business and it ranked one of the top two or three optics companies of its type in the world. Although its reputation for skilled lens making had not diminished since then, the instrument side of the company had come to dominate sales once again in the 80s and 90s.

UPM product range

UPM's product range on the optical side included lenses for inspection systems which were used mainly in the manufacture of microchips. These lenses were sold both to the inspection system manufacturers and to the chip manufacturers themselves. They were very high-precision lenses, however, most of the company's optical products were specialist photographic and cinema lenses. In addition about 15 per cent of the company's optical work was concerned with the development and manufacture of 'one or two off' extremely high-precision lenses for defense contracts, specialist scientific instrumentation, and other optical companies. The group's instrument product range consisted largely of electromechanical assemblies with an increasing emphasis on software based recording, display and diagnostic abilities. This move towards more software-based products had led the instrument side of the business towards accepting some customised orders. The growth of this part of the instrumentation had resulted in a special development unit being set up: the Customer Services Unit (CSU) who modified, customised or adapted products for those customers who required an unusual product. Often CSU's work involved incorporating the company's products into larger systems for a customer.

In 1995, United Photonics Corporation had set up its first non-North American facility just outside Kuala Lumpur in Malaysia. United Photonics Malaysia Sdn Bhd (UPM) had started by manufacturing subassemblies for Photonics instrumentation products, but soon had developed a laboratory for the modification of United Photonics products for customers throughout the Asian region. This part of the Malaysian business was headed by T.S. Lim, a Malaysian engineer who had taken his post-graduate qualifications at Stanford and three years ago moved back to his native KL to head up the Malaysian outpost of the CSU, reporting directly to Bob Brierly, the Vice-President of Development, who ran the main CSU in Detroit. Over the last three years, T.S. Lim and his small team of engineers had gained quite a reputation for innovative development. Bob Brierly was delighted with their enthusiasm. *'Those guys really do know how to make things happen. They are giving us all a run for our money.'*

The Laz-skan project

The idea for Laz-skan had come out of a project which T.S. Lim's CSU had been involved with in 2004. At that time the CSU had successfully installed a high-precision Photonics lens into a character-recognition system for a large clearing bank. The enhanced capability which the lens and software modifications had given had enabled the bank to scan documents even when they were not correctly aligned. This had led to CSU proposing the development of a 'vision metrology' device that could optically scan a product at some point in the manufacturing process, and check the accuracy of up to 20 individual dimensions. The geometry of the product to be scanned, the dimensions to be gauged, and the tolerances to be allowed, could all be programmed into the control-logic of the device. The T.S. Lim team were convinced that the idea could have considerable potential. The proposal, which the CSU team had called the Laz-skan project, was put forward to Bob Brierly in August 2004. Brierly both saw the potential value of the idea and was again impressed by the CSU team's enthusiasm. *'To be frank, it was their evident enthusiasm that influenced me as much as anything. Remember that the Malaysian CSU had only been existence for two years at this time – they were a group of keen but relatively young engineers. Yet their proposal was well thought out and, on reflection, seemed to have considerable potential.'*

In November 2004, Lim and his team were allocated funds (outside the normal budget cycle) to investigate the feasibility of the Laz-skan idea. Lim was given one further engineer and a technician, and a three-month deadline to report to the board. In this time he was expected to overcome any fundamental technical problems, assess the feasibility of successfully developing the concept into a working prototype, and plan the development task that would lead to the prototype stage.

The Lim investigation

T.S. Lim, even at the start of his investigation, had some firm views as to the appropriate 'architecture' for the Laz-skan project. By 'architecture' he meant the major elements of the system, their functions, and how they related to each other. The Laz-skan system architecture would consider five major sub-systems: the lens and lens mounting, the vision support system, the display system, the control logic software and the documentation.

T.S. Lim's first task, once the system's overall architecture was set, was to decide whether the various components in the major sub-systems would be developed in-house, developed by outside specialist companies from UPM's specifications, or bought in as standard units and if necessary modified in-house. Lim and his colleagues made these decisions themselves, while recognising that a more consultative process might have been preferable. *'I am fully aware that ideally we should have made more use of the expertise within the company to decide how units were to be developed. But within the time available we just did not have the time to explain the product concept, explain the choices, and wait for already busy people to come up with a recommendation. Also there was the security aspect to think of. I'm sure our employees are to be trusted but the more people who know about the project, the more chance there is for leaks. Anyway, we did not see our decisions as final. For example, if we decided that a component was to be bought in and modified for the prototype building stage it does not mean that we can't change our minds and develop a better component in-house at a later stage.'* By February 2005, TS's small team had satisfied themselves that the system could be built to achieve their original technical performance targets. Their final task before reporting to Brierly would be to devise a feasible development plan.

Planning the Laz-skan development

As a planning aid the team drew up a network diagram for all the major activities within the project from its start through to completion, when the project would be handed over to Manufacturing Operations. This is shown in Figure 15.16 and the complete list of all events in the diagram is shown in Table 15.4. The duration of all the activities in the project were estimated either by T.S. Lim or (more often) by him consulting a more experienced engineer back in Detroit. While he was reasonably confident in the estimates, he was keen to stress that they were just that – estimates.

Two draughting conventions on these networks need explanation. The three figures in brackets by each activity arrow represent the 'optimistic', 'most likely', and 'pessimistic' times (in weeks) respectively. The left side figure in the event circles indicates the earliest time the event could take place and the figure in the right side of the circles indicates the latest time the event could take place without delaying the whole project. Dotted lines represent 'dummy' activities. These are nominal activities which have no time associated with them and are there either to maintain the logic of the network or for drafting convenience.

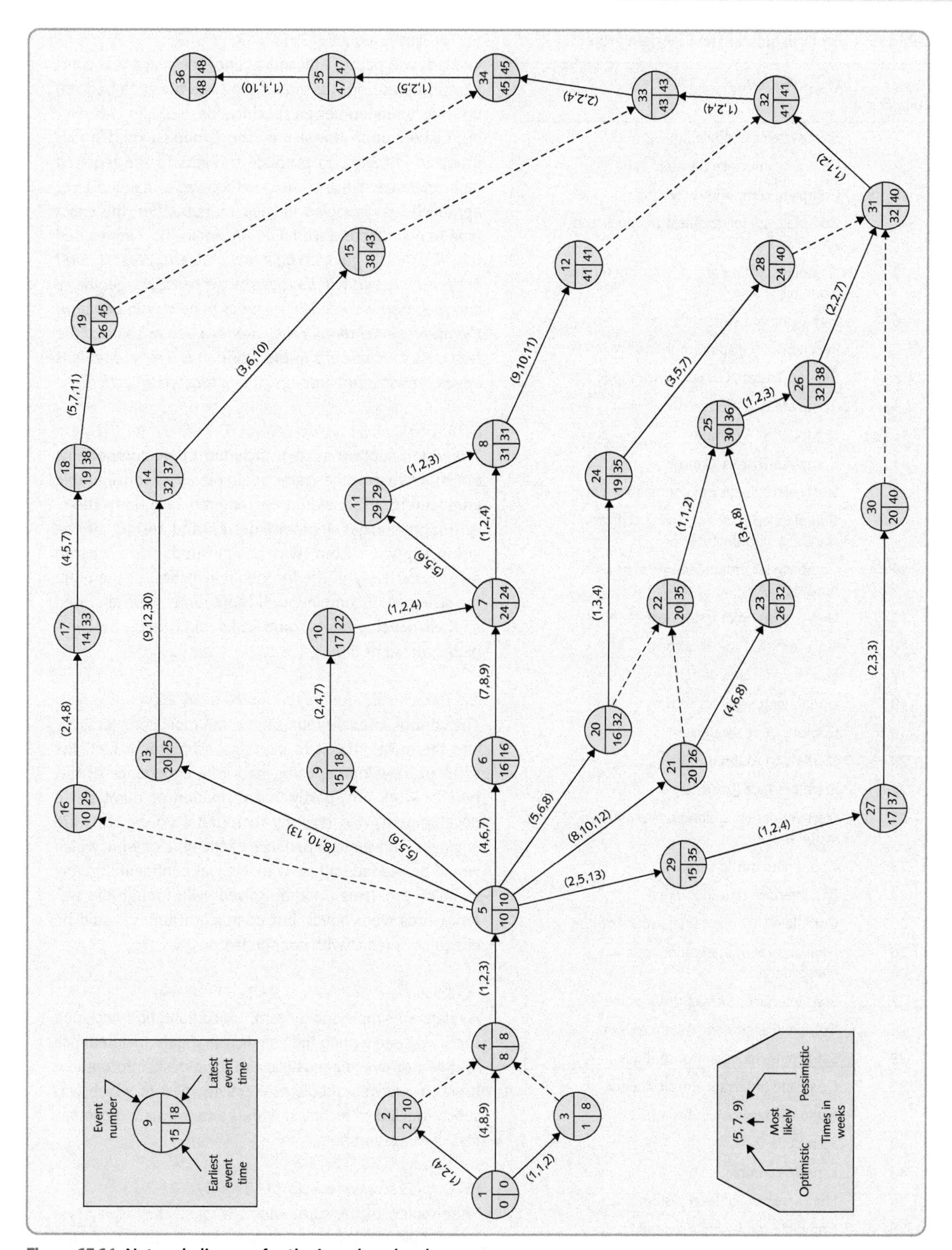

Figure 15.16 Network diagram for the Laz-skan development

Table 15.4 Event listing for the Laz-skan project

Event number	*Event description*
1	Start systems engineering
2	Complete interface transient tests
3	Complete compatibility testing
4	Complete overall architecture block and simulation
5	Complete costing and purchasing tender planning
6	End alignment system design
7	Receive S/T/G, start synch mods
8	Receive Triscan/G, start synch mods
9	Complete B/A mods
10	Complete S/T/G mods
11	Complete Triscan/G mods
12	Start laser sub-system compatibility tests
13	Complete optic design and specification, start lens manufacture
14	Complete lens manufacture, start lens housing S/A
15	Lens S/A complete, start tests
16	Start technical specifications
17	Start help routine design
18	Update engineering mods
19	Complete doc sequence
20	Start vision routines
21	Start interface (tmsic) tests
22	Start system integration compatibility routines
23	Coordinate trinsic tests
24	End interface development
25	Complete alignment integration routine
26	Final alignment integration data consolidation
27	Start interface (tmnsic) programming
28	Complete alignment system routines
29	Start tmnsic comparator routines
30	Complete (interface) trinsic coding
31	Begin all logic system tests
32	Start cycle tests
33	Lens S/A complete
34	Start assembly of total system
35	Complete total system assembly
36	Complete final tests and dispatch

(1) The lens (events 5–13–14–15)

The lens was particularly critical since the shape was complex and precision was vital if the system was to perform up to its intended design specification. T.S. Lim was relying heavily upon the skill of the Group's expert optics group in Pittsburg to produce the lens to the required high tolerance. Since what in effect was a trial and error approach was involved in their manufacture, the exact time to manufacture would be uncertain. T.S. Lim realised this. *'The lens is going to be a real problem. We just don't know how easy it will be to make the particular geometry and precision we need. The optics people won't commit themselves even though they are regarded as some of the best optics technicians in the world. It is a relief that lens development is not amongst the 'critical path' activities.'*

(2) Vision support system (events 6–7–8–12, 9–5, 11)

The vision support system included many components which were commercially available, but considerable engineering effort would be required to modify them. Although the development design and testing of the vision support system was complicated, there was no great uncertainty in the individual activities, or therefore the schedule of completion. If more funds were allocated to their development, some tasks might even be completed ahead of time.

(3) The control software (events 20 to 26, 28)

The control software represented the most complex task, and the most difficult to plan and estimate. In fact, the software development unit had little experience of this type of work but (partly in anticipation of this type of development) had recently recruited a young software engineer with some experience of the type of work which would be needed for Laz-skan. He was confident that any technical problems could be solved even though the system needs were novel, but completion times would be difficult to predict with confidence.

(4) Documentation (events 5–16–17–18–19)

A relatively simple sub-system, 'documentation' included specifying and writing the technical manuals, maintenance routines, on-line diagnostics, and 'help desk' information. It was a relatively predictable activity, part of which was subcontracted to technical writers and translation companies in Kuala Lumpur.

(5) Display system (events 29–27–30)

The simplest of the sub-systems to plan, the display system, would need to be manufactured entirely outside the company and tested and calibrated on receipt.

Market prospects

In parallel with T.S. Lim's technical investigation, sales and marketing had been asked to estimate the market potential of Laz-skan. In a very short time, the Laz-skan project had aroused considerable enthusiasm within the function, to the extent that Halim Ramli, the Asian Marketing Vice President, had taken personal charge of the market study. The major conclusions from this investigation were:

1 The global market for Laz-skan type systems was unlikely to be less than 50 systems per year in 2008, climbing to more than 200 per year by 2012.
2 The volume of the market in financial terms was more difficult to predict, but each system sold was likely to represent around US$300,000 of turnover.
3 Some customisation of the system would be needed for most customers. This would mean greater emphasis on commissioning and post-installation service than was necessary for UPM's existing products.
4 Timing the launch of Laz-skan would be important. Two 'windows of opportunity' were critical. The first and most important was the major world trade show in Geneva in April 2006. This show, held every two years, was the most prominent show-case for new products such as Laz-skan. The second related to the development cycles of the original equipment manufacturers who would be the major customers for Laz-skan. Critical decisions would be taken in the autumn of 2006. If Laz-skan was to be incorporated into these companies products it would have to be available from October 2006.

The Laz-skan go ahead

At the end of February 2005, UPM considered both the Lim and the Ramli reports. In addition, estimates of Laz-skan's manufacturing costs had been sought from George Hudson, the Head of Instrument Development. His estimates indicated that Laz-skan's operating contribution would be far higher than the company's existing products. The board approved the immediate commencement of the Laz-skan development through to prototype stage, with an initial development budget of US$4.5m. The objective of the project was to *'build three prototype Laz-skan systems to be 'up and running' for April 2006'.*

The decision to go ahead was unanimous. Exactly how the project was to be managed provoked far more discussion. The Laz-skan project posed several problems. First, engineers had little experience of working on such a major project. Second, the crucial deadline for the first batch of prototypes meant that some activities might have to be been accelerated, an expensive process that would need careful judgement. A very brief investigation into which activities could be accelerated had identified those where acceleration definitely would be possible and the likely cost of acceleration (see Table 15.5). Finally, no one could agree either whether there should be a single project leader, or which function he or she should come from, or how senior the project leader should be. Anuar Kamaruddin knew that these decisions could affect the success of the project, and possibly the company, for years to come.

Table 15.5 Acceleration opportunities for Laz-skan

Activity	*Acceleration cost (US$/ week)*	*Likely maximum activity time, with acceleration (weeks)*	*Normal most likely time (weeks)*
5–6	23,400	3	6
5–9	10,500	2	5
5–13	25,000	8	10
20–24	5,000	2	3
24–28	11,700	3	5
33–34	19,500	1	2

QUESTIONS

1 Who do you think should manage the Laz-skan development project?
2 What are the major dangers and difficulties that will be faced by the development team as they manage the projects towards its completion?
3 What can they do about these dangers and difficulties?

APPLYING THE PRINCIPLES

Hints

Some of these exercises can be answered by reading the chapter. Other will require some general knowledge of business activity and some might require an element of investigation. Hints on how they can all be answered are to be found in the eText at **www.pearsoned.co.uk/slack.**

1 The activities, their durations and precedences for designing, writing and installing a bespoke computer database are shown in the table below. Draw a Gantt chart and a network diagram for the project and calculate the fastest time in which the operation might be completed.

Bespoke computer database activities

Activity	*Duration (weeks)*	*Activities that must be completed before it can start*
1 Contract negotiation	1	–
2 Discussions with main users	2	1
3 Review of current documentation	5	1
4 Review of current systems	6	2
5 Systems analysis (a)	4	3, 4
6 Systems analysis (b)	7	5
7 Programming	12	5
8 Testing (prelim)	2	7
9 Existing system review report	1	3, 4
10 System proposal report	2	5, 9
11 Documentation preparation	19	5, 8
12 Implementation	7	7, 11
13 System test	3	12
14 Debugging	4	12
15 Manual preparation	5	11

2 Identify a project of which you have been part (for example moving apartments, a holiday, dramatic production, revision for an examination, etc.).

(a) Who were the stakeholders in this project?

(b) What was the overall project objective (especially in terms of the relative importance of cost, quality and time)?

(c) Were there any resource constraints?

(d) Looking back, how could you have managed the project better?

3 The Channel Tunnel project was the largest construction project ever undertaken in Europe and the biggest single investment in transport anywhere in the world. The project, which was funded by the private sector, made provision for a 55-year concession for the owners to design, build and run the operation. The Eurotunnel Group awarded the contract to design and build the tunnel to TML (Trans-Manche Link), a consortium of ten French and British construction companies. For the project managers it was a formidable undertaking. The sheer scale of the project was daunting in itself. The volume of rubble removed from the tunnel increased the size of Britain by the equivalent to 68 football fields. Two main railway tunnels, split by a service/access tunnel, each 7.6 metres in diameter, run 40 metres below the sea bed. In

total there are in excess of 150 kilometres of tunnel. The whole project was never going to be a straightforward management task. During the early negotiations, political uncertainty surrounded the commitment of both governments, and in the planning phase geological issues had to be investigated by a complex series of tests. Even the financing of the project was complex. It required investment by over 200 banks and finance houses, as well as over half a million shareholders. Furthermore, the technical problems posed by the drilling itself and, more importantly, in the commissioning of the tracks and systems within the tunnel needed to be overcome. Yet in spite of some delays and cost overruns, the project ranks as one of the most impressive of the twentieth century.

(a) What factors made the Channel Tunnel a particularly complex project and how might these have been dealt with?

(b) What factors contributed to 'uncertainty' in the project and how might these factors have been dealt with?

(c) Look on the internet to see what has happened to the channel tunnel since it was built. How does this affect your view on the project of building it.

4 Identify your favourite sporting team (Manchester United, the Toulon rugby team, or if you are not a sporting person, choose any team you have heard of). What kind of projects do you think they need to manage? For example, merchandising, sponsorship, etc. What do you think are the key issues in making a success of managing each of these different types of project?

5 Visit the websites of some companies that have developed computer-based project management software (for example, www.primavera.com, www.welcome.com, www.microsoft.com, or just put 'project management software' in a search engine). What appear to be the common elements in the software packages on offer from these companies? Develop a method that could be used by any operation to choose different types of software.

Notes on chapter

1 Reiss, G. (1996) *Programme Management Demystified*, E. & F.N. Spon, London.

2 Slack, A. (2005) 'Popping the Millau Cork' (2004), translated and adapted from *Le Figaro Entreprises*, Mercredi, 15 Decembre.

3 Source: organisation website.

4 Weiss, J.W. and Wysocki, R.K. (1992) *Five-Phase Project Management: A Practical Planning and Implementation Guide*, Addison-Wesley.

TAKING IT FURTHER

There are hundreds of books on project management. They range from the introductory to the very detailed, and from the managerial to the highly mathematical. Here are two general (as opposed to mathematical) books which are worth looking at.

Maylor, H. (2010) *Project Management* (4th edn), Financial Times Prentice Hall.

Meredith, J.R. and Mantel, S. (2009) *Project Management: A Managerial Approach*, John Wiley.

USEFUL WEBSITES

www.opsman.org *Definitions, links and opinions on operations and process management.*

http://apm.org.uk *The UK Association for Project Management. Contains a description of what professionals consider to be the body of knowledge of project management.*

http://pmi.org *The Project Management Institute's home page. An American association for professionals. Insights into professional practice.*

http://ipma.ch *The International Project Management Association, based in Zurich. Some definitions and links.*

Interactive quiz

For further resources including examples, animated diagrams, self-test questions, Excel spreadsheets and video materials please explore the eText on the companion website at **www.pearsoned.co.uk/slack**.

Index